MicroRNAs in Cancer Translational Research

William C.S. Cho
Editor

MicroRNAs in Cancer Translational Research

Springer

Editor
Dr. William C.S. Cho
Queen Elizabeth Hospital
Department of Clinical Oncology
30 Gascoigne Road
Hong Kong SAR
PR China
chocs@ha.org.hk

ISBN 978-94-007-0297-4 e-ISBN 978-94-007-0298-1
DOI 10.1007/978-94-007-0298-1
Springer Dordrecht Heidelberg London New York

Printed on acid-free paper

Springer is part of Springer Science+Business Media (www.springer.com)

Preface

MicroRNAs (miRNAs) have a revolutionary impact on cancer research over recent years. These small non-coding RNAs emerge as important players in tumorigenesis, leading to a paradigm shift in oncology. The rapid discovery of many miRNA targets and their associated pathways has contributed to the development of miRNA-based therapeutics for cancer. These exciting findings reveal the potential of miRNA studies translating from bench to bedside applications. Thorough understanding of miRNAs in cancer translational research is apparently a prerequisite for all professionals involved and interested in this area. MiRNA is a cutting-edge topic in the scientific and medical fields, the identification of cancer-related miRNAs and their state-of-the-art treatment approaches proceed at a fast pace. These stimulating advances drove me to compile this unique book entitled MicroRNAs in Cancer Translational Research.

Hong Kong, China
January 2011

William C.S. Cho

Contents

1 **MicroRNAs in Cancer (An Overview)** 1
Manuela Ferracin, George A. Calin, and Massimo Negrini

2 **The Role of MicroRNAs in Lung Cancer Development, Progression, and Metastasis** 73
Don L. Gibbons, Zain H. Rizvi, and Jonathan M. Kurie

3 **MicroRNAs in Breast Cancer** 91
Suvi-Katri Leivonen

4 **MicroRNAs in Colorectal Cancer** 107
Ondrej Slaby, Marek Svoboda, Jaroslav Michalek, and Rostislav Vyzula

5 **MicroRNAs in Gastric Cancer** 135
Hidekazu Suzuki, Yoshimasa Saito, and Toshifumi Hibi

6 **MicroRNAs in Prostate Cancer: A Possible Role as Novel Biomarkers and Therapeutic Targets?** 145
Paolo Gandellini, Marco Folini, and Nadia Zaffaroni

7 **MicroRNAs in Hepatocellular Carcinoma** 163
Junfang Ji and Xin Wei Wang

8 **MicroRNAs in Cervical Carcinoma** 189
Jeong-Won Lee, Byoung-Gie Kim, and Duk-Soo Bae

9 **MicroRNAs in Esophageal Cancer** 201
Jian Gu and Xifeng Wu

10 **MicroRNAs in Bladder Cancer** 223
Michael B. Williams and Liana Adam

11 **MicroRNAs in Lymphoma** 239
Alba Navarro López and Luis Hernández Pous

12 **MicroRNAs in Leukemia** 269
Florian Kuchenbauer, Johannes Bloehdorn, Lars Bullinger, and Thorsten Zenz

13 MicroRNAs in Pancreatic Cancer: Potential Interests as Biomarkers and Therapeutic Tools 287
Pierre Cordelier and Jérôme Torrisani

14 MicroRNAs in Epithelial Ovarian Cancer 309
Chunsheng Li, Yi Feng, George Coukos, and Lin Zhang

15 MicroRNAs in Brain Tumors 343
Tara Spence, Johnny Nguyen, Eric Bouffet, and Annie Huang

16 MicroRNAs and Cancer Stem Cells 373
Zuoren Yu and Richard G. Pestell

17 MicroRNAs in Cancer Invasion and Metastasis 389
Ashhar S. Ali, Shadan Ali, Aamir Ahmad, Philip A. Philip, and Fazlul H. Sarkar

18 MicroRNAs in Predicting Radiotherapy and Chemotherapy Response 415
Emily J. Noonan, Robert F. Place, and Long-Cheng Li

19 Role of MicroRNAs in Anti-cancer Drug Resistance 449
Erik A.C. Wiemer

20 MicroRNAs and Cancer Metabolism 485
Ping Gao

21 MicroRNAs as Blood-based Biomarkers of Cancer 499
Xi Chen, Junfeng Zhang, Ke Zen, and Chen-Yu Zhang

22 RNAi-based Approaches to the Treatment of Brain Tumors 533
Saroj P. Mathupala, Sandeep Mittal, Murali Guthikonda, and Andrew E. Sloan

Index 551

Contributors

Liana Adam Department of Urology, MD Anderson Cancer Center, The University of Texas, Houston, TX 77030, USA, ladam@mdanderson.org

Aamir Ahmad Department of Pathology, Karmanos Cancer Institute, Wayne State University, Detroit, MI 48201, USA, ahmada@karmanos.org

Ashhar S. Ali Department of Pathology, Karmanos Cancer Institute, Wayne State University, Detroit, MI 48201, USA, ashhar.s.ali@gmail.com

Shadan Ali Division of Hematology and Oncology, Karmanos Cancer Institute, Wayne State University, Detroit, MI 48201, USA, alis@karmanos.org

Duk-Soo Bae Department of Obstetrics and Gynecology, Samsung Medical Center, Sungkyunkwan University School of Medicine, Gangnam-gu, Seoul 135-710, Korea, ds123.bae@samsung.com

Johannes Bloehdorn Department of Internal Medicine III, University of Ulm, 89081 Ulm, Germany, johannes.bloehdorn@uniklinik-ulm.de

Eric Bouffet Pediatric Brain Tumour Program, Division of Hematology Oncology, SickKids Hospital, Toronto, ON, Canada; Department of Pediatrics, University of Toronto, Toronto, ON, Canada M5G 1X8, eric.bouffet@sickkids.ca

Lars Bullinger Department of Internal Medicine III, University of Ulm, 89081 Ulm, Germany, lars.bullinger@uniklinik-ulm.de

George A. Calin Department of Experimental Therapeutics and Cancer Genetics, MD Anderson Cancer Center, University of Texas, Houston, TX 77030, USA, gcalin@mdanderson.org

Xi Chen State Key Laboratory of Pharmaceutical Biotechnology, School of Life Sciences, Jiangsu Diabetes Center, Nanjing University, Nanjing, Jiangsu 210093, PR China, sunnycissy@163.com

Pierre Cordelier Inserm U858, I2MR Institut de Médecine Moléculaire de Rangueil, BP 84225, 31 432 Toulouse Cedex 4, France, pierre.cordelier@inserm

George Coukos Department of Obstetrics and Gynecology, Abramson Family Cancer Research Institute, Center for Research on Early Detection and Cure of

Ovarian Cancer, University of Pennsylvania, Philadelphia, PA 19104, USA, gcks@mail.med.upenn.edu

Yi Feng Abramson Family Cancer Research Institute, University of Pennsylvania, Philadelphia, PA 19104, USA, fengyi2@mail.med.upenn.edu

Manuela Ferracin Department of Experimental and Diagnostic Medicine, University of Ferrara, Ferrara 44121, Italy, manuela.ferracin@unife.it

Marco Folini Department of Experimental Oncology and Molecular Medicine, Fondazione IRCCS Istituto Nazionale dei Tumori, Milan, Italy, marco.folini@istitutotumori.mi.it

Paolo Gandellini Molecular Pharmacology Unit, Department of Experimental Oncology and Molecular Medicine, Fondazione IRCCS Istituto Nazionale dei Tumori, 20133 Milan, Italy, paolo.gandellini@istitutotumori.mi.it

Ping Gao Division of Hematology, Department of Medicine, Johns Hopkins University School of Medicine, Baltimore, MD, USA; School of Life Sciences, University of Science and Technology of China, Hefei, Anhui 230027, PR China, pgao2@jhmi.edu; pgao2@ustc.edu.cn

Don L. Gibbons Department of Thoracic, Head and Neck Medical Oncology and Dept. of Molecular & Cellular Oncology, MD Anderson Cancer Center, The University of Texas, Houston, TX 77030, USA, dlgibbon@mdanderson.org

Jian Gu Department of Epidemiology, MD Anderson Cancer Center, The University of Texas, Houston, TX 77030, USA, jiangu@mdanderson.org

Murali Guthikonda Department of Neurological Surgery, Karmanos Cancer Institute, Wayne State University School of Medicine, Detroit, MI 48201, USA, mguthikonda@med.wayne.edu

Luis Hernández Pous Molecular Pathology Department, Facultad de medicina, Institut d'Investigacions Biomèdiques August Pi i Sunyer (IDIBAPS), 08036 Barcelona, Spain, hernan@clinic.ub.es

Toshifumi Hibi Department of Internal Medicine, Keio University School of Medicine, Tokyo 160-8582, Japan, thibi@sc.itc.keio.ac.jp

Annie Huang Pediatric Brain Tumour Program, Program in Cell Biology, Division of Hematology Oncology, Department of Pediatrics, Sonia and Arthur Labatt Brain Tumour Research Centre, Hospital for Sick Children, Toronto, ON, Canada, annie.huang@sickkids.ca

Junfang Ji Liver Carcinogenesis Section, Laboratory of Human Carcinogenesis, National Cancer Institute, Center for Cancer Research, Bethesda, MD 20892-4258, USA, jijun@mail.nig.gov

Byoung-Gie Kim Department of Obstetrics and Gynecology, Samsung Medical Center, Sungkyunkwan University School of Medicine, Gangnam-gu, Seoul 135-710, Korea, bksong.kim@samsung.com

Florian Kuchenbauer Department of Internal Medicine III, University of Ulm, 89081 Ulm, Germany, florian.kuchenbauer@uniklinik-ulm.de

Jonathan M. Kurie Department of Thoracic, Head and Neck Medical Oncology, MD Anderson Cancer Center, The University of Texas, Houston, TX 77030, USA, jkurie@mdanderson.org

Jeong-Won Lee Department of Obstetrics and Gynecology, Samsung Medical Center, Sungkyunkwan University School of Medicine, Gangnam-gu, Seoul 135-710 Korea, garden.lee@samsung.com

Suvi-Katri Leivonen VTT Medical Biotechnology, FI-20521 Turku, Finland, suvi-katri.leivonen@vtt.fi

Chunsheng Li Center for Research on Early Detection and Cure of Ovarian Cancer, University of Pennsylvania, Philadelphia, PA 19104, USA, lich@mail.med.upenn.edu

Long-Cheng Li Department of Urology, Helen Diller Comprehensive Cancer Center, University of California, San Francisco, CA, USA, LiLC@urology.ucsf.edu

Saroj P. Mathupala Department of Neurological Surgery, Karmanos Cancer Institute, Wayne State University School of Medicine, Detroit, MI 48201, USA, smathupala@med.wayne.edu

Jaroslav Michalek Faculty of Medicine, University Cell Immunotherapy Center, Masaryk University, 625 00 Brno, Czech Republic, jmichalek@fnbrno.cz

Sandeep Mittal Department of Neurological Surgery, Karmanos Cancer Institute, Wayne State University School of Medicine, Detroit, MI 48201, USA, smittal@med.wayne.edu

Alba Navarro López Molecular Pathology Department, Institut d'Investigacions Biomèdiques August Pi i Sunyer (IDIBAPS), Barcelona, Spain, anavarrl@clinic.ub.es

Massimo Negrini Dipartimento di Medicina Sperimentale e Diagnostica, Università di Ferrara, Ferrara 44121, Italy, ngm@unife.it

Johnny Nguyen Program in Cell Biology, Sonia and Arthur Labatt Brain Tumour Research Centre, SickKids Hospital, Toronto, ON, Canada M5G 1X8; Department of Laboratory Medicine and Pathobiology, University of Toronto, Toronto, ON, Canada, jj.nguyen@utoronto.ca

Emily J. Noonan Department of Medicine and Department of Hematology, Center for Molecular Biology in Medicine, Stanford University School of Medicine, Veterans Affairs Palo Alto Health Care System, Palo Alto, CA, USA, emily.noonan@gmail.com

Richard G. Pestell Departments of Cancer Biology and Medical Oncology, Kimmel Cancer Center, Thomas Jefferson University, Philadelphia, PA 19107, USA, Richard.pestell@jefferson.edu

Philip A. Philip Division of Hematology and Oncology, Karmanos Cancer Institute, Wayne State University, Detroit, MI 48201, USA, philipp@karmanos.org

Robert F. Place Department of Urology, Helen Diller Comprehensive Cancer Center, University of California, San Francisco, CA, USA, place.robert@gmail.com

Zain H. Rizvi Department of Thoracic, Head and Neck Medical Oncology, MD Anderson Cancer Center, The University of Texas, Houston, TX 77030, USA; Howard Hughes Medical Institute, Chevy Chase, MD 20815-6789, USA, zrizvi@mdanderson.org

Yoshimasa Saito Division of Gastroenterology and Hepatology, Department of Internal Medicine, Keio University School of Medicine, Tokyo 160-8582, Japan, yoshimasa.saito@gmail.com

Fazlul H. Sarkar Department of Pathology, School of Medicine, Karmanos Cancer Institute, Wayne State University, 740 Hudson Webber Cancer Research Center, Detroit, MI 48201, USA, fsarkar@med.wayne.edu

Ondrej Slaby Department of Comprehensive Cancer Care, Masaryk Memorial Cancer Institute, 656 53 Brno, Czech Republic, slaby@mou.cz

Andrew E. Sloan Department of Neurological Surgery, Case Comprehensive Cancer Center, Case Medical Center, University Hospitals, Cleveland, OH 44106, USA, Andrew.Sloan@UHhospitals.org

Tara Spence Program in Cell Biology, Sonia and Arthur Labatt Brain Tumour Research Centre, SickKids Hospital, Toronto, ON, Canada M5G 1X8; Department of Laboratory Medicine and Pathobiology, University of Toronto, Toronto, ON, Canada, Tara.Spence@utoronto.ca

Hidekazu Suzuki Division of Gastroenterology and Hepatology, Department of Internal Medicine, Keio University School of Medicine, Tokyo 160-8582, Japan, hsuzuki@sc.itc.keio.ac.jp

Marek Svoboda Department of Comprehensive Cancer Care, Masaryk Memorial Cancer Institute, 656 53 Brno, Czech Republic, msvoboda@mou.cz

Jérôme Torrisani INSERM UMR 1037, Cancer Research Center of Toulouse, Université Paul Sabatier, CHU Rangueil Bat L3, 31432 Toulouse Cedex 4, France, jerome.torrisani@inserm.fr

Rostislav Vyzula Department of Comprehensive Cancer Care, Masaryk Memorial Cancer Institute, 656 53 Brno, Czech Republic, vyzula@mou.cz

Xin Wei Wang Liver Carcinogenesis Section, Laboratory of Human Carcinogenesis, National Cancer Institute, Center of Cancer Research, Bethesda, MD 20892-4258, USA, xw3u@nih.gov

Erik A.C. Wiemer Department of Medical Oncology, Josephine Nefkens Institute, Erasmus Medical Center, 3015 GE Rotterdam, The Netherlands, e.wiemer@erasmusmc.nl

Michael B. Williams Department of Urology, MD Anderson Cancer Center, The University of Texas, Houston, TX 77030, USA, mbwillia95@gmail.com

Xifeng Wu Department of Epidemiology, MD Anderson Cancer Center, The University of Texas, Houston, TX 77030, USA, xwu@mdanderson.org

Zuoren Yu Department of Cancer Biology, Kimmel Cancer Center, Thomas Jefferson University, Philadelphia, PA 19107, USA, zuoren.yu@jefferson.edu

Nadia Zaffaroni Department of Experimental Oncology and Molecular Medicine, Fondazione IRCCS Istituto Nazionale dei Tumori, Milan, Italy, nadia.zaffaroni@istitutotumori.mi.it

Ke Zen State Key Laboratory of Pharmaceutical Biotechnology, School of Life Sciences, Jiangsu Diabetes Center, Nanjing University, Nanjing, Jiangsu 210093, PR China, kzen@nju.edu.cn

Thorsten Zenz Department of Internal Medicine III, University of Ulm, 89081 Ulm, Germany, thorsten.zenz@uniklinik-ulm.de

Chen-Yu Zhang State Key Laboratory of Pharmaceutical Biotechnology, School of Life Sciences, Jiangsu Diabetes Center, Nanjing University, Nanjing, Jiangsu 210093, PR China, cyzhang@nju.edu.cn

Junfeng Zhang State Key Laboratory of Pharmaceutical Biotechnology, School of Life Sciences, Jiangsu Diabetes Center, Nanjing University, Nanjing, Jiangsu 210093, PR China, jfzhang@nju.edu.cn

Lin Zhang Department of Obstetrics and Gynecology, Center for Research on Early Detection and Cure of Ovarian Cancer, University of Pennsylvania School of Medicine, Philadelphia, PA 19104, USA, linzhang@mail.med.upenn.edu

Chapter 1
MicroRNAs in Cancer (An Overview)

Manuela Ferracin, George A. Calin, and Massimo Negrini

Abstract The expression of microRNAs (miRNAs) is deregulated in human cancer, with some miRNAs consistently up- or down-regulated in more than one type of neoplasm. The demonstration that aberrantly expressed miRNAs can affect the function of known oncogenes and tumor suppressor genes established molecular links with pathways implicated in malignant transformation. Cell cycle progression, loss of differentiation, increased survival, invasion, and metastasis were shown to be all under the influence of miRNAs, thereby implicating that miRNAs can themselves act as oncogenes or tumor suppressor genes. Besides increasing our knowledge on the molecular basis of cancer, accumulating evidences indicate that miRNA expression profiling has the potential of being translated into clinical applications. Analysis of cancer tissues revealed that miRNAs could be molecular markers useful for cancer classification, prognostic stratification, and drug-response prediction. MiRNAs also emerged as circulating markers, which may become valuable for early diagnosis and follow-up investigations. If we consider that studies on miRNAs in cancer therapy have already produced important results, in just few years, miRNAs have had a great impact in all cancer areas. Whether this will translate into important clinical applications is still too early to say.

1.1 Introduction

MicroRNAs (miRNAs) are a class of 19–24 nucleotides long, philogenetically conserved, non-coding RNAs. Discovered in *C. elegans* in 1993 and soon after in every plant and animal, today the miRNA registry (miRBase) accounts for 15,172 miRNAs, among which 1,048 are human (Release 16, September 2010). MiRNAs act by fine tuning gene expression through a post-transcriptional mechanism. Each miRNA accomplishes its molecular function by guiding the RISC (RNA-induced silencing complex) complex towards target mRNAs (Bartel 2004; Winter et al.

M. Negrini (✉)
Dipartimento di Medicina Sperimentale e Diagnostica, Università di Ferrara, Ferrara 44121, Italy
e-mail: ngm@unife.it

W.C.S. Cho (ed.), *MicroRNAs in Cancer Translational Research*,
DOI 10.1007/978-94-007-0298-1_1, © Springer Science+Business Media B.V. 2011

2009), thereby causing translation inhibition and/or cleavage of target mRNA. Each miRNA can potentially target several mRNAs and can directly or indirectly control the expression of thousands of proteins (Lim et al. 2005), suggesting that the modulation of even a single miRNA may have substantial biological consequences. Recent reviews summarize the mechanisms of miRNA ways of action (Bartel 2009; Brodersen and Voinnet 2009). Although the main mechanism of miRNA action remains the down-regulation of target genes, in some conditions it has been demonstrated that miRNAs may also act by activating translation (Vasudevan et al. 2007) or even by modulating transcription (Dharap et al. 2009; Han et al. 2007a; Janowski et al. 2007; Khraiwesh et al. 2010; Li et al. 2006; Place et al. 2008).

With this knowledge, it seems now obvious that aberrant expression of miRNAs is involved in cancer as well as other human diseases (Negrini et al. 2007). However, the first evidence of miRNA involvement in human cancer was unexpected: by examining a recurring deletion at chromosome 13q14 in the search for a tumor suppressor gene involved in chronic lymphocytic leukemia (CLL), Calin and colleagues (2002) found that the smallest minimal common region of deletion encoded two miRNAs, *miR-15a* and *miR-16-1*, which suggested their role as tumor suppressor genes in CLL. Subsequent investigations confirmed the involvement of these miRNAs in the pathogenesis of human CLL. The putative tumor suppressive role of *miR-15a* and *miR-16-1* was supported by the demonstration that *miR-15a* and *miR-16-1* could induce apoptosis and suppress tumorigenicity in the leukemic cell line MEG-01 and by the demonstration that the anti-apoptotic oncogene BCL2 is a gene target that is up-regulated by their concomitant loss (Calin et al. 2008; Cimmino et al. 2005). The discovery of a germ-line point mutation that results in reduced levels of mature *miR-15a* and *miR-16-1* in two CLL patients (Calin et al. 2005) and a similar mutation in the New Zealand Black (NZB) mouse strain, which is predisposed to the development of a CLL-like disease, further strengthened the idea. Conclusive proofs were generated by LEU2/*miR-15/-16* gene knock-out studies (Klein et al. 2010; Lichter 2010): deletion of this region causes the development of an indolent B cell clonal lymphoproliferative disorder, which represents an accurate model for human CLL. Additional studies (Bonci et al. 2008) extended the tumor suppressive function of *miR-15a/16-1* to other human cancers, such as prostate, and proved that the *miR-15a* and *miR-16-1* cluster can also target CCND1 (encoding cyclin D1) and WNT3A, which promotes tumorigenic features such as survival, proliferation and invasion.

Other examples of miRNAs acting as tumor suppressors are members of the *let-7* family (Esquela-Kerscher et al. 2008; Kumar et al. 2008; Legesse-Miller et al. 2009; Slack 2009). It has been shown that *let-7* is frequently down-regulated in human cancer, leading to the up-regulation of the proto-oncogenes *RAS* (Johnson et al. 2005), *HMGA2* (Lee and Dutta 2007; Mayr et al. 2007; Park et al. 2007; Peng et al. 2008), *MYC* (Sampson et al. 2007), integrin beta 3 (Muller and Bosserhoff 2008), the oncofetal gene *IMP-1* (Boyerinas et al. 2008) and the miRNA maturation RNAse *dicer* (Forman et al. 2008; Tokumaru et al. 2008). *Let-7b* was shown to down-regulate the expression of cyclins D1, D3, and A, and cyclin-dependent kinase (Cdk) 4, in melanoma cells (Schultz et al. 2008).

Importantly, *let-7* seems to be involved in the regulation of "stemness": it was reported that cancer stem cells or cancer-initiating cells exhibited a reduced level of *let-7* miRNAs, which increases upon differentiation. In the same tumor-initiating cells, silencing *H-RAS* could reduce self renewal without affecting differentiation, while silencing *HMGA2* enhanced differentiation without affecting self renewal. The authors concluded that, by silencing its targets, *let-7* regulates multiple stem cell-like properties (Droge and Davey 2008; Yu et al. 2007a). Another link to this stem-regulating property comes from studies on *LIN28* (Chang et al. 2009; Hagan et al. 2009; Heo et al. 2008; Lehrbach et al. 2009; Newman et al. 2008; Rybak et al. 2008). *LIN28*, in combination with *OCT4*, *SOX2*, *NANOG* factors, has been involved in reprogramming fibroblasts back to a pluripotent condition. In mammalian cells, Lin28 and Lin28b act as RNA binding proteins able to associate with the terminal loop of the precursors of *let-7* miRNAs family and block their processing into mature miRNAs (Heo et al. 2008; Newman et al. 2008; Piskounova et al. 2008; Rybak et al. 2008; Viswanathan et al. 2008). Since Lin28 and Lin28b are often over-expressed in human cancer (Guo et al. 2006; Lu et al. 2009; Viswanathan et al. 2009; West et al. 2009), this mechanism explains the down-modulation of *let-7* and the consequent downstream effects.

In addition to tumor suppressors, miRNAs may also act as oncogenes when over-expressed. The *miR-17-92* family is one of the most studied examples. This family includes fourteen homologous miRNAs, which are encoded by three gene clusters on chromosomes 7, 13 and X (Mendell 2008; Tanzer and Stadler 2004). The cluster on chromosome 13 is amplified in human B cell lymphomas (Ota et al. 2004), which leads to increased expression of various miRNA members. Interestingly, enforced expression of the *miR-17-92* cluster acts together with *MYC* to accelerate tumor development in a mouse B cell lymphoma model (He et al. 2005b), thus acting as an oncogene. It has been reported that transcription of this cluster is also induced by *MYC* itself; oncogene members of the *miR-17-92* cluster might therefore act as *MYC* effectors. The up-regulation of members of this large miRNA family protects cells from apoptosis by inhibiting the expression of E2F, p21 and Bim (Kan et al. 2009; Petrocca et al. 2008).

In sum, by inhibiting oncogenes or functioning as their effectors, miRNAs could themselves act as tumor suppressors or oncogenes (Fig. 1.1).

1.2 Mechanisms of MiRNA Deregulation in Human Cancer

Deregulation of miRNAs expression emerges as the main mechanism that triggers their loss or gain of function in cancer cells (Fig. 1.1). As described above, the activation of oncogenic transcription factors, such as *MYC*, represents an important mechanism for altering miRNA expression (O'Donnell et al. 2005). Similarly to *c-MYC*, *MYCN* can up-regulate the *miR-17-92* cluster as well as other miRNAs, including *miR-221*, which is up-regulated in several types of cancer (Schulte et al. 2008). In addition to up-regulate the oncogenic *miR-17-92* cluster, c-Myc regulates a large set of miRNAs that, unexpectedly, become repressed in consequence of

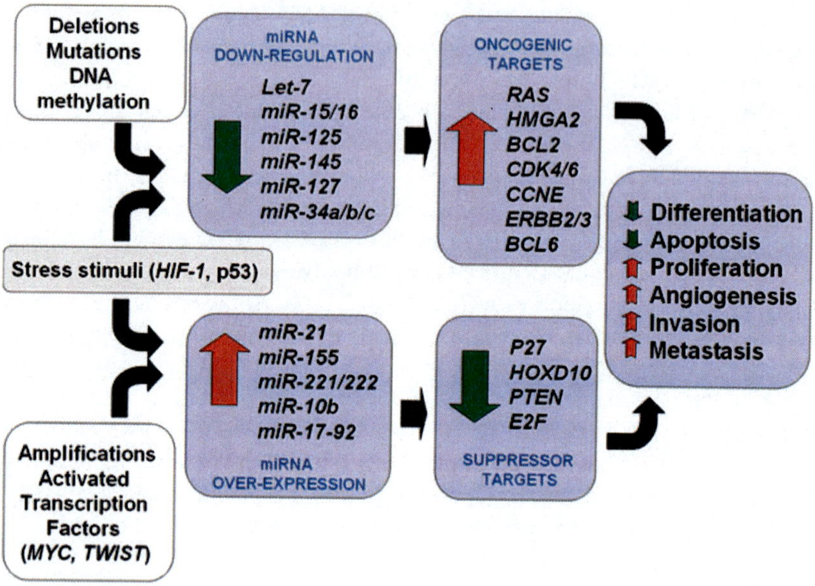

Fig. 1.1 MiRNAs may act as oncogenes and tumor suppressor genes by repressing protein-coding genes with tumor suppressor or oncogenic functions, respectively. Deregulation of miRNAs in cancer cells may occur by genetic (deletions, amplifications, point mutations) or epigenetic (DNA methylation) changes, by aberrant transcription factors expression (Myc, Twist, p53, Hif-1), and by abnormal responsiveness to various stimuli (differentiation, proliferation, hypoxia or other stress stimuli). Abnormally high or abnormally persistent, or lack of miRNA expression may ultimately affect the expression of target genes. As a result, cells may not differentiate or undergo apoptosis, or they may increase their proliferation rate, motility and invasiveness, properties that represent hallmarks of cancer

Myc activation. Ectopic expression of repressed miRNAs diminishes the tumorigenic potential of lymphoma cells, thus proving that the miRNAs repressed by Myc also contribute to tumorigenesis (Chang et al. 2008).

Additional examples of transcription factors regulating miRNAs involved in tumorigenesis are represented by the invasion and metastasis-associated factor Twist, which promotes transcription of *miR-10b* (Ma et al. 2007); the tumor suppressor p53, which is altered in a large fraction of human neoplasms, induces the transcription of *miR-34a*, *miR-34b* and *miR-34c* (Chang et al. 2007; Corney et al. 2007; He et al. 2007; Raver-Shapira et al. 2007; Tarasov et al. 2007; Tazawa et al. 2007); the hypoxia-inducible factor-1 (*HIF-1*) is responsible for the activation of a number of miRNAs in response to hypoxic environment (Kulshreshtha et al. 2007a). Thus, either physiologically induced or aberrantly expressed, transcription factors are largely responsible for miRNA deregulation in human cancer (Fig. 1.1).

Hypoxia, a feature of the neoplastic microenvironment that leads to acidosis and toxic effects, requires genetic or adaptive cellular changes. Hypoxic environment induces specific changes in miRNA expression profile. Indeed, a group of miRNAs, including *miR-21, 23, 24, 26, 27, 30, 93, 103, 106, 107, 181, 192, 195, 210 and 213*, is up-regulated under hypoxic conditions. To assess a biological function of hypoxia-induced miRNAs, the effect on apoptosis was measured. At least for *miR-26*, *-107* and *-210*, caspase inhibition was detected, indicating that target effectors of these miRNAs have a role in protecting cells from apoptosis. Twenty of the 23 commonly up-regulated miRNAs in cancer are induced by hypoxia (Kulshreshtha et al. 2007b), which suggests that the response to this stress is one of the main mechanisms affecting miRNA expression in cancer cells. Interestingly, *miR-21*, one of the miRNAs that is induced by hypoxia and up-regulated in cancer, possesses anti-apoptotic properties; its expression may therefore represent an adaptation to a hypoxic environment that favors cancer cell survival.

Several recent studies have shown that expression of the *miR-34* family is induced by DNA damage and oncogenic stress in a p53-dependent manner (Bommer et al. 2007; Chang et al. 2007; Corney et al. 2007; He et al. 2007; Raver-Shapira et al. 2007; Tarasov et al. 2007; Tazawa et al. 2007) (Fig. 1.2). P53, a tumor suppressor transcription factor at the cross-road of a variety of signalling pathways, is post-transcriptionally activated by DNA damage, oxidative stress, or activation of oncogenes, and functions to safeguard cellular integrity (Vousden and Lane 2007). P53 activation leads to the transcription of genes, whose products in turn induce cell cycle arrest, which can be transient or permanent (senescence), or promote apoptosis in cases where stress cannot be overcome. Biological outcome is largely dependent on cellular background. *miR-34s* form an evolutionarily conserved miRNA family, with three members in vertebrate genomes (*miR-34a*, *miR-34b*, and *miR-34c*), organized in two separate loci: *miR-34a*, located at chromosome 1p16, is encoded by its own transcript, whereas *miR-34b* and *miR-34c,* at 11q23, share a common primary transcript (Bommer et al. 2007). Through various induction stimuli, such as DNA damage and oncogenic activation, *miR-34a* and *miR-34b*/*miR-34c* loci were shown to be directly regulated by interaction of p53. Thus, the identification of the *miR-34* family as p53 targets expands the repertoire of p53-regulated genes to

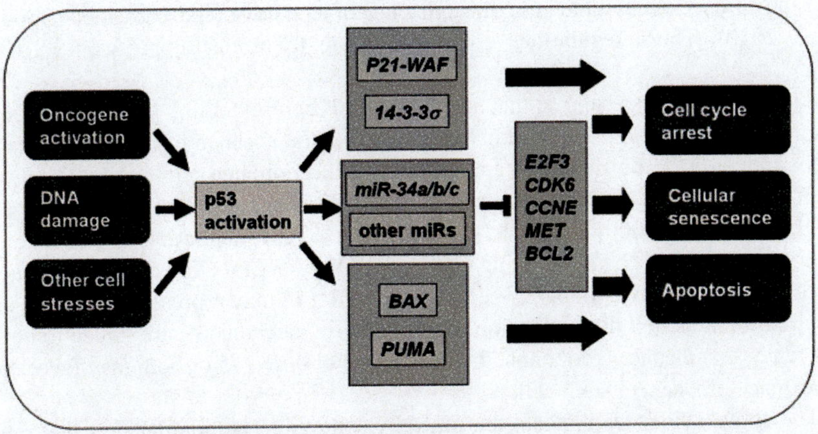

Fig. 1.2 MiRNAs of the *miR-34* family are p53 downstream effectors. In response to various stress stimuli, p53 is activated to promote transcription of a variety of downstream effectors, which are responsible for the final outcomes. *MiR-34a/b/c* are directly induced through a p53 responsive element present in their promoters. Because several effectors downstream to p53 results down-regulated, *miR-34s* and/or other miRNAs appear to be responsible for these effects

include miRNAs. Importantly, biological responses to ectopic expression of *miR-34* included senescence, cell cycle arrest or apoptosis depending on cellular model. As cell cycle arrest and apoptosis are common endpoints of p53 activation, *miR-34* genes may represent important mediators of tumor suppression by p53 (Fig. 1.2). Achieving these different outcomes may depend on the spectrum of *miR-34* regulatory targets that are expressed in a given cell type. Various gene targets compatible with the observed biological effects were identified. For examples, cyclin E2, cyclin-dependent kinase 4 and hepatocyte growth factor MET, were down-regulated by *miR-34* expression. It is interesting to note that p53 has been reported to induce the down-regulation of cyclin-dependent kinases (CDK4) and cyclins (Cyclin E2), which may contribute to p53-induced cell cycle arrest (Spurgers et al. 2006), suggesting that down-regulation of these proteins could be mediated by induction of *miR-34*.

Although the importance of *miR-34* as p53 effector is proven, additional miRNAs are also modulated by p53. Notably, quantitative analyses indicated that 34 miRNAs were significantly induced by p53, whereas 16 miRNAs were repressed (Tarasov et al. 2007). Some of these differentially regulated miRNAs were connected to cancer: among the induced miRNAs were *miR-15/16*, which target the oncogene product Bcl2, and *let-7*, which down-regulates *RAS* and *HMGA2* (Calin and Croce 2006a; Johnson et al. 2005; Lee and Dutta 2007; Mayr et al. 2007). Among the miRNAs repressed by p53 was *miR-221*, which is known to down-regulate the CDK inhibitor p27 (Galardi et al. 2007; le Sage et al. 2007).

More recently, the relationship between p53 and *miR-145* emerged. It has been shown that p53 can induce the expression of mature *miR-145* by a transcriptional mechanism (Sachdeva et al. 2009; Spizzo et al. 2010). It seems therefore possible that one mechanism for the loss of expression of *miR-145* in most human neoplasms is linked to the loss of p53 activity, which is common in human cancer. On the other hand, it has also been demonstrated that ectopic expression of *miR-145* can induce p53 activation, thereby indicating the existence of a positive feed-back loop (Spizzo et al. 2010). These regulations could contribute to tumor suppression activity of p53.

These findings also suggest that *miR-145* could represent a link between p53 and "stemness". Indeed, different studies revealed that the absence of p53 is important in the induction of pluripotent stem cells from differentiated adult cells (Cicalese et al. 2009; Hong et al. 2009; Kawamura et al. 2009; Li et al. 2009a; Marion et al. 2009; Utikal et al. 2009) and *miR-145* was shown to repress pluripotency in human embryonic stem cells by modulating *OCT4*, *SOX2*, and *KLF4* (Xu et al. 2009b), suggesting that the positive loop between p53 and *miR-145* may represent a mechanism required for stem cell differentiation. Because these mechanisms are linked to both self-renewal and cancer programs, the loss of p53 or *miR-145* or both may favor the appearance of a cancer stem cell.

The above evidences indicate that miRNA expression is modulated at the transcriptional level in response to either physiological stimuli, such as environmental changes or differentiation signals, or to aberrantly expressed transcription factors. Changes in transcriptional programs will affect miRNA expression and, in turn, all downstream gene targets. Thus, by dissecting the involvement of miRNAs in cancer-associated pathways various aspects of cell physiopathology should be clarified.

In addition to transcriptional regulation of miRNA, a post-transcriptional level of mature miRNA control is also emerging. In fact, as mentioned above, it was discovered that Lin28 and Lin28b act as RNA binding proteins able to associate with the terminal loop of the precursors of *let-7* family miRNAs and block their processing into mature miRNAs (Heo et al. 2008; Newman et al. 2008; Piskounova et al. 2008; Rybak et al. 2008; Viswanathan et al. 2008). Because *LIN-28* is over-expressed in human cancer, this finding suggests a mechanism for down-regulating *Let-7* in cancer cells (Lu et al. 2009; Viswanathan et al. 2009; West et al. 2009). In addition and in analogy with what described for *miR-145*/p53 loop, a link with stem cell emerges also in this case. Indeed, *LIN-28*, in a cocktail of *OCT4, SOX2, NANOG* factors, can reprogram fibroblasts back to a pluripotent condition (Yu et al. 2007b), thereby suggesting that repression of *let-7* may be important in establishing a pluripotent state (Boyerinas et al. 2008; Droge and Davey 2008; Yu et al. 2007a), again further suggesting that perturbation of physiological stem cell processes may promote the appearance of a cancer stem cell.

Although several are the examples that indicate that abnormal miRNA expression in cancer is caused by transcriptional and post-transcriptional mechanisms instead of genetic alterations, genomic aberrations may also affect miRNA expression. MiRNA up-regulation has been associated with genomic amplification (He et al. 2005b; O'Donnell et al. 2005). Examples of up-regulated miRNAs in consequence of genomic amplification include the previously described clusters *miR-17-92* at chromosome 13q31 and *miR-106b-25* at chromosome 7q22 (Kan et al. 2009; Ota et al. 2004; Rinaldi et al. 2007; Tagawa and Seto 2005). The *miR-106b-25* polycistron is activated by genomic amplification in digestive tract neoplasms and the *miR-17-92* in malignant lymphomas. Another example of up-regulated miRNA is *miR-21*, a gene located at chromosome 17q23 in a chromosomal region frequently amplified in human cancer (Ehrbrecht et al. 2006; Sinclair et al. 2003), which is up-regulated in several human cancers (Chan et al. 2005; Ciafre et al. 2005; Iorio et al. 2005). MiRNA down-regulation has been associated with chromosomal deletions, point mutations and aberrant promoter methylation (Calin et al. 2002; 2004a, 2005; Datta et al. 2008; Han et al. 2007b; Lujambio et al. 2008; Saito et al. 2006; Toyota et al. 2008) (Fig. 1.1). Using thousands of tumor samples, Volinia et al. (2010) reported regions of amplification or deletion that most likely can affect miRNA expression.

1.3 Numerous MiRNAs are Deregulated in Human Cancer

Evidence now indicates that the involvement of miRNAs in cancer is extensive. Direct evidence for this has been provided by genome-wide expression studies. Microarray technology and techniques based on quantitative PCR (qPCR) helped to rapidly establish miRNA expression profiles of normal versus neoplastic tissues.

Every examined neoplasm was shown to have a specific miRNA expression profile, able to distinguish neoplastic from normal counterpart (Calin and Croce 2006b; Croce 2009; Esquela-Kerscher and Slack 2006).

Two large expression profiling studies have been reported. In a large profiling analysis of 334 leukemias and solid cancers, Lu et al. (2005) found that miRNA-expression profiles classify human cancers based on developmental lineage and differentiation state of the tumor. This study also revealed a globally decreased miRNA expression in tumors with respect to their normal counterpart. Volinia et al. (2006) conducted a large-scale miRNome analysis on 540 samples representing six solid cancers (lung, breast, stomach, prostate, colon and pancreas) and corresponding normal tissues and have shown the existence of a tumor-specific miRNA signature, 43 deregulated miRNAs (26 up- and 17 down-regulated).

The above studies were supported by a number of investigations on individual types of neoplasms. All the studies revealed the existence of differences in miRNA expression in neoplastic versus normal tissues in all human solid tumors and haematologic malignancies (Akao et al. 2007; Amaral et al. 2009; Avissar et al. 2009; Bandi et al. 2009; Bloomston et al. 2007; Bonci et al. 2008; Bottoni et al. 2007; Budhu et al. 2008; Busacca et al. 2010; Calin et al. 2004b, 2005; Careccia et al. 2009; Cervigne et al. 2009; Chan et al. 2005; Childs et al. 2009; Ciafre et al. 2005; Cimmino et al. 2005; Costinean et al. 2006; Datta et al. 2008; Davoren et al. 2008; Dyrskjot et al. 2009; Eis et al. 2005; Esquela-Kerscher et al. 2008; Fabbri et al. 2007; Feber et al. 2008; Felicetti et al. 2008; Ferretti et al. 2009; Friedman et al. 2009; Fulci et al. 2007; Galardi et al. 2007; Garzon et al. 2008a, b; Gebeshuber et al. 2009; Gottardo et al. 2007; Gramantieri et al. 2007; Guled et al. 2009; Guo et al. 2008; He et al. 2005a; Henson et al. 2009; Huang et al. 2008a, b; Ichimi et al. 2009; Inamura et al. 2007; Iorio et al. 2005; 2007; Ji et al. 2009b; Jiang et al. 2008; Jung et al. 2009; Kalscheuer et al. 2008; Kozaki et al. 2008; Kutay et al. 2006; Lee et al. 2007; 2008a; Lerner et al. 2009; Li et al. 2008; 2009c; Liu et al. 2004; 2009a, c; Lodygin et al. 2008; Lui et al. 2007; Magrelli et al. 2009; Markou et al. 2008; Martinez and Busslinger 2007; Mattie et al. 2006; Mendell 2008; Meng et al. 2007a; Michael et al. 2003; Motoyama et al. 2009; Mott et al. 2007; Murakami et al. 2006; Nam et al. 2008; Nikiforova et al. 2008; Pallante et al. 2006; Pekarsky et al. 2006; Porkka et al. 2007; Rai et al. 2008; Roccaro et al. 2009; Rokhlin et al. 2008; Roldo et al. 2006; Sampson et al. 2007; Saydam et al. 2009; Schaefer et al. 2009; Sengupta et al. 2008; Strillacci et al. 2009; Su et al. 2009; Subramanian et al. 2008; Szafranska et al. 2007; Taulli et al. 2009; Tetzlaff et al. 2007; Tong et al. 2009; Toyota et al. 2008; Varambally et al. 2008; Veerla et al. 2009; Ventura et al. 2008; Veronese et al. 2010; Voorhoeve et al. 2006; Wang et al. 2007; 2008c; Weber et al. 2006; Welch et al. 2007; Wong et al. 2008a, b; Xiao et al. 2008; Yanaihara et al. 2006; Yu et al. 2007a; Zanette et al. 2007; Zenz et al. 2009).

In 2010, Volinia and colleagues (2010) reported the identification of the most significantly deregulated miRNAs in cancer from the analysis of 2,532 cancer samples (31 types of solid cancers) versus 806 corresponding normal samples. The study revealed the miRNAs whose deregulation is shared among several types of neoplasms and, of course, did not reveal tumor-specific deregulated miRNAs. With the exception of some miRNAs, this study essentially confirmed the results of earlier studies and pointed the attention to some critical miRNAs functioning as hubs, focal points of connections for multiple molecular pathways.

Cumulatively, about 120 up-regulated miRNAs and about 120 down-regulated miRNAs were evidenced in at least one of the above reports. Overall, these studies proved that each neoplasm exhibits a distinct miRNA signature that differs from that of the other neoplasms and of each normal tissue counterpart. Moreover, they established that several miRNAs are recurrently deregulated in human cancer (Tables 1.1 and 1.2) and, in most cases, deregulation was consistently univocal, namely up-regulation or down-regulation, suggesting their involvement in tumorigenic processes shared among the different types of cancers, although examples of miRNAs deregulated in specific neoplasms also exist. A paradigmatic example of the latter type is *miR-122*, a liver specific miRNA, which is down-regulated in hepatocellular carcinoma only (Girard et al. 2008; Gramantieri et al. 2007; Kutay et al. 2006; Lin et al. 2008; Sarasin-Filipowicz et al. 2009).

There are also examples of miRNAs, whose deregulation is not univocal in all cancers (Ferdin et al. 2010). Although the reason for it is not fully clear at this stage, it seems possible that miRNA function may vary because of tissue specific expression of miRNAs and their targets. Hence, if the networks of miRNAs:targets interactions vary in the different tissues, it is plausible that miRNAs can act as either oncogenes or tumor suppressors, depending on the context.

A database (smirnaDB) containing expression information for human, mouse, rat, zebrafish, worm and fruitfly small RNAs (mostly miRNAs), with the possibility of visualizing cluster analysis and downloading normalized data is available at http://www.mirz.unibas.ch/cloningprofiles/. Primary data of published reports can be downloaded from public databases, such as GEO (http://www.ncbi.nlm.nih.gov/geo/) or ArrayExpress (http://www.ebi.ac.uk/microarray-as/ae/). Together with functional studies that have connected deregulated miRNAs to human cancer, it is now clear that the expression profiling work has indeed identified miRNAs relevant in cancer pathogenesis.

1.4 MiRNAs and Their Targets

The molecular function of mature miRNAs is to guide the protein complex RISC (RNA-induced silencing complex) toward regions of partial complementarity, usually in the 3′UTR of target mRNAs, and triggers either RNA degradation or inhibition of translation depending on the degree of complementarity between the miRNA and its target (Tang 2005). Hence, the consistent deregulation of several miRNAs in cancer will also affect expression of their gene targets: identification and validation of targets of the deregulated miRNAs is essential to understand the molecular and biological functions that are influenced as the function of each miRNA is dictated by its own targets.

Table 1.1 Most commonly down-regulated microRNAs in cancer

MicroRNA family	MicroRNA	Rank[a]	MicroRNA cluster	Genome coordinates	Host gene	Associated cancer[b]
let-7	*let-7a-1*	28	*let-7a-1/let-7f-1/let-7d*	9: 96938239-96938318 [+]	Intergenic	Ovary, prostate, lung, liver ca
	let-7d	5	*let-7a-1/let-7f-1/let-7d*	9: 96941116-96941202 [+]	Intergenic	Breast, lung, ovary, prostate, colon, liver ca, hematopoietic malignancies
	let-7f-1	31	*let-7a-1/let-7f-1/let-7d*	9: 96938629-96938715 [+]	Intergenic	Breast, liver, thyroid ca
	let-7a-3	29	*let-7a-3/let-7b*	22: 46508629-46508702 [+]	Sense MIRLET7B-201, exon 5	Breast, lung, prostate, liver ca
	let-7b	30	*let-7a-3/let-7b*	22: 46509566-46509648 [+]	Sense MIRLET7B-201, exon 5	Ovary, prostate, liver ca
	let-7f-2	31	*let-7f-2/miR-98*	X: 53584153-53584235 [–]	Sense HUWE1, intron	Breast, liver, thyroid ca
	let-7a-2	20	*miR-125b-1/let7a-2/ miR-100*	11: 122017230-122017301 [–]	Intergenic	Breast, lung, prostate, liver, lung ca
	let-7c	21	*miR-99a/let-7c/miR-125b-2*	21: 17912148-17912231 [+]	Sense, C21 or f34, intron	Ovary, prostate, liver ca
miR-101	*miR-101*	22	–	1: 65524117-65524191 [–]	Intergenic	Breast, ovary, prostate, colon, liver, lung ca, hematopoietic malignancies
miR-122	*miR-122*	23	–	18: 56118306-56118390 [+]	Intergenic	Liver ca
miR-125	*miR-125b-1*	3	*miR-125b-1/let7a-2/ miR-100*	11: 121970465-121970552 [–]	Intergenic	Breast, ovary, prostate, tongue, liver ca
	miR-125b-2	8	*miR-99a/let-7c/miR-125b-2*	21: 17962557-17962645 [+]	Sense C21 or f34, intron	Breast, ovary, prostate, tongue ca
	miR-125a	6	*miR-99b/let-7e/miR-125a*	19: 52196507-52196592 [+]	Intergenic	Breast, ovary, prostate, liver, lung ca
miR-126	*miR-126*	11	–	9: 139565054-139565138 [+]	Sense EGFL7, intron	Breast, ovary, liver, lung ca

Table 1.1 (continued)

MicroRNA family	MicroRNA	Rank[a]	MicroRNA cluster	Genome coordinates	Host gene	Associated cancer[b]
miR-128	*miR-128a*	12	–	2: 136422967-136423048 [+]	Sense R3HDM1. intron	Multiple, breast, prostate, colon ca, pituitary tumor
miR-130	*miR-130a*	34	–	11: 57408671-57408759 [+]	Intergenic	Breast, liver, lung ca
miR-132	*miR-212*	41	*miR-212/miR-132*	17: 1953565-1953674 [–]	Intergenic	Stomach, colon, pancreas ca
miR-133	*miR-133a*	13	*miR-1-2/miR-133a-1*	18: 19405659-19405746 [–]	Antisense MIB1, intron	Multiple, ovary, tongue ca
miR-136	*miR-136*	35	*miR-665/miR-441/miR-433/miR-127/miR-432/miR-136*	14: 101351039-101351120 [+]	Intergenic	Stomach, colon, liver ca
miR-138	*miR-138-2*	14	–	16: 56892430-56892513 [+]	Intergenic	Tongue, stomach, colon, pancreas, thyroid ca
miR-139	*miR-139*	36	–	11: 72326107-72326174 [–]	Sense, PDE2A, intron	Multiple, tongue, liver ca
miR-140	*miR-140*	37	–	16: 69966984-69967083 [+]	Sense WWP2, intron	Ovary, lung, thyroid ca
miR-143	*miR-143*	2	*miR-143/miR-145*	5: 148808481-148808586 [+]	Intergenic	Multiple, breast, ovary, prostate, colon, liver, lung ca
miR-145	*miR-145*	1	*miR-143/miR-145*	5: 148810209-148810296 [+]	Intergenic	Multiple, breast, ovary, prostate, colon, liver, lung ca
miR-15	*miR-195*	15	*miR-497/miR-195*	17: 6920934-6921020 [–]	Sense AC027763, intron	Prostate, tongue, liver ca
miR-181	*miR-181c*	39	*miR-181c/miR-181d*	19: 13985513-13985622 [+]	Intergenic	Liver, lung ca, glioblastoma
miR-199	*miR-199a-1-5p*	4	–	19: 10928102-10928172 [–]	Antisense DNM2, intron	Ovary, prostate, colon, liver, lung ca

Table 1.1 (continued)

MicroRNA family	MicroRNA	Rank[a]	MicroRNA cluster	Genome coordinates	Host gene	Associated cancer[b]
	miR-199a-2-5p	16	*miR-199a-2/miR-3124/miR-214*	1: 172113675-172113784 [–]	Antisense DNM3, intron	Ovary, tongue, liver ca
	miR-199b	24	*miR-3154/miR-199b*	9: 131007000-131007109 [–]	Antisense DNM1, intron	Ovary, liver, lung ca
miR-203	*miR-203*	26	–	14: 104583742-104583851 [+]	Intergenic	Multiple
miR-204	*miR-204*	17	–	9: 73424891-73425000 [–]	Sense TRPM3, intron	Multiple, breast, lung ca
miR-205	*miR-205*	40	–	1: 209605478-209605587 [+]	Sense CTA-55I10, exon	Multiple
miR-214	*miR-214*	9	*miR-199a-2/miR-3124/miR-214*	1: 172107938-172108047 [–]	Antisense DNM3, intron	Multiple, ovary, liver ca
miR-218	*miR-218-2*	10	–	5: 168195151-168195260 [–]	Sense SLIT3, intron	Multiple, prostate, stomach, colon, lung ca
miR-219	*miR-219-1*	42	–	6: 33175612-33175721 [+]	Intergenic	Tongue, lung, thyroid ca
miR-223	*miR-223*	43	–	X: 65238712-65238821 [+]	Intergenic	Liver ca
miR-26	*miR-26a*	18	–	3: 38010895-38010971 [+]	Sense CTDSPL, intron	Ovary, prostate, liver, lung, thyroid ca
miR-8	*miR-200b*	25	*miR-200b/miR-200a/miR-429*	1: 1102484-1102578 [+]	Intergenic	Ovary, liver ca
miR-9	*miR-9-1*	19	–	1: 156390133-156390221 [–]	Sense C1 or f61, intron	Multiple, ovary, lung, thyroid ca
	miR-9-2	19	–	5: 87962671-87962757 [–]	Intergenic	Ovary, lung, thyroid ca
	miR-9-3	19	–	15: 89911248-89911337 [+]	Intergenic	Ovary, lung, thyroid ca
miR-99	*miR-100*	7	*miR-125b-1/let7a-2/miR-100*	11: 122022937-122023016 [–]	Intergenic	Ovary, prostate, tongue ca
	miR-99a	27	*miR-99a/let-7c/miR-125b-2*	21: 17911409-17911489 [+]	Sense C21 or f34, intron	Ovary, prostate, tongue ca

[a]Ranking score has been empirically determined on the basis of the number of reports combined with the numbers of tumors reported to be associated with a specific miRNA deregulation.

[b]The list derives from results reported in papers indicated in the text; the term "multiple" refers to the association with multiple tumor types, reported in the paper from Volinia et al. (2010).

Table 1.2 Most commonly up-regulated microRNAs in cancer

MicroRNA family	MicroRNA	Rank[a]	MicroRNA cluster	Genome coordinates	Host gene	Associated cancer[b]
miR-1	*miR-206*	45	*miR-206/miR-133b*	6: 52009147-52009232 [+]	Intergenic	Breast, prostate ca
miR-103	*miR-107*	13	–	10: 91352504-91352584 [–]	Sense PANK1, intron	Multiple, breast, stomach, colon, pancreas ca
	miR-103-1	9	*miR-103-1as/miR-103-1*	5: 167987901-167987978 [–]	Sense PANK3, intron	Multiple, breast, stomach, colon, pancreas, bladder ca
	miR-103-2	17	*miR-103-2/miR-103-2as*	20: 3898141-3898218 [+]	Sense PANK2, intron	Multiple, prostate, colon ca
miR-125	*miR-125b-1*	28	*miR-125b-1/let7a-2/miR-100*	11: 121970465-121970552 [–]	Intergenic	Stomach, pancreas, thyroid ca, glioblastoma
	miR-125b-2	39	*miR-99a/let-7c/miR-125b-2*	21: 17962557-17962645 [+]	Intergenic	Stomach, pancreas, thyroid ca
miR-128	*miR-128b*	29	–	3: 35785968-35786051 [+]	Sense ARPP-21, intron	Breast, pancreas, lung ca
miR-146	*miR-146a*	14	–	5: 159912359-159912457 [+]	Intergenic	Multiple, prostate, colon, lung, thyroid ca
miR-155	*miR-155*	3	–	21: 26946292-26946356 [+]	Sense MIRHG2, exon 4	Breast, colon, tongue, lung, thyroid ca, lymphoma, CLL
miR-17	*miR-106a*	18	*miR-106a/miR-18b/miR-20b/miR-19b-2/miR-19a-2/miR-92a-2/miR-363*	X: 133304228-133304308 [–]	Intergenic	Multiple, colon, lung ca
	miR-93-1	25	*miR-106b/miR-25/miR-93-1*	7: 99691391-99691470 [–]	Sense MCM7, intron	Multiple, breast, ovary, prostate ca
	miR-17-5p	10	*miR-17/miR-18a/miR-19a/miR-20a/miR-19b-1/miR-92a-1*	13: 92002859-92002942 [+]	Sense MIRHG1, intron	Multiple, prostate, tongue, lung, bladder ca

Table 1.2 (continued)

MicroRNA family	MicroRNA	Rank[a]	MicroRNA cluster	Genome coordinates	Host gene	Associated cancer[b]
	miR-20a	20	*miR-17/miR-18a/miR-19a/miR-20a/miR-19b-1/miR-92a-1*	13: 92003319-92003389 [+]	Sense MIRHG1, intron	Multiple, ovary, prostate, lung ca
miR-181	*miR-181a*	40	*miR-181a-1/miR-181b-1*	1: 198828173-198828282 [–]	Sense RP11-31E23, intron	Multiple, breast, tongue, liver, thyroid ca
	miR-181b-1	4	*miR-181a-1/miR-181b-1*	1: 198828002-198828111 [–]	Sense RP11-31E23, intron	Multiple, breast, prostate, pancreas, liver, thyroid ca
miR-191	*miR-191*	5	*miR-191/miR-425*	3: 49058051-49058142 [–]	Sense DALRD3, intron	Breast, prostate, stomach, colon, pancreas, lung ca
miR-192	*miR-215*	46	*miR-194-1/miR-215*	1: 220291195-220291304 [–]	Antisense IARS2, intron	Prostate, stomach, pancreas ca
	miR-192	41	*miR-194-2/miR-192*	11: 64658609-64658718 [–]	Intergenic	Stomach, liver, lung ca
miR-197	*miR-197*	42	–	1: 110141515-110141589 [+]	Intergenic	Tongue, lung ca
miR-202	*miR-202*	44	–	10: 135061015-135061124 [–]	Antisense RP13-49I15, intron	Breast, prostate, thyroid ca
miR-205	*miR-205*	31	–	1: 209605478-209605587 [+]	Sense CTA-55I10, exon	Lung, bladder ca
miR-21	*miR-21*	1	–	17: 57918627-57918698 [+]	Intergenic	Multiple, breast, ovarian, tongue, liver, lung, colon, pancreas, prostate, stomach, thyroid ca, glioblastoma, AML, CLL
miR-210	*miR-210*	6	–	11: 568089-568198 [–]	Intergenic	Breast, prostate, liver, lung ca

Table 1.2 (continued)

MicroRNA family	MicroRNA	Rank[a]	MicroRNA cluster	Genome coordinates	Host gene	Associated cancer[b]
miR-221	*miR-221*	2	*miR-221/miR-222*	X: 45605585-45605694 [–]	Intergenic	Stomach, colon, pancreas, liver, bladder, thyroid prostate ca, glioblastoma, melanoma, CLL
	miR-222	11	*miR-221/miR-222*	X: 45606421-45606530 [–]	Intergenic	Stomach, colon, pancreas, liver, bladder, thyroid ca
miR-223	*miR-223*	33	–	X: 65238712-65238821 [+]	Intergenic	Stomach, colon, pancreas, bladder, ca
miR-224	*miR-224*	7	–	X: 151127050-151127130 [–]	Sense GABRE, intron	Pancreas, liver, thyroid ca
miR-23	*miR-23a*	22	*miR-23a/miR-27a/miR-24-2*	19: 13947401-13947473 [–]	Intergenic	Breast, ovarian, prostate, lung, bladder ca
	miR-23b	34	*miR-23b/miR-27b/miR-3074/miR-24-1*	9: 97847490-97847586 [+]	Sense C9 or f3, intron	Ovary, colon, pancreas, bladder ca
miR-24	*miR-24-2*	35	*miR-23a/miR-27a/miR-24-2*	19: 13947101-13947173 [–]	Intergenic	Stomach, pancreas, lung, thyroid ca
miR-25	*miR-92a-2*	12	*miR-106a/miR-18b/miR-20b/miR-19b-2/miR-19a-2/miR-92a-2/miR-363*	X: 133303568-133303642 [–]	Intergenic	Multiple, prostate, stomach, colon, pancreas ca
	miR-25	15	*miR-106b/miR-25/miR-93-1*	7: 99691183-99691266 [–]	Sense MCM7, intron	Multiple, prostate, stomach, pancreas ca, glioblastoma
	miR-92a-1	16	*miR-17/miR-18a/miR-19a/miR-20a/miR-19b-1/miR-92a-1*	13: 92003568-92003645 [+]	Sense MIRHG1, intron	Multiple, stomach, colon, pancreas ca
miR-26	*miR-26b*	47	–	2: 219267369-219267445 [+]	Sense CTDSP1, intron	Colon, pancreas, bladder ca

Table 1.2 (continued)

MicroRNA family	MicroRNA	Rank[a]	MicroRNA cluster	Genome coordinates	Host gene	Associated cancer[b]
miR-29	*miR-29a*	23	*miR-29b-1/miR-29a*	7: 130561506-130561569 [–]	Antisense AC016831, intron	Breast, prostate, pancreas, thyroid ca
	miR-29b-1	48	*miR-29b-1/miR-29a*	7: 130562218-130562298 [–]	Antisense AC016831, intron	Breast, prostate, thyroid ca
	miR-29b-2	49	*miR-29b-2/miR-29c*	1: 207975788-207975868 [–]	Intergenic	Prostate, pancreas, thyroid ca
	miR-29c	36	*miR-29b-2/miR-29c*	1: 207975197-207975284 [–]	Intergenic	Breast, pancreas, thyroid ca
miR-30	*miR-30c*	50	*miR-30e/miR-30c-1*	1: 41222956-41223044 [+]	Sense NFYC, intron	Multiple, prostate, pancreas ca
miR-31	*miR-31*	51	–	9: 21512114-21512184 [–]	Sense RP11-354P17, intron	Tongue, liver ca
miR-34	*miR-34a*	8	–	1: 9211727-9211836 [–]	Intergenic	Multiple, breast, tongue, liver, thyroid, ca
miR-345	*miR-345*	52	–	14: 100774196-100774293 [+]	Intergenic	Multiple
miR-373	*miR-373*	37	*miR-371/miR-372/miR-373*	19: 54291959-54292027 [+]	Intergenic	Breast, prostate, liver, germinal centre ca
miR-483	*miR-483-3p*	24	–	11: 2155364-2155439 [–]	Sense IGF2, intron	Nephroblastoma, liver, colon, breast ca
miR-8	*miR-200a*	43	*miR-200b/miR-200a/miR-429*	1: 1103243-1103332 [+]	Intergenic	Ovary, thyroid ca

AML, acute myeloid leukemia; CLL, chronic lymphocytic leukemia.

[a]Ranking score has been empirically determined on the basis of the number of reports combined with the numbers of tumors reported to be associated with a specific miRNA deregulation.

[b]The list derives from results reported in papers indicated in the text; the term "multiple" refers to the association with multiple tumor types, reported in the paper from Volinia et al. (2010).

In animal cells, post-transcriptional regulation by miRNA requires a mRNA sequence that is perfectly complementary to the "seed sequence" (positions 2–7 of the mature miRNA). Various online algorithms (http://www.microrna.org/; http://www.targetscan.org/; http://pictar.bio.nyu.edu/) have been developed for predicting miRNA targets interactions. These methods are not only based on the presence of nucleotides exhibiting homology with "seed sequence" on the 3′UTR of potential targets, but also on phylogenetic conservation of these target sites. Several additional algorithms have also been developed, which may be used to improve and refine target prediction (Bandyopadhyay and Mitra 2009; Barbato et al. 2009; Hausser et al. 2009; Kim et al. 2006; Kruger and Rehmsmeier 2006; Maragkakis et al. 2009; Ragan et al. 2009; Ruan et al. 2008; Wang and Li 2009; Wang 2008; Wang and El Naqa 2008). The performance of miRNA target prediction programs was recently assessed by Alexiou et al. (2009) by using some available proteomic and gene expression datasets. Based on predictive algorithms, it can be concluded that each miRNA may potentially regulate hundreds of target mRNAs (Lewis et al. 2005) and it seems plausible that most, if not all, mRNAs are post-transcriptionally regulated by miRNAs.

The demonstration that several gene targets are indeed modulated by a single miRNA has been experimentally proven either by large scale proteomic studies (Baek et al. 2008; Selbach et al. 2008) as well as gene expression investigations (Lim et al. 2005; Selbach et al. 2008). These studies have shown that a single miRNA can directly down-regulate the production of hundreds of proteins either by repression of translation and mRNA degradation. They also confirmed that the seed sequence in the 3′UTR of target mRNA as the main motive of miRNA-mediated regulation of protein production. The seed correlated with both mRNA degradation and translational repression. These studies indicated that both proteomic and transcriptomic studies, together with the more traditional western blot analyses and luciferase assays, may be used to experimentally validate predicted targets of each miRNA.

A list of validated targets is available on the Tarbase online database (Papadopoulos et al. 2009; Sethupathy et al. 2006). Gene targets modulated by commonly deregulated miRNAs in cancer are shown in Table 1.3. Information in Table 1.3 demonstrates that cancer miRNAs can alter the expression of oncogenic and tumor suppressor proteins. In this manner, cancer-associated biological pathways become unbalanced: both early events in tumorigenesis, such as abnormal cell growth, cell death and differentiation, and late events, such as angiogenesis, invasion of surrounding tissues and metastasis to distant sites (Fig. 1.3) have been in fact associated with aberrant miRNA expression. Various reviews have summarized these molecular and biological connections (Cho 2010a; Gramantieri et al. 2008).

1.5 MiRNA Expression Profiles as Cancer Classifiers

In addition to the acquisition of new elements for understanding the molecular basis of human cancer, the interest in studying miRNA profiles attracted several investigators for two additional reasons: first, the microRNome comprises a

Table 1.3 Gene targets of commonly deregulated MicroRNAs in cancer

MicroRNA family	Cluster	MicroRNA	MicroRNA expression in cancer	Validated target (gene symbol)	Validated target (gene name)	Association with cancer or other biological processes	References
let-7			Down	STAT-3	Signal transducer and activator of transcription 3	Cholangiocarcinoma	Meng et al. (2007b)
			Down	DCLK1	Doublecortin-like kinase 1, DCAMKL-1	Colon ca	Sureban et al. (2009)
			Down	BCL2L1	BCL2-like 1/BCL-XL	Hepatocellular ca	Shimizu et al. (2010)
			Down	COL1A2	Collagen, type I, alpha 2	Hepatocellular ca	Ji et al. (2010a)
			Down	KRAS	v-Ki-ras2 Kirsten rat sarcoma viral oncogene homolog	Lung ca	Johnson et al. (2005)
			Down	NRAS	Neuroblastoma RAS viral (v-ras) oncogene homolog	Lung ca	Johnson et al. (2005)
			Down	MYC	v-myc myelocytomatosis viral oncogene homolog	Lymphoma	Sampson et al. (2007)
			Down	PRDM1	PR domain containing 1, with ZNF domain, Blimp-1	Lymphoma	Nie et al. (2008)
	let-7a-1/let-7f-1/let-7d	*let-7a-1*	Down	CCNA	Cyclins A	Melanoma	Schultz et al. (2008)
	miR-125b-1/let7a-2/miR-100	*let-7a-2*	Down	CCND1	Cyclins D1	Melanoma	Schultz et al. (2008)
	let-7a-3/let-7b	*let-7a-3*	Down	CCND3	Cyclins D3	Melanoma	Schultz et al. (2008)

Table 1.3 (continued)

MicroRNA family	Cluster	MicroRNA	MicroRNA expression in cancer	Validated target (gene symbol)	Validated target (gene name)	Association with cancer or other biological processes	References
	let-7a-3/let-7b	*let-7b*	Down	CDK4	Cyclin-dependent kinase 4	Melanoma	Schultz et al. (2008)
	miR-99a/let-7c/miR-125b-2	*let-7c*	Down	ITGB3	Integrin beta 3	Melanoma	Muller and Bosserhoff (2008)
	let-7a-1/let-7f-1/let-7d	*let-7d*	Down	HMGA2	High mobility group AT-hook 2	Oncogenic transformation; early cancer; leiomyoma, embryonic development	Boyerinas et al. (2008); Lee and Dutta (2007); Mayr et al. (2007); Park et al. (2007); Peng et al. (2008)
	let-7a-1/let-7f-1/let-7d	*let-7f-1*	Down	KLK10	Kallikrein 10	Ovarian ca	Chow et al. (2008); White et al. (2010)
	let-7f-2/miR-98	*let-7f-2*	Down	CCND2	Cyclins D2	Prostate ca	Dong et al. (2010)
	–	*let-7g*	Down	E2F2	E2F transcription factor 2	Prostate ca	Dong et al. (2010)
				Cdc34	Cell division cycle 34 homolog	Cell cycle control	Legesse-Miller et al. (2009)
				CASP3	Caspase-3	Apoptosis	Tsang and Kwok (2008)
				DICER1	Dicer	Auto regulatory loop	Forman et al. (2008); Tokumaru et al. (2008)
				IGF2BP1	Insulin-like growth factor 2 mRNA binding protein 1, IMP1	Embryonic development	Boyerinas et al. (2008)
				Trim71	Tripartite motif-containing 71, LIN41	Embryonic survival and neural tube closure	Maller Schulman et al. (2008)

Table 1.3 (continued)

MicroRNA family	Cluster	MicroRNA	MicroRNA expression in cancer	Validated target (gene symbol)	Validated target (gene name)	Association with cancer or other biological processes	References
				TLR4	Toll-like receptor 4	Immune response	Chen et al. (2007)
				HNF1A	Hepatic nuclear factor 4 alpha	Mesenchymal stem cells	Koh et al. (2010)
				NR2E1	Nuclear receptor subfamily 2, group E, member 1, TLX	Neural stem cell proliferation and differentiation	Zhao et al. (2010a)
				NKIRAS2	NFKB inhibitor interacting Ras-like 2	NF-kappaB signaling	Murphy et al. (2010)
				KLK6	Kallikrein 6	Regulatory mechanism	Chow et al. (2008)
miR-8	*miR-200b/miR-200a/miR-429*	*miR-200a*	Down	ZEB1	Zinc finger E-box binding homeobox 1	EMT phenotype, control of CDH1 expression	Burk et al. (2008); Gregory et al. (2008); Korpal et al. (2008); Park et al. (2008)
			Down	ZEB2	Zinc finger E-box binding homeobox 2	EMT phenotype, control of CDH1 expression	Bracken et al. (2008); Christoffersen et al. (2007); Xia et al. (2010)
		miR-200b	Down	ZEB1	Zinc finger E-box binding homeobox 1	EMT phenotype, control of CDH1 expression	Bracken et al. (2008); Hurteau et al. (2007)
			Down	ZEB2	Zinc finger E-box binding homeobox 2	EMT phenotype, control of CDH1 expression	Gregory et al. (2008); Korpal et al. (2008); Park et al. (2008)
				CTNNB1	Beta catenin		Xia et al. (2010)
miR-9	–	*miR-9-1/2/3*	Down	NFKB1	Nuclear factor of kappa light polypeptide gene enhancer in B cells 1, NF-kappa-B	Gastric ca	Wan et al. (2010)

Table 1.3 (continued)

MicroRNA family	Cluster	MicroRNA	MicroRNA expression in cancer	Validated target (gene symbol)	Validated target (gene name)	Association with cancer or other biological processes	References
			Down	PRDM1	PR domain containing 1, with ZNF domain, Blimp-1	Lymphoma	Nie et al. (2008)
			Down	CDH1	E-cadherin	Metastasis	Ma et al. (2010)
			Down	CDH1	E-cadherin	Hepatocellular ca	
				NR2E1	Nuclear receptor subfamily 2, group E, member 1, TLX	Stem cell fate determination	Zhao et al. (2009)
miR-17	*miR-17/miR-18a/miR-19a/miR-20a/miR-19b-1/miR-92a-1*	*miR-17-5p*	Down	CCND1	Cyclin D1	Breast ca	Yu et al. (2008b)
			Up	NCOA3	Nuclear receptor coactivator 3, AIB1	Breast ca	Hossain et al. (2006)
			Up	E2f1	E2F transcription factor 1	Lymphoma	O'Donnell et al. (2005)
			Up	BCL2L11	BCL2-like 11, BIM	Lymphoma, Raji	Inomata et al. (2009)
			Up	CDKN1A	Cyclin dependent kinase inhibitor 1A, p21	Lymphoma, SUDHL4	Inomata et al. (2009)
				RBL2	Retinoblastoma-like 2 (p130)	Adipocyte differentiation	Wang et al. (2008b)
				Mapk14	Mitogen-activated protein kinase 14	Lung epithelial branching morphogenesis	Carraro et al. (2009)

Table 1.3 (continued)

MicroRNA family	Cluster	MicroRNA	MicroRNA expression in cancer	Validated target (gene symbol)	Validated target (gene name)	Association with cancer or other biological processes	References
				Stat3	Signal transducer and activator of transcription 3	Lung epithelial branching morphogenesis	Carraro et al. (2009)
				CCL1	Chemokine (C-C motif) ligand 1	miRNA-targetome method	Tan et al. (2009)
				DNAJC27	DnaJ (Hsp40) homolog, subfamily C, member 27 (RBJ)	miRNA-targetome method	Tan et al. (2009)
				FBXO31	F-box protein 31	miRNA-targetome method	Tan et al. (2009)
				GPR137B	G protein-coUpled receptor 137B	miRNA-targetome method	Tan et al. (2009)
				NPAT	Nuclear protein, ataxia-telangiectasia locus	miRNA-targetome method	Tan et al. (2009)
				OBFC2A	Oligonucleotide/oligosaccharide-binding fold containing 2A	miRNA-targetome method	Tan et al. (2009)
				RAB12	RAB12, member RAS oncogene family	miRNA-targetome method	Tan et al. (2009)
				YES1	v-yes-1 Yamaguchi sarcoma viral oncogene homolog 1	miRNA-targetome method	Tan et al. (2009)
				ZNFX1	Zinc finger, NFX1-type containing 1	miRNA-targetome method	Tan et al. (2009)
				FN1	Fibronectin	Transgenic *miR-17*	Shan et al. (2009)

Table 1.3 (continued)

MicroRNA family	Cluster	MicroRNA	MicroRNA expression in cancer	Validated target (gene symbol)	Validated target (gene name)	Association with cancer or other biological processes	References
				FNDC3A	Fibronectin type-III domain containing 3A	Transgenic *miR-17*	Shan et al. (2009)
		miR-18a	Up	ESR1	Estrogen receptor 1 (alpha)	Hepatocellular ca	Liu et al. (2009b)
			Up	ESR1	Estrogen receptor 1 (alpha)	Neuroblastoma	Loven et al. (2010)
				CTGF	Connective tissue growth factor, CCN2	Chondrocytic phenotype	Ohgawara et al. (2009)
				NR3C1	Nuclear receptor subfamily 3, group C, member 1 (glucocorticoid receptor)		Vreugdenhil et al. (2009)
		miR-20a	Down	CCND1	Cyclin D1	Breast ca	Yu et al. (2008b)
			Up	HIF1A	Hypoxia-inducible factor-1, alpha subunit (basic helix-loop-helix transcription factor)	Lung ca	Taguchi et al. (2008)
			Up	Stat3	Signal transducer and activator of transcription 3	Lung epithelial branching morphogenesis	Carraro et al. (2009)
			Up	CDKN1A	Cyclin dependent kinase inhibitor 1A, p21	Lymphoma	Inomata et al. (2009)
			Up	E2f1	E2F transcription factor 1	Lymphoma	O'Donnell et al. (2005)

Table 1.3 (continued)

MicroRNA family	Cluster	MicroRNA	MicroRNA expression in cancer	Validated target (gene symbol)	Validated target (gene name)	Association with cancer or other biological processes	References
				MAPK14	Mitogen-activated protein kinase 14	Lung epithelial branching morphogenesis	Carraro et al. (2009)
				ZBTB7A	Zinc finger and BTB domain containing 7A, LRF	Senescence in mouse embryo fibroblasts	Poliseno et al. (2008)
				E2F1	E2F transcription factor 1		Sylvestre et al. (2007)
				E2F2	E2F transcription factor 2		Sylvestre et al. (2007)
				E2F3	E2F transcription factor 3		Sylvestre et al. (2007)
	miR-106a/miR-18b/miR-20b/miR-19b-2/miR-19a-2/miR-92a-2/miR-363	*miR-20b*		HIF1A	Hypoxia-inducible factor-1, alpha subunit (basic helix-loop-helix transcription factor)	Breast ca, MCF7	Cascio et al. (2010)
				STAT3	Signal transducer and activator of transcription 3	Breast ca, MCF7	Cascio et al. (2010)
		miR-106a	Up	CDKN1A	Cyclin dependent kinase inhibitor 1A, p21	Rescue from Ras(G12V)-induced senescence	Borgdorff et al. (2010)
				IL10	Interleukin 10	Expression regulation	Sharma et al. (2009)

Table 1.3 (continued)

MicroRNA family	Cluster	MicroRNA	MicroRNA expression in cancer	Validated target (gene symbol)	Validated target (gene name)	Association with cancer or other biological processes	References
				MAPK14	Mitogen-activated protein kinase 14	Lung epithelial branching morphogenesis	Carraro et al. (2009)
				STAT3	Signal transducer and activator of transcription 3	Lung epithelial branching morphogenesis	Carraro et al. (2009)
				CDKN1A	Cyclin dependent kinase inhibitor 1A, p21		Ivanovska et al. (2008)
	miR-106b/miR-25/miR-93-1	*miR-93*	Up	BCL2L11	BCL2-like 11, BIM	Gastric ca, TGF beta resistance	Petrocca et al. (2008)
			Up	CDKN1A	Cyclin dependent kinase inhibitor 1A, p21	Gastric ca, TGF beta resistance	Petrocca et al. (2008)
		miR-106b	Up	BCL2L11	BCL2-like 11, BIM	Gastric ca, TGF beta resistance	Petrocca et al. (2008)
			Up	CDKN1A	Cyclin dependent kinase inhibitor 1A, p21	Gastric ca, TGF beta resistance	Petrocca et al. (2008)
miR-19	*miR-17/miR-18a/miR-19a/miR-20a/miR-19b-1/miR-92a-1*	*miR-19a*	Up	ESR1	Estrogen receptor 1 (alpha)	Neuroblastoma	Loven et al. (2010)
				BCL2L11	BCL2-like 11, BIM	Notch-induced T cell	Mavrakis et al. (2010)
				Ppp2r5e	Protein phosphatase 2, regulatory subunit B', epsilon isoform	Notch-induced T cell	Mavrakis et al. (2010)

Table 1.3 (continued)

MicroRNA family	Cluster	MicroRNA	MicroRNA expression in cancer	Validated target (gene symbol)	Validated target (gene name)	Association with cancer or other biological processes	References
				Prkaa1	Protein kinase, AMP-activated, alpha 1 catalytic subunit	Notch-induced T cell	Mavrakis et al. (2010)
				Pten	Phosphatase and tensin homolog	Notch-induced T cell	Mavrakis et al. (2010)
				CCND1	Cyclin D1		Qin et al. (2010)
				PTEN	Phosphatase and tensin homolog		Pezzolesi et al. (2008)
miR-21	–	*miR-21*	Up	FasL	Fas ligand	Anti-apoptosis	Sayed et al. (2010)
			Up	SERPINB5	Serpin peptidase inhibitor, clade B (ovalbumin), member 5 (maspin)	Breast ca	Zhu et al. (2008)
			Up	PDCD4	Programmed cell death 4 (neoplastic transformation inhibitor)	Breast ca, cervical ca, cholangioca, colon ca, gastric ca, glioblastoma	Asangani et al. (2008); Chen et al. (2008c); Lu et al. (2008); Motoyama et al. (2010); Selaru et al. (2009); Yao et al. (2009); Zhu et al. (2008)
			Up	TIMP3	Tissue inhibitor of metalloproteinase 3	Breast ca, cholangioca, glioblastoma	Gabriely et al. (2008); Selaru et al. (2009); Song et al. (2010)
			Up	SPRY2	Sprouty homolog 2 (Drosophila)	Cellular outgrowth	Sayed et al. (2008)

Table 1.3 (continued)

MicroRNA family	Cluster	MicroRNA	MicroRNA expression in cancer	Validated target (gene symbol)	Validated target (gene name)	Association with cancer or other biological processes	References
			Up	LRRFIP1	Leucine rich repeat (in FLII) interacting protein 1	Glioblastoma	Li et al. (2009f)
			Up	RECK	Reversion-inducing-cysteine-rich protein with kazal motifs	Glioblastoma	Gabriely et al. (2008)
			Up	PTEN	Phosphatase and tensin homolog	Hepatocellular ca, NSCLC	Meng et al. (2007a); Zhang et al. (2010)
			Up	BTG2	B cell translocation gene 2, anti-proliferative	Laryngeal ca	Liu et al. (2009a)
			Up	Peli1	Pellino homolog 1 (Drosophila)	Liver rigeneration	Marquez et al. (2010)
			Up	HNRPK	Heterogeneous nuclear ribonucleoprotein K	P53 and TGF beta paths	Papagiannakopoulos et al. (2008)
			Up	TP63	Tumor protein p63	P53 and TGF beta paths	Papagiannakopoulos et al. (2008)
			Up	MARCKS	Myristoylated alanine-rich protein kinase C substrate	Prostate ca	Li et al. (2009e)
			Up	TPM1	Tropomyosin 1		Zhu et al. (2007)
				PDCD4	Programmed cell death 4 (neoplastic transformation inhibitor)	Protection in myocardial injusry	Cheng et al. (2009)

Table 1.3 (continued)

MicroRNA family	Cluster	MicroRNA	MicroRNA expression in cancer	Validated target (gene symbol)	Validated target (gene name)	Association with cancer or other biological processes	References
miR-23	*miR-23b/miR-27b/miR-(3074)/miR-24-1*	*miR-23b*	Up	PLAU	Plasminogen activator, urokinase	Hepatocellular ca	Salvi et al. (2009)
			Up	MET	Met proto-oncogene (hepatocyte growth factor receptor)	Hepatocellular ca	Salvi et al. (2009)
				Hes1	Hairy/enhancer of split protein	Neuronal differentiation	Kimura et al. (2004)
				SMAD	Smad 3, 4, or 5	Hepatic stem cell	Rogler et al. (2009)
miR-24	*miR-23a/miR-27a/miR-24-2*	*miR-24-2*	Down	AURKB	Aurora kinase B	Up-regulated during terminal differentiation of multiple lineages, inhibits cell-cycle progression	Lal et al. (2009a)
			Down	CCNA2	Cyclin A2	Up-regulated during terminal differentiation of multiple lineages, inhibits cell-cycle progression	Lal et al. (2009a)

Table 1.3 (continued)

MicroRNA family	Cluster	MicroRNA	MicroRNA expression in cancer	Validated target (gene symbol)	Validated target (gene name)	Association with cancer or other biological processes	References
			Down	CDK1	Cyclin-dependent kinase 1, CDC2	Up-regulated during terminal differentiation of multiple lineages, inhibits cell-cycle progression	Lal et al. (2009a)
			Down	CDK4	Cyclin-dependent kinase 4	Up-regulated during terminal differentiation of multiple lineages, inhibits cell-cycle progression	Lal et al. (2009a)
			Down	E2F2	E2F transcription factor 2	Up-regulated during terminal differentiation of multiple lineages, inhibits cell-cycle progression	Lal et al. (2009a)
			Down	FEN7		Up-regulated during terminal differentiation of multiple lineages, inhibits cell-cycle progression	Lal et al. (2009a)

Table 1.3 (continued)

MicroRNA family	Cluster	MicroRNA	MicroRNA expression in cancer	Validated target (gene symbol)	Validated target (gene name)	Association with cancer or other biological processes	References
			Down	MYC	v-myc myelocytomatosis viral oncogene homolog	Up-regulated during terminal differentiation of multiple lineages, inhibits cell-cycle progression	Lal et al. (2009a)
			Up	H2AFX	H2A histone family, member X		Lal et al. (2009b)
			Up	CDKN2A	Cyclin-dependent kinase inhibitor 2A (melanoma, p16, inhibits CDK4)		Lal et al. (2008)
				ACVR1B	Activin type I receptor, ALK4	Inhibits erythropoiesis	Wang et al. (2008a)
miR-25	*miR-106b/miR-25/miR-93-1*	*miR-25*	Up	BCL2L11	BCL2-like 11, BIM	Esophageal ca	Kan et al. (2009)
			Up	CDKN1A	Cyclin dependent kinase inhibitor 1A, p21	Esophageal ca	Kan et al. (2009)
			Up	PTEN	Phosphatase and tensin homolog	Prostate ca	Poliseno et al. (2010)
	miR-17/miR-18/miR-19a/miR-20a/miR-19b-1/miR-92a-1	*miR-92a-1*	Up	TP63	Tumor protein p63		Manni et al. (2009)
			Up	VHL	von Hippel-Lindau tumor suppressor	CLL	Ghosh et al. (2009)

Table 1.3 (continued)

MicroRNA family	Cluster	MicroRNA	MicroRNA expression in cancer	Validated target (gene symbol)	Validated target (gene name)	Association with cancer or other biological processes	References
				SLC12A5	Solute carrier family 12 (potassium/chloride transporter), member 5, KCC2	Cerebellar granule neurons	Barbato et al. (2010)
	miR-106a/miR-18b/miR-20b/miR-19b-2/miR-19a-2/miR-92a-2/miR-363	*miR-92a-2*	Up	TP63	Tumor protein p63		Manni et al. (2009)
			Up	VHL	von Hippel-Lindau tumor suppressor	CLL	Ghosh et al. (2009)
				SLC12A5	Solute carrier family 12 (potassium/chloride transporter), member 5, KCC2	Cerebellar granule neurons	Barbato et al. (2010)
miR-26	–	*miR-26a*	Down	EZH2	Enhancer of zeste homolog 2	Myc regulation	Sander et al. (2008); Sander et al. (2009); Wong and Tellam (2008)
				CACNA1C	L-type voltage-gated calcium channel alpha1C subunit		Shi et al. (2009)
				PTGS2	Prostaglandin-endoperoxide synthase 2, cyclooxygenase-2		Ji et al. (2010b)

Table 1.3 (continued)

MicroRNA family	Cluster	MicroRNA	MicroRNA expression in cancer	Validated target (gene symbol)	Validated target (gene name)	Association with cancer or other biological processes	References
miR-29	*miR-29b-1/miR-29a*	*miR-29a*	Down	BCL2	B cell CLL/lymphoma 2	Hepatocellular ca	Xiong et al. (2010)
			Down	CDK6	Cyclin-dependent kinase 6	Lymphoma	Zhao et al. (2010b)
			Down	MCL1	Myeloid cell leukemia sequence 1 (BCL2-related)	Hepatocellular ca	Xiong et al. (2010)
			Up	CD276	CD276 molecule, B7-H3	Immune escape by solid tumors	Xu et al. (2009a)
				CDC42	Cell division cycle 42 (GTP binding protein, 25 kDa)		Park et al. (2009)
				ZFP36	Zinc finger protein 36, C3H type (tristetraprolin)	Epithelial-mesenchymal transition (EMT) and metastasis	Gebeshuber et al. (2009)
				PMP22	Peripheral myelin protein 22		Verrier et al. (2009)
				PIK3R1	Phosphoinositide-3-kinase, regulatory subunit 1 (p85 alpha)		Park et al. (2009)
			Down	Sp1	Sp1 transcription factor	AML	Liu et al. (2010)
				CDC42	Cell division cycle 42 (GTP binding protein, 25 kDa)		Park et al. (2009)
				Col1A1	Collagen, type I, alpha 1		Ogawa et al. (2010)

Table 1.3 (continued)

MicroRNA family	Cluster	MicroRNA	MicroRNA expression in cancer	Validated target (gene symbol)	Validated target (gene name)	Association with cancer or other biological processes	References
	miR-29b-1/miR-29a	*miR-29b1*		DNMT3A	DNA (cytosine-5-)-methyltransferase 3 alpha	de novo DNA methylation	Garzon et al. (2009); Fabbri et al. (2007)
	miR-29b-2/miR-29c	*miR-29b2*		DNMT3B	DNA (cytosine-5-)-methyltransferase 3 beta	de novo DNA methylation	Garzon et al. (2009); Fabbri et al. (2007)
				MCL1	Myeloid cell leukemia sequence 1 (BCL2-related)	Apoptosis	Mott et al. (2007)
				TCL1	T cell lymphoma breakpoint 1		Anastasiadou et al. (2010); Pekarsky et al. (2006)
				PIK3R1	Phosphoinositide-3-kinase, regulatory subunit 1 (p85 alpha)		Park et al. (2009)
	miR-29b-2/miR-29c	*miR-29c*	Down	CDK6	Cyclin-dependent kinase 6	Lymphoma	Zhao et al. (2010b)
			Down	LAMC1	Laminin gamma1	Nasopharyngeal ca	SengUpta et al. (2008)
			Down	COL...	Collagens	Nasopharyngeal ca	SengUpta et al. (2008)
				CDC42	Cell division cycle 42 (GTP binding protein, 25 kDa)		Park et al. (2009)
				PIK3R1	Phosphoinositide-3-kinase, regulatory subunit 1 (p85 alpha)		Park et al. (2009)

Table 1.3 (continued)

MicroRNA family	Cluster	MicroRNA	MicroRNA expression in cancer	Validated target (gene symbol)	Validated target (gene name)	Association with cancer or other biological processes	References
miR-99	*miR-125b-1/let7a-2/miR-100*	*miR-100*	Down	MTOR	Mechanistic target of rapamycin (serine/threonine kinase)	Ovarian ca	Nagaraja et al. (2010)
miR-101	–	*miR-101*	Down	Plk1	Polo-like kinase 1	Nasopharingeal ca	Shi et al. (2010)
			Down	COX-2	Cyclooxygenase-2	Colon ca	Strillacci et al. (2009)
			Down	FOS	FBJ osteosarcoma oncogene	Hepatocellular ca	Li et al. (2009d)
				ATXN1	Ataxin 1	SCA1 pathogenesis	Lee et al. (2008b)
				EZH2	Enhancer of zeste homolog 2	Modulates the cancer epigenome	Friedman et al. (2009); Varambally et al. (2008)
miR-103	–	*miR-107*	Down	ARNT	Aryl hydrocarbon receptor nuclear translocator, alias HIF1 beta	Mediate p53 regulation of hypoxic signaling	Yamakuchi et al. (2010)
			Down	CDK6	Cyclin-dependent kinase 6	Pancreatic ca	Lee et al. (2009b)
			Down	PLAG1	Pleiomorphic adenoma gene 1	CLL	Pallasch et al. (2009)
miR-122	–	*miR-122*	Down	BCL2L2	Bcl2-like 2, BCL-W	Hepatocellular ca	Lin et al. (2008)
			Down	CCNG1	Cyclin G1	Hepatocellular ca	Gramantieri et al. (2007)
				CAT1	Cationic amino acid transporter		Chang et al. (2004)
miR-125	*miR-99b/let-7e/miR-125a*	*miR-125a*	Down	BMF	Bcl2 modifying factor	Glioma	Xia et al. (2009b)

Table 1.3 (continued)

MicroRNA family	Cluster	MicroRNA	MicroRNA expression in cancer	Validated target (gene symbol)	Validated target (gene name)	Association with cancer or other biological processes	References
			Down	ERBB2	v-erb-b2 erythroblastic leukemia viral oncogene homolog 2	Breast ca	Scott et al. (2007)
			Down	ERBB3	v-erb-b2 erythroblastic leukemia viral oncogene homolog 3	Breast ca	Scott et al. (2007)
	miR-125b-1/let7a-2/miR-100	*miR-125b-1*		BMPR1B	Bone morphogenetic protein receptor, type 1B	Target site variant in breast ca	Saetrom et al. (2009)
	miR-99a/let-7c/miR-125b-2	*miR-125b-2*	Down	ERBB2	v-erb-b2 erythroblastic leukemia viral oncogene homolog 2	Breast ca	Scott et al. (2007)
			Down	ERBB3	v-erb-b2 erythroblastic leukemia viral oncogene homolog 3	Breast ca	Scott et al. (2007)
miR-126	–	*miR-126*	Down	EGFL7	EGF-like-domain, multiple 7	NSCLC	Sun et al. (2010)
			Down	IRS1	Insulin receptor substrate 1	Breast ca	Zhang et al. (2008)
				HOXA9	Homeobox A9	Differentiation	Shen et al. (2008)
miR-128	–		Down	AFF1	AF4/FMR2 family, member 1	AML	Kotani et al. (2009)
			Down	MLL	Myeloid/lymphoid or mixed-lineage leukemia (trithorax homolog, Drosophila)	AML	Kotani et al. (2009)

Table 1.3 (continued)

MicroRNA family	Cluster	MicroRNA	MicroRNA expression in cancer	Validated target (gene symbol)	Validated target (gene name)	Association with cancer or other biological processes	References
			Down	Bmi1	Bmi1 polycomb ring finger oncogene	Glioma	Godlewski et al. (2008)
		miR-128-1	Down	E2F3	E2F transcription factor 3	Glioma	Zhang et al. (2009c)
		miR-128-2	Down	DCX	Doublecortin	Neuroblastoma	Evangelisti et al. (2009)
			Down	RELN	Reelin	Neuroblastoma	Evangelisti et al. (2009)
			Down	EGFR	Epidermal growth factor receptor	Non small cell lung ca	Weiss et al. (2008)
			Up	TGFBR1	Transforming growth factor, beta receptor 1	Resistance to endocrine therapy agents in breast ca	Masri et al. (2010)
				ADORA2B	Adenosine A2b receptor, A2BAR		Kolachala et al. (2010)
miR-133	*miR-1-2/miR-133a-1*	*miR-133a*	Down	FSCN1	Fascin homolog 1, actin-bundling protein	Bladder ca; esophageal ca	Chiyomaru et al. (2010); Kano et al. (2010)
			Down	BCL2L2	Bcl2-like 2, BCL-W	Lung ca	Crawford et al. (2009)
			Down	MCL1	Myeloid cell leukemia sequence 1 (BCL2-related)	Lung ca	Crawford et al. (2009)
				CASP9	Caspase 9	Heart physiology	Xu et al. (2007)
miR-143	*miR-143/miR-145*	*miR-143*	Down	DNMT3A	DNA methyltransferases 3A	Colon ca	Ng et al. (2009b)
			Down	KRAS	v-Ki-ras2 Kirsten rat sarcoma viral oncogene homolog	Colon ca	Chen et al. (2009)
			Down	FNDC3B	Fibronectin type III domain containing 3B	Hepatocellular ca	Zhang et al. (2009b)

Table 1.3 (continued)

MicroRNA family	Cluster	MicroRNA	MicroRNA expression in cancer	Validated target (gene symbol)	Validated target (gene name)	Association with cancer or other biological processes	References
				MAPK7	Mitogen-activated protein kinase 7, ERK5	Adypocyte differentiation	Esau et al. (2004)
				MAPK7	Mitogen-activated protein kinase 7, ERK5	Fas-mediated apoptosis	Akao et al. (2009)
				ACE	Angiotensin-converting enzyme	Smooth muscle phenotype	Boettger et al. (2009)
miR-145	*miR-143/miR-145*	*miR-145*	Down	FSCN1	Fascin homolog 1, actin-bundling protein	Bladder ca	Chiyomaru et al. (2010)
			Down	ESR1	Estrogen receptor 1, alpha	Breast ca	Spizzo et al. (2010)
			Down	MUC1	Mucin 1	Breast ca	Sachdeva and Mo (2010)
			Down	RTKN	Rhotekin	Breast ca	Wang et al. (2009)
			Down	IRS1	Insulin receptor substrate 1	Colon ca	Shi et al. (2007)
			Down	STAT1	Signal transducer and activator of transcription 1	Colon ca	Gregersen et al. (2010)
			Down	YES	v-yes-1 Yamaguchi sarcoma viral oncogene homolog 1	Colon ca	Gregersen et al. (2010)
			Down	FSCN1	Fascin homolog 1, actin-bundling protein	Esophageal squamous cell ca	Kano et al. (2010)

Table 1.3 (continued)

MicroRNA family	Cluster	MicroRNA	MicroRNA expression in cancer	Validated target (gene symbol)	Validated target (gene name)	Association with cancer or other biological processes	References
			Down	MYC	v-myc myelocytomatosis viral oncogene homolog	P53 path	Sachdeva et al. (2009)
			Down	KLF4	KrUppel-like factor 4	Stemness	Xu et al. (2009b)
			Down	POU5F1	POU class 5 homeobox 1, OCT4	Stemness	Xu et al. (2009b)
			Down	SOX2	SRY (sex determining region Y)-box 2	Stemness	Xu et al. (2009b)
				ELK1	ELK1, member of ETS oncogene family	Smooth muscle cell fate	Cordes et al. (2009)
				KLF4	KrUppel-like factor 4	Smooth muscle cell fate	Cordes et al. (2009)
				MYOCD	Myocardin	Smooth muscle cell fate	Cordes et al. (2009)
				ACE	Angiotensin-converting enzyme	Smooth muscle phenotype	Boettger et al. (2009)
miR-146	–	*miR-146a*	Down	MMP16	Matrix metallopeptidase 16 (membrane-inserted)	Glioblastoma	Xia et al. (2009a)
			Up	CXCR4	Chemokine (C-X-C motif) receptor 4	Kaposi's sarcoma	Punj et al. (2010)
				Numb	Numb homolog (Drosophila)	Proliferation and differentiation of muscle stem cells	Kuang et al. (2009)
				CCL8	Chemokine (C-C motif) ligand 8, MCP-2		Rom et al. (2010)
miR-155	–	*miR-155*	Up	FOXO3	Forkhead box O3	Breast ca	Kong et al. (2010)

Table 1.3 (continued)

MicroRNA family	Cluster	MicroRNA	MicroRNA expression in cancer	Validated target (gene symbol)	Validated target (gene name)	Association with cancer or other biological processes	References
			Up	SOCS1	Suppressor of cytokine signaling 1	Breast ca	Jiang et al. (2010)
			Up	MLH1	mutL homolog 1, colon cancer, nonpolyposis type 2	Colon ca, inactivation of mismatch repair (MMR)	Valeri et al. (2010)
			Up	MSH2	mutS homolog 2, colon cancer, nonpolyposis type 1	Colon ca, inactivation of mismatch repair (MMR)	Valeri et al. (2010)
			Up	MSH6	mutS homolog 6	Colon ca, inactivation of mismatch repair (MMR)	Valeri et al. (2010)
			Up	CEBPB	CCAAT/enhancer binding protein (C/EBP), beta	Leukemia, regulators of the interleukin-6 signaling	Costinean et al. (2009)
			Up	INPP5D	Inositol polyphosphate-5-phosphatase, 145 kDa, SHIP	Leukemia, regulators of the interleukin-6 signaling	Costinean et al. (2009)
			Up	SMAD5	SMAD family member 5	Lymphoma, escape TGF-beta's growth-inhibitory effects	Rai et al. (2010)
				INPP5D	Inositol polyphosphate-5-phosphatase, 145 kDa, SHIP	Immune cell development	O'Connell et al. (2009)
				JARID2	Jumonji, AT rich interactive domain 2	Cell survival	Bolisetty et al. (2009)

Table 1.3 (continued)

MicroRNA family	Cluster	MicroRNA	MicroRNA expression in cancer	Validated target (gene symbol)	Validated target (gene name)	Association with cancer or other biological processes	References
				RHOA	Ras homolog gene family, member A	Epithelial cell plasticity	Kong et al. (2008)
				AID	Activation-induced cytidine deaminase	Ig somatic hypermutation and class-switch recombination	Dorsett et al. (2008); Teng et al. (2008)
				Ets-1	v-ets erythroblastosis virus E26 oncogene homolog 1	Megakaryopoiesis	Romania et al. (2008)
				Meis1	Meis homeobox 1	Megakaryopoiesis	Romania et al. (2008)
				MyD88	Myeloid differentiation primary response gene (88)	Negative regulation of *Helicobacter pylori*-induced inflammation	Tang et al. (2010)
				IFNGR1	Interferon gamma receptor 1 (alpha chain)	T cell differentiation	Banerjee et al. (2010)
				AGTR1	Angiotensin II receptor, type 1, AT1R		Martin et al. (2006)
				BMP	Bone Morphogenetic Protein		Yin et al. (2010)
miR-181	*miR-181a-1/miR-181b-1*	*miR-181b-1*	Down	PLAG1	Pleiomorphic adenoma gene 1	CLL	Pallasch et al. (2009)
			Down	TCL1	T cell lymphoma breakpoint 1	CLL	Pekarsky et al. (2006)
			Down	BCL2	B cell leukemia/lymphoma 2	Gastric ca	Zhu et al. (2010b)

Table 1.3 (continued)

MicroRNA family	Cluster	MicroRNA	MicroRNA expression in cancer	Validated target (gene symbol)	Validated target (gene name)	Association with cancer or other biological processes	References
			Up	CDX2	Caudal type homeobox transcription factor 2	Hepatocellular ca	Ji et al. (2009b)
			Up	GATA6	GATA binding protein 6	Hepatocellular ca	Ji et al. (2009b)
			Up	NLK	Nemo-like kinase	Hepatocellular ca	Ji et al. (2009b)
			Up	TIMP3	Tissue inhibitor of metalloprotease 3	Hepatocellular ca	Wang et al. (2010)
				AID	Activation-induced cytidine deaminase	Ig somatic hypermutation and class switch recombination	de Yebenes et al. (2008)
				HOXA11	Homeobox A11	Myoblast differentiation	Naguibneva et al. (2006)
miR-199	–	*miR-199a-1-5p*	Down	KRT7	Keratin 7	Bladder ca	Ichimi et al. (2009)
	miR-199a2/miR-(3120)/miR-214	*miR-199a-2-5p*	Down	KRT7	Keratin 7	Bladder ca	Ichimi et al. (2009)
	miR-(3154)/miR-199b	*miR-199b*	Down	SET	SET nuclear oncogene, protein phosphatase 2A inhibitor	Chriocarcinoma	Chao et al. (2010)
			Down	IKBKB	Inhibitor of kappaB kinase beta, IKK beta	Ovarian ca	Chen et al. (2008a)
			Down	MAPK1	Mitogen-activated protein kinase 1, ERK2		Kim et al. (2008)
			Down	MET	Met proto-oncogene (hepatocyte growth factor receptor)		Kim et al. (2008)

Table 1.3 (continued)

MicroRNA family	Cluster	MicroRNA	MicroRNA expression in cancer	Validated target (gene symbol)	Validated target (gene name)	Association with cancer or other biological processes	References
				HES1	Hairy and enhancer of split 1	Cancer stem cell	Garzia et al. (2009)
				Smad1	SMAD family member 1	Chondrogenesis	Lin et al. (2009)
				HIF1A	Hypoxia-inducible factor-1, alpha subunit	Hypoxia	Rane et al. (2009)
				MET	Met proto-oncogene (hepatocyte growth factor receptor)	Invasive growth	Migliore et al. (2008)
miR-203	–	*miR-203*	Down	TP63	Tumor protein p63	Epithelial differentiation, stemness repression	Lena et al. (2008); Melar-New and Laimins (2010); Yi et al. (2008)
miR-205	–	*miR-205*	Down	ERBB3	v-erb-b2 erythroblastic leukemia viral oncogene homolog 3, HER3	Breast ca	(Iorio et al. (2009))
			Down	PRKCE	Protein kinase C, epsilon	Prostate ca	Gandellini et al. (2009)
			Down	SIP1	Survival of motor neuron protein interacting protein 1	Regulation of epithelial to mesenchymal transition	Gregory et al. (2008)
			Down	ZEB1	Zinc finger E-box binding homeobox 1	Regulation of epithelial to mesenchymal transition	Gregory et al. (2008)
			Down	Lrp1	LDL receptor-related protein 1	Tumor cell migration	Song and Bu (2009)

Table 1.3 (continued)

MicroRNA family	Cluster	MicroRNA	MicroRNA expression in cancer	Validated target (gene symbol)	Validated target (gene name)	Association with cancer or other biological processes	References
			Up	PTEN	Phosphatase and tensin homolog	Stem-cell-enriched cell populations	Greene et al. (2010)
			Up	ZEB1	Zinc finger E-box binding homeobox 1	Stem-cell-enriched cell populations	Greene et al. (2010)
			Up	ZEB2	Zinc finger E-box binding homeobox 2	Stem-cell-enriched cell populations	Greene et al. (2010)
				MED1	Mediator complex subunit 1	Trophoblast	Mouillet et al. (2010)
miR-210	–	*miR-210*	Up	MNT	MAX binding protein, myc antagonist		Zhang et al. (2009d)
			Up	EFNA3	Ephrin-A3	Response to hypoxia	Fasanaro et al. (2008)
			Up	CASP8AP2	Caspase-8 associated protein 2	Stem cell survival	Kim et al. (2009)
				AcvR1b	Activin A receptor, type IB	Osteoblastic differentiation	Mizuno et al. (2009)
miR-218	–	*miR-218-2*	Down	NFKB1	Nuclear factor of kappa light polypeptide gene enhancer in B cells 1	Gastric ca	Gao et al. (2010)
			Down	ROBO1	Roundabout, axon guidance receptor, homolog 1	Gastric ca	Tie et al. (2010)
				ONECUT1	One cut homeobox 1, HNF6		Simion et al. (2010)
				ONECUT2	One cut homeobox 2, OC2		Simion et al. (2010)
miR-221	*miR-221/miR-222*		Up	BMF	Bcl2 modifying factor	Hepatocellular ca	Gramantieri et al. (2009)

Table 1.3 (continued)

MicroRNA family	Cluster	MicroRNA	MicroRNA expression in cancer	Validated target (gene symbol)	Validated target (gene name)	Association with cancer or other biological processes	References
			Up	CDKN1B	Cyclin-dependent kinase inhibitor 1B, p27kip1	Breast ca, tamoxifen resistance; CLL; glioblastoma; hepatocellular ca; myelodysplastic syndrome; prostate ca; thyroid ca	Fornari et al. (2008); Frenquelli et al. (2010); Galardi et al. (2007); Gillies and Lorimer (2007); Hussein et al. (2010); le Sage et al. (2007); Miller et al. (2008); Visone et al. (2007); Zhang et al. (2009a)
			Up	CDKN1C	Cyclin-dependent kinase inhibitor 1C, p57kip2	Hepatocellular ca	Fornari et al. (2008)
			Up	ESR1	Estrogen receptor 1 (alpha)	Breast ca	Zhao et al. (2008)
		miR-221	Up	ICAM1	Intercellular adhesion molecule 1	Interferon-gamma path	Hu et al. (2010)
		miR-222	Up	KIT	v-kit Hardy-Zuckerman 4 feline sarcoma viral oncogene homolog	Melanoma	Igoucheva and Alexeev (2009)
			Up	PTEN	Phosphatase and tensin homolog	TRAIL resistance	Garofalo et al. (2010)
			Up	TIMP3	Tissue inhibitor of metalloproteinase 3	TRAIL resistance	Garofalo et al. (2010)
			Up	MET	Met proto-oncogene (hepatocyte growth factor receptor)	Breast ca	Di Leva et al. (2010)

Table 1.3 (continued)

MicroRNA family	Cluster	MicroRNA	MicroRNA expression in cancer	Validated target (gene symbol)	Validated target (gene name)	Association with cancer or other biological processes	References
			Up	FOXO3	Forkhead box O3	Breast ca	Di Leva et al. (2010)
miR-224	–	*miR-224*	Up	CD40	CD40 molecule, TNF receptor superfamily member 5	Pancreatic ca	Mees et al. (2009)
			Up	CDC42	Cell division cycle 42 (GTP binding protein, 25 kDa)	Breast ca, inhibition of Ubc9-mediated invasion	Zhu et al. (2010a)
			Up	CXCR4	Chemokine (C-X-C motif) receptor 4	Breast ca, inhibition of Ubc9-mediated invasion	Zhu et al. (2010a)
			Up	KLK10	Kallicrein 10	Ovarian ca	White et al. (2010)
			Up	Smad4	SMAD family member 4	TFG-beta-mediated granulosa cell proliferation	Yao et al. (2010)
			Up	API5	Apoptosis inhibitor-5	Hepatocellular ca	Wang et al. (2008c)
miR-373	*miR-371/miR-372/miR-373*	*miR-373*	Up	CD44	CD44 molecule	Breast ca metastasis	Huang et al. (2008a)
			Up	LATS2	LATS, large tumor suppressor, homolog 2	Esophageal ca, testicular ca	Lee et al. (2009a); Voorhoeve et al. (2006)

AML, acute myeloid leukemia; CLL, chronic lymphocytic leukemia.

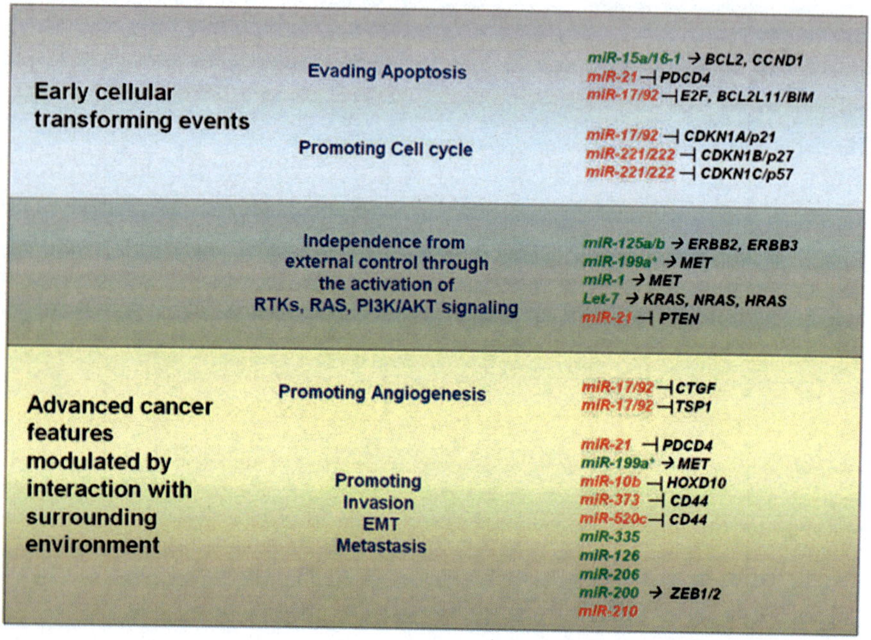

Fig. 1.3 MiRNA deregulation affects both early and late events in tumorigenesis. By upsetting oncogene and tumor suppressors, miRNA deregulation can promote both early and late cancer events. MiRNAs indicated in *red* are over-expressed and miRNAs indicated in *green* are down-regulated in human cancer

considerably smaller number of elements than the mRNome, thereby simplifying the procedures for global expression assessment and validation; second, miRNAs are more stable than mRNAs, thereby allowing the use, for instance, of RNA isolated from formalin-fixed paraffin-embedded (FFPE) samples for microarray or qRT-PCR analyses (Doleshal et al. 2008; Li et al. 2007; Xi et al. 2007).

A diagnostic application of miRNA profiling has been the subclassification of metastases. MiRNA profiling has been applied to the identification of tissue of origin of metastases (Rosenfeld et al. 2008; Rosenwald et al. 2010), which may become important to the subclassification of unknown primary cancers (UPC). UPCs represent the 3–5% of newly diagnosed cancers and constitute a disease with poor prognosis, partly because of the lack of complete diagnostic information. The attempts to determine the primary site by using mRNA expression profiling led to assignments that were discrepant from those derived from autopsies and results did not lead to patients benefits (Pentheroudakis et al. 2007). By examining the global miRNA profile of 253 samples from 22 different tumor tissues, Rosenfeld and colleagues (2008) were able to create a classifier based on 48 miRNAs that correctly predicted the tissue of origin in ~85% of the cases. The classifier reached an 81% sensitivity also if focusing only on metastases prediction. The possibility of using a small number of miRNAs to determine the unknown primary site with a high confidence could become a helpful tool for pathologists.

Another potentially useful diagnostic application derives from the molecular classification of cancer based on the combination of miRNA with gene expression and/or proteomic profiles in the same samples. Few attempts to combine miRNA and mRNA datasets to increase the quality and quantity of information obtainable from microarray analysis were published: Lanza et al. (2007) investigated a colorectal cancer dataset with or without microsatellite instability, and has shown that the mRNA/miRNA combination could achieve a better performance in tumor subtype classification; Lionetti et al. (2009) have used this approach to decipher the role of miRNAs dysregulated in multiple myeloma, and Fulci et al. (2009) examine the combined miRNA/mRNA profile in the context of B- and T-lineage acute lymphoblastic leukemia.

The above applications represent just few examples of the possibilities offered by miRNA expression profiling. The potentials are actually broader than those offered by mRNA profiling because of the possibility of using the huge sets of well-characterized FFPE tumor samples and, most recently, the possibility of testing miRNA profiling in serum/plasma samples.

1.6 MiRNAs as Prognostic Predictors

Because of the important role of miRNAs in multiple pathways, it seemed possible that the aberrant expression of specific miRNAs could reveal distinct clinicopathologic subgroups of cancer. Indeed, this has been demonstrated in several instances. One important field of application of miRNA expression profiles was the use of microarray and/or qRT-PCR expression profiles to predict clinical outcomes or patient survival or response to therapy, just to mention the most obvious examples.

The approach, which is based on the analysis of a training set of samples and validation of results through an independent set of samples, indicates how useful a panel of genes can be in determining the a priori chance for a correct prediction. Several studies have indeed demonstrated the potential usefulness of miRNAs when their expression could be associated with prognostic endpoints, such as overall survival and relapse-free survival, through the use Kaplan-Meier and uni- or multi-variate regression analyses.

Among the studies, Calin et al. (2005) used the microarray profile of 94 CLLs to identify a panel of 9 miRNAs (*miR-181a, miR-155, miR-146, miR-24-2, miR-23b, miR-23a, miR-222, miR-221, miR-29c*) whose expression was associated to different intervals from diagnosis to the beginning of treatment, which is given when the disease becomes more aggressive.

In the classification of non-Hodgkin's lymphoma, diffuse large B cell lymphomas (DLBCLs) with an activated B cell phenotype exhibited a higher level of *miR-155* than DLBCLs with the germinal center phenotype. Because patients with activated

B cell-type DLBCL have a poorer clinical prognosis, quantification of this miRNA may be clinically useful (Eis et al. 2005).

In solid tumors, Bloomston and colleagues (2007) were able to identify 6 miRNAs linked to short or long-term survival in pancreatic adenocarcinoma; they found also that the expression level of *miR-196a-2* was able to predict patients' survival, since higher miRNA levels mark the poor survivors group.

A study by Budhu et al. (2008) revealed a 20-miRNA signature associated with venous invasion of hepatocellular carcinoma (HCC) ($p = 0.002$). Significantly, the same signature could also correlate with disease-free and overall survival ($p = 0.022$). The results by Budhu et al. (2008) may be particularly useful to classify patients with HCC at early stage, which may provide a more rational approach to treatment intervention. In HCC, up-regulation of *miR-221* and down-regulation of *miR-122* were associated with shorter time to recurrence (Fornari et al. 2009; Gramantieri et al. 2009).

In colon adenocarcinoma, Schetter et al. (2009) identified *miR-21* up-regulated in tumors and associated with poor survival in two independent cohorts of patients (recruited in US and China); *miR-21* was shown to be significant also in multivariate regression analysis.

MiRNAs related to survival in lung cancer have been identified in four different investigations with partially overlapping results: reduced levels of *let-7* were found in patients with shorter survival time after resection (Takamizawa et al. 2004; Yanaihara et al. 2006). The same miRNA was used by Yu and co-workers (2008a), in combination with 4 more miRNAs (*miR-221, miR-137, miR-372 and miR-182**), to create a risk-score formula able to predict NSCLC patient relapse and survival in an independent test set of patients. Recently, low levels of *miR-34a* have been as well highlighted as prognostic marker in NSCLC (Gallardo et al. 2009), either alone or in combination with p53 mutational status information. *MiR-34a* was also found down-regulated in CLLs with p53 dysfunctions (Mraz et al. 2009; Zenz et al. 2009), a group of CLL with poor prognosis.

A large multicenter investigation by Mathè et al. (2009) on esophageal cancer, including both adenocarcinoma and squamous cell carcinoma (SCC), revealed a significant association of lower levels of *miR-375* in Barrett's-linked adenocarcinoma patients with worse prognosis (in term of overall survival) and of higher levels of *miR-21* in SCC non-cancerous tissue (but surprisingly not in SCCs themselves) with poor survival. Higher levels of *miR-21* were found to be independent poor prognosis predictor also in tongue squamous cell carcinoma (Li et al. 2009b).

Finally, higher levels of *miR-15b* were associated with poor survival and poor recurrence in melanoma (Satzger et al. 2009) while, surprisingly, high *miR-21* expression, detected by qRT-PCR in sera of diffuse large B cell lymphoma patients, was associated with a longer relapse-free survival but not with a better overall survival (Lawrie et al. 2008).

Breast cancer metastatic process has been associated with over-expression of *miR-10b* (Ma et al. 2007) and with loss of expression of *miR-126* and *miR-335* (Tavazoie et al. 2008).

As a whole, the above studies support a promising role of miRNA in prognosis stratification. Even if frequently limited by the size of the population used or the lack of validation studies, they proved the possibility of using miRNA expression data to guide clinical decisions.

1.7 MiRNAs as Drug Response Predictors/Modulators

Another important clinical information is related to individual drug responsiveness. Chemotherapy is often used as treatment for cancer patients. However, cancer patients may have a different susceptibility to anti-cancer drugs, due to their genetic and epigenetic background, or cancerous cells may become resistant during tumor progression.

Does miRNAs exert a role in this context or can be useful in predicting differential susceptibility? To answer this question, Salter and co-workers (2008) examined the full mRNA and miRNA profile on the NCI-60 panel of cell line to identify signatures linked to sensitivity to Paclitaxel, 5-Fluorouracil, Adriamycin and Cyclophosphamide (TFAC): a miRNA signature linked to each drug response was identified. Then, they assessed the performance of the mRNA signature on a cohort of 133 breast cancer patients, treated with the TFAC regimen and they compared the prediction results with that obtained by using traditional markers of chemosensitivity (ER, PR, HER2 status and Topoisomerase IIA levels). They suggested the importance of integrating information derived from miRNA profile with those currently used markers.

To date, several studies have focused on the identification of miRNAs linked to the acquisition of a resistant phenotype in cancer cell lines. The investigation by Blower et al. (2008) in three cancer cell lines established the influence of 3 cancer-related miRNAs (*let-7i, miR-16 and miR-21*) in drug sensitivity. By testing 14 different anti-cancer compounds, alone or in couples, with increasingly levels of the above-mentioned miRNAs they were able to see that increasing levels of *miR-21* reduce the efficacy of near half the tested compounds.

In a previous work by Meng et al. (2006), it was demonstrated that inhibition of *miR-21* and *miR-200b* increase the sensitivity of cholangiocarcinoma cells to gemcitabine. *MiR-21* is one of the most frequently over-expressed miRNAs in cancer and, as previously underlined, it is also a very promising prognostic predictor; this result give a possible explanation to how the miRNA could influence the course of the disease.

A handful of studies were instead performed in vivo, on cancer patients, to investigate the individual chemoresponsiveness. An association between low *miR-34* levels and fludarabine resistance in CLL, independently from 17p deletion, was found by Zenz et al. (2009) in CLL patients and was associated with the apoptosis resistance and to an impaired response to DNA damage. In line with this report, two recent studies on the role of fludarabine resistance in CLL indicated a critical role of *miR-21*, *miR-221/222* and a dysfunctional p53 pathway (Ferracin et al. 2010; Moussay et al. 2010).

Yang et al. (2008) identified *miR-214*, a miRNA up-regulated in ovarian cancer, as responsible of cisplatin resistance through its action on PTEN/Akt pathway. Results were supported by strong in vitro evidences and indeed, patients with chemoresistant or recurrent tumors displayed higher levels of *miR-214*.

Svoboda et al. (2008) examined the miRNA profile of near 30 patients with colorectal cancer either before and 2 weeks after chemoradiotherapy with capecitabine. They found that *miR-125b* and *miR-137* were up-regulated after therapy in the majority of patients and in particular in not-responding patients.

Ji et al. (2009a) investigated the role of *miR-26* in hepatocellular carcinoma (HCC). The authors profiled *miR-26* expression in three cohorts of HCC patients (total of 455 patients). They found that tumors had reduced levels of *miR-26*, if compared to match non-cancerous tissue, and that tumors with lower miRNA levels had shorter overall survival but showed a better response to interferon therapy. These observations seem to be very promising in the identification of patients who could benefit of an adjuvant therapy. Importantly, an animal model for liver cancer revealed a striking susceptibility to *miR-26* replacement therapy (Kota et al. 2009), suggesting that *miR-26* is not only a prognostic marker, but its loss is involved in hepatocarcinogenesis and *miR-26* may itself be potentially used in cancer therapy.

To what concern the acquisition of a resistant phenotype, Mishra et al. (2007) discovered that a SNP (829 C>T mutation) in human dihydropholate reductase 3′UTR was responsible of methotrexate resistance in cancer cells due to the lack of inhibition by *miR-24*, whose binding site is located near the polymorphism. They were the first to establish a strong link between miRNA-related SNPs and drug resistance. This interesting result has opened the way to the so called "miRNA pharmacogenomics" (Mishra et al. 2008; Mishra and Bertino 2009) providing a new strong challenge for cancer pharmacologic research.

The identification of patients that can benefit of chemotherapeutic agents in terms of global quality of life is of great importance in oncology, also because ineffective chemotherapy may increase patients' mortality. An extension of the number of studies on miRNA profile application in patient therapy response prediction is necessary.

Among the miRNAs mentioned in the previous sessions, *miR-21* always emerges. It is not only over-expressed in most of human cancers, where it appears to play a key role in the regulation of apoptosis and invasion, but it appears to confer poor prognosis and drug resistance. More importantly, anti-*miR-21* oligonucleotides may sensitize cells and play a role in modulating drug response, suggesting that *miR-21* may not only represent a useful diagnostic and prognostic marker, but it may also represent a very attractive therapeutic target.

1.8 MiRNAs as Cancer Biomarkers

An important goal of current studies is the early diagnosis of cancer patients, to significantly reduce morbidity and mortality rates associated with the disease and to improve the prognosis of diagnosed patients. For many years the oncologic research have sought for useful biomarkers among proteins detectable in tumor specimens or circulating in patients' blood. Now, miRNAs could be added to the panel of potential biomarkers thanks to their high stability and the availability of assays able to quantify their level.

A work by Mitchell et al. (2008) revealed not only that miRNAs are present in human serum or plasma in a highly stable form, but also that the levels of specific miRNAs (in this case *miR-141*) are significantly different in plasma of prostate cancer patients compared to healthy donors. Few other papers have been published on this topic till now. Chen et al. (2008b) characterized the miRNAs species in serum of lung cancer, colorectal cancer and diabetes patients by extensive sequencing, compared to healthy subjects, revealing considerable differences. Other papers examined the miRNAs circulating in plasma or serum through a different approach: first, a large panel of miRNAs was examined in a small set of colorectal cancers (Ng et al. 2009a), ovarian cancers (Resnick et al. 2009), and acute myeloid leukemias (Tanaka et al. 2009) either by microarray or real time qRT-PCR and compared to healthy controls; then, a subgroup of selected miRNAs was validated in a larger cohort of patients and controls.

Interestingly, *miR-92* was identified as useful prognostic biomarker in each of these investigations, although detected as up-regulated in the examined solid tumors while down-regulated in blood neoplasm. Several other miRNAs were identified as biomarkers for ovarian cancer, but the dataset was considerably smaller than in the other studies. Encouraging results emerged also from the analysis of diffuse large B cell lymphoma, where higher levels of *miR-155*, *miR-21* and *miR-210* were detected in serum of patients than in controls (Lawrie et al. 2008), while elevated levels of *miR-500* were detected in HCC patients' sera and returned low after HCC surgical resection (Yamamoto et al. 2009).

Although the above results should be taken carefully at this time since they are to be validated in large prospective studies, these findings are very promising and suggest that the use of serum/plasma miRNA markers may become useful in a clinical setting.

1.9 Conclusions and Future Perspectives

The past 5 years have seen an explosion of studies on miRNAs. In particular, studies on the involvement of miRNAs in cancer are rapidly moving toward translational applications (Cho 2010b).

The demonstration of the higher stability of miRNAs compared to other RNA classes, has opened the possibility to use large sets of well-characterized FFPE tumor tissues, which may lead to improved classifications, prognostic stratifications and drug response. An early application of this opportunity was the use of miRNA expression profiles for the identification of tissue of origin of metastases.

Moreover, the demonstration of the stability of miRNAs in serum/plasma has made available a completely new set of biomarkers that could outperform present protein biomarkers for early diagnosis and overall management of cancer patients.

Finally, the demonstration of the key role played by miRNAs in tumor initiation and progression provided the ground for the development of anti-miRNA oligonucleotides or miRNAs themselves as therapeutic molecules. The demonstration of their efficacy in vivo has opened the way to the development of a new class of molecularly-targeted anti-cancer drugs.

It is clear that the above issues represent opportunities rather than established applications. Further research in these areas will certainly be conducted in the years to come to turn opportunities into real facts.

References

Akao Y, Nakagawa Y, Iio A, et al. Role of microRNA-143 in fas-mediated apoptosis in human T-cell leukemia jurkat cells. Leuk Res. 2009;33:1530–8.

Akao Y, Nakagawa Y, Kitade Y, et al. Downregulation of microRNAs-143 and -145 in B-cell malignancies. Cancer Sci. 2007;98:1914–20.

Alexiou P, Maragkakis M, Papadopoulos GL, et al. Lost in translation: an assessment and perspective for computational microRNA target identification. Bioinformatics. 2009;25(23): 3049–55.

Amaral FC, Torres N, Saggioro F, et al. MicroRNAs differentially expressed in ACTH-secreting pituitary tumors. J Clin Endocrinol Metab. 2009;94:320–3.

Anastasiadou E, Boccellato F, Vincenti S, et al. Epstein-Barr virus encoded LMP1 downregulates TCL1 oncogene through miR-29b. Oncogene. 2010;29:1316–28.

Asangani IA, Rasheed SA, Nikolova DA, et al. MicroRNA-21 (miR-21) post-transcriptionally downregulates tumor suppressor PDCD4 and stimulates invasion, intravasation and metastasis in colorectal cancer. Oncogene. 2008;27:2128–36.

Avissar M, Christensen BC, Kelsey KT, et al. MicroRNA expression ratio is predictive of head and neck squamous cell carcinoma. Clin Cancer Res. 2009;15:2850–5.

Baek D, Villen J, Shin C, et al. The impact of microRNAs on protein output. Nature. 2008; 455:64–71.

Bandi N, Zbinden S, Gugger M, et al. MiR-15a and miR-16 are implicated in cell cycle regulation in a RB-dependent manner and are frequently deleted or down-regulated in non-small cell lung cancer. Cancer Res. 2009;69:5553–9.

Bandyopadhyay S, Mitra R. TARGetMINER: MicroRNA target prediction with systematic identification of tissue specific negative examples. Bioinformatics. 2009;25:2625–31.

Banerjee A, Schambach F, DeJong CS, et al. Micro-RNA-155 inhibits IFN-gamma signaling in CD4+ T cells. Eur J Immunol. 2010;40:225–31.

Barbato C, Arisi I, Frizzo ME, et al. Computational challenges in miRNA target predictions: to be or not to be a true target? J Biomed Biotechnol. 2009;2009:803069.

Barbato C, Ruberti F, Pieri M, et al. MicroRNA-92 modulates K(+) Cl(−) co-transporter KCC2 expression in cerebellar granule neurons. J Neurochem. 2010;113:591–600.

Bartel DP. MicroRNAs: genomics, biogenesis, mechanism, and function. Cell. 2004;116:281–97.

Bartel DP. MicroRNAs: target recognition and regulatory functions. Cell. 2009;136:215–33.

Bloomston M, Frankel WL, Petrocca F, et al. MicroRNA expression patterns to differentiate pancreatic adenocarcinoma from normal pancreas and chronic pancreatitis. JAMA. 2007;297: 1901–8.

Blower PE, Chung JH, Verducci JS, et al. MicroRNAs modulate the chemosensitivity of tumor cells. Mol Cancer Ther. 2008;7:1–9.

Boettger T, Beetz N, Kostin S, et al. Acquisition of the contractile phenotype by murine arterial smooth muscle cells depends on the MiR143/145 gene cluster. J Clin Invest. 2009;119: 2634–47.

Bolisetty MT, Dy G, Tam W, et al. Reticuloendotheliosis virus strain T induces miR-155 which targets JARID2 and promotes cell survival. J Virol. 2009;83(23):12009–17.

Bommer GT, Gerin I, Feng Y, et al. P53-mediated activation of miRNA34 candidate tumor-suppressor genes. Curr Biol. 2007;17:1298–307.

Bonci D, Coppola V, Musumeci M, et al. The miR-15a-miR-16-1 cluster controls prostate cancer by targeting multiple oncogenic activities. Nat Med. 2008;14:1271–7.

Borgdorff V, Lleonart ME, Bishop CL, et al. Multiple microRNAs rescue from Ras-induced senescence by inhibiting p21(WAF1/Cip1). Oncogene. 2010;29:2262–71.

Bottoni A, Zatelli MC, Ferracin M, et al. Identification of differentially expressed microRNAs by microarray: a possible role for microRNA genes in pituitary adenomas. J Cell Physiol. 2007;210:370–7.

Boyerinas B, Park SM, Shomron N, et al. Identification of let-7-regulated oncofetal genes. Cancer Res. 2008;68:2587–91.

Bracken CP, Gregory PA, Kolesnikoff N, et al. A double-negative feedback loop between ZEB1-SIP1 and the microRNA-200 family regulates epithelial-mesenchymal transition. Cancer Res. 2008;68:7846–54.

Brodersen P, Voinnet O. Revisiting the principles of microRNA target recognition and mode of action. Nat Rev Mol Cell Biol. 2009;10:141–8.

Budhu A, Jia HL, Forgues M, et al. Identification of metastasis-related microRNAs in hepatocellular carcinoma. Hepatology. 2008;47:897–907.

Burk U, Schubert J, Wellner U, et al. A reciprocal repression between ZEB1 and members of the miR-200 family promotes EMT and invasion in cancer cells. EMBO Rep. 2008;9:582–9.

Busacca S, Germano S, De Cecco L, et al. MicroRNA signature of malignant mesothelioma with potential diagnostic and prognostic implications. Am J Respir Cell Mol Biol. 2010; 42:312–9.

Calin GA, Cimmino A, Fabbri M, et al. MiR-15a and miR-16-1 cluster functions in human leukemia. Proc Natl Acad Sci USA. 2008;105:5166–71.

Calin GA, Croce CM. Genomics of chronic lymphocytic leukemia microRNAs as new players with clinical significance. Semin Oncol. 2006a;33:167–73.

Calin GA, Croce CM. MicroRNA signatures in human cancers. Nat Rev Cancer. 2006b;6:857–66.

Calin GA, Dumitru CD, Shimizu M, et al. Frequent deletions and down-regulation of micro-RNA genes miR15 and miR16 at 13q14 in chronic lymphocytic leukemia. Proc Natl Acad Sci USA. 2002;99:15524–9.

Calin GA, Ferracin M, Cimmino A, et al. A microRNA signature associated with prognosis and progression in chronic lymphocytic leukemia. N Engl J Med. 2005;353:1793–801.

Calin GA, Liu CG, Sevignani C, et al. MicroRNA profiling reveals distinct signatures in B cell chronic lymphocytic leukemias. Proc Natl Acad Sci USA. 2004a;101:11755–60.

Calin GA, Sevignani C, Dumitru CD, et al. Human microRNA genes are frequently located at fragile sites and genomic regions involved in cancers. Proc Natl Acad Sci USA. 2004b;101: 2999–3004.

Careccia S, Mainardi S, Pelosi A, et al. A restricted signature of miRNAs distinguishes APL blasts from normal promyelocytes. Oncogene. 2009;28(45):4034–40.

Carraro G, El-Hashash A, Guidolin D, et al. MiR-17 family of microRNAs controls FGF10-mediated embryonic lung epithelial branching morphogenesis through MAPK14 and STAT3 regulation of E-cadherin distribution. Dev Biol. 2009;333:238–50.

Cascio S, D'Andrea A, Ferla R, et al. MiR-20b modulates VEGF expression by targeting HIF-1 alpha and STAT3 in MCF-7 breast cancer cells. J Cell Physiol. 2010;224:242–9.

Cervigne NK, Reis PP, Machado J, et al. Identification of a microRNA signature associated with progression of leukoplakia to oral carcinoma. Hum Mol Genet. 2009;18:4818–29.

Chan JA, Krichevsky AM, Kosik KS. MicroRNA-21 is an antiapoptotic factor in human glioblastoma cells. Cancer Res. 2005;65:6029–33.

Chang J, Nicolas E, Marks D, et al. MiR-122, a mammalian liver-specific microRNA, is processed from HCR mRNA and may downregulate the high affinity cationic amino acid transporter CAT-1. RNA Biol. 2004;1:106–13.

Chang TC, Wentzel EA, Kent OA, et al. Transactivation of miR-34a by p53 broadly influences gene expression and promotes apoptosis. Mol Cell. 2007;26:745–52.
Chang TC, Yu D, Lee YS, et al. Widespread microRNA repression by myc contributes to tumorigenesis. Nat Genet. 2008;40:43–50.
Chang TC, Zeitels LR, Hwang HW, et al. Lin-28b transactivation is necessary for myc-mediated let-7 repression and proliferation. Proc Natl Acad Sci USA. 2009;106:3384–9.
Chao A, Tsai CL, Wei PC, et al. Decreased expression of microRNA-199b increases protein levels of SET (protein phosphatase 2a inhibitor) in human choriocarcinoma. Cancer Lett. 2010;291:99–107.
Chen R, Alvero AB, Silasi DA, et al. Regulation of IKKbeta by miR-199a affects NF-kappaB activity in ovarian cancer cells. Oncogene. 2008a;27:4712–23.
Chen X, Ba Y, Ma L, et al. Characterization of microRNAs in serum: a novel class of biomarkers for diagnosis of cancer and other diseases. Cell Res. 2008b;18:997–1006.
Chen X, Guo X, Zhang H, et al. Role of miR-143 targeting KRAS in colorectal tumorigenesis. Oncogene. 2009;28:1385–92.
Chen XM, Splinter PL, O'Hara SP, et al. A cellular micro-RNA, let-7i, regulates toll-like receptor 4 expression and contributes to cholangiocyte immune responses against cryptosporidium parvum infection. J Biol Chem. 2007;282:28929–38.
Chen Y, Liu W, Chao T, et al. MicroRNA-21 down-regulates the expression of tumor suppressor PDCD4 in human glioblastoma cell T98G. Cancer Lett. 2008c;272:197–205.
Cheng Y, Liu X, Zhang S, et al. MicroRNA-21 protects against the $H_{(2)}O_{(2)}$-induced injury on cardiac myocytes via its target gene PDCD4. J Mol Cell Cardiol. 2009;47:5–14.
Childs G, Fazzari M, Kung G, et al. Low-level expression of microRNAs let-7d and miR-205 are prognostic markers of head and neck squamous cell carcinoma. Am J Pathol. 2009;174:736–45.
Chiyomaru T, Enokida H, Tatarano S, et al. MiR-145 and miR-133a function as tumour suppressors and directly regulate FSCN1 expression in bladder cancer. Br J Cancer. 2010;102:883–91.
Cho WC. MicroRNAs in cancer – from research to therapy. Biochim Biophys Acta. 2010a;1805:209–17.
Cho WC. MicroRNAs: potential biomarkers for cancer diagnosis, prognosis and targets for therapy. Int J Biochem Cell Biol. 2010b;42:1273–81.
Chow TF, Crow M, Earle T, et al. Kallikreins as microRNA targets: an in silico and experimental-based analysis. Biol Chem. 2008;389:731–8.
Christoffersen NR, Silahtaroglu A, Orom UA, et al. MiR-200b mediates post-transcriptional repression of ZFHX1B. RNA. 2007;13:1172–8.
Ciafre SA, Galardi S, Mangiola A, et al. Extensive modulation of a set of microRNAs in primary glioblastoma. Biochem Biophys Res Commun. 2005;334:1351–8.
Cicalese A, Bonizzi G, Pasi CE, et al. The tumor suppressor p53 regulates polarity of self-renewing divisions in mammary stem cells. Cell. 2009;138:1083–95.
Cimmino A, Calin GA, Fabbri M, et al. MiR-15 and miR-16 induce apoptosis by targeting BCL2. Proc Natl Acad Sci USA. 2005;102:13944–9.
Cordes KR, Sheehy NT, White MP, et al. MiR-145 and miR-143 regulate smooth muscle cell fate and plasticity. Nature. 2009;460:705–10.
Corney DC, Flesken-Nikitin A, Godwin AK, et al. MicroRNA-34b and MicroRNA-34c are targets of p53 and cooperate in control of cell proliferation and adhesion-independent growth. Cancer Res. 2007;67:8433–8.
Costinean S, Sandhu SK, Pedersen IM, et al. Src homology 2 domain-containing inositol-5-phosphatase and CCAAT enhancer-binding protein beta are targeted by miR-155 in B cells of Emicro-miR-155 transgenic mice. Blood. 2009;114:1374–82.
Costinean S, Zanesi N, Pekarsky Y, et al. Pre-B cell proliferation and lymphoblastic leukemia/high-grade lymphoma in E(mu)-miR155 transgenic mice. Proc Natl Acad Sci USA. 2006;103:7024–9.

Crawford M, Batte K, Yu L, et al. MicroRNA 133b targets pro-survival molecules MCL-1 and BCL2L2 in lung cancer. Biochem Biophys Res Commun. 2009;388:483–9.

Croce CM. Causes and consequences of microRNA dysregulation in cancer. Nat Rev Genet. 2009;10:704–14.

Datta J, Kutay H, Nasser MW, et al. Methylation mediated silencing of microRNA-1 gene and its role in hepatocellular carcinogenesis. Cancer Res. 2008;68:5049–58.

Davoren PA, McNeill RE, Lowery AJ, et al. Identification of suitable endogenous control genes for microRNA gene expression analysis in human breast cancer. BMC Mol Biol. 2008;9:76.

de Yebenes VG, Belver L, Pisano DG, et al. MiR-181b negatively regulates activation-induced cytidine deaminase in B cells. J Exp Med. 2008;205:2199–206.

Dharap A, Bowen K, Place R, et al. Transient focal ischemia induces extensive temporal changes in rat cerebral microRNAome. J Cereb Blood Flow Metab. 2009;29:675–87.

Di Leva G, Gasparini P, Piovan C, et al. MicroRNA cluster 221-222 and estrogen receptor {alpha} interactions in breast cancer. J Natl Cancer Inst. 2010;102(10):706–21.

Doleshal M, Magotra AA, Choudhury B, et al. Evaluation and validation of total RNA extraction methods for microRNA expression analyses in formalin-fixed, paraffin-embedded tissues. J Mol Diagn. 2008;10:203–11.

Dong Q, Meng P, Wang T, et al. MicroRNA let-7a inhibits proliferation of human prostate cancer cells in vitro and in vivo by targeting E2F2 and CCND2. PLoS One. 2010;5:e10147.

Dorsett Y, McBride KM, Jankovic M, et al. MicroRNA-155 suppresses activation-induced cytidine deaminase-mediated myc-IgH translocation. Immunity. 2008;28:630–8.

Droge P, Davey CA. Do cells let-7 determine stemness? Cell Stem Cell. 2008;2:8–9.

Dyrskjot L, Ostenfeld MS, Bramsen JB, et al. Genomic profiling of microRNAs in bladder cancer: miR-129 is associated with poor outcome and promotes cell death in vitro. Cancer Res. 2009;69:4851–60.

Ehrbrecht A, Muller U, Wolter M, et al. Comprehensive genomic analysis of desmoplastic medulloblastomas: identification of novel amplified genes and separate evaluation of the different histological components. J Pathol. 2006;208:554–63.

Eis PS, Tam W, Sun L, et al. Accumulation of miR-155 and BIC RNA in human B cell lymphomas. Proc Natl Acad Sci USA. 2005;102:3627–32.

Esau C, Kang X, Peralta E, et al. MicroRNA-143 regulates adipocyte differentiation. J Biol Chem. 2004;279:52361–5.

Esquela-Kerscher A, Slack FJ. Oncomirs – microRNAs with a role in cancer. Nat Rev Cancer. 2006;6:259–69.

Esquela-Kerscher A, Trang P, Wiggins JF, et al. The let-7 microRNA reduces tumor growth in mouse models of lung cancer. Cell Cycle. 2008;7:759–64.

Evangelisti C, Florian MC, Massimi I, et al. MiR-128 up-regulation inhibits reelin and DCX expression and reduces neuroblastoma cell motility and invasiveness. FASEB J. 2009;23: 4276–87.

Fabbri M, Garzon R, Cimmino A, et al. MicroRNA-29 family reverts aberrant methylation in lung cancer by targeting DNA methyltransferases 3a and 3b. Proc Natl Acad Sci USA. 2007;104:15805–10.

Fasanaro P, D'Alessandra Y, Di Stefano V, et al. MicroRNA-210 modulates endothelial cell response to hypoxia and inhibits the receptor tyrosine kinase ligand ephrin-A3. J Biol Chem. 2008;283:15878–83.

Feber A, Xi L, Luketich JD, et al. MicroRNA expression profiles of esophageal cancer. J Thorac Cardiovasc Surg. 2008;135:255–60, discussion 260.

Felicetti F, Errico MC, Bottero L, et al. The promyelocytic leukemia zinc finger-microRNA-221/-222 pathway controls melanoma progression through multiple oncogenic mechanisms. Cancer Res. 2008;68:2745–54.

Ferdin J, Kunej T, Calin GA. Non-coding RNAs: identification of cancer-associated microRNAs by gene profiling. Technol Cancer Res Treat. 2010;9:123–38.

Ferracin M, Zagatti B, Rizzotto L, et al. MicroRNAs involvement in fludarabine refractory chronic lymphocytic leukemia. Mol Cancer. 2010; 9:123.
Ferretti E, De Smaele E, Po A, et al. MicroRNA profiling in human medulloblastoma. Int J Cancer. 2009;124:568–77.
Forman JJ, Legesse-Miller A, Coller HA. A search for conserved sequences in coding regions reveals that the let-7 microRNA targets dicer within its coding sequence. Proc Natl Acad Sci USA. 2008;105:14879–84.
Fornari F, Gramantieri L, Ferracin M, et al. MiR-221 controls CDKN1C/p57 and CDKN1B/p27 expression in human hepatocellular carcinoma. Oncogene. 2008;27:5651–61.
Fornari F, Gramantieri L, Giovannini C, et al. MiR-122/cyclin G1 interaction modulates p53 activity and affects doxorubicin sensitivity of human hepatocarcinoma cells. Cancer Res. 2009;69:5761–7.
Frenquelli M, Muzio M, Scielzo C, et al. MicroRNA and proliferation control in chronic lymphocytic leukemia: functional relationship between miR-221/222 cluster and p27. Blood. 2010;115(19):3949–59.
Friedman JM, Liang G, Liu CC, et al. The putative tumor suppressor microRNA-101 modulates the cancer epigenome by repressing the polycomb group protein EZH2. Cancer Res. 2009;69:2623–9.
Fulci V, Chiaretti S, Goldoni M, et al. Quantitative technologies establish a novel microRNA profile of chronic lymphocytic leukemia. Blood. 2007;109:4944–51.
Fulci V, Colombo T, Chiaretti S, et al. Characterization of B- and T-lineage acute lymphoblastic leukemia by integrated analysis of MicroRNA and mRNA expression profiles. Genes Chromosomes Cancer. 2009;48:1069–82.
Gabriely G, Wurdinger T, Kesari S, et al. MicroRNA 21 promotes glioma invasion by targeting matrix metalloproteinase regulators. Mol Cell Biol. 2008;28:5369–80.
Galardi S, Mercatelli N, Giorda E, et al. MiR-221 and miR-222 expression affects the proliferation potential of human prostate carcinoma cell lines by targeting p27kip1. J Biol Chem. 2007;282:23716–24.
Gallardo E, Navarro A, Vinolas N, et al. MiR-34a as a prognostic marker of relapse in surgically resected non-small-cell lung cancer. Carcinogenesis. 2009;30:1903–9.
Gandellini P, Folini M, Longoni N, et al. MiR-205 exerts tumor-suppressive functions in human prostate through down-regulation of protein kinase cepsilon. Cancer Res. 2009;69:2287–95.
Gao C, Zhang Z, Liu W, et al. Reduced microRNA-218 expression is associated with high nuclear factor kappa B activation in gastric cancer. Cancer. 2010;116:41–9.
Garofalo M, Condorelli GL, Croce CM, et al. MicroRNAs as regulators of death receptors signaling. Cell Death Differ. 2010;17:200–8.
Garzia L, Andolfo I, Cusanelli E, et al. MicroRNA-199b-5p impairs cancer stem cells through negative regulation of HES1 in medulloblastoma. PLoS One. 2009;4:e4998.
Garzon R, Garofalo M, Martelli MP, et al. Distinctive microRNA signature of acute myeloid leukemia bearing cytoplasmic mutated nucleophosmin. Proc Natl Acad Sci USA. 2008a;105:3945–50.
Garzon R, Liu S, Fabbri M, et al. MicroRNA-29b induces global DNA hypomethylation and tumor suppressor gene reexpression in acute myeloid leukemia by targeting directly DNMT3A and 3b and indirectly DNMT1. Blood. 2009;113:6411–8.
Garzon R, Volinia S, Liu CG, et al. MicroRNA signatures associated with cytogenetics and prognosis in acute myeloid leukemia. Blood. 2008b;111:3183–9.
Gebeshuber CA, Zatloukal K, Martinez J. MiR-29a suppresses tristetraprolin, which is a regulator of epithelial polarity and metastasis. EMBO Rep. 2009;10:400–5.
Ghosh AK, Shanafelt TD, Cimmino A, et al. Aberrant regulation of pVHL levels by microRNA promotes the HIF/VEGF axis in CLL B cells. Blood. 2009;113:5568–74.
Gillies JK, Lorimer IA. Regulation of p27Kip1 by miRNA 221/222 in glioblastoma. Cell Cycle. 2007;6:2005–9.

Girard M, Jacquemin E, Munnich A, et al. MiR-122, a paradigm for the role of microRNAs in the liver. J Hepatol. 2008;48:648–56.
Godlewski J, Nowicki MO, Bronisz A, et al. Targeting of the Bmi-1 oncogene/stem cell renewal factor by microRNA-128 inhibits glioma proliferation and self-renewal. Cancer Res. 2008;68:9125–30.
Gottardo F, Liu CG, Ferracin M, et al. Micro-RNA profiling in kidney and bladder cancers. Urol Oncol. 2007;25:387–92.
Gramantieri L, Ferracin M, Fornari F, et al. Cyclin G1 is a target of miR-122a, a microRNA frequently down-regulated in human hepatocellular carcinoma. Cancer Res. 2007;67:6092–9.
Gramantieri L, Fornari F, Callegari E, et al. MicroRNA involvement in hepatocellular carcinoma. J Cell Mol Med. 2008;12:2189–204.
Gramantieri L, Fornari F, Ferracin M, et al. MicroRNA-221 targets bmf in hepatocellular carcinoma and correlates with tumor multifocality. Clin Cancer Res. 2009;15:5073–81.
Greene SB, Gunaratne PH, Hammond SM, et al. A putative role for microRNA-205 in mammary epithelial cell progenitors. J Cell Sci. 2010;123:606–18.
Gregersen LH, Jacobsen AB, Frankel LB, et al. MicroRNA-145 targets YES and STAT1 in colon cancer cells. PLoS One. 2010;5:e8836.
Gregory PA, Bert AG, Paterson EL, et al. The miR-200 family and miR-205 regulate epithelial to mesenchymal transition by targeting ZEB1 and SIP1. Nat Cell Biol. 2008;10:593–601.
Guled M, Lahti L, Lindholm PM, et al. CDKN2A, NF2, and JUN are dysregulated among other genes by miRNAs in malignant mesothelioma -A miRNA microarray analysis. Genes Chromosomes Cancer. 2009;48:615–23.
Guo Y, Chen Y, Ito H, et al. Identification and characterization of lin-28 homolog B (LIN28B) in human hepatocellular carcinoma. Gene. 2006;384:51–61.
Guo Y, Chen Z, Zhang L, et al. Distinctive microRNA profiles relating to patient survival in esophageal squamous cell carcinoma. Cancer Res. 2008;68:26–33.
Hagan JP, Piskounova E, Gregory RI. Lin28 recruits the TUTase Zcchc11 to inhibit let-7 maturation in mouse embryonic stem cells. Nat Struct Mol Biol. 2009;16:1021–5.
Han J, Kim D, Morris KV. Promoter-associated RNA is required for RNA-directed transcriptional gene silencing in human cells. Proc Natl Acad Sci USA. 2007a;104:12422–7.
Han L, Witmer PD, Casey E, et al. DNA methylation regulates MicroRNA expression. Cancer Biol Ther. 2007b;6:1284–8.
Hausser J, Berninger P, Rodak C, et al. MirZ: an integrated microRNA expression atlas and target prediction resource. Nucleic Acids Res. 2009;37:W266–72.
He H, Jazdzewski K, Li W, et al. The role of microRNA genes in papillary thyroid carcinoma. Proc Natl Acad Sci USA. 2005a;102:19075–80.
He L, He X, Lowe SW, et al. MicroRNAs join the p53 network–another piece in the tumour-suppression puzzle. Nat Rev Cancer. 2007;7:819–22.
He L, Thomson JM, Hemann MT, et al. A microRNA polycistron as a potential human oncogene. Nature. 2005b;435:828–33.
Henson BJ, Bhattacharjee S, O'Dee DM, et al. Decreased expression of miR-125b and miR-100 in oral cancer cells contributes to malignancy. Genes Chromosomes Cancer. 2009;48:569–82.
Heo I, Joo C, Cho J, et al. Lin28 mediates the terminal uridylation of let-7 precursor microRNA. Mol Cell. 2008;32:276–84.
Hong H, Takahashi K, Ichisaka T, et al. Suppression of induced pluripotent stem cell generation by the p53-p21 pathway. Nature. 2009;460:1132–5.
Hossain A, Kuo MT, Saunders GF MiR-17-5P regulates breast cancer cell proliferation by inhibiting translation of AIB1 mRNA. Mol Cell Biol. 2006;26:8191–201.
Hu G, Gong AY, Liu J, et al. MiR-221 suppresses ICAM-1 translation and regulates interferon-gamma-induced ICAM-1 expression in human cholangiocytes. Am J Physiol Gastrointest Liver Physiol. 2010;298:G542–50
Huang Q, Gumireddy K, Schrier M, et al. The microRNAs miR-373 and miR-520c promote tumour invasion and metastasis. Nat Cell Biol. 2008a;10:202–10.

Huang YS, Dai Y, Yu XF, et al. Microarray analysis of microRNA expression in hepatocellular carcinoma and non-tumorous tissues without viral hepatitis. J Gastroenterol Hepatol. 2008b;23:87–94.

Hurteau GJ, Carlson JA, Spivack SD, et al. Overexpression of the microRNA hsa-miR-200c leads to reduced expression of transcription factor 8 and increased expression of E-cadherin. Cancer Res. 2007;67:7972–6.

Hussein K, Theophile K, Busche G, et al. Significant inverse correlation of microRNA-150/MYB and microRNA-222/p27 in myelodysplastic syndrome. Leuk Res. 2010;34:328–34.

Ichimi T, Enokida H, Okuno Y, et al. Identification of novel microRNA targets based on microRNA signatures in bladder cancer. Int J Cancer. 2009;125:345–52.

Igoucheva O, Alexeev V. MicroRNA-dependent regulation of cKit in cutaneous melanoma. Biochem Biophys Res Commun. 2009;379:790–4.

Inamura K, Togashi Y, Nomura K, et al. Let-7 microRNA expression is reduced in bronchioloalveolar carcinoma, a non-invasive carcinoma, and is not correlated with prognosis. Lung Cancer. 2007;58:392–6.

Inomata M, Tagawa H, Guo YM, et al. MicroRNA-17-92 down-regulates expression of distinct targets in different B-cell lymphoma subtypes. Blood. 2009;113:396–402.

Iorio MV, Casalini P, Piovan C, et al. MicroRNA-205 regulates HER3 in human breast cancer. Cancer Res. 2009;69:2195–200.

Iorio MV, Ferracin M, Liu CG, et al. MicroRNA gene expression deregulation in human breast cancer. Cancer Res. 2005;65:7065–70.

Iorio MV, Visone R, Di Leva G, et al. MicroRNA signatures in human ovarian cancer. Cancer Res. 2007;67:8699–707.

Ivanovska I, Ball AS, Diaz RL, et al. MicroRNAs in the miR-106b family regulate p21/CDKN1A and promote cell cycle progression. Mol Cell Biol. 2008;28:2167–74.

Janowski BA, Younger ST, Hardy DB, et al. Activating gene expression in mammalian cells with promoter-targeted duplex RNAs. Nat Chem Biol. 2007;3:166–73.

Ji J, Shi J, Budhu A, et al. MicroRNA expression, survival, and response to interferon in liver cancer. N Engl J Med. 2009a;361:1437–47.

Ji J, Yamashita T, Budhu A, et al. Identification of microRNA-181 by genome-wide screening as a critical player in EpCAM-positive hepatic cancer stem cells. Hepatology. 2009b;50:472–80.

Ji J, Zhao L, Budhu A, et al. Let-7g targets collagen type I alpha2 and inhibits cell migration in hepatocellular carcinoma. J Hepatol. 2010a;52(5):690–7.

Ji Y, He Y, Liu L, et al. MiRNA-26b regulates the expression of cyclooxygenase-2 in desferrioxamine-treated CNE cells. FEBS Lett. 2010b;584:961–7.

Jiang J, Gusev Y, Aderca I, et al. Association of microRNA expression in hepatocellular carcinomas with hepatitis infection, cirrhosis, and patient survival. Clin Cancer Res. 2008;14:419–27.

Jiang S, Zhang HW, Lu MH, et al. MicroRNA-155 functions as an oncomir in breast cancer by targeting the suppressor of cytokine signaling 1 gene. Cancer Res. 2010;70:3119–27.

Johnson SM, Grosshans H, Shingara J, et al. RAS is regulated by the let-7 microRNA family. Cell. 2005;120:635–47.

Jung M, Mollenkopf HJ, Grimm C, et al. MicroRNA profiling of clear cell renal cell cancer identifies a robust signature to define renal malignancy. J Cell Mol Med. 2009;13(9B): 3918–28.

Kalscheuer S, Zhang X, Zeng Y, et al. Differential expression of microRNAs in early-stage neoplastic transformation in the lungs of F344 rats chronically treated with the tobacco carcinogen 4-(methylnitrosamino)-1-(3-pyridyl)-1-butanone. Carcinogenesis. 2008;29:2394–9.

Kan T, Sato F, Ito T, et al. The miR-106b-25 polycistron, activated by genomic amplification, functions as an oncogene by suppressing p21 and bim. Gastroenterology. 2009;136:1689–700.

Kano M, Seki N, Kikkawa N, et al. MiR-145, miR-133a and miR-133b: tumor suppressive miRNAs target FSCN1 in esophageal squamous cell carcinoma. Int J Cancer. 2010;127:2804–14.

Kawamura T, Suzuki J, Wang YV, et al. Linking the p53 tumour suppressor pathway to somatic cell reprogramming. Nature. 2009;460:1140–4.

Khraiwesh B, Arif MA, Seumel GI, et al. Transcriptional control of gene expression by microRNAs. Cell. 2010;140:111–22.

Kim HW, Haider HK, Jiang S, et al. Ischemic preconditioning augments survival of stem cells via miR-210 expression by targeting caspase-8-associated protein 2. J Biol Chem. 2009;284(48):33161–8.

Kim S, Lee UJ, Kim MN, et al. MicroRNA miR-199a* regulates the MET proto-oncogene and the downstream extracellular signal-regulated kinase 2 (ERK2). J Biol Chem. 2008;283:18158–66.

Kim SK, Nam JW, Rhee JK, et al. MiTarget: microRNA target gene prediction using a support vector machine. BMC Bioinformatics. 2006;7:411.

Kimura H, Kawasaki H, Taira K Mouse microRNA-23b regulates expression of hes1 gene in P19 cells. Nucleic Acids Symp Ser (Oxf). 2004;48:213–4.

Klein U, Lia M, Crespo M, et al. The DLEU2/miR-15a/16-1 cluster controls B cell proliferation and its deletion leads to chronic lymphocytic leukemia. Cancer Cell. 2010;17:28–40.

Koh W, Sheng CT, Tan B, et al. Analysis of deep sequencing microRNA expression profile from human embryonic stem cells derived mesenchymal stem cells reveals possible role of let-7 microRNA family in downstream targeting of hepatic nuclear factor 4 alpha. BMC Genomics. 2010;11(Suppl 1):S6.

Kolachala VL, Wang L, Obertone TS, et al. Adenosine 2b receptor expression is post-transcriptionally regulated by microRNA. J Biol Chem. 2010;285(24):18184–90.

Kong W, He L, Coppola M, et al. MicroRNA-155 regulates cell survival, growth and chemosensitivity by targeting FOXO3a in breast cancer. J Biol Chem. 2010;285:17869–79.

Kong W, Yang H, He L, et al. MicroRNA-155 is regulated by the transforming growth factor beta/smad pathway and contributes to epithelial cell plasticity by targeting RhoA. Mol Cell Biol. 2008;28:6773–84.

Korpal M, Lee ES, Hu G, et al. The miR-200 family inhibits epithelial-mesenchymal transition and cancer cell migration by direct targeting of E-cadherin transcriptional repressors ZEB1 and ZEB2. J Biol Chem. 2008;283:14910–4.

Kota J, Chivukula RR, O'Donnell KA, et al. Therapeutic microRNA delivery suppresses tumorigenesis in a murine liver cancer model. Cell. 2009;137:1005–17.

Kotani A, Ha D, Hsieh J, et al. MiR-128b is a potent glucocorticoid sensitizer in MLL-AF4 acute lymphocytic leukemia cells and exerts cooperative effects with miR-221. Blood. 2009;114(19):4169–78.

Kozaki K, Imoto I, Mogi S, et al. Exploration of tumor-suppressive microRNAs silenced by DNA hypermethylation in oral cancer. Cancer Res. 2008;68:2094–105.

Kruger J, Rehmsmeier M RNAhybrid: microRNA target prediction easy, fast and flexible. Nucleic Acids Res. 2006;34:W451–4.

Kuang W, Tan J, Duan Y, et al. Cyclic stretch induced miR-146a upregulation delays C2C12 myogenic differentiation through inhibition of numb. Biochem Biophys Res Commun. 2009;378:259–63.

Kulshreshtha R, Ferracin M, Negrini M, et al. Regulation of microRNA expression: the hypoxic component. Cell Cycle. 2007a;6:1426–31.

Kulshreshtha R, Ferracin M, Wojcik SE, et al. A microRNA signature of hypoxia. Mol Cell Biol. 2007b;27:1859–67.

Kumar MS, Erkeland SJ, Pester RE, et al. Suppression of non-small cell lung tumor development by the let-7 microRNA family. Proc Natl Acad Sci USA. 2008;105:3903–8.

Kutay H, Bai S, Datta J, et al. Downregulation of miR-122 in the rodent and human hepatocellular carcinomas. J Cell Biochem. 2006;99:671–8.

Lal A, Kim HH, Abdelmohsen K, et al. P16(INK4a) translation suppressed by miR-24. PLoS One. 2008;3:e1864.

Lal A, Navarro F, Maher CA, et al. MiR-24 inhibits cell proliferation by targeting E2F2, MYC, and other cell-cycle genes via binding to "seedless" 3′UTR microRNA recognition elements. Mol Cell. 2009a;35:610–25.

Lal A, Pan Y, Navarro F, et al. MiR-24-mediated downregulation of H2AX suppresses DNA repair in terminally differentiated blood cells. Nat Struct Mol Biol. 2009b;16:492–8.
Lanza G, Ferracin M, Gafa R, et al. MRNA/microRNA gene expression profile in microsatellite unstable colorectal cancer. Mol Cancer. 2007;6:54.
Lawrie CH, Gal S, Dunlop HM, et al. Detection of elevated levels of tumour-associated microRNAs in serum of patients with diffuse large B-cell lymphoma. Br J Haematol. 2008;141: 672–5.
le Sage C, Nagel R, Egan DA, et al. Regulation of the p27(Kip1) tumor suppressor by miR-221 and miR-222 promotes cancer cell proliferation. EMBO J. 2007;26:3699–708.
Lee EJ, Gusev Y, Jiang J, et al. Expression profiling identifies microRNA signature in pancreatic cancer. Int J Cancer. 2007;120:1046–54.
Lee JW, Choi CH, Choi JJ, et al. Altered MicroRNA expression in cervical carcinomas. Clin Cancer Res. 2008a;14:2535–42.
Lee KH, Goan YG, Hsiao M, et al. MicroRNA-373 (miR-373) post-transcriptionally regulates large tumor suppressor, homolog 2 (LATS2) and stimulates proliferation in human esophageal cancer. Exp Cell Res. 2009a;315:2529–38.
Lee KH, Lotterman C, Karikari C, et al. Epigenetic silencing of microRNA miR-107 regulates cyclin-dependent kinase 6 expression in pancreatic cancer. Pancreatology. 2009b;9:293–301.
Lee Y, Samaco RC, Gatchel JR, et al. MiR-19, miR-101 and miR-130 co-regulate ATXN1 levels to potentially modulate SCA1 pathogenesis. Nat Neurosci. 2008b;11:1137–9.
Lee YS, Dutta A. The tumor suppressor microRNA let-7 represses the HMGA2 oncogene. Genes Dev. 2007;21:1025–30.
Legesse-Miller A, Elemento O, Pfau SJ, et al. Let-7 overexpression leads to an increased fraction of cells in G2/M, direct down-regulation of CDC34, and stabilization of wee1 kinase in primary fibroblasts. J Biol Chem. 2009;284:6605–9.
Lehrbach NJ, Armisen J, Lightfoot HL, et al. LIN-28 and the poly(U) polymerase PUP-2 regulate let-7 microRNA processing in *caenorhabditis elegans*. Nat Struct Mol Biol. 2009;16: 1016–20.
Lena AM, Shalom-Feuerstein R Rivetti di val cervo P, et al. MiR-203 represses 'stemness' by repressing DeltaNp63. Cell Death Differ. 2008;15:1187–95.
Lerner M, Harada M, Loven J, et al. DLEU2, frequently deleted in malignancy, functions as a critical host gene of the cell cycle inhibitory microRNAs miR-15a and miR-16-1. Exp Cell Res. 2009;315:2941–52.
Lewis BP, Burge CB, Bartel DP. Conserved seed pairing, often flanked by adenosines, indicates that thousands of human genes are microRNA targets. Cell. 2005;120:15–20.
Li H, Collado M, Villasante A, et al. The ink4/arf locus is a barrier for iPS cell reprogramming. Nature. 2009a;460:1136–9.
Li J, Huang H, Sun L, et al. MiR-21 indicates poor prognosis in tongue squamous cell carcinomas as an apoptosis inhibitor. Clin Cancer Res. 2009b;15:3998–4008.
Li J, Smyth P, Flavin R, et al. Comparison of miRNA expression patterns using total RNA extracted from matched samples of formalin-fixed paraffin-embedded (FFPE) cells and snap frozen cells. BMC Biotechnol. 2007;7:36.
Li LC, Okino ST, Zhao H, et al. Small dsRNAs induce transcriptional activation in human cells. Proc Natl Acad Sci USA. 2006;103:17337–42.
Li N, Fu H, Tie Y, et al. MiR-34a inhibits migration and invasion by down-regulation of c-met expression in human hepatocellular carcinoma cells. Cancer Lett. 2009c;275:44–53.
Li S, Fu H, Wang Y, et al. MicroRNA-101 regulates expression of the v-fos FBJ murine osteosarcoma viral oncogene homolog (FOS) oncogene in human hepatocellular carcinoma. Hepatology. 2009d;49:1194–202.
Li T, Li D, Sha J, et al. MicroRNA-21 directly targets MARCKS and promotes apoptosis resistance and invasion in prostate cancer cells. Biochem Biophys Res Commun. 2009e;383:280–5.
Li W, Xie L, He X, et al. Diagnostic and prognostic implications of microRNAs in human hepatocellular carcinoma. Int J Cancer. 2008;123:1616–22.

Li Y, Li W, Yang Y, et al. MicroRNA-21 targets LRRFIP1 and contributes to VM-26 resistance in glioblastoma multiforme. Brain Res. 2009f;1286:13–8.

Lichter P. All you need is a miR-acle: the role of nontranslated RNAs in the suppression of B cell chronic lymphocytic leukemia. Cancer Cell. 2010;17:3–4.

Lim LP, Lau NC, Garrett-Engele P, et al. Microarray analysis shows that some microRNAs downregulate large numbers of target mRNAs. Nature. 2005;433:769–73.

Lin CJ, Gong HY, Tseng HC, et al. MiR-122 targets an anti-apoptotic gene, bcl-w, in human hepatocellular carcinoma cell lines. Biochem Biophys Res Commun. 2008;375: 315–20.

Lin EA, Kong L, Bai XH, et al. MiR-199a, a bone morphogenic protein 2-responsive MicroRNA, regulates chondrogenesis via direct targeting to smad1. J Biol Chem. 2009;284:11326–35.

Lionetti M, Biasiolo M, Agnelli L, et al. Identification of microRNA expression patterns and definition of a microRNA/mRNA regulatory network in distinct molecular groups of multiple myeloma. Blood. 2009;114:e20–6.

Liu CG, Calin GA, Meloon B, et al. An oligonucleotide microchip for genome-wide microRNA profiling in human and mouse tissues. Proc Natl Acad Sci USA. 2004;101:9740–4.

Liu M, Wu H, Liu T, et al. Regulation of the cell cycle gene, BTG2, by miR-21 in human laryngeal carcinoma. Cell Res. 2009a;19:828–37.

Liu S, Wu LC, Pang J, et al. Sp1/NFkappaB/HDAC/miR-29b regulatory network in KIT-driven myeloid leukemia. Cancer Cell. 2010;17:333–47.

Liu WH, Yeh SH, Lu CC, et al. MicroRNA-18a prevents estrogen receptor-alpha expression, promoting proliferation of hepatocellular carcinoma cells. Gastroenterology. 2009b;136:683–93.

Liu X, Jiang L, Wang A, et al. MicroRNA-138 suppresses invasion and promotes apoptosis in head and neck squamous cell carcinoma cell lines. Cancer Lett. 2009c;286:217–22.

Lodygin D, Tarasov V, Epanchintsev A, et al. Inactivation of miR-34a by aberrant CpG methylation in multiple types of cancer. Cell Cycle. 2008;7:2591–600.

Loven J, Zinin N, Wahlstrom T, et al. MYCN-regulated microRNAs repress estrogen receptor-alpha (ESR1) expression and neuronal differentiation in human neuroblastoma. Proc Natl Acad Sci USA. 2010;107:1553–8.

Lu J, Getz G, Miska EA, et al. MicroRNA expression profiles classify human cancers. Nature. 2005;435:834–8.

Lu L, Katsaros D, Shaverdashvili K, et al. Pluripotent factor lin-28 and its homologue lin-28b in epithelial ovarian cancer and their associations with disease outcomes and expression of let-7a and IGF-II. Eur J Cancer. 2009;45:2212–8.

Lu Z, Liu M, Stribinskis V, et al. MicroRNA-21 promotes cell transformation by targeting the programmed cell death 4 gene. Oncogene. 2008;27:4373–9.

Lui WO, Pourmand N, Patterson BK, et al. Patterns of known and novel small RNAs in human cervical cancer. Cancer Res. 2007;67:6031–43.

Lujambio A, Calin GA, Villanueva A, et al. A microRNA DNA methylation signature for human cancer metastasis. Proc Natl Acad Sci USA. 2008;105:13556–61.

Ma L, Teruya-Feldstein J, Weinberg RA. Tumour invasion and metastasis initiated by microRNA-10b in breast cancer. Nature. 2007;449:682–8.

Ma L, Young J, Prabhala H, et al. MiR-9, a MYC/MYCN-activated microRNA, regulates E-cadherin and cancer metastasis. Nat Cell Biol. 2010;12:247–56.

Magrelli A, Azzalin G, Salvatore M, et al. Altered microRNA expression patterns in hepatoblastoma patients. Transl Oncol. 2009;2:157–63.

Maller Schulman BR, Liang X, Stahlhut C, et al. The let-7 microRNA target gene, mlin41/trim71 is required for mouse embryonic survival and neural tube closure. Cell Cycle. 2008;7:3935–42.

Manni I, Artuso S, Careccia S, et al. The microRNA miR-92 increases proliferation of myeloid cells and by targeting p63 modulates the abundance of its isoforms. FASEB J. 2009;23: 3957–66.

Maragkakis M, Reczko M, Simossis VA, et al. DIANA-microT web server: elucidating microRNA functions through target prediction. Nucleic Acids Res. 2009;37:W273–6.

Marion RM, Strati K, Li H, et al. A p53-mediated DNA damage response limits reprogramming to ensure iPS cell genomic integrity. Nature. 2009;460:1149–53.

Markou A, Tsaroucha EG, Kaklamanis L, et al. Prognostic value of mature microRNA-21 and microRNA-205 overexpression in non-small cell lung cancer by quantitative real-time RT-PCR. Clin Chem. 2008;54:1696–704.

Marquez RT, Wendlandt E, Galle CS, et al. MicroRNA-21 is upregulated during the proliferative phase of liver regeneration, targets pellino-1, and inhibits NF-kappaB signaling. Am J Physiol Gastrointest Liver Physiol. 2010;298:G535–41.

Martin MM, Lee EJ, Buckenberger JA, et al. MicroRNA-155 regulates human angiotensin II type 1 receptor expression in fibroblasts. J Biol Chem. 2006;281:18277–84.

Martinez J, Busslinger M Life beyond cleavage: the case of Ago2 and hematopoiesis. Genes Dev. 2007;21:1983–8.

Masri S, Liu Z, Phung S, et al. The role of microRNA-128a in regulating TGF beta signaling in letrozole-resistant breast cancer cells. Breast Cancer Res Treat. 2010;124:89–99.

Mathe EA, Nguyen GH, Bowman ED, et al. MicroRNA expression in squamous cell carcinoma and adenocarcinoma of the esophagus: associations with survival. Clin Cancer Res. 2009;15: 6192–200.

Mattie MD, Benz CC, Bowers J, et al. Optimized high-throughput microRNA expression profiling provides novel biomarker assessment of clinical prostate and breast cancer biopsies. Mol Cancer. 2006;5:24.

Mavrakis KJ, Wolfe AL, Oricchio E, et al. Genome-wide RNA-mediated interference screen identifies miR-19 targets in notch-induced T-cell acute lymphoblastic leukaemia. Nat Cell Biol. 2010;12:372–9.

Mayr C, Hemann MT, Bartel DP. Disrupting the pairing between let-7 and hmga2 enhances oncogenic transformation. Science. 2007;315:1576–9.

Mees ST, Mardin WA, Sielker S, et al. Involvement of CD40 targeting miR-224 and miR-486 on the progression of pancreatic ductal adenocarcinomas. Ann Surg Oncol. 2009;16:2339–50.

Melar-New M, Laimins LA. Human papillomaviruses modulate expression of microRNA 203 upon epithelial differentiation to control levels of p63 proteins. J Virol. 2010;84:5212–21.

Mendell JT. MiRiad roles for the miR-17-92 cluster in development and disease. Cell. 2008;133:217–22.

Meng F, Henson R, Lang M, et al. Involvement of human micro-RNA in growth and response to chemotherapy in human cholangiocarcinoma cell lines. Gastroenterology. 2006;130: 2113–29.

Meng F, Henson R, Wehbe-Janek H, et al. MicroRNA-21 regulates expression of the PTEN tumor suppressor gene in human hepatocellular cancer. Gastroenterology. 2007a;133:647–58.

Meng F, Henson R, Wehbe-Janek H, et al. The MicroRNA let-7a modulates interleukin-6-dependent STAT-3 survival signaling in malignant human cholangiocytes. J Biol Chem. 2007b;282:8256–64.

Michael MZ, O'Connor SM, van Holst Pellekaan NG, et al. Reduced accumulation of specific microRNAs in colorectal neoplasia. Mol Cancer Res. 2003;1:882–91.

Migliore C, Petrelli A, Ghiso E, et al. MicroRNAs impair MET-mediated invasive growth. Cancer Res. 2008;68:10128–36.

Miller TE, Ghoshal K, Ramaswamy B, et al. MicroRNA-221/222 confers tamoxifen resistance in breast cancer by targeting p27Kip1. J Biol Chem. 2008;283:29897–903.

Mishra PJ, Banerjee D, Bertino JR. MiRSNPs or MiR-polymorphisms, new players in microRNA mediated regulation of the cell: introducing microRNA pharmacogenomics. Cell Cycle. 2008;7:853–8.

Mishra PJ, Bertino JR. MicroRNA polymorphisms: the future of pharmacogenomics, molecular epidemiology and individualized medicine. Pharmacogenomics. 2009;10:399–416.

Mishra PJ, Humeniuk R, Longo-Sorbello GS, et al. A miR-24 microRNA binding-site polymorphism in dihydrofolate reductase gene leads to methotrexate resistance. Proc Natl Acad Sci USA. 2007;104:13513–8.

Mitchell PS, Parkin RK, Kroh EM, et al. Circulating microRNAs as stable blood-based markers for cancer detection. Proc Natl Acad Sci USA. 2008;105:10513–8.

Mizuno Y, Tokuzawa Y, Ninomiya Y, et al. MiR-210 promotes osteoblastic differentiation through inhibition of AcvR1b. FEBS Lett. 2009;583:2263–8.

Motoyama K, Inoue H, Mimori K, et al. Clinicopathological and prognostic significance of PDCD4 and microRNA-21 in human gastric cancer. Int J Oncol. 2010;36:1089–95.

Motoyama K, Inoue H, Takatsuno Y, et al. Over- and under-expressed microRNAs in human colorectal cancer. Int J Oncol. 2009;34:1069–75.

Mott JL, Kobayashi S, Bronk SF, et al. MiR-29 regulates mcl-1 protein expression and apoptosis. Oncogene. 2007;26:6133–40.

Mouillet JF, Chu T, Nelson DM, et al. MiR-205 silences MED1 in hypoxic primary human trophoblasts. FASEB J. 2010;24(6):2030–9.

Moussay E, Palissot V, Vallar L, et al. Determination of genes and microRNAs involved in the resistance to fludarabine in vivo in chronic lymphocytic leukemia. Mol Cancer. 2010;9:115.

Mraz M, Malinova K, Kotaskova J, et al. MiR-34a, miR-29c and miR-17-5p are downregulated in CLL patients with TP53 abnormalities. Leukemia. 2009;23:1159–63.

Muller DW, Bosserhoff AK. Integrin beta 3 expression is regulated by let-7a miRNA in malignant melanoma. Oncogene. 2008;27:6698–706.

Murakami Y, Yasuda T, Saigo K, et al. Comprehensive analysis of microRNA expression patterns in hepatocellular carcinoma and non-tumorous tissues. Oncogene. 2006;25:2537–45.

Murphy AJ, Guyre PM, Pioli PA. Estradiol suppresses NF-kappaB activation through coordinated regulation of let-7a and miR-125b in primary human macrophages. J Immunol. 2010;184:5029–37.

Nagaraja AK, Creighton CJ, Yu Z, et al. A link between mir-100 and FRAP1/mTOR in clear cell ovarian cancer. Mol Endocrinol. 2010;24:447–63.

Naguibneva I, Ameyar-Zazoua M, Polesskaya A, et al. The microRNA miR-181 targets the homeobox protein hox-A11 during mammalian myoblast differentiation. Nat Cell Biol. 2006;8:278–84.

Nam EJ, Yoon H, Kim SW, et al. MicroRNA expression profiles in serous ovarian carcinoma. Clin Cancer Res. 2008;14:2690–5.

Negrini M, Ferracin M, Sabbioni S, et al. MicroRNAs in human cancer: from research to therapy. J Cell Sci. 2007;120:1833–40.

Newman MA, Thomson JM, Hammond SM. Lin-28 interaction with the let-7 precursor loop mediates regulated microRNA processing. RNA. 2008;14:1539–49.

Ng EK, Chong WW, Jin H, et al. Differential expression of microRNAs in plasma of patients with colorectal cancer: a potential marker for colorectal cancer screening. Gut. 2009a;58: 1375–81.

Ng EK, Tsang WP, Ng SS, et al. MicroRNA-143 targets DNA methyltransferases 3a in colorectal cancer. Br J Cancer. 2009b;101:699–706.

Nie K, Gomez M, Landgraf P, et al. MicroRNA-mediated down-regulation of PRDM1/blimp-1 in Hodgkin/Reed-Sternberg cells: a potential pathogenetic lesion in Hodgkin lymphomas. Am J Pathol. 2008;173:242–52.

Nikiforova MN, Tseng GC, Steward D, et al. MicroRNA expression profiling of thyroid tumors: biological significance and diagnostic utility. J Clin Endocrinol Metab. 2008;93:1600–8.

O'Connell RM, Chaudhuri AA, Rao DS, et al. Inositol phosphatase SHIP1 is a primary target of miR-155. Proc Natl Acad Sci USA. 2009;106:7113–8.

O'Donnell KA, Wentzel EA, Zeller KI, et al. C-myc-regulated microRNAs modulate E2F1 expression. Nature. 2005;435:839–43.

Ogawa T, Iizuka M, Sekiya Y, et al. Suppression of type I collagen production by microRNA-29b in cultured human stellate cells. Biochem Biophys Res Commun. 2010;391:316–21.

Ohgawara T, Kubota S, Kawaki H, et al. Regulation of chondrocytic phenotype by microRNA 18a: involvement of Ccn2/Ctgf as a major target gene. FEBS Lett. 2009;583: 1006–10.

Ota A, Tagawa H, Karnan S, et al. Identification and characterization of a novel gene, C13orf25, as a target for 13q31-q32 amplification in malignant lymphoma. Cancer Res. 2004;64:3087–95.
Pallante P, Visone R, Ferracin M, et al. MicroRNA deregulation in human thyroid papillary carcinomas. Endocr Relat Cancer. 2006;13:497–508.
Pallasch CP, Patz M, Park YJ, et al. MiRNA deregulation by epigenetic silencing disrupts suppression of the oncogene PLAG1 in chronic lymphocytic leukemia. Blood. 2009;114(15):3255–64.
Papadopoulos GL, Reczko M, Simossis VA, et al. The database of experimentally supported targets: a functional update of TarBASE. Nucleic Acids Res. 2009;37:D155–8.
Papagiannakopoulos T, Shapiro A, Kosik KS. MicroRNA-21 Targets a network of key tumor-suppressive pathways in glioblastoma cells. Cancer Res. 2008;68:8164–72.
Park SM, Gaur AB, Lengyel E, et al. The miR-200 family determines the epithelial phenotype of cancer cells by targeting the E-cadherin repressors ZEB1 and ZEB2. Genes Dev. 2008;22: 894–907.
Park SM, Shell S, Radjabi AR, et al. Let-7 prevents early cancer progression by suppressing expression of the embryonic gene HMGA2. Cell Cycle. 2007;6:2585–90.
Park SY, Lee JH, Ha M, et al. MiR-29 miRNAs activate p53 by targeting p85 alpha and CDC42. Nat Struct Mol Biol. 2009;16:23–9.
Pekarsky Y, Santanam U, Cimmino A, et al. Tcl1 expression in chronic lymphocytic leukemia is regulated by miR-29 and miR-181. Cancer Res. 2006;66:11590–3.
Peng Y, Laser J, Shi G, et al. Antiproliferative effects by let-7 repression of high-mobility group A2 in uterine leiomyoma. Mol Cancer Res. 2008;6:663–73.
Pentheroudakis G, Golfinopoulos V, Pavlidis N. Switching benchmarks in cancer of unknown primary: from autopsy to microarray. Eur J Cancer. 2007;43:2026–36.
Petrocca F, Visone R, Onelli MR, et al. E2F1-regulated microRNAs impair TGF beta-dependent cell-cycle arrest and apoptosis in gastric cancer. Cancer Cell. 2008;13:272–86.
Pezzolesi MG, Platzer P, Waite KA, et al. Differential expression of PTEN-targeting microRNAs miR-19a and miR-21 in cowden syndrome. Am J Hum Genet. 2008;82:1141–9.
Piskounova E, Viswanathan SR, Janas M, et al. Determinants of microRNA processing inhibition by the developmentally regulated RNA-binding protein lin28. J Biol Chem. 2008;283:21310–4.
Place RF, Li LC, Pookot D, et al. MicroRNA-373 induces expression of genes with complementary promoter sequences. Proc Natl Acad Sci USA. 2008;105:1608–13.
Poliseno L, Pitto L, Simili M, et al. The proto-oncogene LRF is under post-transcriptional control of MiR-20a: implications for senescence. PLoS One. 2008;3:e2542.
Poliseno L, Salmena L, Riccardi L, et al. Identification of the miR-106b~25 microRNA cluster as a proto-oncogenic PTEN-targeting intron that cooperates with its host gene MCM7 in transformation. Sci Signal. 2010;3:ra29.
Porkka KP, Pfeiffer MJ, Waltering KK, et al. MicroRNA expression profiling in prostate cancer. Cancer Res. 2007;67:6130–5.
Punj V, Matta H, Schamus S, et al. Kaposi's sarcoma-associated herpesvirus-encoded viral FLICE inhibitory protein (vFLIP) K13 suppresses CXCR4 expression by upregulating miR-146a. Oncogene. 2010;29:1835–44.
Qin X, Wang X, Wang Y, et al. MicroRNA-19a mediates the suppressive effect of laminar flow on cyclin D1 expression in human umbilical vein endothelial cells. Proc Natl Acad Sci USA. 2010;107:3240–4.
Ragan C, Cloonan N, Grimmond SM, et al. Transcriptome-wide prediction of miRNA targets in human and mouse using FASTH. PLoS One. 2009;4:e5745.
Rai D, Karanti S, Jung I, et al. Coordinated expression of microRNA-155 and predicted target genes in diffuse large B-cell lymphoma. Cancer Genet Cytogenet. 2008;181:8–15.
Rai D, Kim SW, McKeller MR, et al. Targeting of SMAD5 links microRNA-155 to the TGF-beta pathway and lymphomagenesis. Proc Natl Acad Sci USA. 2010;107:3111–6.
Rane S, He M, Sayed D, et al. Downregulation of miR-199a derepresses hypoxia-inducible factor-1alpha and sirtuin 1 and recapitulates hypoxia preconditioning in cardiac myocytes. Circ Res. 2009;104:879–6.

Raver-Shapira N, Marciano E, Meiri E, et al. Transcriptional activation of miR-34a contributes to p53-mediated apoptosis. Mol Cell. 2007;26:731–43.

Resnick KE, Alder H, Hagan JP, et al. The detection of differentially expressed microRNAs from the serum of ovarian cancer patients using a novel real-time PCR platform. Gynecol Oncol. 2009;112:55–9.

Rinaldi A, Poretti G, Kwee I, et al. Concomitant MYC and microRNA cluster miR-17-92 (C13orf25) amplification in human mantle cell lymphoma. Leuk Lymphoma. 2007;48:410–2.

Roccaro AM, Sacco A, Thompson B, et al. MicroRNAs 15a and 16 regulate tumor proliferation in multiple myeloma. Blood. 2009;113:6669–80.

Rogler CE, Levoci L, Ader T, et al. MicroRNA-23b cluster microRNAs regulate transforming growth factor-beta/bone morphogenetic protein signaling and liver stem cell differentiation by targeting smads. Hepatology. 2009;50:575–84.

Rokhlin OW, Scheinker VS, Taghiyev AF, et al. MicroRNA-34 mediates AR-dependent p53-induced apoptosis in prostate cancer. Cancer Biol Ther. 2008;7:1288–96.

Roldo C, Missiaglia E, Hagan JP, et al. MicroRNA expression abnormalities in pancreatic endocrine and acinar tumors are associated with distinctive pathologic features and clinical behavior. J Clin Oncol. 2006;24:4677–84.

Rom S, Rom I, Passiatore G, et al. CCL8/MCP-2 is a target for miR-146a in HIV-1-infected human microglial cells. FASEB J. 2010;24(7):2292–300.

Romania P, Lulli V, Pelosi E, et al. MicroRNA 155 modulates megakaryopoiesis at progenitor and precursor level by targeting ETS-1 and MEIS1 transcription factors. Br J Haematol. 2008;143:570–80.

Rosenfeld N, Aharonov R, Meiri E, et al. MicroRNAs accurately identify cancer tissue origin. Nat Biotechnol. 2008;26:462–9.

Rosenwald S, Gilad S, Benjamin S, et al. Validation of a microRNA-based qRT-PCR test for accurate identification of tumor tissue origin. Mod Pathol. 2010;23(6):814–23.

Ruan J, Chen H, Kurgan L, et al. HuMiTar: a sequence-based method for prediction of human microRNA targets. Algorithms Mol Biol. 2008;3:16.

Rybak A, Fuchs H, Smirnova L, et al. A feedback loop comprising lin-28 and let-7 controls pre-let-7 maturation during neural stem-cell commitment. Nat Cell Biol. 2008;10: 987–93.

Sachdeva M, Mo YY. MicroRNA-145 Suppresses cell invasion and metastasis by directly targeting mucin 1. Cancer Res. 2010;70:378–87.

Sachdeva M, Zhu S, Wu F, et al. P53 represses c-myc through induction of the tumor suppressor miR-145. Proc Natl Acad Sci USA. 2009;106:3207–12.

Saetrom P, Biesinger J, Li SM, et al. A risk variant in an miR-125b binding site in BMPR1B is associated with breast cancer pathogenesis. Cancer Res. 2009;69:7459–65.

Saito Y, Liang G, Egger G, et al. Specific activation of microRNA-127 with downregulation of the proto-oncogene BCL6 by chromatin-modifying drugs in human cancer cells. Cancer Cell. 2006;9:435–43.

Salter KH, Acharya CR, Walters KS, et al. An integrated approach to the prediction of chemotherapeutic response in patients with breast cancer. PLoS One. 2008;3:e1908.

Salvi A, Sabelli C, Moncini S, et al. MicroRNA-23b mediates urokinase and c-Met downmodulation and a decreased migration of human hepatocellular carcinoma cells. FEBS J. 2009;276:2966–82.

Sampson VB, Rong NH, Han J, et al. MicroRNA let-7a down-regulates MYC and reverts MYC-induced growth in burkitt lymphoma cells. Cancer Res. 2007;67:9762–70.

Sander S, Bullinger L, Klapproth K, et al. MYC stimulates EZH2 expression by repression of its negative regulator miR-26a. Blood. 2008;112:4202–12.

Sander S, Bullinger L, Wirth T. Repressing the repressor: a new mode of MYC action in lymphomagenesis. Cell Cycle. 2009;8:556–9.

Sarasin-Filipowicz M, Krol J, Markiewicz I, et al. Decreased levels of microRNA miR-122 in individuals with hepatitis C responding poorly to interferon therapy. Nat Med. 2009;15:31–3.

Satzger I, Mattern A, Kuettler U, et al. MicroRNA-15b represents an independent prognostic parameter and is correlated with tumor cell proliferation and apoptosis in malignant melanoma. Int J Cancer. 2009;126:2553–62.

Saydam O, Shen Y, Wurdinger T, et al. Down-regulated microRNA-200a in meningiomas promotes tumor growth by reducing E-cadherin and activating the wnt/{beta} catenin signaling pathway. Mol Cell Biol. 2009;29:5923–40.

Sayed D, He M, Hong C, et al. MicroRNA-21 is a downstream effector of AKT that mediates its antiapoptotic effects via suppression of fas ligand. J Biol Chem. 2010;285(26): 20281–90.

Sayed D, Rane S, Lypowy J, et al. MicroRNA-21 targets Sprouty2 and promotes cellular outgrowths. Mol Biol Cell. 2008;19:3272–82.

Schaefer A, Jung M, Mollenkopf HJ, et al. Diagnostic and prognostic implications of microRNA profiling in prostate carcinoma. Int J Cancer. 2009;126(5):1166–76.

Schetter AJ, Nguyen GH, Bowman ED, et al. Association of inflammation-related and microRNA gene expression with cancer-specific mortality of colon adenocarcinoma. Clin Cancer Res. 2009;15:5878–87.

Schulte JH, Horn S, Otto T, et al. MYCN regulates oncogenic MicroRNAs in neuroblastoma. Int J Cancer. 2008;122:699–704.

Schultz J, Lorenz P, Gross G, et al. MicroRNA let-7b targets important cell cycle molecules in malignant melanoma cells and interferes with anchorage-independent growth. Cell Res. 2008;18:549–57.

Scott GK, Goga A, Bhaumik D, et al. Coordinate suppression of ERBB2 and ERBB3 by enforced expression of micro-RNA miR-125a or miR-125b. J Biol Chem. 2007;282:1479–86.

Selaru FM, Olaru AV, Kan T, et al. MicroRNA-21 is overexpressed in human cholangiocarcinoma and regulates programmed cell death 4 and tissue inhibitor of metalloproteinase 3. Hepatology. 2009;49:1595–601.

Selbach M, Schwanhausser B, Thierfelder N, et al. Widespread changes in protein synthesis induced by microRNAs. Nature. 2008;455:58–63.

Sengupta S, den Boon JA, Chen IH, et al. MicroRNA 29c is down-regulated in nasopharyngeal carcinomas, up-regulating mRNAs encoding extracellular matrix proteins. Proc Natl Acad Sci USA. 2008;105:5874–8.

Sethupathy P, Corda B, Hatzigeorgiou AG. TarBASE: A comprehensive database of experimentally supported animal microRNA targets. RNA. 2006;12:192–7.

Shan SW, Lee DY, Deng Z, et al. MicroRNA miR-17 retards tissue growth and represses fibronectin expression. Nat Cell Biol. 2009;11:1031–8.

Sharma A, Kumar M, Aich J, et al. Posttranscriptional regulation of interleukin-10 expression by hsa-miR-106a. Proc Natl Acad Sci USA. 2009;106:5761–6.

Shen WF, Hu YL, Uttarwar L, et al. MicroRNA-126 regulates HOXA9 by binding to the homeobox. Mol Cell Biol. 2008;28:4609–19.

Shi B, Sepp-Lorenzino L, Prisco M, et al. MicroRNA 145 targets the insulin receptor substrate-1 and inhibits the growth of colon cancer cells. J Biol Chem. 2007;282:32582–90.

Shi L, Ko ML, Ko GY. Rhythmic expression of microRNA-26a regulates the L-type voltage-gated calcium channel alpha1c subunit in chicken cone photoreceptors. J Biol Chem. 2009;284:25791–803.

Shi W, Alajez NM, Bastianutto C, et al. Significance of Plk1 regulation by miR-100 in human nasopharyngeal cancer. Int J Cancer. 2010;126:2036–48.

Shimizu S, Takehara T, Hikita H, et al. The let-7 family of microRNAs inhibits bcl-xL expression and potentiates sorafenib-induced apoptosis in human hepatocellular carcinoma. J Hepatol. 2010;52(5):698–704.

Simion A, Laudadio I, Prevot PP, et al. MiR-495 and miR-218 regulate the expression of the onecut transcription factors HNF-6 and OC-2. Biochem Biophys Res Commun. 2010;391:293–8.

Sinclair CS, Rowley M, Naderi A, et al. The 17q23 amplicon and breast cancer. Breast Cancer Res Treat. 2003;78:313–22.

Slack F. Let-7 microRNA reduces tumor growth. Cell Cycle. 2009;8:1823.

Song B, Wang C, Liu J, et al. MicroRNA-21 regulates breast cancer invasion partly by targeting tissue inhibitor of metalloproteinase 3 expression. J Exp Clin Cancer Res. 2010;29:29.

Song H, Bu G. MicroRNA-205 Inhibits tumor cell migration through down-regulating the expression of the LDL receptor-related protein 1. Biochem Biophys Res Commun. 2009;388: 400–5.

Spizzo R, Nicoloso MS, Lupini L, et al. MiR-145 participates with TP53 in a death-promoting regulatory loop and targets estrogen receptor-alpha in human breast cancer cells. Cell Death Differ. 2010;17:246–54.

Spurgers KB, Gold DL, Coombes KR, et al. Identification of cell cycle regulatory genes as principal targets of p53-mediated transcriptional repression. J Biol Chem. 2006;281:25134–42.

Strillacci A, Griffoni C, Sansone P, et al. MiR-101 downregulation is involved in cyclooxygenase-2 overexpression in human colon cancer cells. Exp Cell Res. 2009;315:1439–47.

Su H, Yang JR, Xu T, et al. MicroRNA-101, down-regulated in hepatocellular carcinoma, promotes apoptosis and suppresses tumorigenicity. Cancer Res. 2009;69:1135–42.

Subramanian S, Lui WO, Lee CH, et al. MicroRNA expression signature of human sarcomas. Oncogene. 2008;27:2015–26.

Sun Y, Bai Y, Zhang F, et al. MiR-126 inhibits non-small cell lung cancer cells proliferation by targeting EGFL7. Biochem Biophys Res Commun. 2010;391:1483–9.

Sureban SM, May R, Ramalingam S, et al. Selective blockade of DCAMKL-1 results in tumor growth arrest by a let-7a MicroRNA-dependent mechanism. Gastroenterology. 2009;137:649–659, 59e1–2.

Svoboda M, Izakovicova Holla L, Sefr R, et al. Micro-RNAs miR125b and miR137 are frequently upregulated in response to capecitabine chemoradiotherapy of rectal cancer. Int J Oncol. 2008;33:541–7.

Sylvestre Y, De Guire V, Querido E, et al. An E2F/miR-20a autoregulatory feedback loop. J Biol Chem. 2007;282:2135–43.

Szafranska AE, Davison TS, John J, et al. MicroRNA expression alterations are linked to tumorigenesis and non-neoplastic processes in pancreatic ductal adenocarcinoma. Oncogene. 2007;26:4442–52.

Tagawa H, Seto M. A microRNA cluster as a target of genomic amplification in malignant lymphoma. Leukemia. 2005;19:2013–6.

Taguchi A, Yanagisawa K, Tanaka M, et al. Identification of hypoxia-inducible factor-1 alpha as a novel target for miR-17-92 microRNA cluster. Cancer Res. 2008;68:5540–5.

Takamizawa J, Konishi H, Yanagisawa K, et al. Reduced expression of the let-7 microRNAs in human lung cancers in association with shortened postoperative survival. Cancer Res. 2004;64:3753–6.

Tan LP, Seinen E, Duns G, et al. A high throughput experimental approach to identify miRNA targets in human cells. Nucleic Acids Res. 2009;37:e137.

Tanaka M, Oikawa K, Takanashi M, et al. Down-regulation of miR-92 in human plasma is a novel marker for acute leukemia patients. PLoS One. 2009;4:e5532.

Tang B, Xiao B, Liu Z, et al. Identification of MyD88 as a novel target of miR-155, involved in negative regulation of *Helicobacter pylori*-induced inflammation. FEBS Lett. 2010;584: 1481–6.

Tang G. SiRNA and miRNA: an insight into RISCs. Trends Biochem Sci. 2005;30:106–14.

Tanzer A, Stadler PF. Molecular evolution of a microRNA cluster. J Mol Biol. 2004;339:327–35.

Tarasov V, Jung P, Verdoodt B, et al. Differential regulation of microRNAs by p53 revealed by massively parallel sequencing: miR-34a is a p53 target that induces apoptosis and G1-arrest. Cell Cycle. 2007;6:1586–93.

Taulli R, Bersani F, Foglizzo V, et al. The muscle-specific microRNA miR-206 blocks human rhabdomyosarcoma growth in xenotransplanted mice by promoting myogenic differentiation. J Clin Invest. 2009;119:2366–78.

Tavazoie SF, Alarcon C, Oskarsson T, et al. Endogenous human microRNAs that suppress breast cancer metastasis. Nature. 2008;451:147–52.

Tazawa H, Tsuchiya N, Izumiya M, et al. Tumor-suppressive miR-34a induces senescence-like growth arrest through modulation of the E2F pathway in human colon cancer cells. Proc Natl Acad Sci USA. 2007;104:15472–7.

Teng G, Hakimpour P, Landgraf P, et al. MicroRNA-155 is a negative regulator of activation-induced cytidine deaminase. Immunity. 2008;28:621–9.

Tetzlaff MT, Liu A, Xu X, et al. Differential expression of miRNAs in papillary thyroid carcinoma compared to multinodular goiter using formalin fixed paraffin embedded tissues. Endocr Pathol. 2007;18:163–73.

Tie J, Pan Y, Zhao L, et al. MiR-218 inhibits invasion and metastasis of gastric cancer by targeting the robo1 receptor. PLoS Genet. 2010;6:e1000879.

Tokumaru S, Suzuki M, Yamada H, et al. Let-7 regulates dicer expression and constitutes a negative feedback loop. Carcinogenesis. 2008;29:2073–7.

Tong AW, Fulgham P, Jay C, et al. MicroRNA profile analysis of human prostate cancers. Cancer Gene Ther. 2009;16:206–16.

Toyota M, Suzuki H, Sasaki Y, et al. Epigenetic silencing of microRNA-34b/c and B-cell translocation gene 4 is associated with CpG island methylation in colorectal cancer. Cancer Res. 2008;68:4123–32.

Tsang WP, Kwok TT. Let-7a microRNA suppresses therapeutics-induced cancer cell death by targeting caspase-3. Apoptosis. 2008;13:1215–22.

Utikal J, Polo JM, Stadtfeld M, et al. Immortalization eliminates a roadblock during cellular reprogramming into iPS cells. Nature. 2009;460:1145–8.

Valeri N, Gasparini P, Fabbri M, et al. Modulation of mismatch repair and genomic stability by miR-155. Proc Natl Acad Sci USA. 2010;107:6982–7.

Varambally S, Cao Q, Mani RS, et al. Genomic loss of microRNA-101 leads to overexpression of histone methyltransferase EZH2 in cancer. Science. 2008;322:1695–9.

Vasudevan S, Tong Y, Steitz JA. Switching from repression to activation: microRNAs can up-regulate translation. Science. 2007;318:1931–4.

Veerla S, Lindgren D, Kvist A, et al. MiRNA expression in urothelial carcinomas: important roles of miR-10a, miR-222, miR-125b, miR-7 and miR-452 for tumor stage and metastasis, and frequent homozygous losses of miR-31. Int J Cancer. 2009;124:2236–42.

Ventura A, Young AG, Winslow MM, et al. Targeted deletion reveals essential and overlapping functions of the miR-17 through 92 family of miRNA clusters. Cell. 2008;132:875–86.

Veronese A, Lupini L, Consiglio J, et al. Oncogenic role of miR-483-3p at the IGF2/483 locus. Cancer Res. 2010;70:3140–9.

Verrier JD, Lau P, Hudson L, et al. Peripheral myelin protein 22 is regulated post-transcriptionally by miRNA-29a. Glia. 2009;57:1265–79.

Visone R, Russo L, Pallante P, et al. MicroRNAs (miR)-221 and miR-222, both overexpressed in human thyroid papillary carcinomas, regulate p27Kip1 protein levels and cell cycle. Endocr Relat Cancer. 2007;14:791–8.

Viswanathan SR, Daley GQ, Gregory RI. Selective blockade of microRNA processing by lin28. Science. 2008;320:97–100.

Viswanathan SR, Powers JT, Einhorn W, et al. Lin28 promotes transformation and is associated with advanced human malignancies. Nat Genet. 2009;41:843–8.

Volinia S, Calin GA, Liu CG, et al. A microRNA expression signature of human solid tumors defines cancer gene targets. Proc Natl Acad Sci USA. 2006;103:2257–61.

Volinia S, Galasso M, Costinean S, et al. Reprogramming of miRNA networks in cancer and leukemia. Genome Res. 2010;20:589–99.

Voorhoeve PM, le Sage C, Schrier M, et al. A genetic screen implicates miRNA-372 and miRNA-373 as oncogenes in testicular germ cell tumors. Cell. 2006;124:1169–81.

Vousden KH, Lane DP. P53 in health and disease. Nat Rev Mol Cell Biol. 2007;8:275–83.

Vreugdenhil E, Verissimo CS, Mariman R, et al. MicroRNA 18 and 124a down-regulate the glucocorticoid receptor: implications for glucocorticoid responsiveness in the brain. Endocrinology. 2009;150:2220–8.

Wan HY, Guo LM, Liu T, et al. Regulation of the transcription factor NF-kappaB1 by microRNA-9 in human gastric adenocarcinoma. Mol Cancer. 2010;9:16.

Wang B, Hsu SH, Majumder S, et al. TGF beta-mediated upregulation of hepatic miR-181b promotes hepatocarcinogenesis by targeting TIMP3. Oncogene. 2010;29:1787–97.

Wang H, Li WH. Increasing MicroRNA target prediction confidence by the relative R(2) method. J Theor Biol. 2009;259:793–8.

Wang Q, Huang Z, Xue H, et al. MicroRNA miR-24 inhibits erythropoiesis by targeting activin type I receptor ALK4. Blood. 2008a;111:588–95.

Wang Q, Li YC, Wang J, et al. MiR-17-92 cluster accelerates adipocyte differentiation by negatively regulating tumor-suppressor RB2/p130. Proc Natl Acad Sci USA. 2008b;105:2889–94.

Wang S, Bian C, Yang Z, et al. MiR-145 inhibits breast cancer cell growth through RTKN. Int J Oncol. 2009;34:1461–6.

Wang T, Zhang X, Obijuru L, et al. A micro-RNA signature associated with race, tumor size, and target gene activity in human uterine leiomyomas. Genes Chromosomes Cancer. 2007;46: 336–47.

Wang X. MiRDB: a microRNA target prediction and functional annotation database with a wiki interface. RNA. 2008;14:1012–7.

Wang X, El Naqa IM. Prediction of both conserved and nonconserved microRNA targets in animals. Bioinformatics. 2008;24:325–32.

Wang Y, Lee AT, Ma JZ, et al. Profiling microRNA expression in hepatocellular carcinoma reveals microRNA-224 up-regulation and apoptosis inhibitor-5 as a microRNA-224-specific target. J Biol Chem. 2008c;283:13205–15.

Weber F, Teresi RE, Broelsch CE, et al. A limited set of human MicroRNA is deregulated in follicular thyroid carcinoma. J Clin Endocrinol Metab. 2006;91:3584–91.

Weiss GJ, Bemis LT, Nakajima E, et al. EGFR regulation by microRNA in lung cancer: correlation with clinical response and survival to gefitinib and EGFR expression in cell lines. Ann Oncol. 2008;19:1053–9.

Welch C, Chen Y, Stallings RL. MicroRNA-34a functions as a potential tumor suppressor by inducing apoptosis in neuroblastoma cells. Oncogene. 2007;26:5017–22.

West JA, Viswanathan SR, Yabuuchi A, et al. A role for lin28 in primordial germ-cell development and germ-cell malignancy. Nature. 2009;460:909–13.

White NM, Chow TF, Mejia-Guerrero S, et al. Three dysregulated miRNAs control kallikrein 10 expression and cell proliferation in ovarian cancer. Br J Cancer. 2010;102:1244–53.

Winter J, Jung S, Keller S, et al. Many roads to maturity: microRNA biogenesis pathways and their regulation. Nat Cell Biol. 2009;11:228–34.

Wong CF, Tellam RL. MicroRNA-26a targets the histone methyltransferase enhancer of Zeste homolog 2 during myogenesis. J Biol Chem. 2008;283:9836–43.

Wong TS, Liu XB, Chung-Wai Ho A, et al. Identification of pyruvate kinase type M2 as potential oncoprotein in squamous cell carcinoma of tongue through microRNA profiling. Int J Cancer. 2008a;123:251–7.

Wong TS, Liu XB, Wong BY, et al. Mature miR-184 as potential oncogenic microRNA of squamous cell carcinoma of tongue. Clin Cancer Res. 2008b;14:2588–92.

Xi Y, Nakajima G, Gavin E, et al. Systematic analysis of microRNA expression of RNA extracted from fresh frozen and formalin-fixed paraffin-embedded samples. RNA. 2007;13:1668–74.

Xia H, Ng SS, Jiang S, et al. MiR-200a-mediated downregulation of ZEB2 and CTNNB1 differentially inhibits nasopharyngeal carcinoma cell growth, migration and invasion. Biochem Biophys Res Commun. 2010;391:535–41.

Xia H, Qi Y, Ng SS, et al. MicroRNA-146b inhibits glioma cell migration and invasion by targeting MMPs. Brain Res. 2009a;1269:158–65.

Xia HF, He TZ, Liu CM, et al. MiR-125b expression affects the proliferation and apoptosis of human glioma cells by targeting Bmf. Cell Physiol Biochem. 2009b;23:347–58.

Xiao C, Srinivasan L, Calado DP, et al. Lymphoproliferative disease and autoimmunity in mice with increased miR-17-92 expression in lymphocytes. Nat Immunol. 2008;9:405–14.

Xiong Y, Fang JH, Yun JP, et al. Effects of microRNA-29 on apoptosis, tumorigenicity, and prognosis of hepatocellular carcinoma. Hepatology. 2010;51:836–45.

Xu C, Lu Y, Pan Z, et al. The muscle-specific microRNAs miR-1 and miR-133 produce opposing effects on apoptosis by targeting HSP60, HSP70 and caspase-9 in cardiomyocytes. J Cell Sci. 2007;120:3045–52.

Xu H, Cheung IY, Guo HF, et al. MicroRNA miR-29 modulates expression of immunoinhibitory molecule B7-H3: potential implications for immune based therapy of human solid tumors. Cancer Res. 2009a;69:6275–81.

Xu N, Papagiannakopoulos T, Pan G, et al. MicroRNA-145 regulates OCT4, SOX2, and KLF4 and represses pluripotency in human embryonic stem cells. Cell. 2009b;137:647–58.

Yamakuchi M, Lotterman CD, Bao C, et al. P53-induced microRNA-107 inhibits HIF-1 and tumor angiogenesis. Proc Natl Acad Sci USA. 2010;107:6334–9.

Yamamoto Y, Kosaka N, Tanaka M, et al. MicroRNA-500 as a potential diagnostic marker for hepatocellular carcinoma. Biomarkers. 2009;14(7):529–38.

Yanaihara N, Caplen N, Bowman E, et al. Unique microRNA molecular profiles in lung cancer diagnosis and prognosis. Cancer Cell. 2006;9:189–98.

Yang H, Kong W, He L, et al. MicroRNA expression profiling in human ovarian cancer: miR-214 induces cell survival and cisplatin resistance by targeting PTEN. Cancer Res. 2008;68:425–33.

Yao G, Yin M, Lian J, et al. MicroRNA-224 is involved in transforming growth factor-beta-mediated mouse granulosa cell proliferation and granulosa cell function by targeting Smad4. Mol Endocrinol. 2010;24:540–51.

Yao Q, Xu H, Zhang QQ, et al. MicroRNA-21 promotes cell proliferation and down-regulates the expression of programmed cell death 4 (PDCD4) in HeLa cervical carcinoma cells. Biochem Biophys Res Commun. 2009;388:539–42.

Yi R, Poy MN, Stoffel M, et al. A skin microRNA promotes differentiation by repressing 'stemness'. Nature. 2008;452:225–9.

Yin Q, Wang X, Fewell C, et al. MiR-155 inhibits bone morphogenetic protein (BMP) signaling and BMP mediated Epstein-Barr virus reactivation. J Virol. 2010;84(13):6318–27.

Yu F, Yao H, Zhu P, et al. Let-7 regulates self renewal and tumorigenicity of breast cancer cells. Cell. 2007a;131:1109–23.

Yu J, Vodyanik MA, Smuga-Otto K, et al. Induced pluripotent stem cell lines derived from human somatic cells. Science. 2007b;318:1917–20.

Yu SL, Chen HY, Chang GC, et al. MicroRNA signature predicts survival and relapse in lung cancer. Cancer Cell. 2008a;13:48–57.

Yu Z, Wang C, Wang M, et al. A cyclin D1/microRNA 17/20 regulatory feedback loop in control of breast cancer cell proliferation. J Cell Biol. 2008b;182:509–17.

Zanette DL, Rivadavia F, Molfetta GA, et al. MiRNA expression profiles in chronic lymphocytic and acute lymphocytic leukemia. Braz J Med Biol Res. 2007;40:1435–40.

Zenz T, Mohr J, Eldering E, et al. MiR-34a as part of the resistance network in chronic lymphocytic leukemia. Blood. 2009;113:3801–8.

Zhang C, Kang C, You Y, et al. Co-suppression of miR-221/222 cluster suppresses human glioma cell growth by targeting p27Kip1 in vitro and in vivo. Int J Oncol. 2009a;34: 1653–60.

Zhang J, Du YY, Lin YF, et al. The cell growth suppressor, miR-126, targets IRS-1. Biochem Biophys Res Commun. 2008;377:136–40.

Zhang JG, Wang JJ, Zhao F, et al. MicroRNA-21 (miR-21) represses tumor suppressor PTEN and promotes growth and invasion in non-small cell lung cancer (NSCLC). Clin Chim Acta. 2010;411:846–52.

Zhang X, Liu S, Hu T, et al. Up-regulated microRNA-143 transcribed by nuclear factor kappa B enhances hepatocarcinoma metastasis by repressing fibronectin expression. Hepatology. 2009b;50:490–9.

Zhang Y, Chao T, Li R, et al. MicroRNA-128 inhibits glioma cells proliferation by targeting transcription factor E2F3a. J Mol Med. 2009c;87:43–51.

Zhang Z, Sun H, Dai H, et al. MicroRNA miR-210 modulates cellular response to hypoxia through the MYC antagonist MNT. Cell Cycle. 2009d;8:2756–68.

Zhao C, Sun G, Li S, et al. MicroRNA let-7b regulates neural stem cell proliferation and differentiation by targeting nuclear receptor TLX signaling. Proc Natl Acad Sci USA. 2010a;107:1876–81.

Zhao C, Sun G, Li S, et al. A feedback regulatory loop involving microRNA-9 and nuclear receptor TLX in neural stem cell fate determination. Nat Struct Mol Biol. 2009;16:365–71.

Zhao JJ, Lin J, Lwin T, et al. MicroRNA expression profile and identification of miR-29 as a prognostic marker and pathogenetic factor by targeting CDK6 in mantle cell lymphoma. Blood. 2010b;115:2630–9.

Zhao JJ, Lin J, Yang H, et al. MicroRNA-221/222 negatively regulates estrogen receptor alpha and is associated with tamoxifen resistance in breast cancer. J Biol Chem. 2008;283:31079–86.

Zhu S, Sachdeva M, Wu F, et al. Ubc9 promotes breast cell invasion and metastasis in a sumoylation-independent manner. Oncogene. 2010a;29:1763–72.

Zhu S, Si ML, Wu H, et al. MicroRNA-21 targets the tumor suppressor gene tropomyosin 1 (TPM1). J Biol Chem. 2007;282:14328–36.

Zhu S, Wu H, Wu F, et al. MicroRNA-21 targets tumor suppressor genes in invasion and metastasis. Cell Res. 2008;18:350–9.

Zhu W, Shan X, Wang T, et al. MiR-181b modulates multidrug resistance by targeting BCL2 in human cancer cell lines. Int J Cancer. 2010b;127:2520–9.

Chapter 2
The Role of MicroRNAs in Lung Cancer Development, Progression, and Metastasis

Don L. Gibbons, Zain H. Rizvi, and Jonathan M. Kurie

Abstract Lung cancer is the leading cause of cancer death globally. Although molecularly targeted agents have made small advances in the treatment options for patients, the overall 5-year survival rate has changed little in the past several decades, necessitating a greater understanding of the biology driving tumor progression and metastasis. MicroRNAs (miRNAs) are a relatively recently discovered class of non-protein coding RNAs that modulate extremely important cellular functions via their post-transcriptional regulation of mRNAs. Recent evidence from multiple tumor types and model systems implicates miRNA dysregulation as a common mechanism of tumorigenesis and progression. This represents a rapidly emerging and changing field with new biological connections and applications being reported each month, which provide unique insights into miRNA functions and potential new approaches for diagnosis and therapy.

2.1 Introduction

Lung cancer is the leading cause of cancer-related death in the United States and worldwide (Jemal et al. 2010; Kamangar et al. 2006). In the year 2010, it is estimated that approximately 222,520 new cases will be diagnosed and 157,300 deaths will occur in the United States alone. Only 16% of patients diagnosed with lung cancer are alive at 5 years (Jemal et al. 2010; Kamangar et al. 2006), because at the time of diagnosis more than 70% of patients are found to have advanced disease that is not amenable to curative therapy (Goldstraw et al. 2007) and among the remaining patients who undergo surgical resection with curative intent for early-stage disease, there is a high rate of recurrence. Given the generally poor prognosis of patients with this disease, better treatments are needed based upon the molecular events driving tumor progression and metastasis.

J.M. Kurie (✉)
Department of Thoracic, Head and Neck Medical Oncology, MD Anderson Cancer Center, The University of Texas, Houston, TX 77030, USA
e-mail: jkurie@mdanderson.org

W.C.S. Cho (ed.), *MicroRNAs in Cancer Translational Research*,
DOI 10.1007/978-94-007-0298-1_2,

Lung cancer is broadly divided into non-small cell lung cancer (NSCLC), which arises from epithelial cells of the airways and accounts for 85% of cases, and small cell lung cancer (SCLC), which is a neuroendocrine tumor that comprises the remainder. Non-small cell lung cancer is further divided into several major histologic subtypes, including adenocarcinoma, squamous cell carcinoma, and large cell. Although significant work has gone into identifying and studying the oncogenes responsible for tumorigenesis and progression in NSCLC and SCLC, there has been limited success in translating these findings into the therapeutic setting or in using this information to better stratify patient risk and select treatments.

MicroRNAs (miRNAs) are a class of highly conserved non-coding RNAs, 19–25 nucleotides in length, that regulate gene expression by recognition and binding of the 3′ untranslated regions (3′UTRs) of mRNAs. Mature miRNAs form stem-loop structures with sequences partially complementary to their target mRNAs, which alter mRNA stability or the efficiency of translation. Based upon recent re-evaluation of the miRNAs in the mouse genome by high-throughput sequencing, there are ~500 confidently identified murine miRNAs, ~300 of which are conserved in mammals (Chiang et al. 2010), and it is estimated that approximately one-third of the genome is regulated by miRNAs (Lewis et al. 2005). Interestingly, since each miRNA can regulate the expression of large sets of targets and any one gene can be regulated by multiple miRNAs, a rich network of regulatory fine-tuning is at play that can go awry in pathologic states such as cancer. Given this powerful biology, various miRNAs have been implicated as either tumor suppressors or oncogenes ("oncomirs") in many different tumor types. Genomically miRNAs are frequently found to be at fragile sites in the human genome (Calin et al. 2004), but there are myriad additional mechanisms by which the miRNAs can become dysregulated in cancer. This chapter will outline our understanding of miRNA biology in lung cancer development and metastasis.

2.2 MiRNA Profiling of Lung Cancer and Clinical Application

MiRNAs have the ability to post-transcriptionally regulate large sets of genes, and as such their expression levels should faithfully represent the overall biologic state of tumor cells. In fact, miRNA signatures segregate samples of different tumor types better than mRNA signatures (Lu et al. 2005), suggesting that they may be useful biomarkers of the unique underlying set of molecular events driving tumorigenesis and metastasis (Cho 2010). This idea has raised significant research interest in identification of oncogenic drivers of tumor progression, and in the development of miRNA signatures that would be clinically prognostic of outcome in early-stage lung cancer. Such signatures could be useful for stratification of patient risk for recurrence, helping to inform the decision of who should receive adjuvant treatment after surgery, or predictive of which tumors will respond to particular chemotherapeutic agents. Several different approaches have been utilized in these efforts, including comparison of tumor samples to matched non-cancerous lung tissue with

Table 2.1 MicroRNAs implicated in NSCLC based upon tumor profiling

Representative microRNAs in signature	Measured outcome	References
miR-221, let-7a, miR-137, miR-372, miR-182	Overall and disease-free survival	Yu et al. (2008)
miR-155, miR-17-3p, miR-106a, miR-93, let-7a-2, miR-145, let-7b, miR-21	Overall survival	Yanaihara et al. (2006)
miR-200b, miR-30c-1, miR-510 miR-630, miR-657, miR-146b-3p, miR-124, miR-585, miR-708	Overall and recurrence-free survival	Patnaik et al. (2009)
miR-34a	Post-surgical recurrence	Gallardo et al. (2009)
miR-146b, miR-155	Overall survival in squamous cell carcinoma	Raponi et al. (2009)
miR-34b, miR-34c, miR-449	Diagnostic discrimination and early cancer detection	Liang (2008)
miR-486, miR-30d, miR-1, miR-499 (serum samples)	Overall survival	Hu et al. (2010)
miR-221, let-7a, miR-137, miR-372, miR-182	Overall and disease-free survival	Yu et al. (2008)

the aim of identifying pathways driving tumorigenesis, and the use of miR profiling across a large panel of tumors to derive multi-miR signatures for classification of patient risk for recurrence from this heterogeneous group (Table 2.1).

Profiling of miRNA expression in 104 NSCLC tumors versus matching non-cancerous lung tissue demonstrated that 43 miRs were differentially expressed, including 15 up-regulated (e.g. *miR-21*, *miR-191*, *miR-155*, and *miR-17-3p*) and 28 down-regulated (e.g. *let-7a-2* and *miR-145*) (Yanaihara et al. 2006). The authors specifically demonstrated that high *miR-155* and low *let-7a-2* levels correlated with survival in multiple independent sets of patients with adenocarcinoma. Many of these changes were similar in a separate study of multiple solid tumor types versus normal tissue, which included 123 NSCLC cases, along with breast, colon, gastric, endocrine pancreatic, and prostate cancers (Volinia et al. 2006). These data demonstrate the large number of differences seen in tumors versus normal tissue, likely representing both the biological heterogeneity of NSCLC and the different stages of cancer being analyzed. Despite this complexity, these analyses have clearly identified multiple interesting miRNAs involved in tumorigenesis and progression.

An alternative approach to miRNA profiling of tumor tissues is to measure miR levels in relation to outcome, without explicit comparison to normal or uninvolved tissue. This approach has principally been applied in lung cancer to address the question of whether miRNA signatures can be developed to stratify early-stage patients (those who have undergone surgical resection) into high or low risk for early recurrence or metastasis. By using either microarray or quantitative real-time PCR, multiple studies have demonstrated the ability to stratify patient risk for recurrence or survival based upon multi-miR classifiers, and often independently of the clinical staging (Patnaik et al. 2009; Yu et al. 2008). One study approached the specific question of whether there is a miR signature in squamous cell carcinoma of the lung (Raponi et al. 2009). Interestingly, this study identified several miRNA families that have also been found to be altered in other studies of NSCLC, in which there was strong representation of adenocarcinoma samples (Table 2.1). An alternative approach infers the miRNA networks involved from the changes in mRNA expression data (Liang 2008). Using this methodology, reduced expression of *miR-34* family members was identified in multiple datasets. Currently it is unclear how to use these signatures clinically, and whether they are true biomarkers of the disease, linked with the underlying pathways driving the tumor biology, or simply epiphenomena. However, as our understanding of the miRNA/mRNA target biology advances, this approach may be clinically useful in understanding how to assign risk for recurrence in individual patients.

A final way in which tumor miR profiling data may become clinically useful is in the selection of chemotherapeutic agents for patients. A recent study correlated measured drug sensitivity across the panel of drugs in the NCI Developmental Therapeutics Program drug screen with both miRNA and mRNA expression in the NCI-60 panel of cell lines (Liu et al. 2010a). They demonstrated strong associations between the sensitivity to drugs with known mechanisms of action and the patterns of miRNA and mRNA expression. This type of analysis might provide a miRNA-based molecular pharmacologic classification for tumors that would be helpful in either trial design or personalized selection of drugs for individual patients.

In thinking about how to incorporate these predictive or prognostic signatures into standard clinical practice or investigational trial design, certain practical considerations must be kept in mind. MiRNAs may be ideal for these types of studies, especially in comparison to mRNA-based studies, owing to their relative stability against degradation in formalin-fixed paraffin-embedded tissues. However, there is currently no gold standard methodology for the evaluation of miRNA levels from samples. Some investigators have reported poor overall correlation between high-throughput microarray-based techniques and quantitative real-time reverse transcription-PCR (qRT-PCR), which is more commonly used in clinical laboratories (Koshiol et al. 2010; Liu et al. 2010b). Alternate methods, such as cloning or in situ hybridization are technically challenging, time consuming and/or expensive for adoption by clinical laboratories. As the availability of high-throughput sequencing technology continues to spread, with the attendant decrease in cost, this may well become an appropriate standard that could be adopted by clinical labs. Finally, consideration must be given to the fact that diagnostic biopsy specimens, either from bronchoscopy or from fine needle biopsies, frequently provide relatively small amounts of tissue. These may be insufficient for standard pathologic analysis of stained slides, newer molecular tests such as epidermal growth factor receptor (*EGFR*) or *KRAS* mutational status, and miRNA analysis. One potential solution to this issue would be the measurement of miRNAs from patient serum or plasma, which would provide the added clinical advantage of being able to measure repeatedly during the course of treatment or in post-treatment surveillance. Several studies have reported the differential expression of miRNAs in patient serum that correlates with disease or outcome, including one study in a group of 303 early-stage NSCLC patients (Chen et al. 2008; Ng et al. 2009; Hu et al. 2010). Using high-throughput sequencing on miRNAs derived from serum in a cohort of Chinese patients versus healthy donor controls, a four-miRNA signature (*miR-486, miR-30d, miR-1,* and *miR-499*) was derived that was prognostic for patients with a shorter median survival and increased risk for death (Hu et al. 2010). Such studies provide a starting point for trials of how to incorporate these potential biomarkers into clinical decision-making, but much work still remains to be done before circulating miRNA levels become a standard clinical tool in the diagnosis and management of cancer patients.

2.3 Pathogenesis

Lung tumorigenesis is frequently related to changes induced in the epithelium of the airway by tobacco or carcinogen exposure. Recent studies to identify the link between tobacco exposure and changes in miRNA expression have been performed in rodents (both mouse and rat) and in chronic smokers (Izzotti et al. 2009a, b; Schembri et al. 2009). In all three studies it was observed that the majority of changes involved down-regulation of miRNAs, which for the rodent studies correlated strongly with the changes in mRNA and protein documented from prior work using the same model system (Izzotti et al. 2009a, b). Additionally, measurement of the miRNA levels proved to be an extremely sensitive marker of tobacco exposure, producing down-regulation in 126 out of 484 measured miRNAs (26%). Comparison of the results from the mouse and rat studies found involvement of the same miR pathways, with 13 of 15 miRs down-regulated at least twofold in the mouse lungs also significantly down-regulated in rat (including the *let-7* family members, *miR-34b*, *miR-30b*, *miR-30c*, and *miR-125*). Although the overlap between the rodent samples and human were more limited (e.g. *miR-30*, *miR-99*, and *miR-125*), there are likely differences related to the pathology of cells analyzed (mixed cell population of whole animal lungs versus relatively pure human bronchial epithelial cell population), chronicity of exposure (4 week exposure in animals versus chronic human smokers with an average of 18.8 pack years) and species. The human data also clearly indicates that changes in a relatively small number of miRs could account for a large percentage of the documented smoking-associated mRNA changes (Schembri et al. 2009).

Smoking is the single most important risk factor for the development of lung cancer. However, 10–15% of cases of NSCLC occur in never smokers, corresponding to approximately 20,000 deaths annually and making this subcategory one of the top 10 causes of cancer mortality (Samet et al. 2009). This is an incompletely understood class of patients that has received increasing focus in recent years, especially given the findings that never-smokers have a higher incidence of activating mutation in the *EGFR* and are therefore more likely to benefit from the approved tyrosine kinase inhibitors (Engelman and Janne 2008). To assess the role of miRNA changes in this unique group of patients, one study compared the global miRNA expression profile from tumors in smokers versus never-smokers. Among the differences identified, one of the most significant changes was the increase in levels of *miR-21*, which strongly correlated with mutation in *EGFR* (Seike et al. 2009). The authors further demonstrated that EGFR activation drives *miR-21* expression and that antisense targeting of *miR-21* enhanced the apoptotic response induced by EGFR tyrosine kinase inhibition. A separate study by Cho and coworkers found similar results for the levels of *miR-21* in a small panel of non-smoking patients with lung adenocarcinoma, along with changes in several other miRs, including *miR-145*, *miR-126**, *miR-182*, *miR-183*, and *miR-210* (Cho et al. 2009).

2.4 Regulation of Known Oncogenes in Lung Cancer

2.4.1 Let-7, RAS, c-Myc and HMGA2

The *RAS* proto-oncogene family plays a central role in the growth factor receptor signaling pathways and is found to have an activating mutation in many epithelial tumor types. Approximately 30% of human NSCLC cases have mutation of the *KRAS* gene, frequently associated with a history of tobacco exposure. In genetic mouse models an activating mutation in *KRAS* produces lung adenocarcinoma, the most common histologic subtype of lung cancer, with differing propensities to invade and metastasize (Fisher et al. 2001; Jackson et al. 2001; 2005; Johnson et al. 2001; Liu et al. 2000; Olive et al. 2004; Zheng et al. 2007). One of the first miRNAs to be identified and one of the best studied to date is *let-7*, which was originally identified as a gene responsible for regulating temporally-specific developmental changes in *C. elegans* (Reinhart et al. 2000). The *let-7* family contains at least nine members and is highly conserved across species (Pasquinelli et al. 2000). In *C. elegans* it was demonstrated that *let-7* regulates *let-60/RAS* and inversely correlates with the altered *RAS* levels in human NSCLC tumors, providing clear evidence that miRNAs can act as tumor suppressors (Johnson et al. 2005) (Table 2.2). Several studies also demonstrated that loss of *let-7* expression in surgically resected tumor specimens of NSCLC is prognostic of survival, regardless of the pathologic staging (Takamizawa et al. 2004; Yanaihara et al. 2006). Using multiple in vivo models of NSCLC development, re-expression of *let-7* suppressed tumorigenesis in a manner that was largely dependent upon modulation of the Ras levels (Kumar et al. 2008). In addition, *let-7* has been demonstrated to down-regulate *MYC* and revert *MYC*-induced growth in Burkitt's lymphoma cells (Sampson et al. 2007), a finding that may have relevance to NSCLC and which further demonstrates how the loss in expression of a single miRNA can activate multiple cooperative oncogenic pathways.

As mentioned in the section on miRNA changes in response to tobacco exposure, *let-7* suppression can occur with relatively acute exposure to tobacco smoke, but multiple potential mechanisms have been proposed based upon analysis of human cancer cell lines, tissue specimens, and developmental studies (Boyerinas et al.

Table 2.2 Dysregulated microRNAs in lung cancer and their validated targets

MicroRNA	Target gene(s)	References
let-7	*RAS, Myc, HMGA2, EGFR*	Johnson et al. (2005); Sampson et al. (2007); Shell et al. (2007); Webster et al. (2009)
miR-128b	*EGFR*	Weiss et al. (2008)
miR-21	*PTEN*	Talotta et al. (2009)
miR-17-92	*PTEN, E2F1-3, CDK4, BIM*	Matsubara et al. (2007); O'Donnell et al. (2005); Ventura et al. (2008)
miR-93, miR-98, miR-197, miR-378	*FUS1*	Du et al. (2009a); Lee et al. (2007)
miR-34	*CDK4, CCNE2, pRB*	He et al. (2007a)
miR-200	*ZEB1, ZEB2*	Burk et al. (2008); Gibbons et al. (2009a); Gregory et al. (2008); Park et al. (2008)

2010). One of the most intriguing mechanisms was described in a study of single nucleotide polymorphisms (SNPs) in the 3′UTR of *KRAS*, which found that the prevalence of a variant allele containing an altered *let-7* complementary site was higher in a cohort of patients with NSCLC (20.3%) than in the general population (7.6% in a population of European descent) (Chin et al. 2008). Interestingly, the variant allele predicted increased risk for NSCLC in patients with a moderate smoking history (< 40 pack years) in multiple independent patient cohorts. This data suggests a synergy in miR regulation of oncogenes between the host genetic background and exposure history up to an exposure threshold, past which the exposure becomes the dominant tumorigenic factor.

Finally, from analyses of the NCI-60 cell line panel and ovarian cancer patient samples, it was demonstrated that *let-7* loss correlated with the expression of markers for less differentiated tumors, such as *HMGA2* (Shell et al. 2007). Analysis of survival data from the patient samples clearly indicated that tumors with loss of *let-7* were more likely to metastasize or be poorly responsive to treatment, producing striking differences in patient survival. *HMGA2* is also an established epithelial-mesenchymal transition-inducing gene (Thuault et al. 2006), which would provide a reasonable link between the observed histological differences and a mechanism of tumor progression.

2.4.2 EGFR

The epidermal growth factor receptor, EGFR (ErbB1/Her-1), is a member of the ErbB family of growth factor receptor tyrosine kinases that is frequently found to be activated by mutation or amplification in epithelial malignancies. In lung cancer it is mutated in approximately 10% of patients in the United States and 30% of Asian patients, but is also frequently over-expressed at the protein level and/or has increased gene copy number (Gazdar 2010). Overall, blockade of EGFR activation has been the focus of tremendous efforts to develop oral tyrosine kinase inhibitors, e.g. erlotinib and gefitinib. Currently such medications are approved for use in NSCLC patients who have failed prior chemotherapy, and their use has been found to be of greatest benefit in patients whose tumors carry an activating *EGFR* mutation.

Multiple miRNAs have been implicated in the regulation of EGFR, including *miR-128b*, *let-7* and *miR-21* (as discussed in the previous section on pathogenesis). *MiR-128b* exists on chromosome 3p22, which is a region frequently deleted in lung cancer. A recent study evaluated whether *miR-128b* levels in human NSCLC cell lines regulated *EGFR* levels and demonstrated that *miR-128b* directly recognizes the 3′UTR of *EGFR* (Weiss et al. 2008). In an analysis of patient samples, the same group also showed that loss-of-heterozygosity at the locus containing *miR-128b* was frequent and strongly correlated with improved disease control with gefitinib treatment, producing improved overall survival (23.4 vs. 10.5 months, $p = 0.02$). These findings suggest that the loss of *miR-128b* may be useful as a predictive marker of response to treatment with EGFR inhibitors. Similarly, *let-7* was found to regulate EGFR mRNA and protein levels in multiple different cancer cell lines, including lung, and to subsequently reduce signaling through the Akt pathway and decrease cell viability by an apoptosis-independent process (Webster et al. 2009).

2.4.3 p53 and MiR-34

The *miR-34* family (*miR-34a*, *miR-34b*, and *miR-34c*) is frequently decreased in expression in solid tumors, including NSCLC (He et al. 2007b). The two genomic loci encoding the three family members each have a p53 binding site in their promoter, and their expression is induced in a p53-dependent fashion by oncogenic stress or DNA damage (He et al. 2007a), demonstrating that *miR-34* is an effector in the p53 tumor suppressor network. In a cohort of 70 patients who underwent surgical resection for NSCLC, *miR-34a* and *miR-34b* levels were significantly repressed versus paired normal tissue, and mutations in *p53* were much more frequent in cases with low *miR-34a* expression (Gallardo et al. 2009). The expression level of *miR-34* in the patient samples was independently prognostic for relapse, while the combination of *p53* mutational status and *miR-34* expression level was an even more powerful prognosticator. This study also confirmed prior observations that *miR-34* expression is frequently regulated by the methylation status of the promoter region (Lodygin et al. 2008).

2.4.4 Fus-1 and the 3p21.3 Deletion

The 3p21.3 region in the human genome has been associated with inhibition of tumor growth and progression, and within this locus the *FUS1* gene (or tumor suppressor candidate 2, *TUSC2*) is a tumor suppressor that is lost in expression in

90–100% of cases (Zabarovsky et al. 2002). It is hypothesized that hemizygous deletion, coupled with additional epigenetic regulation, may account for the complete loss of expression of this region in lung tumors. In fact, it was recently demonstrated that expression levels were reduced or absent in 82% of non-small cell and 100% of small cell lung cancer specimens (Prudkin et al. 2008). This gene has been recently shown to be under the regulation of four different miRNAs, *miR-93*, *miR-98*, *miR-197*, and *miR-378* (Du et al. 2009a; Lee et al. 2007).

2.4.5 MiR-17-92

The *miR-17-92* cluster (also called oncomir-1) contains six miRNAs located at 13q31.3, a region that is frequently amplified in lymphoma and solid tumors (Mendell 2008), with concordant up-regulation of expression in many solid tumors, including lung (Volinia et al. 2006). Additionally, expression of the cluster is regulated by *c-Myc* and subsequently targets the transcription factor *E2F1* (O'Donnell et al. 2005). This cluster is therefore proposed to act as a regulator of tumor cell growth/proliferation and apoptosis, dysregulation of which produces a phenotype of hyperproliferation. In fact, when the *miR-17-92* locus was expressed from an early time point during development in a transgenic murine model, the animals developed hyperplasia of the lung epithelium, along with a block in epithelial cell differentiation, producing few primitive alveoli in the distal airways (Lu et al. 2007). Conversely, mice with a homozygous knockout of the *miR-17-92 locus* have significant hypoplasia of the lung, along with a ventricular septal defect, which produced 100% neonatal lethality (Ventura et al. 2008). Besides *E2F1-3*, the *miR-17-92* cluster modulates other downstream targets such as *PTEN*, *CDK4*, and *BIM* (He et al. 2007a; Ventura et al. 2008; Xiao et al. 2008), providing multiple potential mechanistic explanations for a proliferative phenotype. Finally, targeting of *miR-17-5p* and *miR-20a* in lung cancer cell lines over-expressing the *miR-17-92* locus induced apoptosis (Matsubara et al. 2007).

2.4.6 MiR-155

Despite the association of high *miR-155* levels with poor outcome in NSCLC patients, it is currently unclear how *miR-155* is involved in NSCLC pathogenesis. *MiR-155* has been postulated to provide a pivotal link between chronic inflammatory states and cancer development (Tili et al. 2009). It has also been demonstrated to play an important role in the cellular reactivation of oncogenic viruses such as Epstein-Barr virus (EBV) and in modulating the anti-tumor effects of the bone morphogenetic protein pathway signaling (Yin et al. 2010). Most recently *miR-155* has been shown to regulate components of the mismatch repair machinery in colon cancer (including *MLH1*, *MSH2*, and *MSH6*), with high expression resulting in a mutator phenotype and genomic instability (Valeri et al. 2010). Each of these biologic functions could conceivably play a role in lung cancer biology, but further studies will be necessary to elucidate the particular mechanisms at work.

With increasing frequency, other miRNAs are being identified as regulators of various oncogenic functions in human samples, cell lines and mouse models of non-small cell lung cancer, adding to the growing list of potential ways in which miRs regulate tumor initiation or progression. Some examples include *miR-31*-mediated down-regulation of the tumor suppressors LATS2 and PPP2R2A (Liu et al. 2010b) and the loss of regulation of the anti-apoptotic protein PED/PEA-15 by *miR-212* (Incoronato et al. 2010). This later report highlights the fact that some of the changes in miR expression may directly affect the sensitivity of malignant cells to therapeutic intervention, which would clearly be useful in formulating treatment plans. It will be exciting to monitor the increasing list of roles for miRNA function in lung tumors and how this information can be incorporated into ways to personalize treatment for patients.

2.5 Invasion and Metastasis Progression

Tumorigenesis and metastasis are two inter-locking, multi-step processes. Because of their ability to simultaneously modulate many targets, miRNA dysregulation could certainly affect many of the independent steps necessary in the transformation to a tumorigenic and metastatic cell. Using the well-characterized RIP-Tag2 murine model of pancreatic neuroendocrine carcinoma, Hanahan's group recently documented the miRNA changes associated with each of the discrete steps in carcinogenesis, from normal islets to hyperplasia, development of the angiogenic switch, followed by encapsulated tumor development, then invasive carcinoma and metastasis (Olson et al. 2009). This work highlighted that the observed miRNA changes correlate closely with the hallmark capabilities acquired by tumor cells during progression, and that pharmacologic anti-angiogenic therapy could alter the angiogenesis signature while invoking some of the changes observed in the metastatic signature. Several groups have also reported elegant mechanistic work in human breast cancer and murine model systems, demonstrating the role of multiple miRs during progression and metastasis, including *miR-10b*, *miR-31*, *miR-126*, and *miR-335* (Ma et al. 2007; Tavazoie et al. 2008; Valastyan et al. 2009). However, in lung cancer much less is known about the role of miRNAs in invasion and metastasis.

Work by our group has demonstrated that a mutant p53 allele ($p53^{R172H\Delta G}$) confers metastatic potential to lung adenocarcinomas arising in mice due to a latent, somatically-activated $Kras^{G12D}$ allele (Zheng et al. 2007). mRNA expression profiling of metastatic versus matched lung tumors from this model revealed a signature of differentially expressed genes, that when applied to patient cohorts identified a subset of early-stage lung cancer patients with poor prognosis (Gibbons et al. 2009b). Subsequent work with this model identified epithelial-mesenchymal transition (EMT) as a critical step in metastasis, regulated by the expression level of the *miR-200* family (including *miR-141*, *miR-200a*, *miR-200b*, *miR-200c*, and *miR-429*) and the EMT-inducing transcription factor ZEB1 (Gibbons et al. 2009a). Human NSCLC cell lines also displayed a strong correlation between the *miR-200* family levels and the EMT markers, suggesting this mechanism as a potential driver of the biology of these cells. As noted in the section on miRNA profiling of human NSCLC specimens, *miR-200* family members are part of a multi-miRNA signature defining high risk for recurrence after resection (Patnaik et al. 2009). Our work also demonstrated that the *miR-200*/ZEB1 balance is under epigenetic regulation from interactions between the tumor cells and their microenvironment, including cell-matrix interactions and interactions with morphogens such as TGFβ (Gibbons et al. 2009a). It has also been reported that expression of the *miR-200* family in murine and human cells (including normal human mammary epithelial cells, breast cancer cell lines and prostate cancer cell lines) is regulated by DNA methylation of the promoter (Vrba et al. 2010).

The association between altered *miR-200* levels and EMT or metastasis has been demonstrated in other epithelial tumor types, including breast, ovarian and gastric (Du et al. 2009b; Gregory et al. 2008; Hu et al. 2009; Park et al. 2008). Additionally, several studies have demonstrated the role of the *miR-200* family in regulation or maintenance of normal and tumor stem-cell features in breast, colon and pancreatic cancer (Shimono et al. 2009; Wellner et al. 2009). These data support the concept that EMT links the acquisition of stem cell features with metastasis, suggesting that during metastasis tumor cells might acquire certain properties of progenitor cells, either transiently or permanently.

2.6 Therapeutics

Because miRNAs control many known oncogenes, while also acting on their own as either oncogenes or tumor suppressors, greater insight into their biology also carries great promise for this class of molecules as potential therapeutic agents or targets. Techniques to directly target their expression or modulate their levels in tumor cells are an intense area of investigation. In general the inhibitors are single-stranded oligonucleotides complementary to the mature miRNAs, chemically modified with phosphorothioate, 2′-*O*-methyl or locked nucleic acid (LNA) substitutions to improve their resistance to nuclease-mediated degradation (Krutzfeldt et al. 2007). Targeted tissue delivery of either antisense oligonucleotides complementary to specific mature miRNAs, or of precursor or mature miRNAs to replace loss of expression, is still a significant challenge and has not been translated to the clinical cancer setting. However, significant work is being conducted in pre-clinical animal models to demonstrate the in vivo biology of the miRNAs during the complex progression of tumors and to test the ability to target these processes.

In human and murine models of lung adenocarcinoma, expression of the *let-7* family is frequently suppressed. Exogenous in vivo delivery by inhalation of adenoviral or lentiviral vectors expressing *let-7* genes at the time of mutant *KRAS* activation was able to halt tumor progression (Esquela-Kerscher et al. 2008; Kumar et al. 2008), while delivery after the establishment of tumors halted tumor proliferation, producing a significant reduction in tumor burden after only a few weeks of treatment (Trang et al. 2010). One advantage in the treatment of lung cancer versus other solid tumor types may be the ability to deliver these reagents by inhalation or tracheal instillation, providing an appropriate therapeutic effect to the primary tumors, without the potential systemic side effects. Obviously this route of administration would not provide any advantage in patients who have developed metastatic disease.

Similarly, an adeno-associated virus-mediated delivery system was used in a murine model of hepatocellular carcinoma to re-express *miR-26a*. Even with preexisting disease, re-expression of *miR-26a* had significant effect to induce apoptosis and halt tumor progression in the treated animals (Kota et al. 2009). Of equal importance, there was no evidence of systemic or liver toxicity, despite the high levels of expression of the exogenously-delivered miRNA vector.

Metastasis prevention or targeting of occult tumor metastases, rather than primary tumor growth, is emerging as a potential design for adjuvant trials of targeted biological agents. In a recent study on the *miR-10b*, which is frequently up-regulated in metastatic tumors, Ma and colleagues demonstrated selective inhibition of the metastatic process upon targeting of orthotopic mammary tumors by the systemic administration of an antagomir to *miR-10b* (Ma 2010). Antagomirs are antisense oligonucleotides containing a 2′-*O*-methyl on the ribose moieties, partial replacement of the phosphodiester backbone with phosphorothioate bonds, and conjugation of a cholesterol moiety to the 3′ end (Krutzfeldt et al. 2007). Despite a pronounced treatment effect on the development of metastases, there was no effect on the growth of primary tumors. Interestingly, this effect was due to suppression of the early steps in metastasis, rather than the later stages of the metastatic cascade such as colonization. Again, this study was notable for no evidence of systemic toxicity in the animals.

Currently there are no human trials targeting over-expression or loss of miRNA function in cancer, but the clinical application and the necessary technology are progressing in another patient setting. In chronic hepatitis C virus (HCV) infection, viral replication is dependent upon the cellular host factor *miR-122* and in in vitro experiments can be readily suppressed by inhibition of *miR-122* expression. Following the systemic administration of a LNA-anti-miR recognizing *miR-122* in healthy African green monkeys, the *miR-122* levels were repressed with concordant decrease of the serum cholesterol levels, demonstrating delivery of the anti-miR to the liver, but without evidence of toxicity (Elmen 2008). In chimpanzees chronically infected with HCV, use of the agent (now termed SPC3649) decreased serum HCV RNA levels by 2.3 orders of magnitude, with no evidence for the emergence of viral resistance to the therapy (Lanford 2010). The company developing this LNA-anti-miR, Santaris Pharma, currently has two Phase I clinical trials underway in healthy volunteers and reportedly plans to open a Phase II clinical trial in late 2010 for patients with HCV infection. The progress of this agent through clinical development will be closely monitored by those in the field of cancer biology and trial development, for as the biology of miRNAs in cancer becomes more thoroughly understood, there will be increasing efforts to translate those findings into the clinical trial setting for cancer patients.

2.7 Conclusions

The current revolution in non-coding RNAs, especially in the field of miRNAs, provides tremendous opportunity for cancer biologists to further define and refine our understanding of the basis for carcinogenesis and tumor progression. This discovery and enhanced understanding may by itself produce additional insights into better clinical decision making, patient risk stratification and potential therapeutic options. However, the greater promise of miRNAs is in their pleiotrophic biology, which may provide for uniquely innovative strategies to target the spectrum of heterogeneity in cancer that arises from the cumulative changes in multiple pathways. Finally, these insights will likely provide some unifying strategies against multiple cancers, as many of the specifics outlined in this chapter probably have counterparts in the biology of other epithelial and non-epithelial tumor types.

Acknowledgments This work was supported by NIH R01 grants CA132608 and CA117965 (J.M.K.), a Howard Hughes Medical Institute Medical Scholar Fellowship to Z.H.R., an American Society for Clinical Oncology Young Investigator Award, an International Association for the Study of Lung Cancer Fellowship, and NIH 1K08CA151651 and NCI 5 T32 CA009666 (D.L.G.).

References

Boyerinas B, Park SM, Hau A, et al. The role of let-7 in cell differentiation and cancer. Endocr Relat Cancer. 2010;17:F19–36.

Burk U, Schubert J, Wellner U, et al. A reciprocal repression between ZEB1 and members of the miR-200 family promotes EMT and invasion in cancer cells. EMBO Rep. 2008;9:582–9.

Calin GA, Sevignani C, Dumitru CD, et al. Human microRNA genes are frequently located at fragile sites and genomic regions involved in cancers. Proc Natl Acad Sci USA. 2004;101:2999–3004.

Chen X, Ba Y, Ma L, et al. Characterization of microRNAs in serum: a novel class of biomarkers for diagnosis of cancer and other diseases. Cell Res. 2008;18:997–1006.

Chiang HR, Schoenfeld LW, Ruby JG, et al. Mammalian microRNAs: experimental evaluation of novel and previously annotated genes. Genes Dev. 2010;24:992–1009.

Chin LJ, Ratner E, Leng S, et al. A SNP in a let-7 microRNA complementary site in the KRAS 3′ untranslated region increases non-small cell lung cancer risk. Cancer Res. 2008;68:8535–40.

Cho WC. MicroRNAs: potential biomarkers for cancer diagnosis, prognosis and targets for therapy. Int J Biochem Cell Biol. 2010;42:1273–81.

Cho WC, Chow AS, Au JS. Restoration of tumour suppressor hsa-miR-145 inhibits cancer cell growth in lung adenocarcinoma patients with epidermal growth factor receptor mutation. Eur J Cancer. 2009;45:2197–206.

Du L, Schageman JJ, Subauste MC, et al. MiR-93, miR-98, and miR-197 regulate expression of tumor suppressor gene fus1. Mol Cancer Res. 2009a;7:1234–43.

Du Y, Xu Y, Ding L, et al. Down-regulation of miR-141 in gastric cancer and its involvement in cell growth. J Gastroenterol. 2009b;44:556–61.

Elmén J, Lindow M, Schütz S, et al. LNA-mediated microRNA silencing in non-human primates. Nature. 2008;452:896–9.

Engelman JA, Janne PA. Mechanisms of acquired resistance to epidermal growth factor receptor tyrosine kinase inhibitors in non-small cell lung cancer. Clin Cancer Res. 2008;14: 2895–99.

Esquela-Kerscher A, Trang P, Wiggins JF, et al. The let-7 microRNA reduces tumor growth in mouse models of lung cancer. Cell Cycle. 2008;7:759–64.

Fisher GH, Wellen SL, Klimstra D, et al. Induction and apoptotic regression of lung adenocarcinomas by regulation of a k-ras transgene in the presence and absence of tumor suppressor genes. Genes Dev. 2001;15:3249–62.

Gallardo E, Navarro A, Vinolas N, et al. MiR-34a as a prognostic marker of relapse in surgically resected non-small-cell lung cancer. Carcinogenesis. 2009;30:1903–9.

Gazdar AF. Epidermal growth factor receptor inhibition in lung cancer: the evolving role of individualized therapy. Cancer Metastasis Rev. 2010;29:37–48.

Gibbons DL, Lin W, Creighton CJ, et al. Contextual extracellular cues promote tumor cell EMT and metastasis by regulating miR-200 family expression. Genes Dev. 2009a;23: 2140–51.

Gibbons DL, Lin W, Creighton CJ, et al. Expression signatures of metastatic capacity in a genetic mouse model of lung adenocarcinoma. PLoS One. 2009b;4:e5401.

Goldstraw P, Crowley J, Chansky K, et al. The IASLC lung cancer staging project: proposals for the revision of the TNM stage groupings in the forthcoming (seventh) edition of the TNM classification of malignant tumours. J Thorac Oncol. 2007;2:706–14.

Gregory PA, Bert AG, Paterson EL, et al. The miR-200 family and miR-205 regulate epithelial to mesenchymal transition by targeting ZEB1 and sip1. Nat Cell Biol. 2008;10:593–601.

He L, He X, Lim LP, et al. A microRNA component of the p53 tumour suppressor network. Nature. 2007a;447:1130–4.

He L, He X, Lowe SW, et al. MicroRNAs join the p53 network–another piece in the tumour-suppression puzzle. Nat Rev Cancer. 2007b;7:819–22.

Hu Z, Chen X, Zhao Y, et al. Serum microRNA signatures identified in a genome-wide serum microRNA expression profiling predict survival of non-small-cell lung cancer. J Clin Oncol. 2010;28:1721–6.

Hu X, Macdonald DM, Huettner PC, et al. A miR-200 microRNA cluster as prognostic marker in advanced ovarian cancer. Gynecol Oncol. 2009;114:457–64.

Incoronato M, Garofalo M, Urso L, et al. MiR-212 increases tumor necrosis factor-related apoptosis-inducing ligand sensitivity in non-small cell lung cancer by targeting the antiapoptotic protein ped. Cancer Res. 2010;70:3638–46.

Izzotti A, Calin GA, Arrigo P, et al. Downregulation of microRNA expression in the lungs of rats exposed to cigarette smoke. FASEB J. 2009a;23:806–12.

Izzotti A, Calin GA, Steele VE, et al. Relationships of microRNA expression in mouse lung with age and exposure to cigarette smoke and light. FASEB J. 2009b;23:3243–50.

Jackson EL, Olive KP, Tuveson DA, et al. The differential effects of mutant p53 alleles on advanced murine lung cancer. Cancer Res. 2005;65:10280–8.

Jackson EL, Willis N, Mercer K, et al. Analysis of lung tumor initiation and progression using conditional expression of oncogenic k-ras. Genes Dev. 2001;15:3243–8.

Jemal A, Siegel R, Xu J, et al. Cancer statistics, 2010. CA Cancer J Clin. 2010;60:277–300.

Johnson SM, Grosshans H, Shingara J, et al. Ras is regulated by the let-7 microRNA family. Cell. 2005;120:635–47.

Johnson L, Mercer K, Greenbaum D, et al. Somatic activation of the k-ras oncogene causes early onset lung cancer in mice. Nature. 2001;410:1111–6.

Kamangar F, Dores GM, Anderson WF. Patterns of cancer incidence, mortality, and prevalence across five continents: defining priorities to reduce cancer disparities in different geographic regions of the world. J Clin Oncol. 2006;24:2137–50.

Koshiol J, Wang E, Zhao Y, et al. Strengths and limitations of laboratory procedures for microRNA detection. Cancer Epidemiol Biomarkers Prev. 2010;19:907–11.

Kota J, Chivukula RR, O'Donnell KA, et al. Therapeutic microRNA delivery suppresses tumorigenesis in a murine liver cancer model. Cell. 2009;137:1005–17.

Krutzfeldt J, Kuwajima S, Braich R, et al. Specificity, duplex degradation and subcellular localization of antagomirs. Nucleic Acids Res. 2007;35:2885–92.

Kumar MS, Erkeland SJ, Pester RE, et al. Suppression of non-small cell lung tumor development by the let-7 microRNA family. Proc Natl Acad Sci USA. 2008;105:3903–8.

Lanford RE, Hildebrandt-Eriksen ES, Petri A, et al. Therapeutic silencing of microRNA-122 in primates with chronic hepatitis C virus infection. Science. 2010;327:198–201.

Lee DY, Deng Z, Wang CH, et al. MicroRNA-378 promotes cell survival, tumor growth, and angiogenesis by targeting SuFu and Fus-1 expression. Proc Natl Acad Sci USA. 2007;104: 20350–5.

Lewis BP, Burge CB, Bartel DP. Conserved seed pairing, often flanked by adenosines, indicates that thousands of human genes are microRNA targets. Cell. 2005;120:15–20.

Liang Y. An expression meta-analysis of predicted microRNA targets identifies a diagnostic signature for lung cancer. BMC Med Genomics. 2008;1:61.

Liu H, D'Andrade P, Fulmer-Smentek S, et al. mRNA and microRNA expression profiles of the NCI-60 integrated with drug activities. Mol Cancer Ther. 2010a;9:1080–91.

Liu G, McDonnell TJ, Montes de Oca Luna R, et al. High metastatic potential in mice inheriting a targeted p53 missense mutation. Proc Natl Acad Sci USA. 2000;97:4174–9.

Liu X, Sempere LF, Ouyang H, et al. MicroRNA-31 functions as an oncogenic microRNA in mouse and human lung cancer cells by repressing specific tumor suppressors. J Clin Invest. 2010b;120:1298–309.

Lodygin D, Tarasov V, Epanchintsev A, et al. Inactivation of miR-34a by aberrant CPG methylation in multiple types of cancer. Cell Cycle. 2008;7:2591–600.

Lu J, Getz G, Miska EA, et al. MicroRNA expression profiles classify human cancers. Nature. 2005;435:834–8.

Lu Y, Thomson JM, Wong HY, et al. Transgenic over-expression of the microRNA miR-17-92 cluster promotes proliferation and inhibits differentiation of lung epithelial progenitor cells. Dev Biol. 2007;310:442–53.

Ma L, Reinhardt F, Pan E, et al. Therapeutic silencing of miR-10b inhibits metastasis in a mouse mammary tumor model. Nat Biotechnol. 2010;28:341–7.

Ma L, Teruya-Feldstein J, Weinberg RA. Tumour invasion and metastasis initiated by microRNA-10b in breast cancer. Nature. 2007;449:682–8.

Matsubara H, Takeuchi T, Nishikawa E, et al. Apoptosis induction by antisense oligonucleotides against miR-17-5p and miR-20a in lung cancers overexpressing miR-17-92. Oncogene. 2007;26:6099–105.

Mendell JT. Miriad roles for the miR-17-92 cluster in development and disease. Cell. 2008;133:217–22.

Ng EK, Chong WW, Jin H, et al. Differential expression of microRNAs in plasma of patients with colorectal cancer: a potential marker for colorectal cancer screening. Gut. 2009;58: 1375–81.

Olive KP, Tuveson DA, Ruhe ZC, et al. Mutant p53 gain of function in two mouse models of li-fraumeni syndrome. Cell. 2004;119:847–60.

Olson P, Lu J, Zhang H, et al. MicroRNA dynamics in the stages of tumorigenesis correlate with hallmark capabilities of cancer. Genes Dev. 2009;23:2152–65.

O'Donnell KA, Wentzel EA, Zeller KI, et al. C-myc-regulated microRNAs modulate E2F1 expression. Nature. 2005;435:839–43.

Park SM, Gaur AB, Lengyel E, et al. The miR-200 family determines the epithelial phenotype of cancer cells by targeting the E-cadherin repressors ZEB1 and ZEB2. Genes Dev. 2008;22: 894–907.

Pasquinelli AE, Reinhart BJ, Slack F, et al. Conservation of the sequence and temporal expression of let-7 heterochronic regulatory RNA. Nature. 2000;408:86–9.

Patnaik SK, Kannisto E, Knudsen S, et al. Evaluation of microRNA expression profiles that may predict recurrence of localized stage I non-small cell lung cancer after surgical resection. Cancer Res. 2009;70:36–45.

Prudkin L, Behrens C, Liu DD, et al. Loss and reduction of fus1 protein expression is a frequent phenomenon in the pathogenesis of lung cancer. Clin Cancer Res. 2008;14:41–7.
Raponi M, Dossey L, Jatkoe T, et al. MicroRNA classifiers for predicting prognosis of squamous cell lung cancer. Cancer Res. 2009;69:5776–83.
Reinhart BJ, Slack FJ, Basson M, et al. The 21-nucleotide let-7 RNA regulates developmental timing in *caenorhabditis elegans*. Nature. 2000;403:901–6.
Samet JM, Avila-Tang E, Boffetta P, et al. Lung cancer in never smokers: clinical epidemiology and environmental risk factors. Clin Cancer Res. 2009;15:5626–45.
Sampson VB, Rong NH, Han J, et al. MicroRNA let-7a down-regulates myc and reverts myc-induced growth in burkitt lymphoma cells. Cancer Res. 2007;67:9762–70.
Schembri F, Sridhar S, Perdomo C, et al. MicroRNAs as modulators of smoking-induced gene expression changes in human airway epithelium. Proc Natl Acad Sci USA. 2009;106:2319–24.
Seike M, Goto A, Okano T, et al. MiR-21 is an EGFR-regulated anti-apoptotic factor in lung cancer in never-smokers. Proc Natl Acad Sci USA. 2009;106:12085–90.
Shell S, Park SM, Radjabi AR, et al. Let-7 expression defines two differentiation stages of cancer. Proc Natl Acad Sci USA. 2007;104:11400–5.
Shimono Y, Zabala M, Cho RW, et al. Downregulation of miRNA-200c links breast cancer stem cells with normal stem cells. Cell. 2009;138:592–603.
Takamizawa J, Konishi H, Yanagisawa K, et al. Reduced expression of the let-7 microRNAs in human lung cancers in association with shortened postoperative survival. Cancer Res. 2004;64:3753–6.
Talotta F, Cimmino A, Matarazzo MR, et al. An autoregulatory loop mediated by miR-21 and pdcd4 controls the ap-1 activity in ras transformation. Oncogene. 2009;28:73–84.
Tavazoie SF, Alarcon C, Oskarsson T, et al. Endogenous human microRNAs that suppress breast cancer metastasis. Nature. 2008;451:147–52.
Thuault S, Valcourt U, Petersen M, et al. Transforming growth factor-beta employs HMGA2 to elicit epithelial-mesenchymal transition. J Cell Biol. 2006;174:175–83.
Tili E, Croce CM, Michaille JJ. MiR-155: on the crosstalk between inflammation and cancer. Int Rev Immunol. 2009;28:264–84.
Trang P, Medina PP, Wiggins JF, et al. Regression of murine lung tumors by the let-7 microRNA. Oncogene. 2010;29:1580–7.
Valastyan S, Reinhardt F, Benaich N, et al. A pleiotropically acting microRNA, miR-31, inhibits breast cancer metastasis. Cell. 2009;137:1032–46.
Valeri N, Gasparini P, Fabbri M, et al. Modulation of mismatch repair and genomic stability by miR-155. Proc Natl Acad Sci USA. 2010;107:6982–7.
Ventura A, Young AG, Winslow MM, et al. Targeted deletion reveals essential and overlapping functions of the miR-17 through 92 family of miRNA clusters. Cell. 2008;132:875–6.
Volinia S, Calin GA, Liu CG, et al. A microRNA expression signature of human solid tumors defines cancer gene targets. Proc Natl Acad Sci USA. 2006;103:2257–61.
Vrba L, Jensen TJ, Garbe JC, et al. Role for DNA methylation in the regulation of miR-200c and miR-141 expression in normal and cancer cells. PLoS One. 2010;5:e8697.
Webster RJ, Giles KM, Price KJ, et al. Regulation of epidermal growth factor receptor signaling in human cancer cells by microRNA-7. J Biol Chem. 2009;284:5731–41.
Weiss GJ, Bemis LT, Nakajima E, Sugita M, et al. EGFR regulation by microRNA in lung cancer: correlation with clinical response and survival to gefitinib and EGFR expression in cell lines. Ann Oncol. 2008;19:1053–9.
Wellner U, Schubert J, Burk UC, et al. The EMT-activator ZEB1 promotes tumorigenicity by repressing stemness-inhibiting microRNAs. Nat Cell Biol. 2009;11:1487–95.
Xiao C, Srinivasan L, Calado DP, et al. Lymphoproliferative disease and autoimmunity in mice with increased miR-17-92 expression in lymphocytes. Nat Immunol. 2008;9:405–14.
Yanaihara N, Caplen N, Bowman E, et al. Unique microRNA molecular profiles in lung cancer diagnosis and prognosis. Cancer Cell. 2006;9:189–98.

Yin Q, Wang X, Fewell C, et al. MiR-155 inhibits bone morphogenetic protein (BMP) signaling and BMP mediated Epstein-Barr virus reactivation. J Virol. 2010;84:6318–27.
Yu SL, Chen HY, Chang GC, et al. MicroRNA signature predicts survival and relapse in lung cancer. Cancer Cell. 2008;13:48–57.
Zabarovsky ER, Lerman MI, Minna JD. Tumor suppressor genes on chromosome 3p involved in the pathogenesis of lung and other cancers. Oncogene. 2002;21:6915–35.
Zheng S, El-Naggar AK, Kim ES, et al. A genetic mouse model for metastatic lung cancer with gender differences in survival. Oncogene. 2007;26:6896–904.

Chapter 3
MicroRNAs in Breast Cancer

Suvi-Katri Leivonen

Abstract Breast cancer is a complex, phenotypically diverse genetic disease, which involves changes in gene expression and structure. Despite advances in the detection and therapy of breast cancer, it is the leading cause of cancer deaths among women worldwide. Therefore, novel diagnostic and prognostic indicators, as well as potential therapeutic targets are needed. Recently, several studies have shown an involvement for the microRNAs (miRNAs) in breast cancer. MiRNA expression profiling studies have identified sets of miRNAs which are deregulated in breast cancer, and which can separate tumor from normal tissue. Functional studies have uncovered the role of miRNAs in breast cancer as both tumor suppressors and oncogenes. Furthermore, miRNAs have been suggested to play a critical role in regulating breast cancer metastasis. The purpose of this chapter is to present an overview of our current understanding of miRNAs in breast cancer with specific emphasis on miRNAs as potential therapeutic targets.

3.1 Deregulated Expression of MiRNAs in Breast Cancer

Breast cancer is a heterogeneous disease that results from the accumulation of various genetic defects. Traditionally, mRNA expression profiling has been used to molecularly characterize and classify breast tumors and to associate them with clinical and pathological factors as well as in prediction of prognosis (Sørlie 2004). The classification of breast cancer has mainly been based on the expression of certain receptors, such as the estrogen receptor (ER), progesterone receptor (PR), or human epidermal growth factor receptor 2 (HER2 or ErbB2).

MicroRNA (miRNA) expression profiling presents an attractive approach to further dissect and characterize breast cancer subtypes. MiRNA expression profiles can be used to differentiate tumor tissue from surrounding normal tissue for tumor classification and for prognostication. In a miRNA microarray analysis,

S.-K. Leivonen (✉)
VTT Medical Biotechnology, FI-20521 Turku, Finland
e-mail: suvi-katri.leivonen@vtt.fi

W.C.S. Cho (ed.), *MicroRNAs in Cancer Translational Research*,
DOI 10.1007/978-94-007-0298-1_3, © Springer Science+Business Media B.V. 2011

where 245 miRNAs were profiled from 76 breast tumor and 10 normal specimens, 29 miRNAs were identified as being deregulated in breast cancer as compared to normal breast tissue (Iorio et al. 2005). The most significantly deregulated miRNAs were *miR-125b, miR-145*, and *miR-10b*, which were down-regulated, and *miR-21* and *miR-155*, which were up-regulated. Additionally, miRNAs were shown to be differentially expressed between various biopathologic features of breast cancer. For instance, *miR-30* family was significantly down-regulated in ER and PR negative tumors, and *let-7* family was down-regulated in samples with either lymph node metastasis or higher proliferation index. In a subsequent study by Blenkiron and colleagues (2007), a set of 309 miRNAs were profiled in human primary breast tumors, normal breast samples and cell lines, and 31 miRNAs were identified to be associated with the molecular subtype, ER status or grade. Additional studies have identified several miRNAs to correlate with the breast cancer molecular markers ERα, PR, and HER2 (Lowery et al. 2009; Mattie et al. 2006). miRNA expression analyses have also linked four miRNAs, *miR-7, miR-128a, miR-210*, and *miR-516-3p*, with the aggressiveness of lymph node-negative, ER-positive breast cancer (Foekens et al. 2008). In addition, the spatial distribution of miRNA expression within breast cancer tissue has been studied by in situ hybridization method (Sempere et al. 2007). This study revealed that the expression of *miR-145* and *miR-205* was restricted to the myoepithelial/basal cells of normal mammary ducts and lobules, but remarkably reduced or completely eliminated in matching tumor specimens. Conversely, expression of *let-7a, miR-21, miR-141*, and *miR-214* was detected at varying levels predominantly within luminal epithelial cells in normal tissue. *MiR-21* expression was frequently increased, whereas *let-7a* expression was decreased in malignant cells.

Although there are discrepancies between the different studies described above, there is some overlap between the miRNAs identified. For instance, *miR-21* and *miR-155* are clearly over-expressed in breast cancer, whereas *miR-145* and *let-7* are down-regulated (Iorio et al. 2005; Sempere et al. 2007; Volinia et al. 2006). The inconsistence between the separate studies could be partly explained by the application of different microarray platforms or analytical tools.

The molecular mechanisms of miRNA deregulation in breast cancer are still largely unknown. Generally, miRNAs tend to be located in cancer-associated genomic regions or fragile sites (Calin et al. 2004). There is evidence, that 73% of miRNA genes in breast cancer are located in regions with DNA copy number abnormalities (Zhang et al. 2006). For instance, the polycistron cluster *miR-17-92* is located at the c13orf25 locus on chromosome 13q31, and this locus is known to undergo loss of heterozygosity in breast cancer (Negrini et al. 1995). Similarly, *miR-125b*, which is under-expressed in breast cancer, is located at chromosome 11q23-24, one of the regions most frequently deleted in breast tumors (Negrini et al. 1995). Differential miRNA expression could also be caused by sequence variations, such as mutations or single nucleotide polymorphisms (SNPs), as is the case with *miR-27a* (Yang et al. 2010). Additionally, various miRNA genes, including *miR-9-1, miR-124a-3, miR-148, miR-152*, and *miR-663*, are affected by epigenetic inactivation due to aberrant hypermethylation and this is suggested to be an early and frequent event in breast cancer development (Lehmann et al. 2008).

3.2 Biological Functions and Targets of MiRNAs in Breast Cancer

Several functional studies have linked miRNAs as tumor suppressors or oncogenes in breast cancer via the protein targets they regulate (Table 3.1). Selected tumor suppressive and oncogenic miRNAs are discussed in detail below. Additionally, miRNAs influencing estrogen receptor, an important target in breast cancer, are discussed.

3.2.1 Tumor Suppressive Functions of MiRNAs

MiRNAs may participate in the initial stage of breast tumorigenesis by regulating the behavior of breast cancer stem cells (CSC). CSCs are rare, slow-dividing, and self-renewing cells, which may be responsible for the initiation, progression, metastasis, resistance to therapy and subsequent recurrence of tumors (Al-Hajj 2007; Al-Hajj et al. 2003). Isolated breast CSCs exhibit a $CD44^{+}CD24^{-/\mathrm{low}}$ phenotype and can grow into so-called mammospheres in suspension culture, which is a measure of self-renewal capability. Yu et al. (2007) provided evidence for negative regulation of stemness by *let-7* family of miRNAs in breast cancer CSCs. The expression of 52 miRNAs was analyzed in mammosphere-derived CSCs, and the most significantly down-regulated miRNA family was *let-7*. Validated *let-7* targets include high mobility group AT hook-2 (*HMGA2*) and *RAS* (Johnson et al. 2005; Lee and Dutta 2007; Park et al. 2007), and the expression of both proteins was inversely correlated with *let-7* expression in breast CSCs. Forced expression of *let-7* in CSCs reduced their proliferative capacity, their ability to form mammospheres in culture, and their growth and metastasis in mouse models.

Another study identified *miR-200c* as a stemness regulator in breast CSCs (Shimono et al. 2009). Expression of BMI1 polycomb ring finger oncogene (*BMI1*), a known regulator of stem cell self-renewal, was modulated by *miR-200c*. *MiR-200c* inhibited the clonal expansion of breast cancer cells and suppressed the growth of embryonal carcinoma cells in vitro. Most importantly, *miR-200c* strongly suppressed the ability of normal mammary stem cells to form mammary ducts and tumor formation driven by human breast CSCs in vivo (Shimono et al. 2009).

The miRNA expression studies indicated that miRNAs consistently down-regulated in breast cancer are putative tumor suppressors. In the expression studies, *miR-205*, *miR-125b*, and *miR-145* were identified as significantly down-regulated miRNAs in breast cancer (Iorio et al. 2005; Sempere et al. 2007). Subsequent studies have indeed provided functional evidence for the tumor suppressive role for these miRNAs (Iorio et al. 2009; Scott et al. 2007; Sachdeva and Mo 2010; Spizzo et al. 2010). The functional analyses demonstrated that restoration of *miR-145* expression resulted in inhibition of proliferation and induction of apoptosis in several breast cancer cell lines (Spizzo et al. 2010). This was, at least partly, due to induction in *p53* activation and down-regulation of ERα. Moreover, there is evidence, that *miR-145* not only suppresses proliferation, but also inhibits invasion of breast cancer cells by targeting mucin 1 (*MUC1*) (Sachdeva and Mo 2010).

Table 3.1 Tumor suppressive and oncogenic microRNAs in breast cancer

MiRNA	Expression in BCa	Function	Validated target gene in BCa	References
Tumor suppressive				
miR-17-5p	Low in BCa cells	Cell cycle	*CCND1, AIB1, IL8*	Hossain et al. (2006); Yu et al. (2008)
miR-206	Down-regulated in ER+ tumors, low in metastatic BCa cells	Cell proliferation, ER pathway, inhibition of metastasis	*ESR1*	Adams et al. (2007); Kondo et al. (2008); Tavazoie et al. (2008); Leivonen et al. (2009)
miR-193b	–	ER pathway, cell proliferation, invasion	*ESR1, PLAU*	Leivonen et al. (2009); Li et al. (2009)
miR-18a/b	Down-regulated in ER+ tumors	ER pathway	*ESR1*	Leivonen et al. (2009)
miR-145	Down-regulated	Cell proliferation, invasion	*ESR1, MUC1*	Iorio et al. (2005); Sempere et al. (2007); Sachdeva et al. (2010); Spizzo et al. (2010)
let-7	Down-regulated	Regulation of stem cells	*HMGA2, RAS*	Johnson et al. (2005); Lee and Dutta (2007); Yu et al. (2007)
miR-125a/b	Down-regulated	ErbB2 signalling	*ERBB2, ERBB3*	Iorio et al. (2005); Scott et al. (2007)
miR-205	Down-regulated	ErbB2 signalling, EMT	*ERBB3, VEGF-A*	Gregory et al. (2008); Iorio et al. (2009); Sempere et al. (2007)
miR-200a/b/c	Down-regulated in invasive, mesenchymal BCa cells	EMT, regulation of stem cells	*ZEB1, ZEB2, BMI1*	Gregory et al. (2008); Hurteau et al. (2009); Korpal et al. (2008); Shimono et al. (2009); Tryndyak et al. (2010)
miR-126	Low in metastatic BCa cells	Metastasis and tumor growth inhibition	–	Tavazoie et al. (2008)
miR-335	Low in metastatic BCa cells	Inhibition of metastasis	*SOX4, TNC, PTPRN2, MERTK, COL1A1*	Tavazoie et al. (2008)
miR-31	Low in metastatic BCa cells	Inhibition of metastasis	*RHOA, RDX, ITGA5, MMP16, FDZ, MRIP*	Valastyan et al. (2009a, b)

Table 3.1 (continued)

MiRNA	Expression in BCa	Function	Validated target gene in BCa	References
Oncogenic				
miR-21	Up-regulated	Migration, invasion	*TPM1, PDCD4, SERPINB5, BCL2*	Frankel et al. (2008); Iorio et al. (2005); Si et al. 2007; Zhu et al. (2007)
miR-27a	Up-regulated in BCa cells	Proliferation, cell cycle, angiogenesis	*ZBTB10, FOXO1, MYT1, BIRC5*	Guttilla and White (2009); Li et al. (2010); Mertens-Talcott et al. (2007)
miR-155	Up-regulated	Proliferation, tumor growth	*FOXO3A, RHOA, SOCS1*	Iorio et al. (2005); Jiang et al. (2010); Kong et al. (2008); (2010)
miR-9	Up-regulated	Metastasis	*CDH1*	Ma et al. (2010b)
miR-10b	Up-regulated in metastatic BCa cells	Metastasis	*HOXD10, TIAM1*	Ma et al. (2007); (2010a)
miR-373/520c	Up-regulated in lymph node positive BCa	Metastasis	*CD44*	Huang et al. (2008)
miR-221/222	Down-regulated in ER+ tumors	Tamoxifen resistance	*ESR1, p27*	Di Leva et al. (2010); Miller et al. (2008); Zhao et al. (2008)

BCa, Breast cancer; EMT, Epithelial-mesenchymal transition; ER, Estrogen receptor.

MiR-205 is able to interfere with the proliferative pathway mediated by the HER receptor family by directly targeting *ERBB3* (Iorio et al. 2009). *MiR-205* improved the responsiveness of breast cancer cells to tyrosine kinase inhibitor treatment, thus raising the possibility that *miR-205* could be used as a new biomarker for response to specific therapies. Additional miRNAs influencing HER receptor family are *miR-125a* and *miR-125b*, which also are down-regulated in breast cancer. Scott et al. (2007) showed that retroviral over-expression of *miR-125a* and *miR-125b* in *ERBB2*-dependent human breast cancer cell line SKBR3 resulted in the down-regulation of *ERBB2* and *ERBB3* mRNA and protein levels, suppression of anchorage-dependent growth, and inhibition of motility and invasion. Additionally, *ERBB2* downstream signalling via the ERK1/2 and Akt pathways was inhibited. *ERBB2*–over-expressing breast tumors exhibit aggressive growth and unpredictable response to therapy (Hynes and MacDonald 2009). Therefore, enhanced understanding of the regulation of *ERBB2* expression has the potential to greatly improve the management of these aggressive breast tumors.

A putative tumor suppressor is also *miR-17-5p*, which regulates amplified in breast cancer 1 (*AIB1*), a co-activator for nuclear receptors, such as ERα (Hossain et al. 2006). *AIB1* is over-expressed in primary breast cancer and can accelerate breast cancer cell proliferation by amplifying the mitogenic effect of estrogen through an increase in the transactivation functions of ERα. *MiR-17-5p*-mediated *AIB1* down-regulation resulted in inhibition of ER-dependent as well as ER-independent gene expression and retardation of proliferation of breast cancer cells. *MiR-17-5p* also completely abrogated the insulin-like growth factor 1-mediated anchorage-independent growth of breast cancer cells. The loss of heterozygosity of *miR-17-5p* genomic location has been found in different types of cancer, including breast cancer, and expression of *miR-17-5p* is low in breast cancer cell lines (Hossain et al. 2006).

MiR-17-5p belongs to a highly conserved, polycistronic miRNA cluster *miR-17-92*, which encodes six miRNAs: *miR-17*, *miR-18a*, *miR-19a*, *miR-20a*, *miR-19b-1*, and *miR-92-1* (Mendell 2008). Yu et al. (2008) described the function of this cluster as a negative regulator of cell cycle and proliferation of human breast cancer cells, which directly regulates cyclin D1 (*CCND1*). An inverse correlation between the levels of *miR-17-5p/miR-20a* miRNAs and cyclin D1 abundance was found in human breast tumors and cell lines. Interestingly, cyclin D1 could bind to the *miR-17/20* cluster promoter regulatory region and induce *miR-17-5p* and *miR-20a* expression in vivo. In turn, *miR-17/20* limited the proliferative function of cyclin D1, thus linking expression of this specific miRNA cluster to the regulation of oncogenesis.

3.2.2 Oncogenic MiRNAs

One of the most highly expressed miRNAs in human breast cancer is *miR-21*, which has been shown to be oncogenic by several studies. Inhibition of *miR-21* function by anti-miR constructs promoted apoptosis and inhibited breast cancer cell growth and survival both in vitro and in vivo (Si et al. 2007). The growth of tumor xenografts in a mouse model was inhibited by *miR-21* knockdown, probably via down-regulation of the anti-apoptotic protein Bcl-2. Thereafter, proteomics and microarray profiling studies have identified tropomyosin 1 (*TPM1*), maspin (*SERPINB5*), and the tumor suppressor gene programmed cell death-4 (*PDCD4*) as direct targets of *miR-21* in breast cancer (Frankel et al. 2008; Zhu et al. 2007; 2008). Down-regulation of these tumor suppressor genes by *miR-21* leads to enhanced tumor cell growth, migration, and invasion. Recently, the oncogenic receptor tyrosine kinase HER2 was shown to initiate signalling that leads in *miR-21* up-regulation, resulting in down-regulation of *PDCD4* and increased invasion capacity of the tumor cells (Huang et al. 2009).

Initially, Iorio et al. (2005) identified *miR-155* as significantly up-regulated in breast cancer as compared to normal breast tissue, suggesting that it may have oncogenic activities. Kong et al. (2008) demonstrated that *miR-155* is regulated by the transforming growth factor-β (TGF-β)/Smad pathway and plays a role in mammary epithelial cell plasticity through the targeting of Ras homolog gene family, member A (*RHOA*). *MiR-155* also induces cell survival and plays an important role in chemoresistance in breast cancer cells, and its anti-apoptotic function is mediated by direct inhibition of *FOXO3A* (Kong et al. 2010). Furthermore, suppressor of cytokine signaling 1 (*SOCS1*) has been verified to be a target of *miR-155* in breast cancer cells (Jiang et al. 2010). *SOCS1* is a tumor suppressor that normally functions as a negative feedback regulator of Janus-activated kinase (JAK)/signal transducer and activator of transcription (STAT) signalling. Over-expression of *miR-155* in breast cancer cells promoted cell proliferation, colony formation, and xenograft tumor growth through the repression of *SOCS1*. Finally, up-regulation of *miR-155* in breast cancer contributed to the post-transcriptional silencing of *SOCS1* and led to constitutive activation of *STAT3* in breast cancer cells. Collectively, these results suggest that *miR-155* is an oncomir for breast cancer.

Another oncogenic miRNA in breast cancer is *miR-27a*, which is over-expressed in breast cancer cell lines (Mertens-Talcott et al. 2007). *MiR-27a* down-regulates the zinc finger *ZBTB10* gene, a putative suppressor of specificity proteins (Sp). Over-expression of Sp in tumors contributes to the proliferative and angiogenic phenotype of breast cancer cells. Knock-down of *miR-27a* in MDA-MB-231 cells resulted in increased expression of *ZBTB10* and decreased expression of Sp proteins, accompanied by inhibition of Sp-dependent survival and angiogenic genes, including survivin (*BIRC5*), vascular endothelial growth factor (*VEGF*) and VEGF receptor 1 (*VEGFR1*) (Mertens-Talcott et al. 2007). *MiR-27a* was also shown to influence cell cycle progression by targeting myelin transcription factor 1 (*MYT1*), which inhibits the percentage of G2/M by inactivating Cdc2 kinase. In another study, *miR-27a* was shown to target the tumor suppressor *FOXO1*, thus facilitating breast cancer cell growth and cell cycle (Guttilla and White 2009). *MiR-27a* also indirectly regulates estrogen-responsiveness in breast cancer cells through suppression of *ZBTB10*, thereby enhancing the expression of estrogen receptor ERα (Li et al. 2010).

A central regulator of cell fate is the "genome guardian" *p53*, which induces several cellular responses, including cell cycle arrest, apoptosis, and senescence (Junttila and Evan 2009). MiRNAs have been linked to the *p53* network in many cancers. Brosh and colleagues (2008) identified a novel miRNA signature that is transcriptionally repressed by *p53* in breast cancers. These miRNAs included *miR-155* and the *miR-106b/93/25*, *miR-106a-92*, and *miR-17-92* polycistrons, which were repressed by *p53* through inhibition of *E2F1*. *E2F1* transcriptionally controlled these polycistrons, and together they regulated a mutual set of target genes, which are considered as anti-proliferative cell-cycle regulators. Consistent with *p53* function, the three polycistrons and *miR-155* are considered as oncogenic miRNAs and are over-expressed in diverse types of tumors, including breast cancer (Eis et al. 2005; He et al. 2005; Volinia et al. 2006; Yanaihara et al. 2006).

3.2.3 MiRNAs in Estrogen Receptor Signalling

Estrogen receptor status is important for the development, progression, and treatment of breast cancer, and it has become a major target for the treatment of this disease. Most human breast cancers are initially positive for ER, and their growth is stimulated by estrogens and inhibited by anti-estrogens. There are two distinct human estrogen receptors, ERα and ERβ, which are encoded by separate genes *ESR1* and *ESR2*, respectively (Cheskis et al. 2007; Hewitt and Korach 2003). ERα is an inducible transcription factor and primary mediator of the mitogenic activities of estrogen, and its levels are tightly controlled in a tissue- and development-specific manner. Generally, estrogen target genes have different sensitivities to ERα, and their transcriptional responses are often rate limited by the cellular receptor concentrations (Fowler et al. 2006). Therefore, it is important that the ERα level is strictly controlled to ensure appropriate target gene expression.

MiRNAs participate in estrogen signalling by regulating estrogen-responsive genes and pathways. Estrogen stimulation modulates the expression of several miRNAs (Cicatiello et al. 2010). There is a widespread repression of miRNAs in breast cancer cell lines upon estrogen stimulation, and this affects the estrogen-dependent cell growth in vivo (Maillot et al. 2009). The most potent miRNAs repressing cell growth in Maillot and colleagues' study were *miR-26a* and *miR-181a*, which regulated progesterone receptor and many other genes associated with cell proliferation. Furthermore, expression of several estrogen-down-regulated miRNAs was up-regulated by anti-estrogen treatment in breast tumors, suggesting the potential use of miRNAs as markers of response to anti-estrogens (Maillot et al. 2009).

Direct binding of ERα in the promoter region of *miR-21* has been verified by chromatin immunoprecipitation assays, providing evidence that ERα participates in the transcriptional regulation of miRNA genes (Bhat-Nakshatri et al. 2009). Estrogen-bound ERα has been shown to inhibit the maturation of miRNAs by associating with the Drosha complex and preventing the conversion of pri-miRNAs into pre-miRNAs, indicating that ERα regulates miRNA biogenesis (Yamagata et al. 2009). Estrogen stimulation can also up-regulate the expression of a small subset of miRNAs, including the *miR-17-92* and *miR-106a-363* clusters, and this is mediated by direct binding of *c-Myc* in the promoter region of the miRNA clusters (Castellano et al. 2009).

The first miRNA shown to directly target ERα was *miR-206*, which binds to the 3′UTR region of *ERα* and which is down-regulated by estrogen (Adams et al. 2007). The expression of *miR-206* was also shown to negatively correlate with ERα in clinical breast tumor samples (Kondo et al. 2008). Subsequently, 21 miRNAs were identified as being able to down-regulate ERα by a protein lysate microarray analysis (Leivonen et al. 2009). Among them were *miR-18a*, *miR-18b*, *miR-193b*, *miR-206*, and *miR-302c*, which directly targeted *ERα* and inhibited the estrogen-dependent proliferation and cell cycle of breast cancer cells. These miRNAs also targeted the ERα signalling pathway in general and down-regulated many estrogen-responsive genes, the effects being largely overlapping with that of silencing ERα by a small interfering RNA (siRNA). Furthermore, *miR-18a* and *miR-18b* levels were significantly higher in vivo in ERα-negative breast cancer, suggesting a role for these miRNAs in the development of ERα negative breast cancers (Leivonen et al. 2009).

MiRNAs also influence the anti-estrogen (e.g. Tamoxifen) response in breast cancer. ER is the main predictor of response to anti-estrogen therapy and *miR-221* and *miR-222* were demonstrated to negatively regulate ERα and contribute to the Tamoxifen resistance in breast cancer (Zhao et al. 2008). *MiR-221/222* were highly expressed in Tamoxifen-resistant breast cancer cells, and their expression correlated with *ERBB2* expression and a more aggressive tumor phenotype (Miller et al. 2008). Di Leva et al. (2010) showed that the enforced expression of ERα in ER-negative cells suppressed *miR-221/222* expression, whereas the elimination of ERα from ER-positive cells induced *miR-221/222* expression, suggesting a novel function of ERα as a negative regulator of the expression of these miRNAs. This molecular circuitry composed of *miR-221/222* and ERα may provide a basis for understanding how activation of miRNA expression can induce ER-positive breast tumors to become ER-negative, a frequent observation with profound impact on clinical outcomes in recurrent breast cancer.

3.3 MiRNAs in Breast Cancer Invasion and Metastasis

Metastatic breast cancer is the most advanced stage of breast cancer. The knowledge on the role of miRNAs as negative or positive regulators of invasion and metastasis has evolved recently, and multiple miRNAs regulate the breast cancer metastasis, as summarized in Fig. 3.1. Initially, the contribution of miRNAs in the development of metastases was discovered by Ma et al. (2007), who demonstrated that *miR-10b* regulates the initiation of invasion and metastasis in breast cancer. *MiR-10b* is highly expressed in metastatic breast cancer cells, and over-expression of *miR-10b* in otherwise non-invasive cancer cells was able to initiate invasion. A metastasis-promoting transcription factor Twist directly up-regulated *miR-10b* expression. *MiR-10b* targeted homeobox D10 (*HOXD10*), the down-regulation of which led to over-expression of a pro-metastatic gene *RHOC*. Another study identified T lymphoma and metastasis 1 (*TIAM1*) as an additional target for *miR-10b* (Moriarty et al. 2010). *MiR-10b* appears to be a promising target in inhibiting breast cancer metastasis, since therapeutic silencing of *miR-10b* by antagomirs inhibited

Fig. 3.1 Summary of microRNAs positively or negatively regulating breast cancer invasion and metastasis

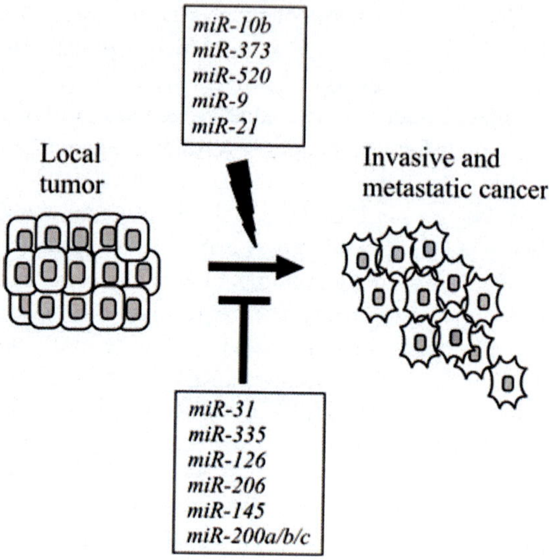

metastasis in a mouse mammary tumor model (Ma et al. 2010a). In addition to *miR-10b*, *miR-373* and *miR-520c* have been shown to be able to stimulate breast cancer cell migration and invasion, at least partly via suppression of *CD44*, a cell surface receptor for hyaluronan (Huang et al. 2008).

E-cadherin (CDH1) is a tumor suppressor protein that is used as a prognostic marker for breast cancer (Baranwal and Alahari 2009). It is an important factor in maintaining adherents junctions between adjacent epithelial cells, and its inactivation promotes epithelial-mesenchymal transition (EMT) and metastasis by enabling the dissociation of carcinoma cells from one another. In addition, E-cadherin loss liberates β-catenin, which may move into the nucleus and activate pro-metastatic genes. Ma et al. (2010b) demonstrated that *miR-9* is a pro-metastatic miRNA, which regulates E-cadherin expression and breast cancer metastasis. *MiR-9*, shown to be up-regulated in breast cancer cells by Iorio et al. (2005), directly targeted E-cadherin, leading to increased cell motility and invasiveness of breast cancer cells. *MiR-9*-mediated E-cadherin down-regulation resulted in the activation of β-catenin signalling, and in up-regulation of VEGF expression, which led to increased tumor angiogenesis. The pro-metastatic function of *miR-9* was further confirmed by over-expressing it in otherwise non-metastatic breast tumor cells, which enabled these cells to form pulmonary micrometastases in vivo in mice. Ma et al. (2010b) further showed that the expression of *miR-9* was activated by Myc and MycN binding to the *miR-9-3* locus. Furthermore, in human cancers, *miR-9* levels correlated with *MYCN* amplification, tumor grade and metastatic status.

EMT is a widespread, developmental program that regulates cell migration in many tissues and organs and is associated with normal and malignant mammary stem cell function (Lee et al. 2006). Malignant breast tumor cells are known to re-activate embryonic programs such as EMT to obtain a selective advantage such as enhanced motility and invasiveness. There are several studies demonstrating that the *miR-200* family (*miR-200a*, *miR-200b*, *miR-200c*, *miR-141*, and *miR-429*) inhibits EMT and the initiating steps of metastasis by maintaining the epithelial phenotype of cells. Hurteau et al. (2009) reported that *miR-200c* targets Zinc finger E-box-binding homeobox 1 (*ZEB1*), a transcription factor repressing E-cadherin, in breast cancer cell lines and showed that ectopic expression of *miR-200c* enhanced E-cadherin expression and promoted an epithelial-like morphology. Subsequently, other reports have linked the *miR-200* family with the epithelial phenotype, EMT, and its inverse process, mesenchymal-epithelial transition (MET) (Gregory et al. 2008; Korpal et al. 2008; Park et al. 2008; Tryndyak et al. 2010). The expression of *miR-200* family is lost in invasive breast cancer cell lines with mesenchymal phenotype (Gregory et al. 2008), and there is a significant correlation between the expression of *miR-200* and the E-cadherin/Vimentin ratio across all NCI60 cells, suggesting that the *miR-200* family is an universal regulator of the epithelial phenotype of cancer cells (Park et al. 2008). Furthermore, the expression of *miR-200* family is significantly lower in sarcomatoid metaplastic breast tumors as compared to the more epithelial ductal breast carcinomas (Gregory et al. 2008). Inhibition of endogenous *miR-200* expression levels was sufficient to induce EMT, whereas ectopic expression of *miR-200* induced MET in normal and cancer cell lines through direct targeting of *ZEB1* and *ZEB2* (Gregory et al. 2008; Korpal et al. 2008). This led in increase in E-cadherin expression and in reduced motility and aggressiveness of breast carcinoma cells.

Tavazoie et al. (2008) identified a set of eight miRNAs (*miR-335*, *miR-199a**, *miR-122a*, *miR-126*, *miR-206*, *miR-203*, *miR-489*, and *miR-127*) which had lower expression in metastatic breast cancer cells as compared to their non-metastatic counterparts. They further showed that restoration of *miR-335*, *miR-206*, or *miR-126* suppressed metastasis of breast cancer cells to the lung and bone in vivo in mice. *MiR-126* suppressed the overall tumor growth and proliferation, whereas *miR-335* and *miR-206* regulated migration and morphology. *MiR-335* targeted a set of metastasis-associated genes, whose collective expression in a large cohort of human tumors was associated with risk of distal metastasis. The genes regulated by *miR-335* included SRY-box containing gene 4 (*SOX4*), a migration-associated transcription factor, and tenascin C (*TNC*), an extracellular matrix component. Furthermore, expression of *miR-126* and *miR-335* was lost in the majority of primary breast tumors from patients who relapse, and the loss of expression of either miRNA was associated with poor distal metastasis-free survival. This suggests that the loss of these miRNAs could be used as a prognostic indicator for breast cancer metastasis.

Similarly, the expression of *miR-31* negatively correlates with breast cancer metastasis (Valastyan et al. 2009b). *MiR-31* does not inhibit primary tumor growth, but it significantly reduces local invasion and motility, as well as later steps of metastasis, such as metastatic colonization. Thus, it intervenes in multiple distinct stages of the invasion-metastasis cascade. *MiR-31* directly targeted many metastasis-promoting genes, such as frizzled3 (*FZD3*), integrin a5 (*ITGA5*), myosin phosphatase-Rho interacting protein (*M-RIP*), matrix metallopeptidase 16 (*MMP16*), radixin (*RDX*), and *RHOA* (Valastyan et al. 2009b). Interestingly, concurrent re-expression of *ITGA5*, *RDX*, and *RHOA* abrogated *miR-31*-imposed metastasis suppression (Valastyan et al. 2009a), indicating that these three genes were the main mediators of *miR-31* effects.

3.4 MiRNAs in Breast Cancer Therapeutics

The capacity of miRNA expression profiles to classify breast tumors according to clinicopathologic variables currently used to predict disease progression highlights the potential of miRNA signatures as novel prognostic indicators which may contribute to the improved selection of patients for therapy. MiRNA expression based clustering of breast cancer tumors according to their distinct molecular subclasses (Sørlie et al. 2001) could further characterize the molecular basis underlying these subtypes, perhaps define more precise subsets of breast cancer, and provide opportunities for the identification of novel targets that can be exploited for targeted therapy.

The functional studies described above have established that individual miRNAs play important roles in proliferation, invasion and metastasis of breast cancer. These observations suggest that miRNAs have the potential to provide novel therapeutic strategies for treatment of breast cancer. The advantage of miRNAs over siRNA/shRNA is their ability to affect multiple targets, thus regulating several genes on a same pathway and a whole network of interacting molecules (Tsang et al. 2010). Inhibition of oncomirs, such as *miR-21* or *miR-10b*, in breast tumors may suppress tumor growth and metastasis (Ma et al. 2010a; Si et al. 2007). Conversely, the induction of tumor suppressor miRNA expression, the so-called "miRNA replacement therapy" (Kota et al. 2009), may result in the prevention of progression or even shrinking of breast tumors. This has been shown in principle by an artificial miRNA against a metastasis associated gene, CXC chemokine receptor-4 (*CXCR4*), which inhibited the invasive and metastatic potential of breast tumor cells (Liang et al. 2007). In addition, miRNAs could be used to sensitize tumor cells to anti-cancer therapy, such as anti-estrogen or Trastuzumab therapy, which is an exciting prospect for patients exhibiting a poor response to primary chemotherapy or treatment.

The biggest problem in miRNA therapeutics is the delivery to the target tissue. A number of strategies for the delivery of miRNA precursors or antagomirs could be applied to the miRNA-based therapy. Chemically modified miRNAs can be delivered to the cells by using suitable lipid-based transfection reagents, liposomes, or cholesterol-conjugates, which have been successfully used in miRNA in vivo studies (Elmen et al. 2008; Krutzfeldt et al. 2005). In addition, ligands for specific cell surface receptors, such as *ERBB2*, capable of being internalized can be conjugated to the miRNA-oligonucleotides, thereby facilitating both cellular uptake and cell type–specific delivery (Juliano et al. 2008). An alternative means of triggering miRNA over-expression is through promoter-driven miRNA expression cassettes encoding miRNA precursors, which are processed as mature, active miRNAs in the target cells (Gentner et al. 2009; McLaughlin et al. 2007).

3.5 Conclusions and Future Perspectives

The involvement of miRNAs in the initiation and progression of breast cancer has great potential for new developments in current diagnostic and therapeutic strategies in the future management of breast cancer. MiRNAs and their targets seem to form complex regulatory networks, but as the functional roles of miRNAs in cancer biology are further uncovered, it is conceivable that there will be improved methods of stratifying and subclassifying breast cancers. This will eventually lead to tailored and individualized therapy. There is apparently great demand for further intensive research into the identification of novel miRNAs, the elucidation of their targets, and determining their functional effects, so as to improve our knowledge of the roles of miRNAs in breast cancer and to expose their potential as therapeutic agents.

References

Adams BD, Furneaux H, White BA. The micro-ribonucleic acid (miRNA) miR-206 targets the human estrogen receptor-alpha (ERa) and represses ERa messenger RNA and protein expression in breast cancer cell lines. Mol Endocrinol. 2007;21:1132–47.

Al-Hajj M. Cancer stem cells and oncology therapeutics. Curr Opin Oncol. 2007;19:61–4.

Al-Hajj M, Wicha MS, Benito-Hernandez A, et al. Prospective identification of tumorigenic breast cancer cells. Proc Natl Acad Sci USA. 2003;100:3983–8.

Baranwal S, Alahari SK. Molecular mechanisms controlling E-cadherin expression in breast cancer. Biochem Biophys Res Commun. 2009;384:6–11.

Bhat-Nakshatri P, Wang G, Collins NR, et al. Estradiol-regulated microRNAs control estradiol response in breast cancer cells. Nucleic Acids Res. 2009;37:4850–61.

Blenkiron C, Goldstein LD, Thorne NP, et al. MicroRNA expression profiling of human breast cancer identifies new markers of tumor subtype. Genome Biol. 2007;8:R214.

Brosh R, Shalgi R, Liran A, et al. p53-Repressed miRNAs are involved with E2F in a feed-forward loop promoting proliferation. Mol Syst Biol. 2008;4:229.

Calin GA, Sevignani C, Dumitru CD, et al. Human microRNA genes are frequently located at fragile sites and genomic regions involved in cancers. Proc Natl Acad Sci USA. 2004;101: 2999–3004.

Castellano L, Giamas G, Jacob J, et al. The estrogen receptor-alpha-induced microRNA signature regulates itself and its transcriptional response. Proc Natl Acad Sci USA. 2009;106: 15732–7.

Cheskis BJ, Greger JG, Nagpal S, et al. Signaling by estrogens. J Cell Physiol. 2007;213:610–7.

Cicatiello L, Mutarelli M, Grober OM, et al. Estrogen receptor α controls a gene network in luminal-like breast cancer cells comprising multiple transcription factors and MicroRNAs. Am J Pathol. 2010;176:2113–30.

Di Leva G, Gasparini P, Piovan C, et al. MicroRNA cluster 221-222 and Estrogen receptor α interactions in breast cancer. J Natl Cancer Inst. 2010;102:706–21.

Eis PS, Tam W, Sun L, et al. Accumulation of miR-155 and BIC RNA in human B cell lymphomas. Proc Natl Acad Sci USA. 2005;102:3627–32.

Elmen J, Lindow M, Schutz S, et al. LNA-mediated microRNA silencing in non-human primates. Nature. 2008;452:896–9.

Foekens JA, Sieuwerts AM, Smid M, et al. Four miRNAs associated with aggressiveness of lymph node-negative, estrogen receptor-positive human breast cancer. Proc Natl Acad Sci USA. 2008;105:13021–6.

Fowler AM, Solodin NM, Valley CC, et al. Altered target gene regulation controlled by estrogen receptor-alpha concentration. Mol Endocrinol. 2006;20:291–301.

Frankel LB, Christoffersen NR, Jacobsen A, et al. Programmed cell death 4 (PDCD4) is an important functional target of the microRNA miR-21 in breast cancer cells. J Biol Chem. 2008;283:1026–33.
Gentner B, Schira G, Giustacchini A, et al. Stable knockdown of microRNA in vivo by lentiviral vectors. Nat Methods. 2009;6:63–6.
Gregory PA, Bert AG, Paterson EL, et al. The miR-200 family and miR-205 regulate epithelial to mesenchymal transition by targeting ZEB1 and SIP1. Nat Cell Biol. 2008;10: 593–601.
Guttilla IK, White BA. Coordinate regulation of FOXO1 by miR-27a, miR-96, and miR-182 in breast cancer cells. J Biol Chem. 2009;284:23204–16.
He L, Thomson JM, Hemann MT, et al. A microRNA polycistron as a potential human oncogene. Nature. 2005;435:828–33.
Hewitt SC, Korach KS. Oestrogen receptor knockout mice: roles for oestrogen receptors a and b in reproductive tissues. Reproduction. 2003;125:143–9.
Hossain A, Kuo MT, Saunders GF. MiR-17-5p regulates breast cancer cell proliferation by inhibiting translation of AIB1 mRNA. Mol Cell Biol. 2006;26:8191–201.
Huang Q, Gumireddy K, Schrier M, et al. The microRNAs miR-373 and miR-520c promote tumour invasion and metastasis. Nat Cell Biol. 2008;10:202–10.
Huang TH, Wu F, Loeb GB, et al. Up-regulation of miR-21 by HER2/neu signaling promotes cell invasion. J Biol Chem. 2009;284:18515–24.
Hurteau GJ, Carlson JA, Roos E, et al. Stable expression of miR-200c alone is sufficient to regulate TCF8 (ZEB1) and restore E-cadherin expression. Cell Cycle. 2009;8:2064–9.
Hynes NE, MacDonald G. ErbB receptors and signaling pathways in cancer. Curr Opin Cell Biol. 2009;21:177–84.
Iorio MV, Casalini P, Piovan C, et al. MicroRNA-205 regulates HER3 in human breast cancer. Cancer Res. 2009;69:2195–200.
Iorio MV, Ferracin M, Liu CG, et al. MicroRNA gene expression deregulation in human breast cancer. Cancer Res. 2005;65:7065–70.
Jiang S, Zhang HW, Lu MH, et al. MicroRNA-155 functions as an oncomir in breast cancer by targeting the suppressor of cytokine signaling 1 gene. Cancer Res. 2010;70:3119–27.
Johnson SM, Grosshans H, Shingara J, et al. RAS is regulated by the let-7 microRNA family. Cell. 2005;120:635–47.
Juliano R, Alam MR, Dixit V, et al. Mechanisms and strategies for effective delivery of antisense and siRNA oligonucleotides. Nucleic Acids Res. 2008;36:4158–71.
Junttila MR, Evan GI. p53–a Jack of all trades but master of none. Nat Rev Cancer. 2009;9: 821–9.
Kondo N, Toyama T, Sugiura H, et al. MiR-206 expression is down-regulated in estrogen receptor a-positive human breast cancer. Cancer Res. 2008;68:5004–8.
Kong W, He L, Coppola M, et al. MicroRNA-155 regulates cell survival, growth and chemosensitivity by targeting FOXO3a in breast cancer. J Biol Chem. 2010;285:17869–79.
Kong W, Yang H, He L, et al. MicroRNA-155 is regulated by the transforming growth factor beta/Smad pathway and contributes to epithelial cell plasticity by targeting RhoA. Mol Cell Biol. 2008;28:6773–84.
Korpal M, Lee ES, Hu G, et al. The miR-200 family inhibits epithelial-mesenchymal transition and cancer cell migration by direct targeting of E-cadherin transcriptional repressors ZEB1 and ZEB2. J Biol Chem. 2008;283:14910–4.
Kota J, Chivukula RR, O'Donnell KA, et al. Therapeutic microRNA delivery suppresses tumorigenesis in a murine liver cancer model. Cell. 2009;137:1005–17.
Krutzfeldt J, Rajewsky N, Braich R, et al. Silencing of microRNAs in vivo with 'antagomirs'. Nature. 2005;438:685–9.
Lee JM, Dedhar S, Kalluri R, et al. The epithelial-mesenchymal transition: new insights in signaling, development, and disease. J Cell Biol. 2006;172:973–81.
Lee YS, Dutta A. The tumor suppressor microRNA let-7 represses the HMGA2 oncogene. Genes Dev. 2007;21:1025–30.

Lehmann U, Hasemeier B, Christgen M, et al. Epigenetic inactivation of microRNA gene hsa-mir-9-1 in human breast cancer. J Pathol. 2008;214:17–24.

Leivonen S-K, Mäkelä R, Östling P, et al. Protein lysate microarray analysis to identify microRNAs regulating estrogen receptor signaling in breast cancer cell lines. Oncogene. 2009;28: 3926–36.

Li X, Mertens-Talcott SU, Zhang S, et al. MicroRNA-27a indirectly regulates estrogen receptor α expression and hormone responsiveness in MCF-7 breast cancer cells. Endocrinology. 2010;151:2462–73.

Li X, Yan P, Shao, Z. Downregulation of miR-193b contributes to enhance urokinase-type plasminogen activator (uPA) expression and tumor progression and invasion in human breast cancer. Oncogene. 2009;28:3937–48.

Liang Z, Wu H, Reddy S, et al. Blockade of invasion and metastasis of breast cancer cells via targeting CXCR4 with an artificial microRNA. Biochem Biophys Res Commun. 2007;363: 542–6.

Lowery AJ, Miller N, Devaney A, et al. MicroRNA signatures predict oestrogen receptor, progesterone receptor and HER2/neu receptor status in breast cancer. Breast Cancer Res. 2009;11:R27.

Ma L, Reinhardt F, Pan E, et al. Therapeutic silencing of miR-10b inhibits metastasis in a mouse mammary tumor model. Nat Biotechnol. 2010a;28:341–7.

Ma L, Teruya-Feldstein J, Weinberg RA. Tumour invasion and metastasis initiated by microRNA-10b in breast cancer. Nature. 2007;449:682–8.

Ma L, Young J, Prabhala H, et al. miR-9, a MYC/MYCN-activated microRNA, regulates E-cadherin and cancer metastasis. Nat Cell Biol. 2010b;12:247–56.

Maillot G, Lacroix-Triki M, Pierredon S, et al. Widespread estrogen-dependent repression of micrornas involved in breast tumor cell growth. Cancer Res. 2009;69:8332–40.

Mattie MD, Benz CC, Bowers J, et al. Optimized high-throughput microRNA expression profiling provides novel biomarker assessment of clinical prostate and breast cancer biopsies. Mol Cancer. 2006;5:24.

McLaughlin J, Cheng D, Singer O, et al. Sustained suppression of Bcr-Abl-driven lymphoid leukemia by microRNA mimics. Proc Natl Acad Sci USA. 2007;104:20501–6.

Mendell JT. miRiad roles for the miR-17-92 cluster in development and disease. Cell. 2008;133:217–22.

Mertens-Talcott SU, Chintharlapalli S, Li X, et al. The oncogenic microRNA-27a targets genes that regulate specificity protein transcription factors and the G2-M checkpoint in MDA-MB-231 breast cancer cells. Cancer Res. 2007;67:11001–11.

Miller TE, Ghoshal K, Ramaswamy B, et al. MicroRNA-221/222 confers tamoxifen resistance in breast cancer by targeting p27Kip1. J Biol Chem. 2008;283:29897–903.

Moriarty CH, Pursell B, Mercurio AM. miR-10b targets Tiam1: implications for Rac activation and carcinoma migration. J Biol Chem. 2010;285:20541–6.

Negrini M, Rasio D, Hampton GM, et al. Definition and refinement of chromosome 11 regions of loss of heterozygosity in breast cancer: identification of a new region at 11q23.3. Cancer Res. 1995;55:3003–7.

Park SM, Gaur AB, Lengyel E, et al. The miR-200 family determines the epithelial phenotype of cancer cells by targeting the E-cadherin repressors ZEB1 and ZEB2. Genes Dev. 2008;22: 894–907.

Park SM, Shell S, Radjabi AR, et al. Let-7 prevents early cancer progression by suppressing expression of the embryonic gene HMGA2. Cell Cycle. 2007;6:2585–90.

Sachdeva M, Mo YY. MicroRNA-145 suppresses cell invasion and metastasis by directly targeting mucin 1. Cancer Res. 2010;70:378–87.

Scott GK, Goga A, Bhaumik D, et al. Coordinate suppression of ERBB2 and ERBB3 by enforced expression of micro-RNA miR-125a or miR-125b. J Biol Chem. 2007;282:1479–86.

Sempere LF, Christensen M, Silahtaroglu A, et al. Altered MicroRNA expression confined to specific epithelial cell subpopulations in breast cancer. Cancer Res. 2007;67: 11612–20.

Shimono Y, Zabala M, Cho RW, et al. Downregulation of miRNA-200c links breast cancer stem cells with normal stem cells. Cell. 2009;138:592–603.

Si ML, Zhu S, Wu H, et al. miR-21-mediated tumor growth. Oncogene. 2007;26:2799–803.

Sørlie T. Molecular portraits of breast cancer: tumour subtypes as distinct disease entities. Eur J Cancer. 2004;40:2667–75.

Sørlie T, Perou CM, Tibshirani R, et al. Gene expression patterns of breast carcinomas distinguish tumor subclasses with clinical implications. Proc Natl Acad Sci USA. 2001;98:10869–74.

Spizzo R, Nicoloso MS, Lupini L, et al. miR-145 participates with TP53 in a death-promoting regulatory loop and targets estrogen receptor-alpha in human breast cancer cells. Cell Death Differ. 2010;17:246–54.

Tavazoie SF, Alarcon C, Oskarsson T, et al. Endogenous human microRNAs that suppress breast cancer metastasis. Nature. 2008;451:147–52.

Tryndyak VP, Beland FA, Pogribny IP. E-cadherin transcriptional down-regulation by epigenetic and microRNA-200 family alterations is related to mesenchymal and drug-resistant phenotypes in human breast cancer cells. Int J Cancer. 2010;126:2575–83.

Tsang JS, Ebert MS, van Oudenaarden A. Genome-wide dissection of microRNA functions and cotargeting networks using gene set signatures. Mol Cell. 2010;38:140–53.

Valastyan S, Benaich N, Chang A, et al. Concomitant suppression of three target genes can explain the impact of a microRNA on metastasis. Genes Dev. 2009a;23:2592–7.

Valastyan S, Reinhardt F, Benaich N, et al. A pleiotropically acting microRNA, miR-31, inhibits breast cancer metastasis. Cell. 2009b;137:1032–46.

Volinia S, Calin GA, Liu CG, et al. A microRNA expression signature of human solid tumors defines cancer gene targets. Proc Natl Acad Sci USA. 2006;103:2257–61.

Yamagata K, Fujiyama S, Ito S, et al. Maturation of microRNA is hormonally regulated by a nuclear receptor. Mol Cell. 2009;36:340–7.

Yanaihara N, Caplen N, Bowman E, et al. Unique microRNA molecular profiles in lung cancer diagnosis and prognosis. Cancer Cell. 2006;9:189–98.

Yang R, Schlehe B, Hemminki K, et al. A genetic variant in the pre-miR-27a oncogene is associated with a reduced familial breast cancer risk. Breast Cancer Res Treat. 2010;121:693–702.

Yu Z, Wang C, Wang M, et al. A cyclin D1/microRNA 17/20 regulatory feedback loop in control of breast cancer cell proliferation. J Cell Biol. 2008;182:509–17.

Yu F, Yao H, Zhu P, et al. let-7 regulates self renewal and tumorigenicity of breast cancer cells. Cell. 2007;131:1109–23.

Zhang L, Huang J, Yang N, et al. microRNAs exhibit high frequency genomic alterations in human cancer. Proc Natl Acad Sci USA. 2006;103:9136–41.

Zhao JJ, Lin J, Yang H, et al. MicroRNA-221/222 negatively regulates ERa and associates with tamoxifen resistance in breast cancer. J Biol Chem. 2008;283:31079–86.

Zhu S, Si ML, Wu H, et al. MicroRNA-21 targets the tumor suppressor gene tropomyosin 1 (TPM1). J Biol Chem. 2007;282:14328–36.

Zhu S, Wu H, Wu F, et al. MicroRNA-21 targets tumor suppressor genes in invasion and metastasis. Cell Res. 2008;18:350–9.

Chapter 4
MicroRNAs in Colorectal Cancer

Ondrej Slaby, Marek Svoboda, Jaroslav Michalek, and Rostislav Vyzula

Abstract Colorectal cancer (CRC) represents one of the most frequent causes of death for cancer. CRC has been recently defined as the third most common cancer. MicroRNAs (miRNAs) are a class of small interfering RNAs frequently involved in the pathogenesis of cancer. Polymorphisms within miRNAs binding regions have been described as new risk factors for CRC. Several genome-wide profiling studies have identified miRNAs deregulated in CRC tissue. A number of experimental studies on these miRNAs revealed insight into miRNA-mediated regulatory links to well-known oncogenic and tumor suppressor signaling pathways. Several investigations have also described the ability of specific miRNA expression profiles to predict prognosis and therapy response in CRC patients. In this chapter, we focus on the most significant findings of original studies on miRNAs involvement in CRC pathogenesis, focusing also on the potential of cancer-related miRNAs as biomarkers for diagnosis, prognosis, prediction, and therapeutical targets.

4.1 Introduction

Deregulation of microRNAs (miRNAs) can affect carcinogenesis if their mRNA targets are encoded by oncogenes or tumor suppressor genes. Both, over-expression and silencing or switching off of specific miRNAs, have been described in the carcinogenesis of colorectal cancer (CRC). Up-regulation of mature miRNA may occur as a consequence of transcriptional activation or amplification of the miRNA encoding gene, whereas silencing or reduced expression may result from deletion of a particular chromosomal region, epigenetic silencing, or defects in their biogenesis (Rossi et al. 2010).

In principle, two approaches are applied today to reveal the connection between miRNAs and CRC. On one hand, miRNAs seem to regulate many known oncogenic

O. Slaby (✉)
Department of Comprehensive Cancer Care, Masaryk Memorial Cancer Institute, 656 53 Brno, Czech Republic
e-mail: slaby@mou.cz

W.C.S. Cho (ed.), *MicroRNAs in Cancer Translational Research*,
DOI 10.1007/978-94-007-0298-1_4, © Springer Science+Business Media B.V. 2011

and tumor suppressor pathways involved in the pathogenesis of CRC. Many proteins involved in key signaling pathways of CRC, such as members of the Wnt/β-catenin and phosphatidylinositol 3-kinase (PI3K) pathways, KRAS, p53, extracellular matrix regulators as well as epithelial-mesenchymal transition (EMT) transcription factors (Fearon and Vogelstein 1990), are altered and seem to be affected by miRNA regulation in CRC (Fig. 4.1). Analyses of these miRNAs in mechanistic studies are crucial to better understanding CRC pathogenesis (Aslam et al. 2009; Faber et al. 2009) and may provide exciting therapeutic opportunities (Cho 2010a). Findings from this area are discussed in the first part of this chapter. On the other hand, expression profiles of hundreds of miRNAs have been shown to have at least the same potential for identification of biomarkers as profiling of their mRNA or protein counterparts. This allows predicting prognosis and therapy response as well as

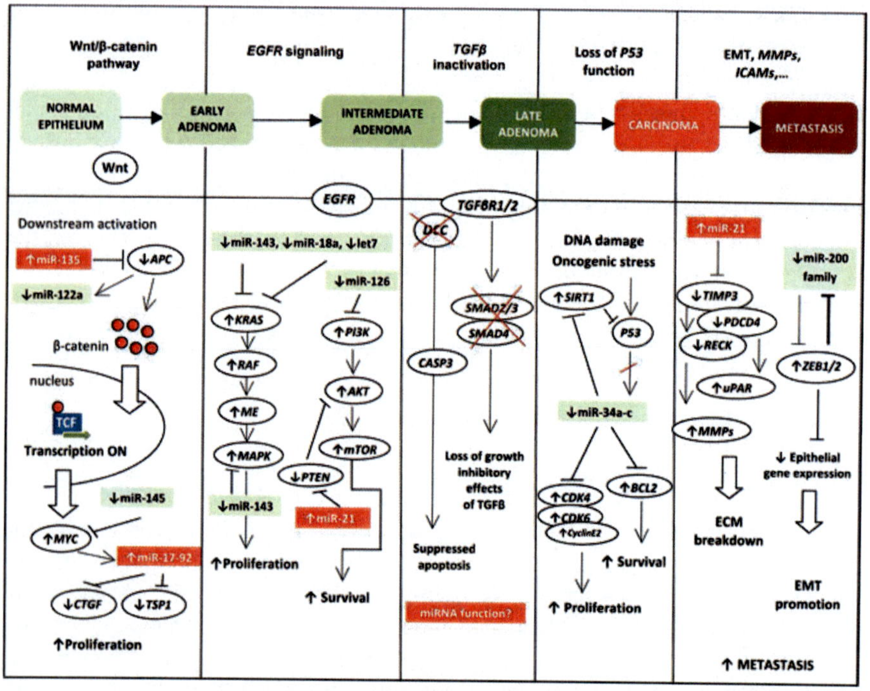

Fig. 4.1 MicroRNAs involvement in Vogelstein's model of colorectal cancer development. (APC – adenomatous polyposis coli, CTGF – connective tissue growth factor, TSP1 – thrombospondin 1, EGFR – epidermal growth factor receptor, mTOR – mammalian target of rapamycin, PTEN – phosphatase and tensin homolog, DCC – deleted in colorectal carcinoma, TGFβR1/2 – transforming growth factor, beta receptor 1/2, CASP3 – caspase 3, SIRT1 – sirtuin 1, CDK4,6 – cyclin-dependent kinase 4,6, ECM – extracellular matrix, EMT – epithelial–mesenchymal transition, ICAMs – intercellular adhesive molecule*s*, TIMP3 – tissue inhibitor of metalloproteinase 3, PDCD4 – programmed cell death 4, RECK – reversion-inducing-cysteine-rich protein with kazal motifs, uPAR – plasminogen activator, urokinase receptor, MMPs – matrix metallopeptidases, ZEB1/2 – zinc-finger E-box binding homeobox 1) (Slaby et al. 2009)

distinguishing certain disease entities, including CRC, as discussed in the second part that focuses on miRNA expression profiling (Cho 2010b; Slaby et al. 2009).

4.2 MiRNAs in Colorectal Cancer Signaling

CRC development has been linked to the progressive acquisition of mutations in genes with a crucial role in cell growth, proliferation and programmed cell death As shown in many different studies, miRNAs might perfectly fit and integrate this model initially postulated by Vogelstein by controlling several pathways involved in CRC development (Valeri et al. 2009).

4.2.1 Wnt/β-Catenin Pathway

The Wnt/β-catenin pathway plays a pivotal role in an early colorectal tumor development. Inactivation of the adenomatous polyposis coli (*APC*) gene is a major initiating event in colorectal carcinogenesis occurring in more than 60% of colorectal adenomas and carcinomas and leading to stimulation of the Wnt pathway via free β-catenin (Fearon and Vogelstein 1990). According to a recent study by Nagel et al. (2008), miRNAs represent a novel mechanism for *APC* regulation in CRC. *MiR-135a* and *miR-135b* decrease translation of the *APC* transcript in vitro. Of note, *miR-135a* and *miR-135b* were also found to be up-regulated in vivo in colorectal adenomas and carcinomas and correlated with low *APC* levels (Nagel et al. 2008). These observations suggest that alteration in the *miR-135* family can be one of the early events in CRC's molecular pathogenesis. On other hand, restoration of *APC* function in colorectal cancer cells led to the deregulation of several cancer-related miRNAs, such as *miR-122a* which was recognized as the liver-specific microRNA. *MiR-122a* was down-regulated in gastrointestinal cancer cell lines as well as primary carcinoma tissues. Inhibition of *miR-122a* could reverse wild-type *APC*-induced growth inhibition of gastrointestinal cancer cells while *miR-122a* mimic inhibited cell growth (Wang et al. 2009).

4.2.2 EGFR Signaling (KRAS and PI3K Pathways)

The epidermal growth factor receptor (EGFR) signaling significantly participates in promotion and progression of broad spectrum of solid tumors, and members of EGFR pathway constitute promising targets for anti-cancer therapy. Two different studies showed that *miR-7* and *miR-128* are able to down-regulate *EGFR* and its downstream pathways in breast, glioblastoma and lung cancer cell lines and in lung cancer patients (Webster et al. 2009; Weiss et al. 2008). Even though *miR-128* and *miR-7* have been shown to be deregulated in CRC (Motoyama et al. 2009; Ng et al. 2009) no study support a role for their down-regulation in *EGFR* over-expression. Stimulation of the EGFR and, subsequently, KRAS signaling lead to the activation of numerous signal transduction molecules initiating a cascade of downstream effectors that mediate tumor growth, survival, angiogenesis and metastasis (Ciardiello and Tortora 2008). *KRAS* oncogene has been reported to be a direct target of the *let-7* miRNA family (Johnson et al. 2005). When *let-7* low-expressing DLD-1 colon cancer cells were transfected with *let-7a-1* precursor, significant growth suppression with concurrent decrease of the *KRAS* protein levels was observed while the levels of both of their mRNAs remained almost unchanged (Akao et al. 2006). Another miRNA associated with *KRAS* regulation in CRC is *miR-143* (Chen et al. 2009). *KRAS* was deduced to be a *miR-143* target not only by computational prediction but also by noting the inverse correlation between *miR-143* and KRAS protein level in clinical samples. *KRAS* expression in vitro was significantly abolished by treatment with *miR-143* precursor, whereas *miR-143* inhibitor increased the KRAS protein level. Moreover, constitutive phosphorylation of MAPK was efficiently blocked through inhibition of *KRAS* expression by *miR-143* (Chen et al. 2009). Recently, *miR-18a* was observed to directly regulate *KRAS* but not *N-* and *HRAS* levels in the colon adenocarcinoma HT-29 cells (Tsang et al. 2009).

Another central signaling pathway downstream from EGFR and important in CRC development is the phosphatidylinositol 3-kinase (PI3K) pathway. MiRNAs arrays-based studies revealed a ubiquitous loss of *miR-126* expression in CRC cell lines when compared to normal human colon epithelium. Reconstitution of *miR-126* resulted in a significant growth reduction (Guo et al. 2008). As a direct target of *miR-126,* the p85β regulatory subunit involved in stabilizing and propagating the PI3K signal was mechanistically proven. Furthermore, this p85β reduction mediated by *miR-126* was accompanied by a substantial reduction in phosphorylated Akt levels in the cancer cells, suggesting an impairment in PI3K signaling. In a series of matched normal colon and primary colon tumor tissues, each of the tumors demonstrated *miR-126* down-regulation together with an increase in the p85β protein level (Guo et al. 2008). Another important regulatory component of the PI3K pathway, the tumor suppressor gene *PTEN*, is strongly repressed by *miR-21*, which was demonstrated on a hepatocellular carcinoma model (Meng et al. 2007). However, *miR-21* is the miRNA most frequently up-regulated in CRC (Schetter et al. 2008; Slaby et al. 2007; Krichevsky and Gabriely 2009). It seems that suppression of *PTEN* controlled by *miR-21* is associated with augmentation of PI3K signaling and progression of CRC.

4.2.3 p53 Pathway

A well-known tumor suppressor gene, *p53* is mutated in about 50–75% of all CRCs and many other human tumors. *p53* responds to DNA damage or deregulation of mitogenic oncogenes through the induction of cell cycle checkpoints, apoptosis, or cellular senescence (Fearon and Vogelstein 1990). Although *p53* is clearly a transcriptional activator, numerous reports have indicated that *p53* also represses the expression of specific genes either directly or indirectly (Bailey et al. 2010; He et al. 2007). The manner in which this is achieved was obscure, with both transcriptional and post-transcriptional suppression as possible mechanisms. In the latter case, the discovery of an extensive regulatory network of miRNAs offered the possibility that p53-mediated control of miRNA expression could allow it to act indirectly to repress target gene expression at the post-transcriptional level. Recently, several groups have unraveled important aspects of the connection between *p53* and the miRNA network (Hermeking 2007).The conserved *miR-34a-c* family was found to be direct transcriptional targets of *p53*. MiRNA expression patterns were analyzed in wild-type $p53^{+/+}$ and $p53^{-/-}$ mutant HCT-116 colon cancer cell lines after treatment with DNA damaging agents. Several miRNAs were induced in the wild type but not in the $p53^{-/-}$ mutant cells, thus suggesting a p53-mediated expression. *MiR-34a* showed the strongest induction. Expression of *miR-34a* was sufficient to induce apoptosis through p53-dependent and independent mechanisms. *MiR-34a*-responsive genes are highly enriched for those that regulate cell-cycle progression, cellular proliferation, apoptosis, DNA repair and angiogenesis (Chang et al. 2007). By experimentally over-expressing *miR-34a*, *p53* effects like cell-cycle arrest and apoptosis could be achieved. p53's connection to the *miR-34* family was successfully evaluated also on a model of lung carcinoma cells harboring regulated *p53* alleles (Bommer et al. 2007) and $p53^{+/+}$ and $p53^{-/-}$ mouse embryo fibroblasts (Corney et al. 2007). *MiR-34a* induction leads to dramatic reprogramming of gene expression. Among the down-regulated targets of the *miR-34* family were well-characterized *p53* targets like *CDK4/6*, cyclin E2, *E2F5*, *BIRC3* and *Bcl-2*. Notably, these effects were nearly identical irrespective of whether *miR-34-a*, *miR-34-b* or *miR-34-c* was introduced. Others have identified SIRT1, a regulator of apoptosis in response to cellular stress, as an additional target of *miR-34a*. Interestingly, the suppression of *SIRT1* by *miR-34a* resulted in apoptosis in wild-type colon cancer cells but not in $p53^{-/-}$ mutants. This suggests a positive feedback loop between *p53* and *miR-34* (Yamakuchi et al. 2008).

Decreased levels of the *miR-34* family have been found in many tumors, including CRC. *MiR-34a* expression was found to be down-regulated in 9 of 25 CRCs (Tazawa et al. 2007). Loss of 1p36, the genomic interval harboring *miR-34a*, is common in diverse human cancers (Chang et al. 2007) but one of the other mechanisms responsible for decrease of *miR-34* family expression levels seems to be CpG island hypermethylation. *MiR-34a* promoter methylation was reported in 3 of 23 cases of colon cancer (Lodygin et al. 2008). *MiR-34b/c* were found to be epigenetically silenced in 9 of 9 cell lines examined and in 101 of 111 primary CRC tumors, but they were not in normal colonic epithelium. After treatment with demethylating agents, *miR-34b/c* expression was restored. That resulted in inhibition of tumor motility and metastasis formation (Toyota et al. 2008). The high frequency of their methylation in CRC and their contribution to the *p53* network imply that *miR-34a-c* function as important tumor suppressors which can be lost during CRC development (Faber et al. 2009).

Like *p53*, members of the *miR-34* family can be considered as tumor suppressors to date, making them potential candidates for causing cancer by way of their inactivation.

4.2.4 IGF Signaling

Insulin receptor substrate-1 (IRS-1) plays an important role in cell growth and cell proliferation. IRS-1, especially when activated by the type 1 insulin-like growth factor receptor (IGF-IR), sends an unambiguous mitogenic, anti-apoptotic, and anti-differentiation signal. IRS-1 levels are often increased in cases of human cancer and are low or even absent in differentiating cells (Pechlivanis et al. 2007). *MiR-145* has been proposed as a tumor suppressor and it had been shown previously that *miR-145* targets the 3′ untranslational region (3′UTR) of *IRS-1* and dramatically inhibits the growth of colon cancer cells (Shi et al. 2007). More recently, *IGF-IR* was proven to be another direct target of *miR-145*. It was shown that an *IRS-1* lacking its 3′UTR is no longer down-regulated by *miR-145* and rescues colon cancer cells from growth inhibition induced by *miR-145*. An *IGF-IR* resistant to *miR-145* (again by elimination of its 3′UTR) was not down-regulated by *miR-145* but failed to rescue colon cancer cells from growth inhibition. These data indicate that *miR-145* plays a significant role in IGF signaling in cancer pathogenesis (La Rocca et al. 2009).

4.2.5 E2F Family and Cell Cycle Regulation

The *miR-17-92* cluster encodes six miRNAs *(miR-17, miR-18a, miR-19a, miR-20a, miR-19b-1*, and *miR-92-1)* located on chromosome 13q31.3. The human genomic locus encoding these miRNAs undergoes amplification in several types of lymphoma and solid tumors. The *miR-17-92* cluster seems to be tightly linked to the functions of the E2F family of transcription factors, which are critical regulators of the cell cycle and apoptosis. E2F1, E2F2, and E2F3-activating E2Fs that induce expression of genes driving the progression from the G1 into the S phase – were among the first experimentally verified targets of the *miR-17-92* cluster. Moreover, both *E2F1* and *E2F3* can directly activate transcription of these miRNAs, thereby establishing a negative feedback loop (Woods et al. 2007). The pro-tumorigenic activity of the *miR-17-92* cluster also involves cell-non-autonomous functions that include induction of angiogenesis in solid tumors. Using a mouse model of colon cancer, Dews et al. (2006) demonstrated that the angiogenic activity of *c-Myc* is due at least in part to downstream activation of the *miR-17-92* cluster. The anti-angiogenic factors thrombospondin-1 (*TSP1*) and connective tissue growth factor (*CTGF*) are negatively regulated by these miRNAs, which are potently induced by *c-Myc* in this model. Robust vascularization of tumors can be induced by expression of either *c-Myc* or the *miR-17-92* cluster (Dews et al. 2006).

4.2.6 MET Signaling

Hu et al. (2010) provided evidence that *miR-133b* regulated tumor cell proliferation and apoptosis through modulation of the MET signaling pathway. In the CRC cell lines SW-620 and HT-29, ectopic expression of *miR-133b* potently affected tumor cell proliferation and apoptosis in vitro and in vivo by direct targeting of the receptor tyrosine kinase *MET*. Transfection of SW-620 and HT-29 cells with *miR-133b* significantly suppressed a luciferase-reporter containing the *MET*-3′UTR.

4.2.7 Cyclooxygenase-2

Cyclooxygenase-2 (COX-2) catalyzes the production of prostaglandin E2, one of the most important products of the arachidonate metabolism. Over-expressed COX-2 strongly contributes to the growth and invasiveness of tumor cells in patients with CRC (Strillacci et al. 2009). It has been demonstrated that COX-2 up-regulation depends upon various cellular pathways involving both transcriptional and post-transcriptional regulations. An inverse correlation was reported between *COX-2* and *miR-101* expression in CRC cell lines. It was demonstrated in vitro that the direct translational inhibition of *COX-2* mRNA is mediated by *miR-101*. Moreover, this correlation was supported by data collected ex vivo, in which colon cancer tissues and liver metastases derived from CRC patients were analyzed. Impairment of *miR-101* levels could represent one of the leading causes of *COX-2* over-expression in CRC cells (Strillacci et al. 2009; Young 2010).

4.2.8 DNA Reparation

DNA reparation impairment has also been linked to CRC development and progression. A link between microenvironmental factors, microRNAs and DNA repair system has been highlighted by Crosby et al. (2009) in breast and cervical cancer cell lines. *miR-210* and *miR-373* expression can be induced by the hypoxia-inducible factor-1 alpha (*HIF-1α*). This transcription factor can bind to microRNA gene promoters and induce their expression. *MiR-210* and *miR-373* in turn down-regulate two major key players in DNA repair system *RAD52* and *RAD23B* respectively implicated in homology-dependent repair (HDR) and nucleotide excision repair (NER). Although this interaction has not been shown in CRC yet, it might contribute to colon tumorigenesis.

4.2.9 Extracellular Matrix Breakdown and Epithelial-mesenchymal Transition

The extracellular matrix (ECM) and its remodeling play a crucial role in the development of blood supply and interaction with the mesenchymal stroma upon which tumor cells grow. ECM remodeling is one of the necessary conditions of tumor growth, survival, invasiveness, and metastasizing. The key enzymes, among the many involved in ECM breakdown, are proteinases, and among these are the urokinase plasminogen activator (uPA) cascade and the matrix metalloproteinases

(MMPs) (Takayama et al. 2006). Substantial data indicate that *miR-21* is significantly elevated in CRC and in many other tumors of various origins (Krichevsky and Gabriely 2009). Based upon a glioblastoma model, it was described that *miR-21* regulates multiple genes associated with cellular motility and ECM remodeling. These included the *RECK* and *TIMP3* genes which are suppressors of malignancy and inhibitors of *MMPs* (Gabriely et al. 2008). Specific inhibition of *miR-21* with antisense oligonucleotides leads to elevated levels of *RECK* and *TIMP3* and therefore reduces MMP activities. Although these observations originate from a glioblastoma model, up-regulation of *miR-21* in CRC cells has been shown to increase their migratory and invasive abilities and *miR-21* seems to act, in this case, in a similar manner (Gabriely et al. 2008). Furthermore, *miR-21* was shown to act on *PDCD4*, a tumor suppressor gene that is an independent prognostic factor in resected CRC and a new negative regulator of intravasation through the invasion-related urokinase receptor gene (*uPAR*) (Asangani et al. 2008). Silencing of *miR-21* by anti-*miR-21* resulted in increased levels of PDCD4 in colorectal cell lines and decreased invasion in a chicken-embryo-metastasis assay. In addition, 22 resected human tumors showed higher *miR-21* expression than did the corresponding normal mucosa and decreased amounts of PDCD4 protein while mRNA levels were unchanged. These results argue for *miR-21*'s having an important function in the pathogenesis of CRC, as it also shows an inverse correlation with survival.

EMT is the conversion of an epithelial cell into a mesenchymal cell. Morphologically, EMT is characterized by a decrease of E-cadherin, loss of cell adhesion, and increased cell motility leading to promotion of metastatic behavior of cancer cells (including CRC) (Natalwala et al. 2008). The transcriptional repressor zinc-finger E-box binding homeobox 1 (*ZEB1*) is a crucial inducer of EMT in various human tumors, and it recently was shown to promote invasion and metastasis of tumor cells. The functional links to EMT come from members of the *miR-200* family (*miR-200a, miR-200b, miR-200c, miR-141*, and *miR-429*). *ZEB1* directly suppresses transcription of *miRNA-200* family members *miR-141* and *miR-200c*, which strongly activate epithelial differentiation in pancreatic, colorectal and breast cancer cells (Burk et al. 2008). Notably, the EMT activators transforming growth factor β2 and *ZEB1* are the predominant targets down-regulated by these miRNAs. These results indicate that *ZEB1* triggers a miRNA-mediated feedforward loop that stabilizes EMT and promotes the invasion of cancer cells. Alternatively, depending on the environmental trigger, this loop might switch and induce epithelial differentiation, thereby explaining strong intratumoral heterogeneity. A recent study associated the expression of *let-7* with two differentiation stages of a panel of cell lines (with epithelial and a mesenchymal gene signatures) and linked *let-7* to EMT (Shell et al. 2007).

4.2.10 Alterations in MiRNA Processing Machinery

The idea of an altered miRNA processing was first postulated by Michael et al. (2003) who identified 28 microRNA sequences that included three completely new miRNAs (*miR-320*, *miR-321*, and *miR-200c*), seven mouse-specific miRNAs and two miRNAs (*miR-143* and *miR-145*) that were consistently down-regulated in all cancer specimens by comparative analysis of CRC tissue and normal colonic mucosa. In the attempt to find an explanation for the deregulation of these miRNAs they analyzed the expression of the RISC complex associated genes, *DICER* and *eIF2C2*, by real-time PCR analysis in a subset of matched tissue-RNA samples. No discrepancies in gene expression between normal and tumoral tissues were found hypothesising that other factors might regulate microRNA expression in CRC (Valeri et al. 2009).

More recently frameshift mutations in *TARBP2* were identified (TAR RNA-binding protein 2, an essential element of the *DICER* machinery) as very frequent in sporadic and hereditary CRCs associated with microsatellite instability. These mutations can cause decrease in TRBP protein expression, destabilization of DICER1 protein and defect in the processing of miRNAs, providing a possible explanation for miRNAs deregulation in colorectal cancers (Melo et al. 2009). Conversely *SND1*, another important component of the RISC complex has been reported as up-regulated gene in human colon cancers. *SND1* mRNA was frequently up-regulated in human and mice cancers as well as in aberrant crypt foci and can down-regulate APC protein expression by post-transcriptional regulation. Based on these findings, authors speculated that *SND1* may control gene expression of *APC* or other cancer-related genes through the regulation of miRNA-induced translational repression (Tsuchiya et al. 2007).

4.2.11 Others

In the study of Arndt et al. (2009), SW620 colon cancer cells were stably transduced with *miR-143* or *miR-145* expression vectors and analyzed in vitro for cell proliferation, cell differentiation and anchorage-independent growth. Signaling pathways associated with differentially expressed miRNAs were identified using a gene set enrichment analysis. Interestingly, *miR-143* and *miR-145* appeared to function in opposing manners to either inhibit or augment cell proliferation in a metastatic CRC model. The pathways targeted by *miR-143* and *miR-145* showed no significant overlap. Furthermore, gene expression analysis of metastatic versus non-metastatic isogenic cell lines indicated that *miR-145* targets involved in cell cycle and neuregulin pathways were significantly down-regulated in the metastatic context.

Gregersen et al. (2010) employed a different microarray based approach to identify *miR-145* targets. Based on seed site enrichment analyses and unbiased word analyses, a significant enrichment of miRNA binding sites in the 3′UTRs of transcripts down-regulated upon miRNA over-expression was found. Gene Ontology analysis showed an over-representation of genes involved in cell death, cellular growth and proliferation, cell cycle, gene expression and cancer. A number of the identified *miR-145* targets have previously been implicated in cancer, *YES* and *STAT1* were verified in vitro as direct targets.

4.3 Single Nucleotide Polymorphisms (SNPs) and MiRNAs: Risk Factors for CRC

The binding of miRNA to mRNA is critical for regulating the mRNA level and protein expression. This binding can be affected, however, by SNPs that can occur in the miRNA target site and can either abolish existing binding sites or create illegitimate binding sites. Therefore, SNPs within miRNA binding sites can have differing effects on gene and protein expression and represent another type of genetic variability that can influence the risk of certain human diseases, including CRC (Fig. 4.2). Various approaches have been used to predict and identify functional SNPs within miRNA binding sites and their biological relevance is beginning to be evaluated in large case–control studies (Chen et al. 2008). SNPs may occur also at the level of the pri-miRNA, pre-miRNA or the mature miRNA sequence (miR-polymorphisms).

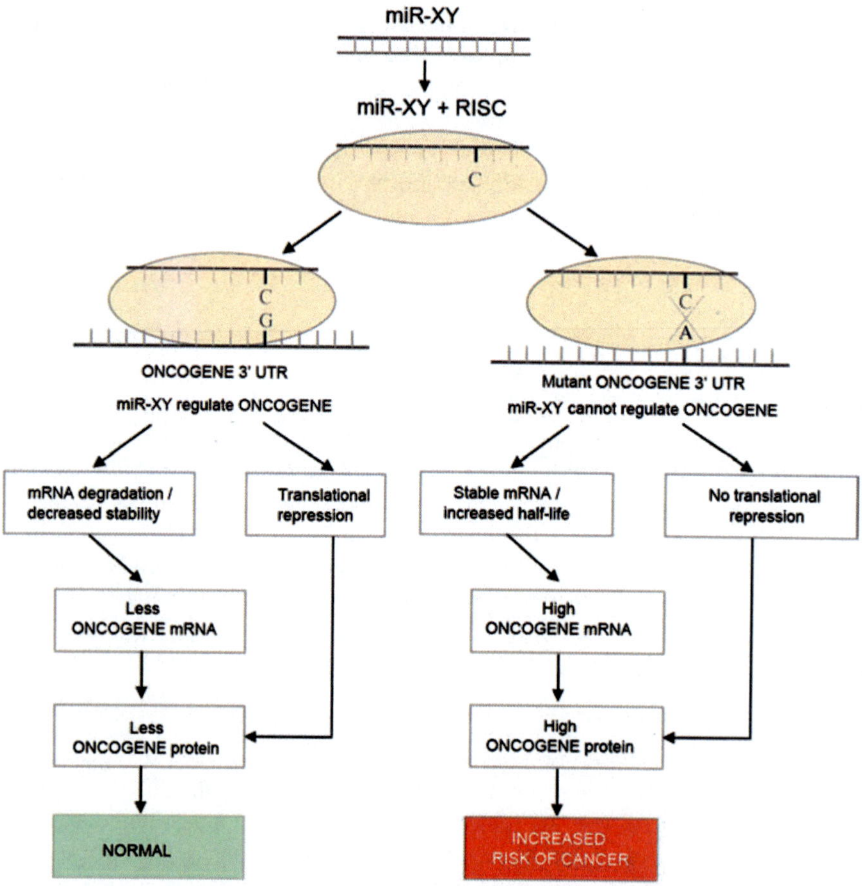

Fig. 4.2 Association of single nucleotide polymorphisms within miRNA binding sites and the risk of colorectal cancer

Such polymorphisms may be functional as to the biogenesis and action of the mature miRNA (Mishra and Bertino 2009).

Regarding CRC, out of eight candidates predicted by computer simulation, the two genes for *CD86*, a costimulatory ligand on lymphocytes, and for the insulin receptor carry an SNP that are significantly associated with the risk of sporadic CRC (odds ratios 2.74 and 1.94, respectively) (Landi et al. 2008). However, the biological relevance of these SNPs has not yet been confirmed by functional in vitro studies. Lee et al. (2010) analyzed the 40 SNPs of miRNA-related genes and their impact on the prognosis in the group 462 Korean CRC patients. None of tested SNPs was found to be an independent prognostic marker in CRC.

4.4 Plasma and Serum MiRNAs: Diagnosis and Monitoring of CRC

Circulating nucleic acids (CNAs) offer unique opportunities for an early diagnosis of CRC. Dysregulated expression of miRNAs in various tissues has been associated with a variety of human cancers. More recently, miRNAs' occurrence in the serum and plasma of humans has been repeatedly observed. The levels of miRNAs in serum are more stable, reproducible, and consistent among individuals of the same species than are other CNAs (Chen et al. 2008). The detection of serum miRNAs have been tested in prostate cancer, ovarian cancer and CRC patients as possible early diagnostic biomarkers (Mitchell et al. 2008; Ng et al. 2009; Resnick et al. 2009).

In a study by Chen et al. (2008), CRC patients had a significantly different serum miRNA profile compared to healthy subjects (HS). In all cases, 69 miRNAs were detected in the CRC serum but not in HS. It is of interest to note that CRC patients shared a large number of serum miRNAs (e.g. *miR-134, miR-146a, miR-221, miR-222, miR-23a*, etc.) with lung cancer patients. Pearson correlation further indicated that the levels of miRNAs in serum from lung cancer patients and CRC patients were consistent, suggesting that there are some "common" tumor-related miRNAs in serum (Chen et al. 2008). Differentially expressed miRNAs in the plasma of patients with CRC have been also reported (Ng et al. 2009). Expression pattern of 30 miRNAs (*miR-17-3p, miR-92, miR-135b, miR-222, miR-95*, etc.) in the plasma of patients with CRC were analyzed by real-time PCR expression profiling. Both *miR-17-3p* and *miR-92* were significantly elevated ($p<0.0005$). The plasma levels of these miRNAs were significantly reduced after surgery in 10 patients with CRC ($p<0.05$). Further validation with an independent set of plasma samples ($n = 180$) indicated that *miR-92* differentiates CRC not only from normal subjects but also from gastric cancer and inflammatory bowel disease. This marker yielded an ROC (receiver operating characteristic) curve area of 88.5%. In discriminating CRC from control subjects, the sensitivity was 89% and the specificity was 70%. *MiR-92* has reasonable sensitivity for CRC detection and compares favorably with the fecal occult blood test (Ng et al. 2009). Huang et al. (2010) measured the levels of 12 miRNAs (*miR-134, -146a, -17-3p, -181d, -191, -221, -222, -223, -25, -29a,*

-320a, and *-92a*) in plasma samples from patients with advanced colorectal neoplasia (carcinomas and advanced adenomas) and healthy controls using real-time PCR. Authors found that plasma *miR-29a* and *miR-92a* have significant diagnostic value for advanced neoplasia. *miR-29a* yielded an AUC (the areas under the ROC curve) of 0.844 and *miR-92a* yielded an AUC of 0.838 in discriminating CRC from controls. More importantly, these 2 miRNAs also could discriminate advanced adenomas from controls and yielded an AUC of 0.769 for *miR-29a* and 0.749 for *miR-92a*. Combined ROC analyses using these 2 miRNAs revealed an elevated AUC of 0.883 with 83.0% sensitivity and 84.7% specificity in discriminating CRC, and AUC of 0.773 with 73.0% sensitivity and 79.7% specificity in discriminating advanced adenomas. These data suggest that plasma *miR-29a* and *miR-92a* have strong potential as novel non-invasive biomarkers for early detection of CRC.

More recently, feasibility of fecal miRNAs as biomarkers for colorectal neoplasia screening was evaluated. miRNA expression profiles from stool of 29 patients showed higher expression of *miR-21* and *miR-106a* in patients with adenomas and CRCs compared with individuals free of colorectal neoplasia (Link et al. 2010).

4.5 MiRNA Expression Profiles of CRC Tissue

Alterations in miRNA expression profiles have been successively detected in many types of human tumors (Garzon et al. 2006). The causes of the widespread differential expression of miRNA genes between malignant and normal cells can be explained by the gene location in cancer-associated regions, alterations in the miRNA processing machinery, and epigenetic mechanisms (Garzon et al. 2009). In reports on various cancer samples, generally lower miRNA levels were identified in tumors in comparison with normal tissue and, lower miRNA levels in poorly differentiated tumors compared to well-differentiated tumors in tissue samples (Lu et al. 2005) as well as in cell lines (Gaur et al. 2007). Studies focusing on miRNA expression profiling in CRC are summarized in Table 4.1, and some of these studies are described in detail below.

In 2003, Michael et al. (2003) published the first such study. Using cloning technology followed by northern blotting, he observed consistently reduced accumulation of the specific mature *miR-143* and *miR-145* in the adenomatous and carcinoma stages of colorectal neoplasia. The same blots, however, displayed consistent levels of the ~70-bp pre-*miR-143* in each of the cell lines. The authors concluded that the levels of mature *miR-143* in these cells were controlled post-transcriptionally. These data suggested that abnormal processing might affect miRNAs expression in colon cancer cells.

Bandres et al. (2006) examined by real-time PCR the expression of 156 mature miRNAs in colorectal tumors and adjacent non-neoplastic tissues from patients and CRC cell lines. This permitted them to identify a group of 13 miRNAs whose expression is significantly altered in this type of tumor. The most significantly deregulated miRNAs were *miR-31*, *miR-96*, *miR-135b*,

Table 4.1 Summary of studies focusing on microRNA expression profiling in colorectal cancer (Slaby et al. 2009)

Study	Design	Tumors (*n*)	Normal[a] (*n*)	MicroRNAs examined (*n*)	Technology	MicroRNAs deregulated in tumors[b]
Michael et al. (2003)	Tumor vs. normal	14	4	28	Cloning	*miR-143, miR-145, miR-21*
Bandres et al. (2006)	Tumor vs. normal, TNM staging correlation	12	12	156	RT-PCR	*miR-31, miR-96, miR-133b, miR-135b, miR-145, miR-183*
Cummins et al. (2006)	MiRNAs identification	4	2	*x*	MiRAGE	133 novel microRNAs
Volinia et al. (2006)	Tumor vs. normal	46	8	245	MicroRNA microarray	*miR-24-1, miR-29b-2, miR-20a, miR-10a, miR-32, miR-203, miR-106a, miR-17-5p, miR-30c, miR-223, miR-126, miR-128b, miR-21, miR-24-2, miR-99b, miR-155, miR-213, miR-150, miR-107, miR-191, miR-221, miR-9-3*
Xi et al. (2007)	Tumor vs. normal, correlation with survival, p53 mutation status	24	24	10	RT-PCR	*miR-15b, miR-181b, miR-191, miR-200c*
Lanza et al. (2007)	MSS tumors vs. MSI tumors	23 MSS 16 MSI	0	230	MicroRNA microarray	Comparison with colon epithelium was not analyzed
Slaby et al. (2007)	Tumor vs. normal, clinicopathological parameters correlation	29	6	5	RT-PCR	*miR-143, miR-145, miR-21, miR-31*
Nakajima et al. (2006)	Tumor vs. normal, chemoresponse to S-1	21	21	5	RT-PCR	*let-7g, miR-181b, miR-200c*

Table 4.1 (continued)

Study	Design	Tumors (*n*)	Normal[a] (*n*)	MicroRNAs examined (*n*)	Technology	MicroRNAs deregulated in tumors[b]
Monzo et al. (2008)	Tumor vs. normal and embryonic colon	22	22	156	RT-PCR	Stage I ($n = 28$; *miR-106a, miR-125b, miR-145, miR-17-5p, miR-200c, miR-21, miR-34a*); Stage II ($n = 64$; *miR-106a, miR-128a, miR-15a, miR-15b, miR-17-5p, miR-181a-c, let-7g, miR-200a-c, miR-21, miR-31, miR-34a, miR-92*)
Schetter et al. (2008)	Tumor vs. normal, correlation with survival, TNM staging	84 (113)[c]	84 (113)[c]	389	MicroRNA microarray	*miR-20a, miR-21, miR-106a, miR-181b, miR-203*
Schepeler et al. (2008)	Tumor vs. normal, correlation with recurrence, MSS vs. MSI	49	19	315	MicroRNA microarray	*miR-145, miR-455, miR-484, miR-101, miR-30b, miR-26b, miR-20a, miR-510, miR-92, miR-513, miR-526c, miR-320, miR-527, miR-432, miR-492, miR-200a, miR-191, miR-302a, miR-512-5p*
Motoyama et al. (2009)	Tumor vs. normal, correlation with survival	4 (69)[c]	4 (69)[c]	455	MicroRNA microarray	*miR-31, miR-183, miR-17-5p, miR-18a, miR-20a, miR-92, miR-143, miR-145*

Table 4.1 (continued)

Study	Design	Tumors (*n*)	Normal[a] (*n*)	MicroRNAs examined (*n*)	Technology	MicroRNAs deregulated in tumors[b]
Ng et al. (2009)	Tumor, plasma vs. control	5 (25)[c]	5 (20)[c]	96	RT-PCR	*miR-18a, miR-223, miR-224, miR-135b, miR-95, miR-106a, miR-19a+b, miR-17-3p, miR-20a, miR-92, miR-221, miR-106b, miR-222*, etc.; Markers up-regulated in both plasma and tumors are given in bold type.
Sarver et al. (2009)	Tumor vs. normal, MSS vs. MS	80	28	735	MicroRNA microarray	The most significant: *miR-135b, miR-96, miR-182, miR-183, miR-1, miR-133a*
Earle et al. (2010)	Tumor vs. normal, MSS vs. MSI	55	55	24	RT-PCR	*miR-183, miR-31, miR-20, miR-25, miR-92, miR-93, miR-17, miR-135a, miR-203, miR-133b, miR-223, miR-192, miR-215, miR-26b, miR-143, miR-145, miR-191, miR-196a, miR-16, let-7a*
Wang et al. (2010)	Tumor vs. normal, TNM staging correlation	6	6	723	MicroRNA microarray	*miR-106b, miR-135b, miR-18a, miR-18b, miR-196b, miR-19a, miR-224, miR-335, miR-424, miR-20a*, miR-301b, miR-374a, miR-378, miR-378**

MiRAGE, miRNA serial analysis of gene expression; MSI, microsatellite instability; MSS, microsatellite stability; RT-PCR, reverse transcription-polymerase chain reaction.

[a]Normal colonic epithelium.

[b]Only microRNAs deregulated in CRC in comparison to normal colon epithelium are mentioned.

[c]Numbers of patients in the validation cohorts.

miR-183 (up-regulated in tumors and CRC cell-lines), and *miR-133b, miR-145* (down-regulated). In addition, the expression level of *miR-31* was positively correlated with the stage of CRC tumor. These results, achieved through a standardized real-time PCR method, suggest that miRNA expression profile could have relevance to the biological and clinical behavior of colorectal neoplasia.

Velculescu's group developed an experimental approach called miRNA serial analysis of gene expression (miRAGE) and used it to perform one of the largest experimental analyses of human miRNAs. Sequence analysis of 273,966 small RNA tags from human colorectal cells allowed them to identify 200 known mature miRNAs, 133 novel miRNA candidates, and 112 previously uncharacterized miRNA forms. To aid in evaluating the candidate miRNAs, they disrupted the Dicer locus in three human CRC cell lines and examined known and novel miRNAs in these cells. This study indicates that the human genome contains many more miRNAs than currently identified (Cummins et al. 2006).

From a large-scale analysis of miRNA expression profiles on 540 samples of solid cancers, including CRC, Volinia et al. (2006) identified a solid cancer miRNA signature composed by a large portion of over-expressed miRNAs. Among these miRNAs were some with well characterized cancer associations, such as *miR-17-5p, miR-20a*, *miR-21*, *miR-92*, *miR-106a*, and *miR-155*. A microarray-based approach for analysis of miRNA expression profiles in CRC was successfully applied also by Motoyama et al. (2009).

In another profiling study, Lanza et al. (2007) evaluated the expression of miRNAs and mRNAs in CRC samples characterized by microsatellite stability (MSS) or by high levels of microsatellite instability (MSI-H). Their analysis of miRNA expression profiles of MSI-H ($n = 16$) and MSS CRCs ($n = 23$) identified 14 differentially expressed miRNAs, while their analysis of messenger RNA expression profiles in these tumors identified 451 differentially expressed genes. Consequently, a smaller selected signature of best predictors of microsatellite status was generated: 27 genes, including 8 miRNAs, were identified as predictors. Further cluster analysis using just these 27 miRNAs and mRNAs also perfectly separated the two tumor classes. Cluster analyses run using either the mRNAs or the miRNAs independently did not perform as well in discriminating the tumor types. Therefore, the combined miRNA/mRNA fingerprint worked as the best discriminator for MSS versus MSI-H. In the study of Earle et al. (2010), relative expression levels of *miR-92, -223, -155, -196a, -31,* and *-26b* were significantly different among MSI subgroups (including low MSI and hereditary non-polyposis colorectal cancer (HNPCC) syndrome), and *miR-31* and *miR-223* were over-expressed in CRC of patients with HNPCC-associated cancer. To identify miRNAs that are differentially expressed in CRC and CRC subtypes, Sarver et al. (2009) carried out highly expression profiling of 735 miRNAs on samples obtained from a statistically powerful set of tumors ($n = 80$) and normal colon tissue ($n = 28$). Tumor specimens showed highly significant and large fold change differential expression of the levels of 39 miRNAs including *miR-135b, miR-96, miR-182, miR-183, miR-1*, and *miR-133a*, relative to normal colon tissue. Significant differences were observed in 6 miRNAs: decreased levels in MSS relative to MSI-H tumors included *miR-552, miR-592, miR-181c*, and *miR-196*. *MiR-625* and *miR-31* exhibited increased levels in MSI-H relative to MSS tumors.

Monzo et al. (2008) assessed the expression of mature miRNAs in human embryonic colon tissue, as well as in CRC and paired normal colon tissue. Overlapping miRNA expression was detected between embryonic colonic mucosa and CRC. The *miR-17-92* cluster and its target, *E2F1*, exhibit a similar pattern of expression in human colon development and in colonic carcinogenesis – regulating cell proliferation in both cases. Authors of this study conclude that miRNA pathways play a major role in both embryonic development and neoplastic transformation of the colonic epithelium.

From a diagnostic point of view, miRNA expression profiles might also contribute significantly to the further determination of the tissue origin of the cancer of unknown primary sites. Cancer of unknown primary (CUP) is usually a very aggressive disease with a poor prognosis. Identifying of colorectal cancer among adenocarcinoma of unknown primary site may improve prognosis of these patients by giving them a chance for modern anti-cancer targeted therapy. Two recent studies examined metastases of unknown primary tumors with miRNA microarrays for their potential to identify the tissue of origin. After establishing a miRNA classifier ($n = 68$, 11 tumor types, 217 miRNAs), 12 out of 17 poorly differentiated tumors were accurately classified by miRNA profiling (Lu et al. 2005). A second publication reported an overall accuracy of 90% in classifying more than 400 malignant tumor samples of 22 tissue origins based on a set of 48 miRNAs (Rosenfeld et al. 2008).

A recent study on lymph node metastases of several malignant tumors, including CRC, identified three specific miRNAs (*miR-148a*, *miR-34b/c*, and *miR-9*), specifically down-regulated by CpG island hyper-methylation (Lujambio et al. 2008).

To offer new approach for preventing and controlling lymphatic metastasis in colon cancer, Huang et al. (2009) compared the miRNA expression profiles (723 human microRNA probes) of normal colonic epithelium from the two CRC patient groups; those with confirmed lymph node metastasis ($n = 3$), and those without detectable lymph node metastasis ($n = 3$). Two microRNA (*hsa-miR-129**, *hsa-miR-137*) were differentially expressed in the lymph node positive group compared with the lymph node negative group. After validation through real-time PCR method, *hsa-miR-137* expression was significantly up-regulated nearly 6.6-fold in lymph node positive specimens ($p = 0.036$).

In general, there are several advantages of using miRNA expression profiling instead of its mRNA counterpart for biomarker identification and also for routine diagnostics. As a consequence of the fact that miRNAs target mRNAs with an imperfect sequence complementarity, a single miRNA can regulate the expression of more than 100 mRNAs simultaneously (Esquela-Kerscher and Slack 2006). This might explain why microarrays of 217 miRNAs have much higher information content than 16,000 mRNAs in distinguishing different tissues and tumors (Faber et al. 2009). It is relatively easier to discover reliable biomarkers from the approximately hundreds of miRNA candidates discovered to date than from over 40,000 genes. A further advantage is that, due to their small size and stem-loop structure, miRNAs are relatively more stable and less subjected to degradation during fixation and sample processing. One recent examination compared the miRNA-expression profiles from fresh frozen versus formalin-fixed paraffin-embedded (FFPE) CRC tissues. A good correlation coefficient of 0.86–0.89 was observed. Worthy of note is that differing formalin fixation times – inevitable in a routine pathology lab – did not significantly influence the expression of miRNAs in 40 CRC specimens (Xi et al. 2007). This can also be of benefit for large retrospective studies based on archived FFPE samples. Furthermore, miRNAs can be visualized at the cellular and subcellular levels by conventional as well as fluorescence in situ hybridization (Nuovo et al. 2008).

4.6 MiRNAs in CRC Prognosis and Prediction

Accumulating evidence shows that miRNA expression patterns are unique to certain cancers and have potential to be used as prognostic and predictive factors in clinical routine. Xi et al. (2007) performed Kaplan-Meier analysis for CRC patients with International Union Against Cancer (UICC) stages I-IV and found that tumors expressing high levels of *miR-200c*, recently connected to EMT, are correlated with poorer prognosis, regardless of tumor stage. These investigators also found that *p53* mutation, commonly found in CRC, is strongly associated with greater than twofold *miR-200c* over-expression. Chen et al. (2010) identified siginificant decrese of *miR-148a* and *miR-152* expression levels in CRC (both $p<0.001$) tissue in comparison to their matched non-tumoral tissues. Authors further observed correlation of low expression levels of *miR-152* and *miR-148a* with increased tumor size ($p = 0.004$ and 0.018, respectively) and advanced pT stage ($p = 0.002$ and 0.023, respectively).

MiR-21 is up-regulated in many solid tumors, including CRC (Krichevsky and Gabriely 2009). Recently, we have found that *miR-21* over-expression shows a strong correlation with the established prognostic factors as nodal stage, metastatic disease and UICC stage (Slaby et al. 2007). Kulda et al. (2010) correlated *miR-21* and *miR-143* expression to disease-free interval (DFI) ($p = 0.0026$ and $p = 0.0191$, respectively). There was shorter DFI in patients with a higher expression of *miR-21* and, surprisingly, also in patients with a higher expression of *miR-143*, which is a putative tumor suppressor. Using class comparison analysis, Shetter et al. (2008) later found that 37 miRNAs were differentially expressed in tumors of CRC patients. From this group, *miR-20a*, *miR-21*, *miR-106a*, *miR-181b*, and *miR-203* were found by Cox regression analysis to be associated also with poor survival and were selected for validation. Validation was performed by measuring miRNAs' expression using real-time PCR in tumor and paired non-tumor tissues in the validation cohort. In the validation set, only high expression of *miR-21* was significantly associated with poor prognosis, and this association was independent of age, sex, and tumor location. Multivariate analysis further revealed that high *miR-21* expression in tumors was associated with poor survival, independent of the tumor stage. In patients who received adjuvant therapy, high *miR-21* expression indicated a poor response to therapy (Schetter et al. 2008).

Schepeler et al. (2008) found that miRNAs were associated with tumor microsatellite status in stage II colon cancer. The predictive molecular signature was composed of only four miRNAs (*miR-142-3p*, *miR-212*, *miR-151*, and *miR-144*).

Furthermore, a biomarker based on miRNA expression profiles could predict recurrence of disease with an overall performance accuracy of 81%, thus, indicating a potential role for miRNAs in determining tumor aggressiveness. Kaplan-Meier survival curves showed that patients who had stage II CRC tumors with high expression of *miR-320* or *miR-498* had significantly shorter progression-free survival than did patients whose tumors showed low expression. These miRNAs were correlated with the probability of progression-free survival also by multivariate analysis. Although these results are promising, larger studies will be needed to prove whether miRNAs really have significant potential to extend prognostic information based on the recent standard diagnostic procedures.

Another important question for management of CRC patients is the possibility of predicting therapy response. Nakajima et al. (2006) evaluated the significance of five mature miRNAs in tumors of CRC patients treated with 5-fluorouracil (5-FU)-based anti-metabolite S-1. They identified *let-7g* and *miR-181b* as significant indicators for chemoresponse to S-1-based chemotherapy.

A study published by Rossi et al. (2007) reported a suggestive pattern of miRNAs rearrangement in HT-29 and HCT-116 human colon cancer cell lines after exposure to 5-FU, a classical anti-metabolite in broad clinical use. At clinically relevant concentrations, the drug up-regulated or down-regulated in vitro the expression of 19 and 3 miRNAs, respectively, by a factor of not less than two-fold. In some instances, 5-FU up-regulated miRNAs that are already over-expressed in tumor tissue, including, for example, *miR-21* (Rossi et al. 2007).

In other instances, by contrast, the drug influenced the expression of miRNAs in a direction that is opposite to that induced by neoplastic transformation. A typical example is provided by *miR-200b*, which is up-regulated in various tumors but down-regulated by treatment with 5-FU. Interestingly, it is known that *miR-200b* targets mRNA that codes for a protein tyrosine phosphatase (*PTPN12*) which inactivates products of oncogenes, such as *ABL*, *SRC*, or *KRAS* (Schepeler et al. 2008).

Another study evaluated changes in miRNA expression profiles as a response to therapy, focusing on the effects of capecitabine chemoradiotherapy on rectal tumors in vivo (Svoboda et al. 2008). Tumor microexcisions were taken before starting a therapy and, again, after a 2-week therapy. The extent of tumor response to the therapy was investigated microscopically by an experienced pathologist according to Mandard's tumor regression criteria. In this study, many miRNAs (*miR-10a*, *miR-21*, *miR-145*, *miR-212*, *miR-339*, *miR-361*) responded to capecitabine chemoradiotherapy in individual tumor samples. In most samples, however, only two miRNAs, *miR-125b* and *miR-137*, showed significant increase in expression levels after 2-week therapy.

Zhou et al. (2010) determined how 5-FU and oxaliplatin (L-OHP) modify the expression profiles (856 human miRNA probes) of miRNAs in HCT-8 and HCT-116 colon cancer cells. Fifty-six up- and 50 down-regulations of miRNA expression with statistical significance were identified in colon cancer cells following exposure to 5-FU or L-OHP compared to matched control cells. Expression levels of *miR-197*, *miR-191*, *miR-92a*, *miR-93*, *miR-222*, and *miR-1826* were significantly down-regulated in both cell lines after the treatment of one drug or in one cell line following exposure to either drug.

HCT116 human colorectal cancer cells were used to investigate the biological and potential chemosensitizing role of *miR-143* in the study of Borralho et al. (2009). Transient *miR-143* over-expression resulted in an approximate 60% reduction in cell viability. In addition, stable *miR-143* over-expressing cells were selected with G418 and exposed to 5-FU. Increased stable expression of *miR-143* was associated with decreased viability and increased cell death after exposure to 5-FU. These changes were associated with increased nuclear fragmentation and caspase -3, -8 and -9 activities. In addition, extracellular-regulated protein kinase 5, nuclear factor-kappaB and *Bcl-2* protein expression were down-regulated by *miR-143*, and further reduced by exposure to 5-FU.

MiR-215, through the suppression of denticleless protein homolog (*DTL*), a cell cycle-regulated nuclear and centrosome protein, induces decreased colon cancer cell proliferation by causing G2-arrest, thereby leading to an increase in their chemoresistance to the chemotherapeutic agents, methotrexate and Tomudex (Song et al. 2010).

Boni et al. (2010) investigated associations between 18 polymorphisms in both miRNA-containing genomic regions (primary and precursor miRNA) and in genes related to miRNA biogenesis with clinical outcome in 61 metastatic colorectal cancer (mCRC) patients treated with 5-FU and irinotecan (CPT-11). A significant association with tumor response and time to progression (TTP) was found for SNP rs7372209 in pri-*miR26a-1* ($p = 0.041$ and $p = 0.017$, respectively). The genotypes CC and CT were favorable when compared with the TT variant genotype. SNP rs1834306, located in the pri-*miR-100* gene, significantly correlated with a longer TTP ($p = 0.04$). In the miRNA-biogenesis pathway, a trend was identified between SNP rs11077 in the exportin-5 gene and disease control rate ($p = 0.076$).

There is increasing evidence that the *let-7* miRNA exerts an effect as a tumor suppressor by targeting the *KRAS* mRNA. The *let-7* complementary site (LCS6) T > G variant in the *KRAS* 3′UTR weakens *let-7* binding. Graziano et al. (2010) analyzed whether the LCS6 variant may be clinically relevant to patients with metastatic colorectal cancer (MCRC) treated with anti-EGFR therapy. LCS6 genotypes and *KRAS/BRAF* mutations were determined in the tumor DNA of patients with MCRC who underwent salvage cetuximab-irinotecan therapy. There were 25% G-allele (T/G+G/G) carriers and 75% T/T genotype carriers. G-allele carriers were significantly more frequent in the *KRAS* mutation group than in patients with *KRAS* wild type ($p = 0.004$). In the patients without *BRAF* V600E mutation, overall survival (OS) and progression-free survival (PFS) times were compared between carriers of the LCS6 G-allele genotypes and carriers of the wild-type T/T genotype. LCS6 G-allele carriers showed worse OS ($p = 0.001$) and PFS ($p = 0.004$) than T/T genotype carriers (confirmed in the multivariate model including the *KRAS* status). In the exploratory analysis of the unresponsive patients with *KRAS* mutation, LCS6 G-allele carriers showed adverse OS and PFS times (Graziano et al. 2010).

Despite these results, more studies are needed that will examine the effects of chemotherapeutic agents on the miRNA expression profiles and their possible usage for predicting therapy response in CRC patients.

4.7 MiRNAs as Potential Therapeutic Targets in CRC

Since miRNAs constitute a robust network for gene regulation, they possess a great potential as both, a novel class of therapeutic targets and a powerful intervention tool (Fig. 4.3). The biosynthesis, maturation and activity of miRNAs can be manipulated with various oligonucleotides that encode the sequences complementary to mature miRNAs (Aslam et al. 2009). Over-expression of miRNAs can be induced either by using synthetic miRNA mimics or chemically modified oligonucleotides. Conversely, miRNAs can be silenced by antisense miRNA oligonucleotides (AMOs), locked nucleic acids (LNAs) and "antagomirs" (synthetic analogues of miRNAs). Cross-sensitivity with endogenous miRNAs and lack of specificity for cancer cells can cause non-specific side effects during miRNA modulation therapy.

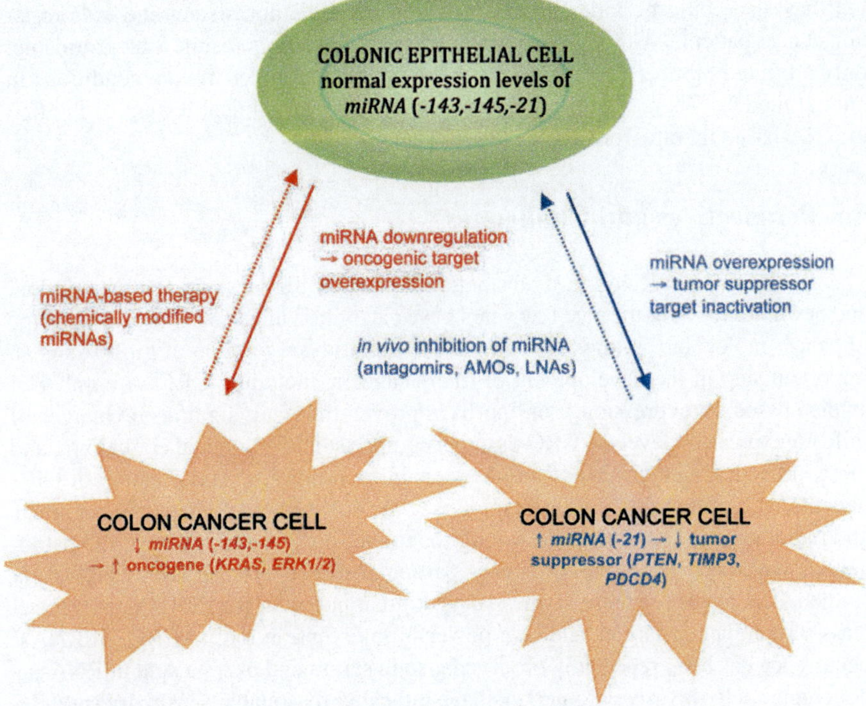

Fig. 4.3 MicroRNAs as therapeutic targets in colorectal cancer

However, the use of an effective delivery system and less toxic synthetic anti-miRNA oligonucleotides may minimize such side effects (Krutzfeldt et al. 2005; Zhang and Farwell 2008). Gene therapies may be designed to treat CRC and to block the progression of precursor lesions by manipulating the tumor suppressive or oncogenic miRNAs. Such manipulation may control the tumor growth rate and have potential as a new therapy for both early and advanced cancers (Calin and Croce 2007; Tong and Nemunaitis 2008).

Studies have revealed that inhibition of *miR-21* and *miR-17-92* activity is associated with reduced tumor growth, invasion, angiogenesis and metastasis (Krichevsky and Gabriely 2009; Dews et al. 2006). Moreover, over-expression of *miR-21* is associated with low sensitivity and a poor response to chemotherapy, and its inhibition may improve the response to chemotherapy. On the other hand, restoration of *miR-145* expression has been associated with inhibition of tumor cells growth via down-regulation of IRS-1. Expression levels of *miR-451*, down-regulated in tumor tissues of CRC patients, were increased in vitro and caused reduced cell proliferation and increased sensitivity to radiotherapy (La Rocca et al. 2009). These miRNAs present examples of miRNAs validated as oncogenes or tumor suppressors in CRC and thus of potential candidates for miRNA-based targeted CRC therapy. Targeting such miRNAs may help to not only prevent the recurrence of disease in high-risk tumors in UICC stage II and control the growth of advanced metastatic tumors, but they also could provide another possibility for chemoresistant and radioresistant cancer patients. Although experimental miRNA therapy results look promising, only a limited number of studies have been conducted under in vivo conditions in animal models. There is still a long way to go to reach clinical testing of the first miRNA-based therapy for CRC in the future.

4.8 Perspectives and Challenges

The discovery of miRNAs has substantially changed the view on gene regulation, and new findings over the past few years have catapulted miRNAs to the center stage of cancer molecular biology. It is now evident that dysregulation of miRNAs is an important step in the development of many cancers, including CRC. A number of studies based on expression profiling have proven there are significant changes of miRNA expression levels in CRC tissue in comparison to colorectal epithelium, and these have identified groups of miRNAs enabling prognostic stratification of CRC patients and prediction of their responses to selected chemotherapeutic regimens and radiotherapy. To improve knowledge as to the roles of miRNAs in CRC pathogenetic pathways, functional effects of particular miRNAs have been successfully studied. The results of these studies suggest great potential for miRNAs as a novel class of therapeutic targets and as a powerful intervention tool in CRC. MiRNAs' occurrence has been repeatedly observed also in serum and plasma, and miRNAs as novel minimally invasive biomarkers have indicated reasonable sensitivity for CRC detection and compare favorably with the fecal occult blood test. Figure 4.4 summarizes how miRNAs may enter the clinical management of CRC patients in the near future.

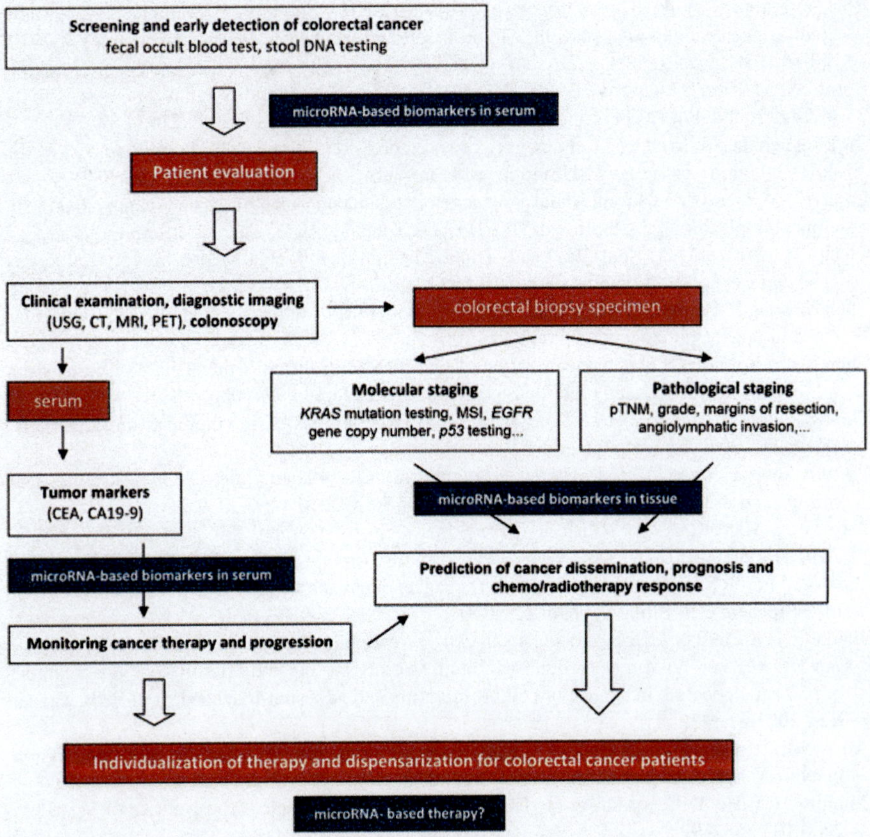

Fig. 4.4 The potential usage of microRNAs in the clinical management of the colorectal cancer patients (Slaby et al. 2009)

References

Akao Y, Nakagawa Y, Naoe T. Let-7 microRNA functions as a potential growth suppressor in human colon cancer cells. Biol Pharm Bull. 2006;29:903–6.

Arndt GM, Dossey L, Cullen LM, et al. Characterization of global microRNA expression reveals oncogenic potential of miR-145 in metastatic colorectal cancer. BMC Cancer. 2009;9:374.

Asangani IA, Rasheed SA, Nikolova DA, et al. MicroRNA-21 (miR-21) post-transcriptionally downregulates tumor suppressor Pdcd4 and stimulates invasion, intravasation and metastasis in colorectal cancer. Oncogene. 2008;27:2128–36.

Aslam MI, Taylor K, Pringle JH, et al. MicroRNAs are novel biomarkers of colorectal cancer. Br J Surg. 2009;96:702–10.

Bailey SG, Sanchez-Elsner T, Stephanou A, et al. Regulating the genome surveillance system: miRNAs and the p53 super family. Apoptosis. 2010;15:541–52.

Bandres E, Cubedo E, Agirre X, et al. Identification by Real-time PCR of 13 mature microRNAs differentially expressed in colorectal cancer and non-tumoral tissues. Mol Cancer. 2006;5:29.

Bommer GT, Gerin I, Feng Y, et al. p53-mediated activation of miRNA34 candidate tumor-suppressor genes. Curr Biol. 2007;17:1298–307.

Boni V, Zarate R, Villa JC, et al. Role of primary miRNA polymorphic variants in metastatic colon cancer patients treated with 5-fluorouracil and irinotecan. Pharmacogenomics J. 2010. doi:10.1038/tpj.2010.58.

Borralho PM, Kren BT, Castro RE, et al. MicroRNA-143 reduces viability and increases sensitivity to 5-fluorouracil in HCT116 human colorectal cancer cells. FEBS J. 2009;276:6689–700.

Burk U, Schubert J, Wellner U, et al. A reciprocal repression between ZEB1 and members of the miR-200 family promotes EMT and invasion in cancer cells. EMBO Rep. 2008;9:582–9.

Calin GA, Croce CM. Chromosomal rearrangements and microRNAs: a new cancer link with clinical implications. J Clin Invest. 2007;117:2059–66.

Chang TC, Wentzel EA, Kent OA, et al. Transactivation of miR-34a by p53 broadly influences gene expression and promotes apoptosis. Mol Cell. 2007;26:745–52.

Chen K, Song F, Calin GA, et al. Polymorphisms in microRNA targets: a gold mine for molecular epidemiology. Carcinogenesis. 2008;29:1306–11.

Chen X, Ba Y, Ma L, et al. Characterization of microRNAs in serum: a novel class of biomarkers for diagnosis of cancer and other diseases. Cell Res. 2008;18:997–1006.

Chen X, Guo X, Zhang H, et al. Role of miR-143 targeting KRAS in colorectal tumorigenesis. Oncogene. 2009;28:1385–92.

Chen Y, Song Y, Wang Z, et al. Altered expression of miR-148a and miR-152 in gastrointestinal cancers and its clinical significance. J Gastrointest Surg. 2010;14:1170–9.

Cho WC. MicroRNAs in cancer – from research to therapy. Biochim Biophys Acta. 2010a;1805:209–17.

Cho WC. MicroRNAs: Potential biomarkers for cancer diagnosis, prognosis and targets for therapy. Int J Biochem Cell Biol. 2010b;42:1273–81.

Ciardiello F, Tortora G. EGFR antagonists in cancer treatment. N Engl J Med. 2008;358:1160–74.

Corney DC, Flesken-Nikitin A, Godwin AK, et al. MicroRNA-34b and MicroRNA-34c are targets of p53 and cooperate in control of cell proliferation and adhesion-independent growth. Cancer Res. 2007;67:8433–8.

Crosby ME, Kulshreshtha R, Ivan M, et al. MicroRNA regulation of DNA repair gene expression in hypoxic stress. Cancer Res. 2009;69:1221–9.

Cummins JM, He Y, Leary RJ, et al. The colorectal microRNAome. Proc Natl Acad Sci USA. 2006;103:3687–92.

Dews M, Homayouni A, Yu D, et al. Augmentation of tumor angiogenesis by a Myc-activated microRNA cluster. Nat Genet. 2006;38:1060–5.

Earle JS, Luthra R, Romans A, et al. Association of microRNA expression with microsatellite instability status in colorectal adenocarcinoma. J Mol Diagn. 2010;12:433–40.

Esquela-Kerscher A, Slack FJ. Oncomir – microRNAs with a role in cancer. Nat Rev Cancer. 2006;6:259–69.

Faber C, Kirchner T, Hlubek F. The impact of microRNAs on colorectal cancer. Virchows Arch. 2009;454:359–67.

Fearon ER, Vogelstein B. A genetic model for colorectal tumorigenesis. Cell. 1990;61:759–67.

Gabriely G, Wurdinger T, Kesari S, et al. MicroRNA 21 promotes glioma invasion by targeting matrix metalloproteinase regulators. Mol Cell Biol. 2008;28:5369–80.

Garzon R, Calin GA, Croce CM. MicroRNAs in cancer. Annu Rev Med. 2009;60:167–79.

Garzon R, Fabbri M, Cimmino A, et al. MicroRNA expression and function in cancer. Trends Mol Med. 2006;12:580–7.

Gaur A, Jewell DA, Liang Y, et al. Characterization of microRNA expression levels and their biological correlates in human cancer cell lines. Cancer Res. 2007;67:2456–68.

Graziano F, Canestrari E, Loupakis F, et al. Genetic modulation of the Let-7 microRNA binding to KRAS 3′-untranslated region and survival of metastatic colorectal cancer patients treated with salvage cetuximab-irinotecan. Pharmacogenomics J. 2010;10:458–64.

Gregersen LH, Jacobsen AB, Frankel LB, et al. MicroRNA-145 targets YES and STAT1 in colon cancer cells. PLoS One. 2010;5:e8836.

Guo C, Sah JF, Beard L, et al. The noncoding RNA, miR-126, suppresses the growth of neoplastic cells by targeting phosphatidylinositol 3-kinase signaling and is frequently lost in colon cancers. Genes Chromosomes Cancer. 2008;47:939–46.

He L, He X, Lowe SW, Hannon GJ. MicroRNAs join the p53 network – another piece in the tumour-suppression puzzle. Nat Rev Cancer. 2007;7:819–22.

Hermeking H. p53 enters the microRNA world. Cancer Cell. 2007;12:414–8.

Hu G, Chen D, Li XM, et al. MiR-133b regulates the MET proto-oncogene and inhibits the growth of colorectal cancer cells in vitro and in vivo. Cancer Biol Ther. 2010;10:2.

Huang Z, Huang D, Ni S, et al. Plasma microRNAs are promising novel biomarkers for early detection of colorectal cancer. Int J Cancer. 2010;127:118–26.

Huang ZM, Yang J, Shen XY, et al. MicroRNA expression profile in non-cancerous colonic tissue associated with lymph node metastasis of colon cancer. J Dig Dis. 2009;10:188–94.

Johnson SM, Grosshans H, Shingara J, et al. RAS is regulated by the let-7 microRNA family. Cell. 2005;120:635–47.

Krichevsky AM, Gabriely G. MiR-21: a small multi-faceted RNA. J Cell Mol Med. 2009;13: 39–53.

Krutzfeldt J, Rajewsky N, Braich R, et al. Silencing of microRNAs in vivo with 'antagomirs'. Nature. 2005;438:685–9.

Kulda V, Pesta M, Topolcan O, et al. Relevance of miR-21 and miR-143 expression in tissue samples of colorectal carcinoma and its liver metastases. Cancer Genet Cytogenet. 2010;200:154–60.

La Rocca G, Badin M, Shi B, et al. Mechanism of growth inhibition by microRNA 145: the role of the IGF-I receptor signaling pathway. J Cell Physiol. 2009;220:485–91.

Landi D, Gemignani F, Naccarati A, et al. Polymorphisms within micro-RNA-binding sites and risk of sporadic colorectal cancer. Carcinogenesis. 2008;29:579–84.

Lanza G, Ferracin M, Gafa R, et al. mRNA/microRNA gene expression profile in microsatellite unstable colorectal cancer. Mol Cancer. 2007;6:54.

Lee HC, Kim JG, Chae YS, et al. Prognostic impact of microRNA-related gene polymorphisms on survival of patients with colorectal cancer. J Cancer Res Clin Oncol. 2010;136:1073–8.

Link A, Balaguer F, Shen Y, et al. Fecal MicroRNAs as novel biomarkers for colon cancer screening. Cancer Epidemiol Biomarkers Prev. 2010;19:1766–74.

Lodygin D, Tarasov V, Epanchintsev A, et al. Inactivation of miR-34a by aberrant CpG methylation in multiple types of cancer. Cell Cycle. 2008;7:2591–600.

Lu J, Getz G, Miska EA, et al. MicroRNA expression profiles classify human cancers. Nature. 2005;435:834–8.

Lujambio A, Calin GA, Villanueva A, et al. A microRNA DNA methylation signature for human cancer metastasis. Proc Natl Acad Sci USA. 2008;105:13556–61.

Melo SA, Ropero S, Moutinho C, et al. A TARBP2 mutation in human cancer impairs microRNA processing and DICER1 function. Nat Genet. 2009;41:365–70.

Meng F, Henson R, Wehbe-Janek H, et al. MicroRNA-21 regulates expression of the PTEN tumor suppressor gene in human hepatocellular cancer. Gastroenterology. 2007;133:647–58.

Michael MZ, O'Connor SM, van Holst Pellekaan NG, et al. Reduced accumulation of specific microRNAs in colorectal neoplasia. Mol Cancer Res. 2003;1:882–91.

Mishra PJ, Bertino JR. MicroRNA polymorphisms: the future of pharmacogenomics, molecular epidemiology and individualized medicine. Pharmacogenomics. 2009;10:399–416.

Mitchell PS, Parkin RK, Kroh EM, et al. Circulating microRNAs as stable blood-based markers for cancer detection. Proc Natl Acad Sci USA. 2008;105:10513–8.

Monzo M, Navarro A, Bandres E, et al. Overlapping expression of microRNAs in human embryonic colon and colorectal cancer. Cell Res. 2008;18:823–33.

Motoyama K, Inoue H, Takatsuno Y, et al. Over- and under-expressed microRNAs in human colorectal cancer. Int J Oncol. 2009;34:1069–75.

Nagel R, le Sage C, Diosdado B, et al. Regulation of the adenomatous polyposis coli gene by the miR-135 family in colorectal cancer. Cancer Res. 2008;68:5795–802.

Nakajima G, Hayashi K, Xi Y, et al. Non-coding microRNAs hsa-let-7g and hsa-miR-181b are associated with chemoresponse to S-1 in colon cancer. Cancer Genomics Proteomics. 2006;3:317–24.
Natalwala A, Spychal R, Tselepis C. Epithelial-mesenchymal transition mediated tumourigenesis in the gastrointestinal tract. World J Gastroenterol. 2008;14:3792–7.
Ng EK, Chong WW, Jin H, et al. Differential expression of microRNAs in plasma of patients with colorectal cancer: a potential marker for colorectal cancer screening. Gut. 2009;58:1375–81.
Nuovo GJ. In situ detection of precursor and mature microRNAs in paraffin embedded, formalin fixed tissues and cell preparations. Methods. 2008;44:39–46.
Pechlivanis S, Pardini B, Bermejo JL, et al. Insulin pathway related genes and risk of colorectal cancer: INSR promoter polymorphism shows a protective effect. Endocr Relat Cancer. 2007;14:733–40.
Resnick KE, Alder H, Hagan JP, et al. The detection of differentially expressed microRNAs from the serum of ovarian cancer patients using a novel real-time PCR platform. Gynecol Oncol. 2009;112:55–9.
Rosenfeld N, Aharonov R, Meiri E, et al. MicroRNAs accurately identify cancer tissue origin. Nat Biotechnol. 2008;26:462–9.
Rossi L, Bonmassar E, Faraoni I. Modification of miR gene expression pattern in human colon cancer cells following exposure to 5-fluorouracil in vitro. Pharmacol Res. 2007;56:248–53.
Rossi S, Kopetz S, Davuluri R, et al. MicroRNAs, ultraconserved genes and colorectal cancers. Int J Biochem Cell Biol. 2010;42:1291–7.
Sarver AL, French AJ, Borralho PM, et al. Human colon cancer profiles show differential microRNA expression depending on mismatch repair status and are characteristic of undifferentiated proliferative states. BMC Cancer. 2009;9:401.
Schepeler T, Reinert JT, Ostenfeld MS, et al. Diagnostic and prognostic microRNAs in stage II colon cancer. Cancer Res. 2008;68:6416–24.
Schetter AJ, Leung SY, Sohn JJ, et al. MicroRNA expression profiles associated with prognosis and therapeutic outcome in colon adenocarcinoma. JAMA. 2008;299:425–36.
Shell S, Park SM, Radjabi AR, et al. Let-7 expression defines two differentiation stages of cancer. Proc Natl Acad Sci USA. 2007;104:11400–5.
Shi B, Sepp-Lorenzino L, Prisco M, et al. MicroRNA 145 targets the insulin receptor substrate-1 and inhibits the growth of colon cancer cells. J Biol Chem. 2007;282:32582–90.
Slaby O, Svoboda M, Fabian P, et al. Altered expression of miR-21, miR-31, miR-143 and miR-145 is related to clinicopathologic features of colorectal cancer. Oncology. 2007;72:397–402.
Slaby O, Svoboda M, Michalek J, et al. MicroRNAs in colorectal cancer: translation of molecular biology into clinical application. Mol Cancer. 2009;14;8:102.
Song B, Wang Y, Titmus Ma, et al. Molecular mechanism of chemoresistance by *mir-215* in osteosarcoma and colon cancer cells. Mol Cancer. 2010;9:96.
Strillacci A, Griffoni C, Sansone P, et al. MiR-101 downregulation is involved in cyclooxygenase-2 overexpression in human colon cancer cells. Exp Cell Res. 2009;315:1439–47.
Svoboda M, Izakovicova Holla L, Sefr R, et al. Micro-RNAs miR125b and miR137 are frequently upregulated in response to capecitabine chemoradiotherapy of rectal cancer. Int J Oncol. 2008;3:541–7.
Takayama T, Miyanishi K, Hayashi T, et al. Colorectal cancer: genetics of development and metastasis. J Gastroenterol. 2006;41:185–92.
Tazawa H, Tsuchiya N, Izumiya M, et al. Tumor-suppressive miR-34a induces senescence-like growth arrest through modulation of the E2F pathway in human colon cancer cells. Proc Natl Acad Sci USA. 2007;104:15472–7.
Tong AW, Nemunaitis J. Modulation of miRNA activity in human cancer: a new paradigm for cancer gene therapy? Cancer Gene Ther. 2008;15:341–55.
Toyota M, Suzuki H, Sasaki Y, et al. Epigenetic silencing of microRNA-34b/c and B-cell translocation gene 4 is associated with CpG island methylation in colorectal cancer. Cancer Res. 2008;68:4123–32.

Tsang WP, Kwok TT. The miR-18a* microRNA functions as a potential tumor suppressor by targeting on K-Ras. Carcinogenesis. 2009;30:953–9.

Tsuchiya N, Ochiai M, Nakashima K, et al. SND1, a component of RNA-induced silencing complex, is up-regulated in human colon cancers and implicated in early stage colon carcinogenesis. Cancer Res. 2007;67:9568–76.

Valeri N, Croce CM, Fabbri M. Pathogenetic and clinical relevance of microRNAs in colorectal cancer. Cancer Genomics Proteomics. 2009;6:195–204.

Volinia S, Calin GA, Liu CG, et al. A microRNA expression signature of human solid tumors defines cancer gene targets. Proc Natl Acad Sci USA. 2006;103:2257–61.

Wang X, Lam EK, Zhang J, et al. MicroRNA-122a functions as a novel tumor suppressor downstream of adenomatous polyposis coli in gastrointestinal cancers. Biochem Biophys Res Commun. 2009;387:376–80.

Wang YX, Zhang XY, Zhang BF, et al. Initial study of microRNA expression profiles of colonic cancer without lymph node metastasis. J Dig Dis. 2010;11:50–4.

Webster RJ, Giles KM, Price KJ. Regulation of epidermal growth factor receptor signaling in human cancer cells by microRNA-7. J Biol Chem. 2009;284:5731–41.

Weiss GJ, Bemis LT, Nakajima E, et al. EGFR regulation by microRNA in lung cancer: correlation with clinical response and survival to gefitinib and EGFR expression in cell lines. Ann Oncol. 2008;19:1053–9.

Woods K, Thomson JM, Hammond SM. Direct regulation of an oncogenic micro-RNA cluster by E2F transcription factors. J Biol Chem. 2007;282:2130–4.

Xi Y, Nakajima G, Gavin E, et al. Systematic analysis of microRNA expression of RNA extracted from fresh frozen and formalin-fixed paraffin-embedded samples. RNA. 2007;13:1668–74.

Yamakuchi M, Ferlito M, Lowenstein CJ. MiR-34a repression of SIRT1 regulates apoptosis. Proc Natl Acad Sci USA. 2008;105:13421–6.

Young LE, Dixon DA. Posttranscriptional Regulation of Cyclooxygenase 2 Expression in Colorectal Cancer. Curr Colorectal Cancer Rep. 2010;6:60–7.

Zhang B, Farwell MA. MicroRNAs: a new emerging class of players for disease diagnostics and gene therapy. J Cell Mol Med. 2008;12:3–21.

Zhou J, Zhou Y, Yin B, et al. 5-Fluorouracil and oxaliplatin modify the expression profiles of microRNAs in human colon cancer cells in vitro. Oncol Rep. 2010;23:121–8.

Chapter 5
MicroRNAs in Gastric Cancer

Hidekazu Suzuki, Yoshimasa Saito, and Toshifumi Hibi

Abstract MicroRNAs (miRNAs) are small non-coding RNAs that function as endogenous silencers of numerous target genes. Hundreds of human miRNAs have been identified in the human genome and are expressed in a tissue-specific manner, playing important roles in cell proliferation, apoptosis, and differentiation. Links between miRNAs and gastric cancer have become increasingly apparent, and the aberrant expression of miRNAs may contribute to the development and progression of human gastric cancer. Recent studies have shown that some miRNAs play roles as tumor suppressors or oncogenes in gastric cancer. For example, *miR-21* (one of the major oncogenic miRNAs) is activated by *Helicobacter pylori* infection and is increased in gastric cancer. On the contrary, the *miR-29* family is down-regulated in the malignancy, suggesting that it is a potential tumor suppressor of gastric cancer. MiRNA expression profiling may be a powerful clinical tool for diagnosis and the regulation of miRNA expression could be a novel therapeutic strategy for the disease. This chapter summarizes the biological importance of miRNAs in human gastric cancer.

5.1 MicroRNAs

MicroRNAs (miRNAs) are ~22 nucleotide (nt) non-coding RNAs that can post-transcriptionally down-regulate the expression of various target genes. Currently, ~1,000 human miRNAs have been identified in the human genome, and each miRNA potentially controls hundreds of target genes. MiRNAs are expressed in a tissue-specific manner and play important roles in cell proliferation, apoptosis, and differentiation (He and Hannon 2004). Moreover, recent studies have shown a link between the aberrant expression of miRNAs and the development of cancer (Cho 2007; 2010a, 2010b).

H. Suzuki (✉)
Division of Gastroenterology and Hepatology, Department of Internal Medicine, Keio University School of Medicine, Tokyo 160-8582, Japan
e-mail: hsuzuki@sc.itc.keio.ac.jp

W.C.S. Cho (ed.), *MicroRNAs in Cancer Translational Research*,
DOI 10.1007/978-94-007-0298-1_5,

5.2 Target Genes of MiRNAs

The identification of miRNA target genes is critical to determining the functions of miRNAs. Recent studies have indicated that a single miRNA may regulate more than 200 target genes. This target pool is enriched with genes involved in transcriptional regulation and signal transduction. Several databases of human miRNA target predictions according to different algorithms have been established, such as miRanda, miRBase, PicTar and TargetScan (Griffiths-Jones et al. 2006; John et al. 2004; Krek et al. 2005; Lewis et al. 2003). Conserved perfect 6–8 base pair (bp) matches between the 5′ end of the mature miRNA and the 3′UTR of the predicted target mRNA (called "seed" matches) are generally believed to be the most reliable way of determining miRNA targets. However, experimental confirmation using protein expression and luciferase reporter assays for the miRNA targets is necessary to identify the target genes of miRNAs accurately.

5.3 Regulatory Mechanisms of MiRNA Expression

Since miRNAs can have large-scale effects through the regulation of a variety of genes during mammalian development and carcinogenesis, an understanding of the regulatory mechanisms controlling miRNA expression, which is quite limited at present, is important. Several reports have described the binding of transcription factors to the promoter regions of specific miRNA genes that in turn activate the transcription of pri-miRNAs, resulting in the increased expression of mature miRNAs. For example, *c-Myc* binds to the regulatory region of the *miR-17-92* cluster, and the increased expression of *c-Myc* leads to the activation of the miRNAs in this cluster. On the other hand, numerous human miRNAs have been shown to reside within the intronic regions of either coding or non-coding transcription units. Intronic miRNAs are believed to be coordinately expressed with their host gene *mRNAs*, and the expression of intronic miRNAs is thought to depend on the regulation of their host genes (Rodriguez et al. 2004). The location of miRNA genes is also an important factor in the regulation of miRNA expression. Many miRNAs are located in cancer-associated genomic regions that are frequently involved in chromosomal abnormalities, such as a loss of heterozygosity (LOH), amplification, and breakpoints (Calin et al. 2004). Chromosomal abnormalities during carcinogenesis could lead to the widespread differential expressions of miRNAs in human cancer cells.

5.4 Aberrant Expression of MiRNAs in Gastric Cancer

Analyses of miRNA expression profiles have shown the aberrant expressions of many miRNAs that are correlated with tumorigenesis, tumor progression, and the prognosis of various hematological and solid tumors. Ueda et al. (2010) have shown a distinct connection between the aberrant expressions of miRNAs and gastric cancer. They analyzed 353 gastric cancer samples using a miRNA microarray and demonstrated that miRNAs are expressed differentially in gastric cancers, that the histological subtypes are characterized by specific miRNA signatures, and that unique miRNAs are associated with the progression and prognosis of gastric cancer.

Some miRNA genes are over-expressed in gastric cancers, indicating that they may have roles as oncogenes accelerating the development of gastrointestinal cancer (Table 5.1).

MiR-21 is a major oncogenic miRNA that is over-expressed in various human malignancies including lung cancer, breast cancer and glioblastoma, resulting in the suppression of target tumor suppressor genes. A recent study showed that *miR-21* expression was increased in advanced gastric cancer and that it was activated by *Helicobacter pylori* (*H. pylori*) (Guo et al. 2008; Motoyama et al. 2010; Zhang et al. 2008). Although many clinical and basic studies have suggested a link between *H. pylori* infection and gastric cancer (Suzuki et al. 2007; 2009), the molecular mechanisms underlying *H. pylori*-associated gastric carcinogenesis are still unknown. Thus, a large-scale study of *miR-21* expression in clinical gastric cancer specimens is warranted.

The *miR-17-92* cluster, located on chromosome 13q31, is activated by the oncogene *c-Myc,* which is an important regulator of cell growth and is often mutated or amplified in human cancers. The *miR-17-92* cluster is expressed at high levels in

Table 5.1 MicroRNAs up-regulated or down-regulated in gastric cancer

MicroRNAs	Target genes	References
MicroRNAs up-regulated in gastric cancer		
miR-17-92 cluster	*E2F1*	O'Donnell et al. (2005); Ueda et al. (2010)
miR-21	*RECK*	Guo et al. (2008); Zhang et al. (2008); Motoyama et al. (2010)
miR-106a	*RB-1*	Volinia et al. (2006)
miR-106b-25 cluster	*p21, Bim, E2F1*	Petrocca et al. (2008)
miR-650	*ING4*	Zhang et al. (2010a)
miR-130b	*RUNX3*	Lai et al. (2010)
MicroRNAs down-regulated in gastric cancer		
miR-15b, miR-16	*BCL2*	Xia et al. (2008)
miR-29 family	*MCL1, p85, CDC42*	Mott et al. (2007); Park et al. (2009)
miR-375	*PDK1, 14-3-3zeta*	Tsukamoto et al. (2010)
miR-200	*ZEB1, ZEB2*	Shinozaki et al. (2010)
miR-143, miR-145	*DNMT3A, FNDC3B, EGFR*	Takagi et al. (2009); Cho (2010b)
miR-218	*ECOP*	Gao et al. (2010)
miR-9, miR-433	*GRB2, RAB34*	Luo et al. (2009)
miR-31	Unknown	Zhang et al. (2010b)
miR-141	Unknown	Du et al. (2009)
miR-148a, miR-152	Unknown	Chen et al. (2010)

various human malignancies, including gastric cancer (Ueda et al. 2010). *c-Myc* up-regulates the transcription factor *E2F1*, and O'Donnell et al. found that two miRNAs in the *miR-17-92* cluster, namely, *miR-17-5p* and *miR-20a*, target *E2F1* (O'Donnell et al. 2005). Their work showed the existence of a negative feedback loop involving *c-Myc, E2F1,* and *miR-17-5p* and *miR-20a.*

The *miR-106b-25* cluster, which is up-regulated in human gastric cancers and activated by *E2F1*, impairs the TGF-β tumor suppressor pathway by interfering with the expression of *p21* and *Bim* (Petrocca et al. 2008). *MiR-106a* is also up-regulated in colon cancer and targets *RB-1* (Volinia et al. 2006).

The over-expression of *miR-650* in gastric cancer may promote the proliferation and growth of cancer cells, at least partially through the direct targeting of *ING4* (Zhang et al. 2010a). *MiR-130b* expression was significantly higher, compared to matched normal tissue, and was inversely associated with *RUNX3* hypermethylation (Lai et al. 2010).

On the other hand, recent studies have shown that some miRNAs are down-regulated in gastric cancers, which suggests that they may also function as tumor suppressors. Shinozaki et al. (2010) have reported that *miR-200* is down-regulated in EBV-associated gastric cancer.

MiR-15b and *miR-16*, which are down-regulated in human gastric cancer cells, have been shown to play a role in the development of multidrug resistance (MDR) through the modulation of apoptosis by targeting *BCL2* (Xia et al. 2008).

Recent studies have shown that *miR-29b* and *miR-29c* are down-regulated in gastric cancer (Ueda et al. 2010). Mott et al. (2007) have demonstrated that Mcl-1, encoding an anti-apoptotic Bcl2 family protein, is one of the targets of *miR-29* family and that *miR-29* family regulate apoptosis by targeting *Mcl-1*. The members of the *miR-29* family have also been reported to activate *p53* by targeting *p85* and *CDC42* and inducing apoptosis in a p53-dependent manner (Park et al. 2009). Thus, *miR-29* miRNAs are down-regulated in gastric cancers, resulting in the suppressed expression of the *Mcl-1* oncogene, indicating that *miR-29c* functions as a tumor suppressor in gastric epithelial cells.

MiR-375 was down-regulated in gastric carcinoma cells and inhibited the expression of *PDK1* and *14-3-3zeta* as their direct targets (Tsukamoto et al. 2010). The expression levels of *miR-143* and *miR-145* were decreased in most of the human gastric cancers that were examined, as previously reported in colon tumors (Takagi et al. 2009). *MiR-218* expression was significantly reduced in gastric cancer tissues, in *H. pylori*-infected gastric mucosa, and in *H. pylori*-infected AGS cells (Gao et al. 2010).

In addition, *miR-9*, *miR-31*, *miR-141, miR-148a*, *miR-152* and *miR-433* were reported to be down-regulated in gastric cancer cells (Du et al. 2009; Luo et al. 2009; Zhang et al. 2010b; Chen et al. 2010; Wan et al. 2010).

These findings indicate that miRNAs have critical roles in human carcinogenesis and that the aberrant expression of miRNAs may contribute to the initiation and progression of human gastric cancer (Fig. 5.1).

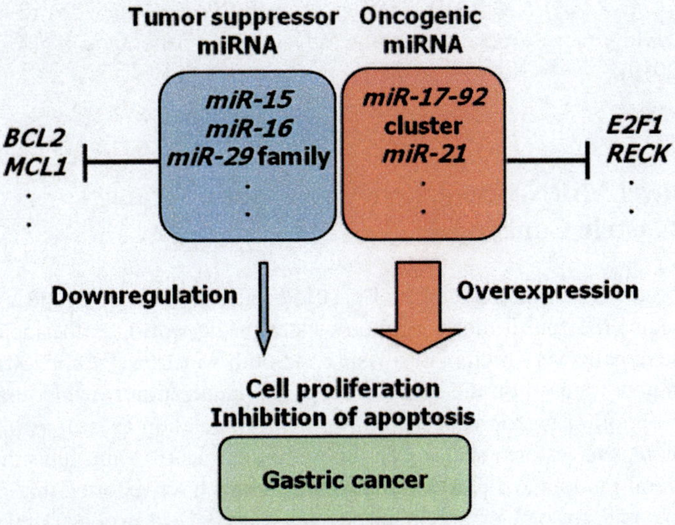

Fig. 5.1 Overexpression of oncogenic miRNAs and downregulation of tumor suppressor miRNAs would link to the development of gastric cancer

5.5 MiRNA Profile as a Diagnostic Tool and a Predictor of the Prognosis of Patients with Gastric Cancer

Lu et al. (2005) reported that miRNA expression profiles can be used to classify the developmental lineages and differentiation stages of tumors. Interestingly, miRNA expression profiles are more accurate for tumor classification than conventional mRNA profiling. Furthermore, recent studies have demonstrated the association of miRNA expression signatures with prognostic factors and disease progression in chronic lymphocytic leukemia and lung cancer (Calin et al. 2005; Volinia et al. 2006). Yanaihara et al. (2006) showed that a low expression level of *let-7a-2* was correlated with poor survival in patients with lung adenocarcinoma, indicating that *let-7* may have a role as a tumor suppressor. On the other hand, the over-expression of *miR-155*, which is a potential oncogenic miRNA, was correlated with poor survival in cases with lung cancer. Schetter et al. have shown that a high expression level of *miR-21* is associated with poor survival and a poor therapeutic outcome in patients with colon cancer (Schetter et al. 2008). These findings indicate that miRNA expression profiling might be a powerful clinical tool for determining the diagnosis and prognosis of patients with gastric cancer, especially in cases that are difficult to diagnose pathologically.

Recent studies have suggested that the detection of miRNA in peripheral blood may be a novel tool for monitoring circulating tumor cells in patients with gastric cancers (Tsujiura et al. 2010; Zhou et al. 2010). In addition, seven miRNA signatures (*miR-10b, miR-21, miR-223, miR-338, let-7a, miR-30a-5p, miR-126*) were closely associated with relapse-free and overall survival among patients with gastric cancer (Li et al. 2010).

5.6 A Novel MiRNA-mediated Therapeutic Strategy for Gastric Cancer

Despite the decline in the incidence of gastric cancer in Western countries, gastric cancer remains the fourth most common cancer in the world, as the incidence of this disease remains very high in East Asia and South America (Parkin 2001). Much research is now focused on the effect of *H. pylori* eradication (Suzuki et al. 2007) on the prevention of gastric cancer in the general population as well as in patients with preneoplastic lesions, such as chronic atrophic gastritis and intestinal metaplasia. Several randomized placebo-controlled studies have reported that *H. pylori* eradication could, to some extent, induce a reversal of gastric atrophy (Kuipers et al. 2004; Leung et al. 2004; Ley et al. 2004; Mera et al. 2005; Schenk et al. 2000); the effect of bacterial eradication on gastric cancer prevention is, however, less evident (Leung et al. 2004; Mera et al. 2005; Schetter et al. 2008). In all of these studies (Kuipers et al. 2004; Leung et al. 2004; Ley et al. 2004; Mera et al. 2005; Schenk et al. 2000; Wong et al. 2004), no significant difference in the incidence of gastric cancer was observed between the *H. pylori*-eradicated and placebo groups during the first 4–12 years after treatment. The outstanding observation in all these studies, however, was that the gastric cancers that developed after *H. pylori* eradication were confined to those subjects who already had atrophic gastritis and intestinal metaplasia, suggesting that the major cancer-preventive effect of *H. pylori* eradication could be expected to occur in subjects without such pre-cancerous conditions. In other words, cohorts with pre-neoplastic lesions may already have passed the point of no return and need to be treated further in addition to *H. pylori* eradication. Maekita et al. (2006) have demonstrated that *H. pylori* infection potently induces the methylation of the promoter regions of some genes (Ushijima et al. 2006). Long-term colonization by *H. pylori* might induce epigenetic modifications of gastric mucosal genes, including the promoters of tumor suppressor miRNAs, which cannot be completely reversed simply by bacterial eradication. Epigenetic therapy using DNA methylation inhibitors and histone deacetylase (HDAC) inhibitors has been reported to show clinical promise for the treatment of human malignancies. According to a recent study conducted by us (Saito et al. 2009), an miRNA microarray analysis showed that most miRNAs activated by 5-Aza-CdR (5-aza-2′-deoxycytidine) and PBA (4-phenylbutyric acid) in gastric cancer cells are located at Alu repeats on chromosome 19, and analyses of chromatin modification showed that DNA demethylation and HDAC inhibition at Alu repeats activate silenced *miR-512-5p* via RNA polymerase II. In addition, the activation of *miR-512-5p* by epigenetic treatment induces the suppression of *Mcl-1*, resulting in the apoptosis of gastric cancer cells, suggesting that chromatin remodeling at Alu repeats may play a critical role in the regulation of miRNA expression and that the epigenetic activation of silenced Alu-associated miRNAs could be a novel therapeutic approach for gastric cancer.

The distinct connection between the aberrant expressions of miRNAs and the development of gastric cancer suggests that miRNAs can be potential therapeutic targets. A recent study showed that chemically engineered oligonucleotides, termed "antagomirs", that were found to be specific inhibitors of endogenous miRNAs in mice (Krutzfeldt et al. 2005), could be used to silence oncogenic miRNAs, such as the *miR-17-92* cluster, as a strategy for cancer therapy. On the other hand, if tumor suppressor miRNAs are down-regulated or silenced in cancer cells, the transfection or the activation of these miRNAs may be beneficial as an anti-cancer therapy.

5.7 Conclusion and Perspectives

MiRNA expression profiling may be a powerful clinical tool for the diagnosis of gastric cancer, and the regulation of miRNA expression could be a novel therapeutic strategy for gastric cancer. Since the relationship between miRNAs and gastric cancer has only just begun to be understood and the number of miRNA genes identified is increasing, a large number of therapeutic targets may exist for gastric cancer. Further studies are necessary to investigate whether miRNA-mediated therapy may be an effective strategy for the management of gastric cancer.

References

Calin GA, Ferracin M, Cimmino A, et al. A MicroRNA signature associated with prognosis and progression in chronic lymphocytic leukemia. N Engl J Med. 2005;353:1793–801.

Calin GA, Liu CG, Sevignani C, et al. MicroRNA profiling reveals distinct signatures in B cell chronic lymphocytic leukemias. Proc Natl Acad Sci USA. 2004;101:11755–60.

Chen Y, Song Y, Wang Z, et al. Altered expression of MiR-148a and MiR-152 in gastrointestinal cancers and its clinical significance. J Gastrointest Surg. 2010;14:1170–9.

Cho WC. MicroRNAs in cancer – from research to therapy. Biochim Biophys Acta. 2010a;1805:209–17.

Cho WC. MicroRNAs: potential biomarkers for cancer diagnosis, prognosis and targets for therapy. Int J Biochem Cell Biol. 2010b;42:1273–81.

Cho WC. OncomiRs: the discovery and progress of microRNAs in cancers. Mol Cancer. 2007;6:60.

Du Y, Xu Y, Ding L, et al. Down-regulation of miR-141 in gastric cancer and its involvement in cell growth. J Gastroenterol. 2009;44:556–61.

Gao C, Zhang Z, Liu W, et al. Reduced microRNA-218 expression is associated with high nuclear factor kappa B activation in gastric cancer. Cancer. 2010;116:41–9.

Griffiths-Jones S, Grocock RJ, van Dongen S, et al. miRBase:microRNA sequences, targets and gene nomenclature. Nucleic Acids Res. 2006;34:D140–4.

Guo Y, Chen Z, Zhang L, et al. Distinctive microRNA profiles relating to patient survival in esophageal squamous cell carcinoma. Cancer Res. 2008;68:26–33.

He L, Hannon GJ. MicroRNAs: small RNAs with a big role in gene regulation. Nat Rev Genet. 2004;5:522–31.

John B, Enright AJ, Aravin A, et al. Human MicroRNA targets. PLoS Biol. 2004;2:e363.

Krek A, Grun D, Poy MN, et al. Combinatorial microRNA target predictions. Nat Genet. 2005;37:495–500.

Krutzfeldt J, Rajewsky N, Braich R, et al. Silencing of microRNAs in vivo with 'antagomirs'. Nature. 2005;438:685–9.

Kuipers EJ, Nelis GF, Klinkenberg-Knol EC, et al. Cure of *Helicobacter pylori* infection in patients with reflux oesophagitis treated with long term omeprazole reverses gastritis without exacerbation of reflux disease: results of a randomised controlled trial. Gut. 2004;53:12–20.

Lai KW, Koh KX, Loh M, et al. MicroRNA-130b regulates the tumour suppressor RUNX3 in gastric cancer. Eur J Cancer. 2010;46:1456–63.

Leung WK, Lin SR, Ching JY, et al. Factors predicting progression of gastric intestinal metaplasia: results of a randomised trial on *Helicobacter pylori* eradication. Gut. 2004;53:1244–9.

Lewis BP, Shih IH, Jones-Rhoades MW, et al. Prediction of mammalian microRNA targets. Cell. 2003;115:787–98.

Ley C, Mohar A, Guarner J, et al. *Helicobacter pylori* eradication and gastric preneoplastic conditions: a randomized, double-blind, placebo-controlled trial. Cancer Epidemiol Biomarkers Prev. 2004;13:4–10.

Li X, Zhang Y, Ding J, et al. Survival prediction of gastric cancer by a seven-microRNA signature. Gut. 2010;59:579–85.

Lu J, Getz G, Miska EA, et al. MicroRNA expression profiles classify human cancers. Nature. 2005;435:834–8.

Luo H, Zhang H, Zhang Z, et al. Down-regulated miR-9 and miR-433 in human gastric carcinoma. J Exp Clin Cancer Res. 2009;28:82.

Maekita T, Nakazawa K, Mihara M, et al. High levels of aberrant DNA methylation in *Helicobacter pylori*-infected gastric mucosae and its possible association with gastric cancer risk. Clin Cancer Res. 2006;12:989–95.

Mera R, Fontham ET, Bravo LE, et al. Long term follow up of patients treated for *Helicobacter pylori* infection. Gut. 2005;54:1536–40.

Motoyama K, Inoue H, Mimori K, et al. Clinicopathological and prognostic significance of PDCD4 and microRNA-21 in human gastric cancer. Int J Oncol. 2010;36:1089–95.

Mott JL, Kobayashi S, Bronk SF, et al. mir-29 regulates Mcl-1 protein expression and apoptosis. Oncogene. 2007;26:6133–40.

O'Donnell KA, Wentzel EA, Zeller KI, et al. c-Myc-regulated microRNAs modulate E2F1 expression. Nature. 2005;435:839–43.

Park SY, Lee JH, Ha M, et al. miR-29 miRNAs activate p53 by targeting p85 alpha and CDC42. Nat Struct Mol Biol. 2009;16:23–9.

Parkin DM. Global cancer statistics in the year 2000. Lancet Oncol. 2001;2:533–43.

Petrocca F, Visone R, Onelli MR, et al. E2F1-regulated microRNAs impair TGFbeta-dependent cell-cycle arrest and apoptosis in gastric cancer. Cancer Cell. 2008;13:272–86.

Rodriguez A, Griffiths-Jones S, Ashurst JL, et al. Identification of mammalian microRNA host genes and transcription units. Genome Res. 2004;14:1902–10.

Saito Y, Suzuki H, Tsugawa H, et al. Chromatin remodeling at Alu repeats by epigenetic treatment activates silenced microRNA-512-5p with downregulation of Mcl-1 in human gastric cancer cells. Oncogene. 2009;28:2738–44.

Schenk BE, Kuipers EJ, Nelis GF, et al. Effect of *Helicobacter pylori* eradication on chronic gastritis during omeprazole therapy. Gut. 2000;46:615–21.

Schetter AJ, Leung SY, Sohn JJ, et al. MicroRNA expression profiles associated with prognosis and therapeutic outcome in colon adenocarcinoma. JAMA. 2008;299:425–36.

Shinozaki A, Sakatani T, Ushiku T, et al. Downregulation of microRNA-200 in EBV-associated gastric carcinoma. Cancer Res. 2010;70:4719–27.

Suzuki H, Hibi T, Marshall BJ. *Helicobacter pylori*: present status and future prospects in Japan. J Gastroenterol. 2007;42:1–15.

Suzuki H, Iwasaki E, Hibi T. *Helicobacter pylori* and gastric cancer. Gastric Cancer. 2009;12: 79–87.

Takagi T, Iio A, Nakagawa Y, et al. Decreased expression of microRNA-143 and -145 in human gastric cancers. Oncology. 2009;77:12–21.

Tsujiura M, Ichikawa D, Komatsu S, et al. Circulating microRNAs in plasma of patients with gastric cancers. Br J Cancer. 2010;102:1174–9.

Tsukamoto Y, Nakada C, Noguchi T, et al. MicroRNA-375 is downregulated in gastric carcinomas and regulates cell survival by targeting PDK1 and 14-3-3zeta. Cancer Res. 2010;70:2339–49.

Ueda T, Volinia S, Okumura H, et al. Relation between microRNA expression and progression and prognosis of gastric cancer: a microRNA expression analysis. Lancet Oncol. 2010;11:136–46.

Ushijima T, Nakajima T, Maekita T. DNA methylation as a marker for the past and future. J Gastroenterol. 2006;41:401–7.

Volinia S, Calin GA, Liu CG, et al. A microRNA expression signature of human solid tumors defines cancer gene targets. Proc Natl Acad Sci USA. 2006;103:2257–61.

Wan HY, Guo LM, Liu T, et al. Regulation of the transcription factor NF-kappaB1 by microRNA-9 in human gastric adenocarcinoma. Mol Cancer. 2010;9:16.

Wong BC, Lam SK, Wong WM, et al. *Helicobacter pylori* eradication to prevent gastric cancer in a high-risk region of China: a randomized controlled trial. JAMA. 2004;291:187–94.

Xia L, Zhang D, Du R, et al. MiR-15b and miR-16 modulate multidrug resistance by targeting BCL2 in human gastric cancer cells. Int J Cancer. 2008;123:372–9.

Yanaihara N, Caplen N, Bowman E, et al. Unique microRNA molecular profiles in lung cancer diagnosis and prognosis. Cancer Cell. 2006;9:189–98.

Zhang X, Zhu W, Zhang J, et al. MicroRNA-650 targets ING4 to promote gastric cancer tumorigenicity. Biochem Biophys Res Commun. 2010a;395:275–80.

Zhang Y, Guo J, Li D, et al. Down-regulation of miR-31 expression in gastric cancer tissues and its clinical significance. Med Oncol. 2010b;27:685–9.

Zhang Z, Li Z, Gao C, et al. miR-21 plays a pivotal role in gastric cancer pathogenesis and progression. Lab Invest. 2008;88:1358–66.

Zhou H, Guo JM, Lou YR, et al. Detection of circulating tumor cells in peripheral blood from patients with gastric cancer using microRNA as a marker. J Mol Med. 2010;88:709–17.

Chapter 6
MicroRNAs in Prostate Cancer: A Possible Role as Novel Biomarkers and Therapeutic Targets?

Paolo Gandellini, Marco Folini, and Nadia Zaffaroni

Abstract Eradication of advanced prostate cancer still represents an unsolved clinical problem, making the development of alternative treatment approaches highly desirable. Understanding the molecular alterations that distinguish the non-progressive from progressive disease will allow the identification of novel biomarkers for improved staging and prognostication, and will also provide mechanistic information to facilitate treatment selection and design of novel therapeutic approaches. MicroRNAs (miRNAs) are small non-coding RNAs that negatively regulate gene expression. Recent findings indicate that miRNAs are deregulated in human tumors, suggesting a potential role for these molecules in the pathogenesis of cancer. Thus far, only a limited number of studies have investigated miRNA expression in prostate cancer. Results indicate that miRNA expression profiles may distinguish carcinoma from non-neoplastic specimens and further classify tumors according to androgen dependence. In addition, a prognostic significance has been attributed to specific miRNAs as predictors of clinical recurrence following radical prostatectomy. These findings, together with the documented possibility to detect cancer-related miRNAs in blood and core biopsies, open a window on the possibility to utilize them as novel biomarkers. For a handful of miRNAs, functional investigation has also been pursued in prostate cancer experimental models to establish the rationale for the development of miRNA-based therapies. However, a better understanding of the role exerted by specific miRNAs in the onset and progression of prostate cancer is needed, as is a precise definition of their targets relevant to the disease, before translating these molecules into the clinical setting.

P. Gandellini (✉)
Molecular Pharmacology Unit, Department of Experimental Oncology and Molecular Medicine, Fondazione IRCCS Istituto Nazionale dei Tumori, 20133 Milan, Italy
e-mail: paolo.gandellini@istitutotumori.mi.it

W.C.S. Cho (ed.), *MicroRNAs in Cancer Translational Research*,
DOI 10.1007/978-94-007-0298-1_6, © Springer Science+Business Media B.V. 2011

6.1 Prostate Cancer

6.1.1 Incidence and Screening

Prostate cancer (PCa) is the most frequently diagnosed, non-dermatologic malignancy in developed countries (Jemal et al. 2010). About 217,730 new cases were diagnosed in 2010 in the United States. It is estimated that at least a third of men over 50 and almost all men over 80 have small hotbeds of prostate adenocarcinoma. Despite a stage migration effect (largely because of prostate-specific antigen [PSA] screening) and decreased mortality rate, PCa is still the second leading cause of cancer-related death in men following lung cancer (Jemal et al. 2010).

As anticipated, early detection is currently based on the measurement of serum levels of PSA/KLK3, a serine protease belonging to the family of glandular kallikrein-related peptidases, which is synthesized by prostate tissue and functions to liquefy seminal fluid. Only a small fraction of PSA leaks into the circulatory system in physiological conditions. The onset of PCa can increase the release of PSA into blood, thus making its serum levels a commonly used marker for detecting and monitoring prostate tumors. However, increased PSA levels cannot be considered a specific tumor marker, given that they can be associated with other benign prostate diseases, such as urinary infections or benign prostatic hyperplasia (BPH).

Programs of screening for PSA have been undertaken to reduce mortality, but prospective randomized trials have not yet provided conclusive evidence of their efficacy (Andriole et al. 2009; Schroeder et al. 2009). The European Randomized Study of Screening for PCa showed that PSA-based screening reduced the rate of death from PCa by 20% (Schroeder et al. 2009), though associated with a high risk of over-diagnosis. However, The Prostate, Lung, Colorectal, and Ovarian Cancer Screening Trial reported that, after 7–10 years of follow-up, the death rate from PCa was very low and did not differ significantly between men who received annual screening and those who underwent usual care (Andriole et al. 2009).

6.1.2 Diagnosis

At present, diagnosis of PCa is based on examination of histopathological specimens from the organ, usually obtained by transrectal or transperineal ultrasound-guided core biopsies or sometimes from transurethral resection or adenomectomy.

Prostate adenocarcinoma is graded using the Gleason pattern score, and its local staging (T category) is attributed according to findings of digital rectal examination. The Gleason score is the sum of two scores, each varying from 1 (well-differentiated) to 5 (undifferentiated), representative of the most prevalent and the second most prevalent differentiation grade within the tumor. To be counted, a pattern needs to occupy more than 5% of the biopsy specimen. The stage of the tumor is instead classified as non-palpable (T1), palpable but confined to the prostate (T2), palpable and extending beyond the prostate (T3) or palpable and invading adjacent structures (T4). T3 and T4 tumors are referred to as locally advanced.

As no forms of imaging accurately estimate the extent of the tumor and its location within the prostate or in the area surrounding it, nomograms based on serum PSA concentration, grade of the biopsy and stage currently represent the standard device to establish the patient's risk of having organ-confined disease versus microscopic extracapsular extension, seminal vesicle penetration or lymph node involvement (Damber and Aus 2008). Definitive grading and staging of the disease can only be assessed on surgical specimens after prostatectomy. Usually, local-regional lymph nodes are removed to ascertain dissemination of the disease. When a tumor is confined to the gland, it is referred to as N0 and local-regionally disseminated disease as N1. Metastatic disease, which usually spreads to bone, is classified as M1.

6.1.3 Current Therapies

The treatment options for localized PCa are radical prostatectomy and radiotherapy, both of which can be administered in different variations. These strategies seem to have similar efficacy, despite different side effects in terms of frequency of erectile dysfunctions and urinary incontinence. The choice for each treatment should be based on the patient's preference, life expectancy and tumor characteristics. Usually, treatments with a curative intent are administered to men with poorly differentiated tumors and/or with a long life expectancy, whereas active monitoring is recommended for low-risk or elderly patients, based on the evidence that many PCa are indolent and progress slowly. After surgery or radiotherapy, evaluation of PSA serum levels is used to monitor eventual relapse of the disease, as an increase in PSA represents the first sign of failure after treatment (biochemical failure). For patients who are at high risk for systemic recurrence, androgen ablation therapy is recommended. Such a strategy is mainly aimed at counteracting hormone production and function, thus reducing cancer cell proliferation. Androgen ablation in PCa patients is mainly pursued by: (i) castration (testosterone elimination), through surgical removal of testicles (orchiectomy) or administration of gonadotropin-releasing hormone agonists (i.e. goserelin); or (ii) administration of androgen-receptor antagonists (i.e. flutamide, bicalutamide, nilutamide). The same treatment is usually administered, with a palliative intent, to patients with distant metastases at diagnosis, whose prognosis is poor. However, the efficacy of androgen deprivation is time-limited, and most patients undergo disease progression and develop a hormone-refractory PCa. Androgen escape is strongly linked to the high rate of PCa mortality, as cytotoxic chemotherapy seems to be generally ineffective against hormone-refractory disease. However, recent clinical trials have shown that docetaxel, alone or in combination with estramustine, might improve the survival of men with advanced PCa (Damber and Aus 2008).

Given this scenario, it is clear that eradication of advanced PCa still represents an unsolved clinical problem, making the development of alternative treatment approaches highly desirable. Understanding the molecular alterations that distinguish progressive from non-progressive disease will allow the identification of novel biomarkers for improved staging and determining prognosis and also for providing mechanistic information to facilitate treatment selection and design of novel therapeutic approaches.

Over the past years, it has become clear that alterations in microRNA (miRNA) expression are a widespread phenomenon in human cancer, thus suggesting a role for these molecules in tumorigenesis (Calin and Croce 2006). In some instances, the expressions of selected miRNAs or miRNA signatures were found to correlate with diverse clinico-pathological parameters and to predict patient clinical outcome or response to treatment (Calin and Croce 2006). Overall, these findings have highlighted the potential of miRNAs as new diagnostic or prognostic biomarkers. In addition, evidence that specific miRNAs are endowed with oncogenic or tumor-suppressive functions has emerged from several studies conducted in experimental models, pointing to a possible role as novel targets or tools for anti-cancer therapy. However, it should be noted that definitive evidence linking miRNAs with the development of cancer is scarce and that further investigation is warranted to ascertain whether changes in the expression levels of specific miRNAs are actually causative rather than a consequence of the disease.

6.2 MiRNA Expression Studies in Prostate Cancer: Towards MiRNA-based Diagnostic and Prognostic Biomarkers

6.2.1 Diagnostic Value of MiRNAs in Prostate Cancer

Attempts to investigate deregulation of miRNAs in PCa, thus validating specific miRNAs as novel diagnostic markers for the disease, have been mainly based on the analysis of miRNA expression profiles in PCa versus normal/benign tissues.

Genome-wide miRNA profiling data from seven independent studies are available. Initially, Lu et al. (2005) analyzed the expression of 217 miRNAs in different types of human cancer, including PCa specimens (6 tumors and 8 normal tissues) and showed a general down-regulation of miRNAs in tumors compared with normal tissues. A subsequent large-scale microarray analysis conducted on different tumor histotypes, including PCa specimens (56 carcinomas and 7 normal tissues from non-cancer individuals), identified a common miRNA signature for solid tumors that was mainly composed of over-expressed miRNAs (Volinia et al. 2006). A PCa-specific miRNA signature was first obtained by Porkka et al. (2007) on 9 PCa and 4 BPH samples. Differential expression of 51 miRNAs (37 down- and 14 up-regulated) was found between cancer and non-malignant tissues. Hierarchical clustering of tumors by their miRNA expression was shown to accurately separate PCa from BPH specimens and further classify the tumors according to their androgen dependence (hormone-naïve versus hormone-refractory). Ozen et al. (2008), who analyzed the expression of 480 miRNAs in 16 clinically localized PCa and 10 BPH samples, found widespread down-regulation of miRNAs in cancer, including *miR-125b*, *miR-145*, and *let-7c*, for which expression data were confirmed by quantitative reverse transcription-polymerase chain reaction (qRT-PCR). A genome-wide expression profiling of miRNAs and *mRNAs*, performed by Ambs et al. (2008) in 60 primary PCa samples and 16 surrounding non-tumor tissues, showed that key components of miRNA processing and several genes carrying miRNA coding sequences in their introns, such as *miR-106b-25* cluster and its host gene *MCM7*, were significantly up-regulated in tumors. In a further study carried out in a series of 57 primary PCa, the same research group found a differential expression of 19 miRNAs in tumors with and without perineural invasion, with high-level expression of *miR-224* in perineural cancer cells (Prueitt et al. 2008).

Tong et al. (2009) carried out the first paired analysis of miRNA expression on microdissected malignant and uninvolved areas of 40 prostatectomy specimens. Five miRNAs (*miR-23b*, *miR-100*, *miR-145*, *miR-221*, and *miR-222*) were found to be significantly down-regulated in malignant compared with normal tissues. Very recently, Schaefer et al. (2010) analyzed miRNA expression in matched tumor and adjacent normal tissues obtained from 76 patients and showed down-regulation of *miR-16*, *miR-31*, *miR-125b*, *miR-145*, *miR-149*, *miR-181b*, *miR-184*, *miR-205*, *miR-221*, and *miR-222*, and up-regulation of *miR-96*, *miR-182*, *miR-182**, *miR-183*, and *miR-375* in cancer specimens. Expression of *miR-31*, *miR-96*, and *miR-205* correlated with Gleason score, whereas expression of *miR-125b*, *miR-205*, and *miR-222* correlated with pathological tumor stage. In addition, *miR-205* was found to be the best discriminating miRNA, providing a correct overall classification of 72%.

Besides genome-wide profiling studies, a deep sequencing approach has been recently used to identify miRNAs with a differential expression in PCa compared to normal tissue. Szczyrba et al. (2010) sequenced small RNAs from two pooled miRNA cDNA libraries of each normal and primary PCa samples, with each library containing five pooled samples from normal and cancer tissue, respectively. They found a total of 260 of the presently > 1,000 known human miRNAs expressed in the normal and tumor-derived cDNA libraries. In addition, 16 and 17 miRNAs were respectively at least 1.5-fold up- and down-regulated in PCa compared to normal tissue.

6.2.2 MiRNAs and Prostate Cancer Progression

Additional studies were aimed at the identification of miRNAs involved in PCa progression by comparing the miRNA profile of primary tumors and metastatic sites. Leite et al. (2010) analyzed by qRT-PCR the expression of 14 miRNAs in tumor lesions obtained from 18 patients who underwent radical prostatectomy for treatment of localized high-grade PCa (with a mean Gleason score 8.6, all staged pT3) and 4 patients with metastatic, androgen-independent PCa. They found that *let-7c*, *miR-100*, and *miR-218* were significantly down-regulated in metastatic carcinomas compared to localized PCa lesions, thus suggesting that reduced expression levels of these miRNAs could be implicated in the metastatic process of PCa. A comparative analysis of *miR-101* expression in benign prostate tissue (15 cases), primary PCa (16 cases) and metastatic PCa (33 cases) revealed that *miR-101* expression decreases during cancer progression mainly due to loss of one or both of its two genomic loci (Varambally et al. 2008). Similarly, an analysis of PCa (134 cases) and benign prostate (83 cases) tissue samples showed significantly decreased *miR-145* in PCa, particularly from patients experiencing tumor progression, in terms of PSA elevation, local progression or metastasis (Chen et al. 2010).

6.2.3 Prognostic Value of MiRNAs in Prostate Cancer

The first evidence of a possible prognostic relevance of miRNAs in PCa was obtained by Tong et al. (2009) in a series of samples from 40 patients who underwent radical prostatectomy. Through the analysis of miRNA expression profiles, they found that individuals with elevation of PSA within 2 years of surgery displayed a distinct expression profile of 16 miRNAs compared with non-relapsing patients, suggesting that aberrant miRNA expression features might reflect a tendency for early biochemical failure. More recently, a qRT-PCR analysis carried out by Spahn et al. (2010) in tumor specimens from 92 patients who underwent radical prostatectomy showed that *miR-221* was down-regulated in samples from patients who experienced local recurrence or distant metastases. In addition, when the sample population was divided into groups with either low or high *miR-221* expression levels, Kaplan-Meier survival estimates predicted a significant difference between groups in 10-year recurrence-free survival (87.2 and 53.2% for patients with high and low *miR-221* expression, respectively). Moreover, Cox proportional hazard models confirmed that dichotomized and continuous *miR-221* expression levels were significant predictors of clinical recurrence in both univariate and multivariate analyses in the presence of PSA levels, pathological stage and Gleason score as covariates. Schaefer et al. (2010) evaluated the prognostic significance of a panel of miRNAs in a series of radical prostatectomy specimens from 75 PCa patients. For analysis, miRNA expression levels were dichotomized by the median, and it was found that the recurrence-free interval was significantly decreased in patients with high *miR-96* expression in their tumor samples. In addition, Cox regression analysis showed that *miR-96* remained a prognostic indicator alone with Gleason score in the prediction model for disease recurrence. This prognostic information was then confirmed in an independent tumor sample set consisting of 79 radical prostatectomy specimens (Schaefer et al. 2010). In a recent study, *miR-34c* was found to be down-regulated in PCa and inversely correlated to the aggressiveness of the tumor, WHO grade, PSA levels and occurrence of metastases. Furthermore, Kaplan-Meier analysis of patient survival, based on *miR-34c* expression levels categorized into low (< 50th percentile) and high (> 50th percentile) expression, significantly divided the patients into high-risk and low-risk patients (Hagman et al. 2010). Chen et al. (2010) showed that decreased *miR-145* levels were significant negative prognostic factors for both disease-specific and progression-free survival in PCa patients. In the same study, a Cox proportion regression model incorporating classic clinico-pathologic parameters (Gleason score, PSA level, and tumor stage) showed that the expression of *miR-145* was also an independent favorable prognostic factor for progression-free survival (Chen et al. 2010).

6.2.4 MiRNAs in Circulating Blood

Current challenges in the management of PCa include a continuing search for sensitive markers detectable by minimally invasive approaches that can be exploited to identify early neoplastic changes, thus facilitating diagnosis of PCa at an early stage, as well as to monitor disease progression and response to treatment. Recent studies have shown that tumor-derived miRNAs are present in biological fluids in remarkably stable forms and are protected from endogenous ribonuclease activity. In addition, such miRNAs were shown to be present in circulating blood at levels sufficient to be measurable as biomarkers for the detection of tumors (Cortez and Calin 2009). The potential to use miRNAs as diagnostic biomarkers in PCa has been thus far investigated in two studies. Specifically, in a case-control cohort of serum samples collected from 25 individuals with metastatic PCa and 25 healthy age-matched male control individuals, Mitchell et al. (2008) found a marked differential expression (46-fold over-expressed) of *miR-141* in PCa compared with the control pool. Consistent with results from the analysis of pooled samples, when analyzed in individual samples, serum levels of *miR-141* were, in general, substantially higher in cases compared with controls. Furthermore, serum levels of *miR-141* were able to detect individuals with cancer with a high accuracy (60% sensitivity and 100% specificity). Successively, Lodes et al. (2009) identified *miR-16*, *miR-92a*, *miR-103*, *miR-107*, *miR-197*, *miR-34b*, *miR-328*, *miR-485-3p*, *miR-486-5p*, *miR-92b*, *miR-574-3p*, *miR-636*, *miR-640*, *miR-766*, and *miR-885-5p* as over-expressed in the serum from six advanced PCa patients with respect to eight normal controls. A slightly elevated signal for *miR-141* in PCa patients compared to normal donors' sera was also shown.

6.2.5 Are We Still Far from MiRNA-based Biomarkers for Prostate Cancer?

Overall, the studies conducted so far convey the notion that miRNAs are aberrantly expressed in PCa compared with normal tissue, thus indicating them as eligible novel markers for PCa diagnosis and prognosis. Nonetheless, a consensus on which miRNAs are specific biomarkers for PCa has not been yet achieved, owing to the large inconsistencies between the collected data (Gandellini et al. 2009a). It is not surprising that miRNA expression profiles published thus far are inconsistent, given that profiling technology is still in its infancy. Results obtained with different measurement platforms may not perfectly correlate. Hence, validation of data by an independent technique, such as northern blotting, RNase protection assay or qRT-PCR, is strongly recommended when translating from high-throughput screening to single candidate miRNAs to be developed as potential biomarkers. Discrepancies in miRNA expression data could also depend on diverse experimental variables, including the selection of the case series (in terms of tumor stage, previous treatment of the patient), the selection (use of BPH, non-neoplastic tissue surrounding the tumor, or prostate tissue from non-cancer individuals as controls; use of bulky

Table 6.1 Comparison of microRNA expression in prostate cancer versus non-neoplastic tissues from the seven profiling studies available in the literature

Author	Lu	Volinia	Porkka	Ozen	Ambs	Tong	Schaefer	Szczyrba
Technique	Bead-based hybridization	Oligo microarray	Oligo microarray	Oligo microarray	Oligo microarray	Bead-based hybridization	Oligo microarray	Deep sequencing
miR-100	↓	=	↓	↓	=	↓	=	↓
miR-125b	=	=	↓	↓	=	↓	↓	↓
miR-145	=	=	↓	↓	↓	↓	↓	↓
miR-16	↓	↑	↓	↓	=	=	↓	=
miR-205	=	=	↓	↓	↓	↓	↓	↓
miR-21	=	↑	=	↓	=	n.a.	=	↑
miR-221	↓	=	↓	↓	↓	↓	↓	↓
miR-222	↓	=	↓	↓	=	↓	↓	↓
miR-23b	↓	=	↓	↓	=	↓	=	=
miR-32	↓	↑	=	n.a.	↑	n.a.	=	↑

↓, Down-regulated; ↑, Up-regulated; =, Not differentially expressed; n.a., Not available.

or microdissected tissues) and preservation of specimens (frozen versus formalin-fixed paraffin-embedded) and the technique of RNA preparation (use of total RNA or purified small RNA samples). In this regard, the achievement of a global consensus on the techniques and controls to be used as well as the conduction of studies encompassing larger case series of paired neoplastic and normal tissues for PCa patients, with associated pathobiological and clinical information, are strongly warranted. Special attention should be paid to the selection of tissue specimens, which appears particularly relevant because of the heterogeneous and multifocal nature of the malignancy. Specifically, eligibility should be determined by an expert pathologist through evaluation of the percentage of tumor or normal epithelial cells within the sample, as well as that of contaminating stromal cells.

Despite the discrepancies in the results of the studies published thus far, it could be tentatively hypothesized that *miR-100*, *miR-125b*, *miR-145*, *miR-205*, *miR-221*, and *miR-222* may represent, in the future, the most promising candidate markers for PCa, as their decreased expression in cancer specimens has been reported in most of the expression profiling studies available in the literature (Table 6.1) and confirmed by an independent technique, such as qRT-PCR (Gandellini et al. 2009a).

6.3 Functional Investigation of MiRNAs in Prostate Cancer: Towards the Development of MiRNA-based Therapies

Oncogenic or tumor-suppressive functions in different human cancers have been experimentally documented for several miRNAs based on loss- or gain-of-function approaches. The function of over-expressed oncogenic miRNAs has been successfully antagonized in experimental models using chemically modified antisense oligonucleotides, such as 2′-*O*-methyl-antisense oligonucleotides or locked nucleic acids (LNA). The latter, which exhibit the highest affinity and specificity for complementary target RNA and show properties of cellular uptake and biodistribution adequate for use in vivo (Stenvang et al. 2008), proved to efficiently inhibit the expression of specific miRNAs (Elmen et al. 2008). Similarly, intravenous injection of antagomirs, 2′-*O*-methyl-antisense oligonucleotides conjugated with a cholesterol moiety at the 3′ end, produced a long-lasting inhibition of specific miRNAs in mice (Krutzfeld et al. 2005). In contrast, restoration of expression of down-modulated tumor-suppressive miRNAs is usually achieved using synthetic double-stranded RNA molecules resembling precursor miRNAs or expression vectors carrying miRNA genes packed into lentiviral systems (Bonci et al. 2008; Gandellini et al. 2009b). Overall, these strategies, besides being powerful approaches for functionally validating miRNAs relevant for a specific disease, constitute the prerequisite to develop potential miRNA-based therapies.

Thus far, the functional activity of only a handful of miRNAs has been experimentally modeled in the context of PCa, and a small number of miRNAs has been characterized as potential oncogenes or tumor suppressors, based on the consequences of the modulation of their expression on the phenotype of experimental

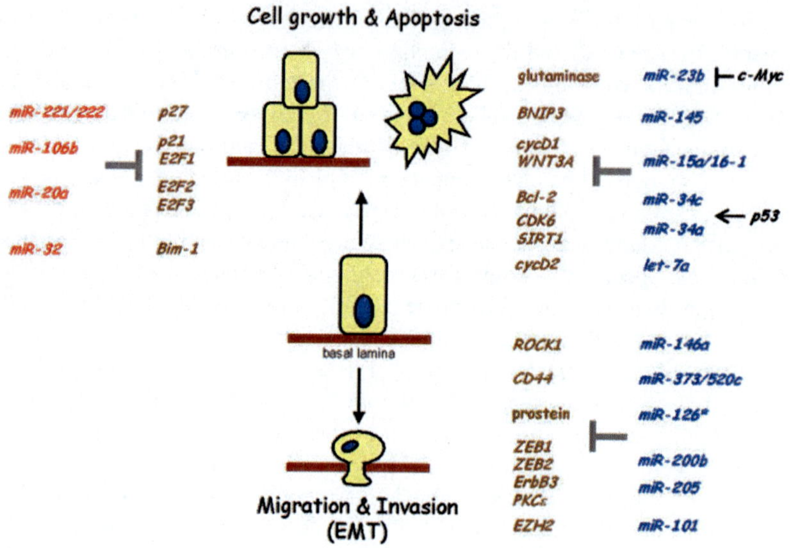

Fig. 6.1 Summary of microRNAs thought to play an oncogenic or tumor-suppressive function in prostate cancer. MicroRNAs are defined as oncogenic or tumor suppressive on the basis of their functional target genes in a given tissue, which may be tumor-suppressor genes or oncogenes. For example, *miR-205* may act as a tumor suppressor by inhibiting expression of the transcriptional repressors of E-cadherin *ZEB1* and *ZEB2*, as well as that of the PKCε. Gandellini et al. (2009b) showed that *miR-205* can actually revert epithelial-mesenchymal transition (EMT) in prostate cancer cells, thus reducing cell migration and invasion. Likewise, *miR-106b* may act as an oncogene by targeting the cell cycle inhibitor $p21^{Waf1}$. When *miR-106b* is over-expressed, as in prostate cancer, $p21^{Waf1}$ would be suppressed and cell growth enhanced (Fig. 6.1 modified from Gandellini et al. 2010)

models of the disease (Fig. 6.1). Specifically, the oncogenic role of selected miRNAs has been mainly associated to their ability to enhance the proliferative potential of PCa cells through the suppression of cell-cycle-dependent kinase inhibitors $p21^{Waf1}$ and $p27^{Kip1}$ (Ambs et al. 2008; Galardi et al. 2007), as well as members of the E2F family of transcription factors (Sylvestre et al. 2007).

A putative oncogenic function was proposed for the *miR-106b-25* cluster and *miR-32* (Ambs et al. 2008), which were shown to directly repress the tumor-suppressors *E2F1*, *p21*, and *Bim* in PCa cell lines. In addition, infection of 22Rv1 PCa cells with a lentivirus encoding the *miR-106b-25* cluster significantly inhibited caspase-3/7 activation induced by doxorubicin and etoposide, thus confirming an anti-apoptotic role for these miRNAs. The intronic *miR-106b-25* cluster was also found to cooperate with its host gene *MCM7* in cellular transformation both in vitro and in vivo, so that the concomitant overexpression of *MCM7* and the miRNA cluster triggered prostatic intraepithelial neoplasia in transgenic mice (Poliseno et al. 2010).

A member of another well-known oncogenic miRNA cluster (*miR-17-92*), namely *miR-20a*, was reported to decrease apoptosis in PC-3 cells through the repression of E2F2 and E2F3, whereas its inhibition by an antisense oligonucleotide resulted in increased cell death after doxorubicin treatment (Sylvestre et al. 2007).

PCa-related miRNAs have more often been described as tumor suppressors, which is not surprising given that most deregulated miRNAs in PCa seem to be down-regulated (Gandellini et al. 2009a). The tumor-suppressive role of miRNAs in PCa has been generally ascribed to their ability to interfere with cell migration and invasion or, alternatively, to their pro-apoptotic functions. Transfection of *miR-126** into LNCaP androgen-dependent PCa cells strongly reduced the translation of prostein (*SLC45A3*), a recently identified prostate-specific antigen that presumably sustains the motility and invasiveness of PCa (Musiyenko et al. 2008). Similar effects were associated with the ectopic expression of *miR-373* and *miR-520c*, which target the adhesion receptor CD44 (Yang et al. 2009a). *MiR-146a* was shown to suppress the Rho-associated protein kinase ROCK1 in PC-3 androgen-independent PCa cells, consequently markedly reducing cell proliferation, invasion, and metastasis to human bone marrow endothelial cell monolayers (Lin et al. 2008). A number of groups have ascribed pro-apoptotic functions to *miR-34a*, a target of the tumor suppressor p53. In this context, ectopic *miR-34a* expression was reported to induce cell cycle arrest and senescence, inhibit cell growth, and attenuate resistance to the chemotherapeutic agent camptothecin by inducing apoptosis in PC-3 cells through the repression of the deacetylase sirtuin and cyclin-dependent kinase 6 (*CDK6*) (Fujita et al. 2008; Lodygin et al. 2008). Ectopic expression of another member of the *miR-34* family, *miR-34c*, was shown to decrease cellular proliferation rate and increase apoptosis by negatively regulating the oncogenes *E2F3* and *BCL-2*. In contrast, blocking *miR-34c* in vitro increased cell growth (Hagman et al. 2010). *E2F2* and *CCND2* were demonstrated to be direct targets of *let-7a*, which inhibited cell proliferation in PC-3 and LNCaP cells by inducing a cell cycle arrest at the G1/S phase (Dong et al. 2010).

Growth-modulatory functions have also been reported for *miR-23b*, *miR-145*, and *miR-100*, as their ectopic expression in LNCaP cells significantly reduced the proliferative potential (Tong et al. 2009). However, an extensive investigation of the oncogenes targeted by these miRNAs in PCa is still lacking. As regards *miR-23b*, it has been recently shown that it is transcriptionally repressed by c-myc in PC-3 cells, thus resulting in the up-modulation of its target mitochondrial glutaminase (Gao et al. 2009). Enhanced glutamine catabolism, as well as altered glucose metabolism (referred to as the Warburg effect (Deberardinis et al. 2008)), confers to cancer cells the capability to grow at a higher rate than normal cells and also to survive in hypoxic conditions. Presumably, restoration of *miR-23b* expression in PCa cells might counteract pro-survival metabolic alterations. More recently, BCL2/adenovirus E1B 19-kDa interacting protein 3 (*BNIP3*), which is over-expressed in various tumors including PCa, was demonstrated to be a direct target of *miR-145*. Artificial over-expression of this miRNA by using adenoviral vectors in PC-3 and DU145 androgen-independent PCa cells significantly down-regulated *BNIP3*, reduced cell growth and increased cell death (Chen et al. 2010).

Varambally et al. (2008) demonstrated that the expression and function of EZH2, a mammalian histone methyltransferase that contributes to epigenetic silencing of target genes and regulates the survival and metastatic potential of PCa cells (Varambally et al. 2002), are inhibited by *miR-101*, whose expression in PCa decreases during progression mainly owing to loss of one or both of its two genomic loci. Similarly, deletion of chromosomal region 13q14 in advanced PCa results in significantly decreased expression levels of *miR-15a* and *miR-16-1*, which were shown to act as tumor suppressors in chronic lymphocytic leukemia by targeting the oncogene *Bcl-2* (Cimmino et al. 2005). Bonci et al. (2008) demonstrated that, in human prostate, the *miR-15a-miR-16-1* cluster can control cell survival, proliferation and invasion by suppressing cyclin D1 and WNT3A. Efficient reconstitution of *miR-15a* and *miR-16-1* expression was achieved by injecting a lentiviral vector expressing the miRNA precursor into LNCaP human PCa xenografts in mice and resulted in marked tumor regression. In addition, the delivery of antagomirs specific for *miR-15a* and *miR-16* was able to induce hyperplasia in normal mouse prostate and promote the growth and invasiveness of untransformed RWPE-1 prostate cells, which became tumorigenic in immunodeficient NOD-SCID mice (Bonci et al. 2008). More recently, injection of *miR-16* with atelocollagen *via* the tail vein was also shown to significantly inhibit the growth of prostate tumors in bone, likely due to the suppression of cyclin-dependent kinases *CDK1* and *CDK2* (Takeshita et al. 2010).

Metastasis is a multistep process that begins when cancer cells break away from their neighboring epithelial and stromal cells and invade through the basement membrane. Cancer cells can acquire motility and invasiveness through a process termed epithelial-mesenchymal transition (EMT), a hallmark of which is the functional loss of E-cadherin. Notably, in most cancers of epithelial origin including PCa, E-cadherin is lost concomitantly with progression towards malignancy, causing the resulting tumors to exhibit an essentially mesenchymal phenotype. It has been recently shown that members of the *miR-200* family are significantly down-regulated in EMT induced by platelet-derived growth factor-D in PC-3 cells and that restoring *miR-200b* expression could reverse this phenotype by suppressing ZEB1, ZEB2 and snail2, which function as transcriptional repressors of E-cadherin (Kong et al. 2009). In human prostate, however, a major role in regulating EMT has been ascribed to *miR-205*, whose expression is reduced in PCa compared to normal tissues (Gandellini et al. 2009b), as well as in a variety of cell lines derived from human tumors. Ectopic expression of *miR-205* in PCa cells was shown to induce cell rearrangements consistent with a mesenchymal-epithelial transition, such as up-regulation of E-cadherin and reduction of cell locomotion and invasion. In addition, *miR-205* suppressed several oncogenes involved in disease progression, such as those encoding interleukin-6, caveolin-1 and EZH2 (Gandellini et al. 2009b). Gandellini et al. suggested that the tumor-suppressive functions of this miRNA could be ascribable to the concurrent repression of specific predicted *miR-205* targets, namely N-chimaerin, ErbB3, E2F1, E2F5, ZEB2 and protein kinase Cε (PKCε). Surprisingly, PKCε appeared to play a direct role in EMT. It is conceivable that reduction of *miR-205* expression levels could represent an oncogenic event that drives the progression towards a cell phenotype with enhanced invasive properties, which presumably favor metastasis. Consistent with this hypothesis, metastasis-derived PCa cell lines express almost undetectable levels of *miR-205,* and tumors from patients with lymph node dissemination of the disease are characterized by lower *miR-205* expression than node-negative patients (Gandellini et al. 2009b).

Given that *miR-205* down-regulation has been reported also in other cancers of epithelial origin, such as breast carcinoma (Wu and Mo 2009), therapies based on the restoration of this miRNA could be investigated for a variety of tumors for which a tumor-suppressive function of *miR-205* has been documented. However, it is not always possible to generalize this concept. In some instances, miRNAs described as therapeutic targets or tools in a specific tumor histotype may not be equally valuable for others. For example, *miR-21* has been proposed as an oncogenic miRNA as well as a therapeutic target in different human cancers (Krichevsky and Gabriely 2009). Nonetheless, Folini et al. showed that *miR-21* knockdown is not sufficient per se to affect the proliferative and invasive potential or the chemosensitivity and radiosensitivity profiles of PCa cells (Folini et al. 2010). In addition, *miR-21* was not differentially expressed in carcinomas and matched normal tissues obtained from patients subjected to radical prostatectomy (Folini et al. 2010). In general, the potential role of a given miRNA as a therapeutic target should always be characterized in the context of the specific disease entity. In addition, the use of experimental models that often do not perfectly recapitulate the features of clinical tumors represents a limiting factor in the definition of the actual role of specific miRNAs in PCa. Even if endowed with oncogenic or tumor-suppressive roles in PCa cell lines, some miRNAs might not be crucial for the pathogenesis of clinical PCa. For example, oncogenic functions have been described in experimental models for *miR-125b* and *miR-221/222* (Galardi et al. 2007; Shi et al. 2007), and an over-expression of these miRNAs could be expected in PCa. Surprisingly, several reports have instead indicated that *miR-125b* and *miR-221/222* are more often down-regulated in PCa than in non-neoplastic tissues (Table 6.1; Gandellini et al. 2009a). For this reason, an extensive characterization of both the expression and the function of candidate miRNAs in a given disease is warranted before translating these molecules into therapy.

6.4 Future Implications for Personalized Prostate Cancer Care

Current understanding of the role exerted by specific miRNAs in the initiation and progression of PCa is still at an early stage, and only scanty evidence that causatively links deregulation of an miRNA to disease development has been obtained. In addition, a consensus on which miRNAs are specific biomarkers for PCa has not yet been reached owing to only partially overlapping of the results collected thus far. However, evidence that miRNAs are highly stable in formalin-fixed tissues (Li et al. 2007) and that their expression levels can be directly measured not only in prostatectomies but also in core biopsies (Mattie et al. 2006) strongly suggests their potential to contribute to the diagnosis of PCa and, eventually, to classify high- and low-risk patients. Specifically, the use of miRNA-related information could permit the pathologist to provide a more accurate grading of the tumor. In addition, miRNAs could be integrated in new prognostic tools to improve the predictive power. In fact, a more precise stratification of patients according to their risk would assist the physician in choosing the best care.

The potential to use miRNAs as biomarkers is also supported by their stability in biological fluids (Mitchell et al. 2008). The possibility to detect cancer-related miRNAs in blood represents the rationale for setting up non-invasive systems for early PCa detection and disease monitoring. This issue could be of potential relevance for promoting active surveillance strategies as well as for monitoring disease recurrence after radical treatment, together with traditional tools such as PSA determination in patients with low-risk disease. Hence, once a clearer overview of the pathological significance of each alteration in miRNA expression is reached and the statistical and clinical value of any deregulation is confirmed in appropriate patient cohorts, selected PCa-related miRNAs might be appealing candidates as new diagnostic and prognostic markers.

In addition, a detailed understanding of the role exerted by specific miRNAs in the development and progression of PCa will be instrumental for the development of miRNA-based therapeutics. The possibility to use miRNA agonists or antagonists, in order to restore or inhibit the function of down-regulated onco-suppressive miRNAs and up-regulated oncogenic miRNAs, respectively, has been already successfully demonstrated in experimental models of human cancers (Gandellini et al. 2009a). However, a crucial factor for the development of successful miRNA-based treatment modalities is the efficient in vivo delivery of miRNA modulators. In this context, reconstitution of *miR-15a* and *miR-16-1* expression has been effectively achieved by injecting a lentiviral vector expressing the miRNA precursor into prostate tumor xenografts in mice (Bonci et al. 2008). In addition, a proof-of-concept of the antagomir treatment has been successfully demonstrated, as tumor growth in a neuroblastoma mouse model was abolished by the injection of antago*miR-17-5p* (Fontana et al. 2008). To the best of our knowledge, there are no ongoing miRNA-based clinical trials in cancer patients. However, a safety study (Phase I) in healthy male volunteers (NCT00688012, ClinicalTrial.gov) has been recently carried out by Santaris Pharma (Hoersholm, Denmark) with SPC3649, a LNA oligomer targeting *miR-122* as a potential new approach for the treatment of hepatitis C infection, but the results have not been published yet. This kind of studies will ascertain the feasibility and assess eventual toxicity of therapies involving miRNAs as targets or tools. In addition, the use of systems displaying carrier-defined specificity (e.g., cell specific immunoliposomes targeting antigens highly expressed on the PCa cell membrane) might represent a suitable approach for controlled release of therapeutic molecules to relevant tissues, thereby avoiding toxic side effects. Prostate-specific membrane antigen, which is specifically expressed on prostate epithelial cells and strongly up-regulated in PCa, has been proven to be a suitable surface molecule to drive PCa-specific delivery of small RNAs (McNamara et al. 2006).

Since miRNAs have the potential to modulate a cohort of cancer-relevant gene networks, they might become therapeutically relevant in a "one-hit multi-target" context against PCa. However, the multispecific nature of miRNA-based gene regulation is both a strength and a weakness, as interfering with miRNA expression might give rise to unpredictable off-target effects thus making imperative the precise definition of the key targets of a given miRNA in PCa cells. In this context, in vivo knock-out or knock-in models should be helpful to understand the pathways that are affected by miRNA manipulations. In addition, gene expression profiling (Creighton et al. 2008) and proteomic analysis (Yang et al. 2009b) associated to loss- and gain-of-function studies in in vivo or in vitro models might contribute to a better understanding of miRNA-controlled cellular networks in a given disease and identify biologically relevant targets.

In the therapeutic setting, interference with PCa-specific miRNAs could be exploited not only to produce a direct anti-cancer effect but also to improve the response of tumor cells to conventional anti-cancer agents, since it has been recently suggested that specific miRNAs may contribute to chemoresistance in different human cancer cell lines and that the modulation of their expression increases cell sensitivity to cytotoxic drugs with different action mechanisms as well as to ionizing radiation (Weidhaas et al. 2007). For instance, transfection with *miR-148a* increased the sensitivity of PC-3 cells to paclitaxel as well as attenuated paclitaxel-resistance of hormone-refractory, drug-resistant PC-3PR cells, in part by regulating the expression of mitogen- and stress-activated protein kinase *MSK1* (Fujita et al. 2010). Altered expression of specific miRNAs could also provide information about sensitivity or resistance of individual tumors to different treatments before starting therapy (response prediction). Again, changes in the expression of specific miRNAs during therapy might offer a tool for monitoring treatment success (response control).

Although a considerable investment in new research is required before a possible use of miRNAs as novel biomarkers and new therapeutic targets or intervention tools can be formulated, the rapid expansion of the field, together with the promising results obtained in other tumor types, suggest an important impact of these molecules for the management of PCa patients in the future.

References

Ambs S, Prueitt RL, Yi M, et al. Genomic profiling of microRNA and messenger RNA reveals deregulated microRNA expression in prostate cancer. Cancer Res. 2008;68:6162–70.

Andriole GL, Crawford ED, Grubb RL 3rd, et al. PLCO project team. Mortality results from a randomized prostate-cancer screening trial. N Engl J Med. 2009;360:1310–9.

Bonci D, Coppola V, Musumeci M, et al. The miR-15a-miR-16-1 cluster controls prostate cancer by targeting multiple oncogenic activities. Nat Med. 2008;14:1271–7.

Calin GA, Croce CM. MicroRNA signatures in human cancers. Nat Rev Cancer. 2006;6:857–66.

Chen X, Gong J, Zeng H, et al. MicroRNA145 targets BNIP3 and suppresses prostate cancer progression. Cancer Res. 2010;70:2728–38.

Cimmino A, Calin GA, Fabbri M, Iorio MV, et al. MiR-15 and miR-16 induce apoptosis by targeting BCL2. Proc Natl Acad Sci USA. 2005;102:13944–9.

Cortez MA, Calin GA. MicroRNA identification in plasma and serum: a new tool to diagnose and monitor diseases. Expert Opin Biol Ther. 2009;9:703–11.

Creighton CJ, Nagaraja AK, Hanash SM, et al. A bioinformatics tool for linking gene expression profiling results with public databases of microRNA target predictions. RNA. 2008;14:2290–6.

Damber JE, Aus G. Prostate cancer. Lancet. 2008;371:1710–21.

Deberardinis RJ, Sayed N, Ditsworth D, et al. Brick by brick: metabolism and tumor cell growth. Curr Opin Genet Dev. 2008;18:54–61.

Dong Q, Meng P, Wang T, et al. MicroRNA let-7a inhibits proliferation of human prostate cancer cells in vitro and in vivo by targeting E2F2 and CCND2. PLoS One. 2010;5:e10147.

Elmen J, Thonberg H, Ljungberg K, et al. Locked nucleic acid (LNA) mediated improvements in siRNA stability and functionality. Nucleic Acids Res. 2005;33:439–47.

Folini M, Gandellini P, Longoni N, et al. MiR-21: an oncomir on strike in prostate cancer. Mol Cancer. 2010;9:12.

Fontana L, Fiori ME, Albini S, et al. Antagomir-17-5pabolishes the growth of therapy-resistant neuroblastoma through p21 and BIM. PLoS One. 2008;3:e2236.

Fujita Y, Kojima K, Hamada N, et al. Effects of miR-34a on cell growth and chemoresistance in prostate cancer PC3 cells. Biochem Biophys Res Commun. 2008;377:114–9.

Fujita Y, Kojima K, Ohhashi R, et al. MiR-148a attenuates paclitaxel-resistance of hormone-refractory, drug-resistant prostate cancer PC3 cells by regulating MSK1 expression. J Biol Chem. 2010;285:19076–84.

Galardi S, Mercatelli N, Giorda E, et al. MiR-221 and miR-222 expression affects the proliferation potential of human prostate carcinoma cell lines by targeting p27Kip1. J Biol Chem. 2007;282:23716–24.

Gandellini P, Folini M, Longoni N, et al. MiR-205 exerts tumor-suppressive functions in human prostate through down-regulation of protein kinase Cepsilon. Cancer Res. 2009b;69: 2287–95.

Gandellini P, Folini M, Zaffaroni N. Towards the definition of prostate cancer-related microRNAs: where are we now? Trends Mol Med. 2009a;15:381–90.

Gandellini P, Folini M, Zaffaroni N. Emerging role of microRNAs in prostate cancer: implications for personalized medicine. Discov Med. 2010;9:212–8.

Gao P, Tchernyshyov I, Chang TC, et al. c-Myc suppression of miR-23a/b enhances mitochondrial glutaminase expression and glutamine metabolism. Nature. 2009;458:762–5.

Hagman Z, Larne O, Edsjö A, et al. MiR-34c is down regulated in prostate cancer and exerts tumor suppressive functions. Int J Cancer. 2010;127:2768–76.

Jemal A, Siegel R, Xu J, et al. Cancer statistics, 2010. CA Cancer J Clin. 2010;60:277–300.

Kong D, Li Y, Wang Z, Banerjee et al. MiR-200 regulates PDGF-D-mediated epithelial-mesenchymal transition, adhesion, and invasion of prostate cancer cells. Stem Cells 2009;27:1712–21.

Krichevsky AM, Gabriely G. MiR-21: a small multi-faceted RNA. J Cell Mol Med. 2009;13: 39–53.

Krützfeldt J, Rajewsky N, Braich R, et al. Silencing of microRNAs in vivo with 'antagomirs'. Nature. 2005;438:685–9.

Leite KR, Sousa-Canavez JM, Reis ST, et al. Change in expression of miR-let7c, miR-100, and miR-218 from high grade localized prostate cancer to metastasis. Urol Oncol. 2010. doi:10.1016/j.urolonc.2009.02.002.

Li J, Smyth P, Flavin R, et al. Comparison of miRNA expression patterns using total RNA extracted from matched samples of formalin-fixed paraffin-embedded (FFPE) cells and snap frozen cells. BMC Biotechnol. 2007;7:36.

Lin SL, Chiang A, Chang D, et al. Loss of miR-146a function in hormone-refractory prostate cancer. RNA. 2008;14:417–24.

Lodes MJ, Caraballo M, Suciu D, et al. Detection of cancer with serum miRNAs on an oligonucleotide microarray. PLoS One. 2009;4:e6229.

Lodygin D, Tarasov V, Epanchintsev A, et al. Inactivation of miR-34a by aberrant CpG methylation in multiple types of cancer. Cell Cycle. 2008;7:2591–600.

Lu J, Getz G, Miska EA, et al. MicroRNA expression profiles classify human cancers. Nature. 2005;435:834–8.

Mattie MD, Benz CC, Bowers J, et al. Optimized high-throughput microRNA expression profiling provides novel biomarker assessment of clinical prostate and breast cancer biopsies. Mol Cancer. 2006;5:24.

McNamara JO 2nd, Andrechek ER, Wang Y, et al. Cell type-specific delivery of siRNAs with aptamer-siRNA chimeras. Nat Biotechnol. 2006;24:1005–15.

Mitchell PS, Parkin RK, Kroh EM, et al. Circulating microRNAs as stable blood-based markers for cancer detection. Proc Natl Acad Sci USA. 2008;105:10513–8.

Musiyenko A, Bitko V, Barik S. Ectopic expression of miR-126*, an intronic product of the vascular endothelial EGF-like 7 gene, regulates prostein translation and invasiveness of prostate cancer LNCaP cells. J Mol Med. 2008;86:313–22.

Ozen M, Creighton CJ, Ozdemir M, et al. Widespread deregulation of microRNA expression in human prostate cancer. Oncogene. 2008;27:1788–93.

Poliseno L, Salmena L, Riccardi L, et al. Identification of the miR-106b~25 microRNA cluster as a proto-oncogenic PTEN-targeting intron that cooperates with its host gene MCM7 in transformation. Sci Signal. 2010;3:ra29.

Porkka KP, Pfeiffer MJ, Waltering KK, et al. MicroRNA expression profiling in prostate cancer. Cancer Res. 2007;67:6130–5.

Prueitt RL, Yi M, Hudson RS, et al. Expression of microRNAs and protein-coding genes associated with perineural invasion in prostate cancer. Prostate. 2008;68:1152–64.

Schaefer A, Jung M, Mollenkopf HJ, et al. Diagnostic and prognostic implications of microRNA profiling in prostate carcinoma. Int J Cancer. 2010;126:1166–76.

Schröder FH, Hugosson J, Roobol MJ, et al. ERSPC investigators. Screening and prostate-cancer mortality in a randomized European study. N Engl J Med. 2009;360: 1320–8.

Shi XB, Xue L, Yang J, et al. An androgen-regulated miRNA suppresses Bak1 expression and induces androgen-independent growth of prostate cancer cells. Proc Natl Acad Sci USA. 2007;104:19983–8.

Spahn M, Kneitz S, Scholz CJ, et al. Expression of microRNA-221 is progressively reduced in aggressive prostate cancer and metastasis and predicts clinical recurrence. Int J Cancer. 2010; 127:394–403.

Stenvang J, Silahtaroglu AN, Lindow M, et al. The utility of LNA in microRNA-based cancer diagnostics and therapeutics. Semin Cancer Biol. 2008;8:89–102.

Sylvestre Y, De Guire V, Querido E, et al. An E2F/miR-20a autoregulatory feedback loop. J Biol Chem. 2007;282:2135–43.

Szczyrba J, Löprich E, Wach S, et al. The microRNA profile of prostate carcinoma obtained by deep sequencing. Mol Cancer Res. 2010;8:529–38.

Takeshita F, Patrawala L, Osaki M, et al. Systemic delivery of synthetic microRNA-16 inhibits the growth of metastatic prostate tumors via downregulation of multiple cell-cycle genes. Mol Ther. 2010;18:181–7.

Tong AW, Fulgham P, Jay C, et al. MicroRNA profile analysis of human prostate cancers. Cancer Gene Ther. 2009;16:206–16.

Varambally S, Cao Q, Mani RS, et al. Genomic loss of microRNA-101 leads to overexpression of histone methyltransferase EZH2 in cancer. Science. 2008;322:1695–9.

Varambally S, Dhanasekaran SM, Zhou M, et al. The polycomb group protein EZH2 is involved in progression of prostate cancer. Nature. 2002;419:624–9.

Volinia S, Calin GA, Liu CG, et al. A microRNA expression signature of human solid tumors defines cancer gene targets. Proc Natl Acad Sci USA. 2006;103:2257–61.

Weidhaas JB, Babar I, Nallur SM, et al. MicroRNAs as potential agents to alter resistance to cytotoxic anticancer therapy. Cancer Res. 2007;67:11111–6.
Wu H, Mo YY. Targeting miR-205 in breast cancer. Expert Opin Ther Targets. 2009;13:1439–48.
Yang K, Handorean AM, Iczkowski KA. MicroRNAs 373 and 520c are downregulated in prostate cancer, suppress CD44 translation and enhance invasion of prostate cancer cells in vitro. Int J Clin Exp Pathol. 2009;2:361–9.
Yang Y, Chaerkady R, Beer MA, et al. Identification of miR-21 targets in breast cancer cells using a quantitative proteomic approach. Proteomics. 2009;9:1374–84.

Chapter 7
MicroRNAs in Hepatocellular Carcinoma

Junfang Ji and Xin Wei Wang

Abstract Hepatocellular carcinoma (HCC) is a major type of liver cancer. It is the third leading cause of cancer-related death worldwide. MicroRNAs (miRNAs), a class of small non-coding RNAs, are aberrantly expressed in HCC. The genomic instability, transcriptionally regulation, and epigenetic alteration have been identified to contribute to the abnormal expression of miRNAs in HCC. Moreover, deregulation of miRNAs not only functionally enables liver tumorigenesis, but also promotes tumor progression. Further, aberrant expression of certain miRNAs is correlated with clinical features of HCC, indicating their potential to serve as diagnostic and prognostic biomarkers of HCC. Several miRNAs have been validated in independent cohorts as HCC biomarkers, paving the way for developing clinically useful platforms to aid in diagnosis and prognosis for patients, and to assist in HCC patient stratification with the potential for personalized adjuvant therapy. In addition, several oncogenic and tumor suppressive miRNAs are being exploited as possible therapeutic targets in HCC. In this chapter, recent studies on miRNA and HCC, particularly those that are clinically relevant, are discussed.

7.1 Introduction

7.1.1 Hepatocellular Carcinoma

Primary liver cancer, predominantly hepatocellular carcinoma (HCC), is the third most deadly and fifth most common malignancy in the world. Although it is predominant in sub-Saharan Africa and Eastern Asia, its incidence has been increasing in the West over the last 20 years (El-Serag and Mason 1999; El-Serag et al. 2003). Potentially curative therapies such as surgical resection and liver transplantation are only available to 5–15% of HCC patients based on the evaluation of their liver function and tumor burden (Budhu and Wang 2006; El Serag et al. 2008). Moreover,

X.W. Wang (✉)
Liver Carcinogenesis Section, Laboratory of Human Carcinogenesis, National Cancer Institute, Center for Cancer Research, Bethesda, MD 20892-4258, USA
e-mail: xw3u@nih.gov

W.C.S. Cho (ed.), *MicroRNAs in Cancer Translational Research*,
DOI 10.1007/978-94-007-0298-1_7, © Springer Science+Business Media B.V. 2011

recurrence and metastasis occur commonly after curative therapy and post-operative 5-year survival is only 30–40% (Blum 2005).

HCC normally develops as a consequence of underlying chronic liver disease. 80% of HCCs are associated with cirrhotic liver caused by chronic infection of hepatitis B virus (HBV) and hepatitis C virus (HCV), excessive alcohol intake, exposure to dietary hepatocarcinogenesis like aflatoxins or other chronic conditions (Parikh and Hyman 2007). HCCs are clinically heterogeneous and have numerous genetic alterations (Thorgeirsson and Grisham 2002). Our understanding of HCC has been improved by the studies on molecular profiling to identify changes in gene expression that are associated with particular phenotypes, such as HCC subtypes, recurrence or metastasis (Boyault et al. 2007; Budhu et al. 2006; Budhu et al. 2008; Hoshida et al. 2008; Lee et al. 2006; Yamashita et al. 2008; Yamashita et al. 2009; Ye et al. 2003). More recently, a growing body of evidence show that microRNAs are aberrantly expressed in cancer including HCC, and are associated with HCC phenotypes. Future HCC diagnosis, prognosis and therapeutic intervention can take advantage of these new findings.

7.1.2 MicroRNAs

MicroRNAs (miRNAs), a class of small (21–23 nucleotides in length), non-coding, regulatory RNA molecules, were first discovered in 1993. Up to date, thousands of miRNAs have been identified in a wide variety of species including animals, plants, and viruses. It is estimated that there are about 1,000 miRNA genes in the human genome (http://microrna.sanger.ac.uk; Release 15: May 2010) and approximately 600 miRNA genes have been identified.

7.1.2.1 Biogenesis of MiRNAs

MiRNA genes are encoded in intergenic chromosomal regions or within introns of protein coding genes. Generally they are transcribed by RNA polymerase II or III into longer primary-miRNAs (pri-miRNAs) with a characteristic of stem loop structure. These transcripts are subsequently processed in nucleus to small (approximately 70 nucleotides), stem-looped, hairpin-like miRNA precursors (pre-miRNAs) by the Drosha (nuclease)-DGCR8 (double-stranded RNA binding protein) microprocessor complex (Borchert, Lanier and Davidson 2006; Lee et al. 2003; Lee et al. 2004). Alternatively, some miRNAs are derived from intronic stem-loops, known as mirtrons, which are processed by the splicesome to pre-miRNAs (Berezikov et al. 2007; Ruby et al. 2007). Through an Exportin-5/Ran complex, pre-miRNAs are exported to the cytoplasm where they are further processed by Dicer-TRBP (double-stranded RNA binding protein) into an imperfect duplex (19–23 nucleotides) consisting of mature miRNAs and a complementary fragment with similar size (miRNAs*) (Lee et al. 2003; Lund et al. 2004; Yi et al. 2003). The mature miRNA strand is loaded onto the RNA-induced silencing complex (RISC) while miRNA* is separated from the duplex and is degraded (Hutvagner and Zamore 2002; Lee et al. 2002). Mature miRNA in RISC base pairs its partially complementary target mRNA within a 3′ untranslated region (3′UTR) of mRNA and regulates its stability or translation (Hutvagner and Zamore 2002). Specifically, only nucleotides 2–8 in the 5′ terminus of the miRNA, called the "seed sequence", is critical for target recognition (Brennecke et al. 2005; Lewis, Burge and Bartel 2005; Lewis et al. 2003). Target recognition by miRNA through imperfect base pairing enables miRNAs to target multiple genes simultaneously. The discovery of miRNA has opened a new era of understanding in post-transcriptional regulation of genes. One miRNA can target multiple mRNA targets due to imperfect base pairing with its targets and it is estimated that as many as 90% of human genes may be regulated by miRNAs.

7.1.2.2 Detection of MiRNA Expression and Function

The development of new technologies offers opportunities to examine the expression and function of miRNAs in cellular biological processes and pathological processes. Publications accumulating at a rapid pace have shown that miRNAs act as key regulatory molecules that control a wide variety of biological processes such as development, differentiation, cell motility, apoptosis and cell cycle, and their deregulation is related to many diseases including solid and hematological malignancies, diabetes and diseases of the nervous system.

A number of different techniques have been developed to examine miRNA expression. Microarray technology is a powerful high-throughput tool to monitor the expression of thousands of miRNAs at once (Liu et al. 2004). Quantitative reverse transcription-polymerase chain reaction (qRT-PCR) is another reliable and highly sensitive technique for miRNA detection, which is simple and robust, and only requires very small amounts of input total RNA. Standard northern blotting is proposed for detecting and validating candidate miRNA. In addition, miRNAs can be detected using in situ hybridization. Most recently, miRNAs can be profiled by the next generation sequencing technologies that allow the measurement of their absolute abundance as well as the discovery of novel miRNAs.

The elucidation of miRNA targets can be achieved bioinformatically by a number of available free programs, such as miRBase (www.microRNA.org), or Targetscan (www.targetscan.org/). It should be noted that these programs only predict putative targets using sets of algorithms based on the principle of the miRNA seed regions and their complementary sites at the 3′UTRs of mRNA sequences. To explore the miRNAs' function, it is important to confirm these predictions using miRNA target validation techniques, which are convincingly achieved by several assays. First, endogenous miRNAs can be silenced by their inhibitors or exogenously they can be introduced into cells, which could lead to increased or decreased levels of their putative targets, respectively. Second, miRNA regulating targets can be mimicking using a luciferase reporter with an inserted 3′UTR of the targeting mRNA. Many methods can be used for altering miRNA expression in vitro and in vivo, such as the vector based miRNA expression systems or antagomir (Krutzfeldt et al. 2005; Voorhoeve et al. 2006). A direct alteration of miRNA expression can be achieved by transfecting cells with the miRNA precursor sequence (for miRNA over-expression) or the antisense of miRNA (for miRNA silencing).

7.2 Expression and Function of MiRNAs in HCC

A high proportion of miRNA genes are encoded at or near fragile genomic sites which are prone to breakage and rearrangement in cancer cells (Calin et al. 2004). Deregulation of miRNA expression is associated with human malignancies including HCC (Ji and Wang 2009). Some miRNAs function as tumor suppressor genes or oncogenes; these are collectively termed "oncomirs".

7.2.1 MiRNAs are Aberrantly Expressed in HCCs

The creation of miRNA high-throughput analysis accelerates the identification of aberrant-expressed miRNAs in tumor. Differential expression of many miRNAs has been found in HCC tissues when compared to normal tissues including adjacent non-tumor liver tissues, cirrhotic and normal liver tissues. Moreover, distinct miRNA profiles have been identified in different HCC subtypes such as cancer stem cell (CSC) like subtype vs. mature hepatocyte subtype, metastatic subtype vs. non-metastatic subtype, viral vs. non-viral HCC, or HBV-HCC subtype vs. HCV-HCC subtype (Table 7.1).

The first miRNA profiling was done in 24 HCCs and 22 non-HCC tissues and eight miRNAs were revealed to be differentially expressed (Murakami et al. 2006). Subsequently, several other groups also identified miRNA signatures that differentiate cancerous from adjacent non-cancerous liver tissues (Chung et al. 2010; Huang et al. 2008b; Jiang et al. 2008; Li et al. 2008b; Meng et al. 2007; Wang et al. 2008). Two groups developed miRNA signatures by using HCC tissues and normal liver tissues (Su et al. 2009; Yang et al. 2010). The differential miRNA profiles are also identified between cirrhosis liver and HCC tissues (Gramantieri et al. 2007; Wong et al. 2010). Moreover, differentially expressed miRNAs with chromosome amplification or deletion were recently reported (Ding et al. 2010). Notably, each study comes up with a unique profile, sharing only few miRNAs in common, which may be due to differences in ethnicity and/or techniques used.

Remarkably, a global profiling over 600 available miRNAs in a large HCC cohort, including 241 HCC cases, revealed that as much as one third of them were significantly altered in HCC (Budhu et al. 2008). However, the expression of these miRNAs was heterogeneous. These results indicate that deregulation of certain miRNAs is associated with a unique tumor biology. Accordingly, miRNA profiling was compared between different HCC subgroups. A 20-miRNA signature including *miR-122*, *let-7g* and *miR-338* was first identified to be differentially expressed in HCCs with metastasis and non-metastatic HCCs (Budhu et al. 2008). Other miRNAs

Table 7.1 Hepatocellular carcinoma (HCC)-associated microRNA signatures

Relevant microRNAs	Description	References
8-miRNA signature (*miR-18, miR-199a*, miR-199a*)	Differ in tumor vs. non-tumor	Murakami et al. (2006)
15-miRNA signature (*miR-21, miR-34a, miR-92-1*)	Differ in tumor vs. non-tumor	Meng et al. (2007)
40-miRNA signature	Differ in tumor vs. non-tumor	Huang et al. (2008b)
69-miRNA signature	Differ in tumor vs. non-tumor	Li et al. (2008b)
22-miRNA signature	Differ in tumor vs. non-tumor	Wang et al. (2008)
16-miRNA signature (*miR-135a, miR-33, miR-221*)	Differ in tumor vs. non-tumor	Jiang et al. (2008)
31 miRNAs (*let-7a, miR-122, miR-338*)	Differ in tumor vs. non-tumor	Huang et al. (2009)
21 miRNAs (*miR-15b, miR-105, miR-339*)	Differ in tumor vs. non-tumor	Chung et al. (2010)
22 miRNAs (*miR-151-5p, miR-96, miR-486*)	Amplification or deletion in tumor vs. non-tumor	Ding et al. (2010)
29-miRNA signature	Differ in HCC vs. normal liver	Su et al. (2009)
14 miRNAs (*miR-143, miR-210, miR-602*)	Differ in HBV-HCC, chronic HBV hepatitis, HBC-cirrhotic and normal para-tumorous liver vs. normal liver	Yang et al. (2010)
35-miRNA signature	Differ in HCC vs. cirrhosis liver	Gramantieri et al. (2007)
40 miRNAs (*miR-222, miR-301, miR-199**)	Differ in tumor vs. adjacent cirrhotic liver tissues	Wong et al. (2010)
29-miRNA signature	Differ in tumor vs. non-tumor from HCV-HCC	Varnholt et al. (2008)
20-miRNA signature (*miR-30c-1, miR-148a, miR-34a*)	Associate with metastasis	Budhu et al. (2008)
20-miRNA signature (*miR-181a-1, miR-181b-1, miR-181b-2*)	Differ in HpSC-HCC vs. MH-HCC	Ji et al. (2009b)
19 miRNAs (*miR-211, miR-134, miR-105*)	Differ in HBV-HCC vs. HCV-HCC	Ura et al. (2008)
31 miRNAs	Associate with HCC staging	Ura et al. (2008)
6 miRNAs (*miR-200c, miR-21, miR-203*)	Differ in benign tumor vs. malignant HCC	Ladeiro et al. (2008)

including *miR-181*, *let-7g*, and *miR-221* are differentially expressed between a subtype with cancer stem cell features and a subtype with mature hepatocyte features (Ji et al. 2009b). Other studies have revealed that there are 6 miRNAs including *miR-200c*, *miR-21*, and *miR-203* with differential expression between malignant HCC and benign HCC. Hepatitis B and C viruses are etiological agents for the development of HCC, and miRNAs that are unique to certain virus-related HCC are identified. By comparing HCV-HCC tissues and adjacent non-HCC tissues, Varnholt et al. (2008) found 29 differentially expressed miRNAs. Ura et al. (2008) revealed 19 miRNAs to be differentially expressed between HBV-HCC and HCV-HCC.

Interestingly, most of these miRNAs are also HCC-associated, indicating dual roles of miRNAs in viral replication and HCC development.

It is important to note that several different groups consistently showed that *let-7s*, *miR-122*, *miR-26*, and *miR-101* are significantly down-regulated while *miR-221*, *miR-181*, and cluster *17-92* are significantly up-regulated in HCC compared to non-cancerous liver tissues. The consistent alteration of these miRNAs in different studies suggests that they are good candidate HCC biomarkers. In addition, the deregulation of certain miRNAs is also identified by some individual studies. For example, *miR-30d*, *miR-195*, *miR-155*, and *miR-143* are increased in HCC while *miR-29* and *miR-375* are decreased in HCC (Table 7.2).

7.2.2 The Mechanism of the MiRNAs Deregulation in HCCs

Recent studies have reported the involvement of both genetic and epigenetic mechanisms of miRNA deregulation in HCC. Genetic mechanisms are usually correlated with genomic alteration including deletion, amplification, single nucleotide polymorphism and mutation of miRNAs, and transcriptional alteration. Epigenetic alteration in cancer, such as methylation, also leads to aberrant expression of miRNAs.

7.2.2.1 Genomic Level

Genome instability is the hallmark of human malignancies including HCC. These chromosomal aberrations lead to critical deregulations in gene expression, structure and function, which are associated with cancer. By analyzing miRNA loci in the common recurrent chromosome aberration regions of HCC, 22 miRNAs are subject to chromosomal gain or loss at the genomic level, and 13 of them were transcriptionally deregulated in HCC, i.e. nine up-regulated miRNAs (*miR-9-1, miR-548a-1, miR-550-2, miR-96, miR-548d-1, miR-151, miR-423, miR-365-2*, and *miR-21*) and four down-regulated miRNAs (*miR-153-2, miR-383, miR-486*, and *miR-138-2*) (Ding et al. 2010).

Mutation of miRNAs in HCC was also found through sequencing analysis of 59 miRNAs with abnormal expression levels in HCC. Sequencing analysis was done in 55 HCCs and only a total of four mutation sites in three miRNAs were found in four HCCs, which were *miR-106b* (2 mutation sites), *miR-192* (1 site), and *let-7a-2* (1 site) (Yang et al. 2008). Although many other studies demonstrate the down-regulation of these miRNAs, it is unknown whether the reduced level of these miRNAs in HCC is due to these mutations.

In addition, a G>C polymorphism was also identified in the stem region opposite to the mature *miR-146a* sequence. HCCs with G-allele display an increased production of mature *miR-146a* when compared to C-allelic cases (Xu et al. 2008). Moreover, HCC cases with *miR-196a2* rs11614913 T-allele was significantly more frequent in male HCC patients with lymphatic metastasis than HCC with C-allele (Qi et al. 2010).

Table 7.2 Hepatocellular carcinoma (HCC)-associated microRNAs

MicroRNAs	Status in HCC	Targets	Associated function	References
miR-122	Down (T vs. NT)	*CCNG1, Bcl-w, ADAM10, SRF, Igf1R, ADAM17*	Inhibit cell growth; promote apoptosis; suppress the metastasis	Bai et al. (2009); Coulouarn et al. (2009); Gramantieri et al. (2007); Lin et al. (2008); Ma et al. (2010); Tsai et al. (2009); Wu et al. (2009)
let-7s	Down (T vs. NT)	*c-myc, lin-28, COL1A2, bcl-xl*	Inhibit cell growth, migration and invasion	Guo et al. (2006); Ji et al. (2010); Lan et al. (2011); Shimizu et al. (2010); Viswanathan et al. (2009)
miR-221	Up (T vs. NT)	*CDKN1B, CDKN1C, DDIT4, Bmf*	Promote tumorigenesis and cell growth, inhibit apoptosis	Fornari et al. (2008); Gramantieri et al. (2009); Pineau et al. (2010)
miR-222	Up (T vs. NT)	*PPP2R2A*	Promote cell motility	Wong et al. (2008, 2010)
miR-181s	Up (HpSC-HCC vs. MH-HCC)	*CDX2, GATA6, NLK, TIMP3*	Promote HCC stemness, cell growth, migration and invasion	Ji et al. (2009b); Wang et al. (2010)
miR-26	Down (T vs. NT)	*CCND2, CCNE2*	Inhibit cell growth, promote apoptosis	Ji et al. (2009a); Kota et al. (2009)
miR-101	Down (T vs. NT)	*Mcl-1, FOS*	Promote apoptosis; inhibit cell growth, migration and invasion	Li et al. (2009b); Su et al. (2009)
miR-17-92	Up (T vs. NT)	*E2F1*	Induce proliferation and anchorage-independent growth	Connolly et al. (2008); Li et al. (2009c)
miR-21	Up (T vs. NT)	*PTEN*	Promote cell proliferation, migration and invasion	Connolly et al. (2008); Meng et al. (2007)
miR-224	Up (T vs. NT)	*API-5*	Inhibit apoptosis and promote cell growth	Wang et al. (2008)
miR-125b	Down (T vs. NT)	–	Inhibit cell growth	Li et al. (2008b)
miR-223	Down (T vs. NT)	*STMN1*	Inhibit cell viability	Wong et al. (2008)
miR-23a~27~24	Up (T vs. NT)	–	Promote cell growth and inhibit apoptosis	Huang et al. (2008a)
miR-106b-25	Up (T vs. NT)	*E2F1, Bim*	Promote cell proliferation	Li et al. (2009c)
miR-18a	UP (female T/NT vs. male T/NT)	*ESR1*	Promote cell growth	Liu et al. (2009)
miR-34a	Down (T vs. NT)	*c-Met*	Inhibit cell migration and invasion	Li et al. (2009a)

Table 7.2 (continued)

MicroRNAs	Status in HCC	Targets	Associated function	References
miR-30d	Up (T vs. NT)	*GNAI2*	Promote tumor invasion and metastasis	Yao et al. (2010)
miR-29	Down (T vs. NT)	*Bcl-2, Mcl-2*	Promote apoptosis	Xiong et al. (2010)
miR-195	Down (T vs. NT)	*CCND1, CDK6, E2F3*	Promote cell growth	Xu et al. (2009)
miR-155	Up (T vs. NT)	*C/EBP beta*	Promote cell growth	Wang et al. (2009)
miR-143	Up (T vs. NT)	*FNDC3B*	Promote tumor invasion and metastasis	Zhang et al. (2009)
miR-23b	–	*uPA, c-met*	Inhibit migration and invasion	Salvi et al. (2009)
miR-602	Up (T vs. NT)	–	Inhibit apoptosis, promote cell growth	Yang et al. (2010)
miR-500	Up (T vs. NT)	–	–	Yamamoto et al. (2009)
miR-375	Down (T vs. NT)	*YAP*	Inhibit proliferation and invasion	Liu, Poon, Luk (2010)
miR-15b	Down (T vs. NT)	*Bcl-w*	Inhibit cell growth, promote apoptosis	Chung et al. (2010)
miR-9	Up (T vs. NT)	–	Promote the cell invasion	Hao-Xiang et al. (2009)
miR-338	Down (T vs. NT)	–	–	Huang et al. (2009)
*miR-126**	Down in alcohol-HCC	–	–	Ladeiro et al. (2008)
miR-96	Up in HBV-HCC	–	–	Ladeiro et al. (2008)
miR-151-5p	Amplification in T	*RhoGDIA*	Promote metastasis	Ding et al. (2010)
miR-106b, miR-192, let-7a-2	Sequence variation	–	–	Yang et al. (2008)
miR-146a	G>C polymorphism	–	Promote cell growth	Xu et al. (2008)
miR-196a2	T>C polymorphism	–	–	Qi et al. (2010)
miR-1	Methylation in T	*Foxp-1, MET*	Inhibit cell growth, promote apoptosis	Datta et al. (2008)
miR-124	Methylation in T	*CDK6, VIM, SMYD3, IQGAP1*	Inhibit cell proliferation	Furuta et al. (2010)
miR-203	Methylation in T	*ABCE1*	Inhibit cell proliferation	Furuta et al. (2010)

7.2.2.2 Transcriptional Level

Due to the similarity of transcriptional mechanism of miRNA and mRNA genes, it is not surprising that miRNAs can be transcriptionally regulated, such as p53 transitionally inducing *miR-34* expression. In HCC, nuclear factor kappa B (NF-κB) signaling was found to induce the expression of *miR-155* through a typical binding to its promoter region (Wang et al. 2009). Meanwhile, the inhibition of this signaling pathway significantly decreased *miR-155* level in HCC cells, indicating the association of up-regulated *miR-155* in HCC with activation of NF-κB signaling. Moreover, NF-κB signaling could also activate *miR-143* expression (Zhang et al. 2009). The level of *miR-143* is dramatically increased in HBV-HCCs but slightly up-regulated in HBV-negative HCCs. It is known that NF-κB transcriptional activity was significantly enhanced through interacting with HBx, indicating that the up-regulation of *miR-143* in HBV-HCC is related to the enhanced NF-κB activity by HBx (Zhang et al. 2009).

It was also shown that a high level of *miR-181* in tumorigenesis is associated with TGF-β signaling and Wnt/β-catenin signaling (Ji et al. 2009b; Wang et al. 2010). The levels of the precursor and mature *miR-181b* were augmented on exposure of hepatic cells to TGF-β and were significantly reduced by small interference RNA-mediated depletion of Smad4, showing the involvement of TGF-β signaling pathway in *miR-181b* expression. Moreover, in HCC cases, the gene expression profiling revealed that *miR-181* up-regulation is positively associated with Wnt/β-catenin signaling activation in HCCs.

7.2.2.3 Epigenetic Level and Others

Epigenetic changes of the miRNA genes in HCC have been described. Datta et al. (2008) first examined the miRNA expression profiles in HCC cell lines after treatment with 5-azacytidine (DNA hypomethylating agent) and/or trichostatin A (histone deacetylase inhibitor). These epigenetic drugs differentially regulate expression of a few miRs, particularly *miR-1-1*. Further, down-regulation of *miR-1* in primary HCCs is validated to be associated with methylation. *MiR-124* and *miR-203* are also found to be silenced in HCC through CpG-island methylation (Furuta et al. 2010).

It has been found that abnormal processing of miRNAs can lead to aberrant expression. Dicer is an endoribonuclease III and cleaves pre-miRNAs into mature miRNAs. The disruption of Dicer results in a loss of mature miRNAs. Dicer deficient hepatocytes exhibit a loss of the expression of all miRNAs, and mice with disrupted Dicer promote hepatocarcinogenesis (Sekine et al. 2009). Moreover, Lin-28 is known as an inhibitor of *let-7* processing, leading to the suppression of mature *let-7s* production in two ways. First, lin28 binds to the loop region of the primary and the precursor of *let-7*, which blocks *let-7* processing at both the level of Drosha and Dicer (Viswanathan, Daley and Gregory 2008). Second, lin28 can induce 3′-terminal uridylation of *let-7* precursors, leading to the failure of Dicer processing and finally degradation (Heo et al. 2008). In metastatic HCCs, *lin-28* is highly expressed and consequently *let-7* expression is significantly down-regulated (Viswanathan et al. 2009).

7.2.3 Function of Aberrant-expressed MiRNAs in HCC

Through base pairing to the 3′UTR of the targeted gene transcripts, miRNAs regulate gene expression by either directing mRNA degradation or repressing mRNA post-transcriptional translation. The expression profiling of miRNAs in HCC reveal that a majority of miRNAs are deregulated in HCC compared to non-tumor liver tissues, and are differentially expressed among different HCC subgroups, suggesting the role of miRNA in HCC pathological processes. Functional studies have revealed that these deregulated miRNAs in HCC promote HCC tumorigenesis and progression.

7.2.3.1 Promote HCC Tumorigenesis

It has been known that miRNAs play a key role in biological processes of cell development and differentiation. Therefore, it is understandable that the deregulation of miRNAs leads to oncogenic transformation.

Several studies have demonstrated that the expression of EpCAM and AFP could predict two different subtypes in HCC, i.e., hepatic stem cell like HCC (HpSC-HCC; $EpCAM^{+}AFP^{+}$) and mature hepatocyte like HCC (MH-HCC; $EpCAM^{-}AFP^{-}$) (Yamashita et al. 2007; 2008; 2009). $EpCAM^{+}AFP^{+}$ HCC cells are hepatic cancer stem cells (HpCSCs) with the abilities to self-renew, differentiate and initiate aggressive tumors in vivo (Yamashita et al. 2009). Further, a 20-miRNA signature was revealed to significantly differentiate HpSC-HCC from MH-HCC by interrogating miRNA expression profiles of 53 HpSC-HCC and 95 MH-HCC clinical specimens (Ji et al. 2009b), indicating a role of this miRNA signature in HCC carcinogenesis. Among them, *miR-181* family members are high-expressed in HpSC-HCC tissues and isolated hepatic cancer stem cells, and they promote the stemness of HCC by targeting CDX2 and GATA6, hepatic transcriptional regulators of differentiation, and NLK, an inhibitor of Wnt/β-catenin signaling (Ji et al. 2009b). These findings define a novel regulatory link between *miR-181s* and human $EpCAM^{+}$ hepatic cancer stem cells, and reveal a negative feedback mechanism that demonstrates how miRNAs play an important role in carcinogenesis by participating in cancer-associated signaling pathways.

Additional studies indicate that the up-regulation of *miR-181* is also found in the early stage of HCC in a feeding choline-deficient and amino acid defined (CDAA) diet mice model (Wang et al. 2010). This model was used to induce HCC and identify the alteration of miRNAs in early stage of HCC development. Thirty miRNAs were significantly altered following tumorigenesis in these mice. Among them, the up-regulation of *miR-181* is found at as early as pre-neoplastic stage of HCC in CDAA diet mouse model. Through targeting *mTIMP3*, *miR-181* was shown to promote cell growth, cell migration and cell invasion. Therefore, up-regulation of *miR-181* at early stages of feeding CDAA diet mice promotes hepatocarcinogenesis (Wang et al. 2010).

Up-regulation of *miR-155*, *miR-221/222*, and *miR-21* and down-regulation of *miR-122*, and *let-7s* was also found at the early stages of hepatic tumorigenesis (Wang et al. 2009). Up-regulation of *miR-155* related to NF-κB signaling pathway may promote cell growth through targeting *C/EBP-β* (Wang et al. 2009). *MiR-122* expression is under the transcriptional control of transcriptional factor HNF1A, HNF3A and HNF3B, and its silencing might partially explain the failed differentiation of HCC cells (Coulouarn et al. 2009).

7.2.3.2 Promote Tumor Progression

MiRNA microarray platforms have been used to reveal miRNA signatures associated with HCC TNM staging and HCC metastasis (Budhu et al. 2008; Ladeiro et al. 2008; Ura et al. 2008), indicating the association between deregulation of miRNAs and HCC progression. Mechanistically, miRNAs may promote HCC progression through three different steps, i.e. increasing cell proliferation, inhibiting cell apoptosis or promoting cell migration and invasion by targeting different genes. Because each miRNA can potentially target multiple mRNAs, it is conceivable that different miRNAs may play similar roles in HCC progression through targeting the same pathways. An example includes *miR-21* and *miR-221/222* that target the phosphoinositide 3-kinase (PI3K)-phosphatase and tensin homolog (PTEN)-v-Akt murine thymoma viral oncogene homolog (Akt)-mammalian target of rapamycin (mTOR) pathway. Conversely, each miRNA may have multiple functions, such as *miR-122* and *let-7*, through targeting different signaling pathways.

7.2.3.2.1 Cell Cycle Related MiRNAs

Most miRNAs with altered expression levels in HCC are involved in promoting the loss of cell cycle control in HCC. These include the up-regulation of *miR-21*, *miR-224*, *miR-23a~27a~24*, *miR-17-92*, *miR-106b-25*, *miR-221*, *miR-146a* and *miR-18a*, and the down-regulation of *miR-122*, *let-7*, *miR-26*, *miR-125b*, *miR-223*, *miR-1*, *miR-195*, and *miR-101* (Table 7.2).

Cell cycle regulation is achieved by two key classes of regulatory molecules, i.e., cyclins and cyclin-dependent kinases (CDKs), as well as two classes of inhibiting factors, i.e. cip/kip family (including p21, p27, and p57) and INK4a/ARF family (p16INK4a and p14Arf). CDKs and cyclins are involved in cell cycle regulation by forming an activated heterodimer. The inhibiting factors bind to cyclin/CDK complexes to cease cell cycle progression and guide cells to apoptosis. Many miRNAs deregulated in HCC could directly target various cell cycle related molecules (Fig. 7.1). *Cyclin G1*, *Cyclin D1*, *Cyclin D2*, and *Cyclin E2* could be directly targeted by *miR-122*, *miR-195*, and *miR-26* (Gramantieri et al. 2007; Kota et al. 2009; Xu et al. 2009). Moreover, *miR-124* and *miR-195* could also target *CDK6*. Consequently, down-regulation of these miRNAs can lead to an increased proliferation of HCC cells. Meanwhile, *CDKN1B/p27* and *CDKN1C/p57*, cyclin-dependent

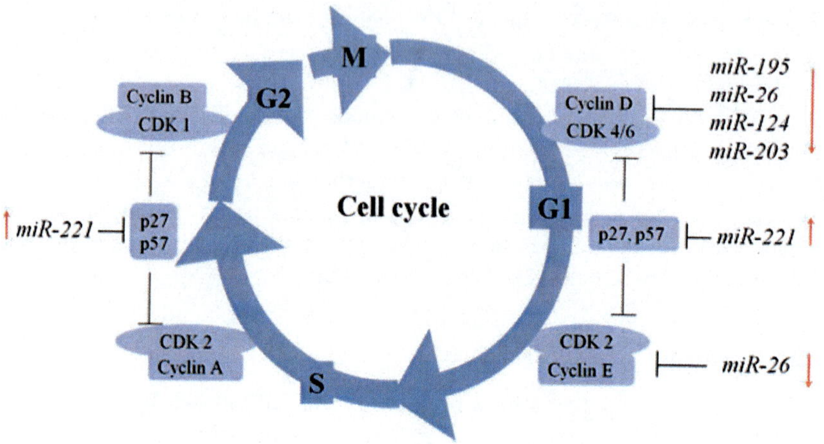

Fig. 7.1 Deregulated microRNAs in hepatocellular carcinoma promote cell proliferation through regulating cell cycle related proteins

kinase inhibitors (CDKIs), are the targets of *miR-221*, which is significantly up-regulated in HCC (Fornari et al. 2008). By controlling these two CDKIs, *miR-221* can promote HCC cell growth.

E2F family members play a major role during G1/S transition through activating the expression of cyclins and CDKs. It has been found that down-regulation of *miR-195* contributes to tumor growth partially through targeting *E2F3* (Xu et al. 2009). Interestingly, *E2F1* is the target of both clusters *miR-17-92* and *miR-106b-25* (Li et al. 2009c). Both clusters are highly expressed in HCC versus non-HCC tissues and stimulate HCC cell growth. At this point, it is not possible to fully ascertain the role of these two clusters in HCC, although it is likely that they only prevent excessively high E2F1 expression.

In addition, silenced *miR-124* and *miR-203* by methylation in HCC could act as proliferation inhibitors through multiple targets such as *CDK6* and *vimentin* (Furuta et al. 2010). Using MALDI-TOF/TOF analysis another study also shows that *miR-34a* can regulate the p53 signaling pathway and cell cycle pathway (Cheng et al. 2010). Interestingly, *miR-18a* was found to be more highly expressed in female HCC than male HCC. Through targeting estrogen receptor, high level of *miR-18a* in HCC contributes to HCC cell proliferation (Liu et al. 2009).

7.2.3.2.2 Apoptosis Related MiRNAs

Certain miRNAs that have an abnormal expression in HCC can inhibit apoptosis of HCC cells. These include up-regulation of *miR-23a~27a~24*, *miR-25*, *miR-221*, *miR-602*, and *miR-224*, down-regulation of *miR-101*, *miR-122*, *miR-1*, *miR-29*, *miR-15b*, *let-7*, and *miR-26* (Chung et al. 2010; Datta et al. 2008; Gramantieri et al. 2009; Huang et al. 2008a; Li et al. 2009b, c; Su et al. 2009; Wang et al. 2008; Xu et al. 2009; Yang et al. 2010).

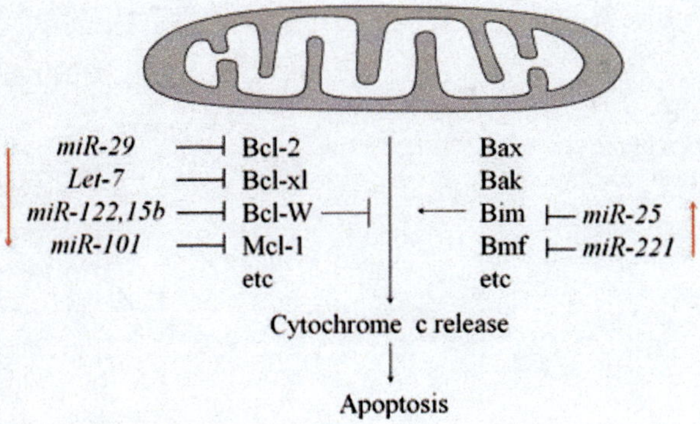

Fig. 7.2 Deregulated microRNAs in hepatocellular carcinoma suppress apoptosis through blocking pro-apoptosis proteins and releasing anti-apoptosis proteins

Apoptosis proteins consist of pro-survival members (Bcl-2, Bcl-xL, Bcl-w, Mcl-1, and A1) and pro-apoptotic members (Bax, Bak, Bok, Bmf, Bim, Bad, Bid, Bik, Puma, Noxa, and Hrk). *Bcl-2* is the direct target of *miR-29*; *Bcl-xl* is the target of *let-7*; Bcl-w is the target of *miR-122* and *miR-15b*, *Mcl-1* is the target of *miR-101* (Chung et al. 2010). All of these miRNAs are down-regulated in HCC, leading to the increased level of pro-survival proteins. Meanwhile, up-regulated *miR-221* and *miR-25* could target pro-apoptosis proteins *Bmf* and *Bim*, respectively (Gramantieri et al. 2009; Li et al. 2009c). Therefore, the deregulated miRNAs changed the balance of pro- and anti-apoptotic genes towards a more anti-apoptotic phenotype, which may promote tumor progression (Fig. 7.2).

Moreover, up-regulation of *miR-602* is associated with low level of apoptosis in HCC cells possibly through regulating Ras-associated domain family member 1A (RASSF1A). The RASSF1A loss is an early event in HCC (Yang et al. 2010).

7.2.3.2.3 Metastasis Related MiRNAs

Metastasis is a complex series of steps in which cancer cells migrate from the original tumor site to other parts of the body via the bloodstream or the lymphatic system. The ability of cell migration, cell invasion and angiogenesis are important for the formation of metastasis nodule. Microarray-based miRNA profiling analyses have revealed a 20-miRNA signature to be associated with HCC metastasis (Budhu et al. 2008). Consistently, up-regulation of *miR-9*, *miR-143*, *miR-30d*, *miR-181s*, *miR-21*, *miR-151-5p*, *miR-221*, and *miR-222*, and the down-regulation of *miR-375*, *miR-101*, *miR-223*, and *let-7* in HCC are functionally associated with HCC metastasis through increasing cell motility (Table 7.2).

Interestingly, among these metastasis-associated miRNAs, three of them, i.e. *miR-21*, *miR-221*, and *miR-222* target Akt signaling pathway (Fig. 7.3). This pathway is involved in many cellular processes including proliferation, differentiation, cell motility, and its abnormal activation is found in cancers including HCC (Chen et al. 2009; Li et al. 2008a; Nakanishi et al. 2005; Steelman et al. 2008). Its core

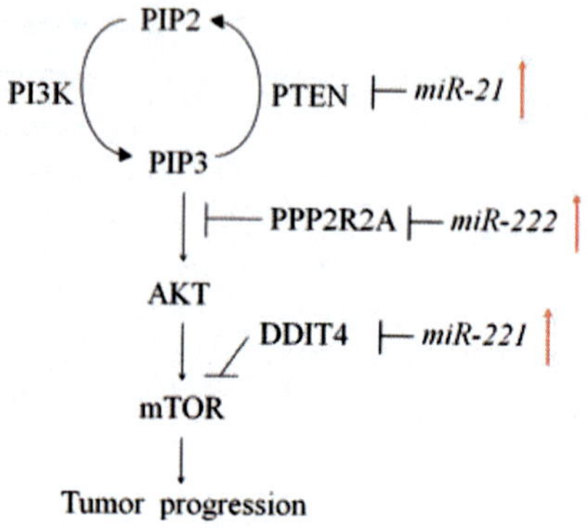

Fig. 7.3 Up-regulation of *miR-21*, *miR-221*, and *miR-222* in hepatocellular carcinoma promotes persistent activation of the PI3K/Akt/mTOR pathway

components include PI3K, PTEN, Akt, and mTOR. PI3K is a member of intracellular lipid kinase family, which catalyzes the generation of phosphatidylinositol-3,4,5- triphosphate (PIP3) from phosphatidylinositol-4,5-triphosphate (PIP2). Akt is recruited to the plasma membrane by PIP3 and is then phosphorylated. Activated Akt translocates to the nucleus and activates mTOR and downstream targets. PTEN negatively regulates the PI3K/Akt/mTOR pathway through dephosphorylating PIP3 back to PIP2 and consequently inhibiting the activation of Akt. Up-regulation of *miR-21* in HCC could target *PTEN*, leading to the activation of this pathway (Meng et al. 2007). Meanwhile, over-expressed *miR-222* could suppress the protein phosphatase 2A subunit B (PPP2R2A), consequently blocking the phosphorylation of Akt (Wong et al. 2010). Moreover, *miR-221* is highly expressed in HCC and targets the DNA damage-inducible transcript 4 (DDIT4), a negative modulator of mTOR pathway (Pineau et al. 2010). Through targeting PTEN-PI3K-Akt-mTOR axis, these miRNAs promote cell migration, invasion and cell proliferation in HCC (Meng et al. 2007; Pineau et al. 2010; Wong et al. 2010). The co-up-regulation of *miR-21/miR-221/miR-222* may act in concert in HCC, which is considered to be crucial for HCC progression.

MiR-34, *miR-23b*, and *miR-1* play tumor suppressor roles through suppressing the oncogenic pathway, hepatocyte growth factor (HGF)/c-Met/MAPK pathway (Salvi et al. 2009). c-Met, a tyrosine kinase receptor, is also known as hepatocyte growth factor receptor. HGF is the only known ligand of c-Met. HGF-stimulated c-met activation triggers phosphorylation of extracellular signal regulated kinases 1 and 2 (ERK1/2), the key factors influencing tumor invasion and migration. *miR-34a* directly targets c-met and reduces its expression at both mRNA and protein levels, leading to a decreased phosphorylation of ERK1/2. These results demonstrate that *miR-34a* is a potent inhibitor of the migrative/invasive capacity of HCC cells (Li et al. 2009a). Moreover, *miR-23b* suppresses the migration and invasion also through targeting *c-Met* (Salvi et al. 2009). *MiR-1* is silenced in HCC through methylation, and reported to regulate cell proliferation and apoptosis through targeting c-met (Datta et al. 2008). Down-regulation of these miRNAs may collectively contribute to HCC metastasis.

Alteration of other signaling pathways by miRNAs in HCC metastasis has also been demonstrated. *MiR-143* and *miR-9* are highly expressed in HCC and promote HCC metastasis (Hao-Xiang et al. 2009; Zhang et al. 2009). *MiR-143* is induced by NF-κB pathway through NF-κB binding sites in its promoter region and targets fibronectin type III domain containing 3B (FNDC3B), which regulates cell motility (Zhang et al. 2009). *MiR-9* promotes cell invasion partially through reducing *E-cadherin* expression (Hao-Xiang et al. 2009). Moreover, the amplified *miRNA-151* on 8q24.3 is correlated with intrahepatic metastasis in HCC through targeting *RhoGDIA*, a metastasis suppressor of HCC (Ding et al. 2010). Both *miR-23b* and *miR-375* are decreased in HCC and act as inhibitors of both cell proliferation and motility (Liu, Poon, Luk 2010; Salvi et al. 2009). *MiR-23b* targets urokinase-type plasminogen activator (uPA), a serine protease that could govern the extracellular matrix turnover, as well as suppress the migration and invasion (Salvi et al. 2009). *MiR-375* suppresses cell proliferation and invasion through targeting yes-associated protein (YAP), a potent oncogenic driver (Liu, Poon, Luk 2010). *MiR-101* inhibits migration and invasion through suppressing the activator protein-1 (AP-1) transcriptional activity by targeting *FOS*, a key component of AP1 transcription factor (Li et al. 2009b).

7.2.3.3 Multiple Functions of *MiR-122* and *Let-7*

MiR-122 is a well-defined liver-specific miRNA and plays important roles in many aspects of liver physiology, such as stress response and lipid metabolism (Esau et al. 2006; Gatfield et al. 2009; Krutzfeldt et al. 2005). The relationship between *miR-122* and HCC was initially established in woodchuck. It was further confirmed that *miR-122* expression was down-regulated in HCC of rodent and human and the down-regulation of *miR-122* associated with HCC metastasis and worse prognosis (Bai et al. 2009; Budhu et al. 2008; Coulouarn et al. 2009; Fornari et al. 2009; Kutay et al. 2006; Tsai et al. 2009; Wu et al. 2009). *Cyclin G1, bcl-W, ADAM17, ADAM10, SRF*, and *Igf1R* are validated as *miR-122* targets since they contain the binding site of *miR-122* at the 3′UTR of their mRNAs (Bai et al. 2009; Coulouarn et al. 2009; Gramantieri et al. 2007; Lin et al. 2008; Ma et al. 2010; Wu et al. 2009). Through these targets, *miR-122* can function as a potential tumor suppressor in three ways as we talked above, i.e. inhibiting cell proliferation, promoting apoptosis and suppressing cell migration and invasion (Bai et al. 2009; Gramantieri et al. 2007; Lin et al. 2008; Tsai et al. 2009; Wu et al. 2009). Moreover, by targeting *cyclin G1*, *miR-122* could increase p53 protein stability and transcriptional activity and reduces invasion capability of HCC-derived cell lines (Fornari et al. 2009). Meanwhile, *miR-122* also plays a positive role in HCC development by stimulating HCV RNA translation (Gramantieri et al. 2007; Jopling, Norman and Sarnow 2006; Kutay et al. 2006; Lin et al. 2008; Randall et al. 2007).

Let-7 is a well-known tumor suppressor in many tumors through targeting oncogenes *Ras*, *Myc*, and *HMGA2*, etc. Consistently, *let-7* family members are widely down-regulated in HCC and its down-regulation is associated with increased tumor growth, HCC metastasis and worse prognosis. In addition to c-myc, other *let-7* targets such as *COL1A2* and *lin28* have recently been identified (Guo et al. 2006; Ji et al. 2010; Lan et al. 2011). These studies revealed that forced expression of *let-7* inhibits cell proliferation through targeting *c-Myc* and consequently up-regulating the tumor suppressor gene *p16* possibly via c-Myc-Bmi-1-p16 regulatory circuit. This circuit is described to be important in simulating epithelial cell growth, in which c-Myc can directly bind to E-box at the Bmi-1 promoter, while Bmi-1 is known as a transcription repressor of p16INK4A. Through inhibiting CDK4 kinase, P16INK4A could potentially suppress the cell growth (Lan et al. 2011). Our recent studies revealed that *COL1A2* is a direct target of *let-7g* and the exogenous collagen I could rescue the migration inhibition by *let-7g* in HCC cells. Further, *COL1A2* is highly expressed in HCCs with metastasis and negatively correlated with *let-7* level, indicating the metastasis suppressor role of *let-7g* (Ji et al. 2010). Interestingly, we could not find any evidence of down-regulation of *c-Myc*, *Ras*, and *HMGA2* in cells over-expressing *let-7g*. Other studies demonstrate that *let-7* could bind to the 3′UTR of *Lin28* and reduce the level of *lin-28*, leading to the malignant phenotype (Guo et al. 2006). It was also found that *c-Myc* could suppress the *let-7* expression. Consistently, it was suggested that an association of lin-28 with HCC malignancy is related to its regulation on *let-7* (Viswanathan et al. 2009). Moreover, lin-28 inhibits the level of *let-7* through interfering with Drosha and Dicer (Heo et al. 2008; Viswanathan, Daley and Gregory 2008). The negative feedback loops of *c-Myc/let-7* and *lin28/let-7* are important in maintaining cellular homeostasis and loss of such feedback loops may contribute to HCC metastasis (Guo et al. 2006; Ji et al. 2010).

7.3 Biomarkers for Diagnosis, Prognosis, and Therapy of HCC

The identification of miRNAs unique to HCC not only provides further insight into our understanding in HCC heterogeneity, but also offers new biomarkers for cancer diagnosis and prognosis. Functional exploration of candidate HCC miRNAs may provide clues for new therapeutic targets in HCC (Table 7.3).

7.3.1 Biomarkers for Diagnosis

HCC typically has a poor prognosis, because it is often diagnosed at an advanced stage. Therefore, HCC biomarkers especially early biomarkers are useful in assisting HCC early diagnosis. Moreover, given that HCC is a clinically and biologically heterogeneous malignancy, miRNAs that are associated with different subgroups would be helpful for clinical HCC diagnosis and consequently HCC therapeutics.

7.3.1.1 HCC Biomarkers

Although many candidate miRNAs are found to be differentially expressed in HCC, only a few miRNAs are further validated for their utilities in predicting HCC using independent cohorts, a practice essential to translate biomarkers into clinic.

Table 7.3 Hepatocellular carcinoma (HCC)-related microRNAs as potential diagnosis and prognosis biomarkers of HCC

Diagnosis/prognosis/therapy	MicroRNAs (top 5)	References
Diagnosis		
HCC markers	*miR-222* and *miR-223*	Wong et al. (2008)
HCC subtypes		
HpSC-HCC vs. MH-HCC	20 *miRNAs* (*miR-181a-1, miR-181a-2, miR-181b-1, miR-181b-2, miR-181c*)	Ji et al. (2009b)
Malignant HCC vs. Benign HCC	6 *miRNAs* (*miR-200c, miR-203, miR-21, miR-224, miR-10b*)	Ladeiro et al. (2008)
Alcohol-HCC vs. HCC	*miR-126**	Ladeiro et al. (2008)
HBV-HCC vs. HCC	*miR-96*	Ladeiro et al. (2008)
Prognosis		
Metastasis	20 *miRNAs* (*miR-30c-1, miR-1-2, miR-34a, miR-148a, miR-124a-2*)	Budhu et al. (2008)
Recurrence	7 *miRNAs* (*miR-15b, miR-17-5-p, miR-92, miR-20a, miR-1a*)	Chung et al. (2010)
Survival	46 *miRNAs* (*miR-338, miR-219-1, miR-206, miR-183, miR-129-2*)	Budhu et al. (2008)
	miR-26	Ji et al. (2009a)
	Let-7	Ji et al. (2010)
Therapy		
Tumorigenesis	Blocking of *miR-181s*	Ji et al. (2009b)
Progression	Delivery of *miR-26*	Kota et al. (2009)
	Delivery of *miR-122*	Bai et al. (2009); Tsai et al. (2009)
Stratification	*miR-26*	Ji et al. (2009a)

It was shown that a combination of *miR-222* and *miR-223* could be served as a HCC diagnosis biomarker. Down-regulation of *miR-223* and up-regulation of *miR-222* in HCC could unequivocally distinguish HCC from adjacent non-tumor liver, irrespective of viral association (Wong et al. 2008). Moreover, *miR-146a* might work as a HCC risk prediction marker. Individuals with a G-allele in *miR-146a* are 2-fold more likely to develop HCC when compared with those with C-allele, indicating the potential of *miR-146a* polymorphism in HCC risk prediction (Xu et al. 2008). Interestingly, the preliminary data show that *miR-500* is highly expressed in tumor and its level is high in serum of HCC patients but dramatically low after HCC resection surgery (Yamamoto et al. 2009). Although further validation is missing, the exploration of HCC-associated miRNAs in serum as a diagnostic biomarker should be encouraged as a less invasive method.

7.3.1.2 HCC Subtype Biomarkers

Due to the heterogeneous tumor pathology and biology, miRNAs can serve as powerful biomarkers for identifying HCC subtypes. One example is the miRNA signature associated with the HpSC-HCC subtype (Ji et al. 2009b). A 20-miRNA signature can discriminate HpSC-HCC from MH-HCC and can predict two HCC subtypes with an overall accuracy of 78%. Among the 20 miRNAs, the conserved *miR-181* family members are validated to be highly up-regulated in HpSC-HCC tissues with cancer stem cell features and in isolated hepatic CSCs, with the use of EpCAM or CD133, two experimentally proven CSC cell surface makers (Ma et al. 2007; Yamashita et al. 2009). Moreover, high levels of *miR-181* family members are also found in embryonic livers and in isolated hepatic stem cells. These data indicate that *miR-181* can serve as a biomarker for hepatic CSC and HCC with hepatic stem cell features, the activities that contribute to HCC metastasis and resistance to chemotherapy. Another study showed that a 19-miRNA signature can differentiate HBV-related HCC from HCV related HCC (Ura et al. 2008).

Moreover, it was shown that the utility of a 6-miRNA signature for distinguishing benign tumors including hepatocellular adenomas and focal nodular hyperplasia from malignant HCC (Ladeiro et al. 2008). HCCs have increased levels of *miR-21*, *miR-10b*, *miR-222*, and *miR-224*, whereas benign tumors show decreased expression of *miR-200c* and *miR-203*. This study also characterized and validated that low level of *miR-126** in HCC with an alcohol consumption subgroup and high level of *miR-96* in HCC with hepatitis B virus infection.

7.3.2 Biomarker for Prognosis

Metastasis and recurrence are major concerns for the long-term survival of post-operative HCC patients. Biomarkers related to these processes could have the clinical prognostic utility.

7.3.2.1 Metastasis and Recurrence

By comparing miRNAs profiling between 29 HCCs with metastasis and 102 non-metastasis/relapse-free HCC cases, we have identified a unique 20-miRNA signature useful to predict primary HCC tissues with venous metastases. This signature is also capable of predicting survival and recurrence of HCC patients with multi-nodular or solitary tumors. Furthermore, This signature is validated in an independent cohort ($n = 110$) of HCC cases including early stage HCC (Budhu et al. 2008). Consistently, other studies have shown that down-regulation of *let-7*, *miR-338*, *miR-34*, and *miR-122*, and up-regulation of *miR-30d*, occurred more frequently in HCCs with metastasis (Coulouarn et al. 2009; Huang et al. 2009; Ji et al. 2010; Li et al. 2009a; Tsai et al. 2009; Yao et al. 2010). These miRNAs are included in the 20-miRNA metastasis signature, suggesting the general utility of this signature in predicting HCC metastasis.

In addition, another study from a small cohort of 25 HCC cases reveal that seven miRNAs (*miR-15b*, *miR-17-5p*, *miR-92*, *miR-20a*, *miR-1a*, *miR-34c*, and *miR-361*) significantly separated the tumors according to their recurrence status with an overall accuracy of 90% (Chung et al. 2010), although independent validation is needed.

7.3.2.2 Survival

Survival associated miRNAs were globally analyzed using a survival risk prediction method with cross validation. In this analysis, patients were assigned to two survival risk groups, i.e. high risk groups and low risk groups, based on a 50% prognostic index cutoff. A value above 50% corresponds to a high hazard of death and consequently a poor predicted survival. This analysis yielded 46 miRNAs associated with survival in a total 241 HCC cases (Budhu et al. 2008).

Given that HCC more frequently occurs in man than women and female HCC cases often have better survival than male cases, miRNA with differential expression in female and male might be used as survival biomarkers. By comparing the miRNA expression between male and female (31 cases for each group), *miR-26* is found to highly express in female non-tumor liver tissues than male. As expected, the low level of this miRNA is significantly associated to short overall survival in three different cohorts including a total of 455 patients (Ji et al. 2009a).

As a well-known tumor suppressor, *let-7* is down-regulated in HCC. The down-regulation of *let-7* could also predict the poor survival of HCC patients, which is validated in two different cohorts (Ji et al. 2010). In addition, Jiang et al identified 19 miRNAs to be associated with patients' outcome and many of the predicted targets of these miRNAs regulate cell division, mitosis, and G1-S transition (Jiang et al. 2008). Several other studies show that down-regulation of *miR-125* (Li et al. 2008b) and *miR-29* (Xiong et al. 2010), and up-regulation of *miR-222* (Wong et al. 2010) are associated with poor survival.

7.3.3 Biomarker for Therapy

The deregulated miRNAs in HCC functionally enable HCC tumorigenesis and promote HCC progression. Our mechanistic studies offer an opportunity to explore miRNAs as targets for cancer therapy.

7.3.3.1 Block Tumor Initiation

Cancer stem cells are thought to be the source of tumor initiation. It has been shown that $EpCAM^+$ CSCs isolated from AFP^+ HCC cases can efficiently form spheroids in vitro and as few as 200 of these cells can initiate tumor growth in NOD/SCID mice (Yamashita et al. 2009). Further, $EpCAM^+$ hepatic CSCs have high levels of *miR-181* and inhibition of *miR-181* in $EpCAM^+$ hepatic CSCs results in a reduction in spheroid formation and tumorigenicity (Ji et al. 2009b). These results support a role for *miR-181* as a therapeutic target in the treatment of hepatic CSCs.

The presence of CSCs in HCC may contribute to cancer relapse due to their drug/radiation resistance. It was also found that EpCAM$^+$ HCC cells are resistant to 5-FU (Yamashita et al. 2009). Therefore, combining chemotherapy (5-FU) and anti-*miR-181* can be a powerful adjuvant HCC therapy to those with HpSC-HCC. Molecular adjuvant therapy might help to eradicate CSCs.

7.3.3.2 Block Tumor Progression

Many miRNAs, such as *miR-122*, *let-7*, and *miR-26*, functionally act as tumor suppressors. Therapeutic delivery of these miRNAs back to HCC cells has been shown to significantly suppress tumor growth in a mouse model. For example, *miR-26* is significantly down-regulated in HCC and low level of *miR-26* is significantly associated with poor prognosis in HCC (Ji et al. 2009a). Moreover, *miR-26* can act as a tumor-suppressor gene due to its role in inhibiting cell proliferation by silencing two cyclins needed for DNA replication. Further, delivery of *miR-26* back to liver cells through adeno-associated virus dramatically blocks *Myc*-induced HCC (Kota et al. 2009). Since *miR-26* is abundantly expressed in most of normal cells, such gene therapy is unlikely to be toxic to normal cells. Furthermore, studies also revealed that HCC cells with forced expression of *miR-122* showed an inhibited in situ tumor formation, tumor growth, metastasis in a xenograft mouse model (Bai et al. 2009; Tsai et al. 2009). HCC cells with forced expression of *miR-195* also showed the inhibited tumor growth in vivo (Xu et al. 2009).

The ability of targeting various miRNAs described above to induce tumor suppression offers a useful therapeutical strategy through manipulating miRNA expression. Because miRNAs potentially suppress hundreds of cellular genes to regulate multiple signaling pathways, it may be difficult for tumors to escape from the effect of miRNA drugs. Consistently, the utility of artificial miRNAs being designed to target osteopontin, which is highly expressed in HCC with metastasis, successfully led to a decreased lung metastasis in mice model (Sun et al. 2008). These studies further demonstrate that patients may benefit from specific miRNA adjuvant therapy through miRNA-guided classification of their specific tumor subtype. The major obstacle to therapies that are based on RNA interference is delivering these oligonucleotides into cells. Because of its filtering role, the liver traps and internalizes both small RNA drugs and gene-therapy viruses, making it as an ideal organ for this new approach.

7.3.3.3 Drug Sensitivity

Down-regulation of *miR-26* specifically occurs in HCC compared to paired noncancerous tissues from a subgroup of patients with poor survival, indicating that HCC cases with reduced *miR-26* expression have a distinct transcriptomic pattern. By comparing gene expression between HCC cases with high level of *miR-26* and those with low level of *miR-26*, NF-κB signaling, interleukin-6 signaling and interferon signaling are highly activated in HCC cases with low level of *miR-26*, indicating the association of *miR-26* expression in HCC with an activated inflammatory pathway. It was then hypothesized that HCC with reduced HCC would respond to anti-inflammatory agent related therapy. Consistently, using liver specimens derived from two independent randomized control trials, it was found that HCC cases with low-level of *miR-26* respond favorably to interferon adjuvant therapy while HCC cases with high level of *miR-26* have no response at all (Ji et al. 2009a). These studies indicate that *miR-26* can be used not only as a molecule target to treat cancer, but also as a biomarker to stratify HCC patients for interferon therapy.

A recent study showed that *miR-122* suppresses cell growth, migration and invasion, and sensitizes HCC cells to sorafenib (Bai et al. 2009). In addition, *let-7* can induce apoptosis when cooperating with Sorafenib, an anti-cancer drug by targeting *Bcl-xL* (Shimizu et al. 2010). Taken together, these research activities are important as they pave the way to enable personalized cancer care.

7.4 Conclusion and Prospective

A growing body of evidence shows that many miRNAs are aberrantly expressed in HCC. These HCC-associated miRNAs not only provide new insights into the molecular basis of HCC. Some miRNAs are involved in HCC carcinogenesis by promoting stemness of CSC and by controlling cell proliferation; others are associated with HCC progression by controlling cell growth, migration and invasion. Moreover, the deregulation of several miRNAs has been validated in an independent cohort, and could serve as new biomarkers for HCC diagnosis, prognosis and therapy. However, it is notable that only a few miRNA signatures or individual miRNAs could potentially be used as these biomarkers, and even for these there is still a long way to go before they can be used in clinics. To achieve this goal, these miRNAs need to be further validated with high accuracy in prospective studies.

HCC is a pathologic and genetically heterogeneous disease. The identification of *miR-181*, HpSC-HCC associated miRNA, has given rise to a potential rational therapeutic approach to eliminate cancer stem cell. The expression of *miR-26* in HCCs could serve as a biomarker to stratifying HCC patients for interferon alpha therapy. However, the understanding of different HCC subtypes is still very limited. Further exploration of these subtype associated miRNAs in HCC would not only improve our understanding in the molecular mechanisms, but also useful in assisting patient stratification for prognosis and/or therapy. Moreover, the circulating miRNAs in sera from HCC patients remains largely unexplored. It has been demonstrated that the circulating miRNAs are stable and can be reliably extracted and assayed in either serum or plasma samples. Comparing with serum proteins, detection of serum miRNAs might be easier for HCC diagnosis and prognosis. The exploration of these biomarkers would allow individually-tailored treatment programs that will significantly improve the survival and health of HCC patients.

In addition, a majority of miRNAs and their related signaling pathways are aberrantly activated in HCC. The strategy of reprogramming miRNA expression of HCC cells through changing the expression of certain miRNAs might provide a potent method for HCC therapy. Reprogramming strategy has been successfully used to drive non-pluripotent cells to pluripotent stem cells through forced expression of a combination of several transcriptional factors such as *Oct3/4*, *Sox2*, *Klf4*, and *c-Myc* (Takahashi and Yamanaka 2006; Takahashi et al. 2007). Similar as transcriptional factors, single miRNAs could also regulate a group of genes expression through a similar binding site in the 3′UTR of target genes. Therefore, it is reasonable and worthwhile to test this miRNA combination approach in reprogramming HCC cells to normal cells or metastatic HCC cells to non-metastatic cells. Such an approach would open new therapeutic avenues for HCC patients and may eventually alter the dismal outcome of HCC patients.

References

Bai S, Nasser MW, Wang B, et al. MicroRNA-122 inhibits tumorigenic properties of hepatocellular carcinoma cells and sensitizes these cells to sorafenib. J Biol Chem. 2009;284:32015–27.

Berezikov E, Chung WJ, Willis J, et al. Mammalian mirtron genes. Mol Cell. 2007;28: 328–36.

Blum HE. Hepatocellular carcinoma: therapy and prevention. World J Gastroenterol. 2005;11:7391–400.

Borchert GM, Lanier W, Davidson BL. RNA polymerase III transcribes human microRNAs. Nat Struct Mol Biol. 2006;13:1097–101.

Boyault S, Rickman DS, de Reynies A, et al. Transcriptome classification of HCC is related to gene alterations and to new therapeutic targets. Hepatology. 2007;45:42–52.

Brennecke J, Stark A, Russell RB, et al. Principles of microRNA-target recognition. PLoS Biol. 2005;3:e85.

Budhu A, Forgues M, Ye QH, et al. Prediction of venous metastases, recurrence and prognosis in hepatocellular carcinoma based on a unique immune response signature of the liver microenvironment. Cancer Cell. 2006;10:99–111.

Budhu A, Jia HL, Forgues M, et al. Identification of metastasis-related microRNAs in hepatocellular carcinoma. Hepatology. 2008;47:897–907.

Budhu A, Wang XW. Human hepatocellular carcinoma: new insights from gene expression profiling. In: Jeffreis LP(ed). New Developments in Cancer Research, pp. 1–32. 2006.

Calin GA, Sevignani C, Dumitru CD, et al. Human microRNA genes are frequently located at fragile sites and genomic regions involved in cancers. Proc Natl Acad Sci USA. 2004;101: 2999–3004.

Chen JS, Wang Q, Fu XH, et al. Involvement of PI3K/PTEN/AKT/mTOR pathway in invasion and metastasis in hepatocellular carcinoma: Association with MMP-9. Hepatol Res. 2009;39: 177–86.

Cheng J, Zhou L, Xie QF, et al. The impact of miR-34a on protein output in hepatocellular carcinoma HepG2 cells. Proteomics. 2010;10:1557–72.

Chung GE, Yoon JH, Myung SJ, et al. High expression of microRNA-15b predicts a low risk of tumor recurrence following curative resection of hepatocellular carcinoma. Oncol Rep. 2010;23:113–19.

Connolly E, Melegari M, Landgraf P, et al. Elevated expression of the miR-17-92 polycistron and miR-21 in hepadnavirus-associated hepatocellular carcinoma contributes to the malignant phenotype. Am J Pathol. 2008;173:856–64.

Coulouarn C, Factor VM, Andersen JB, et al. Loss of miR-122 expression in liver cancer correlates with suppression of the hepatic phenotype and gain of metastatic properties. Oncogene. 2009;28:3526–36.

Datta J, Kutay H, Nasser MW, et al. Methylation mediated silencing of MicroRNA-1 gene and its role in hepatocellular carcinogenesis. Cancer Res. 2008;68:5049–58.

Ding J, Huang S, Wu S, et al. Gain of miR-151 on chromosome 8q24.3 facilitates tumour cell migration and spreading through downregulating RhoGDIA. Nat Cell Biol. 2010;12:390–9.

El Serag HB, Marrero JA, Rudolph L, et al. Diagnosis and treatment of hepatocellular carcinoma. Gastroenterology. 2008;134:1752–63.

El-Serag HB, Davila JA, Petersen NJ, et al. The continuing increase in the incidence of hepatocellular carcinoma in the United States: an update. Ann Intern Med. 2003;139:817–23.

El-Serag HB, Mason AC. Rising incidence of hepatocellular carcinoma in the United States. N Engl J Med. 1999;340:745–50.

Esau C, Davis S, Murray SF, et al. miR-122 regulation of lipid metabolism revealed by in vivo antisense targeting. Cell Metab. 2006;3:87–98.

Fornari F, Gramantieri L, Ferracin M, et al. MiR-221 controls CDKN1C/p57 and CDKN1B/p27 expression in human hepatocellular carcinoma. Oncogene. 2008;27:5651–61.

Fornari F, Gramantieri L, Giovannini C, et al. MiR-122/cyclin G1 interaction modulates p53 activity and affects doxorubicin sensitivity of human hepatocarcinoma cells. Cancer Res. 2009;69:5761–7.

Furuta M, Kozaki KI, Tanaka S, et al. miR-124 and miR-203 are epigenetically silenced tumor-suppressive microRNAs in hepatocellular carcinoma. Carcinogenesis. 2010;31:766–76.

Gatfield D, Le Martelot G, Vejnar CE, et al. Integration of microRNA miR-122 in hepatic circadian gene expression. Genes Dev. 2009;23:1313–26.

Gramantieri L, Ferracin M, Fornari F, et al. Cyclin G1 is a target of miR-122a, a microRNA frequently down-regulated in human hepatocellular carcinoma. Cancer Res. 2007;67:6092–9.

Gramantieri L, Fornari F, Ferracin M, et al. MicroRNA-221 targets Bmf in hepatocellular carcinoma and correlates with tumor multifocality. Clin Cancer Res. 2009;15:5073–81.

Guo Y, Chen Y, Ito H, et al. Identification and characterization of lin-28 homolog B (LIN28B) in human hepatocellular carcinoma. Gene. 2006;384:51–61.

Hao-Xiang T, Qian W, Lian-Zhou C, et al. MicroRNA-9 reduces cell invasion and E-cadherin secretion in SK-Hep-1 cell. Med Oncol. 2009;27(3):654–60.

Heo I, Joo C, Cho J, et al. Lin28 mediates the terminal uridylation of let-7 precursor MicroRNA. Mol Cell. 2008;32:276–84.

Hoshida Y, Villanueva A, Kobayashi M, et al. Gene Expression in Fixed Tissues and Outcome in Hepatocellular Carcinoma. N Engl J Med. 2008;359:1995–2004.

Huang YS, Dai Y, Yu XF, et al. Microarray analysis of microRNA expression in hepatocellular carcinoma and non-tumorous tissues without viral hepatitis. J Gastroenterol Hepatol. 2008b;23:87–94.

Huang S, He X, Ding J, et al. Upregulation of miR-23a approximately 27a approximately 24 decreases transforming growth factor-beta-induced tumor-suppressive activities in human hepatocellular carcinoma cells. Int J Cancer. 2008a;123:972–8.

Huang XH, Wang Q, Chen JS, et al. Bead-based microarray analysis of microRNA expression in hepatocellular carcinoma: miR-338 is downregulated. Hepatol Res. 2009;39:786–94.

Hutvagner G, Zamore PD. A microRNA in a multiple-turnover RNAi enzyme complex. Science. 2002;297:2056–60.

Ji J, Shi J, Budhu A, et al. MicroRNA expression, survival, and response to interferon in liver cancer. N Engl J Med. 2009a;361:1437–47.

Ji J, Wang XW. New kids on the block: Diagnostic and prognostic microRNAs in hepatocellular carcinoma. Cancer Biol Ther. 2009;8.

Ji J, Yamashita T, Budhu A, et al. Identification of microRNA-181 by genome-wide screening as a critical player in EpCAM-positive hepatic cancer stem cells. Hepatology. 2009b;50:472–80.

Ji J, Zhao L, Budhu A, et al. Let-7g targets collagen type I alpha2 and inhibits cell migration in hepatocellular carcinoma. J Hepatol. 2010;52:690–7.

Jiang J, Gusev Y, Aderca I, et al. Association of MicroRNA expression in hepatocellular carcinomas with hepatitis infection, cirrhosis, and patient survival. Clin Cancer Res. 2008;14:419–27.

Jopling CL, Norman KL, Sarnow P. Positive and negative modulation of viral and cellular mRNAs by liver-specific microRNA miR-122. Cold Spring Harb Symp Quant Biol. 2006;71:369–76.

Kota J, Chivukula RR, O'donnell KA, et al. Therapeutic microRNA delivery suppresses tumorigenesis in a murine liver cancer model. Cell. 2009;137:1005–17.

Krutzfeldt J, Rajewsky N, Braich R, et al. Silencing of microRNAs in vivo with 'antagomirs'. Nature. 2005;438:685–9.

Kutay H, Bai S, Datta J, et al. Downregulation of miR-122 in the rodent and human hepatocellular carcinomas. J Cell Biochem. 2006;99:671–8.

Ladeiro Y, Couchy G, Balabaud C, et al. MicroRNA profiling in hepatocellular tumors is associated with clinical features and oncogene/tumor suppressor gene mutations. Hepatology. 2008;47:1955–63.

Lan FF, Wang H, Chen YC, et al. Hsa-let-7g inhibits proliferation of hepatocellular carcinoma Cells by down-regulation of c-Myc and Up-regulation of p16(INK4A). Int J Cancer. 2011;128: 319–31.

Lee Y, Ahn C, Han J, et al. The nuclear RNase III Drosha initiates microRNA processing. Nature. 2003;425:415–9.

Lee JS, Heo J, Libbrecht L, et al. A novel prognostic subtype of human hepatocellular carcinoma derived from hepatic progenitor cells. Nat Med. 2006;12:410–6.

Lee Y, Jeon K, Lee JT, et al. MicroRNA maturation: stepwise processing and subcellular localization. EMBO J. 2002;21:4663–70.

Lee Y, Kim M, Han J, et al. MicroRNA genes are transcribed by RNA polymerase II. EMBO J. 2004;23:4051–60.

Lewis BP, Burge CB, Bartel DP. Conserved seed pairing, often flanked by adenosines, indicates that thousands of human genes are microRNA targets. Cell. 2005;120:15–20.

Lewis BP, Shih IH, Jones-Rhoades MW, et al. Prediction of mammalian microRNA targets. Cell. 2003;115:787–98.

Li N, Fu H, Tie Y, et al. MiR-34a inhibits migration and invasion by down-regulation of c-Met expression in human hepatocellular carcinoma cells. Cancer Lett. 2009a;275:44–53.

Li S, Fu H, Wang Y, et al. MicroRNA-101 regulates expression of the v-fos FBJ murine osteosarcoma viral oncogene homolog (FOS) oncogene in human hepatocellular carcinoma. Hepatology. 2009b;49:1194–202.

Li Y, Tan W, Neo TW, et al. Role of the miR-106b-25 microRNA cluster in hepatocellular carcinoma. Cancer Sci. 2009c;100:1234–42.

Li W, Tan D, Zhang Z, et al. Activation of Akt-mTOR-p70S6K pathway in angiogenesis in hepatocellular carcinoma. Oncol Rep. 2008a;20:713–9.

Li W, Xie L, He X, et al. Diagnostic and prognostic implications of microRNAs in human hepatocellular carcinoma. Int J Cancer. 2008b;123:1616–22.

Lin CJ, Gong HY, Tseng HC, et al. MiR-122 targets an anti-apoptotic gene, Bcl-w, in human hepatocellular carcinoma cell lines. Biochem Biophys Res Commun. 2008;375:315–20.

Liu CG, Calin GA, Meloon B, et al. An oligonucleotide microchip for genome-wide microRNA profiling in human and mouse tissues. Proc Natl Acad Sci USA. 2004;101:9740–4.

Liu AM, Poon RT, Luk JM. MicroRNA-375 targets Hippo-signaling effector YAP in liver cancer and inhibits tumor properties. Biochem Biophys Res Commun. 2010;394:623–7.

Liu WH, Yeh SH, Lu CC, et al. MicroRNA-18a prevents estrogen receptor-alpha expression, promoting proliferation of hepatocellular carcinoma cells. Gastroenterology. 2009;136: 683–93.

Lund E, Guttinger S, Calado A, et al. Nuclear export of microRNA precursors. Science. 2004;303:95–8.

Ma S, Chan KW, Hu L, et al. Identification and characterization of tumorigenic liver cancer stem/progenitor cells. Gastroenterology. 2007;132:2542–56.

Ma L, Liu J, Shen J, et al. Expression of miR-122 mediated by adenoviral vector induces apoptosis and cell cycle arrest of cancer cells. Cancer Biol Ther. 2010;9.

Meng F, Henson R, Wehbe-Janek H, et al. MicroRNA-21 regulates expression of the PTEN tumor suppressor gene in human hepatocellular cancer. Gastroenterology. 2007;133:647–58.

Murakami Y, Yasuda T, Saigo K, et al. Comprehensive analysis of microRNA expression patterns in hepatocellular carcinoma and non-tumorous tissues. Oncogene. 2006;25:2537–45.

Nakanishi K, Sakamoto M, Yamasaki S, et al. Akt phosphorylation is a risk factor for early disease recurrence and poor prognosis in hepatocellular carcinoma. Cancer. 2005;103:307–12.

Parikh S, Hyman D. Hepatocellular cancer: a guide for the internist. Am J Med. 2007;120:194–202.

Pineau P, Volinia S, McJunkin K, et al. MiR-221 overexpression contributes to liver tumorigenesis. Proc Natl Acad Sci USA. 2010;107:264–9.

Qi P, Dou TH, Geng L, et al. Association of a variant in MIR 196A2 with susceptibility to hepatocellular carcinoma in male Chinese patients with chronic hepatitis B virus infection. Hum Immunol. 2010;71:621–6.

Randall G, Panis M, Cooper JD, et al. Cellular cofactors affecting hepatitis C virus infection and replication. Proc Natl Acad Sci USA. 2007;104:12884–9.

Ruby JG, Jan CH, Bartel DP. Intronic microRNA precursors that bypass Drosha processing. Nature. 2007;448:83–6.

Salvi A, Sabelli C, Moncini S, et al. MicroRNA-23b mediates urokinase and c-met down-modulation and a decreased migration of human hepatocellular carcinoma cells. FEBS J. 2009;276:2966–82.

Sekine S, Ogawa R, Ito R, et al. Disruption of Dicer1 induces dysregulated fetal gene expression and promotes hepatocarcinogenesis. Gastroenterology. 2009;36:2304–15.

Shimizu S, Takehara T, Hikita H, et al. The let-7 family of microRNAs inhibits Bcl-xL expression and potentiates sorafenib-induced apoptosis in human hepatocellular carcinoma. J Hepatol. 2010;52:698–704.

Steelman LS, Stadelman KM, Chappell WH, et al. Akt as a therapeutic target in cancer. Expert Opin Ther Targets. 2008;12:1139–65.

Su H, Yang JR, Xu T, et al. MicroRNA-101, down-regulated in hepatocellular carcinoma, promotes apoptosis and suppresses tumorigenicity. Cancer Res. 2009;69:1135–42.

Sun BS, Dong QZ, Ye QH, et al. Lentiviral-mediated miRNA against osteopontin suppresses tumor growth and metastasis of human hepatocellular carcinoma. Hepatology. 2008;48:1834–42.

Takahashi K, Tanabe K, Ohnuki M, et al. Induction of pluripotent stem cells from adult human fibroblasts by defined factors. Cell. 2007;131:861–72.

Takahashi K, Yamanaka S. Induction of pluripotent stem cells from mouse embryonic and adult fibroblast cultures by defined factors. Cell. 2006;126:663–76.

Thorgeirsson SS, Grisham JW. Molecular pathogenesis of human hepatocellular carcinoma. Nat Genet. 2002;31:339–46.

Tsai WC, Hsu PW, Lai TC, et al. MicroRNA-122, a tumor suppressor microRNA that regulates intrahepatic metastasis of hepatocellular carcinoma. Hepatology. 2009;49:1571–82.

Ura S, Honda M, Yamashita T, et al. Differential microRNA expression between hepatitis B and hepatitis C leading disease progression to hepatocellular carcinoma. Hepatology. 2008;49:1098–112.

Varnholt H, Drebber U, Schulze F, et al. MicroRNA gene expression profile of hepatitis C virus-associated hepatocellular carcinoma. Hepatology. 2008;47:1223–32.

Viswanathan SR, Daley GQ, Gregory RI. Selective blockade of microRNA processing by Lin28. Science. 2008;320:97–100.

Viswanathan SR, Powers JT, Einhorn W, et al. Lin28 promotes transformation and is associated with advanced human malignancies. Nat Genet. 2009;41:843–8.

Voorhoeve PM, le Sage C, Schrier M, et al. A genetic screen implicates miRNA-372 and miRNA-373 as oncogenes in testicular germ cell tumors. Cell. 2006;124:1169–81.

Wang B, Hsu SH, Majumder S, et al. TGFbeta-mediated upregulation of hepatic miR-181b promotes hepatocarcinogenesis by targeting TIMP3. Oncogene. 2010;29:1787–97.

Wang Y, Lee AT, Ma JZ, et al. Profiling microRNA expression in hepatocellular carcinoma reveals microRNA-224 up-regulation and apoptosis inhibitor-5 as a microRNA-224-specific target. J Biol Chem 2008;283:13205–15.

Wang B, Majumder S, Nuovo G, et al. Role of microRNA-155 at early stages of hepatocarcinogenesis induced by choline-deficient and amino acid-defined diet in C57BL/6 mice. Hepatology. 2009;50:1152–61.

Wong QW, Ching AK, Chan AW, et al. MiR-222 overexpression confers cell migratory advantages in hepatocellular carcinoma through enhancing AKT signaling. Clin Cancer Res. 2010;16: 867–75.

Wong QW, Lung RW, Law PT, et al. MicroRNA-223 is commonly repressed in hepatocellular carcinoma and potentiates expression of Stathmin1. Gastroenterology. 2008;135:257–69.

Wu X, Wu S, Tong L, et al. MiR-122 affects the viability and apoptosis of hepatocellular carcinoma cells. Scand J Gastroenterol. 2009;44:1332–9.

Xiong Y, Fang JH, Yun JP, et al. Effects of microRNA-29 on apoptosis, tumorigenicity, and prognosis of hepatocellular carcinoma. Hepatology. 2010;51:836–45.

Xu T, Zhu Y, Wei QK, et al. A functional polymorphism in the miR-146a gene is associated with the risk for hepatocellular carcinoma. Carcinogenesis. 2008;29:2126–31.

Xu T, Zhu Y, Xiong Y, et al. MicroRNA-195 suppresses tumorigenicity and regulates G1/S transition of human hepatocellular carcinoma cells. Hepatology. 2009;50:113–21.

Yamamoto Y, Kosaka N, Tanaka M, et al. MicroRNA-500 as a potential diagnostic marker for hepatocellular carcinoma. Biomarkers. 2009;14:529–38.

Yamashita T, Budhu A, Forgues M, et al. Activation of hepatic stem cell marker EpCAM by Wnt-ß-catenin signaling in hepatocellular carcinoma. Cancer Research. 2007;67:10831–9.

Yamashita T, Forgues M, Wang W, et al. EpCAM and alpha-fetoprotein expression defines novel prognostic subtypes of hepatocellular carcinoma. Cancer Res. 2008;68:1451–61.

Yamashita T, Ji J, Budhu A, et al. EpCAM-positive hepatocellular carcinoma cells are tumor-initiating cells with stem/progenitor cell features. Gastroenterology. 2009;136:1012–24.

Yang L, Ma Z, Wang D, et al. MicroRNA-602 regulating tumor suppressive gene RASSF1A is overexpressed in hepatitis B virus-infected liver and hepatocellular carcinoma. Cancer Biol Ther. 2010;9.

Yang J, Zhou F, Xu T, et al. Analysis of sequence variations in 59 microRNAs in hepatocellular carcinomas. Mutat Res. 2008;638:205–9.

Yao J, Liang L, Huang S, et al. MicroRNA-30d promotes tumor invasion and metastasis by targeting Galphai2 in hepatocellular carcinoma. Hepatology. 2010;51:846–56.

Ye QH, Qin LX, Forgues M, et al. Predicting hepatitis B virus-positive metastatic hepatocellular carcinomas using gene expression profiling and supervised machine learning. Nat Med. 2003;9:416–23.

Yi R, Qin Y, Macara IG, et al. Exportin-5 mediates the nuclear export of pre-microRNAs and short hairpin RNAs. Genes Dev. 2003;17:3011–6.

Zhang X, Liu S, Hu T, et al. Up-regulated microRNA-143 transcribed by nuclear factor kappa B enhances hepatocarcinoma metastasis by repressing fibronectin expression. Hepatology. 2009;50:490–9.

Chapter 8
MicroRNAs in Cervical Carcinoma

Jeong-Won Lee, Byoung-Gie Kim, and Duk-Soo Bae

Abstract MicroRNAs (miRNAs) have been shown to have important roles in various cellular processes, such as metabolism, proliferation, apoptosis, differentiation, and development. Recent studies have shown that the expression of miRNAs is deregulated in human cancers in which miRNAs can exert their effect as oncogenes or tumor suppressors. The widespread and comprehensive use of miRNA microarrays has enabled the identification of a number of miRNAs as potential biomarkers for cancer. The pathogenesis of cervical cancer is well-known to involve a multistep process that includes the transformation of normal cervical epithelium to pre-neoplastic cervical intraepithelial neoplasia that is subsequently transformed to invasive cervical cancer. Although the causal relationship between high-risk human papillomavirus (HPV) infection and cervical cancer has been well-documented in epidemiologic and functional studies, HPV infection alone is not sufficient to induce the malignant transformation of HPV-infected cells. Hence, other unidentified genetic alterations, such as miRNAs, are required. The differential expression of miRNAs in cervical cancer cells or tissues compared with normal controls has been reported, and candidate miRNAs functioning as oncogenes (including *miR-21, miR-127, miR-146a, miR-199a*) and tumor suppressors (including *miR-34a, miR-143, miR-145, miR-200a, miR-218*) in cervical cancer carcinogenesis have been suggested. Herein we have reviewed recently reported studies involving miRNAs in cervical cancer cell lines and tissues, and summarized the potential for using miRNAs as biomarkers for diagnosis, prognosis, and therapies for cervical carcinoma.

8.1 Introduction

Cervical cancer is the second most common cancer in women worldwide, with an estimated 12,200 new cases and 4,210 deaths in the US in 2010 (Jemal et al. 2010; Whiteside et al. 2008). Although many advances have recently taken place

D.-S. Bae (✉)
Department of Obstetrics and Gynecology, Samsung Medical Center, Sungkyunkwan University School of Medicine, Gangnam-gu, Seoul 135-710, Korea
e-mail: ds123.bae@samsung.com

W.C.S. Cho (ed.), *MicroRNAs in Cancer Translational Research*,
DOI 10.1007/978-94-007-0298-1_8,

in the diagnosis and management of cervical cancer, this disease is the leading cause of death from cancer in many developing countries where widespread screening by cervical cytology is still unavailable. Cervical cancer is known to result from a multi-step process that involves the transformation of normal cervical epithelium to pre-neoplastic cervical intraepithelial neoplasia that is subsequently transformed to invasive cervical cancer (Kim 2009). The causal relationship between high-risk human papillomavirus (HPV) infection and cervical cancer has been well-documented in epidemiologic and functional studies. High-risk HPVs, such as HPV-16, HPV-18, HPV-31, and HPV-33, have been detected in up to 99.7% of cervical squamous cell carcinomas and 94–100% of cervical adenocarcinomas and adenosquamous carcinomas (Kim 2009; Whiteside et al. 2008). The high-risk HPV oncoproteins, E6 and E7, contribute to cervical carcinogenesis by inactivating the cellular tumor suppressor proteins, p53 and pRb, respectively (Kim 2009; Whiteside et al. 2008). Although high-risk HPVs are associated with cervical cancer, a HPV infection alone is not sufficient to induce the malignant transformation of HPV-infected cervical epithelial cells. Characterization of new molecular markers, such as MicroRNAs (miRNAs), should help to disclose the molecular mechanisms of carcinogenesis for cervical carcinomas. Furthermore, these studies will help develop diagnostic, prognostic, and therapeutic tools for cervical carcinoma.

MiRNAs are a recently discovered class of small, non-coding RNAs that regulate gene expression (Cho 2010a; Trang et al. 2008). Mature miRNAs result from the sequential processing of primary transcripts (pri-miRNAs) mediated by two RNase III enzymes, *Drosha* and *Dicer* (Cho 2010a). Mature 18- to 25-nt-long miRNAs regulate gene expression by catalyzing the cleavage of mRNA or repressing mRNA translation (Cho 2010a; Trang et al. 2008).

There has been a significant update in the miRNAs database, with > 1,000 new hairpin sequences and > 1,700 new mature sequences, according to miRBase release 16.0 (http://microrna.sanger.ac.uk/sequences/). To date, there are few miRNAs in which a physiologic function has been elucidated in vivo and in which targets are known (Lui et al. 2007). Extensive researches in model organisms for miRNAs have revealed that miRNAs are involved in the control of developmental timing, cell proliferation, apoptosis, morphogenesis, fat metabolism, and tumorigenesis, as well as in a variety of patterning processes in plants (Cho 2010a, b).

There is increasing evidence that the expression of miRNA genes is deregulated in human cancer (Croce and Calin 2005; Gregory and Shiekhattar 2005; Cho 2010a). Specific over- or under-expression of miRNA genes has been shown to correlate with particular tumor types (Cummins et al. 2006; He et al. 2005b; Lu et al. 2005; Volinia et al. 2006). Altered miRNA expression profiles have been reported in lung cancer (Hayashita et al. 2005), breast cancer (Iorio et al. 2005), glioblastoma (Chan et al. 2005), hepatocellular carcinoma (Murakami et al. 2006), papillary thyroid carcinoma (He et al. 2005a), and colorectal cancer (Cummins et al. 2006). Moreover, some studies have suggested that miRNAs have the potential as biomarkers for diagnosis, prognosis, and cancer therapeutics in a variety of human cancers (Calin et al. 2005; Trang et al. 2008; Yanaihara et al. 2006). However, the detailed molecular basis of miRNA-mediated gene regulation is not fully understood and their role in tumorigenesis remains largely unknown.

In this chapter, we have reviewed the recently published studies involving miRNAs in cervical cancer. Specifically, miRNA expression profiles of cervical cancer to identify specific miRNAs which are important in carcinogenesis, the roles of miRNAs as prognostic markers or therapeutic targets, and the mechanism of interaction with HPV and miRNAs will be discussed in this context. However, in cervical cancer, compared with other cancers, there have not been many studies involving miRNAs research until now. Thus, there are some limitations in summarizing the roles of miRNAs for tumorigenesis, metastasis, prediction of prognosis, and therapeutic targets in cervical cancer.

8.2 MiRNA Expression in Cervical Carcinoma Cell Lines and Tissues

A miRNA expression profiling study could be a good tool to evaluate a detailed molecular analysis of its expression in the process of cancer development. A high-throughput (e.g. microarray- or quantitative PCR-based) characterization of known miRNA profiles are able to identify distinct signatures in each tumor type. Moreover, certain miRNA signatures are correlated with prognosis and can potentially be used to determine the specific course of cancer treatment (He et al. 2005a; Lu et al. 2005; Shell et al. 2007; Visone et al. 2007; Yanaihara et al. 2006).

Recently, several studies involving miRNA expression profiles in cervical cancer cell lines or tissues using a microarray technique have been reported. Wang et al. (2008) reported that normal and cancerous cervical tissues had abundant expression of *miR-23a, miR-23b, let-7a, let-7c*, and *let-7d,* whereas high expression of *miR-26a, miR-29a, miR-99a, miR-100, miR-125b, miR-143, miR-145, miR-195*, and *miR-199a* was only detected in normal cervical tissues, and high expression of *miR-16, miR-21, miR-205*, and *let-7f* was only detected in cervical cancer tissues. Using clustering analysis, they observed significantly increased expression of 18 miRNAs in cervical cancer and 15 miRNAs in normal cervical tissue. They were able to confirm that cancer tissues had reduced expression of *miR-126* and *miR-424*, and increased expression of *miR-15b, miR-16, miR-146a, miR-155*, and *miR-223,* after individual miRNA levels in each sample were quantified and normalized to *U6* expression. Functional studies showed that both *miR-143* and *miR-145* are suppressive to cell growth. When introduced into cell lines, *miR-146a* was shown to promote cell proliferation. These authors suggested that down-regulation of *miR-143* and *miR-145* and up-regulation of *miR-146a* play a role in cervical carcinogenesis (Wang et al. 2008). Lui et al. (2007) used a direct sequencing method to characterize the profiles of miRNAs and other small RNA segments for six human cervical carcinoma cell lines and five normal cervical samples. Of 166 miRNAs expressed in normal cervix and cancer cell lines, they observed significant variation in expression of 6 miRNAs between the 2 groups. They reported that reduced expression of *miR-143* and increased expression of *miR-21* were reproducibly displayed in cancer samples, suggesting the potential value of these miRNAs as tumor markers. Lee et al. (2008) used a real-time quantitative PCR array method to evaluate miRNA expression in early stage invasive squamous cell carcinomas and normal cervical squamous epithelial specimens. Among the 157 miRNAs analyzed, there was a significant difference in the expression of 70 miRNAs in comparing squamous cell carcinomas and normal epithelial tissues; 68 were up-regulated and 2 were down-regulated. Among the 70 miRNAs, 10 were the most significantly over-expressed in squamous cell carcinomas with fold changes of approximately 100 and a $p < 0.0001$, as follows: *miR-199-s, miR-9, miR-199a*, miR-199a, miR-199b, miR-145, miR-133a, miR-133b, miR-214,* and *miR-127*. By contrast, only 2 of the miRNAs, *miR-149* (2.974-fold change) and *miR-203* (3.704-fold change), showed significant down-regulation (Lee et al. 2008). They suggested that miRNA deregulation may play an important role in the malignant transformation of cervical squamous cells.

8.3 MiRNAs Functioning as Potential Oncogenes or Tumor Suppressors in Cervical Cancer

Recent evidences have shown that miRNA deregulation or mutations correlate with various human cancers, and that miRNAs can function as tumor suppressors and oncogenes (Esquela-Kerscher and Slack 2006). Studies have shown that the over-expression of miRNAs can down-regulate a tumor suppressor or other genes involved in cell growth or differentiation, thereby contributing to tumor formation by stimulating proliferation, angiogenesis, and/or invasion. Thus, miRNAs can act as oncogenes, so called "oncomirs" (Miska 2005). Similarly, the action of miRNAs in cancer is dependent on their targets for mRNA, where they act as oncogenes or tumor suppressors. It is interesting to note that some miRNAs may have dual functions as both tumor suppressors and oncogenes (Gebeshuber et al. 2009), depending on the context. Because mis-regulation of miRNA has been associated with various cancers, the identification of specific regulators of miRNAs will be helpful in developing new therapeutic agents (Shenouda and Alahari 2009).

In cervical cancers, there have been several studies involving miRNAs functioning as oncogenes or tumor suppressors in controlling cancer growth. Yao et al. (2009) reported that the inhibition of *miR-21* in HeLa cells caused profound suppression of cell proliferation, and up-regulated the expression of the tumor suppressor gene, *PDCD4*. They also provided direct evidence that *PDCD4*-3′UTR is a functional target of *miR-21* and that the 18 bp putative target site can function as the sole regulatory element in HeLa cells. These results suggested that *miR-21* may play an oncogenic role in the cellular processes of cervical cancer and may serve as a target for effective therapies.

Although the physiologic function is unknown, *miR-34a* is highly expressed in normal tissues, like the testis, lung, adrenal gland, and spleen (Dutta et al. 2007).

Table 8.1 Proposed microRNAs as oncogenes or tumor suppressor in cervical cancer

MicroRNA	Putative function	Predicted targets	References
miR-21	Oncogene	*PDCD4*	Lui et al. (2007); Yao et al. (2009)
miR-34a	TSG[a]	*p53, Notch1, Jagged1*	Pang et al. (2010); Wang et al. (2009)
miR-127	Oncogene		Lee et al. (2008)
miR-143	TSG	*ERK5*	Lui et al. (2007); Wang et al. (2008)
miR-145	TSG		Wang et al. (2008)
miR-146a	Oncogene		Wang et al. (2008)
miR-199a	Oncogene		Lee et al. (2008)
miR-200a	TSG	*ZEB1, ZEB2*	Hu et al. (2010)
miR-214	TSG	*MEK3, JNK1*	Yang et al. (2009)
miR-218	TSG	*LAMB3*	Martinez et al. (2008); Zhou et al. (2010)

[a]TSG: tumor suppressor gene.

Until now, *miR-34a* is one of the first and best studied miRNAs related to carcinogenesis. Pang et al. reported that *miR-34a* is expressed at various levels in cervical cancer cells (including HeLa, SiHa, C4I, C33a, and CaSki) and computational miRNA target prediction suggested that *Notch1* and *Jagged1* were targets of *miR-34a*. Using functional assays, *miR-34a* was shown to bind to the 3′UTRs of *Notch1* and *Jagged1*. They suggested that *miR-34a* inhibits invasiveness through regulation of the *Notch*-pathway and its downstream matrix degrading enzyme (Pang et al. 2010). This study showed that *miR-34a* has tumor suppressor activity in cervical cancer, and this finding is similar to other cancers. Yang et al. (2009) reported the level of expression of *miR-214* is down-regulated in cervical cancer compared with normal tissues. Over-expression of *miR-214* in HeLa cells significantly inhibits cell proliferation. HeLa cells, which stably over-express *miR-214*, down-regulate the expression of MEK3 and JNK1 at the mRNA and protein levels. Further investigation revealed that *miR-214* regulates the expression of MEK3 and JNK1 by targeting the 3′UTRs of these genes. Collectively, they suggested that *miR-214* negatively regulates HeLa cell proliferation by targeting the non-coding regions of MEK3 and JNK1 mRNAs.

We have summarized the reported miRNAs functioning as oncogenes or tumor suppressors in cervical cancer in Table 8.1.

8.4 Interactions of HPV Infection and MiRNA in Cervical Carcinogenesis

It is well-known that many human viruses (both DNA and RNA) produce their own viral miRNAs in the course of infection (Cai et al. 2005; Pfeffer et al. 2005; Samols et al. 2005; Tang et al. 2008; Umbach et al. 2008). Therefore, it seems that a study for the interacion of HPV infection and miRNAs in cervical cancer would be of value in evaluating the mechanism of HPV-related carcingenesis. For the presence or expression of HPV-encoded-miRNA in cervical cancer, Lui et al. (2007) reported that they did not identify any viral-encoded miRNAs from the HPV-infected cells by molecular cloning. Consistent with this observation, no viral-encoded miRNA was predicted from the HPV-18 genome by a computational method (Pfeffer et al. 2005), and no viral-derived small RNAs were detected during a latent or productive replication cycle of HPV-31 by a cloning approach (Cai et al. 2006).

Although the detection of the HPV genome for cervical cancer cell lines has been reported in several studies (Martinez et al. 2008; Zhou et al. 2010), there have been few studies addressing whether or not viral proteins, including E6 and E7, can regulate the expression of cellular miRNAs. There are several reports regarding the differential expression between HPV-positive and -negative cells and tissues of the cervix. Martinez et al. (2008) reported the differential expression of several miRNAs in HPV-positive cervical cell lines and tissues compared to normal cervical tissues and a HPV-negative cervical carcinoma cell line. They also demonstrated that *miR-218* was specifically under-expressed in HPV-positive cell lines, cervical lesions, and cancer tissues containing HPV-16 DNA compared to both C-33A and the normal cervix. Expression of the *E6* oncogene of high-risk HPV-16, but not that of low-risk HPV-6, reduced *miR-218* expression, and conversely, RNA interference of *E6/E7* oncogenes in a HPV-16-positive cell line increased *miR-218* expression. They also demonstrated that the epithelial cell-specific marker, *LAMB3,* is a target of *miR-218.* Therefore, they concluded that *LAMB3* expression is increased in the presence of the HPV-16 *E6* oncogene and this effect is mediated through *miR-218* (Martinez et al. 2008; Zhou et al. 2010) (Fig. 8.1).

Recently, *miR-34a* was identified as a direct transcriptional target of the cellular transcription factor, *p53* (Chang et al. 2007; He et al. 2007). This transactivation of *miR-34a* expression is triggered by the binding of p53 to a consensus p53 binding site identified in the *miR-34a* promoter region. Since HPV E6 oncoprotein destabilizes p53 during virus infection, one may assume down-regulation of *miR-34a* expression in most cervical cancer tissues with oncogenic HPV infections. A direct correlation between down-regulation of *miR-34a* expression and p53 degradation by HPV oncoprotein E6 in cervical cancer has not been reported. Wang et al. (2009) reported, for the first time, that a viral oncoprotein regulates cellular non-coding gene expression; i.e. the first report of viral regulation of expression of a tumor

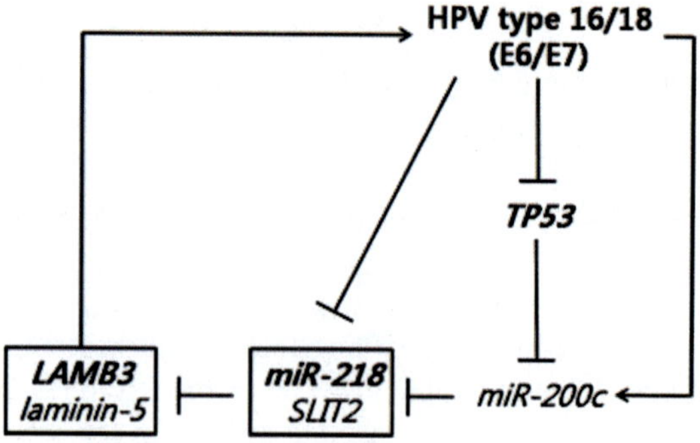

Fig. 8.1 The human papillomavirus-*TP53-miR-218-LAMB3* loop (modified from Zhou et al. 2010)

suppressor miRNA. They have shown that at all stages of pathogenesis induced by high-risk HPV types, E6 destabilization of the tumor suppressor, p53, down-regulates the tumor-suppressive *miR-34a*, leading to elevated expression of cell cycle regulators and cell proliferation.

8.5 MiRNA Predicting Cervical Cancer Prognosis

One continuing concern for oncologists is giving patients the correct prognosis and being able to predict the outcome for certain types of cancer. MiRNA expression signatures have been shown to be promising biomarkers for the classification or prediction of outcome of a wide array of human cancers (Jay et al. 2007; Yu et al. 2007). However, are few studies have been reported for the application of miRNAs to the prognosis for cervical cancer (Lee et al. 2008; Lui et al. 2007; Wang et al. 2008). Hu et al. (2010) reported that using an established PCR-based miRNA assay to analyze 102 cervical cancer samples, they identified *miR-200a* and *miR-9* as 2 miRNAs that could predict patient survival. A logistic regression model was developed based on these two miRNAs and the prognostic value of the model was subsequently validated with independent cervical cancers. They suggested that both *miR-200a* and *miR-9* could play important regulatory roles in cervical cancer control. In particular, *miR-200a* is likely to affect the metastatic potential of cervical cancer cells by coordinated suppression of multiple genes controlling cell motility.

Pelvic or para-aortic lymph node metastasis is the most important prognostic factor in cervical cancer after radical surgery. In order to evaluate lymph node metastasis before treatment, magnetic resonance imaging, computed tomography, and positron emission tomography scans are usually carried out. Although these methods can be useful in recognizing lymph node metastasis, small metastatic lymph nodes often remain undetected and conflicting results have been reported (Kim et al. 2008). Lee et al. (2008) reported the prediction of lymph node metastasis of invasive squamous cell carcinoma using specific miRNA expression. They reported that the expression of *miR-127* was significantly higher in the group of patients with squamous cell carcinomas with lymph node metastasis than in those without metastasis ($p = 0.0061$). They concluded that despite the uncertainty regarding the functional effects of miRNAs, the specific miRNA expression may be used as a prognostic marker for clinical aggressiveness, such as lymph node metastasis in cervical cancer.

8.6 MiRNAs as Potential Cancer Therapeutics

The revealing role of miRNAs functioning as potential oncogenes and tumor suppressors in tumorigenesis has generated great interest in using them as targets for cancer therapies. General therapeutic strategies involving antisense-mediated inhibition of oncogenic miRNAs and miRNA replacements with miRNA mimetics or viral vector-encoded miRNAs would be important (Cho 2010a, b; Trang et al. 2008). In a recent work, Kota et al. (2009) showed that the systemic delivery of a single miRNA can cause tumors from a mouse model of liver cancer to regress. This study illustrated that a single miRNA is powerful enough to have a dramatic suppression of tumor progression. Moreover, the use of miRNA therapy to complement traditional anti-cancer treatments appears to have great potential as it has been reported by several groups that miRNAs can enhance the response and suppress resistance to cytotoxic chemotherapy (Meng et al. 2006; Cochrane et al. 2009) or radiation therapy (Weidhaas et al. 2007).

Recent evidences have shown that various cancer-related miRNAs with potential roles as therapeutic targets for cervical cancer treatment have been reported (Esquela-Kerscher and Slack 2006; Reshmi and Pillai 2008; Cho 2010a, b). Yao et al. (2009) reported that the inhibition of *miR-21* in HeLa cervical cancer cells caused profound suppression of cell proliferation by up-regulating the expression of the tumor suppressor gene, *PDCD4*. Their findings demonstrated that *miR-21* plays the role of an oncogene in the cellular processes of cervical cancer and may serve as a target for effective therapies. Pang et al. reported that *miR-34a* regulates Notch signaling in cervical cell lines. Forced-expression of *miR-34a* inhibited not only *Notch1* and *Jagged1* expression, but also Notch signaling and reduced cell invasion. They proposed that *miR-34a* is a useful target molecule for developing new therapy against cancer (Pang et al. 2010). Lee et al. (2008) reported that *miR-199a* is one of the most significantly over-expressed miRNAs in squamous cell carcinomas compared with normal cervix, and anti-*miR-199a* reduced cervical cancer cell (SiHa and ME-180) growth and increased cisplatin-induced cytotoxicity. This study, for the first time, reported the role of miRNA in chemosensitivity in cervical cancer. They suggested that DNA damage caused by cisplatin might have increased the growth inhibition of anti-*miR-199*a. Hu et al. (2010) reported that *miR-200a* could potentially act as a master suppressor for cervical cancer metastasis. Indeed, cervical cancer cell motility was significantly reduced when *miR-200a* was over-expressed. Thus, one potential tumor control strategy could be the manipulation of *miR-200a* expression. Recent studies indicate that the therapeutic delivery of *miR-26a* and *miR-31* can suppress liver and breast cancers, respectively (Kota et al. 2009; Valastyan et al. 2009). Therefore, it is reasonable to expect that the therapeutic delivery of *miR-200a* could be a promising new treatment strategy for cervical cancer control (Hu et al. 2010).

8.7 Conclusions

Because there are not many studies pertaining to miRNAs in cervical carcinomas, the roles of miRNAs in carcinogenesis associated with HPV infection and the clinical significance, including prognostic factors or therapeutic targets, have been not well established. Although there is still much to be learned concerning the mechanism of miRNAs in cervical cancer, researchers have applied their knowledge to use miRNAs for the diagnosis, prognosis, and therapeutic use for cervical cancer. Recently, many studies involving either anti-miR knockdown or miRNA replacement therapy have moved into animal models with highly encouraging results for cancer therapeutics. To develop an in vivo study for miRNA therapy, the efficient development and delivery of a sufficient amount of anti-miR into the proper target cell is required. In addition, if miRNA-based therapies are to be evaluated in clinical studies, unwanted side effects or toxicities related to miRNA administration may occur in subjects and should be considered.

References

Cai X, Li G, Laimins LA, et al. Human papillomavirus genotype 31 does not express detectable microRNA levels during latent or productive virus replication. J Virol. 2006;80:10890–3.

Cai X, Lu S, Zhang Z, et al. Kaposi's sarcoma-associated herpesvirus expresses an array of viral microRNAs in latently infected cells. Proc Natl Acad Sci USA. 2005;102:5570–5.

Calin GA, Ferracin M, Cimmino A, et al. A MicroRNA signature associated with prognosis and progression in chronic lymphocytic leukemia. N Engl J Med. 2005;353:1793–801.

Chan JA, Krichevsky AM, Kosik KS. MicroRNA-21 is an antiapoptotic factor in human glioblastoma cells. Cancer Res. 2005;65:6029–33.

Chang TC, Wentzel EA, Kent OA, et al. Transactivation of miR-34a by p53 broadly influences gene expression and promotes apoptosis. Mol Cell. 2007;26:745–52.

Cho WC. MicroRNAs in cancer – from research to therapy. Biochim Biophys Acta. 2010a;1805:209–17.

Cho WC. MicroRNAs: Potential biomarkers for cancer diagnosis, prognosis and targets for therapy. Int J Biochem Cell Biol. 2010b;42:1273–81.

Cochrane DR, Spoelstra NS, Howe EN, et al. MicroRNA-200c mitigates invasiveness and restores sensitivity to microtubule-targeting chemotherapeutic agents. Mol Cancer Ther. 2009:8(5);1055–66.

Croce CM, Calin GA. MiRNAs, cancer, and stem cell division. Cell. 2005;122:6–7.

Cummins JM, He Y, Leary RJ, et al. The colorectal microRNAome. Proc Natl Acad Sci USA. 2006;103:3687–92.

Dutta KK, Zhong Y, Liu YT, et al. Association of microRNA-34a overexpression with proliferation is cell type-dependent. Cancer Sci. 2007;98:1845–52.

Esquela-Kerscher A, Slack FJ. Oncomirs – microRNAs with a role in cancer. Nat Rev Cancer. 2006;6:259–69.

Gebeshuber CA, Zatloukal K, Martinez J. MiR-29a suppresses tristetraprolin, which is a regulator of epithelial polarity and metastasis. EMBO Rep. 2009;10:400–5.

Gregory RI, Shiekhattar R. MicroRNA biogenesis and cancer. Cancer Res. 2005;65:3509–12.

Hayashita Y, Osada H, Tatematsu Y, et al. A polycistronic microRNA cluster, miR-17-92, is overexpressed in human lung cancers and enhances cell proliferation. Cancer Res. 2005;65:9628–32.

He L, He X, Lim LP, et al. A microRNA component of the p53 tumour suppressor network. Nature. 2007;447:1130–4.

He H, Jazdzewski K, Li W, et al. The role of microRNA genes in papillary thyroid carcinoma. Proc Natl Acad Sci USA. 2005a;102:19075–80.

He L, Thomson JM, Hemann MT, et al. A microRNA polycistron as a potential human oncogene. Nature. 2005b;435:828–33.

Hu X, Schwarz JK, Lewis JS, Jr., et al. A microRNA expression signature for cervical cancer prognosis. Cancer Res. 2010;70:1441–8.

Iorio MV, Ferracin M, Liu CG, et al. MicroRNA gene expression deregulation in human breast cancer. Cancer Res. 2005;65:7065–70.

Jay C, Nemunaitis J, Chen P, et al. MiRNA profiling for diagnosis and prognosis of human cancer. DNA Cell Biol. 2007;26:293–300.

Jemal A, Siegel R, Xu J, et al. Cancer statistics, 2010. CA Cancer J Clin. 2010;60:277–300.

Kim YT. Current status of cervical cancer and HPV infection in Korea. J Gynecol Oncol. 2009;20:1–7.
Kim TJ, Choi JJ, Kim WY, et al. Gene expression profiling for the prediction of lymph node metastasis in patients with cervical cancer. Cancer Sci. 2008;99:31–8.
Kota J, Chivukula RR, O'Donnell KA, et al. Therapeutic microRNA delivery suppresses tumorigenesis in a murine liver cancer model. Cell. 2009;137:1005–17.
Lee JW, Choi CH, Choi JJ, et al. Altered MicroRNA expression in cervical carcinomas. Clin Cancer Res. 2008;14:2535–42.
Lu J, Getz G, Miska EA, et al. MicroRNA expression profiles classify human cancers. Nature. 2005;435:834–8.
Lui WO, Pourmand N, Patterson BK, et al. Patterns of known and novel small RNAs in human cervical cancer. Cancer Res. 2007;67:6031–43.
Martinez I, Gardiner AS, Board KF, et al. Human papillomavirus type 16 reduces the expression of microRNA-218 in cervical carcinoma cells. Oncogene. 2008;27:2575–82.
Meng F, Henson R, Lang M, et al. Involvement of human micro-RNA in growth and response to chemotherapy in human cholangiocarcinoma cell lines. Gastroenterology. 2006;130:2113–29.
Miska EA. How microRNAs control cell division, differentiation and death. Curr Opin Genet Dev. 2005;15:563–8.
Murakami Y, Yasuda T, Saigo K, et al. Comprehensive analysis of microRNA expression patterns in hepatocellular carcinoma and non-tumorous tissues. Oncogene. 2006;25:2537–45.
Pang RT, Leung CO, Ye TM, et al. MicroRNA-34a suppresses invasion through down-regulation of Notch1 and Jagged1 in cervical carcinoma and choriocarcinoma cells. Carcinogenesis. 2010;31(6);1037–44.
Pfeffer S, Sewer A, Lagos-Quintana M, et al. Identification of microRNAs of the herpesvirus family. Nat Methods. 2005;2:269–76.
Reshmi G, Pillai MR. Beyond HPV: oncomirs as new players in cervical cancer. FEBS Lett. 2008;582:4113–6.
Samols MA, Hu J, Skalsky RL, et al. Cloning and identification of a microRNA cluster within the latency-associated region of Kaposi's sarcoma-associated herpesvirus. J Virol. 2005;79: 9301–5.
Shell S, Park SM, Radjabi AR, et al. Let-7 expression defines two differentiation stages of cancer. Proc Natl Acad Sci USA. 2007;104:11400–5.
Shenouda SK, Alahari SK. MicroRNA function in cancer: oncogene or a tumor suppressor? Cancer Metastasis Rev. 2009;28:369–78.
Tang S, Bertke AS, Patel A, et al. An acutely and latently expressed herpes simplex virus 2 viral microRNA inhibits expression of ICP34.5, a viral neurovirulence factor. Proc Natl Acad Sci USA. 2008;105:10931–6.
Trang P, Weidhaas JB, Slack FJ. MicroRNAs as potential cancer therapeutics. Oncogene. 2008;27 Suppl 2:S52–7.
Umbach JL, Kramer MF, Jurak I, et al. MicroRNAs expressed by herpes simplex virus 1 during latent infection regulate viral mRNAs. Nature. 2008;454:780–3.
Valastyan S, Reinhardt F, Benaich N, et al. A pleiotropically acting microRNA, miR-31, inhibits breast cancer metastasis. Cell. 2009;137:1032–46.
Visone R, Pallante P, Vecchione A, et al. Specific microRNAs are downregulated in human thyroid anaplastic carcinomas. Oncogene. 2007;26:7590–5.
Volinia S, Calin GA, Liu CG, et al. A microRNA expression signature of human solid tumors defines cancer gene targets. Proc Natl Acad Sci USA. 2006;103:2257–61.
Wang X, Tang S, Le SY, et al. Aberrant expression of oncogenic and tumor-suppressive microRNAs in cervical cancer is required for cancer cell growth. PLoS One. 2008;3:e2557.
Wang X, Wang HK, McCoy JP, et al. Oncogenic HPV infection interrupts the expression of tumor-suppressive miR-34a through viral oncoprotein E6. RNA. 2009;15:637–47.
Weidhaas JB, Babar I, Nallur SM, et al. MicroRNAs as potential agents to alter resistance to cytotoxic anticancer therapy. Cancer Res. 2007;67:11111–6.

Whiteside MA, Siegel EM, Unger ER. Human papillomavirus and molecular considerations for cancer risk. Cancer. 2008;113:2981–94.

Yanaihara N, Caplen N, Bowman E, et al. Unique microRNA molecular profiles in lung cancer diagnosis and prognosis. Cancer Cell. 2006;9:189–98.

Yang Z, Chen S, Luan X, et al. MicroRNA-214 is aberrantly expressed in cervical cancers and inhibits the growth of HeLa cells. IUBMB Life. 2009;61:1075–82.

Yao Q, Xu H, Zhang QQ, et al. MicroRNA-21 promotes cell proliferation and down-regulates the expression of programmed cell death 4 (PDCD4) in HeLa cervical carcinoma cells. Biochem Biophys Res Commun. 2009;388:539–42.

Yu SL, Chen HY, Yang PC, et al. Unique MicroRNA signature and clinical outcome of cancers. DNA Cell Biol. 2007;26:283–92.

Zhou X, Chen X, Hu L, et al. Polymorphisms involved in the miR-218-LAMB3 pathway and susceptibility of cervical cancer, a case-control study in Chinese women. Gynecol Oncol. 2010;117:287–90.

Chapter 9
MicroRNAs in Esophageal Cancer

Jian Gu and Xifeng Wu

Abstract Esophageal cancer (EC) is the sixth leading cause of cancer death worldwide. More than 90% of ECs are either esophageal squamous cell carcinomas (ESCC) or adenocarcinomas (EAC). These two histological subtypes share certain risk factors and pathogenic pathways, but mostly have distinct etiology and pathogenesis. The development of EAC generally follows the metaplasia-dysplasia-carcinoma sequence and Barrett's esophagus (BE) is a precursor of EAC. MicroRNAs (miRNAs) are small non-coding RNAs that bind to the 3′ untranslated region of target mRNA and post-transcriptionally regulates gene expression. Several miRNA profiling and candidate miRNA studies have identified changes in miRNA expression that take place during the pathogenesis and progression of EC. *MiR-21* is up-regulated in both ESCC and EAC, whereas a number of other miRNAs are distinctly dysregulated in ESCC or EAC. MiRNA expression profiles can distinguish between BE and EAC. The *miR-106b-25* polycistron is up-regulated progressively at successive stages of BE and EAC. *MiR-196a* is up-regulated in high-grade dysplasia and EAC, and is a potential biomarker of progression from BE to EAC. A few mRNA targets of select miRNAs, including *miR-21, miR-196a, miR-10b, miR-145, miR-133a, miR-133b*, and *miR-106b-25*, have been identified and validated in EC. Several studies have shown the prognostic values of selected miRNAs. Finally, germline genetic variations affecting the binding of miRNA and mRNA have been found to be associated with the risk of EC. A better understanding of the miRNA expression in EC may provide new avenues for the diagnosis, prognosis, and therapy of this deadly disease.

9.1 Introduction

Esophageal cancer is the eighth most common and the sixth most lethal cancer in the world (Garcia et al. 2007). Esophageal cancer has a very poor prognosis because

X. Wu (✉)
Department of Epidemiology, MD Anderson Cancer Center, The University of Texas, Houston, TX 77030, USA
e-mail: xwu@mdanderson.org

W.C.S. Cho (ed.), *MicroRNAs in Cancer Translational Research*,
DOI 10.1007/978-94-007-0298-1_9,

most tumors are diagnosed with advanced diseases and current therapies are largely ineffective for unresectable patients. The overall 5-year survival rate for esophageal cancer remains below 20% (Enzinger and Mayer 2003). This dismal statistics highlights the need for risk prediction, prevention, early diagnosis, better treatment, and improved prediction of prognosis and treatment response. Molecular markers can be highly valuable in each of these steps that will help improve the clinical outcomes of esophageal cancer.

More than 90% of esophageal cancers are either esophageal squamous cell carcinomas (ESCC) or adenocarcinomas (EAC). ESCC forms in squamous cells and most often occurs in the upper and middle part of the esophagus, but can occur anywhere along the esophagus. EAC begins in glandular cells and usually forms in the lower part of the esophagus near the stomach. There are striking geographic differences in the incidences of these two histological subtypes (Hongo et al. 2009). Once a rare tumor representing about 5% of esophageal cancers in the US, EAC is the cancer with the fastest increasing incidence in the US in the past few decades and has replaced ESCC as the most common histological type since the mid 1990s (Pera and Pera 2001). The reason for the dramatic increase of EAC in Western countries is unknown. However, the incidence of EAC remains rare in the Far East and ESCC still dominates in Asian and African countries. ESCC and EAC share a few etiological factors but mostly have distinct etiology. Smoking is a common risk factor for both types, but the effect is much stronger in ESCC than in EAC (Hongo et al. 2009; Pera and Pera 2001). Any factor that causes chronic irritation and inflammation of the esophageal mucosa appears to increase the incidence of ESCC, whereas EAC is associated with Barrett's esophagus (BE), gastro-esophageal reflux disease (GERD), obesity, esophageal ulcer, and frequent use of histamine-H(2) blockers (Pera and Pera 2001).

The common and distinct etiologic risk factors of ESCC and EAC dictate that their carcinogenic processes may involve the same and different molecular pathways. Previous comparative genomic hybridization (CGH) and high-density single nucleotide polymorphism (SNP) array studies have identified many chromosomal regions with high-frequency loss and gain in ESCC and EAC (Nancarrow et al. 2008; Riegman et al. 2001; Li et al. 2008; Gu et al. 2010; Chen et al. 2008a; Hu et al. 2009; Qin et al 2008). Loss of heterogeneity (LOH) of 9p21 and 17p13, and gain of 8q24 are among the most frequent events in both ESCC and EAC, suggesting that *p16*, *p53*, and *Myc* genes are involved in the carcinogenesis of both subtypes. LOH of 13q14, where *Rb* locates, is much more frequent in ESCC than in EAC. LOH of 3p is one of the most common aberrations in both subtypes; however, the LOH region of 3p in EAC and its precursor BE is often small and located within the *FHIT* gene at 3p14.2, the most frequently activated common fragile site (CFS) in human genome (Nancarrow et al. 2008; Li et al. 2008; Gu et al. 2010), whereas the LOH of 3p in ESCC tends to be large and includes several commonly deleted regions (3p26.3, 3p22, 3p21.3 and 3p14.2) (Qin et al. 2008; Zhu et al. 2009). There are extensive focal chromosomal losses in CFS regions in EAC and BE (Nancarrow et al. 2008; Gu et al. 2010), but the involvement of CFS regions in ESCC is not obvious.

Numerous transcriptional profiling and validation studies have identified many differentially expressed genes between tumor and normal tissues of esophagus. Several common pathways and genes were found; for example, the expression of matrix metalloproteinase (*MMP*) family, *CDC25B*, and *LAMC2* genes were up-regulated in both ESCC and EAC compared to normal tissues (Hourihan et al. 2003; Lin et al. 2009; Shou et al. 2008; Su et al. 2003; Wang et al. 2006). However, since transcriptional profiling typically evaluated 20,000–50,000 transcripts, there were often hundreds of differentially expressed genes in the initial screening. Moreover, there were often very few overlaps of the top genes from arrays assessing the same cancer type but conducted by different groups. A recent transcriptional profiling of 41,000 genes in ESCC found that a total of 2,235 genes were differentially regulated in the tumors, of which 881 were significantly up-regulated (Kashyap et al. 2009). It is a daunting task to validate these many genes.

MicroRNAs (miRNAs) are a class of small non-coding endogenous RNAs of 18–25 nucleotides, which was first discovered in *Caenorhabditis elegans* in early 1990s (Lee et al. 1993) and has since been found highly conserved among a wide variety of species (Wheeler et al. 2009). The number of miRNAs keeps increasing. The first release of the miRBase (December 2002) had a total of 210 sequences. The latest version of miRBase (Release 16, September 2010) contained 15,172 entries representing hairpin precursor miRNAs, expressing 17,341 mature miRNA products in 142 species. The number of annotated human miRNA sequences in this version was 1,048. MiRNAs are capable of simultaneous regulation of hundreds of genes through binding to the 3′ untranslated region (3′UTR) of target mRNAs resulting in either mRNA degradation or translation inhibition (Calin and Croce 2006; Lim et al. 2005). It was estimated that miRNAs may regulate one third to as many as two thirds of human and mammalian genes (Friedman et al. 2009; Lewis et al. 2005). The regulation of numerous human genes by miRNAs leads to a variety of physiological functions of miRNAs including development, cell differentiation, proliferation, migration, apoptosis, metabolism, and defense, and pathologic roles of miRNAs in diverse diseases including cancer, cardiovascular disease, stroke, neurodegenerative disease, diabetes, liver disease, kidney disease and infectious disease (Zhang 2009). MiRNAs have attracted particular interest in oncologic research. More than half of miRNA genes are located in cancer-associated genomic regions or in fragile sites in human genome (Calin et al. 2004). Changes of miRNA expression have been observed in almost all solid tumors and hematological malignancies studied (Lu et al. 2005; Calin and Croce 2006). Over-expressed miRNAs in cancers, such as *miR-21* and *miR-17-92* cluster, function as oncogenes and promote cancer development by negatively regulating tumor suppressor genes, whereas under-expressed miRNAs, such as *let-7* family and *miR-34* family miRNAs, function as tumor suppressor genes by regulating oncogenes and genes that control cell differentiation or apoptosis (Garzon et al. 2009). Identification of miRNAs that are altered during the initiation, development, progression, metastasis, and treatment of cancer may not only elucidate the mechanisms of carcinogenesis, but also identify biomarkers of early detection, prognosis and treatment response, and potential targets for prevention and therapy. MiRNAs have been suggested as better biomarkers than mRNAs because of several advantages: first, a single miRNA may regulate hundreds of mRNAs and therefore an array of a few hundreds of miRNAs may contain much more information than an array of tens of thousands of mRNAs, which may explain why an array of 217 microRNAs provided better clustering than an array of ~16,000 mRNAs (Lu et al. 2005); second, the total number of miRNAs is much smaller than that of mRNAs and it is easier to screen and validate biomarkers from a pool of several hundreds of miRNAs than from over 40,000 mRNA transcripts; third, miRNAs are small in size, contain stem-loop structure, and are therefore more stable than mRNAs and less subjected to degradation during sample processing, which is specially beneficial to formalin-fixed paraffin-embedded (FFPE) tissues.

In esophageal cancer, a number of studies using either a candidate miRNA or a genome-wide screening followed by validation approach have been conducted and several miRNAs have been consistently found altered in ESCC and/or EAC, which may become promising biomarkers for detection, diagnosis, prognosis, and/or therapy. In addition, a few miRNA target genes have also been identified that have shed light into the molecular mechanisms of esophageal tumorigenesis. This chapter will provide a thorough review of our current understanding of miRNAs in esophageal cancer.

9.2 MiRNAs in BE and EAC Development

The incidence of EAC in the US has increased about 3-fold in the past 3 decades and currently accounts for approximately three quarters of new esophageal cancer cases in US (Enzinger and Mayer 2003; Lagergren 2005). Most EAC cases arise from BE, a precursor lesion in which the squamous epithelium of the esophagus is replaced by a metaplastic columnar epithelium. BE is estimated to be present in 1–2% of the general population and confers a 30-fold increased risk of developing EAC. The malignant progression of BE follows a generally accepted series of stages, from metaplasia, to low-grade dysplasia (LGD), to high-grade dysplasia (HGD), and finally, to adenocarcinoma. The risk of developing EAC in patients with HGD may be higher than 10% per patient-year (Schnell et al. 2001; Shaheen and Richter 2009). However, this risk has been hard to assess because the grading of dysplasia is subjective and there is relatively high inter-observer variability in its diagnosis. Independent biomarkers thus may improve the assessment of EAC risk among BE patients.

The first study of miRNA expression profile of EAC tissues was reported by Feber et al. (2008), in which they used an Ambion array containing 328 human miRNA probes to profile and compare miRNA expression in 10 EAC, 10 ESCC, 5 BE, 1 HGD, and 9 normal squamous epithelium (NSE) tissues. Unsupervised hierarchic clustering showed that miRNA expression profiles could distinguish different esophageal tissue types and also discriminate malignant from normal esophageal tissue. Furthermore, the NSE and ESCC samples were much more similar to each other than to the EAC samples. Similarly, BE and EAC miRNA expression profiles were more similar to each other than to tissues of squamous origin (ESCC or NSE). The single HGD specimen had a miRNA expression profile similar to the EAC sample. These observations are biologically plausible since EAC is believed to arise from Barrett columnar epithelium and not from squamous epithelium. This study also identified 13 miRNAs that were significantly differentially expressed among EAC, ESCC, and NSE. The miRNAs *miR-194*, *miR-192*, and *miR-200c* were significantly up-regulated in EAC but not in ESCC; *miR-342* was up-regulated in ESCC but not EAC; *miR-21* and *miR-93* were up-regulated, and *miR-205, miR-203, miR-27b, let-7c, miR-100,* and *miR-125b* were down-regulated in both histological subtypes. It also appeared that *miR-194, miR-192,* and *miR-21* were progressively up-regulated from NSE to BE and then to EAC, suggesting that these miRNAs may be valuable biomarkers for the malignant progression of BE patients. In another more detailed study of miRNAs in the context of progression from BE to EAC, Yang et al. (2009a) analyzed the expression patterns of 470 human miRNAs using an Agilent miRNA array in 16 pairs of BE or EAC and normal tissues (including 5 LGD/normal, 5 HGD/normal, and 6 EAC/normal pairs) and then validated select miRNAs in an additional 75 pairs of disease/normal tissues. Unsupervised hierarchical clustering and class comparison analyses showed that that miRNA expression profiles of HGD and EAC tissues were significantly different from their corresponding normal tissues, but not for LGD versus normal tissues. Among the 13 miRNAs that showed differential expression between EAC and normal tissues in the above mentioned study (Feber et al. 2008), 8 (*miR-27b, miR-203, miR-205, let-7c, miR-342, miR-100, miR-21,* and *miR125b*) passed the filtering criteria of this study and all of them exhibited a consistent direction of expression change between the data of Feber and this study, showing the consistency of different array platforms. When the global false discovery rate (FDR) was controlled at 5%, there were no miRNAs that could differentiate LGD from normal tissues. In comparison, the expression of 32 and 39 miRNAs showed significant difference between HGD and normal, and between EAC and normal tissues, respectively. Among these miRNAs, 24 miRNAs exhibited the same trend of expression change in both groups, including 14 up-regulated and 10 down-regulated in disease tissues versus normal tissues. For 9 of the 14 up-regulated miRNAs (*miR-126, miR-143, miR-145, miR-181a, miR-181b, miR-199a, miR-199a*, miR-28,* and *miR-30a-5p*), the ratio of expression level of disease to normal tissues was higher in EAC than in HGD groups. For 7 of the 10 down-regulated miRNAs (*miR-149, miR-203, miR-210, miR-27b, hmiR-513, miR-617,* and *miR-99a*), the ratio of disease to normal expression level was lower in EAC than in HGD groups. These miRNAs need to be validated by quantitative reverse transcription-polymerase chain reaction (qRT-PCR). In the third study that involved miRNA array in BE and EAC, Kan et al. (2009a) first used Agilent's miRNA array to profile a primary normal esophageal epithelial cell, 3 BE-derived cell lines, and an EAC-derived cell line. There were 18 and 35 miRNAs that showed at least 2-fold up- and down-regulation, respectively, in the EAC cell line compared to the normal cell line. They then validated 7 miRNAs in 22 EAC, 24 BE, and 22 normal esophageal (NE) tissues. Notably, 3 members of the *miR-106b-25* polycistron (*miRs-25, -93,* and *-106b*) on chromosome 7q22.1 were gradually up-regulated at successive stages from NE to BE to EAC. Four miRNAs were validated as significantly down-regulated in EAC, including *miR-100, miR-125b, miR-205,* and *miR-19b*. The most recent miRNA profiling of BE and EAC tissues was conducted in Australia (Wijnhoven et al. 2010). Expression profiles of 377 human miRNAs were determined by Ambion's microarray analysis in NSE, normal gastric epithelium, BE, and EAC tissues from 16 individuals. Seven miRNAs (*miR-21, miR-143, miR-145, miR-194, miR-203, miR-205*, and *miR-215*) were chosen for validation using Taqman qRT-PCR in tissues from 32 different individuals. The expression of *miR-21, miR-143, miR-145, miR-194*, and *miR-215* were significantly higher columnar tissues than in squamous tissues, but the expression of *miR-143, miR-145,* and *miR-215* was lower in EAC than in BE tissues, whereas levels of *miR-203* and *miR-205* were significantly lower in BE and EAC tissues than in NSE. In addition, there was a trend of progressive increase of *miR-21* expression from NSE to BE and finally to EAC.

In another large study that focused on tumor tissues including 100 pairs of EAC/normal (32 pairs for training and 68 for validation) and 70 pairs ESCC/normal tissues (44 pairs for training and 26 for validation), Mathe et al. (2009) used a miRNA array chip (version 3, Ohio State University) containing 329 human and 249 mouse miRNA probes to profile miRNA expression in the training set and then used Taqman qRT-PCR to validate the expression of select miRNAs in all samples. In EAC, *miR-21, miR-223, miR-192,* and *miR-194* expression was up-regulated, whereas *miR-203* expression was down-regulated compared to normal tissues. In ESCC, the expression of *miR-21* was elevated and the expression of *miR-375* was reduced. Due to its large sample size of paired tumor/normal tissues, training and validation sets design, and qRT-PCR validation, these findings are not likely to be false positives.

From these five independent array analyses, a number of miRNAs emerged strongly as biomarkers of EAC and/or BE. *MiR-21, miR-192*, and *miR-194* were up-regulated, and *miR-203, miR-205, miR-100, miR-125b, miR-99a*, and *miR-27b* were down-regulated in EAC (Table 9.1). For *miR-21*, there was a progressive increase of expression from NSE to BE and to EAC. For the remaining of these miRNAs, it appears that their up- or down-regulation also starts from BE stage, but whether there is a stepwise trend of alteration from NSE to BE to EAC remains to be confirmed. For many miRNAs that were identified in a single study, they are also likely involved in the carcinogenesis of EAC, but they need to be validated in independent studies.

In addition to these array analyses, there were a few studies that focused on candidate miRNAs. Dijckmeester et al. (2009) showed that the expression of *miR-143* was significantly higher and that of *miR-205* was significantly lower in BE than in NSE, consistent with the aforementioned array analyses. Luthra et al. (2008) reported that *miR-196a* was up-regulated in EAC compared to NSE tissues, and Maru et al. (2009) went one step further to show that *miR-196a* levels are 10- to 100-fold higher in LGD, HGD, and EAC than in NSE tissues and that the levels of *miR-196a* proportionally increase with higher histological grades of dysplasia,

Table 9.1 Summary of studies focusing on microRNA expression in esophageal cancer

Study	Method	Up-regulated microRNAs[a]	Under-regulated microRNAs[a]
Esophageal adenocarcinoma			
Feber et al. (2008)	Array (Ambion)	***miR-21, miR-192, miR-194, miR-93,*** *miR-200c*	***miR-203, miR-205, miR-27b, miR-100, miR-125b,*** *let-7c*
Yang et al. (2009)	Array (Agilent)	*miR-126, mirR-143, miR-145, miR-146a, miR-181a and b, miR-195, miR-199a and b, miR-199a*, miR-28, miR-29c, miR-424, miR-30a-5p*	***miR-203, miR-205, miR-27b, miR-99a****, miR-149, miR-494, miR-221, miR-210, miR-513, miR-617*
Kan et al. (2009)	Array (Agilent) and Taqman	***miR-93,*** *miR-25, miR-106b*	***miR-205, miR-100, miR-125b****, miR-19b*
Mathe et al. (2009)	Array (Ohio State University)	***miR-21, miR-192, mir-194,*** *miR-223*	***miR-203***
Wijnhoven et al. (2010)	Array (Ambion)	***miR-21, miR-194***	***miR-203, miR-205*** **EAC vs. BE:** *miR-145, miR143, miR-215*
Maru et al. (2009)	Taqman	*miR-196a* (progressive from NSE to BE to EAC)	
Dijckmeester et al. (2009)	Taqman	*miR-143* (BE vs. NSE)	***miR-205*** **(BE vs. NSE)**
Esophageal squamous cell carcinoma			
Guo et al. (2008)	Array (custom)	***miR-25, miR-151****, miR-424*	***miR-100, miR-99a****, miR-29c, mmu-miR-140**
Feber et al. (2008)	Array (Ambion)	***miR-21****, miR-93, miR-342*	***miR-203, miR-205, miR-27b, miR-100, miR-125b,*** *let-7c*
Mathe et al. (2009)	Array (Ohio State University)	***miR-21***	*miR-375*
Lee et al. (2009)	Array (Ambion)	*miR-330, miR-340, miR-373, let-7d*	
Ogawa et al. (2009)	Taqman	*miR-20b, miR-34b, miR-34c, miR-129, miR-130b, miR-138* (> 4-fold), ***miR-21, miR-25, miR-151*** and thirteen more (2- to 4-fold)	***miR-133a, miR-133b, miR-145****, miR-139*
Kimura et al. (2010)	Array (Ambion) and Taqman	***miR-21***	***miR-205***
Kano et al. (2010)	Taqman Array Card		***miR-203, miR-100, miR-99a, miR-133a and b, miR-145,*** *miR-143, miR-99b, miR-1, miR-30a-30p, miR-504, miR-139-5p, miR-204, miR-326*
Tian et al. (2010)	Taqman	*miR-10b*	

BE, Barrett's esophagus; EAC, Esophageal adenocarcinoma; NSE, Normal squamous epithelium.
[a]MicroRNAs that are consistent in at least two studies are shown in bold.

suggesting that *miR-196a* alteration is an early event in carcinogenesis of EAC and may be a good potential biomarker of malignant progression in BE.

9.3 MiRNAs in ESCC Development

ESCC is the dominant histological subtypes in Far East and all of the published studies evaluating miRNAs in ESCC have used tissue samples from patients in East Asia (Guo et al. 2008; Ogawa et al. 2009; Mathe et al. 2009; Kano et al. 2010; Kimura et al. 2010; Lee et al. 2009). Guo et al. (2008) published the first miRNA array analysis in esophageal cancer. A custom-designed miRNA array containing 509 mature miRNA sequences from human, rat, and mouse was used to compare miRNA expressions in 31 pairs of ESCC and normal tissues (training set) and 24 paired samples and 1 unpaired ESCC sample (validation set). From a total of 191 evaluable miRNAs, 7 were identified as the best classifier to distinguish ESCC tissues from normal tissues. Among the 7 miRNAs, 3 (*miR-25, miR-424*, and *miR-151*) were over-expressed and 4 (*miR-100, miR-99a, miR-29c*, and *mmu-miR-140**) were under-expressed in ESCC versus normal tissues. It is noteworthy that 4 (*miR-25, miR-151, miR-100*, and *miR-99a*) out of these 7 miRNAs have been validated in at least one of the following studies. Lee et al. (2009) performed microarray analysis for 210 mature human miRNAs (Ambion' array chip) in 5 pairs of ESCC tumors and their adjacent non-tumor esophageal tissues. Four miRNAs, *let-7d, miR-330, miR-340,* and *miR-373*, showed over-expression (> 1.5-fold) in all the 5 ESCC tumors. The sample size of this study was small. Ogawa et al. (2009) quantified the expression of 73 miRNAs by qRT-PCR in 30 primary ESCC specimens and paired normal tissues. They found 6 miRNAs that were up-regulated at least 4-fold (*miR-20b, miR-34b, miR-34c, miR-129, miR-130b,* and *miR-138*), 16 miRNAs were up-regulated 2- to 4-fold (including *miR-21, miR-25,* and *miR-151*), and 4 miRNAs were down-regulated at least 2-fold (*miR-133a, miR-133b, miR-139,* and *miR-145*) in ESCC. Among these differentially expressed miRNAs, *miR-21* has been consistently shown to be up-regulated in ESCC in two other studies (Mathe et al. 2009; Kimura et al. 2010). *MiR-25* and *miR-151* over-expression was consistent with the study of Guo et al. (2008). A most recent study by Kano et al. (2010) used Taqman human miRNA array to determine the expression of 365 mature human miRNAs in 10 matched pairs of ESCC and non-cancerous epithelium and identified 15 down-regulated miRNAs (*miR-375, let-7c, miR-145, miR-143, miR-100, miR-133a, miR-99a, miR-133b, miR-1, miR-30a-3p, miR-504, miR-139-5p, miR-204, miR-203*, and *miR-326*) in ESCC tissues. This group previously identified 7 down-regulated miRNAs in bladder tumors and a comparison of these two cancer sites revealed 4 commonly down-regulated miRNAs (*miR-145, miR-30a-3p, miR-133a,* and *miR-133b*), which is highly consistent with the study of Ogawa et al. (2009) with 3 (*miR-133a, miR-133b,* and *miR-145*) out of 4 down-regulated miRNAs overlapped between these two studies. In addition, two additional down-regulated miRNAs (*miR-100* and *miR-99a*) were consistent with the above mentioned study of Guo et al. (2008). These two miRNAs have also been shown to be down-regulated in EAC tissues (Feber et al. 2008; Kan et al. 2009a; Yang et al. 2009a). Of interesting note, Kimura et al (2010) recently showed that the expression of *miR-205* was high in both malignant and benign squamous epithelia and low in cell lines and tissues other than squamous epithelia, suggesting that *miR-205* may be a specific biomarker of squamous epithelia. This observation is in line with the consistent under-expression of *miR-205* in EAC tissues compared to normal squamous epithelial tissues (Feber et al. 2008; Kan et al. 2009a; Yang et al. 2009a; Wijnhoven et al. 2010). Finally, Tian et al. (2010) reported that *miR-10b* was up-regulated in 95% (38 of 40) of ESCC tissues. In summary, the consistently up-regulated miRNAs in ESCC (at least in two studies) include *miR-21, miR-25,* and *miR-151*, and down-regulated miRNAs include *miR205, miR-203, miR-145, miR-100, miR-99a, miR-27b, miR-125b, miR-133a*, and *miR-133b* (Table 9.1).

9.4 MiRNA as Prognostic Factors

There were only a few studies that evaluated miRNAs as prognostic factors for esophageal cancer, and the results have been inconclusive for a number of reasons, including small sample size, tumor heterogeneity, treatment heterogeneity, and different assay methods. Guo et al. (2008) showed that high expression of human *miR-103* or *miR107* (the mature forms of these two miRNAs were nearly identical) was associated with poor overall survival of Chinese ESCC patients in a training set (31 patients) and a testing set (22 patients). Ogawa et al. (2009) found that high expression of *miR-129* conferred an 18-fold increased risk of death in 30 surgically treated Japanese ESCC patients ($p = 0.031$). An additional 19 patients (testing set) was analyzed for independent validation ($p = 0.011$). These results suggest that *miR-129* may become a biomarker of prognosis and survival in post-operative ESCC patients. Mathe et al. (2009) showed that higher *miR-21* expression in non-cancerous tissue of 69 ESCC patients was associated with poor survival; and low levels of *miR-375* in EAC tissues were strongly associated with worse prognosis. In the largest study so far (158 tissue specimens, 99 from EAC and 59 from ESCC patients), Hu et al. (2011) measured the expression levels of 10 select miRNAs in these tissue specimens by in situ hybridization, which is capable of detecting positive signals at the cellular level. Kaplan-Meier analysis showed that *miR-16-2* expression and *miR-30e* expression was associated with shorter survival in all esophageal cancer patients. In addition, *miR-16-2, miR-30e,* and *miR-200a* expression was associated with shorter survival in EAC patients, but not in ESCC patients. These scattered, heterogeneous findings reflect the major challenges in clinical outcome study of rare cancers due to difficulty in obtaining well-characterized tissue specimens from homogeneously treated patients. Collaborative efforts from investigators in the esophageal cancer community are needed to conduct well-designed, sufficiently powered studies to identify miRNAs as biomarkers for the prognosis and treatment response of esophageal cancer.

9.5 Genes and Molecular Pathways Regulated by MiRNAs in Esophageal Cancer

One of the most challenging tasks in miRNA research is the identification and experimental validation of miRNA targets, more importantly, the mechanistic elucidation of physiological and pathological roles of the specific miRNA and target gene. It is well-known that each miRNA can target hundreds of genes, and the same mRNA can be targeted by multiple miRNAs. The currently available target prediction programs often produce hundreds of potential target genes due to the imperfect complementarity between miRNA and its target, which makes experimental validation extremely important. In this regard, several recent studies have identified and validated a number of target genes for select miRNAs in esophageal cancer; however, compared to other model systems and other cancers, such in vitro studies in esophageal cancer are still scattered (Table 9.2). Future efforts are needed to find target transcripts of these miRNAs in esophageal cancer.

9.5.1 MiRNAs Acting as Oncogenes in Esophageal Cancer

9.5.1.1 *MiR-21*

MiR-21 is a multi-faceted molecule and one of the best studied miRNAs. It is the most commonly up-regulated miRNA in human cancer and is over-expressed in the majority of solid tumors and hematological malignancies that have been analyzed (Krichevsky and Gabriely 2009; Jazbutyte and Thum 2010; Selcuklu et al. 2009). *MiR-21* regulates a plethora of target genes that are involved in cellular survival, apoptosis and cell invasiveness. The consistent *miR-21* over-expression in cancer also points to *miR-21* as a new therapeutic target. A number of tumor-suppressor genes have been validated as targets of *miR-21*, including tropomyosin-1 (*TPM1*), *PTEN*, *maspin*, and programmed cell death 4 (*PDCD4*) (Selcuklu et al. 2009). Hiyoshi et al. (2009) showed that PDCD4 protein levels in ESCC cells had an inverse correlation with *miR-21* expression. Anti-*miR-21*-transfected cells increased PDCD4 protein expression without changing the *PDCD4* mRNA level, but increased the activity of a luciferase reporter containing the 3′UTR of *PDCD4*, supporting that *miR-21* targets *PDCD4* at the post-transcriptional level and regulates cell proliferation and invasion in ESCC cells. *PDCD4* is a tumor suppressor gene regulating multiple proteins in cells, depending on cell type and/or genetic background, at the transcriptional and translational level, which are involved in tumor progression, cell cycle and differentiation. The list of PDCD4-regulatable proteins includes CDKN1A (p21), E-cadherin, uPAR, CA II, dUTPase, Akt, JNK, etc. (Lankat-Buttgereit and Göke 2009). Which one(s) of these downstream targets are involved in the tumorigenesis of ESCC and EAC remain to be determined.

9.5.1.2 *MiR-106-25* Polycistron (*miR-25*, *miR-93*, and *miR-106b*)

Kan et al. (2009a) showed that the *miR-106-25* polycistron, located at chromosome 7q22.1, was up-regulated progressively at successive stages of esophageal

Table 9.2 Aberrantly expressed microRNAs and potential targets in esophageal cancer (EC)

MicroRNAs	Validated target transcripts in other systems[a]	Validated target transcripts in EC[a]	Functions
Oncogenic microRNAs			
miR-21	*PDCD4, PTEN, Cdc25A, TPM1, TM1, TIMP3, Maspin, RECK, TGFBR2, NFIB, BMPRII, LRRF1P1, BTG2, MARCKS, Peli1*	*PDCD4*	Cell cycle control, proliferation, apoptosis, migration
miR-93	*p21, FUS1, TP53INP1, PTEN*	*P21*	Cell cycle control, apoptosis
miR-106b	*p21, PTEN*	*p21*	Cell cycle control, apoptosis
miR-25	*p57, PTEN, MITF*	*Bim*	Cell cycle control, apoptosis, pigmentation
miR-192	*Period* genes *(Per 1, 2, and 3), DHFR, ZEB2*		Circadian clock, cell proliferation, migration
miR-194	*EP300*		Metastasis
miR-196a	*HOX-B7*	*ANXA, KRT5, SPRR2C, S100A9*	Cell proliferation, apoptosis
miR-10b	*Tiam 1, HOXD10*	*KLF4*	Migration and metastasis
miR-373	*LATS2, CD44*	*LATS2*	Proliferation, apoptosis, invasion
Tumor suppressor microRNAs			
miR-145	*c-Myc, IRS-1, MUC1, PPP3CA, CBFB, YES, STAT1, OCT4, SOX2, KLF4, FSCN1*	*FSCN1*	Cell proliferation, invasion, metastasis, stem cell differentiation
miR-205	*ZEB1, ZEB2, LRP1, HER3, VEGF-A, MED1*		Epithelial-mesenchymal transition, cell proliferation, migration, cellular response to hypoxia
miR-203	*TP63, ABL1*		Cell proliferation and differentiation
miR-125b	*Bak1, Bmf, CYP24, ERBB2, ERBB3, IRF-4*		Cell proliferation, differentiation, apoptosis
miR-27b	*A2BAR, PPARg, MMP-13, ST14, CYP1B1*		Cell proliferation
miR-100	*Plk1, mTOR*		Mitosis, cell growth
mIiR-133a	*UCP-2, SRF, CCND2, FSCN1*	*FSCN1*	Cardiomyocyte growth and differentiation
mIiR-133b	*MCL-1, BCL2L2, MET, FSCN1*	*FSCN1*	Cell proliferation
miR-375	*PDK1, 14-3-3zeta*		Cell proliferation, apoptosis

[a]Validated by luciferase reporter assay.

neoplasia, from NSE, to BE, and finally to EAC. In vitro and in vivo experiments showed that *miR-106b-25* polycistron exerted potential proliferative, antiapoptotic, and cell cycle-promoting effects in vitro and tumorigenic activity in vivo. Western blotting and luciferase assays showed that *miR-93* and *miR-106b* targeted *p21* by mRNA degradation, whereas *miR-25* targeted *Bim* (*Bcl-2* interacting mediator of cell death) by translational inhibition without affecting *Bim* mRNA. *p21* is a tumor suppressor gene involved in cell cycle control and apoptosis. It is a *PDCD4*-inducible gene. It is biologically conceivable that *miR-21* targets *PDCD4*, and *miR-93* and *miR-106b* target *p21*. The oncogenic function of these three miRNAs may converge on *p21* and related pathways. *Bim* is a BH3-only proapoptotic gene and is an established tumor suppressor gene in *Myc*-induced oncogenesis (Piñon et al. 2008). Chromosome 8q24 region (containing *Myc* gene) is one of the most commonly amplified region in both EAC and ESCC (Chen et al. 2008a; Gu et al. 2010; Hu et al. 2009; Li et al. 2008; Nancarrow et al. 2008; Qin et al. 2008; Riegman et al. 2001). It is therefore biologically plausible that *miR-25* acts as an oncogene in EAC by inhibiting the tumor suppressor gene *Bim*.

9.5.1.3 *MiR-10b*

MiR-10b has been reported to be over-expressed in several types of cancer, including liver cancer (Ladeiro et al. 2008), metastatic breast cancer (O'Day and Lal 2010), glioma (Sasayama et al. 2009), leukemia (Garzon et al. 2008), and ESCC (Tian et al. 2010). *MiR-10b* was also the first miRNA found to influence the invasion and metastasis of human cancer (breast cancer) (Ma et al. 2007). Similarly in ESCC, Tian et al. (2010) found a significant correlation of *miR-10b* level with cell motility and invasiveness in several human ESCC cell lines. Additionally, they identified *KLF4* (Krüppel-like factor 4) as a direct target of *miR-10b*. The Krüppel-like factor family comprises a family of evolutionarily conserved zinc finger transcription factors that regulate numerous biological processes including proliferation, differentiation, development and apoptosis (McConnell et al. 2007). *KLF4* has tumor suppressor function and the loss of *KLF4* expression has been reported in several human tumors, including colorectal, stomach, esophageal, and bladder cancers (McConnell et al. 2007). Yang et al. (2005) showed that *KLF4* regulate proliferation, apoptosis, and invasion in cultured ESCC cells. Biologically, *KLF4* regulates the expression level of p21 and mediates p53-dependent G1/S cell cycle arrest in response to DNA damage (Yoon et al. 2003). The oncogenic function of *miR-10b* is at least partially mediated by *KLF4* in ESCC.

9.5.1.4 *MiR-196a*

MiR-196a is another miRNA that is over-expressed in several types of cancer as well as pre-malignant esophageal tissues (Hui et al. 2009; Maru et al. 2009; Schimanski et al. 2009; Zhang et al. 2009). The role of *miR-196a* in esophageal cancer was originally discovered in a search of miRBase to identify potential miRNAs that regulate Annexin A1 (*ANXA1*) (Luthra et al. 2008). *ANXA1* is a critical mediator of apoptosis and is implicated in the actions of glucocorticoids, including the inhibition of cell proliferation and regulation of cell migration (Solito et al. 2001; Parente and Solito 2004). Suppression or loss of *ANXA1* expression was consistently reported in esophageal cancers (Hu et al. 2004; Paweletz et al. 2000; Xia et al. 2002). *MiR-196a* showed significant inverse correlation with *ANXA1* mRNA levels in 12 cancer cell lines. A series of in vitro assays clearly validated *ANXA1* as a bona fide target gene of *miR-196a*. *MiR-196a* promoted cell proliferation, anchorage-independent growth and suppressed apoptosis, suggesting its oncogenic potential. Its post-transcriptional inhibition of *ANXA1* expression may be one of the mechanisms. Maru et al. (2009) further identified in silico and validated experimentally a few additional target genes of *miR-196a*, including keratin 5 (*KRT5*), small proline-rich protein 2C (*SPRR2C*), and S100 calcium-binding protein A9 (*S100A9*), all of which are down-regulated during BE progression. It appears that *miR-196a* play an important oncogenic role in the malignant progression of BE via targeting a number of target genes that have tumor suppressor functions (Maru et al. 2009).

9.5.1.5 *MiR-373*

MiR-373 has been suggested as an oncogenic miRNA in testicular germ cell tumors, breast cancer, and prostate cancer and the large tumor suppressor 2 (*LATS2*) is a potential *miR-373* target (Huang et al. 2008; Voorhoeve et al. 2006; Yang et al. 2009b). *LATS2* is a tumor suppressor that binds *MDM2*, activates *p53*, inhibits cell cycle, and (Aylon et al. 2006; 2009). Lee et al. (2009) found that *miR-373* is over-expressed in ESCC cell lines and tumor tissues and its expression was inversely correlated with that of *LATS2*. In vitro assays using a luciferase reporter harboring the 3′UTR of *LATS2* showed that *miR-373* directly targets *LATS2* in esophageal cancer cells. Furthermore, overexpression of *miR-373* enhanced cell proliferation whereas inhibition of *miR-373* by anti-*miR-373* reduced cell proliferation. These findings suggested that over-expression of *miR-373* may inhibit *LATS2* expression resulting in increased cell proliferation which ultimately lead to tumorigenesis of ESCC (Lee et al. 2009)

9.5.1.6 *MiR-194*

MiR-194 is up-regulated in BE and EAC. Hino et al. (2008) showed that *MiR-194* was induced by *HNF-1*, a transcription factor induced in BE and EAC. Mees et al. (2010) showed that *miR-194* was induced during intestinal epithelial cell differentiation and identified *E-300*, a metastasis suppressor gene, as a target of *miR-194*.

9.5.2 *MiRNAs Acting as Tumor Suppressors in Esophageal Cancer*

9.5.2.1 *MiR-145*

MiR-145 has been reported to be down-regulated in a variety of cancers, including cancers of colon, prostate, lung, bladder, liver, pituatory, ovary (Sachdeva and Mo 2010), and pre-cancerous lesions of breast and colon cancer (Michael et al. 2003; Sempere et al. 2007), suggesting loss of *miR-145* is an early event in neoplastic development. A plethora of target genes of *miR-145* have been identified and validated (Gregersen et al. 2010; Ostenfeld et al. 2010; Sachdeva et al. 2009; Xu et al. 2009). In particular, Sachdeva et al. (2009) showed that *c-Myc* is a direct target for *miR-145*. *MiR-145* silences the expression of *c-Myc* and anti-*miR-145* enhances its expression. This specific silencing of *c-Myc* by *miR-145* accounts at least in part for the *miR-145*-mediated inhibition of tumor cell growth both in vitro and in vivo. In addition, *miR-145* is induced by the tumor suppressor *p53* transcriptionally. The blockade of *miR-145* by anti-*miR-145* is able to reverse the *p53*-mediated *c-Myc* repression. Therefore, *miR-145* provides a direct link between *p53* and *c-Myc*. Given that LOH of 17p13 (p53 locus) and amplification of 8q24 (*Myc* locus) are both frequent events in ESCC and EAC, *miR-145* exhibit its tumor suppressor function in esophageal cancer at least partly through the *p53-Myc* network.

MiR-145 is also down-regulated in ESCC tissues (Kano et al. 2010; Ogawa et al. 2009). Moreover, Kano et al. (2010) identified and validated a novel target gene, *FSCN1* (actin-binding protein, Fascin homolog 1), for *miR-145*. An *FSCN1* loss-of-function assay found significant cell growth and invasion inhibition, suggesting *FSCN1* is associated with ESCC carcinogenesis. *FSCN1* is a globular protein that organizes F-actin into well-ordered, tightly packed parallel bundles in cells. In studies of non-small cell lung cancer and gastric adenocarcinoma, high *FSCN1* expression in tumor tissues correlated with poor survival (Hashimoto et al. 2004; Pelosi et al. 2003). In ESCC, *FSCN1* was usually over-expressed in the tumors compared to normal epithelium, and *FSCN1* over-expression was significantly associated with the extent of the tumor, lymph node metastasis, and poor prognosis (Hashimoto et al. 2005). These data suggest that *FSCN1* is involved in the progression of ESCC and inhibition of *FSCN1* may be one of the mechanisms for the potential tumor suppressor function of *miR-145* in ESCC.

9.5.2.2 *MiR-205*

MiR-205 is significantly under-expressed in a number of cancers including breast, prostate, head and neck cancer, and esophageal cancer (Baffa et al 2009; Childs et al. 2009; Gandellini et al. 2009; Wu and Mo 2009). *MiR-205* plays an important role in the regulation of epithelial-mesenchymal transition (EMT), a process critical to tumor progression and metastasis. EMT is characterized by the loss of *E-cadherin* and reduction of cell locomotion and invasion. Gregory et al (2008) showed that *miR-205* directly target the mRNA of the *E-cadherin* transcriptional repressors and EMT-activators *ZEB1* (*TCF8/deltaEF1*) and *ZEB2* (SMAD-interacting protein 1 [*SIP1*]) in breast cell lines. Loss of *miR-205* in tumor cells therefore would result in increased EMT. Gandellini et al. (2009) showed that *miR-205* targeted protein kinase *Cepsilon*, which seemed to play a direct role in regulating EMT in human prostate cells. In fact, the down-regulation of *Cepsilon* led to a cell phenotype similar to that of cells ectopically expressing *miR-205*. Therefore, *miR-205* exerts a tumor-suppressive effect in human prostate by counteracting EMT and reducing cell migration/invasion at least in part through the down-regulation of protein kinase Cepsilon (Gandellini et al. 2009). There have not been reports about the gene targets of *miR-205* in esophageal cancer, but it is expected that it would also target *ZEB1* and *ZEB2* and loss of *miR-205* in esophageal cancer would lead to reduced E-cadherin expression and increased EMT.

9.5.2.3 *MiR-375*

MiR-375 is down-regulated in ESCC (Mathe et al. 2009). Down-regulation of *miR-375* was also reported in liver cancer and gastric cancer (Ladeiro et al. 2008; Tsukamoto et al. 2010). Tsukamoto et al. (2010) found that *miR-375* was the most down-regulated miRNA in gastric cancer. They identified *PDK1* as a direct target of *miR-375*, and inhibition *PDK1* by *miR-375* led to the suppression of Akt phosphorylation. In addition, they identified *14-3-3zeta*, a potent anti-apoptotic gene, as another direct target of *miR-375*. Furthermore, knockdown of either *PDK1* or *14-3-3zeta* in gastric carcinoma cells induced caspase activation, which was also observed in *miR-375*-transfected cells, suggesting that *miR-375* may exert its proapoptotic function, at least in part, through the down-regulation of *PDK1* and *14-3-3zeta* (Tsukamoto et al. 2010). It remains to be seen whether *miR-375* targets these two genes and what are the additional targets in esophageal cancer.

9.5.2.4 *MiR-133a* and *MiR-133b*

MiR-133a and *miR-133b* have been shown under-expressed in ESCC tissues (Kano et al. 2010; Ogawa et al. 2009) and a few cancers including bladder cancer and lung cancer (Chiyomaru et al. 2010; Crawford et al. 2009). Kano et al. (2010) found that *FSCN1* is a common target for *miR-145, miR-133a*, and *miR-133b*. Crawford et al. (2009) identified and validated two members of the *BCL-2* family (*MCL-1* and *BCL2L2*) as direct targets of *miR-133b* in lung cancer cell lines, providing possible biologic explanation for the tumor suppressor function of *miR-133b*. It would be interesting to see whether *miR-133b* also targets these genes in esophageal cancer. *MiR-133a* appears to play an important role in cardiovascular development (Meder et al. 2008). Its role in carcinogenesis and its target genes require further validations.

9.6 Germline Genetic Variations in MiRNA Genes, MiRNA Processing Genes, and MiRNA Binding Sites and the Risk of Esophageal Cancer

The binding of miRNA to the 3′UTR of mRNA of target genes is critical for regulating the mRNA level and/or protein expression. This binding can be affected by genetic variations that affect either miRNA or mRNA binding sites. Single nucleotide polymorphisms (SNPs) in 3 categories of genes may affect the structure and/or function of miRNA-mRNA pairing and are therefore candidates of genetic susceptibility to disease: miRNA-biogenesis pathway genes, precursor and mature miRNA genes, and miRNA target genes. SNPs in these categories of genes can affect the pri-miRNA and pre-miRNA processing, abolish an existing binding site, modulate the affinity of miRNA-mRNA binding, or create abnormal binding sites. Numerous studies have been published evaluating SNPs in these categories and the risks and outcomes of cancers (Ryan et al. 2010). In general, the number of SNPs in pre-miRNA, and particularly in mature miRNA, is very small, suggesting that the functional region of miRNAs is highly conservative and functionally essential. However, SNPs in miRNA target sites are a gold mine for molecular epidemiology (Chen et al. 2008b). For example, Chin et al. (2008) found an increased risk of non-small cell lung cancer in subjects with a SNP in the 3′UTR of *KRAS* gene that interferes the binding of *let-7* miRNA. Nicoloso et al. (2010) recently identified a SNP in *TGFB1* and a SNP in *XRCC1* as target SNPs, whose alleles could modulate gene expression by differential interaction with *miR-187* and *miR-138*, respectively. Genome-wide bioinformatics analysis predicted approximately 64% of transcribed SNPs as target SNPs that can modify (increase/decrease) the binding energy of putative miRNA-mRNA duplexes by > 90%. One study evaluating the SNPs in miRNA related genes with the risk of esophageal cancer has been published (Ye et al. 2008). The most notable finding was a SNP (rs6505162) located in the pre-*miR-423* region, which was associated with a per allele odds ratio (OR) of 0.64 (95% CI 0.51–0.80) under an additive genetic model (p for trend < 0.0001). This association remained significant after correction for multiple comparisons. The biological mechanism of this SNP has not been studied.

9.7 Conclusions

The discovery of miRNAs has revolutionized the basic research of gene regulation and dramatically expanded the scope of biomarker discovery. It is clear that dysregulation of miRNAs plays a pivotal role in the carcinogenesis of the majority of cancers. Multiple genome-wide profiling studies of EAC and ESCC have been performed and several up-regulated and down-regulated miRNAs have been identified. Most of these miRNAs are not specific to esophageal cancer, but showed general dysregulation in other cancer types. However, the target genes may be specific to esophageal cancer. The next critical step is to identify relevant targets for specific miRNAs. Since each miRNA may regulate hundreds of genes, the identification of target genes would likely be a tedious work. Compared to the relatively abundance of studies using miRNAs for molecular classification of normal and tissues of different origins and stages, the study of miRNAs as markers for prognosis, and particularly for treatment response of esophageal cancer, has only been sparsely reported. Future efforts should be made to conduct well-designed, sufficiently powered clinical outcome study. Although having not been tested in esophageal cancer, other potential applications related to miRNA include using circulating miRNAs in serum/plasma as minimally invasive biomarkers for early detection and targeting oncogenic miRNAs (e.g. *miR-21*) or re-introducing tumor suppressor miRNAs (e.g. *miR-145*) for cancer therapy (Kan and Meltzer 2009b; Zhou and Wang 2010). The future clinical application of miRNAs in esophageal cancer is promising.

References

Aylon Y, Michael D, Shmueli A, et al. A positive feedback loop between the p53 and Lats2 tumor suppressors prevents tetraploidization. Genes Dev. 2006;20:2687–700.

Aylon Y, Yabuta N, Besserglick H, et al. Silencing of the Lats2 tumor suppressor overrides a p53-dependent oncogenic stress checkpoint and enables mutant H-Ras-driven cell transformation. Oncogene. 2009;28:4469–79.

Baffa R, Fassan M, Volinia S, et al. MicroRNA expression profiling of human metastatic cancers identifies cancer gene targets. J Pathol. 2009;219:214–21.

Calin GA, Croce CM. MicroRNA signatures in human cancers. Nat Rev Cancer. 2006;6:857–66.

Calin GA, Sevignani C, Dumitru CD, et al. Human microRNA genes are frequently located at fragile sites and genomic regions involved in cancers. Proc Natl Acad Sci USA. 2004;101: 2999–3004.

Chen J, Guo L, Peiffer DA, et al. Genomic profiling of 766 cancer-related genes in archived esophageal normal and carcinoma tissues. Int J Cancer. 2008a;122:2249–54.

Chen K, Song F, Calin GA, et al. Polymorphisms in microRNA targets: a gold mine for molecular epidemiology. Carcinogenesis. 2008b;29:1306–11.

Childs G, Fazzari M, Kung G, et al. Low-level expression of microRNAs let-7d and miR-205 are prognostic markers of head and neck squamous cell carcinoma. Am J Pathol. 2009;174:736–45.

Chin LJ, Ratner E, Leng S, et al. A SNP in a let-7 microRNA complementary site in the KRAS 3′ untranslated region increases non-small cell lung cancer risk. Cancer Res. 2008;68:8535–40.

Chiyomaru T, Enokida H, Tatarano S, et al. miR-145 and miR-133a function as tumour suppressors and directly regulate FSCN1 expression in bladder cancer. Br J Cancer. 2010;102:883–91.

Crawford M, Batte K, Yu L, et al. MicroRNA 133B targets pro-survival molecules MCL-1 and BCL2L2 in lung cancer. Biochem Biophys Res Commun. 2009;388:483–9.

Dijckmeester WA, Wijnhoven BP, Watson DI, et al. MicroRNA-143 and -205 expression in neosquamous esophageal epithelium following Argon plasma ablation of Barrett's esophagus. J Gastrointest Surg. 2009;13:846–53.

Enzinger PC, Mayer RJ. Esophageal cancer. N Engl J Med. 2003;349:2241–52.

Feber A, Xi L, Luketich JD, Pennathur A, et al. MicroRNA expression profiles of esophageal cancer. J Thorac Cardiovasc Surg. 2008;135:255–60

Friedman RC, Farh KK, Burge CB, et al. Most mammalian mRNAs are conserved targets of microRNAs. Genome Res. 2009;19:92–105.

Gandellini P, Folini M, Longoni N, et al. MiR-205 Exerts tumor-suppressive functions in human prostate through down-regulation of protein kinase Cepsilon. Cancer Res. 2009;69: 2287–95.

Garcia M, Jemal A, Ward EM, et al. Global cancer facts & figures 2007. Atlanta, GA:American Cancer Society, 2007.

Garzon R, Calin GA, Croce CM. MicroRNAs in cancer. Annu Rev Med. 2009;60:167–79.

Garzon R, Garofalo M, Martelli MP, et al. Distinctive microRNA signature of acute myeloid leukemia bearing cytoplasmic mutated nucleophosmin. Proc Natl Acad Sci USA. 2008;105:3945–50.

Gregersen LH, Jacobsen AB, Frankel LB, et al. MicroRNA-145 targets YES and STAT1 in colon cancer cells. PLoS One. 2010;5:e8836.

Gregory PA, Bert AG, Paterson EL, et al. The miR-200 family and miR-205 regulate epithelial to mesenchymal transition by targeting ZEB1 and SIP1. Nat Cell Biol. 2008;10:593–601.

Gu J, Ajani JA, Hawk E, et al. Genome-wide catalogue of chromosomal aberrations in Barrett's esophagus and esophageal adenocarcinoma: a high-Density SNP array analysis. Cancer Prev Res (Phila). 2010;3:1176–86.

Guo Y, Chen Z, Zhang L, et al. Distinctive microRNA profiles relating to patient survival in esophageal squamous cell carcinoma. Cancer Res. 2008;68:26–33.

Hashimoto Y, Ito T, Okumura T, et al. Prognostic significance of fascin overexpression in human esophageal squamous cell carcinoma. Clin Cancer Res. 2005;11:2597–605.

Hashimoto Y, Shimada Y, Kawamura J, et al. The prognostic relevance of fascin expression in human gastric carcinoma. Oncology. 2004;67:262–70.

Hino K, Tsuchiya K, Fukao T, et al. Inducible expression of microRNA-194 is regulated by HNF-1alpha during intestinal epithelial cell differentiation. RNA. 2008;14:1433–42.

Hiyoshi Y, Kamohara H, Karashima R, et al. MicroRNA-21 regulates the proliferation and invasion in esophageal squamous cell carcinoma. Clin Cancer Res. 2009;15:1915–22.

Hongo M, Nagasaki Y, Shoji T. Epidemiology of esophageal cancer: orient to occident. Effects of chronology, geography and ethnicity. J Gastroenterol Hepatol. 2009;24:729–35.

Hourihan RN, O'Sullivan GC, Morgan JG. Transcriptional gene expression profiles of oesophageal adenocarcinoma and normal oesophageal tissues. Anticancer Res. 2003;23:161–5.

Hu N, Flaig MJ, Su H, et al. Comprehensive characterization of annexin I alterations in esophageal squamous cell carcinoma. Clin Cancer Res. 2004;10:6013–22.

Hu N, Wang C, Ng D, et al. Genomic characterization of esophageal squamous cell carcinoma from a high-risk population in China. Cancer Res. 2009;69:5908–17.

Hu Y, Correa AM, Hoque A, et al. Prognostic significance of differentially expressed miRNAs in esophageal cancer. Int J Cancer. 2011;128:132–43.

Huang Q, Gumireddy K, Schrier M, et al. The microRNAs miR-373 and miR-520c promote tumour invasion and metastasis. Nat Cell Biol. 2008;10:202–10.

Hui AB, Shi W, Boutros PC, et al. Robust global micro-RNA profiling with formalin-fixed paraffin-embedded breast cancer tissues. Lab Invest. 2009;89:597–606.

Jazbutyte V, Thum T. MicroRNA-21: from cancer to cardiovascular disease. Curr Drug Targets. 2010;11:926–35.

Kan T, Sato F, Ito T, et al. The miR-106b-25 polycistron, activated by genomic amplification, functions as an oncogene by suppressing p21 and Bim. Gastroenterology. 2009a;136:1689–700.

Kan T, Meltzer SJ. MicroRNAs in Barrett's esophagus and esophageal adenocarcinoma. Curr Opin Pharmacol. 2009b;9:727–32.

Kano M, Seki N, Kikkawa N, et al. MiR-145, miR-133a and miR-133b: tumor suppressive miRNAs target FSCN1 in esophageal squamous cell carcinoma. Int J Cancer. 2010;127:2804–14.

Kashyap MK, Marimuthu A, Kishore CJ, et al. Genomewide mRNA profiling of esophageal squamous cell carcinoma for identification of cancer biomarkers. Cancer Biol Ther. 2009; 8:36–46.

Kimura S, Naganuma S, Susuki D, et al. Expression of microRNAs in squamous cell carcinoma of human head and neck and the esophagus: miR-205 and miR-21 are specific markers for HNSCC and ESCC. Oncol Rep. 2010;23:1625–33.

Krichevsky AM, Gabriely G. MiR-21: a small multi-faceted RNA. J Cell Mol Med. 2009;13: 39–53.

Ladeiro Y, Couchy G, Balabaud C, et al. MicroRNA profiling in hepatocellular tumors is associated with clinical features and oncogene/tumor suppressor gene mutations. Hepatology. 2008;47:1955–63.

Lagergren J. Adenocarcinoma of oesophagus: what exactly is the size of the problem and who is at risk? Gut. 2005;54(Suppl 1):i1–5.

Lankat-Buttgereit B, Göke R. The tumour suppressor Pdcd4: recent advances in the elucidation of function and regulation. Biol Cell. 2009;101:309–17.

Lee KH, Goan YG, Hsiao M. MicroRNA-373 (miR-373) post-transcriptionally regulates large tumor suppressor, homolog 2 (LATS2) and stimulates proliferation in human esophageal cancer. Exp Cell Res. 2009;315:2529–38.

Lee RC, Feinbaum RL, Ambros V. The *C. elegans* heterochronic gene lin-4 encodes small RNAs with antisense complementarity to lin-14. Cell. 1993;75:843–54.

Lewis BP, Burge CB, Bartel DP. Conserved seed pairing, often flanked adenosines, indicates that thousands of human genes are microRNA targets. Cell. 2005;120:15–20.

Li X, Galipeau PC, Sanchez CA, et al. Single nucleotide polymorphism-based genome-wide chromosome copy change, loss of heterozygosity, and aneuploidy in Barrett's esophagus neoplastic progression. Cancer Prev Res (Phila) 2008;1:413–23.

Lim LP, Lau NC, Garrett-Engele P, et al. Microarray analysis shows that some microRNAs downregulate large numbers of target mRNAs. Nature. 2005;433:769–73.

Lin DC, Du XL, Wang MR. Protein alterations in ESCC and clinical implications: a review. Dis Esophagus. 2009;22:9–20.

Luthra R, Singh RR, Luthra MG, et al. MicroRNA-196a targets annexin A1: a microRNA-mediated mechanism of annexin A1 downregulation in cancers. Oncogene. 2008;27:6667–78.

Ma L, Teruya-Feldstein J, Weinberg RA. Tumour invasion and metastasis initiated by microRNA-10b in breast cancer. Nature. 2007;449:682–8.

Maru DM, Singh RR, Hannah C, et al. MicroRNA-196a is a potential marker of progression during Barrett's metaplasia-dysplasia-invasive adenocarcinoma sequence in esophagus. Am J Pathol. 2009;174:1940–8.

Mathé EA, Nguyen GH, Bowman ED, et al. MicroRNA expression in squamous cell carcinoma and adenocarcinoma of the esophagus: associations with survival. Clin Cancer Res. 2009;15: 6192–200.

McConnell BB, Ghaleb AM, Nandan MO, et al. The diverse functions of Krüppel-like factors 4 and 5 in epithelial biology and pathobiology. Bioessays. 2007;29:549–57.

Meder B, Katus HA, Rottbauer W. Right into the heart of microRNA-133a. Genes Dev. 2008;22:3227–31.

Mees ST, Mardin WA, Wendel C, et al. EP300–a miRNA-regulated metastasis suppressor gene in ductal adenocarcinomas of the pancreas. Int J Cancer. 2010;126:114–24.

Michael MZ, SM OC. van Holst Pellekaan NG, et al. Reduced accumulation of specific microRNAs in colorectal neoplasia. Mol Cancer Res. 2003;1:882–91.

Nancarrow DJ, Handoko HY, Smithers BM, et al. Genome-wide copy number analysis in esophageal adenocarcinoma using high-density single-nucleotide polymorphism arrays. Cancer Res. 2008;68:4163–72.

Nicoloso MS, Sun H, Spizzo R, et al. Single-nucleotide polymorphisms inside microRNA target sites influence tumor susceptibility. Cancer Res. 2010;70:2789–98.

O'Day E, Lal A. MicroRNAs and their target gene networks in breast cancer. Breast Cancer Res. 2010;12:201.

Ogawa R, Ishiguro H, Kuwabara Y, et al. Expression profiling of micro-RNAs in human esophageal squamous cell carcinoma using RT-PCR. Med Mol Morphol. 2009;42:102–9.

Ostenfeld MS, Bramsen JB, Lamy P, et al. MiR-145 induces caspase-dependent and -independent cell death in urothelial cancer cell lines with targeting of an expression signature present in Ta bladder tumors. Oncogene. 2010;29:1073–84.

Parente L, Solito E. Annexin 1: more than an antiphospholipase protein. Inflamm Res. 2004;53:125–32.

Paweletz CP, Ornstein DK, Roth MJ, et al. Loss of annexin 1 correlates with early onset of tumorigenesis in esophageal and prostate carcinoma. Cancer Res. 2000;60:6293–7.

Pelosi G, Pastorino U, Pasini F, et al. Independent prognostic value of fascin immunoreactivity in stage I no small cell lung cancer. Br J Cancer. 2003;88:537–47.

Pera M, Pera M. Recent changes in the epidemiology of esophageal cancer. Surg Oncol. 2001;10:81–90.

Piñon JD, Labi V, Egle A, et al. Bim and Bmf in tissue homeostasis and malignant disease. Oncogene. 2008;27(Suppl 1):S41–52.

Qin YR, Fu L, Sham PC, et al. Single-nucleotide polymorphism-mass array reveals commonly deleted regions at 3p22 and 3p14.2 associate with poor clinical outcome in esophageal squamous cell carcinoma. Int J Cancer. 2008;123:826–30.

Riegman PH, Vissers KJ, Alers JC, et al. Genomic alterations in malignant transformation of Barrett's esophagus. Cancer Res. 2001;61:3164–70.

Ryan BM, Robles AI, Harris CC. Genetic variation in microRNA networks: the implications for cancer research. Nat Rev Cancer. 2010;10:389–402.

Sachdeva M, Mo YY. MiR-145-mediated suppression of cell growth, invasion and metastasis. Am J Transl Res. 2010;2:170–80.

Sachdeva M, Zhu S, Wu F, et al. p53 represses c-Myc through induction of the tumor suppressor miR-145. Proc Natl Acad Sci USA. 2009;106:3207–12.
Sasayama T, Nishihara M, Kondoh T, et al. MicroRNA-10b is overexpressed in malignant glioma and associated with tumor invasive factors, uPAR and RhoC. Int J Cancer. 2009;125: 1407–13.
Schimanski CC, Frerichs K, Rahman F, et al. High miR-196a levels promote the oncogenic phenotype of colorectal cancer cells. World J Gastroenterol. 2009;15:2089–96.
Schnell TG, Sontag SJ, Chejfec G, et al. Long-term nonsurgical management of Barrett's esophagus with high-grade dysplasia. Gastroenterology. 2001;120:1607–19.
Selcuklu SD, Donoghue MT, Spillane C. MiR-21 as a key regulator of oncogenic processes. Biochem Soc Trans. 2009;37:918–25.
Sempere LF, Christensen M, Silahtaroglu A, et al. Altered microRNA expression confined to specific epithelial cell subpopulations in breast cancer. Cancer Res. 2007;67:11612–20.
Shaheen NJ, Richter JE. Barrett's oesophagus. Lancet. 2009;373:850–61.
Shou JZ, Hu N, Takikita M, et al. Overexpression of CDC25B and LAMC2 mRNA and protein in esophageal squamous cell carcinomas and premalignant lesions in subjects from a high-risk population in China. Cancer Epidemiol Biomarkers Prev. 2008;17:1424–35.
Solito E, de Coupade C, Canaider S, et al. Transfection of annexin 1 in monocytic cells produces a high degree of spontaneous and stimulated apoptosis associated with caspase-3 activation. Br J Pharmacol. 2001;133:217–28.
Su H, Hu N, Shih J, et al. Gene expression analysis of esophageal squamous cell carcinoma reveals consistent molecular profiles related to a family history of upper gastrointestinal cancer. Cancer Res. 2003;63:3872–6.
Tian Y, Luo A, Cai Y, et al. MicroRNA-10b promotes migration and invasion through KLF4 in human esophageal cancer cell lines. J Biol Chem. 2010;285:7986–94.
Tsukamoto Y, Nakada C, Noguchi T, et al. MicroRNA-375 is downregulated in gastric carcinomas and regulates cell survival by targeting PDK1 and 14-3-3zeta. Cancer Res. 2010;70: 2339–49.
Voorhoeve PM, le Sage C, Schrier M, et al. A genetic screen implicates miRNA-372 and miRNA-373 as oncogenes in testicular germ cell tumors. Cell. 2006;124:1169–81.
Wang S, Zhan M, Yin J, et al. Transcriptional profiling suggests that Barrett's metaplasia is an early intermediate stage in esophageal adenocarcinogenesis. Oncogene. 2006;25:3346–56.
Wheeler BM, Heimberg AM, Moy VN, et al. The deep evolution of metazoan microRNAs. Evol Dev. 2009;11:50–68.
Wijnhoven BP, Hussey DJ, Watson DI, et al. MicroRNA profiling of Barrett's oesophagus and oesophageal adenocarcinoma. Br J Surg. 2010;97:853–61.
Wu H, Mo YY. Targeting miR-205 in breast cancer. Expert Opin Ther Targets. 2009;13: 1439–48.
Xia SH, Hu LP, Hu H, et al. Three isoforms of annexin I are preferentially expressed in normal esophageal epithelia but down-regulated in esophageal squamous cell carcinomas. Oncogene. 2002;21:6641–8.
Xu N, Papagiannakopoulos T, Pan G, et al. MicroRNA-145 regulates OCT4, SOX2, and KLF4 and represses pluripotency in human embryonic stem cells. Cell. 2009;137: 647–58.
Yang H, Gu J, Wang KK, et al. MicroRNA expression signatures in Barrett's esophagus and esophageal adenocarcinoma. Clin Cancer Res. 2009a;15:5744–52.
Yang K, Handorean AM, Iczkowski KA. MicroRNAs 373 and 520c are downregulated in prostate cancer, suppress CD44 translation and enhance invasion of prostate cancer cells in vitro. Int J Clin Exp Pathol. 2009b;2:361–9.
Yang Y, Goldstein BG, Chao HH, et al. KLF4 and KLF5 regulate proliferation, apoptosis and invasion in esophageal cancer cells. Cancer Biol Ther. 2005;4:1216–21.
Ye Y, Wang KK, Gu J, et al. Genetic variations in microRNA-related genes are novel susceptibility loci for esophageal cancer risk. Cancer Prev Res (Phila) 2008;1:460–9.

Yoon HS, Chen X, Yang VW. Kruppel-like factor 4 mediates p53-dependent G1/S cell cycle arrest in response to DNA damage. J Biol Chem. 2003;278:2101–5.

Zhang C. Novel functions for small RNA molecules. Curr Opin Mol Ther. 2009;11:641–51.

Zhang Y, Li M, Wang H, et al. Profiling of 95 microRNAs in pancreatic cancer cell lines and surgical specimens by real-time PCR analysis. World J Surg. 2009;33:698–709.

Zhou SL, Wang LD. Circulating microRNAs: novel biomarkers for esophageal cancer. World J Gastroenterol. 2010;16:2348–54.

Zhu C, Qin YR, Xie D, et al. Characterization of tumor suppressive function of P300/CBP-associated factor at frequently deleted region 3p24 in esophageal squamous cell carcinoma. Oncogene. 2009;28:2821–8.

Chapter 10
MicroRNAs in Bladder Cancer

Michael B. Williams and Liana Adam

Abstract Bladder cancer is the fourth most common solid malignancy in men and fifth most common overall with an estimated 70,000 new cases of urothelial carcinoma (UC) and over 14,000 deaths from the disease expected in 2010 in the United States. Although the majority of patients with invasive bladder cancer present without radiographic or clinical evidence of disease beyond the bladder, up to 56% of patients die from the result of occult metastasis not detected by current staging modalities. The potential of microRNAs (miRNAs) as novel tumor markers has been the focus of recent scrutiny because of their tissue specificity, stability, and association with clinical-pathological parameters. Prognostic tools based on conventional clinical and pathologic staging can quantify the risk of death from UC, but their accuracy is imperfect due to the heterogeneous biologic behavior of tumors. Use of biomarkers specific to the tumor and/or patient can provide prognostic utility over that available from routine clinical features. Data have emerged documenting altered systemic miRNAs expression across a spectrum of cancers including urothelial carcinoma of the bladder. Examples include *miR-21* (up-regulated), *miR-200* family (associated with epithelial-mesenchymal transition and *Zeb1/2*), and *miR-145* (apoptosis). Assessing the expression of all known and predicted non-coding RNAs species and contrasting the miRNAs in the circulation of patients with superficial or invasive disease has great potential in determining whether we can identify systemic miRNAs as screening tools for bladder cancer.

10.1 Introduction

Bladder cancer affected nearly 71,000 people in the United States in 2010 and was the fourth most common cancer diagnosis in men. Over fourteen thousand people succumbed to the disease during the same time frame (Jemal et al. 2010). For

---

L. Adam (✉)
Department of Urology, MD Anderson Cancer Center, The University of Texas, Houston, TX 77030, USA
e-mail: ladam@mdanderson.org

W.C.S. Cho (ed.), *MicroRNAs in Cancer Translational Research*,
DOI 10.1007/978-94-007-0298-1_10, © Springer Science+Business Media B.V. 2011

this chapter, bladder cancer is strictly limited to cancer of urothelial origin (also known as transitional cell carcinoma). From a clinical perspective, bladder cancer is traditionally subdivided into either superficial (Ta or T1) or invasive (T2, T3, and T4) subtypes. The two types are quite different in their overall characteristics, with superficial disease having long-recurrence free episodes, while invasive disease often requires multimodal therapies combining surgery as well as chemotherapeutic approaches.

MicroRNA (miRNAs) expression and research in bladder cancer has increased over the last several years, but remains inconclusive with so many new miRNAs being discovered on a monthly basis (Schaefer et al. 2010). This chapter will review the major findings over the last several years related to bladder cancer, as well as discuss new proposed outcomes and mechanisms.

10.2 Historical

Since the discovery by Lee and colleagues (1993) in 1993 of small non-protein encoding RNAs of approximately 22 nucleotides in length, miRNAs have been identified in multiple organisms and tissue types (Chen et al. 2004; Hanke et al. 2010; Lagos-Quintana et al. 2002; Lagos-Quintana et al. 2003). MiRNAs are thought to induce their effect on gene expression based on the amount the miRNA complements the mRNA. If it has a high amount of complement, then the miRNA leads to mRNA degradation; if the complementary sequence is low, however, the miRNA can repress translation of the mRNA (Hutvagner and Zamore 2002).

The first identified miRNA was in *C. elegans* and was labeled *lin-4*, though at the time all that was known was that this particular small RNA was complementary in sequence to *lin-14*. The binding of *lin-4* to the 3′UTR of *lin-14* resulted in decreased mRNA expression suggesting a suppressive role for this small RNA nucleotide (Lee et al. 1993). Identification of miRNAs in humans occurred over seven years later (Pasquinelli et al. 2000). Further, the expression of miRNAs is variable across organ systems, with the highest expression in kidney, lung, and brain, while the lowest expression in heart, thymus, and bone marrow. This finding is a common theme throughout the expression of mRNA with the interaction of miRNAs in varying tissues (Grosshans and Filipowicz 2008; Vasudevan et al. 2007).

10.2.1 Embryonic MiRNA Associations with Bladder Development

In order to shed light on the developmental characteristics of miRNAs in the bladder, Liu and colleagues (2009) evaluated mouse embryo at three time points correlating to the day prior to smooth muscle formation, at smooth muscle formation, and after formation with each time point separated by approximately 24–48 h. Tissue was extracted at each time point from the urothelium and then the smooth muscle with subsequent miRNA characterization via microarray technology and RT-PCR confirmation. Recalling that epithelial cells of the bladder is derived from endoderm and that mesenchymal cells associated with smooth muscle is derived from the urogenital sinus and allantoids (Staack et al. 2005), significant differences are found during each developmental time stage. Of note, the mesenchymal cells of the bladder differentiated into smooth muscle based on signaling from the urothelium that is postulated to occur secondary to a miRNA process (Liu et al. 2009).

Specific miRNA have been identified for varying time points of the developmental process in the bladder (Liu et al. 2009). Of the 187 miRNAs discovered, however, only 16 are known miRNA including *miR-137*, *miR-467*, and *miR-503*. As might be expected, with differentiation some miRNAs increased in expression amount, while others decreased. For example, *miR-137* increased with more differentiated embryonic stages whereas, *miR-503* had a stepwise decrease. Likewise, the expression amounts changed within the same stage of development, but in different tissues. For example, *miR-137* increased in expression when moving from the outermost layer of mesenchymally derived tissue and going toward the epithelial derived tissue (Liu et al. 2009). This finding is characterized due to the extensive amount careful laser dissection performed by the authors, and this finding demonstrates the importance that may be missed by characterization of whole tumors rather than laser dissected ones and the differences that may lie within superficial and invasive cancers.

10.2.2 MiR-200 Family and Epithelial-mesenchymal Transition (EMT)

Numerous papers over the past several years have specifically targeted the five members of the *miR-200* family (*200a*, *200b*, *200c*, *141*, and *429*) and their effects on the EMT. From an embryologic standpoint, this family of miRNAs is more pronounced within differentiated epithelial tissues as opposed to mesenchymal ones (Darnell et al. 2006), and can be organized into one of two clusters based on their base sequences (Park et al. 2008). Further, *miR-200* family members demonstrate an inhibitory role on the mesenchymal associated factors of *Zeb1* and *Zeb2* (Gregory et al. 2008; Park et al. 2008). Specifically, *Zeb1* and *Zeb2* are repressed and EMT prevented when *miR-200* family expression is enforced. Conversely, the loss of *miR-200* expression can lead to elevated *Zeb1* and *Zeb2* causing silencing of *CDH1* and, consequently, increased EMT (Gregory et al. 2008).

Recently, our group published findings in UC cell lines related to the *miR-200* family and its modulatory effects on EMT (Adam et al. 2009). Utilizing nine UC cell lines classified as either epithelial or mesenchymal based on *CDH1* expression, miRNA array screening with RT-PCR validation confirmed expression patterns. Baseline expression of *miR-200c* is elevated in cell lines associated with the epithelial phenotype as compared to those of mesenchymal phenotype and this expression correlates with the expression of *CDH1*. UMUC3, a UC cell line with low baseline expression of *miR-200c*, was transfected with a stable lentivirus construct for *miR-200c* and yielded an approximate 150-fold increased expression of *miR-200c* as compared to the empty vector. Of important significance, the analysis of EMT markers (*CDH1*, *Zeb1*, and *Zeb2*) demonstrates a dramatic increase in *CDH1* with converse decrease in *Zeb1* and *Zeb2* expression for the stable *miR-200c* construct. Confocal microscopy also demonstrates the nuclear down-regulation of *Zeb1* and *Zeb2*. Essentially, an epithelial phenotype can be achieved in a mesenchymal cell line by over-expression of *miR-200c*. Therefore, *miR-200c* is inversely correlated with *Zeb1* and *Zeb2* expression. Further characterization of the 3′UTR of *Zeb1* and *Zeb2* yields five potential binding sites for *miR-200* family members and suggests a molecular mechanism for inhibition of this mesenchymal gene expression. Overall, *miR-200* family members in UC cell lines are associated with epithelial phenotype and are inversely correlative with markers of mesenchymal phenotype.

The exact mechanisms that lead to EMT with respect to loss of *miR-200* expression are just now being characterized. *Zeb1* interacts directly as a repressor of transcription against *miR-200c* and *miR-141* via binding to the E-box promoter regions (Bracken et al. 2008). Further, both the *miR-200* family and *miR-205* are associated with increased CpG islands which suggest a mechanism of inhibition via DNA hypermethylation (Wiklund et al. 2010). Analysis of normal, non-invasive and invasive bladder tumors find confirmation of DNA hypermethylation with silencing of the *miR-200* family and *miR-205* in invasive tumors (Wiklund et al. 2010).

Although it is important to clarify the differences between miRNA expressions in invasive versus non-invasive tumors, one of the more difficult tumors clinically is the T1 UC, where a significant percentage will undergo progression to muscle invasive disease. In this regard, Wiklund and colleagues (2010) evaluated 100 T1 tumors and found loss of *miR-200c* expression was highly correlative to disease progression. This latter finding has important clinical biomarker potential if confirmed as T1 tumors have a varied clinical response and identification of poor risk patients would be extremely useful.

10.2.3 Other Identified MiRNAs

As has previously been discussed, the variability in miRNA expression can be quite high, and this carries over into the cancer phenotype (Lu et al. 2005). As compared to normal tissues, most cancerous tissues have decreased expression of miRNAs (Calin et al. 2002; Lu et al. 2005). However, some combination cancer regimens yield up-regulation of miRNAs. *MiR-127* is one such miRNA where combination treatment with histone deacetylase (HDAC) inhibition and inhibitors of DNA methylation in the T24 UC cell line yields up-regulation. Further, *miR-127* is embedded within a *CpG* island and is not expressed within the T24 invasive mesenchymal phenotype (Saito et al. 2006). Additionally, matched normal and cancerous bladder tissue demonstrate similar findings with *miR-127* over-expression in normal tissues, but non-existent in bladder tumors (Saito et al. 2006).

Utilizing computational analysis, Saito and colleagues identified *BCL6* as a possible target for *miR-127* as over-expression of *miR-127* resulted in decreased protein expression of *BCL6*. Further, when utilizing vectors with mutant and wild-type (wt) *BCL6*, *miR-127* led to decreased translational expression of *BCL6* (Saito et al. 2006). Importantly, this suggests that HDAC inhibition in conjunction with DNA methylation inhibitors can induce expression of miRNAs associated with tumor suppressive actions in cancer cells and suggests further therapeutic targets.

Other tissue derived comparative assessments have also found unique miRNAs and proposals for biomarker evaluations. Utilizing miRNA array expression analysis, 290 unique miRNAs were evaluated in normal and cancerous bladder tissues (Dyrskjot et al. 2009) of which several demonstrated differential expression between normal and cancerous tissue as well as between stages of disease. Of note, *miR-21* was up-regulated in bladder cancer tissue, whereas *miR-143* and *miR-145* were down-regulated. When comparing invasive bladder cancer to superficial disease, *miR-200c* was down-regulated in the invasive subtypes. We have also noted similar findings in our comparison of invasive and superficial bladder tumors for the *miR-200* family (Williams et al. 2010, manuscript in preparation). Further, four miRNAs were identified as being associated with disease progression defined as worsened stage: *miR-133b*, *miR-518c*, *miR-129*, and *miR-29c* (Dyrskjot et al. 2009). These particular markers were then cross-validated resulting in a 63% sensitivity and 66% specificity for disease progression. With in situ hybridization, normal urothelium had elevated *miR-145* and *miR-129*; however, *miR-21* was only detected in carcinoma tissue. Functionally, when the T24 UC cell line was transfected with pre-*miR-129*, the result was a significant anti-proliferative effect that promoted cell death. Knockdown of *miR-129* did not have the counter effect secondary to very low basal levels found within the T24 cell line. Transfection of *miR-21* into the same T24 cell line did not result in phenotypic change. In summary, identification of key miRNAs involved in disease progression have been made, but exact molecular mechanistic identification of why a given tumor progresses as compared to another remains elusive.

The number of tissues evaluated does not limit the number of expression changes found as demonstrated by Gottardo and colleagues (2007). Utilizing an array platform for 245 miRNAs, 27-bladder primary tumor specimens were assessed along with two normal mucosa. Ten miRNAs (*miR-223*, *miR-26b*, *miR-221*, *miR-103-1*, *miR-185*, *miR-23b*, *miR-203*, *miR-17-5p*, *miR-23a*, *miR-205*) were up-regulated in the cancerous tissue as compared to normal bladder mucosa. Twenty of the 25 patients had evaluable stage information, and four of these tumors were Ta. Only *miR-26b* demonstrated a trend for decreasing expression with increasing stage, however, none of the miRNAs were significant based on Stage alone. Speculation was made that due to the known deletion of Chr2q35 in progression from Ta to T1 (Richter et al. 1999), that miRNA regulation of *miR-26b* may also be adversely affected.

A unique concept to evaluating the effects of miRNA expression within invasive and non-invasive cell lines is the comparative ratio. Neely and colleagues (2010) performed expression profiling for 343 miRNAs in 14 UC cell lines and demonstrated 9 differentially expressed miRNAs for invasive and non-invasive phenotypes, including *miR-21*, *miR-205*, and two members of the *miR-200* family (*miR-200a* and *miR-200c*). *MiR-21*, targets caspases and thus prevents apoptosis (Chan et al. 2005), and *miR-205* were targeted based on prior published data demonstrating their association with cancer phenotypes. As might be expected, invasive UC cell

Table 10.1 MicroRNA ratios associated with prognosis in bladder cancer

MicroRNA Ratio	Prognostic application	References
miR-21:miR-205	Increased with increasing stage	Chan et al. (2005)
miR-126:miR-152	Increased with cancer as compared to normal tissues	Hanke et al. (2010)
miR-182:miR-152	Increased with cancer as compared to normal tissues	Hanke et al. (2010)

lines found *miR-21* elevation and decreased expression of *miR-205,* with the converse true for non-invasive UC lines. Upon initial assessment of a small subset of bladder tumors, however, significance was lost due to lack of discriminatory power. Therefore, a ratio approach was applied to this data with invasive lines having a 10-fold higher *miR-21:miR-205* ratio as compared to non-invasive ones. This finding remained significant in a larger cohort of bladder tumors with receiver operator characteristic (ROC) curve analysis being 0.89 for discriminating superficial from invasive bladder tumors and the sensitivity and specificity being 100 and 78%, respectively. This is a very promising biomarker, but needs to be validated in larger cohorts (Table 10.1).

Interestingly, this approach is not limited to utilizing fresh or frozen tumors, as specific techniques can be applied to urine specimens to effectively capture RNA and distinguish cancerous from non-cancerous states via miRNA identification. Of note, Hanke et al. (2010) consistently demonstrated two particular miRNA ratios were higher in bladder cancer patients as compared to normal subjects: *miR-126:miR-152* and *miR-182:miR-152*. It is unclear at this time, however, how the modulation of these miRNAs result in invasive changes (Table 10.1).

Silencing specific miRNAs has been demonstrated to promote cell death. Based on work in melanoma and prostate cancer cell lines, Lu et al. (2009) used the T24 invasive UC cell line and found that *miR-221* was significantly up-regulated and proposed the reversal of this expression might ultimately lead to susceptibility to cell death. Like the prior utilization of combination agents to promote cell death, an antisense *miR-221* was transfected into the T24 cell line and then this UC line was exposed to tumor necrosis factor apoptosis induced ligand (TRAIL) over the ensuing 24 h. *MiR-221* expression was dramatically reduced and, when TRAIL was introduced, apoptosis was achieved at a rate of 50%.

Though most of the studies published have focused on identification of a large number of miRNAs that are differentially expressed as compared to normal, few have focused efforts on the one miRNA. Lin and colleagues (2009) confirmed prior findings of down-regulation of *miR-143* in bladder cancer tissues as compared to normal adjacent urothelium, however no differences were noted between stages of disease. Of the 26 patient tumor samples analyzed, 31% were non-invasive with the remainder muscle-invasive. To further characterize the effects of *miR-143*, the mesenchymal and invasive cell lines, T24 and EJ, were evaluated as to the baseline expression of *miR-143* as well as the end results from transfection with *miR-143*. As compared to control *U6*, T24 and EJ cells had little to no expression of *miR-143*. Once transfected with the mature form of *miR-143*, however, significant growth inhibition occurred suggesting a negative regulation for cell proliferation. Further, *KRAS*, a gene that has been associated with bladder cancer, has several *miR-143* binding sites. When assessing the amount of RAS protein produced by T24 and EJ cells transfected with *miR-143*, the *KRAS* mRNA remained stable, but the protein product was significantly decreased leading to a second methodology for regulation by *miR-143* (Lin et al. 2009). Overall, *miR-143* has decreased expression in bladder cancer and this can yield both lack of growth inhibition as well as negative regulation of the protein product from *KRAS* mRNA.

Evaluation of mechanistic targets of the down-regulated *miR-145* in urothelial carcinoma has also begun to shed some light as to the variability and wide range of possible targets for a given miRNA. Chiyomaru and colleagues performed a microarray analysis of 3 *miR-145* transfected UC cell lines as compared to the normal UC cell lines and performed a genome wide evaluation of possible targets. Ultimately, the focus was on the most down-regulated target, *FSCN1* (Chiyomaru et al. 2010), which is essential in forming filopodia (Vignjevic et al. 2006) and is highly expressed in cells that have increased motility (Pelosi et al. 2003). Identification was made of 2 complementary sites for *miR-145* on *FSCN1* and when transfected with *miR-145*, T24 cell lines had a significant growth inhibition and inhibition of migration as compared to the wt controls, however, matrigel invasion was not prevented by the transfected cell with T24. Of note, the other cell line investigated, BOY, ceased invasion with transfection of *miR-145*. Which raised the possibility for other mechanistic events promoting invasion based on the given cell line. Finally, immunohistochemistry staining for *FSCN1* demonstrated increased intensity with advanced stage and is consistent with prior studies (Tong et al. 2009). These two studies demonstrate the style of mechanistic work that needs assessment in urothelial carcinoma.

10.2.4 Genetic Variability in MiRNA Machinery

Based on previously published findings that the majority of tumors have decreased expression of miRNAs as compared to normal (Lu et al. 2005), this premise was applied to bladder cancer and DNA copy number (Lamy et al. 2006). Evaluation of thirty superficial (Ta or T1) UC tumors based on single nucleotide polymorphism (SNP) and subsequent gain or loss of DNA copy number in those regions was performed. Interestingly, patients with T1 UC with a gain in DNA copy number had decreased expression of miRNAs, whereas, an increased expression of miRNAs has been demonstrated in prostate and colon tumors (Lamy et al. 2006; Volinia et al. 2006).

The effect of miRNA biogenesis in relation to genetic variants in bladder cancer outcome was recently assessed by the MD Anderson Cancer Center group (Yang et al. 2008). In this large case control study, 41 SNPs with potential targeting of miRNA functionality underwent relational assessment to bladder cancer risk. With nearly 1,500 patients evaluated and half with newly diagnosed disease, 7 of the 41 SNPs had at least a borderline statistical increased risk of bladder cancer. Two genes and their regulatory SNPs stand out as prominently increasing the risk of bladder cancer: *GEMIN3* (2.5-fold increased risk) and *GEMIN4* (1.25-fold increased risk) (Yang et al. 2008). Both genes are important in the processing of miRNA precursors (Hutvagner and Zamore 2002) and the proteins made by transcription of these genes interact with the survivor proteins involved with pre-mRNA splicing (Nelson et al. 2004). Beyond this, a decreased bladder cancer risk was also associated with 4 miRNA genes including *miR-423*, *miR-492*, *miR-26a-1*, and *miR-124-1*. In other epithelial cancers, the chromosome responsible for production of *miR-26a-1* (Chr3p21) is deleted (Protopopov et al. 2003) whereas *CpG* island methylation has been attributed to the decreased expression for *miR-124-1* (Lujambio et al. 2007). When the 7 SNPs were then taken together as markers of unfavorable genotypes, the higher the number of unfavorable SNPs led to an increased bladder cancer risk ($p < 0.0001$) and looking at individual numbers of unfavorable genotypes, 3 or more led to an almost 2-fold increased bladder cancer risk (OR 1.92, $p < 0.0001$) (Yang et al. 2008). In summary, this is the first to identify specific genetic variants of miRNA biogenesis machinery and increased risk in bladder cancer.

Another gene adversely regulated in bladder cancer is the tumor suppressor gene, *Rb*. Further, the protein formed by transcription of the *E2F3* oncogene is bound to the *Rb* protein under normal cellular processes. However, with phosphorylation of *Rb*, *E2F3* becomes free to target several different sites with ultimate mitotic promotion. Huang and colleagues (2010) published their results of the interaction of the *miR-125b* with the *E2F3* oncogene. Utilizing miRNA microarray technology, they demonstrated in a prior study *miR-125b* was significantly down-regulated in bladder cancers and was related to cell proliferation in cancer cells (Lee et al. 2005; Lin et al. 2009). Utilizing the tissues from 25 bladder cancer tumors that had never undergone local or systemic treatment prior to operation, they assessed the amount of *miR-125b* and *E2F3* expression along with comparison to eight UC cell lines including T24 and UMUC3 (Huang et al. 2010). *MiR-125b* was decreased across all bladder tumors as compared to adjacent normal tissues, but did not correlate to stage or grade of disease. When stable transfectants of *miR-125b* were placed into mesenchymal UC cell lines (T24 and 5637), cell proliferation was significantly depressed as compared to the wt cells. *E2F3* was identified as a possible target based on several complementary sites for *miR-125b* and, utilizing luciferase reporter, *E2F3* was significantly reduced and protein expression of the two was noted to be inversely correlated ($p < 0.001$). However, mRNA for *E2F3* was not affected by the transfectants, suggesting a role of *miR-125b* for post-translational modification. Finally, *cyclin A2* is inhibited in both T24 and 5687 UC cell lines with induction *miR-125b*. Globally, the repression of the oncogene *E2F3* by *miR-125b* is post-translational and affects cell proliferation by decreasing *cyclin A2* (Huang et al. 2010).

10.3 Prognostic Implications of MiRNAs

Determining the changes between different tumor types allows the ability to evaluate the extent of disease and could lead to determining when progression might occur on a molecular level (Table 10.2). An excellent study evaluating these different tumor

Table 10.2 MicroRNAs in bladder cancer

MicroRNAs	Cancer regulation	Target discovery	Mechanistic discovery	Prognostic discovery	References
miR-124-1	Down-regulated	–	CpG island hyper-methylation	–	Lujambio et al. (2007)
miR-125b	Down-regulated	*E2F3*	Decreased protein expression; Cyclin A2 inhibition	Increased IHC intensity in high grade urothelial carcinoma	Huang et al. (2010)
miR-127	Down-regulated	*BCL6*	Decreases BCL6 protein expression	–	Saito et al. (2006)
miR-129	Down-regulated	–	–	Worse progression and stage	Dyrskjot et al. (2009)
miR-133b	Down-regulated	–	–	Worse progression and stage	Dyrskjot et al. (2009)
miR-137	Down-regulated	–	–	Decreased expression from epithelial tumor to mesenchymal tumor	Liu et al. (2009)
miR-143	Down-regulated	*KRAS, FSCN1*	Decreases protein expression	No differences between stages; IHC of FSCN1 increased with stage	Lin et al. (2009); Tong et al. (2009); Wang et al. (2010)
miR-145	Down-regulated	*CBFB, PPP3CA, CLINT1*	Apoptosis	–	Ostenfeld et al. (2010)
miR-199a	Down-regulated	*KRT7*	Inversely correlated with miRNA	–	Kawakami et al. (2006); Wang et al. (2010)
miR-200a/b/c, 141, 429	Down-regulated	*Zeb1, Zeb2*	EMT prevention with increased expression	Loss of *miR-200c* expression correlated to progression; *miR-141* up-regulated in T1 tumors as compared to ANT	Gregory et al. (2008); Neely et al. (2010); Park et al. (2008); Wiklund et al. (2010); Wszolek et al. (2010); Wang et al. (2010)
miR-205	Up-regulated	–	–	Up-regulated as compared to normal tissue	Gottardo et al. (2007)

Table 10.2 (continued)

MicroRNAs	Cancer regulation	Target discovery	Mechanistic discovery	Prognostic discovery	References
miR-21	Up-regulated/ down-regulated (one study)	*Caspase, PTEN*	Prevention of apoptosis	Only detected in cancerous tissue; more likely to progress in disease	Chan et al. (2005); Dyrskjot et al. (2009); Gottardo et al. (2007); Ichimi et al. (2009); Neely et al. (2010); Wang et al. (2010); Meng et al. (2007)
miR-221	Up-regulated	–	Knockdown of microRNA increased susceptibility to TRAIL	–	Gottardo et al. (2007); Lu et al. (2010)
miR-26a-1	Down-regulated	–	–	Compared to ANT	Wang et al. (2010)
miR-26b	Up-/down-regulated	–	–	Up-regulated as compared to normal tissue; down-regulated with increasing stage	Richer et al. (1999); Gottardo et al. (2007)
miR-29c	Down-regulated	–	–	Disease progression and worsened stage as compared to ANT	Dyrskjot et al. (2009); Wang et al. (2010)
miR-518c	Down-regulated	–	–	Worse disease progression and stage	Dyrskjot et al. (2009)

ANT, Adjacent normal tissue; TRAIL, TNF-related apoptosis inducing ligand.

types was published in this past year, where assessment was made of the following disease states: low-grade non-muscle invasive, high-grade non-muscle invasive, muscle invasive, normal urothelium from patients with bladder cancer, and normal urothelium without evidence of bladder cancer (Catto et al. 2009). RNA was isolated and the expression of 354 known mature miRNAs assessed. Mean follow-up for the entire cohort was just over 3 years (range 0.2–7.34). Of note, patients with bladder cancer had an up-regulation of the adjacent normal urothelium miRNAs as compared to disease free controls and unsupervised cluster analysis demonstrated two branches corresponding to tissue state. Specific miRNAs include the *let-7* family, *miR-492*, *miR-221*, *miR-492*, and *miR-141*. When specifically evaluating urothelial carcinoma tissues only, twelve miRNAs were identified with sensitivity and specificity for identifying cancer of 90–100 and 80–100%, respectively. A similar process was found in regard to cancer tissues where there was up-regulation of miRNA expression from tissues with low-grade tumor to those of invasive disease (Catto et al. 2009).

The gene targets of the up-regulated miRNAs included *RBAK*, *LATS2*, *RAB6c*, and *FGF2*, while down-regulated targets included *SOX4*, *RUNX2*, and *ANGPT1*. Further, *Dicer*, *Drosha*, and *Exportin5* were significantly up-regulated in the normal urothelium adjacent to bladder cancer as compared to the malignant tissue where these three processing genes were down-regulated. Finally, patients with up-regulated *miR-21* and normally expressed *miR-100* and *miR-99a* were more likely to progress to advanced disease (Catto et al. 2009).

In another smaller evaluation of 14 bladder cancers as compared to 5 samples of normal urothelium, seven miRNAs were identified as being down-regulated in bladder cancers (Ichimi et al. 2009). ROC analysis revealed that these seven miRNAs could distinguish normal from cancerous tissues with at least 70% sensitivity and 75% specificity for each miRNA. Of the down-regulated miRNAs, *miR-199a* was significantly reduced in the cancerous state. Further, based on prior research demonstrating the cytokeratin 7 (*KRT7*) as elevated in bladder cancer (Kawakami et al. 2006) and this finding was recapitulated in further study. Of particular interest was the finding that *KRT7* was inversely associated with the amount of each of the 7 miRNAs suggesting *KRT7* as a target of the miRNAs.

Returning to the *miR-200* family, which, as outlined earlier is mechanistically involved in the EMT, while the reduction of *miR-200* yielding increasing expression of the mesenchymal nuclear factors of *Zeb1* and *Zeb2*, clinical application of the expression of *miR-200* has demonstrated similar results in the clinical arena. Utilizing a panel of 57 urothelial carcinoma tumors (45% $\leq$T1) with a mean follow-up of survivors being 92 months, 12 miRNAs were found to be differentially expressed between invasive and non-invasive tumors (Wszolek et al. 2010. In particular, *miR-200c*, *miR-141*, and *miR-30b* demonstrated significantly worse cancer specific survival when they were reduced in expression as compared to normal expression patterns. Five-year survival for *miR-200c* high and low expression was 90 and 36% ($p<0.001$), respectively, with *miR-141* high and low expression of 95 and 31% ($p<0.0001$), and *miR-30b* high and low expression being 100 and 26% ($p<0.0001$). The expression of these three miRNAs had a predictive capacity for distinguishing invasive from non-invasive disease with a sensitivity of 100% and specificity of 96.2%. This results needs to be clarified in a larger cohort of patients (Wszolek et al. 2010).

As depicted earlier, several studies demonstrate significant changes in miRNAs between normal tissues as compared to cancerous ones. Studying adjacent normal tissue with T1 and T2 tumors in 7 patients, Wang and colleagues (2010) demonstrated that T1 and T2 tumors had different miRNA up-regulation and down-regulation between them. In T1 tumors, six known miRNAs were up-regulated (*miR-129, 141, 494, 498, 500*) and 13 miRNAs were down-regulated (*let7a-d, miR-199a, 21, 24, 26a, 29c, 30a-5p, 30c, 30e-5p*). However, in T2 tumors, none of the analyzed miRNAs were up-regulated with 4 miRNAs being down-regulated (*miR-26a, 29c, 30c*, and *30e-5p*). The differences between the T1 and T2 tumors suggest an underlying change initiated in the development of the urothelial cancer. This is depicted in the transition seen in carcinoma in situ tumors where *miR-145* was significantly reduced as compared to adjacent normal tissue (Ostenfeld et al. 2010). Further, identification was made of several gene targets (*CBFB, PPP3CA, CLINT1*) associated with caspase inhibition. Overall, each of the tumor types demonstrate different miRNA expression and may have different mechanistic progression features.

10.4 Perspective and Future Challenges

In this chapter, we have presented several new miRNAs identified as potential differentiating features for bladder cancers. In a few, the mechanistic aspects have been assessed, but many remain strictly as discovery findings of new miRNAs. Examples of these recent discoveries are the *miR-200* family (*Zeb1/2* inhibition and EMT), *miR-21* (apoptosis prevention and *PTEN*), and *miR-143* (*KRT7*). For the mechanisms that have been identified, some demonstrate post-translational protein modification rather than directed gene targeting and repression of mRNA expression (*miR-127* decreased BCL6 protein). Further, miRNAs have been identified as having numerous targets for each individual miRNA (Cho 2010; Nelson and Weiss 2008; Wu et al. 2007). The problem is to determine which of the newly discovered miRNAs should be further assessed for the mechanisms promoting bladder cancer as well with the ultimate goal of developing agents that prevent this transition in the first place. As technology continues to develop, our ability to perform large scale analysis, as demonstrated by microarray technology, will further allow us to refine these characteristics. For the current time, further evaluation of miRNAs that have already demonstrated clinical benefit need to be performed.

10.5 Conclusions

The studies enumerated here suggest that invasive bladder cancer displays panels of miRNAs with tumor-specific profiles. This could aid in discriminating among other subgroups of superficial bladder cancer that are prone to relapse or between different subgroups of invasive bladder cancers that are fatal. These findings are of notable clinical consequence and predict unlimited potential for impacting clinical practice patterns by directing appropriate management of this disease and reducing death from bladder cancer. Understanding the biology of the differentially-expressed miRNAs, combined with the assessment of pattern similarities not only in tumors but also in surrogate tissues or in the systemic circulation (plasma, serum, urine, or saliva of patients) may represent the next step in the development of non-invasive, highly-reliable miRNA-based biomarkers. The use of reliable markers also has the potential for reducing the cost of health care delivery by improving and streamlining surveillance protocols, and by personalizing the therapy and for this, the miRNA signature profiling seems to hold great promise in bladder cancer.

References

Adam L, Zhong M, Choi W, et al. MiR-200 expression regulates epithelial-to-mesenchymal transition in bladder cancer cells and reverses resistance to epidermal growth factor receptor therapy. Clin Cancer Res. 2009;15:5060–72.

Bracken CP, Gregory PA, Kolesnikoff N, et al. A double-negative feedback loop between ZEB1-SIP1 and the microRNA-200 family regulates epithelial-mesenchymal transition. Cancer Res. 2008;68:7846–54.

Calin GA, Dumitru CD, Shimizu M, et al. Frequent deletions and down-regulation of micro-RNA genes miR15 and miR16 at 13q14 in chronic lymphocytic leukemia. Proc Natl Acad Sci USA. 2002;99:15524–9.

Catto JW, Miah S, Owen HC, et al. Distinct microRNA alterations characterize high- and low-grade bladder cancer. Cancer Res. 2009;69:8472–81.

Chan JA, Krichevsky AM, Kosik KS. MicroRNA-21 is an antiapoptotic factor in human glioblastoma cells. Cancer Res. 2005;65:6029–33.

Chen CZ, L, L, Lodish HF, et al. MicroRNAs modulate hematopoietic lineage differentiation. Science. 2004;303:83–6.

Chiyomaru T, Enokida H, Tatarano S, et al. MiR-145 and miR-133a function as tumour suppressors and directly regulate FSCN1 expression in bladder cancer. Br J Cancer. 2010;102: 883–91.

Cho WC. MicroRNAs in cancer – from research to therapy. Biochim Biophys Acta. 2010;1805:209–17.

Darnell DK, Kaur S, Stanislaw S, et al. MicroRNA expression during chick embryo development. Dev Dyn. 2006;235:3156–65.

Dyrskjot L, Ostenfeld MS, Bramsen JB, et al. Genomic profiling of microRNAs in bladder cancer: miR-129 is associated with poor outcome and promotes cell death in vitro. Cancer Res. 2009;69:4851–60.

Gottardo F, Liu CG, Ferracin M, et al. Micro-RNA profiling in kidney and bladder cancers. Urol Oncol. 2007;25:387–92.

Gregory PA, Bert AG, Paterson EL, et al. The miR-200 family and miR-205 regulate epithelial to mesenchymal transition by targeting ZEB1 and SIP1. Nat Cell Biol. 2008;10: 593–601.

Grosshans H, Filipowicz W Molecular biology: the expanding world of small RNAs. Nature. 2008;451:414–6.

Hanke M, Hoefig K, Merz H, et al. A robust methodology to study urine microRNA as tumor marker: microRNA-126 and microRNA-182 are related to urinary bladder cancer. Urol Oncol. 2010;28:655–61

Huang L, Luo J, Cai Q, et al. MicroRNA-125b suppresses the development of bladder cancer by targeting E2F3. Int J Cancer. 2010;28:655–61.

Hutvagner G, Zamore PD. A microRNA in a multiple-turnover RNAi enzyme complex. Science. 2002;297:2056–60.

Ichimi T, Enokida H, Okuno Y, et al. Identification of novel microRNA targets based on microRNA signatures in bladder cancer. Int J Cancer. 2009;125:345–52.

Jemal A, Siegel R, Xu J, et al. Cancer statistics, 2010. CA Cancer J Clin. 2010;60:277–300.

Kawakami K, Enokida H, Tachiwada T, et al. Identification of differentially expressed genes in human bladder cancer through genome-wide gene expression profiling. Oncol Rep. 2006;16:521–31.

Lagos-Quintana M, Rauhut R, Meyer J, et al. New microRNAs from mouse and human. RNA. 2003;9:175–9.

Lagos-Quintana M, Rauhut R, Yalcin A, et al. Identification of tissue-specific microRNAs from mouse. Curr Biol. 2002;12:735–9.

Lamy P, Andersen CL, Dyrskjot L, et al. Are microRNAs located in genomic regions associated with cancer? Br J Cancer. 2006;95:1415–8.

Lee RC, Feinbaum RL and Ambros V. The *C. elegans* heterochronic gene lin-4 encodes small RNAs with antisense complementarity to lin-14. Cell. 1993;75:843–54.

Lee YS, Kim HK, Chung S, et al. Depletion of human micro-RNA miR-125b reveals that it is critical for the proliferation of differentiated cells but not for the down-regulation of putative targets during differentiation. J Biol Chem. 2005;280:16635–41.

Lin T, Dong W, Huang J, et al. MicroRNA-143 as a tumor suppressor for bladder cancer. J Urol. 2009;181:1372–80.

Liu B, Cunha GR, Baskin LS. Differential expression of microRNAs in mouse embryonic bladder. Biochem Biophys Res Commun. 2009;385:528–33.

Lu J, Getz G, Miska EA, et al. MicroRNA expression profiles classify human cancers. Nature. 2005;435:834–8.

Lu Q, Lu C, Zhou GP, et al. MicroRNA-221 silencing predisposed human bladder cancer cells to undergo apoptosis induced by TRAIL. Urol Oncol. 2010;28:635–41.

Lujambio A, Ropero S, Ballestar E, et al. Genetic unmasking of an epigenetically silenced microRNA in human cancer cells. Cancer Res. 2007;67:1424–9.

Meng F, Henson R, Wehbe-Janek H, et al. MicroRNA-21 regulates expression of the PTEN tumor suppressor gene in human hepatocellular cancer. Gastroenterology. 2007;133:647–58.

Neely LA, Rieger-Christ KM, Neto BS, et al. A microRNA expression ratio defining the invasive phenotype in bladder tumors. Urol Oncol. 2010;28:39–48.

Nelson KM, Weiss GJ. MicroRNAs and cancer: past, present, and potential future. Mol Cancer Ther. 2008;7:3655–60.

Nelson PT, Hatzigeorgiou AG, Mourelatos Z. MiRNP:mRNA association in polyribosomes in a human neuronal cell line. RNA. 2004;10:387–94.

Ostenfeld MS, Bramsen JB, Lamy P, et al. MiR-145 induces caspase-dependent and -independent cell death in urothelial cancer cell lines with targeting of an expression signature present in Ta bladder tumors. Oncogene. 2010;29:1073–84.

Park SM, Gaur AB, Lengyel E, et al. The miR-200 family determines the epithelial phenotype of cancer cells by targeting the E-cadherin repressors ZEB1 and ZEB2. Genes Dev. 2008;22: 894–907.

Pasquinelli AE, Reinhart BJ, Slack F, et al. Conservation of the sequence and temporal expression of let-7 heterochronic regulatory RNA. Nature. 2000;408:86–9.

Pelosi G, Pastorino U, Pasini F, et al. Independent prognostic value of fascin immunoreactivity in stage I nonsmall cell lung cancer. Br J Cancer. 2003;88:537–47.

Protopopov A, Kashuba V, Zabarovska VI, et al. An integrated physical and gene map of the 3.5-Mb chromosome 3p21.3 (AP20) region implicated in major human epithelial malignancies. Cancer Res. 2003;63:404–12.

Richter J, Wagner U, Schraml P, et al. Chromosomal imbalances are associated with a high risk of progression in early invasive (pT1) urinary bladder cancer. Cancer Res. 1999;59: 5687–91.

Saito Y, Liang G, Egger G, et al. Specific activation of microRNA-127 with downregulation of the proto-oncogene BCL6 by chromatin-modifying drugs in human cancer cells. Cancer Cell. 2006;9:435–43.

Schaefer A, Jung M, Kristiansen G, et al. MicroRNAs and cancer: current state and future perspectives in urologic oncology. Urol Oncol. 2010;28:4–13.

Staack A, Hayward SW, Baskin LS, et al. Molecular, cellular and developmental biology of urothelium as a basis of bladder regeneration. Differentiation. 2005;73:121–33.

Tong AW, Fulgham P, Jay C, et al. MicroRNA profile analysis of human prostate cancers. Cancer Gene Ther. 2009;16:206–16.

Vasudevan S, Tong Y, Steitz JA. Switching from repression to activation: microRNAs can up-regulate translation. Science. 2007;318:1931–4.

Vignjevic D, Kojima S, Aratyn Y, et al. Role of fascin in filopodial protrusion. J Cell Biol. 2006;174:863–75.

Volinia S, Calin GA, Liu CG, et al. A microRNA expression signature of human solid tumors defines cancer gene targets. Proc Natl Acad Sci USA. 2006;103:2257–61.

Wang G, Zhang H, He H, et al. Up-regulation of microRNA in bladder tumor tissue is not common. Int Urol Nephrol. 2010;42:95–102.

Wiklund ED, Bramsen JB, Hulf T, et al. Coordinated epigenetic repression of the miR-200 family and miR-205 in invasive bladder cancer. Int J Cancer. 2010. doi:10.1002/ijc.25461.

Wszolek MF, Rieger-Christ KM, Kenney PA, et al. A MicroRNA expression profile defining the invasive bladder tumor phenotype. Urol Oncol. 2010. doi:10.1016/j.urolonc.2009.08.024.

Wu W, Sun M, Zou GM, et al. MicroRNA and cancer: current status and prospective. Int J Cancer. 2007;120:953–60.

Yang H, Dinney CP, Ye Y, et al. Evaluation of genetic variants in microRNA-related genes and risk of bladder cancer. Cancer Res. 2008;68:2530–7.

Chapter 11
MicroRNAs in Lymphoma

Alba Navarro López and Luis Hernández Pous

Abstract MicroRNAs (miRNAs) are small regulatory RNAs involved in post-transcriptional gene regulation of molecular pathways related with differentiation, proliferation, and survival of cells. Normal lymphocytes are not an exception, and even relevant molecular pathways involved in the immunologic functions of these cells are also controlled by miRNAs. These important roles make miRNA alterations to be oncogenetic mechanisms with multiple targeting roles in lymphomagenesis. Among miRNA alterations, mutations in mature and pre-miRNA forms have been punctually described, but the most well-known miRNA alteration mechanisms are changes in their expression levels. The origin of these expression abnormalities seems to be diverse regarding the considered particular miRNA and lymphoma entity. Thus, some miRNAs have been described as targets of genomic instability affecting miRNA containing chromosomal loci. On the reverse side, other miRNA expression changes seem to involve epigenetic alterations or other molecular mechanisms not related with direct alterations of their loci. Regardless of their alteration mechanisms, miRNA expression profiles have been shown useful for improvement of clinical management of the patients, through their good performance in the diagnosis discrimination among types and subtypes of lymphomas. In addition to the biological features of these neoplasms, miRNA expression profiles have also proven to help in the identification of patient subsets with different prognosis outcomes. Taking together all these findings, we can foresee an increasing importance of the miRNA-based translational research in practical applications to the lymphoma clinical management and hopefully, even help in achieving more specific treatments.

L. Hernández Pous (✉)
Molecular Pathology Department, Facultad de medicina, Institut d'Investigacions Biomèdiques August Pi i Sunyer (IDIBAPS), 08036 Barcelona, Spain
e-mail: hernan@clinic.ub.es

W.C.S. Cho (ed.), *MicroRNAs in Cancer Translational Research*,
DOI 10.1007/978-94-007-0298-1_11, © Springer Science+Business Media B.V. 2011

11.1 MicroRNAs in the Normal Physiology of Lymphoid Cells

11.1.1 As Regulators of Lymphoid Cell Development in the Primary Lymphoid Organs

Systematic analyses on microRNA (miRNA) expression in mammalian hematopoiesis have revealed distinct profiles in the different related cell types, with specific sets of miRNAs dynamically regulated during T and B cell development (Chen et al. 2004; Chen and Lodish 2005; Ramkissoon et al. 2006; Sonkoly et al. 2008). Even before the stage of the common lymphoid (B/T) progenitor, several miRNAs are involved in the generation of these cells from hematopoietic stem cells in the bone marrow (as *miR-221*, *miR-222*, *miR-196b*, *miR-126*, and *miR-10* family members) through their action on relevant targets as *KIT*, *PLK2*, and several *HOX* family members (O'Connell et al. 2010) (Fig. 11.1).

11.1.2 In B Cells

MiRNAs from the *miR-17-92* cluster are essential for the pro-B to pre-B transition, enhancing the survival of the B cells at this stage by targeting transcripts of the pro-apoptotic protein *BIM*, and the tumor suppressor *PTEN* (Xiao et al. 2008). *MiR-150* has also a role during the B cell development, but blocks the proB-preB transition if its expression is forced at this stage, and involving the targeting on the *MYB* transcription factor (Xiao et al. 2007; Zhou et al. 2007). Constitutive expression of *miR-34a* also have a blocking effect at this stage, as a result of targeting on *FoxP1* transcription factor, that regulates the expression of the relevant immunoglobulin differentiation genes *Rag1* and *Rag2* (O'Connell et al. 2010).

11.1.3 In T Cells

MiR-17-92 cluster expression has been described to be essential for T cell proliferation and survival, sharing some targets as for B cell development (Xiao et al. 2008). *MiR-181a* has been also characterized to regulate relevant signal transduction pathways (*DUSP5*, *DUSP6*, *SHP2*, and *PTPN22* genes involved in the TCR response and $CD4^+$ T cell selection) in the T cell development (Li et al. 2007).

11.2 As Regulators of Immunologic Function of the Mature Lymphoid Cells

As mediators of the acquired immune response, the B and T cell immunological functions are also modulated by miRNA action in these cells (Tsitsiou and Lindsay 2009). Several relevant examples are highlighted below (Fig. 11.1).

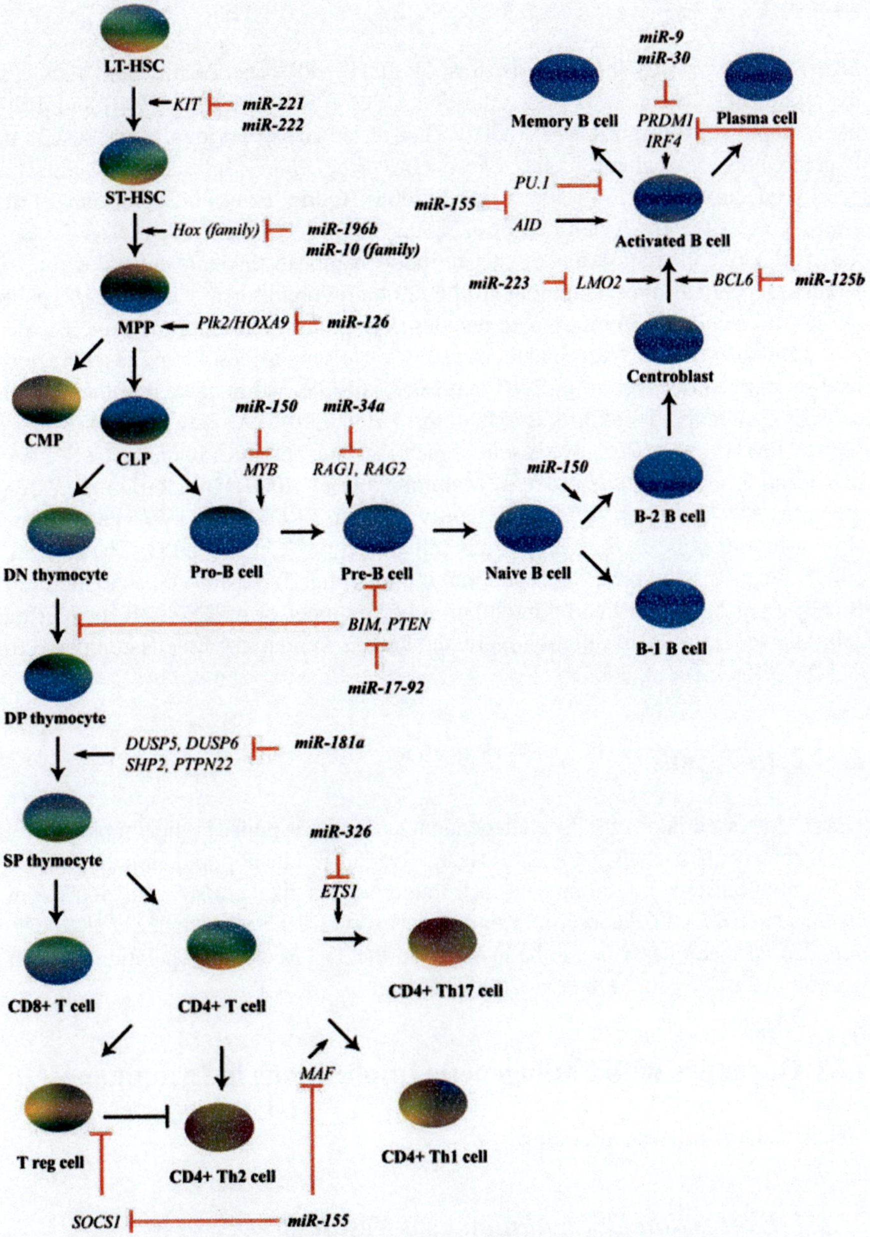

Fig. 11.1 MiRNA-mediated regulation of B and T cell development and function. MiRNAs are involved in the regulation of the expression of target coding genes essential for development and immunologic function of the B and T cells

11.2.1 In B Cells

MiR-155 participates in many aspects of the B cell functionality and thus, its expression in these cells is under the control of B cell receptor, CD40 and Toll-like receptors (Rodriguez et al. 2007; Thai et al. 2007; van den et al. 2003). In B cells, *miR-155* regulate targets like *c-MAF*, *PU.1/Sfpi1*, and activation-induced cytidine deaminase (*AID*) (Dorsett et al. 2008; Rodriguez et al. 2007; Teng et al. 2008; Thai et al. 2007; van den et al. 2003; Vigorito et al. 2007). *AID* is essential in the B cell maturation of the antibody response through the induction of antigen-dependent hypermutations in the immunoglobulin heavy chain (*IgH*) gene during the secondary immunologic response. Thus, differentiation of naïve B cells to centroblasts in the germinal center (GC) of lymphoid follicles is characterized by marked changes in miRNA expression profile, while naïve and memory B cells showed marked similarities (Malumbres et al. 2009). Among these, *miR-125b* repress *PRDM1* and *IRF4*, two essential factors of the post-GC, but not *BCL6*, a key transcription factor of this process (Malumbres et al. 2009). In addition, *miR-223* represses *LMO2*, and *miR-9/miR-30* family regulate *PRDM1/BLIMP1* gene expression, a master gene in the GC-plasma cell transition (Calame 2008; Zhang et al. 2009). Finally, in the post-GC B cell maturation either to plasma cell as to memory B cell, it has been described a modulation of a number of miRNAs. Thus, plasma cells showed 31 miRNAs highly expressed and 27 in memory B cells compared to GC cells (Zhang et al. 2009).

11.2.2 In T Cells

MiR-155 is essential for the T cell dependent antibody response in the lymphoid germinal centers (Thai et al. 2007), and it is involved in T cell subpopulation (T_H1/T_H2) differentiation and T_{reg} homeostasis through targets like *c-MAF* and *SOCS1* in response to CD3/CD28 and FoxP3 signaling (Lu et al. 2009; Rodriguez et al. 2007). *miR-326* has been described to be involved in the T_H17 cell subpopulation through targeting of *ETS1* gene (Du et al. 2009).

11.3 Oncomirs with Pathogenetic Implications in Lymphomas

Mechanisms of miRNA alteration:

11.3.1 Expression Deregulation as a Result of Genomic Instability

11.3.1.1 Involving Gains or Losses of MiRNA Chromosomal Loci

As a deregulation mechanism of miRNA expression levels, imbalances of chromosomal regions have been demonstrated to target miRNAs in several types of neoplasm. In addition, several examples also involving lymphomas are explained below.

MiR-15a/16-1 cluster is located at 13q14.3 locus, the most frequent chromosomic alteration (over 65%) in chronic lymphocytic leukemia (CLL), and this alteration was also present in 50% of mantle cell lymphoma (MCL) and 16–40% of multiple myeloma (MM). In fact, these miRNAs are the most consistent oncogenic targets inside this deletion, included in the genomic locus *DLEU2* (Klein et al. 2010; Lerner et al. 2009), besides from another recently described cancer-related locus (*DLEU7*) in this chromosomic region (Palamarchuk et al. 2010).

MiR-17-92 cluster comprises six precursor miRNAs that are codified inside a non-protein-coding gene *MIR17HG/MIRHG1* (formerly named as *C13orf25*) located at 13q31.3. This region is frequently amplified in several types of lymphomas, including Burkitt lymphoma (BL), diffuse large B cell lymphoma (DLBCL) and MCL (Bea et al. 2009; Ota et al. 2004; Tagawa and Seto 2005), where an associated over-expression of the involved miRNAs has also been found, suggesting that high-level amplification is the main source of oncogenic *miR-17-92* cluster over-expression in these lymphomas (Navarro et al. 2009a; Tagawa et al. 2007) (Fig. 11.2).

Finally, several miRNAs have been described in MM to be associated with loss of heterozygosity, or copy number alterations at their chromosomal loci (Lionetti et al. 2009).

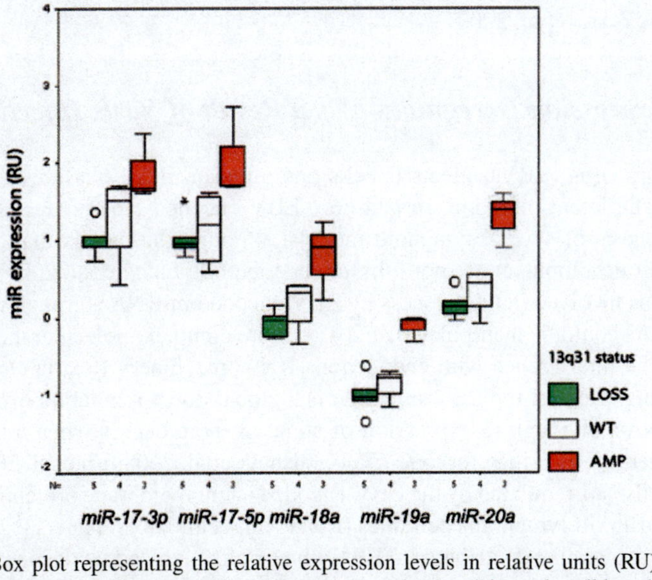

Fig. 11.2 Box plot representing the relative expression levels in relative units (RU) of several miRNAs at 13q31 and their copy number changes in examined nodal mantle cell lymphoma cases. All 5 miRNAs showed a significant correlation between the expression levels quantified by reverse transcriptase-looped quantitative PCR and the gene dosage of this locus according to comparative genomic hybridization results. Loss (LOSS), wild-type (WT), and amplification (AMP). Circles and asterisks, extreme values and outliers, respectively. Reprinted with permission from Navarro et al. (2009a) (corresponding to the original Fig. 4)

11.3.1.2 Involving Epigenetic Mechanisms Acting on MiRNA Genes

A proportion of miRNA promoters seem to be regulated by epigenetic mechanisms and related alterations could lead to oncogenic effects (Guil and Esteller 2009). One example is *miR-203*, that it has been found down-regulated in T cell lymphomas, among other hematological disorders, and targeting the *ABL* oncogene (Bueno et al. 2008). The expression down-regulation mechanism found was the deletion of one allele (at 14q32) and CpG hypermethylation in the promoter of the remaining allele (Bueno et al. 2008). Another example is *miR-124a* that it has been shown to be regulated by CpG methylation in several human tumors, including lymphomas (Lujambio et al. 2007). Other examples are also further explained in the text.

11.3.1.3 Involving Mutation of MiRNA Genes

Mutations in miRNA loci could result in alterations of sequence-specific recognition of their targets. Germ-line and somatic miRNA mutations were found in 12% of sequenced miRNAs and in 15% of CLL patients analyzed. Noticeably, no such mutations could be found in a large series of normal controls, and the majority of patients showing these mutations in the *mR-15a/16-1* locus had a personal or family history of CLL or other hematopoietic and solid tumors, suggesting a possible role of this miRNA cluster in familial CLL (Calin et al. 2005). Nevertheless, mutations of this miRNA have not been found in later studies on sporadic CLL among other neoplasms (Yazici et al. 2009).

11.3.2 Expression Deregulation as a Result of Virus Infection

Epstein-Barr virus (EBV) infects B cells and after an initial phase establishes an asymptomatic latent infection. In addition, EBV encoded factors are related with the oncogenesis of several lymphoid models, although the factors that are conditioning their development are not fully understood (De Falco et al. 2009). MiRNA expression is involved in this process by EBV-encoded miRNAs through which cellular mRNAs could be manipulated, and also more complex molecular mechanisms including the interference with endogenous RNAi machinery (Scaria et al. 2007). The resulting effect of the EBV infection is a global down-regulation of many cellular miRNAs, although re-expression of some of them occurs when the cells are immortalized, as described for *miR-155* (Godshalk et al. 2008; Lu et al. 2008). This miRNA is also up-regulated by the EBV-encoded latent membrane protein 1 (*LMP1*) in infected Burkitt lymphoma cell lines (Rahadiani et al. 2008; Yin et al. 2008), and it has been shown that contributes to EBV immortalization and its oncogenic potential by modulation of the NF-κB signaling (Lu et al. 2008). It also seems to have a cytostatic activity in BL cells. The underlying molecular mechanisms of the growth inhibitory property of *LMP1* seems to involve its ability to down-regulate a major oncogene, *TCL1* in DLBCL and BL cells by inducing *miR-29b* transcript expression (Anastasiadou et al. 2009). The underlying mechanism for this effect involves p38 mitogen-activated protein kinase-activating function of *LMP1*. As *LMP1* is also important for B cell transformation, the final effect on lymphomagenesis may depend on a combination of levels of its expression, lineage and differentiation of the target cells and also of the miRNA expression context of the host cells. Another described miRNA induced by *LMP1* in BL cell lines is *miR-146a*, and the authors proposed that this miRNA should plays a role in EBV latency by modulating innate immune responses to the virus-infected cell (Motsch et al. 2007). As it has been described that *miR-146a* negatively regulates *IRF7*, an activator of the *LMP1* promoter, this miRNA may function as a negative feedback control to modulate the transforming potential of EBV (Cameron et al. 2008; Li et al. 2010). In addition, *miR-21* is induced by EBV infection in human B cells (Mrazek et al. 2007), and it has been found over-expressed in a significant proportion of EBV-positive natural killer (NK)/T cell lymphomas, where targeting over the anti-apoptotic transcript *PTEN* was also demonstrated (Yamanaka et al. 2009). Finally, EBV influence in the expression of a 10 miRNA host subset could be demonstrated in classical Hodgkin's lymphoma (cHL) (Navarro et al. 2008). Regarding EBV-mediated effects on lymphocyte differentiation program, it has been described altered levels of the transcription factors *BLIMP1* and *XBP1* by *miR-127* in EBV-positive BL.

Regarding miRNA influence of other lymphoma-related viruses, the reticuloendotheliosis virus strain T (REV-T) induces B cell lymphomas through up-regulation of *miR-155* resulted in a prosurvival effect by down-regulation of *JARID2/Jumonji*, a cell cycle regulator part of a histone methyltransferase complex (Bolisetty et al. 2009; Yin et al. 2008). In addition, human herpesvirus 8/Kaposi's sarcoma-associated herpesvirus (KSHV) has been characterized as the etiologic agent associated, among others, with primary effusion lymphoma (PEL). This virus encodes orthologue miRNAs for the control of relevant host targets, as KSHV *miR-K12-11* that has a common sequence and targets compared to the cellular *miR-155* (Gottwein et al. 2007; Skalsky et al. 2007).

11.4 As Regulators of Known Oncogenes and Tumor Suppressor Genes

Several examples of both miRNA types are explained below and summarized in Table 11.1.

11.4.1 Tumor Suppressor MiRNAs

MiRNAs encoded by the *miR-15a/16-1* cluster are known to act as tumor suppressors and, in fact, it were the first miRNAs to be found experimentally related to mammalian carcinogenesis. Down-regulation of these miRNAs has been reported in CLL and other neoplasms (Aqeilan et al. 2010). In normal cells, the expression of these miRNAs inhibits cell proliferation, promotes apoptosis of cancer cells, and suppresses tumorigenicity through their suppressive action on

Table 11.1 MicroRNAs with oncogenic or tumor suppressor role in human lymphomas

MicroRNA	Cytoband	Lymphoma[a]	Alteration	Target	Described function
let-7a	9q22.32 (*let-7a1*) 11q24.1 (*let-7a2*) 22q13.31 (*let-7a3*)	DLBCL, cHL	Down-regulation Up-regulation	*PRDM1*, *BLIMP1*	Plasma cell differentiation, oncomir
miR-9 family	1q22 (*miR-9 -1*) 5q14.3 (*miR-9- 2*) 15q25.1 (*miR-9-3*)	DLBCL, cHL	Up-regulation	*PRDM1*, *BLIMP1*	B cell functionality, plasma cell differentiation, oncomir
miR-15a/16 cluster	13q14.3	CLL, MCL, MM	Down-regulation by deletion Point mutations have been also described	*BCL2*, *WT1*, *WNT3A*, *CCND1*, *MCL1*, *ETS*, *RAB9B*, *PDCD6IP*, *AKT3*, *S6*, *MAP3KIP3*	Inhibition proliferation, promotes apoptosis, tumor suppressor microRNA
miR-17-92 cluster (*miR-17*, *miR-18a*, *miR-19a*, *miR-20a*, *miR-19-b1*, *miR-92a-1*)	13q31.3	BL, DLBCL, MCL, MM, BL	Up-regulation (amplification) *c-Myc/E2F* regulated expression	*BIM*, *PTEN*, *c-Myc*, *E2F1*, *E2F3*, *SOCS1*, *CDKN1A*, *CCND1*	Lymphoid cell development (Pro-B to pre-B transition), T cell proliferation and survival, oncomir
miR-21	17q23.1	EBV^+ NK/T, DLBCL, MM, CLL	Up-regulation (induced by EBV infection)	*PTEN*	Oncomir
miR-23a/b	9q22.3	BL, CLL	Up-regulation	*GLS2*	B cell proliferation, tumor suppressor microRNA
miR-25	7q22.1	MM	Up-regulation	*PCAF*	Oncomir
miR-26a	3p22.2 (*miR-26a-1*) 12q14.1 (*miR-26a-2*)	BL	*c-Myc* regulated expression	*EZH2*	Tumor suppressor microRNA

Table 11.1 (continued)

MicroRNA	Cytoband	Lymphoma[a]	Alteration	Target	Described function
miR-29 family (*miR-29-a*, *miR-29-b1/2*, *miR-29-c*)	7q32.3 (*miR-29a/b1*) 1q32.2 (*miR-29c/b2*)	CLL, MCL, SMZL	Down-regulation (7q deletion in SMZL), and *LMP1*-EBV regulated expression	*TCL1, MCL1, CDK6*	Tumor suppressor microRNA
miR-32	9q31.3	MM	Up-regulation	*PCAF*	Oncomir
miR-34 family (*miR-34a*, *miR-34b*, *miR-34c*)	1p36.22 (*miR-34a*) 11q23 (*miR-34*b/c)	CLL, BL	Down-regulation (chromosomal losses, promoter methylation)	*CCND1, FoxP1, CDK4/6, CCNE, E2F3, c-Myc, TCL1*	Pro-B pre-B transition regulation, tumor suppressor microRNA
miR-93	7q22.1	MM	Up-regulation	*PCAF*	Oncomir
miR-106 family	Xq26.2 (*miR-106a*) 7q22.1 (*miR-106b*)	MM, DLBCL, CLL	Up-regulation	*PCAF*	Oncomir
miR-127	14q32.2	EBV$^+$ BL, ABC-DLBCL	Up-regulation (by EBV infection and epigenetic mechanisms) Down-regulation in DLBCL	*BLIMP1, XBP1, BCL6*	Effects on lymphocyte differentiation, oncomir
miR-135a	3q21 (*miR-135a1*) 12q23.1 (*miR-135a2*)	CLL	Down-regulation	*JAK2*	Prosurvival effect, tumor suppressor microRNA
miR-143	5q32	CLL, DLBCL, BL, EBV-transformed cell line	Down-regulation	*ERK5, DNMT3A*	Tumor suppressor microRNA
miR-146a	5q34	BL, CLL	Up-regulation and *LMP1*-EBV-regulated expression	*IRF7*	Oncomir

Table 11.1 (continued)

MicroRNA	Cytoband	Lymphoma[a]	Alteration	Target	Described function
miR-150	19q13.33	CLL, MCL	Up-regulation (CLL) Down-regulation (MCL)	*MYB*	Pro-B pre-B transition regulation
miR-155	21q21.3	ABC-DLBCL, CLL, BL, MCL, cHL, NK/T cell lymphoma, PMBCL	Up-regulation in many but BL (and also by EBV and REV-T infection)	*c-MAF, PU1/sfpil, AID, SOCS1, JARID2, CCND1, SMAD5, SHIP1*	B and T cell functionality, oncomir
miR-181a	1q32.1 (*miR-181a1*) 9q33.3 (*miR-181a2*)	CLL, MM, MCL	Down-regulation (17p deletion) Up-regulation (in MM and CLL trisomy 12)	*DUSP5, DUSP6, SHP2, PTPN22, TCL1, PLAG1, PCAF*	T cell development, oncomir
miR-195	17p13.1	CLL	Up-regulation (in unmutated CLL group)	*CCND1*	Post-transcriptional regulation of *CCND1*
miR-203	14q32	T cell lymphoma	Down-regulation (Deletion of one allele and epigenetic regulation of the promoter)	*ABL*	Oncomir
miR-221/222	Xp11.3	CLL, MM	Down-regulation Up-regulation in MM	*KIT, PLK2, CDKN1B, HOX* family members	Lymphoid cell development, oncomirs

[a]Chronic lymphocytic leukemia (CLL); Diffuse large B cell lymphoma (DLBCL); Mantle cell lymphoma (MCL); Follicular lymphoma (FL); Primary mediastinal B cell lymphoma (PMBCL); Burkitt lymphoma (BL); Classical Hodgkin's lymphoma (cHL); Multiple myeloma (MM); Epstein-Barr virus (EBV); Natural killer T cell lymphoma (NK/T); Splenic marginal zone lymphoma (SMZL).

cancer-related targets like *BCL2*, *WT1*, *WNT3A*, *CCND1*, *MCL1*, *ETS1*, *RAB9B*, and *PDCD6IP* (Aqeilan et al. 2010; Calin et al. 2008). It has been proposed that the high relevance of *miR-15a/16-1* in CLL lymphomagenesis could be mainly resumed by its causal role in *BCL2* over-expression in a proportion of these lymphomas that harbor 13q14 deletions (Cimmino et al. 2005). The experimental evidence showed that the expression of *BCL2* and these miRNAs are inversely correlated in CLL tumors, and in vitro apoptosis could be induced by *BCL2* repression through these miRNAs (Cimmino et al. 2005).

As above mentioned, *CCND1* is a validated *miR-16-1* target. The over-expression of this oncogene is the hallmark of MCL. The t(11;14)(q13:q23) is the primary genetic alteration that induced the *CCND1* over-expression by means of the promoting influence of the *IgH* enhancer onto the *CCND1* promoter. It has been also described in MCL other molecular alterations in additional layers of *CCND1* expression control. One of these mechanisms involves the post-transcriptional regulation of *CCND1* by means of several miRNAs like *miR-19a*, *miR-155*, *miR-503*, *miR-424*, *miR-195*, *miR-34a*, *miR-15a*, and *miR-16-1* (Jiang et al. 2009). It has been described that point mutations and genomic deletions affecting the 3′UTR of *CCND1* resulted in increased proliferation and shorter survival (Wiestner et al. 2007). A *miR-16-1* orientated study has demonstrated in MCL cell lines that these truncated *CCND1* forms lack the *miR-16-1* binding sequences in the *CCND1* 3′UTR (Chen et al. 2008). The proposed oncogenic effect of this alteration is similar to other described examples involving different mechanisms that also generate shorter *CCND1* transcript forms which are associated with a higher mRNA stability (Wiestner et al. 2007).

Another lymphoid neoplasm showing down-regulation of *miR-15a/miR-16* is MM (Roccaro et al. 2009). In this model, the mentioned miRNAs regulate proliferation and induced angiogenesis, and the gene targets that could be related with these miRNAs were *AKT3*, *S6*, *MAP* kinases and the *NF-κB*-activator *MAP3KIP3* (Roccaro et al. 2009). Although, these miRNAs were located in 13q14, a region frequently deleted in MM, their low expression was also detectable in cases without this chromosome imbalance.

Expression levels of *miR-143* and *miR-145* have been found decreased in CLL, DLBCL, EBV-transformed B cell lines, and BL cell lines (Akao et al. 2007). In this study, introduction of precursor forms of these miRNAs into Raji cells resulted in a significant growth inhibition that occurred in a dose-dependent manner. One target gene of *miR-143* was determined to be the extracellular signal-regulated kinase 5 (*ERK5*), involved in mitogen-activated protein kinase proliferation pathways, and also previously found to be targeted by this miRNA in colorectal carcinomas and colon cancer cell lines (Akao et al. 2006; 2007). Another work has showed that DNA methyltransferase 3 (*DNMT3A*) is another target of *miR-143* in colorectal cancer (Ng et al. 2009), suggesting that the down-regulation of *miR-143* could impair the activation of tumor suppressor genes by demethylation. Thus, it could be very interesting to confirm the existence of this oncogenic mechanism in the *miR-143*-down-regulated lymphomas.

Down-regulation of *miR-29* family members (*miR-29-a2*, *miR-29-b2*, and *miR-29c*) have been described associated with the unmutated-*IgH* poor prognosis CLL subgroup (Calin et al. 2005). This association was also confirmed in other CLL series (Fulci et al. 2007; Marton et al. 2008; Stamatopoulos et al. 2009). Two demonstrated oncogenes that are target of several *miR-29* family members are *TCL1* (as also showed for *miR-181b*) and *MCL1*, acting both in PI3K/Akt survival pathways (Mott et al. 2007; Pekarsky et al. 2006). In addition, *TCL1* transgenic mice over-expressing this gene in B cells developed phenotypic changes very similar to that seen in CLL, supporting the pathogenetic role of this gene, and thus of *miR-29b* and *miR-181b* in CLL (Bichi et al. 2002). Expression down-regulation of *miR-29a* and *miR-29b-1* were also described to be associated with 7q32 deletion in splenic marginal zone lymphoma (SMZL). Finally, it has been shown that *miR-29a/b/c* down-regulation in MCL involved their targeting on *CDK6* (Zhao et al. 2010).

MiR-34 family is composed by *miR-34a* and *miR-34b/c* that are tumor suppressor miRNAs induced by *p53* (Hermeking 2009). Thus, *p53* alterations are one of the known *miR-34* family inactivating mechanisms. Demonstrated target genes for these miRNAs included many oncogenes (*CCND1*, *CDK4/6*, *CCNE*, *E2F3*, and *c-Myc* among others) (Hermeking 2010). In addition, chromosomal losses including the coding region of these miRNAs (*miR-34a* at 1p36 and *miR-34b/c* at 11q23) are another possible mechanism of alteration of their expression. In CLL, it has been shown that *miR-34a* is a part of the *p53* response network, thus low levels of this miRNA caused by *p53* abnormalities (17p deletion and point mutations) are associated with an impaired DNA damage response and chemotherapy resistance (Zenz et al. 2009). In addition, some CLL cases also showed low basal *miR-34a* levels in absence of *p53* mutations (Zenz et al. 2009). A possible repression mechanism for these cases could be mediated by promoter hypermethylation. This mechanism has been also previously described in many tumor types, also including non-Hodgkin's lymphomas like extranodal NK/T (44%) as the most frequent non-Hodgkin's lymphoma with this alteration (Chim et al. 2010). As mentioned before, *miR-34b/c* are located at 11q23, a chromosomal region deleted in 10–32% of CLL cases and that it is associated with patients with a poor outcome (Seiler et al. 2006). *ATM* gene has been previously characterized as a relevant target of this chromosomal alteration in CLL. Nevertheless, *miR-34b/c* locus has also been found frequently deleted in 11q- CLL cases (Cardinaud et al. 2009; Lehmann et al. 2008). This is a remarkable observation taking in account that the same authors showed that *miR-34b-5p* regulates the *TCL1* expression that is, as already mentioned, a very relevant gene in CLL pathogenesis (Cardinaud et al. 2009). In addition to gene deletion, it has been also described that expression of pre-*miR-34b/c* is also regulated at methylation level, suggesting an alternative inactivation mechanism of these miRNAs in CLL that deserve further exploration (Toyota et al. 2008).

MiR-135a down-regulation in cHL has been showed to be involved in the *JAK2* expression regulation (Navarro et al. 2009b). Concordantly with the pathogenetic role of this gene in cHL, chromosomal gains and amplifications involving *JAK2* gene have been also described in these lymphomas, leading to the same prosurvival effect of this miRNA alteration.

MiR-26a has been identified to be repressed by *c-Myc* in primary BL cells and in a murine B cell *c-Myc*-over-expressing mouse model (O'Donnell et al. 2005;

Sander et al. 2008). This miRNA has been showed to directly regulate *EZH2* expression (Sander et al. 2008). *EZH2* is member of the Polycomb repressive complex 2 that resulted in epigenetic silencing of genes involved in tumor suppression and differentiation (Sander et al. 2009). Thus, *c-Myc*-dependent repression of this post-transcriptional repressor of *EZH2* constitutes another transforming mechanism of *c-Myc* in BL. This kind of transforming mechanism involving *c-Myc*-dependent repression of miRNA expression has been also described involving *miR-23a/b*, a negative regulator of *GLS2* that codifies for mitochondrial glutaminase, and resulting in a convenient enhancement of energetic metabolism for the proliferating B cells (Gao et al. 2009).

MiR-127 was shown to regulate *BCL6* expression in a BL cell line, suggesting a possible pathogenetic role in lymphomas of germinal center origin (Saito et al. 2006). Interestingly, the expression of this miRNA has described to have clinical impact in DLBCL (Roehle et al. 2008). Epigenetic changes are also described to regulate the expression of this miRNA (Saito et al. 2006). Finally, its expression seem to be modulated in different anatomic locations of DLBCL, being higher in testicular than in central nervous system or nodal DLBCL (Robertus et al. 2009).

11.4.2 Oncomirs

The oncogenic role of *miR-155* was discovered from the precursor transcript corresponding to the *BIC* locus, that was first observed to cooperate with *c-Myc* in chicken B cell lymphomas induced by avian leukosis proviral integrations (Clurman and Hayward 1989). Its over-expression in transgenic lymphoma prone mice accelerates the death rate for malignancy (Costinean et al. 2006). In human B-lymphomas, *miR-155* expression is very reduced in BL (Kluiver et al. 2006), but it is highly over-expressed in cHL and another non-Hodgkin's lymphomas such primary mediastinal B cell lymphoma (PMBCL) and DLBCL of poor prognosis activated B cell (ABC) subtype (Eis et al. 2005; Kluiver et al. 2005). ABC subtypes of DLBCL are characterized by constitutive activation of *NF-κB* pathway. Concordantly, this pathway is also targeted by *miR-155* through *SMAD5* down-regulation, although the detailed mechanism by which this alteration contributes to DLBCL pathogenesis it is not fully understood (Rai et al. 2008; 2010). In addition, *miR-155* is also frequently over-expressed in NK/T cell lymphomas, where this miRNA targets the inositol phosphatase transcript *SHIP1*, that it is involved in Akt signaling, and affecting the apoptotic activity of these cells (Yamanaka et al. 2009). This targeting on *SHIP1* was also described in DLBCL, where it could be demonstrated that high levels of *miR-155* were caused by autocrine stimulation by TNFα (Pedersen et al. 2009). Other potential targets of *miR-155* have been demonstrated, and many of them are relevant either for T and B cell functionality.

It has been hypothesized that *AID* miss-expression could be a source of oncogenic mutations, and the fact that *miR-155* inhibits the expression of *AID* suggested a tumor suppressor activity of this miRNA in B cells. In fact, a relationship of *miR-155* levels with *AID*-induced oncogenic reciprocal chromosomal translocations between *IgH* and *c-Myc* genes has been described (Dorsett et al. 2008). Nevertheless, as *AID* activity is essential for B cell maturation, it seems that the relationship of *miR-155* levels with *AID* or even other targets should be controlled in a certain range to avoid oncogenic effects (Tili et al. 2009). Noticeably, induced *miR-155* levels are higher and transient in physiological situations than those found in lymphomas. The time limited dynamics of *miR-155* up-regulation found in normal lymphocytes is overcame in lymphoid neoplasms probably due to the insufficient miRNA levels to generate a putative inhibitory feedback loop (as the blocking of its own transcription) (Tili et al. 2009).

MiR-17-92 cluster includes six mature miRNAs (*miR-17-5p*, *miR-18a*, *miR-19a*, *miR-20a*, *miR-19b-1, and miR-92-1*). Transfection of the vertebrate-specific *miR-17-miR-19b-1* section showed oncogenic cooperation with *c-Myc* over-expression in transgenic mice models (He et al. 2005; Tagawa et al. 2007). Further experiments have demonstrated that the rationale for this cooperation consisted in the requirement of *miR-17-92* activity to suppress apoptosis in *c-Myc*-driven B cell lymphomas (Mu et al. 2009). Moreover, *miR-19a* and *miR-19b* are necessary and sufficient for this oncogenic prosurvival activity, targeting proapoptotic genes as *PTEN* (Mu et al. 2009). In addition, *BIM* proapoptotic gene has also been showed to be targeted by *miR-19* and *miR-92* in experimental lymphoid mice models and in primary samples of MM (Pichiorri et al. 2008; Xiao et al. 2008). In fact, the relationship of *miR-17-92* cluster with *c-Myc* is very complex. First, *c-Myc* oncogene induces the expression of *miR-17-92* cluster by direct binding to regulatory sequences therein (O'Donnell et al. 2005). Second, *miR-17-5p* and *miR-20* negatively regulate the *c-Myc* -target gene *E2F1*. In addition, *E2F3* and, in a lesser extent, *E2F1* can also activate the transcription of the *miR-17-92* cluster (Sylvestre et al. 2007; Woods et al. 2007), in a complex circuitry with an important role in the control of proliferation and apoptosis (Aguda et al. 2008). Moreover, the alteration of a *miR-17-92* cluster-mediated *E2F1* regulation seem to affect the expression timing of this protein along the cell cycle, and leading to accumulation of DNA double-strand breaks by disruption of the *E2F1*-dependent G1-cell cycle arrest (Pickering et al. 2009). Finally, among other specific targets of the *miR-17-92* components, it could be mentioned that the G1-S checkpoint cyclin-dependent kinase inhibitor *CDKN1A* has been found to be repressed by *miR-17-5p* and *miR-20a* (Cloonan et al. 2008; Pickering et al. 2009). In fact, this target was considered the main oncogenic mechanism underlying the *miR-17-92* over-expression in a MCL cell line (Inomata et al. 2009). *miR-17-5p* and *miR-20a* expression were also found to show oncogenic cooperation with *c-Myc* in primary samples of MCL (Navarro et al. 2009a) (Fig. 11.3). In MM, *miR-19a* and *miR-19b* have been shown to target *SOCS1*, a negative regulator of *IL-6R/STAT3* pathway giving a role of this miRNA within the anti-apoptotic signal of IL-6 in these lymphomas (Pichiorri et al. 2008).

MiR-181a over-expression was found to be associated with the pathogenesis of several lymphoid neoplasms. First, it was described to belong to a miRNA signature over-expressed in a subset of CLL patients with a faster development of clinical features that made need to start treatment, thus suggesting an oncogenic action of *miR-181a* in this lymphoma (Calin et al. 2005). However, the oncogenic zinc-finger transcription factor *PLAG1* was found to be one of the repressed targets

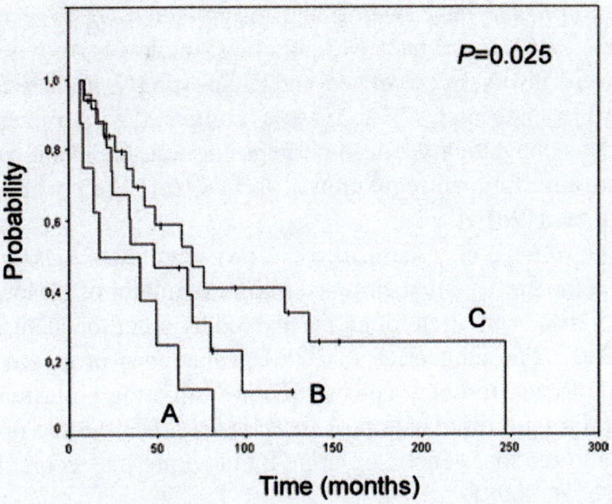

Fig. 11.3 Overall survival of 50 patients with nodal MCL according to the combined expression of *c-Myc* and *miR-17-5p/miR-20a*. Mantle cell lymphoma patients with concomitant high expression of both *c-Myc* and *miR-17-5p/miR-20a* ($n=8$; *line A*) have a significant shorter overall survival than patients with high expression of only one ($n=17$; fail 5; *line B*) or none ($n=25$; fail 10; *line C*) of these factors ($p=0.025$, log-rank test). Reprinted with permission from Navarro et al. (2009a) (corresponding to the original Fig. 5)

of this miRNA in CLL (Pallasch et al. 2009). Marton et al. (2008) also found down-regulation of this miRNA in CLL compared to normal B cells. These results stressed the complex role of this and other miRNAs for which oncogenic and tumor suppressor features seem to coexist, even in a same neoplastic entity. This phenomenon could also be reflecting the different molecular context in which the miRNA expression must be considered. Thus, *miR-181a* expression levels have opposite trends in the progression of CLL depending on the karyotype, as high expression of this miRNA was associated with disease progression in CLL patients with trisomy 12, whereas aggressive CLL forms with 17p deletion corresponded to cases with lower levels of *miR-181a* expression (Visone et al. 2009). Over-expression of this miRNA was also found to be present in MM and oncogenic features could be demonstrated for this alteration (Pichiorri et al. 2008). In this work, a positive regulator of *p53* was established to be an oncogenic target of *miR-181a* together with other as *miR-181b*, *miR-32*, *miR-106b*, *miR-25*, and *miR-93* (Pichiorri et al. 2008). This target was the p300-CBP-associated factor (*PCAF*) that is a histone acetyl-transferase involved in acetylation-dependent *p53* regulation and that, in addition, affect the levels of *p53* through its intrinsic ubiquitination activity controlling *Hdm2* protein levels (Linares et al. 2007; Schiltz and Nakatani 2000). *MiR-181a* expression levels have also been elevated in a distinct biological subset of MCL showing unmutated *IgH* genes, higher proliferation status and higher levels of chromosomal instability (Navarro et al. 2009a). The mechanisms of *miR-181a* expression modulation in all these lymphomas are not fully understood. For instance, CpG methylation status seem to be very different among individual of CLL cases, at least for the *miR-181a2* subtype (Pallasch et al. 2009).

MiR-221 and *miR-222* have been found to regulate *CDKN1B* in CLL in vitro (Frenquelli et al. 2010). In primary CLL samples, the levels of these miRNAs and the *CDKN1B* were inversely correlated and, interestingly, high *miR-221/222* levels were found in lymph nodes/bone marrow compared to peripheral blood CLL samples from the same patients. These findings suggest a relevant role of the miR-dependent post-transcriptional regulation of *CDKN1B* in the proliferation status of CLL (Frenquelli et al. 2010).

MiR-9 and *let-7a* have been described to down-regulate *PRDM1/BLIMP1* gene (Nie et al. 2008), the previously mentioned master regulator of plasma cell differentiation, and that it has been often found inactivated by other mechanisms in DLBCL (Parekh et al. 2007; Pasqualucci et al. 2006). It has been proposed for cHL, that precursors of the Reed-Sternberg cells (RSC) are initiating a plasma cell differentiation program through initial action of *PRDM1/BLIMP1* but the up-regulation of these miRNAs aborted that genetic program by blocking the accumulation of these transcripts (Nie et al. 2008).

11.5 Oncomirs with Biomarker Implications in Lymphomas

11.5.1 Using MiRNA Expression for Lymphoma Diagnosis Improvement

11.5.1.1 In B/T Lymphomas

DLBCL is considered to be a highly heterogeneous disease that includes several subgroups either at biological and pathological levels (Friedberg and Fisher 2008). In addition, FL could also progress to a more aggressive DLBCL form. Gene-expression profiling allows differentiation into 3 major subgroups: ABC-DLBCL, germinal center B cell (GCB) DLBCL, and PMBCL. Several studies have shown that miRNA expression profiling is also different among these entities. Initially, several miRNAs were found to be differentially expressed between DLBCL and normal reactive lymph nodes, and between these tumors and FL (Roehle et al. 2008). The authors conclude that a decision tree using only 4 miRNAs (*miR-330*, *miR-17-5p*, *miR-106a*, and *miR-210*) is enough to differentiate between these three types of lymphoid tissues, although without a comprehensive consideration of the subtypes that exists inside these lymphomas. In this way, a later study also included DLBCL cases transformed from FL, in addition to the de novo DLBCL and other FL cases that did not progressed (Lawrie et al. 2009). Thus, the comparison between DLBCL and FL was then performed only considering de novo DLBCL and non-transformed FL. Twenty-six miRNAs were found to be differentially expressed between these two types of lymphoma, including 6 miRNAs encoded by the *miR-17-92* and/or homologous clusters (Table 11.2). In addition, other miRNAs were also found to be differentially expressed between de novo and transformed DLBCL (14 miRNAs), and between non-transformed FL and the cases of this lymphoma type that underwent

Table 11.2 MicroRNA expression profiles of diagnostic/prognostic value in human lymphomas

Lymphoma type[a]	MicroRNA[b]	Clinical value[c]
BL	*let-7c*	Diagnosis of BL independent of, *c-Myc* translocation
BL	*miR-17-92*	Diagnosis of BL with, *c-Myc* translocation
BL	*miR-34b*	Diagnosis of BL without, *c-Myc* translocation
MM	*let-7e, miR-99b, miR-125a-5p*	In t(4;14) positive TC4 group
MM	*miR-133b*	In t(14;16)/t(14;20) positive TC5 group
MM	*miR-582-5p*	In t(11;14) TC1 group
MM	*miR-181a/b, miR-17-92, miR-106a-92, miR-106b-25, miR-221, miR-222, miR-328, miR-21, miR-32, miR-15a, miR16, miR-193b-365*	Differentially expressed between tumor and normal plasma cells
CLL	*miR-15a/16-1, miR-195, miR-221, miR-23b, miR-155, miR-24-1, miR-146, miR-223, miR-29a-2/b-2/c, miR-150, miR-30d, miR-191*	In poor prognosis compared to good prognosis CLL subgroups
MCL	*miR-150, miR-142-3p/5p, miR-29* family, *miR-124a, miR-155*	In comparison to normal B cells from lymph nodes and peripheral blood
DLBCL	*miR-142, miR-302b, miR-519d*	In de novo DLBCL treated with R-CHOP associated with event free survival
DLBCL	*miR-330, miR-425, miR27a, miR-222, miR-199b*	In de novo DLBCL treated with R-CHOP and associated with poor outcome
DLBCL	*miR-146b-5p, miR-146a, miR-21, miR-155, miR-500, miR-222, miR-363, miR-574-3p, 574-5p*	In ABC-like DLBCL compared to GCB DLBCL
DLBCL	*miR-491, miR-27a, miR-19b, miR-25, miR-18a, miR-636, miR-92, miR-621, miR-526c, miR-766, miR-299-5p, miR-380-3p, miR-129, miR-588*	In transformed DLBCL in comparison with de novo DLBCL
DLBCL	*miR-93, miR-92, miR-106a, miR-20a, miR-19b, miR-18b, miR-200c, miR-638, miR-518a, miR-199a, miR-22, miR-34a, miR-362, miR-206, miR-451, miR-636, miR-27b, miR-199b, miR-27a, miR-24, miR-99a, miR-100, miR-26b, miR-217, miR-634, miR-150*	In de novo DLBCL in comparison with non-transformed FL

Table 11.2 (continued)

Lymphoma type[a]	MicroRNA[b]	Clinical value[c]
PEL	*miR-9, miR-16, miR-17-3p, miR-17-5p, miR-19a, miR-20, miR-23b, miR-26a, miR-29b, miR-30e, miR-34a, miR-103, miR-106, miR-140, miR-142-3p, miR-148a, miR-152, miR-181b, miR-182, miR-186, miR-191, miR-194, miR-210, miR-301, miR-320, miR-331*	In PEL in comparison with other types of DLBCL
FL	*let-7b, let-7i, miR-221, miR-222, miR-223, miR-217*	In transformed FL in comparison with non-transformed FL
NK/T	*miR-34a*	Most frequent lymphoma type showing this alteration
cHL	*miR-138*	Ann Arbor stage I-II
cHL	*miR-20a, miR-21, miR-9, miR-155, miR-16, miR-140, miR-18a, miR-30b, miR-30a-5p, miR-196a, miR-374, miR-186, miR-520a, miR-200a, miR-614*	cHL miRNA signature in malignant Hodgkin and Reed Sternberg cells compared to $CD77^+$progenitor cells
cHL	*miR-21, miR-27a, miR-124a, miR-134, miR-138, miR-147, miR-182, miR-185, miR-198, miR-216, miR-220, miR-302a/b/c, miR-325, miR-335, miR-23b, miR-26b, miR-30b, miR-31, miR-126, miR-135a, miR-183, miR-204, miR-205*	Expression in cHL compared with reactive lymph nodes

[a]Chronic lymphocytic leukemia (CLL); Diffuse large B cell lymphoma (DLBCL), activated B cell (ABC) and germinal center B cell (GCB) subtypes; Mantle cell lymphoma (MCL); Follicular lymphoma (FL); Burkitt lymphoma (BL); Classical Hodgkin's lymphoma (cHL); Multiple myeloma (MM); Natural killer/T cell lymphoma (NK/T); Primary effusion lymphoma (PEL).

[b]Red for up-regulated and green for down-regulated microRNAs in the corresponding comparisons.

[c]Rituximab-cyclophosphamide, hydroxydaunorubicin, oncovin, and prednisone/prednisolone (R-CHOP).

transformation (6 miRNAs) (Table 11.2) (Lawrie et al. 2009). Regarding DLBCL subtypes, current differentiation between GCB and non-GCB subtype without using gene expression profiling has been proposed by immuno-phenotyping with CD10, BCL6, and MUM1 antibodies (Hans et al. 2004), and this classification method was the used in the miR-related studies. Thus, differential expression of 26 miRNAs was found between the immuno-phenotyped groups (as GCB or non-GCB-like) was also found in the above mentioned work (Table 11.2) (Lawrie et al. 2009). However, even less miRNAs seem to be enough to separate these DLBCL subtypes, as has been showed in DLBCL-derived cell lines where 9 miRNAs where differentially expressed between the mentioned DBCL subtypes (Malumbres et al. 2009), including *miR-155* and *miR-21* that were also described to be highly expressed in the ABC-DLBCL subtypes in other previous works with cell lines (Table 11.2) (Lawrie et al. 2007; 2009; Rai et al. 2008). Moreover, these miRNAs could be also detected in the serum of DLBCL patients at significant different levels compared to healthy controls, suggesting that these types of non-invasive analysis are also possible in human lymphomas, or at least in DLBCL (Lawrie et al. 2008).

PEL is a unique type of DLBCL, and it has been shown to express a 26 miRNA signature compared to normal lymphoid tonsils (Table 11.2) (O'Hara et al. 2008). Among them, it could be found B cell lineage and B cell-lymphoma specific miRs. Although some genomic alterations were found to be associated with the miRNA expression changes, no one miRNA locus was consistently lost or amplified in all cases, thus suggesting that the PEL-specific miRNA signature arise from other mechanisms.

The oncogenic hallmark of BL is the *c-Myc* over-expression. As previously mentioned, *c-Myc* induces *miR-17-92* expression and it is involved with this miRNA cluster in a complex regulatory loop that determines the deregulation of *E2F1* normal expression with oncogenic impact in proliferation and apoptosis regulation (Sylvestre et al. 2007). In BL, *c-Myc* over-expression is mainly originated from translocations with immunoglobulin genes, but in a proportion of cases over-expression of this oncogene was present in *c-Myc*-translocation negative BL cases. Interestingly, there was a strong correlation of *miR-34b* down-regulation only in these *c-Myc* translocation-negative cases (Leucci et al. 2008). This result suggested a possible *c-Myc* expression control by this miRNA that was confirmed by in vitro increasing of a synthetic *miR-34b* leading to *c-Myc* expression modulation (Leucci et al. 2008). Noticeably, *miR-34b* down-regulation in the *c-Myc* translocation-negative BL cases was independent of *p53* status, a previously demonstrated transcriptional inducer. This observation suggested that other mechanisms must be causing the *miR-34b* down-regulation in this subset of BL, probably by epigenetic mechanisms, as previously suggested (Toyota et al. 2008). In addition, there was a down-regulation of the *let-7c* miRNA in all BL cases analyzed, either with or without the *c-Myc* translocation, although it is not clear its pathogenetic meaning.

In CLL, it has been described several miRNAs that are differentially expressed in comparison with normal B cells (Calin et al. 2004). Using cloning approaches, *miR-150*, *miR-21*, and *miR-155* were described to be significantly up-regulated and *miR-92* down-regulated in this neoplasm compared to normal $CD19^+$ blood cells (Fulci et al. 2007). A later study using a similar quantification technique also confirmed the over-expression of *miR-155* compared to normal peripheral blood B cells but it could not confirm the remaining mentioned miRNAs (Marton et al. 2008). Another interesting published data about *miR-150* and *miR-155* in CLL is their inverse expression pattern in the proliferation centers of the CLL affected lymph nodes, revealed by in situ hybridization, and with *miR-155* expressed in the proliferation centers whereas *miR-150* is expressed outside these areas (Wang et al. 2008).

In MM, several studies have showed differentially expressed miRNAs between the tumor and normal plasma cells, although with some variations between them. Thus, over-expression of *miR-181a/b* in MM was described in two reports (Pichiorri et al. 2008; Roccaro et al. 2009), similarly to miRNAs of the paralog clusters *miR-17-92*, *miR-106a-92*, and *miR-106b-25* (Pichiorri et al. 2008; Unno et al. 2009). Other miRNAs also showed to be over-expressed are *miR-221*, *miR-222*, *miR-382* (Roccaro et al. 2009), as well as *miR-21* and *miR-32*, among others (Pichiorri et al. 2008). On the other side, *miR-328* (Pichiorri et al. 2008) as well as *miR-15a* and *miR-16* down-regulation were also observed (Roccaro et al. 2009). In addition, other down-regulated miRNAs in MM have been also described (Unno et al. 2009). Finally, the miRNA cluster *miR-193b-365* was identified to be expressed exclusively in MM (Unno et al. 2009).

Also in MM, association of different miRNA expression profiles have been described in relation with cytogenetic tumor subtypes (TC classification (Hideshima et al. 2004)), including over-expression of the clustered *miR-99b*, *let-7e*, and *miR-125a-5p* in the t(4;14) positive TC4 group, over-expression of *miR-133b* in t(14;16)/t(14;20) positive TC5 group, and the *miR-582-5p* over-expression in the t(11;14) TC1 group (Lionetti et al. 2009). Finally, in another work, miRNA expression profiles were also studied in different cytogenetic tumor subtypes of MM, but focusing at their differences regarding which miRNAs are differentially expressed compared to normal plasma cells (Gutierrez et al. 2010).

In MCL, there are two studies characterizing the miRNA expression in comparison to normal B cells. In the study of Zhao et al. the comparison was performed between lymph node MCL samples and peripheral blood normal B cells, and only a few miRNAs were found to be deregulated in more than 50% of cases, including decrease of *miR-150*, *miR-142-3p/5p*, *miR-29 family*, and increase of *miR-124a* and *miR-155* (Zhao et al. 2010). However, a previous study using a selected 86 miRNA set demonstrated significant differences in a longer number of miRNAs either from peripheral blood or lymph node MCL samples compared with $CD5^+$ or $CD5^-$ tonsillar normal B cells (Navarro et al. 2009a). These results showed that, as in other neoplasms, MCL has a characteristic altered miRNA expression profile compared to normal counterparts, although normal or transformed B cells could also show expression profiles affected by the microenvironment of these cells. This later consideration was further extended in the MCL scope, as it was shown that tumoral cells from the same patient showed differential expression of some miRNAs in different anatomic environments (peripheral blood and lymph nodes, Fig. 11.4) (Navarro et al. 2009a). This finding suggests that anatomic source of tumoral samples is a relevant parameter to take in account for consideration of certain miRNAs as potential biomarkers clinically useful, at least in MCL.

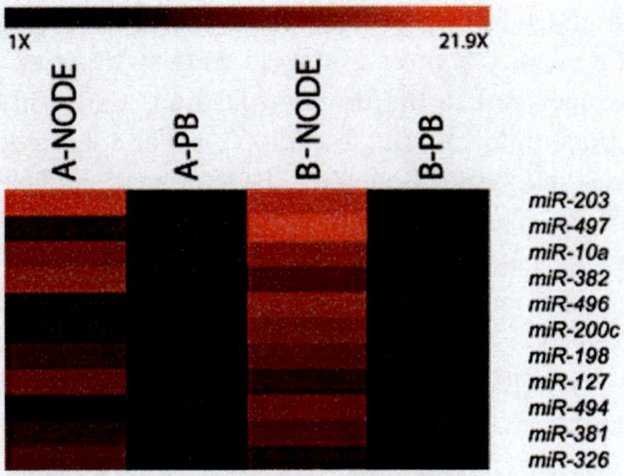

Fig. 11.4 Heat map representation of the significant miRNA expression differences in paired samples of purified tumor cells from simultaneous matched peripheral blood (PB) and tumor lymph nodes (NODE) of two patients (*A* and *B*). Eleven miRNAs were found differentially expressed in the paired significance analysis of microarrays of expression levels between the two tumor cell populations from these two patients. Relative expression was obtained in reference to CD5$^+$ control cells but, additionally for each miRNA, its expression in the nodal samples has been normalized in reference to the peripheral blood values for a more clear representation. Reprinted with permission from Navarro et al. (2009a) (corresponding to the original Fig. 2)

11.5.1.2 In Other Lymphomas

In cHL, miRNA expression profiling allows to differentiate these tumors from reactive lymph nodes through a 25-miRNA signature (Table 11.2) (Navarro et al. 2008). Moreover, major histological subtypes (nodular sclerosis and mixed cellularity) could also be differentiated by the expression of a reduced number of miRNAs (Navarro et al. 2008). Noticeably, this study was performed without tumor cell purification. On the contrary, a study using quantitative PCR on microdissected RSC cells described a signature of 12 over-expressed and 3 down-regulated miRNAs in comparison to normal CD77$^+$ progenitor cells, and showing little overlap with the previous work on whole biopsies (Table 11.2) (Van Vlierberghe et al. 2009). In addition, absolute miRNA abundance was also measured by direct cloning in two HL cell lines, and among the most abundant miRNAs were found *miR-16*, *miR-21*, *miR-155*, and *miR-9*, concordantly with the previous results of Van Vlierberghe et al. (Nie et al. 2008).

11.5.2 Using MiRNA Expression for Lymphoma Prognosis Stratification Improvement

11.5.2.1 In B/T Lymphomas

In DLBCL, the expression of 7 miRNAs was found to be associated with event-free or overall survival, but only *miR-127* expression was associated with both parameters, being the expression of this miRNA lower in poor prognosis cases (Roehle et al. 2008). In another study, de novo DLBCL cases were split into two groups with low (0–2) or high (3–5) international prognostic index (IPI), and 13 miRNAs correctly predicted 85% of the cases (Table 11.2) (Lawrie et al. 2009). Event-free survival (EFS) was analyzed in the de novo DLBCL uniformly treated with the chemotherapy combination of rituximab-cyclophosphamide, hydroxydaunorubicin, oncovin, and prednisone/prednisolone (R-CHOP) and 8 miRNAs were found to be associated with EFS. These including high expression of *miR-142*, *miR-302b*, *miR-519d*, and down-regulation of *miR-330*, *miR-425*, *miR-27a*, *miR-222*, and *miR-199b*, being the last one the most significantly associated with poor outcome. Nevertheless, in a later study using a longer series of R-CHOP uniformly treated DLBCL patients, only high *miR-222* expression (also associated with ABC subtype) showed a significant association with overall survival and progression-free survival (Malumbres et al. 2009). Finally, it must be stressed that the differential expression that has been shown for several miRNAs between ABC and GCB subtypes of DLBCL could also have prognostic relevance as ABC subtypes are more aggressive. Noticeably, this was not the case of *miR-155*, because it was not found related to prognosis of the whole series (Rai et al. 2008; Roehle et al. 2008) and surprisingly, it was associated with a subset of ABC-DLBCL with a less aggressive behavior (Jung and Aguiar 2009).

CLL is characterized by highly variable clinical behavior, with a survival range from 1 to more than 15 years. A plethora of factors have been showed to influence survival of CLL patients (Dal Bo et al. 2009). The expression of several miRNAs has also described to have prognostic value. Thus, loss of 13q14.3, including *miR-16-1* and *miR-15a*, is not only the most frequent genomic alteration of this neoplasm but it is also associated to a good prognosis if it is the sole aberration. Noticeably, the deletion is often interstitial and is homozygous in up to 15% of the cases (Haferlach et al. 2007). In addition, other miRNAs related with prognosis were found by Calin et al. when analyzing a total of 190 miRNAs (Calin et al. 2005). The prognostic signature found consisted in 13 miRNAs that also included the above mentioned miRNAs down-regulated in the group of patients with good prognosis (also characterized by low expression of ZAP-70 and hypermutation of *IgH* genes) (Table 11.2). The fact that *miR-15a/16-1* levels were higher in the poor prognosis group points out that these miRNAs are involved in complex regulatory webs of both oncogenes and tumor suppressor genes and it is difficult to predict the final effect in vivo. Experimental data also support this complex regulatory effects of this miRNA cluster in CLL (Calin et al. 2008). In addition, other miRNAs included in this signature with the same prognosis association that the above mentioned were *miR-195*, *miR-221*, *miR-23b*, *miR-155*, *miR-24-1*, *miR-146*, and *miR-16-2*. On the other hand several miRNAs were down-regulated in the poor prognosis CLL group, including *miR-223* and *miR-29a-2/b-2/c*. In another study, using a different technology for miRNA expression quantification in CLL, lower expression levels of *miR-150*, *miR-223*, and *miR29b/c* were found to be associated with the aggressive patient subgroup (Fulci et al. 2007). The majority of these miRNAs were also found differentially expressed between the two defined CLL prognostic groups in another miRNA cloning study, although also finding *miR-30d* and *miR-191* as differentially expressed in the poor prognosis group (Marton et al. 2008).

In addition, *miR-29c* and *miR-223* were later also showed to be down-regulated in patients with poor prognosis independently of the prognostic marker classification used (Stamatopoulos et al. 2009). Moreover, the expression of these miRNAs were used in an index through which a clear separation of patients into 5 groups with different median treatment free-survival and into 3 groups with different median overall survival was obtained (Stamatopoulos et al. 2009). Noticeably, it has been shown that down-regulation of *miR-29a/b/c* is also associated with poor prognosis in MCL (Zhao et al. 2010).

Another relevant clinical parameter of CLL in which expression of several miRNA has also been associated is the time to initial therapy. Thus, treatment of CLL begins with the development of symptomatic or progressive disease, and patients with related shorter interval have been characterized to show high expression of *miR-181a*, *miR-155*, *miR-146*, *miR-24-2*, *miR-23a/b*, *miR-222*, and *miR-221*, and low expression of *miR-29c*.

Finally, low expression of *miR-15a* and over-expression of *miR-181a/b* were associated with worse prognosis in MM (Roccaro et al. 2009).

11.5.2.2 In Other Lymphomas

In cHL, high expression levels of *miR-138* were found to be related with Ann Arbor stage I-II of these tumors (Navarro et al. 2008), and low expression of *miR-135a* was associated with a higher probability of relapse and shorter disease-free survival (Navarro et al. 2009b).

11.6 Oncomirs as Possible Candidates for Therapeutic Targeting: A Promise of More Specific Lymphoma Therapies?

As we have seen in the previous subheadings, miRNAs are relevant regulatory molecules in the lymphomagenesis. Altered miRNA expression levels through different causal mechanisms constitute the main final pathogenetic effect. In relation with treatment, it is possible that miRNA expression changes could be contributing to the cytotoxic effect of some chemotherapeutics. Thus, inhibition of histone deacetylases induces cell death in CLL cells through up-regulation of *miR-106b* by a characterized molecular pathway controlled by this miRNA (Sampath et al. 2009). Nevertheless, an even more interesting approach is the finding of treatments based on specific expression modulation of relevant miRs, allowing the correction of pathogenic molecular imbalances. In this sense, different molecular systems have been experimentally tested to induce specific miRNA expression changes including locked nucleic acids (LNA) or 2′-*O*-methyl-antisense RNA. These molecules have been used in different tumor models because of their ability to specifically inhibit the miRNA to target by complementary sequence binding (Frieden and Orum 2008). In addition, it is possible to synthesize pre-miRNAs that could be introduced in lymphoma cells to replace low expressed or silenced endogenous miRNAs. Nevertheless, there are at least two issues that should be resolved for a real clinical application: (1) To reach an effective tumor cytotoxicity either alone or combined with existing chemotherapy; and (2) To obtain specific delivering to the tumor cells avoiding side effects on other cell populations, although at enough dose to obtain a physiological effect. Neither of these issues has been resolved yet, including lymphomas. Nevertheless, the known role of miRNAs in apoptosis regulation of the lymphoid neoplasms offers interesting candidates that will be surely targeted in lymphomas as soon as the above mentioned issues will be overcame with new technological improvements.

Acknowledgments The authors were supported (in part) by a grant (RD06/0020/0039) from Red Temática de Investigación Cooperativa en Cáncer (RTICC), Instituto de Salud Carlos III (ISCIII), Spanish Ministry of Science and Innovation and European Regional Development Fund (ERDF). In addition, L. Hernández is supported by FIS and "programa d'estabilització d'investigadors" of Direcció d'Estrategia i Coordinació del Departament de Salut (Generalitat de Catalunya).

References

Aguda BD, Kim Y, Piper-Hunter MG, et al. MicroRNA regulation of a cancer network: consequences of the feedback loops involving miR-17-92, E2F, and Myc. Proc Natl Acad Sci USA. 2008;105:19678–83.

Akao Y, Nakagawa Y, Kitade Y, et al. Downregulation of microRNAs-143 and -145 in B-cell malignancies. Cancer Sci. 2007;98:1914–20.

Akao Y, Nakagawa Y, Naoe T. MicroRNAs 143 and 145 are possible common onco-microRNAs in human cancers. Oncol Rep. 2006;16:845–50.

Anastasiadou E, Boccellato F, Vincenti S, et al. Epstein-Barr virus encoded LMP1 downregulates TCL1 oncogene through miR-29b. Oncogene. 2009;29:1316–28.

Aqeilan RI, Calin GA, Croce CM. MiR-15a and miR-16-1 in cancer: discovery, function and future perspectives. Cell Death Differ. 2010;17:215–20.

Bea S, Salaverria I, Armengol L, et al. Uniparental disomies, homozygous deletions, amplifications, and target genes in mantle cell lymphoma revealed by integrative high-resolution whole-genome profiling. Blood. 2009;113:3059–69.

Bichi R, Shinton SA, Martin ES, et al. Human chronic lymphocytic leukemia modeled in mouse by targeted TCL1 expression. Proc Natl Acad Sci USA. 2002;99:6955–60.

Bolisetty MT, Dy G, Tam W, et al. Reticuloendotheliosis virus strain T induces miR-155, which targets JARID2 and promotes cell survival. J Virol. 2009;83:12009–17.

Bueno MJ, Perez DCI, Gomez DC, et al. Genetic and epigenetic silencing of microRNA-203 enhances ABL1 and BCR-ABL1 oncogene expression. Cancer Cell. 2008;13:496–506.

Calame K. Activation-dependent induction of Blimp-1. Curr Opin Immunol. 2008;20:259–64.

Calin GA, Cimmino A, Fabbri M, et al. MiR-15a and miR-16-1 cluster functions in human leukemia. Proc Natl Acad Sci USA. 2008;105:5166–71.

Calin GA, Ferracin M, Cimmino A, et al. A MicroRNA signature associated with prognosis and progression in chronic lymphocytic leukemia. N Engl J Med. 2005;353:1793–801.

Calin GA, Liu CG, Sevignani C, et al. MicroRNA profiling reveals distinct signatures in B cell chronic lymphocytic leukemias. Proc Natl Acad Sci USA. 2004;101:11755–60.

Cameron JE, Yin Q, Fewell C, et al. Epstein-Barr virus latent membrane protein 1 induces cellular MicroRNA miR-146a, a modulator of lymphocyte signaling pathways. J Virol. 2008;82: 1946–58.

Cardinaud B, Moreilhon C, Marcet B, et al. MiR-34b/miR-34c: a regulator of TCL1 expression in 11q- chronic lymphocytic leukaemia? Leukemia. 2009;23:2174–7.

Chen RW, Bemis LT, Amato CM, et al. Truncation in CCND1 mRNA alters miR-16-1 regulation in mantle cell lymphoma. Blood. 2008;112:822–9.

Chen CZ, Li L, Lodish HF, et al. MicroRNAs modulate hematopoietic lineage differentiation. Science. 2004;303:83–6.

Chen CZ, Lodish HF. MicroRNAs as regulators of mammalian hematopoiesis. Semin Immunol. 2005;17:155–65.

Chim C, Wong K, Qi Y, et al. Epigenetic inactivation of the miR-34a in hematological malignancies. Carcinogenesis. 2010;31:745–50.

Cimmino A, Calin GA, Fabbri M, et al. MiR-15 and miR-16 induce apoptosis by targeting BCL2. Proc Natl Acad Sci USA. 2005;102:13944–9.

Cloonan N, Brown MK, Steptoe AL, et al. The miR-17-5p microRNA is a key regulator of the G1/S phase cell cycle transition. Genome Biol. 2008;9:R127.

Clurman BE, Hayward WS. Multiple proto-oncogene activations in avian leukosis virus-induced lymphomas: evidence for stage-specific events. Mol Cell Biol. 1989;9: 2657–64.

Costinean S, Zanesi N, Pekarsky Y, et al. Pre-B cell proliferation and lymphoblastic leukemia/high-grade lymphoma in E(mu)-miR155 transgenic mice. Proc Natl Acad Sci USA. 2006;103: 7024–9.

Dal Bo M, Bertoni F, Forconi F, et al. Intrinsic and extrinsic factors influencing the clinical course of B-cell chronic lymphocytic leukemia: prognostic markers with pathogenetic relevance. J Transl Med. 2009;7:76.

De Falco G, Antonicelli G, Onnis A, et al. Role of EBV in microRNA dysregulation in Burkitt lymphoma. Semin Cancer Biol. 2009;19:401–6.

Dorsett Y, McBride KM, Jankovic M, et al. MicroRNA-155 suppresses activation-induced cytidine deaminase-mediated Myc-Igh translocation. Immunity. 2008;28:630–8.

Du C, Liu C, Kang J, et al. MicroRNA miR-326 regulates TH-17 differentiation and is associated with the pathogenesis of multiple sclerosis. Nat Immunol. 2009;10:1252–9.

Eis PS, Tam W, Sun L, et al. Accumulation of miR-155 and BIC RNA in human B cell lymphomas. Proc Natl Acad Sci USA. 2005;102:3627–32.

Frenquelli M, Muzio M, Scielzo C, et al. MicroRNA and proliferation control in chronic lymphocytic leukemia: functional relationship between miR-221/222 cluster and p27. Blood. 2010;115:3949–59.

Friedberg JW, Fisher RI. Diffuse large B-cell lymphoma. Hematol Oncol Clin North Am. 2008;22:941–52, ix.

Frieden M, Orum H. Locked nucleic acid holds promise in the treatment of cancer. Curr Pharm Des. 2008;14:1138–42.

Fulci V, Chiaretti S, Goldoni M, et al. Quantitative technologies establish a novel microRNA profile of chronic lymphocytic leukemia. Blood. 2007;109:4944–51.

Gao P, Tchernyshyov I, Chang TC, et al. c-Myc suppression of miR-23a/b enhances mitochondrial glutaminase expression and glutamine metabolism. Nature. 2009;458:762–5.

Godshalk SE, Bhaduri-McIntosh S, Slack FJ. Epstein-Barr virus-mediated dysregulation of human microRNA expression. Cell Cycle. 2008;7:3595–600.

Gottwein E, Mukherjee N, Sachse C, et al. A viral microRNA functions as an orthologue of cellular miR-155. Nature. 2007;450:1096–9.

Guil S, Esteller M. DNA methylomes, histone codes and miRNAs: tying it all together. Int J Biochem Cell Biol. 2009;41:87–95.

Gutierrez NC, Sarasquete ME, Misiewicz-Krzeminska I, et al. Deregulation of microRNA expression in the different genetic subtypes of multiple myeloma and correlation with gene expression profiling. Leukemia. 2010;24:629–37.

Haferlach C, Dicker F, Schnittger S, et al. Comprehensive genetic characterization of CLL: a study on 506 cases analysed with chromosome banding analysis, interphase FISH, IgV(H) status and immunophenotyping. Leukemia. 2007;21:2442–51.

Hans CP, Weisenburger DD, Greiner TC, et al. Confirmation of the molecular classification of diffuse large B-cell lymphoma by immunohistochemistry using a tissue microarray. Blood. 2004;103:275–82.

He L, Thomson JM, Hemann MT, et al. A microRNA polycistron as a potential human oncogene. Nature. 2005;435:828–33.
Hermeking H. MiR-34a and p53. Cell Cycle. 2009;8:1308.
Hermeking H. The miR-34 family in cancer and apoptosis. Cell Death Differ. 2010;17:193–9.
Hideshima T, Bergsagel PL, Kuehl WM, et al. Advances in biology of multiple myeloma: clinical applications. Blood. 2004;104:607–18.
Inomata M, Tagawa H, Guo YM, et al. MicroRNA-17-92 down-regulates expression of distinct targets in different B-cell lymphoma subtypes. Blood. 2009;113:396–402.
Jiang Q, Feng MG, Mo YY. Systematic validation of predicted microRNAs for cyclin D1. BMC Cancer. 2009;9:194.
Jung I, Aguiar RC. MicroRNA-155 expression and outcome in diffuse large B-cell lymphoma. Br J Haematol. 2009;144:138–40.
Klein U, Lia M, Crespo M, et al. The DLEU2/miR-15a/16-1 cluster controls B cell proliferation and its deletion leads to chronic lymphocytic leukemia. Cancer Cell. 2010;17:28–40.
Kluiver J, Haralambieva E, de Jong D, et al. Lack of BIC and microRNA miR-155 expression in primary cases of Burkitt lymphoma. Genes Chromosomes Cancer. 2006;45:147–53.
Kluiver J, Poppema S, de Jong D, et al. BIC and miR-155 are highly expressed in Hodgkin, primary mediastinal and diffuse large B cell lymphomas. J Pathol. 2005;207:243–9.
Lawrie CH, Chi J, Taylor S, et al. Expression of microRNAs in diffuse large B cell lymphoma is associated with immunophenotype, survival and transformation from follicular lymphoma. J Cell Mol Med. 2009;13:1248–60.
Lawrie CH, Gal S, Dunlop HM, et al. Detection of elevated levels of tumour-associated microRNAs in serum of patients with diffuse large B-cell lymphoma. Br J Haematol. 2008;141:672–5.
Lawrie CH, Soneji S, Marafioti T, et al. MicroRNA expression distinguishes between germinal center B cell-like and activated B cell-like subtypes of diffuse large B cell lymphoma. Int J Cancer. 2007;121:1156–61.
Lehmann S, Ogawa S, Raynaud SD, et al. Molecular allelokaryotyping of early-stage, untreated chronic lymphocytic leukemia. Cancer. 2008;112:1296–305.
Lerner M, Harada M, Loven J, et al. DLEU2, frequently deleted in malignancy, functions as a critical host gene of the cell cycle inhibitory microRNAs miR-15a and miR-16-1. Exp Cell Res. 2009;315:2941–52.
Leucci E, Cocco M, Onnis A, et al. MYC translocation-negative classical Burkitt lymphoma cases: an alternative pathogenetic mechanism involving miRNA deregulation. J Pathol. 2008;216:440–50.
Li QJ, Chau J, Ebert PJ, et al. miR-181a is an intrinsic modulator of T cell sensitivity and selection. Cell. 2007;129:147–61.
Li L, Chen XP, Li YJ. MicroRNA-146a and human disease. Scand J Immunol. 2010;71:227–31.
Linares LK, Kiernan R, Triboulet R, et al. Intrinsic ubiquitination activity of PCAF controls the stability of the oncoprotein Hdm2. Nat Cell Biol. 2007;9:331–8.
Lionetti M, Biasiolo M, Agnelli L, et al. Identification of microRNA expression patterns and definition of a microRNA/mRNA regulatory network in distinct molecular groups of multiple myeloma. Blood. 2009;114:e20–6.
Lu LF, Thai TH, Calado DP, et al. Foxp3-dependent microRNA155 confers competitive fitness to regulatory T cells by targeting SOCS1 protein. Immunity. 2009;30:80–91.
Lu F, Weidmer A, Liu CG, et al. Epstein-Barr virus-induced miR-155 attenuates NF-kappaB signaling and stabilizes latent virus persistence. J Virol. 2008;82:10436–43.
Lujambio A, Ropero S, Ballestar E, et al. Genetic unmasking of an epigenetically silenced microRNA in human cancer cells. Cancer Res. 2007;67:1424–9.
Malumbres R, Sarosiek KA, Cubedo E, et al. Differentiation stage-specific expression of microRNAs in B lymphocytes and diffuse large B-cell lymphomas. Blood. 2009;113:3754–64.
Marton S, Garcia MR, Robello C, et al. Small RNAs analysis in CLL reveals a deregulation of miRNA expression and novel miRNA candidates of putative relevance in CLL pathogenesis. Leukemia. 2008;22:330–8.

Motsch N, Pfuhl T, Mrazek J, et al. Epstein-Barr virus-encoded latent membrane protein 1 (LMP1) induces the expression of the cellular microRNA miR-146a. RNA Biol. 2007;4:131–7.

Mott JL, Kobayashi S, Bronk SF, et al. Mir-29 regulates Mcl-1 protein expression and apoptosis. Oncogene. 2007;26:6133–40.

Mrazek J, Kreutmayer SB, Grasser FA, et al. Subtractive hybridization identifies novel differentially expressed ncRNA species in EBV-infected human B cells. Nucleic Acids Res. 2007;35:e73.

Mu P, Han YC, Betel D, et al. Genetic dissection of the miR-17~92 cluster of microRNAs in Myc-induced B-cell lymphomas. Genes Dev. 2009;23:2806–11.

Navarro A, Bea S, Fernandez V, et al. MicroRNA expression, chromosomal alterations, and immunoglobulin variable heavy chain hypermutations in Mantle cell lymphomas. Cancer Res. 2009a;69:7071–8.

Navarro A, Diaz T, Martinez A, et al. Regulation of JAK2 by miR-135a: prognostic impact in classic Hodgkin lymphoma. Blood. 2009b;114:2945–51.

Navarro A, Gaya A, Martinez A, et al. MicroRNA expression profiling in classic Hodgkin lymphoma. Blood. 2008;111:2825–32.

Ng EK, Tsang WP, Ng SS, et al. MicroRNA-143 targets DNA methyltransferases 3A in colorectal cancer. Br J Cancer. 2009;101:699–706.

Nie K, Gomez M, Landgraf P, et al. MicroRNA-mediated down-regulation of PRDM1/Blimp-1 in Hodgkin/Reed-Sternberg cells: a potential pathogenetic lesion in Hodgkin lymphomas. Am J Pathol. 2008;173:242–52.

Ota A, Tagawa H, Karnan S, et al. Identification and characterization of a novel gene, C13orf25, as a target for 13q31-q32 amplification in malignant lymphoma. Cancer Res. 2004;64:3087–95.

O'Connell RM, Rao DS, Chaudhuri AA, et al. Physiological and pathological roles for microRNAs in the immune system. Nat Rev Immunol. 2010;10:111–22.

O'Donnell KA, Wentzel EA, Zeller KI, et al. c-Myc-regulated microRNAs modulate E2F1 expression. Nature. 2005;435:839–43.

O'Hara AJ, Vahrson W, Dittmer DP. Gene alteration and precursor and mature microRNA transcription changes contribute to the miRNA signature of primary effusion lymphoma. Blood. 2008;111:2347–53.

Palamarchuk A, Efanov A, Nazaryan N, et al. 13q14 deletions in CLL involve cooperating tumor suppressors. Blood. 2010;115:3916–22.

Pallasch CP, Patz M, Park YJ, et al. MiRNA deregulation by epigenetic silencing disrupts suppression of the oncogene PLAG1 in chronic lymphocytic leukemia. Blood. 2009;114:3255–64.

Parekh S, Polo JM, Shaknovich R, et al. BCL6 programs lymphoma cells for survival and differentiation through distinct biochemical mechanisms. Blood. 2007;110:2067–74.

Pasqualucci L, Compagno M, Houldsworth J, et al. Inactivation of the PRDM1/BLIMP1 gene in diffuse large B cell lymphoma. J Exp Med. 2006;203:311–7.

Pedersen IM, Otero D, Kao E, et al. Onco-miR-155 targets SHIP1 to promote TNFalpha-dependent growth of B cell lymphomas. EMBO Mol Med. 2009;1:288–95.

Pekarsky Y, Santanam U, Cimmino A, et al. Tcl1 expression in chronic lymphocytic leukemia is regulated by miR-29 and miR-181. Cancer Res. 2006;66:11590–3.

Pichiorri F, Suh SS, Ladetto M, et al. MicroRNAs regulate critical genes associated with multiple myeloma pathogenesis. Proc Natl Acad Sci USA. 2008;105:12885–90.

Pickering MT, Stadler BM, Kowalik TF. MiR-17 and miR-20a temper an E2F1-induced G1 checkpoint to regulate cell cycle progression. Oncogene. 2009;28:140–5.

Rahadiani N, Takakuwa T, Tresnasari K, et al. Latent membrane protein-1 of Epstein-Barr virus induces the expression of B-cell integration cluster, a precursor form of microRNA-155, in B lymphoma cell lines. Biochem Biophys Res Commun. 2008;377:579–83.

Rai D, Karanti S, Jung I, et al. Coordinated expression of microRNA-155 and predicted target genes in diffuse large B-cell lymphoma. Cancer Genet Cytogenet. 2008;181:8–15.

Rai D, Kim SW, McKeller MR, et al. Targeting of SMAD5 links microRNA-155 to the TGF-beta pathway and lymphomagenesis. Proc Natl Acad Sci USA. 2010;107:3111–6.

Ramkissoon SH, Mainwaring LA, Ogasawara Y, et al. Hematopoietic-specific microRNA expression in human cells. Leuk Res. 2006;30:643–7.
Robertus JL, Harms G, Blokzijl T, et al. Specific expression of miR-17-5p and miR-127 in testicular and central nervous system diffuse large B-cell lymphoma. Mod Pathol. 2009;22: 547–55.
Roccaro AM, Sacco A, Thompson B, et al. MicroRNAs 15a and 16 regulate tumor proliferation in multiple myeloma. Blood. 2009;113:6669–80.
Rodriguez A, Vigorito E, Clare S, et al. Requirement of bic/microRNA-155 for normal immune function. Science. 2007;316:608–11.
Roehle A, Hoefig KP, Repsilber D, et al. MicroRNA signatures characterize diffuse large B-cell lymphomas and follicular lymphomas. Br J Haematol. 2008;142:732–44.
Saito Y, Liang G, Egger G, et al. Specific activation of microRNA-127 with downregulation of the proto-oncogene BCL6 by chromatin-modifying drugs in human cancer cells. Cancer Cell. 2006;9:435–43.
Sampath D, Calin GA, Puduvalli VK, et al. Specific activation of microRNA106b enables the p73 apoptotic response in chronic lymphocytic leukemia by targeting the ubiquitin ligase Itch for degradation. Blood. 2009;113:3744–53.
Sander S, Bullinger L, Klapproth K, et al. MYC stimulates EZH2 expression by repression of its negative regulator miR-26a. Blood. 2008;112:4202–12.
Sander S, Bullinger L, Wirth T. Repressing the repressor: a new mode of MYC action in lymphomagenesis. Cell Cycle. 2009;8:556–9.
Scaria V, Hariharan M, Pillai B, et al. Host-virus genome interactions: macro roles for microRNAs. Cell Microbiol. 2007;9:2784–94.
Schiltz RL, Nakatani Y. The PCAF acetylase complex as a potential tumor suppressor. Biochim Biophys Acta. 2000;1470:M37–53.
Seiler T, Dohner H, Stilgenbauer S. Risk stratification in chronic lymphocytic leukemia. Semin Oncol. 2006;33:186–94.
Skalsky RL, Samols MA, Plaisance KB, et al. Kaposi's sarcoma-associated herpesvirus encodes an ortholog of miR-155. J Virol. 2007;81:12836–45.
Sonkoly E, Stahle M, Pivarcsi A. MicroRNAs and immunity: novel players in the regulation of normal immune function and inflammation. Semin Cancer Biol. 2008;18:131–40.
Stamatopoulos B, Meuleman N, Haibe-Kains B, et al. MicroRNA-29c and microRNA-223 down-regulation has in vivo significance in chronic lymphocytic leukemia and improves disease risk stratification. Blood. 2009;113:5237–45.
Sylvestre Y, De G, V, Querido E, et al. An E2F/miR-20a autoregulatory feedback loop. J Biol Chem. 2007;282:2135–43.
Tagawa H, Karube K, Tsuzuki S, et al. Synergistic action of the microRNA-17 polycistron and Myc in aggressive cancer development. Cancer Sci. 2007;98:1482–90.
Tagawa H, Seto M. A microRNA cluster as a target of genomic amplification in malignant lymphoma. Leukemia. 2005;19:2013–6.
Teng G, Hakimpour P, Landgraf P, et al. MicroRNA-155 is a negative regulator of activation-induced cytidine deaminase. Immunity. 2008;28:621–9.
Thai TH, Calado DP, Casola S, et al. Regulation of the germinal center response by microRNA-155. Science. 2007;316:604–8.
Tili E, Croce CM, Michaille JJ. MiR-155: on the crosstalk between inflammation and cancer. Int Rev Immunol. 2009;28:264–84.
Toyota M, Suzuki H, Sasaki Y, et al. Epigenetic silencing of microRNA-34b/c and B-cell translocation gene 4 is associated with CpG island methylation in colorectal cancer. Cancer Res. 2008;68:4123–32.
Tsitsiou E, Lindsay MA. MicroRNAs and the immune response. Curr Opin Pharmacol. 2009;9:514–20.
Unno K, Zhou Y, Zimmerman T, et al. Identification of a novel microRNA cluster miR-193b-365 in multiple myeloma. Leuk Lymphoma. 2009;50:1865–71.

van den BA, Kroesen BJ, Kooistra K, et al. High expression of B-cell receptor inducible gene BIC in all subtypes of Hodgkin lymphoma. Genes Chromosomes Cancer. 2003;37:20–8.

Van Vlierberghe P, De Weer A, Mestdagh P, et al. Comparison of miRNA profiles of microdissected Hodgkin/Reed-Sternberg cells and Hodgkin cell lines versus CD77$^+$ B-cells reveals a distinct subset of differentially expressed miRNAs. Br J Haematol. 2009;147:686–90.

Vigorito E, Perks KL, Abreu-Goodger C, et al. MicroRNA-155 regulates the generation of immunoglobulin class-switched plasma cells. Immunity. 2007;27:847–59.

Visone R, Rassenti LZ, Veronese A, et al. Karyotype-specific microRNA signature in chronic lymphocytic leukemia. Blood. 2009;114:3872–9.

Wang M, Tan LP, Dijkstra MK, et al. miRNA analysis in B-cell chronic lymphocytic leukaemia: proliferation centres characterized by low miR-150 and high BIC/miR-155 expression. J Pathol. 2008;215:13–20.

Wiestner A, Tehrani M, Chiorazzi M, et al. Point mutations and genomic deletions in CCND1 create stable truncated cyclin D1 mRNAs that are associated with increased proliferation rate and shorter survival. Blood. 2007;109:4599–606.

Woods K, Thomson JM, Hammond SM. Direct regulation of an oncogenic micro-RNA cluster by E2F transcription factors. J Biol Chem. 2007;282:2130–4.

Xiao C, Calado DP, Galler G, et al. MiR-150 controls B cell differentiation by targeting the transcription factor c-Myb. Cell. 2007;131:146–59.

Xiao C, Srinivasan L, Calado DP, et al. Lymphoproliferative disease and autoimmunity in mice with increased miR-17-92 expression in lymphocytes. Nat Immunol. 2008;9:405–14.

Yamanaka Y, Tagawa H, Takahashi N, et al. Aberrant overexpression of microRNAs activate AKT signaling via down-regulation of tumor suppressors in natural killer-cell lymphoma/leukemia. Blood. 2009;114:3265–75.

Yazici H, Zipprich J, Peng T, et al. Investigation of the miR16-1 (C > T) + 7 substitution in seven different types of cancer from three ethnic groups. J Oncol. 2009;2009:827532.

Yin Q, McBride J, Fewell C, et al. MicroRNA-155 is an Epstein-Barr virus-induced gene that modulates Epstein-Barr virus-regulated gene expression pathways. J Virol. 2008;82:5295–306.

Zenz T, Mohr J, Eldering E, et al. MiR-34a as part of the resistance network in chronic lymphocytic leukemia. Blood. 2009;113:3801–8.

Zhang J, Jima DD, Jacobs C, et al. Patterns of microRNA expression characterize stages of human B-cell differentiation. Blood. 2009;113:4586–94.

Zhao JJ, Lin J, Lwin T, et al. microRNA expression profile and identification of miR-29 as a prognostic marker and pathogenetic factor by targeting CDK6 in mantle cell lymphoma. Blood. 2010;115:2630–9.

Zhou B, Wang S, Mayr C, et al. miR-150, a microRNA expressed in mature B and T cells, blocks early B cell development when expressed prematurely. Proc Natl Acad Sci USA. 2007;104:7080–5.

Chapter 12
MicroRNAs in Leukemia

Florian Kuchenbauer, Johannes Bloehdorn, Lars Bullinger, and Thorsten Zenz

Abstract Deregulated gene expression plays a crucial role in leukemia and recent evidence suggests that in hematologic malignancies deregulated microRNAs (miRNAs) have the ability to function as both tumor suppressor and oncogene. Microarray-based miRNA expression profiling has been performed for different leukemias and can identify leukemia subtypes and prognostic classification. Specific expression patterns also suggest that miRNAs contribute to leukemogenesis. Indeed, in chronic lymphocytic leukemia (CLL), the putative tumor suppressor *miR-15* and *miR-16* are down-regulated and recent data from animal models suggest that the deregulation of *miR-15* and *miR-16* can cause CLL/lymphoma. Similarly, a growing number of particular miRNAs are associated with clinical course or specific genetic aberrations. In this chapter, we will summarize current knowledge of the role of miRNAs in acute and chronic leukemia.

12.1 Introduction

Myeloid and lymphoid leukemias are hematopoietic stem cell disorders characterized by the clonal expansion of malignant myeloid or lymphoid cells. Over past years our knowledge of these hematologic malignancies has increased tremendously resulting in a classification based on morphology, immunophenotype, genetic and clinical features. Current classifications thus attempt to define biologically and clinically relevant entities. Nevertheless, in many cases the pathogenic event(s) are still unknown, and even well-defined leukemia subgroups (e.g. based on cytogenetics) exhibit considerable heterogeneity. Therefore, an improved understanding of the underlying tumor biology represents a prerequisite for refined molecular taxonomy of myeloid and lymphoid malignancies.

Chronic myeloid leukemia (CML) was the first malignancy associated with a recurrent chromosomal abnormality, the Philadelphia chromosome, which turned

T. Zenz (✉)
Department of Internal Medicine III, University of Ulm, 89081 Ulm, Germany
e-mail: thorsten.zenz@uniklinik-ulm.de

W.C.S. Cho (ed.), *MicroRNAs in Cancer Translational Research*,
DOI 10.1007/978-94-007-0298-1_12, © Springer Science+Business Media B.V. 2011

out to be also the first aberration that could be shown to result from a translocation of genetic material among chromosomes. By forming a fusion gene (*BCR-ABL*) the Philadelphia chromosome translocation t(9;22) is responsible for the pathogenesis of the disease (Chen et al. 2010). With the invention of chromosomal banding techniques, recurrent cytogenetic aberrations were also discovered in acute myeloid leukemia (AML), the most common acute leukemia in adults. Here, the translocation t(15;17) associated acute promyelocytic leukemia (APL) was the first malignancy to be successfully treated with a targeted approach that specifically overcomes the transforming potential of the molecular defect by adding all-*trans* retinoic acid to leukemia therapy (Huang et al. 1988). Likewise, CML has become the first disorder in which a small molecule inhibitor has been designed to specifically target the aberrant tyrosine kinase function responsible for the malignant transformation (Druker et al. 1996).

In addition, during the last decade genomics technologies have provided further "high resolution" insights into the molecular variation underlying the biological and clinical heterogeneity of malignancies. In many ways leukemias proved to be valuable models to demonstrate the utility and promise of the use of DNA microarray technology to study human tumors. For example, based on genome-wide gene expression profiling studies of acute lymphoblastic leukemia (ALL) and AML samples it was first demonstrated that distinct tumor subclasses exhibit characteristic gene expression changes that can be used for tumor class prediction (Golub et al. 1999).

In chronic lymphocytic leukemia (CLL) genomics approaches comparing gene expression profiles of CLL samples with unmutated and mutated *IGHV* led to the discovery of the differentially expressed gene *ZAP70* (Rosenwald et al. 2001; Rosenwald and Staudt 2002), which became the first microarray analysis derived marker that was clinically implemented.

As deregulated gene expression plays a crucial role in leukemias, recent evidence has been provided that in hematologic malignancies deregulated microRNAs (miRNAs) that have the ability to function as both tumor suppressors and oncogenes might also play a crucial role (He et al. 2005). Here, the observations that the putative tumor suppressors *miR-15* and *miR-16* are down-regulated in patients with CLL (Calin et al. 2004; 2005), and that microarray based miRNA expression profiling can identify leukemia subtype specific expression patterns in ALL (Lu et al. 2005) further suggested miRNAs to be important players in leukemogenesis. Similarly, in AML the expression of miRNAs could be correlated with morphology and genomic aberrations (Chen et al. 2010), and in CML the deregulation of the polycistronic *miR-17-92* cluster seems to also be of pathogenic relevance (Venturini et al. 2007).

12.2 Deregulated MiRNA Expression in AML

Initial studies revealed lineage specific expression of miRNAs in hematopoietic differentiation (Chen et al. 2004) (Fig. 12.1), and *miR-223* became one of the best investigated miRNAs in myelopoiesis (and AML) due to its specific expression in differentiated myeloid cells. Profiling miRNA expression in hematopoietic subpopulations as well as in a human APL cell line (NB4) upon differentiation with ATRA revealed *miR-223* to be expressed at low levels in the stem cell compartment with increasing expression throughout myeloid differentiation (Chen et al. 2004). Lentiviral over-expression of *miR-223* in NB4 cells induces myeloid differentiation (Fazi et al. 2005), and in primary AML cells minimal over-expression of *miR-223* (less than 2-fold expression levels) led to similar results (Fazi et al. 2007). This was the first study demonstrating that changes in miRNA expression levels can promote reprogramming of AML cells.

Interestingly, over-expression of *miR-223* and loss of *miR-223* might have distinct effects, as genetic depletion of *miR-223* led to a significant increase of myeloid progenitor cells as well as the number of circulating and bone marrow neutrophils (Johnnidis et al. 2008). Despite its important role in granulopoiesis, broader profiling studies did not connect *miR-223* expression to a particular AML subtype (Isken et al. 2008; Garzon et al. 2008a, 2008b; Jongen-Lavrencic et al. 2008; Debernardi et al. 2007; Marcucci et al. 2008). However, Fazi and colleagues demonstrated that the AML1-ETO oncoprotein, the product of the gene fusion resulting from the t(8;21), induces heterochromatic silencing of *miR-223* expression by recruiting chromatin remodeling enzymes at an AML1-binding site on the *miR-223* gene (Fazi et al. 2007), thereby showing for the first time that epigenetic silencing of a miRNA locus can be associated with the pathogenesis of AML.

12.2.1 Profiling Deregulated MiRNA Expression in AML – Association of MiRNA Profiles with Distinct AML Subtypes

Based on the initial observations and considering the steadily increasing number of newly discovered miRNAs, it was reasonable to perform broader miRNA expression profiling studies in AML. Indeed, the miRNA expression profiles of AML patient samples have been recently addressed by several studies that used different methodological approaches and screened slightly different patient subgroups (Isken et al. 2008; Garzon et al. 2008a, 2008b; Jongen-Lavrencic et al. 2008; Debernardi et al. 2007; Marcucci et al. 2008; Cammarata et al. 2010).

For example, Garzon and colleagues profiled miRNA expression in 240 AML patient samples with predominantly intermediate and poor cytogenetics using DNA microarrays. With this approach, the authors could identify molecular signatures associated with balanced 11q23 translocations, isolated trisomy 8 and *FLT3* mutations (*FLT3*-ITD), implying that miRNA profiles are at least partially driven by cytogenetics (Garzon et al. 2008b). Based on these findings the role of miRNAs in AML carrying *NPM1* and *FLT3-ITD* mutations, the two most frequent molecular aberrations in AML, was further investigated (Garzon et al. 2008). A signature distinguishing mutated *NPM1* from wildtype cases included the up-regulation of *miR-10a*, *miR-10b* as well as *let-7* and *miR-29* family members. Interestingly, an independent study confirmed a correlation of *FLT3-ITD* samples and *miR-155* up-regulation (Garzon et al. 2008a; Jongen-Lavrencic et al. 2008), though FLT3 inhibitor studies showed that the up-regulation of *miR-155* was independent from FLT3 signaling in these cases (Garzon et al. 2008b).

In accordance, a large study exploring the expression of miRNAs by quantitative reverse transcription-polymerase chain reaction (qRT-PCR) in a large AML patient cohort could also associate miRNA expression patterns with cytogenetic and molecular subtypes (Jongen-Lavrencic et al. 2008). Interestingly, for a miRNA based prediction of leukemia classes the necessary number of miRNAs varied drastically between AML subtypes. For example, a class predictor of only 10 miRNAs could reliably predict AML with t(8;21) and a set of 7 miRNAs AML with t(15;17). In contrast, a predictor comprising 72 miRNAs was necessary for AML with inv(16), thereby suggesting that not all cytogenetic aberrations might have a quite unique miRNA expression pattern and indeed some of the heterogeneity may stem from the fact that current subgroups may need to be further subdivided. In addition, miRNA expression levels might mainly be influenced by the differentiation stage of the leukemic cells, and thus e.g. inv(16) might be hard to distinguish form other inv(16)-negative AML cases with an identical morphology.

Similarly, smaller genome-wide miRNA expression studies using bead-based miRNA profiling approaches, microarrays and qRT-PCR confirmed miRNA expression patterns characteristic of cytogenetic subgroups such as t(15;17), t(11q23), t(8;21) and inv(16) (Cammarata et al. 2010; Li et al. 2008), as well as molecular subtypes like AML with *CEBPA* and *NPM1* mutations or deregulated *MN1* expression (Langer et al. 2008). Interestingly, all studies pointed towards the deregulation of miRNAs located in the *HOX* gene cluster, such as *miR-10a/b* and *miR-196a/b*, as well as *miR-221*, *let-7* family members, *miR-155*, *miR-29*, *miR-125b*, members of the *miR-17-92* cluster in AML. This suggests that a defined group of miRNAs might be involved in leukemogenic processes such as impaired differentiation and increased self-renewal.

12.2.2 Prognostic Impact of Altered MiRNA Expression in AML

Being correlated with altered gene expression and cytogenetic and molecular genetics aberrations, miRNA expression signatures have been shown to also confer prognostic information. While the study of Garzon and colleagues could identified two miRNAs, *miR-191* and *miR-199a*, to be significantly correlated with overall and disease-free survival (Garzon et al. 2008b), a large AML cohort by the Rotterdam group (based on qRT-PCR) could not identify any miRNAs that were correlated with overall or event-free survival independent of known factors such as cytogenetics (Jongen-Lavrencic et al. 2008). On the other hand, Marcucci and colleagues could identify a miRNA signature associated with event-free survival studying cytogenetically normal AML samples from adults under the age of 60 years who had high-risk molecular features such as *FLT3*–ITD, a wildtype *NPM1*, or both (Marcucci et al. 2008). However, in the future additional studies are needed to determine the impact of miRNAs as reliable biomarkers for diagnosis as well as prognosis in AML.

12.2.3 MiRNAs of Pathogenic and Functional Relevance in AML

Today, no study has been published yet investigating miRNA signatures in leukemic stem cells. So far, only few in vivo studies demonstrated the potential of a single miRNA to contribute to the development of myeloid neoplasia. Recently, Han and colleagues showed that retroviral over-expression of *miR-29a* can induce AML in mice (Han et al. 2010), and O'Connell and colleagues demonstrated similar findings through over-expression of *miR-155* in primitive hematopoietic cells that led to a myelo-proliferative syndrome (O'Connell et al. 2008). In contrast to miRNAs that can function as proto-oncogenes, there also has been evidence that distinct miRNAs can function as tumor-suppressor in AML. For example, recently Garzon and colleagues highlighted the potential of *miR-29b* as tumor suppressor by inducing apoptosis and reducing tumorgenicity in a xenograft AML model (Garzon et al. 2009a). Furthermore, miRNAs have also been shown to represent both targets and effectors of the epigenetic machinery. Similar to other genes, miRNA expression can be affected by DNA promoter methylation and histone modifications. As mentioned above AML1-ETO can lead to heterochromatic silencing of the *miR-223* genomic region and demethylation can restore *miR-223* expression followed by differentiation of leukemic blasts (Fazi et al. 2007). On the other hand, miRNA expression can impact the epigenetic modifications as e.g. *miR-29b* targets DNA methyltransferases in AML (Garzon et al. 2009b).

With regard to novel biological insights, the list of miRNA-target gene interactions likely involved in leukemogenesis is rapidly growing. Here, further in-depth analyses of the miRNA transcriptome using next generation sequencing will also provide novel insights by being potentially less error-prone than hybridization-based technologies (Kuchenbauer et al. 2008). These novel approaches might significantly add to the understanding how miRNAs contribute to the development of cancer, by revealing novel miRNAs, miRNA isoforms, mutations and absolute sequence counts, thereby highlighting additional miRNAs that might serve as future therapeutic targets in AML.

12.3 MiRNAs in CML

Important contributions to the understanding of CML pathogenesis have been shown for several miRNAs. GEP-studies identified specific miR expression profiles for CML compared to other tumor types as well as malignant hematopoietic cell lines and purified normal human blood cells (Lu et al. 2005; Ramkissoon et al. 2006).

BCR-ABL and c-MYC dependent regulation of miRNAs encoded within the polycistronic *miR-17-92* cluster were identified in CML cell lines (Venturini, 2007). Previous reports had already shown the involvement of this cluster in malignancies of B cell origin (Ota et al. 2004) and a direct regulation by c-MYC has been reported for the *miR-17-92* cluster (He et al. 2005; O'Donnell et al. 2005). In the recent study the authors linked BCR-ABL-driven leukemogenesis with c-Myc dependent regulation of the *miR-17-92* cluster. The link was identified through inhibition of BCR/ABL after treatment with STI-571 (Imatinib) and by using RNAi. Moreover, the authors suggested a BCR/ABL-c-MYC-*miR-17-92* pathway that mediates enhanced miRNA expression in early chronic phase $CD34^+$cells in CML (Venturini et al. 2007).

Agirre and colleagues reported a specific miR profile after analysing expression levels in mononuclear cells from healthy persons and from patients with newly diagnosed CML (Agirre et al. 2008). *MiR-10a*, *miR-150*, and *miR-151* were down-regulated, whereas *miR-96* was up-regulated in CML cells. Down-regulation of *miR-10a* was not dependent on BCR-ABL activity and contributed to the increased cell growth of CML cells. The group identified the upstream stimulatory factor 2 (USF2) as a potential target of *miR-10a* and showed that over-expression of USF2 increases cell growth. There were some indications in the study for clinical relevance: In a group of 85 newly diagnosed patients with CML expression of *miR-10a* was down-regulated in 71% of the patients. Conversely expression of USF2 was up-regulated in 60% of the CML patients and there was an inverse correlation with over-expression of USF2 being significantly associated with decreased expression of *miR-10a*. While direct evidence is lacking, the results indicate that down-regulation of *miR-10a* may increase USF2 and contribute to the increase in cell proliferation of CML (Agirre et al. 2008).

A subsequent study identified miRNAs differentially expressed between resistant and responding samples that might, if confirmed, predict resistance to imatinib in patients with newly diagnosed CML. Potential targets for these miRNAs included membrane transporters which have been implicated in resistance to chemotherapy in general and specific imatinib resistance as well (San Jose-Eneriz et al. 2009).

Soon after the report on the role of the *miR-17-92* cluster, a second microRNA, *miR-203*, involved in the regulation of the ABL/BCR-ABL pathway was identified by analysing a fragile region on the mouse chromosome 12. This region was found to be frequently lost in γ-radiation-induced T cell lymphomas and contained approximately 12% of the known miRNAs.

Down-regulation of this miRNA turned out to be caused not only by deletion but also by silencing through promoter hypermethylation of the upstream region of *miR-203* in a majority of Ph-positive tumors, including B-ALL, primary CML, and cultured CML cell lines. Further analysis using computational prediction, mRNA expression analysis and functional validation identified *miR-203* as a regulator of ABL and BCR-ABL fusion protein levels. These data suggest that miR cannot only be regulated by BCR/ABL but may control oncogene levels as e.g. ABL or BCR/ABL. Reexpression of *miR-203* resulted in inhibition of tumor cell proliferation (Bueno et al. 2008).

CML blast crisis progenitors have been shown to loose their ability for differentiation by suppressed expression of CEBPα, a transcription factor controlling myeloid differentiation. Altered post-transcriptional gene regulation by the translational regulator poly(rC)-binding protein hnRNP E2 has recently been shown to be centrally involved in the pathogenesis of blast crisis in CML. High expression of hnRNPE2 leads to extensive down-regulation of CEBPα protein levels through interaction with the mRNAs 5′ untranslated region (Melo and Barnes 2007; Perrotti et al. 2002; Perrotti and Neviani 2007; Chang et al. 2007).

The role of miRNAs in this process has recently been elucidated. D. Perrotti and colleagues (Eiring et al. 2010) identified *miR-328* which was differentially modulated in a BCR-ABL dose and kinase-dependent manner through the MAPK-hnRNP2 pathway. The group showed decreased levels of *miR-328* in blast crisis CML. *MiR-328* was identified to harbour a C-rich element that resembles the negative regulatory hnRNP E2-binding site contained in the CEBPα intercistronic mRNA region. As hypothesised, *miR-328* turned out to competitively target hnRNP E2 for binding and decreases association of hnRNP E2 to CEBPα. In further experiments it was possible to recover maturation in BCR-ABL positive cells that lost the ability for differentiation by reconstitution of *miR-328* expression with consecutively decreased survival of leukemic blasts mediated through the described mechanisms and by targeting PIM1 survival factor (Eiring et al. 2010).

The interaction with hnRNP E2 was independent of the miRNA's seed sequence but lead to the release of *CEBPα* mRNA from hnRNP E2-mediated translational inhibition. Importantly, these data reveal the ability of a miRNA to control cell fate not only through base pairing with mRNA targets but also through a "decoy" activity that interferes with regulatory proteins.

12.4 MiRNA in CLL

CLL is the most prevalent form of adult leukemia in the Western world showing an estimated incidence about 3.9 per 100,000 people per year. Men are affected two times more often than women and the median age at diagnosis peaks at 72 years (Dores et al. 2007).

CLL has turned out to be a disease with multiple facets in its pathogenic mechanisms including genetic aberrations, antigen drive and microenvironmental interactions (Zenz et al. 2010). Different risk groups can be closely linked to the clinical and genetic features. Identification of recurring genomic aberrations, gene mutations such as *TP53* (Döhner et al. 1995) and *ATM* (Schaffner et al. 1999), the detection of somatic mutations in the variable regions of the immunoglobulin (Ig) heavy chain (*IGHV*) genes (Fais et al. 1998; Hamblin et al. 1999) and the evidence of biased *IGHV* usage and stereotyped B cell receptors (BCRs) (Fais et al. 1998; Tobin et al. 2003; Stamatopoulos et al. 2007) have essentially led to better prognostication and insight into CLL-biology (Zenz et al. 2010).

Chromosomal aberrations occur in approximately 80% of all CLL cases. Most often chromosome 13q14 is affected ($13q^-$: 55%) by monoallelic (76%) or biallelic (24%) deletion followed by 11q deletion ($11q^-$: 18%), trisomy of chromosome 12 (16%) and the 17p deletion ($17p^-$: 7%) (Döhner et al. 2000).

Although research succeeded to map the ataxia teleangiectasia-mutated (*ATM*) gene to the affected minimal consensus regions at the chromosome bands 11q22.3-q23.1 and the tumor suppressor gene *TP53* is inactivated in 80% of cases with deletion of 17p13 respectively, no classical tumor suppressor gene could be located at the minimal deleted region (MDR) at 13q14.3 despite intensive investigations for many years (Bullrich et al. 2001; Corcoran et al. 1998; Migliazza et al. 2001; Stilgenbauer et al. 1998).

Interestingly, the cells of the precursor state of CLL named monoclonal B-lymphocytosis (MBL) also exhibit 13q14 deletions at rates of more than 50% (Rawstron et al. 2008). In addition this characteristic deletion is present in a variety of other cancers with remarkable regularity (DLBCL, mantle cell lymphoma and multiple myeloma as well as mature T cell lymphoma (Rosenwald et al. 1999); Cigudosa et al. 1998; Stilgenbauer et al. 1998) prostate cancer (Dong et al. 2001).

12.4.1 MiR-15a and MiR-16-1

The missing link to the mechanism underlying the tumor suppression at band 13q14 was found in 2002 when it became evident that a cluster of two newly cloned miRNA genes, *miR-15a* and *miR-16-1* (Lagos-Quintana et al. 2001; 2002; Lee et al. 2001; Lau et al. 2001), was exactly located in a 30-kb region within the *DLEU2* gene at the MDR on chromosome 13q14.3 (Calin et al. 2002). Both *miR-15a* and *miR-16-1* were either deleted or down-regulated in most investigated CLL cases (Calin et al. 2002).

These findings were further strengthened after investigators reported that a germline mutation in the primary precursor of *miR-15a* and *miR-16-1* seemed to reduce the expression of this cluster in two CLL patients (Calin et al. 2005). Although mutations in the miR locus appear to be very rare, another study identified a point mutation in the 3′-DNA neighbouring the *miR-16-1* region that has been associated with reduced *miR-16-1* expression in the New Zealand black mouse, a model for indolent, late-onset CLL (Raveche et al. 2007).

A characteristic of CLL is that the clonal expansion of $CD5^+$ B cells is nourished only by a small proliferating cell-pool in CLL. Most CLL cells are locked in the G0/G1 phase of the cell cycle and non-dividing. Accumulation of malignant cells results from resistance to apoptosis as a further feature for CLL. CLL cells show consistent over-expression of the B cell lymphoma 2 (Bcl-2) protein (Kitada et al. 1998) which exerts its anti-apoptotic function by blocking the release of cytochrome c from mitochondria (Cory and Adams 2002). Bcl-2 over-expression can be caused by characteristic chromosomal translocations. In follicular lymphoma with the classical translocation t(14;18)(q32;q21), the *BCL2* gene is brought under the control of the Immunoglobulin heavy chain enhancers, resulting in forced transcription of the gene (Horsman et al. 1995). Similar mechanisms were reported for CLL but account only for approximately 5% of cases in which the *BCL2* gene is over-expressed (Adachi et al. 1990). Therefore other mechanisms explaining BCL-2 deregulation still needed to be found.

The search for the targets of *miR-15a* and *miR-16-1* revealed significant complementarity between parts of the *BCL-2* sequence and the seed region of both miRs. Studying the functional aspect of this finding, both miRs turned out to be inversely correlated with BCL-2 protein levels in CLL and were reported to act as direct

Table 12.1 Selected targets identified for *miR-15a/16-1*

Targets (gene symbol)	Tissue/cell lines	References
Bcl-2, WT1, CDC2, ETS1, JUN, MCL1, MSH2, PDCD6IP, RAB9B, WT1	MEG-01 cell lines	Calin et al. (2008); Cimmino et al. (2005)
CCND1, WNT3A	Prostate cancer	Bonci et al. (2008)
CCND1 (by *miR-16-1*)	Mantle cell lymphoma	Chen et al. (2008)
CCND3, CCNE1, CDK6	HeLa, HepG2, and A549 cell lines	(Liu et al. (2008))
c-MYB	K562 myeloid leukemia cells, human $CD34^+$ cells	Zhao et al. (2009)
CCNE1, CHK1, MCM5, CCND2, CDK6, CDK4, CCND3, C-MYB, BCL, CSE1L, ARL2, IGF1Rβ	CLL I83E95 cells	Klein et al. (2010)

post-transcriptional repressors of BCL-2 in transfection experiments on a MEG-01 cell line. Transfection with wild-type *miR-15/16* lead to an increased rate of apoptosis in transfected MEG-01 cells (Cimmino et al. 2005). Moreover, in in vivo studies in immunocompromised nude mice, tumorigenicity of leukemic xenografts using MEG-01 leukemic cells was significantly suppressed by ectopic expression of *miR-15a/16-1* (Calin et al. 2008).

By applying bioinformatic programs for screening and functional experiments for validation, several proteins involved in the regulation of cell cycle, cell growth and apoptosis have been reported to be directly regulated by *miR-15a/16-1* (Table 12.1).

Further and definite evidence for the pathogenic meaning of the 13q14 MDR and *miR-15a/miR-16-1* in B cell malignancies has been shown recently by generating a mouse model deleting the DLEU2/*miR-15a/16-1* cluster (Klein et al. 2010).

Mice presenting deletions of the MDR or with sole *miR-15a/16-1* deletion developed clonal B cell lymphoproliferation at the age of 15–18 months, including $CD5^+$ MBL, CLL/SLL, and NHL in 42% of $MDR^{-/-}$ and 26% of *miR-15a/16-1*$^{-/-}$ cases (Klein et al. 2010). While these data suggest that the deletion of the *miR-15a/16-1* is sufficient for lymphomagenesis, it also suggests that the non-protein coding DLEU2 is also contributing as the phenotype in these mice was more pronounced. Therefore additional tumor suppressive functions for residual parts of the MDR on 13q14, including *DLEU2* and potentially parts of *DLEU5* in 13q14 deletions exceeding the MDR, can be assumed (Klein et al. 2010).

In the work of Ulf Klein and colleagues, the *miR15a/16-1* cluster has been shown to influence growth, cell-cycle control and/or apoptosis, although the exact regulatory function by which *miR15a/16-1* exert its pathogenic effects in CLL. However the group found limited evidence of the *miR-15a/miR-16-1* to regulate BCL-2 (Cimmino et al. 2005; Fulci et al. 2007; Linsley et al. 2007; Klein et al. 2010; Calin et al. 2008).

Patients with a monoallelic 13q14 deletion seem to express higher *miR-15a/16-1* levels than patients with a biallelic 13q14 deletion (Fulci et al. 2007) and patients with a monoallelic del13q14 show slower lymphocyte growth kinetics than patients with biallelic deletions (Pfeifer et al. 2007), implicating mechanism which regulate the gene-dosage of these miRs. Nonetheless, questions remain as for example the question if and how *miR-15a/16-1* levels could be deregulated in cases without 13q deletion.

There appears to be some functional redundancy among members of the *miR-16* family (Linsley et al. 2007) and there are hints for further, potentially overlapping regulatory mechanisms modulating the expression levels (Zhao et al. 2009) and primary effect and of the *miR-15a/16-1* cluster.

12.4.2 MiRNA Microarray Screens Identifying Signatures and Subgroups for Prognostication

Although distinctive markers have been successfully identified to subgroup different CLL entities, further insights are needed to develop robust prognostic and predictive models. Estimations for the clinical course are currently based on the Binet classification and Rai staging system, age, beta2-microglobulin/thymidine kinase levels and genetic features (Zenz et al. 2010).

An adverse clinical course in CLL-patients has been proven for genetic factors such as unmutated *IGHV* genes associated with high ZAP-70 expression, *V3-21*-gene usage and particularly for genomic aberrations as 17p- and 11q- deletions or *TP53* mutation, which frequently lead to a treatment-refractory course (Zenz et al. 2010).

Microarray technology and global gene expression profiling (GEP) has been a successfully applied method to identify gene-signatures associated with specific biological processes, clinical stages and outcome in different cancer types (Quackenbush 2006). However, GEP performed on CLL samples mostly failed to identify specific expression-patterns (if unsupervised analysis was performed) that could differentiate CLL-subtypes. It was only after supervised analysis based on *IGHV*-mutation status that differentially expressed mRNAs were identified (Rosenwald et al. 2001; Klein et al. 2001; Jelinek et al. 2003; Durig et al. 2003; Stratowa et al. 2001).

MiRNAs profiling studies were quickly initiated using either microarrays (Calin et al. 2005; 2004) or PCR based approaches (Fulci et al. 2007).

Besides revealing distinct signatures that differ between CLL tumor-samples and normal $CD5^+$ B cells, characteristic signatures with respect to *IGHV* mutation status and/or ZAP-70 expression were observed in these early studies again based on supervised analysis (Calin et al. 2004; 2005).

Although the initially reported gene-signatures could not be completely reproduced by subsequent studies (Fulci et al. 2007; Marton et al. 2008), down-regulation of *miR-223* and members of the *miR-29* family were independently reported to be associated with unmutated *IGHV* genes and disease progression (Fulci et al. 2007;

Table 12.2 Selected microRNAs with characteristic signatures and potential prognostic relevance in chronic lymphocytic leukemia

MicroRNA signatures	Reference prognostic markers	Relevance	References
miR-15a, *miR-195*, *miR-221*, *miR-23b*, *miR-155*, *miR-24-1*, *miR-146*, *miR-16-1*, *miR-16-2* (all up)	Unmutated *IGHV*, ZAP-70	Association with adverse markers	Calin et al. (2005)
miR-223, *miR-29a-2*, *miR-29b-2*, *miR-29c* (all down)			
miR-29c, *miR-223*, *miR-150*, *miR-29b* (all down)	Unmutated *IGHV*	Disease progression	Fulci et al. (2007)
miR-181a, *let-7a*, *miR-30d* (all down)	Unmutated *IGHV*	Disease progression	Marton et al. (2008)
miR-155 (up), *miR-29* (down)			
miR-29c, *miR-223*	Significantly associated with Binet stage, sCD23, beta2-M; ZAP-70, LPL *IGHV* mutational status, CD38 expression, and cytogenetics	Disease progression score combining *miR-29c*, *miR-223*, ZAP70, and LPL stratifies survival	Stamatopoulos et al. (2009)
miR-223, *miR-29b*, *miR-29c*, *miR-181* family (all down)	17p deletion	Potential subdivision of CLL with 17p deletion	Visone et al. (2009)
miR151-3p, *miR-29c*, *miR-34a* (all down)	17p deletion		
miR-29b (down), *miR-155* (up)	11q deletion	"Karyotype specific" microRNA	Visone et al. (2009)
miR-640/*miR-148a* (down), *miR-146b-5p*/*miR-146a* (up)	Trisomy 12		
miR-148a (up)	Normal karyotype		
miR-155 (down), *miR-640* (up)	13q deletion		

Marton et al. 2008) (Table 12.2). Further evaluation of these miRNAs in a retrospective study showed that low *miR-223* and *miR-29c* levels were associated with poor prognosis by showing shorter treatment free survival and reduced overall survival. A qPCR-score developed in this study integrated ZAP-70, LPL, *miR-223* and *miR-29c* and was able to discriminate prognostic subgroups of CLL patients (Stamatopoulos et al. 2009).

In a study from Pallasch and colleagues, 50 treatment-naïve CLL patients and peripheral B cells of 14 healthy donors were assessed (Pallasch et al. 2009). In CLL cells, a set of 7 up- and 19 down-regulated miRNAs was identified including the up-regulation of *miR-155*. Among the miRNAs down-regulated in CLL cells, 6 of 10 miRNA promoters examined showed gain of methylation compared with normal B cell controls suggesting that methylation plays an important role in controlling miR expression. Subsequent target prediction of deregulated miRNAs suggested significant binding prediction at the 3′ untranslated region of the pleomorphic adenoma gene 1 (*PLAG1*) oncogene. *PLAG1* was shown to be regulated by *miR-181a*, *miR-181b*, *miR-107*, and *miR-424*.

Another profiling-study identified a karyotype-specific signature consisting of 32 miRNAs of which 9 miRs validated by qRT-PCR were associated with common cytogenetic features (Visone et al. 2009). This study confirmed previous reports of low expression levels for *miR-29b* and *miR-29c* in CLL with 11q deletion (Pekarsky et al. 2006) and the significance of *miR-29c* expression for poor prognostic CLL subgroups (Stamatopoulos et al. 2009). Moreover, it was possible to further subdivide the group of patients with 17p- based on the down-regulation of *miR-223*, *miR-29b*, *miR-29c*, and the *miR-181* family (Visone et al. 2009) (Table 12.2). In a similar study by Rossi et al. the authors were able to show that miRs correlated with 17p status and specifically were down- (*miR-34a*, *miR-181b*, *miR-497*) or up-regulated (*miR-15a*, *miR-21*, *miR-155*) in CLL with 17p deletion. In addition, *miR-21* and *miR-181b* expression were shown to be associated with overall and progression free survival respectively. These data added important information to other recently published studies showing that *miR-34a*, *miR-29c*, *miR-17-5p*, and *miR 151-3p* are differentially expressed based on 17p status (Mraz et al. 2009; Visone et al. 2009; Zenz et al. 2009. To address the mechanisms underlying the deregulation of the miRNAs, the authors have taken advantage of a cell line model where *p53* was silenced to assess if the miRNAs are direct *p53* targets. In accordance with prior studies the authors show evidence that both *miR-34a* and *miR-155* are dependent on *p53* as these were up- (*miR-34a*) and down-regulated (*miR-155*) after silencing of *p53*.

MiR-29c and *miR-181* have been shown to regulate the *TCL1*-oncogene, which is highly expressed in CLL samples with 11q deletions (Pekarsky et al. 2006) or CLL with high ZAP-70 levels and unmutated *IGHV*-genes (Herling et al. 2006). Other studies report regulatory functions of the *miR-29* family by suppression of MCL1, a member of the bcl-2 family with anti-apoptotic function, which is correlated with classic adverse prognostic factors and unfavourable disease course (Pepper et al. 2008). Enforced expression of *miR-29* family members in lung cancer cell lines and AML restores normal patterns of DNA methylation by targeting DNA methyltransferases (DNMTs) and leads to normal DNA methylation patterns with consequent re-expression of silenced tumor suppressor genes (Fabbri et al. 2007; Garzon et al. 2009b).

12.5 Summary

The growing understanding of miRNA deregulation has greatly advanced our understanding of leukemia. While a great deal of mechanistic insight into the precise

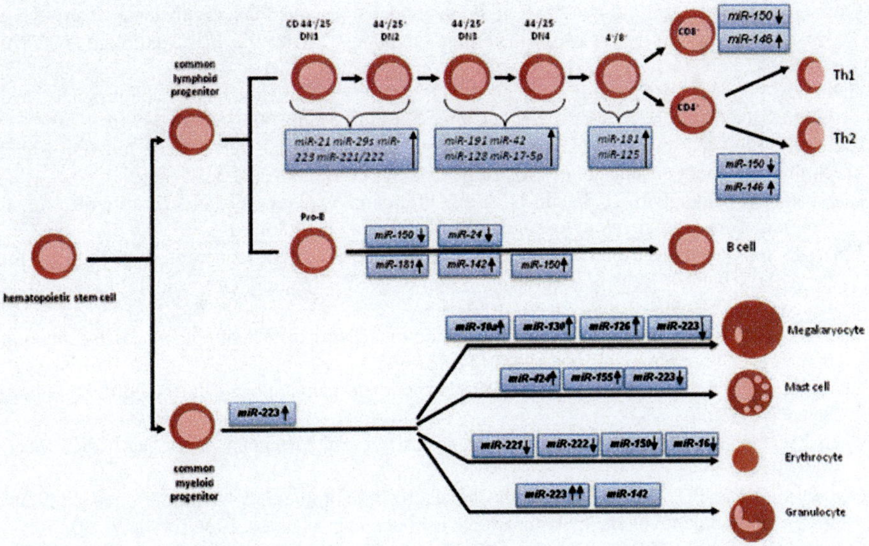

Fig. 12.1 MicroRNAs are involved in the differentiation of hematopoietic progenitors. Multiple miRNAs of this process have also been shown to be deregulated in leukemia (e.g. *miR-155*, *miR29s*, and *miR-223*).

function of deregulated miRNA remains to be discovered, there are a number of miRNAs (e.g. *miR-223*, *miR-34a*, *miR-15a/miR16-1*) with a clearer pathogenetic role in leukemogenesis. The near future is thus likely to yield a translation of these discoveries into targeted treatment approaches in leukemia and other cancers. This translation could relate to the use of the miRNA expression and a prognostic or predictive marker (e.g. *miR-34a* in *TP53* mutant cases) or *miR-15a/miR-16-1* as a therapeutic target in CLL.

References

Adachi M, Tefferi A, Greipp PR, et al. Preferential linkage of bcl-2 to immunoglobulin light chain gene in chronic lymphocytic leukemia. J Exp Med. 1990;171:559–64.

Agirre X, Jimenez-Velasco A, San Jose-Eneriz E, et al. Down-regulation of hsa-miR-10a in chronic myeloid leukemia $CD34^+$ cells increases USF2-mediated cell growth. Mol Cancer Res. 2008;6:1830–40.

Bonci D, Coppola V, Musumeci M, et al. The miR-15a-miR-16-1 cluster controls prostate cancer by targeting multiple oncogenic activities. Nat Med. 2008;14:1271–7.

Bueno MJ, Perez de Castro I, Gomez de Cedron M, et al. Genetic and epigenetic silencing of microRNA-203 enhances ABL1 and BCR-ABL1 oncogene expression. Cancer Cell. 2008;13:496–506.

Bullrich F, Fujii H, Calin G, et al. Characterization of the 13q14 tumor suppressor locus in CLL: identification of ALT1, an alternative splice variant of the LEU2 gene. Cancer Res. 2001;61:6640–8.

Calin GA, Cimmino A, Fabbri M, et al. MiR-15a and miR-16-1 cluster functions in human leukemia. Proc Natl Acad Sci USA. 2008;105:5166–71.

Calin GA, Dumitru CD, Shimizu M, et al. Frequent deletions and down-regulation of micro-RNA genes miR15 and miR16 at 13q14 in chronic lymphocytic leukemia. Proc Natl Acad Sci USA. 2002;99:15524–9.

Calin GA, Ferracin M, Cimmino A, et al. A MicroRNA signature associated with prognosis and progression in chronic lymphocytic leukemia. N Engl J Med. 2005;353:1793–801.

Calin GA, Liu CG, Sevignani C, et al. MicroRNA profiling reveals distinct signatures in B cell chronic lymphocytic leukemias. Proc Natl Acad Sci USA. 2004;101:11755–60.

Cammarata G, Augugliaro L, Salemi D, et al. Differential expression of specific microRNA and their targets in acute myeloid leukemia. Am J Hematol. 2010;85:331–9.

Chang JS, Santhanam R, Trotta R, et al. High levels of the BCR/ABL oncoprotein are required for the MAPK-hnRNP-E2 dependent suppression of C/EBPalpha-driven myeloid differentiation. Blood. 2007;110:994–1003.

Chen RW, Bemis LT, Amato CM, et al. Truncation in CCND1 mRNA alters miR-16-1 regulation in mantle cell lymphoma. Blood. 2008;112:822–9.

Chen CZ, Li L, Lodish HF, et al. MicroRNAs modulate hematopoietic lineage differentiation. Science. 2004;303:83–6.

Chen J, Odenike O, Rowley JD. Leukaemogenesis: more than mutant genes. Nat Rev Cancer. 2010;10:23–36.

Cigudosa JC, Rao PH, Calasanz MJ, et al. Characterization of nonrandom chromosomal gains and losses in multiple myeloma by comparative genomic hybridization. Blood. 1998;91:3007–10.

Cimmino A, Calin GA, Fabbri MI, et al. MiR-15 and miR-16 induce apoptosis by targeting BCL2. Proc Natl Acad Sci USA. 2005;102:13944–9.

Corcoran MM, Rasool O, Liu Y, et al. Detailed molecular delineation of 13q14.3 loss in B-cell chronic lymphocytic leukemia. Blood. 1998;91:1382–90.

Cory S, Adams JM. The Bcl2 family: regulators of the cellular life-or-death switch. Nat Rev Cancer. 2002;2:647–56.

Debernardi S, Skoulakis S, Molloy G, et al. MicroRNA miR-181a correlates with morphological sub-class of acute myeloid leukaemia and the expression of its target genes in global genome-wide analysis. Leukemia. 2007;21:912–16.

Dohner H, Fischer K, Bentz M, et al. p53 gene deletion predicts for poor survival and non-response to therapy with purine analogs in chronic B-cell leukemias. Blood. 1995;85:1580–9.

Dohner H, Stilgenbauer S, Benner A, et al. Genomic aberrations and survival in chronic lymphocytic leukemia. N Engl J Med. 2000;343:1910–6.

Dong JT, Boyd JC, Frierson HF Jr. Loss of heterozygosity at 13q14 and 13q21 in high grade, high stage prostate cancer. Prostate. 2001;49:166–71.

Dores GM, Anderson WF, Curtis RE, et al. Chronic lymphocytic leukaemia and small lymphocytic lymphoma: overview of the descriptive epidemiology. Br J Haematol. 2007;139:809–19.

Druker BJ, Tamura S, Buchdunger E, et al. Effects of a selective inhibitor of the Abl tyrosine kinase on the growth of Bcr-Abl positive cells. Nat Med. 1996;2:561–6.

Durig J, Nuckel H, Huttmann A, et al. Expression of ribosomal and translation-associated genes is correlated with a favorable clinical course in chronic lymphocytic leukemia. Blood. 2003;101:2748–55.

Eiring AM, Harb JG, Neviani P, et al. MiR-328 functions as an RNA decoy to modulate hnRNP E2 regulation of mRNA translation in leukemic blasts. Cell. 2010;140:652–65.

Fabbri M, Garzon R, Cimmino A, et al. MicroRNA-29 family reverts aberrant methylation in lung cancer by targeting DNA methyltransferases 3A and 3B. Proc Natl Acad Sci USA. 2007;104:15805–10.

Fais F, Ghiotto F, Hashimoto S, et al. Chronic lymphocytic leukemia B cells express restricted sets of mutated and unmutated antigen receptors. J Clin Invest. 1998;102:1515–25.

Fazi F, Racanicchi S, Zardo G, et al. Epigenetic Silencing of the Myelopoiesis Regulator microRNA-223 by the AML1/ETO Oncoprotein. Cancer Cell. 2007;12:457–66.

Fazi F, Rosa A, Fatica A, et al. A minicircuitry comprised of microRNA-223 and transcription factors NFI-A and C/EBPalpha regulates human granulopoiesis. Cell. 2005;123:819–31.

Fulci V, Chiaretti S, Goldoni M, et al. Quantitative technologies establish a novel microRNA profile of chronic lymphocytic leukemia. Blood. 2007;109:4944–51.

Garzon R, Garofalo M, Martelli MP, et al. Distinctive microRNA signature of acute myeloid leukemia bearing cytoplasmic mutated nucleophosmin. Proc Natl Acad Sci USA. 2008a;105:3945–50.

Garzon R, Heaphy CE, Havelange V, et al. MicroRNA 29b functions in acute myeloid leukemia. Blood. 2009a;114:5331–41.

Garzon R, Liu S, Fabbri M, et al. MicroRNA-29b induces global DNA hypomethylation and tumor suppressor gene reexpression in acute myeloid leukemia by targeting directly DNMT3A and 3B and indirectly DNMT1. Blood. 2009b;113:6411–8.

Garzon R, Volinia S, Liu CG, et al. MicroRNA signatures associated with cytogenetics and prognosis in acute myeloid leukemia. Blood. 2008b;111:3183–9.

Golub TR, Slonim DK, Tamayo P, et al. Molecular classification of cancer: class discovery and class prediction by gene expression monitoring. Science. 1999;286:531–7.

Hamblin TJ, Davis Z, Gardiner A, et al. Unmutated Ig V(H) genes are associated with a more aggressive form of chronic lymphocytic leukemia. Blood. 1999;94:1848–54.

Han YC, Park CY, Bhagat G, et al. MicroRNA-29a induces aberrant self-renewal capacity in hematopoietic progenitors, biased myeloid development, and acute myeloid leukemia. J Exp Med. 2010;207:475–89.

He L, Thomson JM, Hemann MT, et al. A microRNA polycistron as a potential human oncogene. Nature. 2005;435:828–33.

Herling M, Patel KA, Khalili J, et al. TCL1 shows a regulated expression pattern in chronic lymphocytic leukemia that correlates with molecular subtypes and proliferative state. Leukemia. 2006;20:280–5.

Horsman DE, Gascoyne RD, Coupland RW, et al. Comparison of cytogenetic analysis, southern analysis, and polymerase chain reaction for the detection of t(14; 18) in follicular lymphoma. Am J Clin Pathol. 1995;103:472–8.

Huang ME, Ye YC, Chen SR, et al. Use of all-trans retinoic acid in the treatment of acute promyelocytic leukemia. Blood. 1988;72:567–72.

Isken F, Steffen B, Merk S, et al. Identification of acute myeloid leukaemia associated microRNA expression patterns. Br J Haematol. 2008;140:153–61.

Jelinek DF, Tschumper RC, Stolovitzky GA, et al. Identification of a global gene expression signature of B-chronic lymphocytic leukemia. Mol Cancer Res. 2003;1:346–61.

Johnnidis JB, Harris MH, Wheeler RT, et al. Regulation of progenitor cell proliferation and granulocyte function by microRNA-223. Nature. 2008;451:1125–9.

Jongen-Lavrencic M, Sun SM, Dijkstra MK, et al. MicroRNA expression profiling in relation to the genetic heterogeneity of acute myeloid leukemia. Blood. 2008;111:5078–85.

Kitada S, Andersen J, Akar S, et al. Expression of apoptosis-regulating proteins in chronic lymphocytic leukemia: correlations with in vitro and in vivo chemoresponses. Blood. 1998;91: 3379–89.

Klein U, Lia M, Crespo M, et al. The DLEU2/miR-15a/16-1 cluster controls B cell proliferation and its deletion leads to chronic lymphocytic leukemia. Cancer Cell. 2010;17: 28–40.

Klein U, Tu Y, Stolovitzky GA, et al. Gene expression profiling of B cell chronic lymphocytic leukemia reveals a homogeneous phenotype related to memory B cells. J Exp Med. 2001;194:1625–38.

Kuchenbauer F, Petriv OI, Delaney A, et al. Comprehensive Profiling microRNAs in murine hematopoietic stem cells and lineages using a microfluidics approach. Blood ASH Annual Meeting 2008; Abstract: 2468.

Lagos-Quintana M, Rauhut R, Lendeckel W, et al. Identification of novel genes coding for small expressed RNAs. Science. 2001;294:853–8.

Lagos-Quintana M, Rauhut R, Yalcin A, et al. Identification of tissue-specific microRNAs from mouse. Curr Biol. 2002;12:735–9.

Langer C, Radmacher MD, Ruppert AS, et al. High BAALC expression associates with other molecular prognostic markers, poor outcome and a distinct gene-expression signature in cytogenetically normal acute myeloid leukemia: a Cancer and Leukemia Group B (CALGB) study. Blood. 2008;111:5371–9.

Lau NC, Lim LP, Weinstein EG, et al. An abundant class of tiny RNAs with probable regulatory roles in *Caenorhabditis elegans*. Science. 2001;294:858–62.

Lee RC, Ambros V. An extensive class of small RNAs in *Caenorhabditis elegans*. Science. 2001;294:862–4.

Li Z, Lu J, Sun M, et al. Distinct microRNA expression profiles in acute myeloid leukemia with common translocations. Proc Natl Acad Sci USA. 2008;105:15535–40.

Linsley PS, Schelter J, Burchard J, et al. Transcripts targeted by the microRNA-16 family cooperatively regulate cell cycle progression. Mol Cell Biol. 2007;27:2240–52.

Liu Q, Fu H, Sun F, et al. MiR-16 family induces cell cycle arrest by regulating multiple cell cycle genes. Nucleic Acids Res. 2008;36:5391–404.

Lu J, Getz G, Miska EA, et al. MicroRNA expression profiles classify human cancers. Nature. 2005;435:834–8.

Marcucci G, Radmacher MD, Maharry K, et al. MicroRNA expression in cytogenetically normal acute myeloid leukemia. N Engl J Med. 2008;358:1919–28.

Marton S, Garcia MR, Robello C, et al. Small RNAs analysis in CLL reveals a deregulation of miRNA expression and novel miRNA candidates of putative relevance in CLL pathogenesis. Leukemia. 2008;22:330–8.

Melo JV, Barnes DJ. Chronic myeloid leukaemia as a model of disease evolution in human cancer. Nat Rev Cancer. 2007;7:441–53.

Migliazza A, Bosch F, Komatsu H, et al. Nucleotide sequence, transcription map, and mutation analysis of the 13q14 chromosomal region deleted in B-cell chronic lymphocytic leukemia. Blood. 2001;97:2098–104.

Mraz M, Malinova K, Kotaskova J, et al. MiR-34a, miR-29c and miR-17-5p are downregulated in CLL patients with TP53 abnormalities. Leukemia. 2009;23:1159–63.

Ota A, Tagawa H, Karnan S, et al. Identification and characterization of a novel gene, C13orf25, as a target for 13q31-q32 amplification in malignant lymphoma. Cancer Res. 2004;64:3087–95.

O'Connell RM, Rao DS, Chaudhuri AA, et al. Sustained expression of microRNA-155 in hematopoietic stem cells causes a myeloproliferative disorder. J Exp Med. 2008;205:585–94.

O'Donnell KA, Wentzel EA, Zeller KI, et al. c-Myc-regulated microRNAs modulate E2F1 expression. Nature. 2005;435:839–43.

Pallasch CP, Patz M, Park YJ, et al. miRNA deregulation by epigenetic silencing disrupts suppression of the oncogene PLAG1 in chronic lymphocytic leukemia. Blood. 2009;114: 3255–64.

Pekarsky Y, Santanam U, Cimmino A, et al. Tcl1 expression in chronic lymphocytic leukemia is regulated by miR-29 and miR-181. Cancer Res. 2006;66:11590–3.

Pepper C, Lin TT, Pratt G, et al. Mcl-1 expression has in vitro and in vivo significance in chronic lymphocytic leukemia and is associated with other poor prognostic markers. Blood. 2008;112:3807–17.

Perrotti D, Cesi V, Trotta R, et al. BCR-ABL suppresses C/EBPalpha expression through inhibitory action of hnRNP E2. Nat Genet. 2002;30:48–58.

Perrotti D, Neviani P. From mRNA metabolism to cancer therapy: chronic myelogenous leukemia shows the way. Clin Cancer Res. 2007;13:1638–42.

Pfeifer D, Pantic M, Skatulla I, et al. Genome-wide analysis of DNA copy number changes and LOH in CLL using high-density SNP arrays. Blood. 2007;109:1202–10.

Quackenbush J. Microarray analysis and tumor classification. N Engl J Med. 2006;354:2463–72.

Ramkissoon SH, Mainwaring LA, Ogasawara Y, et al. Hematopoietic-specific microRNA expression in human cells. Leuk Res. 2006;30:643–7.

Raveche ES, Salerno E, Scaglione BJ, et al. Abnormal microRNA-16 locus with synteny to human 13q14 linked to CLL in NZB mice. Blood. 2007;109:5079–86.

Rawstron AC, Bennett FL, O'Connor SJ, et al. Monoclonal B-cell lymphocytosis and chronic lymphocytic leukemia. N Engl J Med. 2008;359:575–83.

Rosenwald A, Alizadeh AA, Widhopf G, et al. Relation of gene expression phenotype to immunoglobulin mutation genotype in B cell chronic lymphocytic leukemia. J Exp Med. 2001;194:1639–47.

Rosenwald A, Ott G, Krumdiek AK, et al. A biological role for deletions in chromosomal band 13q14 in mantle cell and peripheral T-cell lymphomas? Genes Chromosomes Cancer. 1999;26:210–4.

Rosenwald A, Staudt LM. Clinical translation of gene expression profiling in lymphomas and leukemias. Semin Oncol. 2002;29:258–63.

San Jose-Eneriz E, Roman-Gomez J, Jimenez-Velasco A, et al. MicroRNA expression profiling in Imatinib-resistant Chronic Myeloid Leukemia patients without clinically significant ABL1-mutations. Mol Cancer. 2009;8:69.

Schaffner C, Stilgenbauer S, Rappold GA, et al. Somatic ATM mutations indicate a pathogenic role of ATM in B-cell chronic lymphocytic leukemia. Blood. 1999;94:748–53.

Stamatopoulos K, Belessi C, Moreno C, et al. Over 20% of patients with chronic lymphocytic leukemia carry stereotyped receptors: pathogenetic implications and clinical correlations. Blood. 2007;109:259–70.

Stamatopoulos B, Meuleman N, Haibe-Kains B, et al. microRNA-29c and microRNA-223 down-regulation has in vivo significance in chronic lymphocytic leukemia and improves disease risk stratification. Blood. 2009;113:5237–45.

Stilgenbauer S, Nickolenko J, Wilhelm J, et al. Expressed sequences as candidates for a novel tumor suppressor gene at band 13q14 in B-cell chronic lymphocytic leukemia and mantle cell lymphoma. Oncogene. 1998;16:1891–7.

Stratowa C, Loffler G, Lichter P, et al. CDNA microarray gene expression analysis of B-cell chronic lymphocytic leukemia proposes potential new prognostic markers involved in lymphocyte trafficking. Int J Cancer. 2001;91:474–80.

Tobin G, Thunberg U, Johnson A, et al. Chronic lymphocytic leukemias utilizing the VH3-21 gene display highly restricted Vlambda2-14 gene use and homologous CDR3 s: implicating recognition of a common antigen epitope. Blood. 2003;101:4952–7.

Venturini L, Battmer K, Castoldi M, et al. Expression of the miR-17-92 polycistron in chronic myeloid leukemia (CML) $CD34^+$ cells. Blood. 2007;109:4399–405.

Visone R, Rassenti LZ, Veronese A, et al. Karyotype-specific microRNA signature in chronic lymphocytic leukemia. Blood. 2009;114:3872–9.

Zenz T, Mertens D, Kuppers R, et al. From pathogenesis to treatment of chronic lymphocytic leukaemia. Nat Rev Cancer. 2010;10:37–50.

Zenz T, Mohr J, Eldering E, et al. MiR-34a as part of the resistance network in chronic lymphocytic leukemia. Blood. 2009;113:3801–8.

Zhao H, Kalota A, Jin S, et al. The c-myb proto-oncogene and microRNA-15a comprise an active autoregulatory feedback loop in human hematopoietic cells. Blood. 2009;113:505–16.

Chapter 13
MicroRNAs in Pancreatic Cancer: Potential Interests as Biomarkers and Therapeutic Tools

Pierre Cordelier and Jérôme Torrisani

Abstract Pancreatic ductal adenocarcinoma (PDAC) is the fourth cause of death by cancer in Western countries. Its poor prognosis is primarily explained by a lack of early diagnostic markers and efficient therapeutic treatments. PDAC does not appear de novo but rather originates of an accumulation of genetic and epigenetic alterations that leads to an aberrant production of diverse molecules such as RNA and proteins. These altered expression profiles result in a multi-step progression of precursor lesions to invasive PDAC. Therefore, a better understanding of the early genetic and epigenetic alterations occurring in PDAC development is valuable for diagnostic and new therapeutic strategies. MicroRNAs (miRNAs) are small endogenous RNA molecules that function as translation inhibitors of messenger RNA by binding to their 3′ untranslated region. These molecules are tightly involved in the regulation of many physiological processes such as development, proliferation, invasion, and apoptosis among others. Their expressions are profoundly altered in PDAC and are strongly involved in PDAC carcinogenesis. In this chapter, we describe the miRNAs for which the expression is altered in PDAC and PDAC pre-neoplastic lesions. We outline the different molecular mechanisms that lead to altered miRNA expression in PDAC cells as well as the signaling pathways affected in response to altered miRNA expression. Lastly, we review the potential interests of miRNA as biological markers and therapeutic tools for PDAC.

13.1 Introduction

Pancreatic ductal adenocarcinoma (PDAC) is the fourth cause of death by cancer in Western countries with a 5-year survival rate of ~6% (Jemal et al. 2010). This poor prognosis is frequently explained by lack of early symptoms, of early diagnostic markers and of efficient therapeutic treatments. The vast majority of patients with PDAC displays locally or advanced distant metastasis, which renders

J. Torrisani (✉)
INSERM UMR 1037, Cancer Research Center of Toulouse, Université Paul Sabatier, CHU Rangueil Bat L3, 31432 Toulouse Cedex 4, France
e-mail: jerome.torrisani@inserm.fr

W.C.S. Cho (ed.), *MicroRNAs in Cancer Translational Research*,
DOI 10.1007/978-94-007-0298-1_13,

their disease surgically inoperable. Moreover, PDAC is highly resistant to cytotoxic chemotherapy and ionizing radiation (Safioleas and Moulakakis 2004).

PDAC does not develop de novo. It takes origin from an accumulation of genetic and epigenetic alterations in ductal pancreatic cells that result in altered expression of different molecules (RNA, proteins) (Torrisani and Buscail 2002). These altered expression patterns lead to a multi-step progression from pre-neoplastic lesions to invasive PDAC. Pancreatic intraepithelial neoplasia (PanIN) and intraductal papillary mucinous neoplasia (IPMN) are the most common pre-neoplastic lesions of PDAC. Both lesions affect pancreatic ducts. Therefore, a better understanding of the early genetic and epigenetic events governing PDAC development is valuable for diagnostic and new therapeutic strategies.

Many genetic and epigenetic alterations have been described in PDAC and pre-neoplastic lesions. The most frequent genetic mutations in PDAC is mutations of the oncogene *K-RAS* (*Kirsten RAS*). These activating mutations occur in 60–80% of patients with PDAC (Torrisani and Buscail 2002). Moreover, multiple molecular events (chromosomal alterations, point mutations, DNA hypermethylation, etc. . .) are responsible for the inactivation of tumor suppressor genes such *p16*, *Smad/DPC-4 (Mothers against decapentaplegic homolog 4/Deleted in pancreatic cancer 4)*, *TP 53 (TP53)*, among others during PDAC development. These different alterations are also present in variable proportions in pre-neoplastic lesions (Feldmann et al. 2007).

Altered expression of microRNA (miRNA) was described in PDAC and pre-neoplastic lesions. MiRNA expression profiles were demonstrated to provide a precise classification of cancers from different origins (Lu et al. 2005; Volinia et al. 2006). In addition, expression of specific miRNAs can correlate with clinicopathological features of cancers (Lu et al. 2005).

In this chapter, we will list the miRNAs for which the expression is altered in PDAC and in PDAC pre-neoplastic lesions. We will describe the different molecular mechanisms that lead to their altered expression in PDAC cells. The functional consequences of miRNA altered expression will be also described. Finally, we will discuss the potential interests in microRNAs as biological markers and therapeutic tools for PDAC.

13.2 MiRNAs Differentially Expressed in PDAC

MiRNA expression profiles are specific for each cell types, and these profiles are profoundly altered in cancer cells (Calin and Croce 2006). Interestingly, the altered profiles are in fact specific for different tumor cell type. This specificity, therefore, allows for the classification of tumors depending on their organ of origin. qRT-PCR and the development of microarray technologies allow the simultaneous measurement of several hundred miRNAs and for the identification of miRNAs that are differentially expressed in PDAC when compared to normal pancreas or other pancreatic pathologies (Seux et al. 2008).

The first study on miRNA expression in PDAC determined the expression level of 222 miRNA precursors in 28 PDAC samples as compared to 15 adjacent benign tissues and 6 normal pancreases (Lee et al. 2007). This initial study revealed that one hundred miRNA precursors are aberrantly expressed in PDAC and that most of them display an increased expression in the tumors (Table 13.1). *MiR-221*, *miR-100*, *miR-21* and *miR-155* are among the most abundantly expressed in pancreatic tumors. Soon after, by use of miRNA microarray technologies, the expression of 326 miRNAs were measured in 65 microdissected PDAC samples and compared to adjacent pancreatic tissues (Bloomston et al. 2007). This analysis revealed that 30 miRNAs are over-expressed in PDAC whereas three were down-regulated (*miR-148a*, *miR-148b* and *miR-375*) (Table 13.1). When PDAC samples are compared with those from chronic pancreatitis, 15 miRNAs were over-expressed and 8 were under-expressed in PDAC. Meanwhile, Hahn's group established the first pancreatic *miRNAome* by comparing the data from normal pancreas and a reference set of 33 human tissues. This miRNA microarray analysis revealed that most miRNAs have similar expression in both sample sets; and many miRNAs known to be highly expressed in all tissue types such as *miR-16*, *miR-21* and *let-7* family members, are also very abundant in pancreas. Nevertheless, several miRNAs such as *miR-148a*, *miR-200a,b,c*, *miR-216*, *miR-217* and *miR-375* are evidently enriched in pancreas (Szafranska et al. 2007). In parallel, the comparison of 377 miRNA expression of five normal pancreases and eight PDAC revealed that 41 miRNAs are down-regulated and 32 are up-regulated at least twofold (Table 13.1). Among them, the *miR-148a*, *miR-148b*, *miR-216* and *miR-217* are down-regulated more than five folds in PDAC whereas *miR-155* and *miR-196a* are robustly enriched. Similar results were further observed from PDAC biopsy samples (Szafranska et al. 2008). Moreover, the over-expression of *miR-196a* in PDAC was confirmed after microdissection of PDAC, healthy ductal cells and acinar cells.

Following these large scale miRNA analysis, other groups have focused on specific miRNAs in PDAC. Indeed, profiling analysis of 95 miRNA chosen for their potential functions related to cancer biology lead to the identification of eight miRNA that are significantly up-regulated in PDAC when compared to adjacent tissues (*miR-196a*, *miR -190*, *miR-186*, *miR -221*, *miR-222*, *miR-200b*, *miR-15b* and *miR-95*) (Zhang et al. 2009).

MiR-21, one of the most often up-regulated miRNA in human cancer is also commonly over-expressed in the vast majority of miR expression studies on PDAC samples (Dillhoff et al. 2008; Giovannetti et al. 2010; Moriyama et al. 2009; Zhu et al. 2009). Similarly, *miR-155*, *miR-221*, *miR-222*, *miR-196a* are also frequently found over-expressed in PDAC (Bloomston et al. 2007; Lee et al. 2007; Szafranska et al. 2007). Recently, an increased expression of *miR-200a*, *b* and *c* was reported in PDAC and PDAC derived cell lines (Li et al. 2010a; Yu et al. 2010a).

Although most deregulated miRNAs are indeed due to an over-expression in PDAC, a specific subset of miRNAs are down-regulated in tumors. Among them, *miR-148a* is frequently down-regulated in tumor samples (Bloomston et al. 2007; Hanoun et al. 2010; Szafranska et al. 2007). *MiR-34a* for which the transcription is activated by the protein *TP53* was found to be down-regulated in 15/15 PDAC-derived cell lines when compared to hPNE (human Pancreatic Nestion-positive Epithelial) and hPDE (human pancreatic ductal epithelial) cell lines (Chang et al.

Table 13.1 Differentially expressed microRNAs in PDAC and PDAC-derived cell lines

Up-regulated microRNAs	References	Down-regulated microRNAs	References
miR-21	Bloomston et al. (2007); Lee et al. (2007); Dillhoff et al. (2008); Moriyama et al. (2009); Zhu et al. (2009); Giovannetti et al. (2010)	***miR-148a***	Bloomston et al. (2007); Szafranska et al. (2007); Hanoun et al. (2010)
miR-221	Bloomston et al. (2007); Lee et al. (2007); Szafranska et al. (2007); Zhang et al. (2009)	***miR-148b***	Bloomston et al. (2007); Szafranska et al. (2007)
miR-222	Bloomston et al. (2007); Szafranska et al. (2007); Zhang et al. (2009)	***miR-375***	Bloomston et al. (2007); Szafranska et al. (2007)
miR-15b	Lee et al. (2007); Zhang et al. (2009)	***let-7***	Bloomston et al. (2007); Torrisani et al. (2009)
miR-125-b1	Bloomston et al. (2007); Lee et al. (2007)	***miR-34***	Chang et al. (2007); Lodygin et al. (2008)
miR-100	Bloomston et al. (2007); Lee et al. (2007)	*miR-96*	Yu et al. (2010b)
miR-107	Bloomston et al. (2007); Lee et al. (2007)	*miR-20a*	Yan et al. (2010)
miR-143	Bloomston et al. (2007); Szafranska et al. (2007)	*miR-29c*	Szafranska et al. (2007)
miR-146	Bloomston et al. (2007); Szafranska et al. (2007)	*miR-96*	Szafranska et al. (2007)
miR-155	Bloomston et al. (2007); Lee et al. (2007); Szafranska et al. (2007)	*miR-130b*	Szafranska et al. (2007)
miR-181a	Bloomston et al. (2007); Lee et al. (2007)	*miR-139*	Lee et al. (2007)
miR-181c	Bloomston et al. (2007); Lee et al. (2007)	*miR-141*	Szafranska et al. (2007)
miR-196a	Szafranska et al. (2007); Zhang et al. (2009)	*miR-142p*	Lee et al. (2007)
miR-200b	Zhang et al. (2009); Li et al. (2010a)	*miR-146*	Li et al. (2010b)
miR-205	Bloomston et al. (2007); Szafranska et al. (2007)	*miR-216*	Szafranska et al. (2007)
miR-210	Bloomston et al. (2007); Szafranska et al. (2007)	*miR-217*	Szafranska et al. (2007)
miR-223	Bloomston et al. (2007); Szafranska et al. (2007)	*miR-345*	Lee et al. (2007)
let-7f-1	Lee et al. (2007)		
let-7d	Lee et al. (2007)		
miR-10a	Bloomston et al. (2007)		
miR-10b	Bloomston et al. (2007)		
miR-16-1	Lee et al. (2007)		
miR-18a	Szafranska et al. (2007)		

Table 13.1 (continued)

Up-regulated microRNAs	References	Down-regulated microRNAs	References
miR-23a	Bloomston et al. (2007)		
miR-23b	Bloomston et al. (2007)		
miR-24-1	Lee et al. (2007)		
miR-24-2	Lee et al. (2007)		
miR-31	Szafranska et al. (2007)		
miR-92-1	Lee et al. (2007)		
miR-93	Szafranska et al. (2007)		
miR-99	Bloomston et al. (2007)		
miR-95	Zhang et al. (2009)		
miR-99-a1	Bloomston et al. (2007)		
miR-99-a2	Bloomston et al. (2007)		
miR-100-1,2	Bloomston et al. (2007)		
miR-103-2	Bloomston et al. (2007)		
miR-125a	Bloomston et al. (2007)		
miR-145	Szafranska et al. (2007)		
miR-150	Szafranska et al. (2007)		
miR-181b	Bloomston et al. (2007)		
miR-181b-1	Bloomston et al. (2007)		
miR-181b-2	Bloomston et al. (2007)		
miR-181d	Bloomston et al. (2007)		
miR-186	Zhang et al. (2009)		
miR-190	Zhang et al. (2009)		
miR-196b	Szafranska et al. (2007)		
miR-199a-1	Bloomston et al. (2007)		
miR-199a-2	Bloomston et al. (2007)		
miR-203	Szafranska et al. (2007)		
miR-212	Lee et al. (2007)		
miR-213	Bloomston et al. (2007)		

Table 13.1 (continued)

Up-regulated microRNAs	References	Down-regulated microRNAs	References
miR-220	Bloomston et al. (2007)		
miR-224	Szafranska et al. (2007)		
miR-200a	Li et al. (2010a)		
miR-200c	Yu et al. (2010a)		
miR-301	Lee et al. (2007)		
miR-376a	Lee et al. (2007)		
miR-424	Lee et al. (2007)		

MicroRNAs are ranked according to number of citations and numeric order.
MicroRNAs cited more than twice in PDAC and PDAC-derived cell lines are in bold.

2007). Similar results were also observed in MiaPACA-2 and BxPC-3 cell lines (Lodygin et al. 2008). Reduced expression of the *let-7* miRNA family was also reported in PDAC samples (Bloomston et al. 2007; Torrisani et al. 2009). More recently, down-regulation of *miR-146* was measured in PANC-1 when compared to hPDE control cell lines (Li et al. 2010b). Although microarray and qRT-PCR approaches are now usually used, miRNA expression can also be evaluated by in situ hybridization (ISH) or by Northern blot as recently reported the down-regulation of *miR-20a* and *miR-96* in PDAC, respectively (Yan et al. 2010; Yu et al. 2010b).

Altogether, the plethora of studies emerging from numerous scientific groups show a number of miRNA that is undoubtedly deregulated in PDAC samples (Table 13.1). This collection of miRNAs could serve as the basis for future studies.

13.3 MiRNA Expression Profiles in PDAC Precursor Lesions

As described earlier, altered expression of miRNA is commonly encountered in PDAC samples. However, alteration of miRNA expression in pre-neoplastic lesions is less documented. Several precursor lesions of PDAC have been described. These include PanIN, intraductal papillary mucinous neoplasms (IPMN) and mucinous cystic neoplasms (Hruban and Adsay 2009). To date, miRNA expression patterns have been explored only in PanIN and IPMN lesions.

13.3.1 PanIN

PanIN lesions are the most common non-invasive precursor to invasive PDAC. PanINs are microscopic lesions that are categorized into three histological stages (PanIN-1A/B, PanIN-2, and PanIN-3) based on the increasing degrees of architectural and nuclear atypia. Several murine models for pancreatic carcinogenesis have been established to mimic this pancreatic cancer progression (Hingorani et al. 2003; Hruban et al. 2000; Maitra and Hruban 2008). The expression of three miRNAs, *miR-21*, *miR-155* and *miR-221* was measured in microdissected human PanIN lesions (Ryu et al.). A significant over-expression of *miR-155* was measured at stage PanIN-2 when compared to non-neoplastic ductal epithelium and confirmed by ISH. Over-expression of *miR-21* was only observed at stage PanIN-3. No difference in *miR-221* levels was observed between normal epithelium and PanIN lesions. A similar study from our group determined the expression of seven miRNAs (*miR-21*, *miR-221*, *miR-222*, *let-7a*, *miR-29c*, *miR-200* and *miR-205*) in murine and human PanIN (du Rieu et al. 2010). Globally, our study revealed that *miR-21*, *miR-205* and *miR-200* expression parallels PanIN progression in a KRAS (G12D) mouse model (Hruban et al. 2001) (Fig. 13.1). Moreover, this study further demonstrates that the expression of *miR-21*, *miR-221* and *miR-222* increases with human PanIN grade, with a peak of expression in hyperplastic PanIN-2/3 lesions. A recent study from our group demonstrated a down-regulation of the *miRNA-148a* in murine and human microdissected PanIN lesions (Hanoun et al. 2010) (Fig. 13.1).

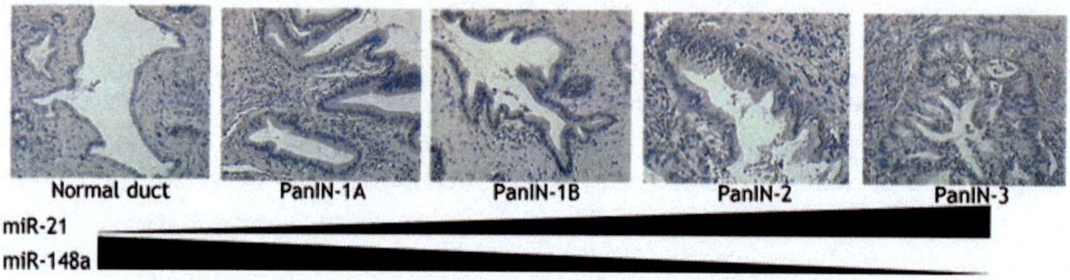

Fig. 13.1 *MiR-21* and *miR-148a* expression in microdissected murine pancreatic intraepithelial neoplasia (PanIN) precursor lesions (du Rieu et al. 2010; Hanoun et al. 2010)

13.3.2 Intraductal Papillary Mucinous Neoplasm (IPMN)

IPMN are large (usually >1 cm) non-invasive mucin-producing epithelial neoplasms that arise in the larger pancreatic ducts (Hruban and Adsay 2009). IPMNs can be a precursor to invasive adenocarcinoma of the pancreas. So far, only one study has evaluated the expression of miRNAs in this type of lesion. Relative expression of 12 miRNAs elevated in PDAC were assessed from 15 non-invasive IPMN (*miR-21*, *miR-155*, *miR-107*, *miR-223*, *miR-181a*, *miR-181c*, *miR-221*, *miR-210*, *miR-16*, *miR-100*, *miR-15a* and *miR-17-5p*) (Habbe et al. 2009). These studies revealed that, except for *miR-15a* and *miR-17-5p*, the remaining 10 miRNAs reach a significant over-expression in IPMN. *MiR-155* and *miR-21* display the highest relative fold changes in these lesions. Their expression was evaluated by ISH in a panel of 64 archival IPMN. *MiR-155* and *miR-21* expression is elevated in 83 and 81% of IPMN compare to normal ducts, respectively.

Altogether, these studies on PDAC precursor lesions clearly demonstrate that alteration of miRNA expression is an early event in PDAC and therefore advocate a role for microRNA in PDAC initiation and progression.

13.4 Molecular Mechanisms Responsible for Altered MiRNA Expression

Several molecular mechanisms responsible for altered miRNA expression in cancers have been described. Some of which were described in PDAC or PDAC-derived cell lines.

13.4.1 Chromosomal Alterations

Chromosomal alterations (deletions, amplifications) are responsible for the loss- or gain- of miRNA expression in cancers. For example, *miR-15* and *miR-16* are located at chromosome 13q14, a region deleted in more than half of B cell chronic lymphocytic leukemias (Calin et al. 2002). Thus far, there are no major chromosomal alterations that lead to miRNA misexpression described in PDAC.

13.4.2 Transcription Factor Defects

Deregulated expression or function of transcription factors occurs frequently in cancers. One example is the transcription factor *c-Myc* which is pathologically activated in human malignancies. The predominant consequence of this activation is a widespread repression of miRNA expression (Chang et al. 2008). Although the protein *TP53* was shown to activate *miR-34a* promoter, the loss of expression/function of this protein cannot completely explain for the loss of *miR-34a* expression observed in PDAC (Chang et al. 2007; Lodygin et al. 2008).

13.4.3 Epigenetic Alterations

DNA hypermethylation, a major epigenetic alteration, is responsible for silencing numerous tumor suppressor genes in cancer (Bird 1996). It was recently shown that this epigenetic alteration can affect miRNA expression (Saito et al. 2006). A number of examples for epigenetic alterations are described in PDAC samples or PDAC-derived cell lines. In fact, the *miR-34* promoter is hypermethylated in

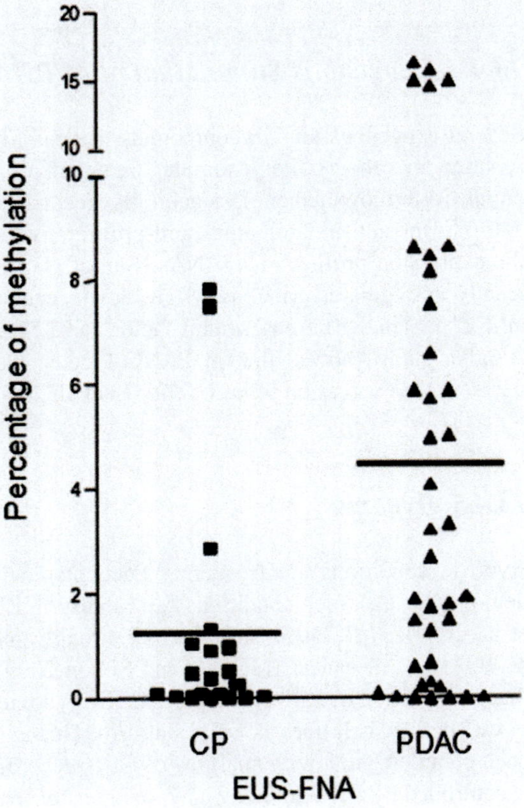

Fig. 13.2 Methylation level of *miR-148a* coding region in endoscopic ultrasound-guided fine needle aspiration (EUS-FNA) biopsies from patients with chronic pancreatitis (PC) and pancreatic ductal adenocarcinoma (PDAC). DNA methylation level of *miR-148a* coding region as determined by quantitative methylation specific PCR (qMS-PCR) from EUS-FNA-extracted DNA of 20 patients with proven PC and 39 patients with proven PDAC (Hanoun et al. 2010)

3 out 19 PDAC-derived cell lines (Lodygin et al. 2008). Moreover, the treatment of MiaPACA-2 and PANC-1 cell lines with the DNA demethylating agent 5-aza-deoxycytidine, and the histone deacetylase inhibitor, Trichostatin A, demonstrated that 14 miRNA are induced by at least two fold. Among them, *miR-107*, *miR-103*, *miR-29a,b* and *miR-320* are up-regulated in both cell lines (Lee et al. 2009). More recently, we have shown that the DNA region encoding *miR-148a* is hypermethylated in 10/14 PDAC samples when compared to non-pathological adjacent tissues. We further demonstrated that this hypermethylation is inversely correlated with *miR-148a* expression (Fig. 13.2) (Hanoun et al. 2010).

Conversely, DNA hypomethylation of genomic sequences can be responsible for the over-expression of multiple genes in PDAC (Sato et al. 2003). Recently, Goggins and colleagues identified two members of *miR-200* family, *miR-200a* and *miR-200b* that are over-expressed in PDAC, and for which the DNA coding region is hypomethylated (Li et al. 2010a).

Histone modifications are also implicated in the regulation of miRNA expression (Scott et al. 2006). In the PDAC-derived BxPC-3 cells, treatment with Trichostatin A, a potent histone deacetylase inhibitor, was shown to down-regulate 24 miRNAs and to up-regulate 5 miRNAs (Zhang et al. 2008). Among them, the *miR-200c* is up-regulated and *miR-21* is down-regulated.

13.4.4 Exogenous Stimulation of MiRNA Expression

Several exogenous stimuli/compounds that modify miRNA expression have been reported in cancers. For example, treatment of PDAC BxPC-3 cells with curcumin (diferuloylmethane), a naturally occurring flavonoid displays anti-cancer, anti-oxidant, anti-inflammatory, anti-proliferative and pro-apoptotic activities, alters the expression profiles of miRNAs (Sun et al. 2008). Eleven miRNAs are significantly up-regulated whereas 18 are down-regulated. Among those up-regulated, *miR-22* targets SP1 transcription factor and ESR1 (estrogen receptor 1) mRNAs. Another study showed that in PANC-1 cells, the isoflavones B-DIM or G2535 decrease the expression of *miR-146a* (Li et al. 2010b).

13.4.5 Hypoxia

Hypoxic conditions are frequently encountered in solid tumors. Recent studies demonstrate that such conditions can modify miRNA expression patterns (Crosby et al. 2009). *MiR-210* is shown to be strongly induced in hypoxic conditions in several types of cancer (Mathew and Simon 2009). Therefore, it is not surprising that an induction of *miR-210* expression by hypoxic condition was also described in PDAC-derived cell lines as well as in vivo (Huang et al. 2009). These findings have been complemented with analyses of *miR-210* promoter sequences showing that it is controlled by *HIF1alpha* by a hypoxia responsive element (HRE).

13.4.6 Genetic Mutations in MiRNA Genes

Even though mutations in the genomic sequences encoding for miRNA would, in theory, not modify their expression level, such mutations were reported in both PDAC tissues and PDAC-derived cell lines (Zhu et al. 2009). Four mutations were identified in *miR-155* and *miR-21* genes. Although the two mutations of *miR-21* did not affect the promoter activity of this gene, findings for a mutation (A-to-G at 29 nt) downstream of pre-*miR-21* resulted in a conformational change of the DNA secondary structure close to the stem reaching into the pre-*miR-21* that led to a relative reduction of the mature *miR-21* expression in vivo.

13.5 Role of MiRNAs in Pancreatic Cancer Development and Metastasis

The consequence of altered miRNA expression in cancer cells is the deregulation of targeted mRNAs. These changes result in an aberrant translation of proteins involved in the control of several cellular processes.

13.5.1 Cell Proliferation and Tumor Growth

The oncogene *K-RAS* is mutated in 60–80% of PDAC and; thus considered as a potential marker and target for PDAC treatment. Interestingly, *let-7* miRNA family targets K-RAS protein expression in human cancer cells (Morris and McManus 2005). We showed that the restoration of *let-7* level in Capan-1 PDAC cells strongly inhibit their proliferation in vitro but we failed to impede tumor growth progression following intra-tumoral gene transfer (Torrisani et al. 2009). Nevertheless, Watanabe et al. (2009) reported that the transfection of *let-7* inhibitors or the *let-7* clusters does not influence the proliferation rate of AsPC-1 or PANC-1 cells that endogenously express this miRNA. *MiR-21* is largely implicated in the control of proliferation in various cancer cell lines. Similarly, it was reported that in PDAC cells, transfection with *miR-21* precursor significantly increases their proliferation (Giovannetti et al. 2010; Moriyama et al. 2009). As mentioned earlier, *miR-34* expression is reduced in PDAC cells (Chang et al. 2007; Lodygin et al. 2008). Lodygin et al. (2008) showed its involvement in the control of cell proliferation. Indeed, they demonstrated that the restoration of *miR-34* in MiaPACA-2 cells suppresses colony formation and decreases the percentage of cells in S phase by directly targeting the cyclin dependent kinase 6 (*CDK6*). These proliferative effects induced by the restoration of *miR-34* were later confirmed by other groups (Ji et al. 2009; Kent et al. 2009). Xu's group further showed that *miR-34* restoration also inhibits tumor-sphere growth in vitro and tumor formation in vivo. Similarly, it was reported that an enforced expression of *miR-107* in PANC-1 and MiaPACA-2 down-regulates in vitro growth and represses the expression of a putative target *CDK 6* (Lee et al. 2009). Using PDAC SU86.86 cells, the stable ectopic expression of *miR-210*, which is highly induced in hypoxic conditions, delays the tumor growth when subcutaneously implanted into nude mice (Huang et al. 2009). Interestingly, *miR-210* expression has little consequence on angiogenesis, proliferation and apoptosis. Nonetheless, the inhibitory effect of *miR-210* on tumor growth initiation is partially rescued by expressing the homeobox protein A1 (*HOXA1*) and fibroblast growth factor receptor-like 1 (*FGFRL1*) coding sequence without 3′UTR. Furthermore, lentiviral over-expression of *miR-20a* was recently reported to result in anti-proliferative effects in vitro and in vivo. These findings were attributed to *miR-20a* functions associated with an increased percentage of cells in G0/G1 phase and with an post-transcriptional regulation of the protein signal transducer and activator of transcription 3 (STAT3) (Yan et al. 2010). More recently, an ectopic over-expression of *miR-200c* was shown to increase PDAC cell proliferation (Yu et al. 2010a). Inversely, the transient and stable over-expression of *miR-96*, which directly targets the oncogene *K-RAS*, inhibits proliferation of MiaPCACA-2 and PANC-1 cells in vitro and in vivo (Yu et al. 2010b).

13.5.2 Epithelial-mesenchymal Transition (EMT), Invasion, and Metastatic Behavior

Invasion and metastasis of carcinoma is promoted by the activation of the embryonic EMT program, which triggers cellular mobility and subsequent dissemination of tumor cells. Several miRNAs have been involved in EMT transition, invasive capacity and metastatic behavior of PDAC cells. Burk et al. (2008) demonstrated that *miR-200c* and *miR-141* are strong inducers of epithelial phenotype. They showed that the over-expression of both miRNAs results in the induction of the epithelial differentiation of PANC-1 cells. Watanabe et al. (2009) demonstrated that the high-mobility group A protein 2 (*HMGA2*), a non-histone chromatin factor, is strongly involved in EMT maintenance in PDAC cells. However, they showed that although the *let-7* family directly targets the 3′UTR of *HMGA1, 2* and *K-RAS* mRNAs, these miRNAs do not inhibit EMT in PDAC cells that have endogenous *let-7* expression. MiRNA expression profiles were compared between gemcitabine-resistant and -sensitive PDAC cell lines. Interestingly, it was reported that the expression of *miR-200b*, *miR-200c*, *let -7b, -7c, -7d* and *-7e* is significantly reduced in gemcitabine-resistant cells that display EMT characteristics (Li et al. 2009). Moreover, transfection of *miR-200* results in the reversal of EMT phenotype leading to epithelial morphology.

13.5.2.1 Invasion

MiR-21 is involved in invasive capacity of PDAC cells. Transfection with the pre-*miR-21* markedly increases PANC-1 cell invasion by indirectly mediating matrix metalloproteinase *MMP-2* and *MMP-9* expression (Moriyama et al. 2009). On the other hand, restoration of *miR-34* expression in PDAC cells was showed to inhibit their invasion (Ji et al. 2009). Low expression of *miR-146a* was reported in PDAC cells when compared with normal pancreatic duct epithelial cells (Li et al. 2010b). The same study further demonstrated that the re-expression of *miR-146a* inhibits the invasive capacity of PDAC cells with concomitant down-regulation of epidermal growth factor (*EGFR*) and the NF-κB regulatory kinase interleukin 1 receptor-associated kinase 1 (*IRAK-1*). In the same way, the transfection of the high invasive PANC-1 cells with the precursor of *miR-29a* reduces their capacity of invasion as determined by use of the Boyden chamber in vitro invasion assay (Muniyappa et al. 2009). More recently, high levels of *miR-200c* as measured in 15 PDAC cell lines was correlated with low invasion ability (Yu et al. 2010a). Similarly, *miR-96* displays inhibitory effects of migration and invasion of PDAC cells as lately reported by Yu et al. (2010b).

13.5.2.2 Metastatic Behavior

MiRNA participate in the metastatic behavior of PDAC cells. Indeed, sixteen human PDAC cell lines were orthopically xenografted in a murine PDAC model and classified in three hierarchical groups according to their metastatic potential. *CD40* was detected as a relevant target for differentially expressed miRNA in the highly metastatic group. A significant over-expression of *CD40*-related *miR-224* and *miR-486* was measured in this group, which was correlated with reduced expression of *CD40* protein levels (Mees et al. 2009). Using the same approach, the authors observed a significant induction of *EP300 (E1A binding protein p300)* targeting miRNA in the high metastatic group (*miR-194*, *miR-200b*, *miR-200c*, and *miR-429*) (Mees et al. 2010). *MiR-200* family members (*miR-141*, *miR-200a, b* and *c*, and *miR-429*) induce epithelial differentiation, thereby suppressing EMT by inhibiting translation of the EMT-activators zinc finger E-box-binding homeobox (*ZEB)1* and *ZEB2* transcription factor. Wellner et al. (2009) by use of PDAC cells showed that *ZEB1* not only promotes tumor cell dissemination but it is also necessary for the tumor initiating capacity. The authors further showed that *ZEB1* represses the expression of *miR-203*, known to inhibit stem-cell identity, and that candidate targets of the *miR-200* family are also stem cell factors; such as sex determining region Y-box 2 (*SOX 2*) and Krüppel like factor 4 (*KLF4*). What is more, *miR-200c*, *miR-203*, and *miR-183* cooperate to suppress expression of stem cell factors. *MiR-10a* was identified as an important mediator of metastasis formation in pancreatic tumor cells (Weiss et al. 2009). The authors observed an increased expression of *miR-10a* in metastatic PDAC. They also demonstrated that *miR-10a* promotes metastatic behavior of PDAC cells and inversely, that repression of *miR-10a* was sufficient to inhibit invasion and metastasis formation. These effects were presumed mediated by the inhibition of the proteins *HOXB1* and *HOXB3*. Over-expression of *miR-200a* and *miR-200b* were shown to regulate smad interacting protein (*SIP*) expression, which is transcriptional repressor of *E-cadherin*. However, even if over-expression of *miR-200a/200b* induces *E-cadherin* in PDAC cell lines, it was incapable of reversing the EMT morphology (Yan et al. 2010). More recently, the over-expression of *miR-20a* in PDAC cell lines was shown to inhibit the protein *STAT-3* translation and consequently reduced the invasive capacity of these cells and metastatic properties of the generated tumors (Yan et al. 2010). By targeting *STAT-3*, *miR-20a* over-expression can suppress *MMP-2* and *VEGF* expression; two important players in EMT.

13.5.3 Apoptosis and Sensitivity to Anti-cancer Agents

MiRNAs also play important roles in the control of apoptotic processes and in the sensitivity of PDAC cells to anti-cancer agents. Several examples for miRNA controlled apoptosis exist. For example, Chang et al. (2007) showed that *miR-34* is an important component of the tumor protein 53 (TP53) transcriptional networks. Re-expression of *miR-34* promotes apoptosis by repressing the expression of the anti-apoptotic protein, B cell lymphoma 2 (*BCL2*). These results were complemented by Xu and colleagues who showed that, in addition to inducing apoptosis, *miR-34* restoration sensitizes PDAC cells to chemotherapy and radiation (Ji et al. 2009). In contrast, Dusetti's group elegantly demonstrated that *miR-155*, which is up-regulated in PDAC (Bloomston et al. 2007; Lee et al. 2007; Szafranska et al. 2007), controls the expression of TP53-induced nuclear protein, a potent pro-apoptotic protein (Gironella et al. 2007). Moreover, Capan-2 cells transfected with an anti-*miR-155* oligonucleotide demonstrated with a significant increase in apoptosis. In the same way, the ectopic expression of *miR-96*, for which expression is decreased in PDAC, induces cell death by apoptosis resulting from a decreased phosphorylation of Bcl-associated death (BAD) protein (Yu et al. 2010b). The transfection with *miR-21* was shown to decrease the sensitivity to gemcitabine treatment of PANC-1 cells (Moriyama et al. 2009). These findings were recently confirmed by demonstrating that the addition of phosphoinositide 3-kinase (PI3K) inhibitors and mammalian target of rapamycin (mTOR) resulted in a decrease of phosphoprotein kinase B/Akt; and prevented *miR-21*-induced resistance to the pro-apoptotic effects of gemcitabine (Giovannetti et al. 2010). Inversely, the transfection of *miR-200b* increases the gemcitabine sensitivity of MiaPACA-2 cells (Li et al. 2009).

13.6 Potential Interests in MiRNAs as Biomarkers for PDAC

Molecular markers are still required to help alleviate PDAC prognosis. Progress must be made for the characterization of large scale diagnostic markers, and for differential markers to distinguish PDAC from other pancreatic diseases such as pseudo-tumorous pancreatitis (Bournet et al. 2009). Also, the ability to predict whether a patient will respond to therapy early in the treatment may be of particular value to stratify and individualize patient treatment strategies. In addition, a validated shorter term surrogate end point would reduce the treatment follow-up period in an attempt to prevent unnecessary disease progression, and ultimately reduce drug development costs through the more rapid completion of randomized controlled Phase III trials. A recent study effectively demonstrated the robust stability of miRNAs, even from human tissues and cell samples for which reliable mRNA analyses are commonly inapplicable (Jung et al. 2010). Therefore, this robust stability confers to miRNAs a great interest as molecular biomarkers for PDAC.

13.6.1 MiRNAs as Diagnostic Marker

Due to their robust stability, the interest in miRNAs as potential diagnostic markers for PDAC was reported in different types of biological samples (resected tumors, biopsies, blood samples).

Several attempts were made to establish a miRNA signature for PDAC. Schmittgen's group was the first to report the expression profile of miRNA in PDAC resected tissues. In this pioneering study, the authors demonstrated that the PAM algorithm based on miRNA expression was able to correctly classified 28/28 tumors, 6/6 normal pancreas and 11 of 15 adjacent benign tissues (Lee et al. 2007). Expression profiles of miRNAs are also very informative not only to discriminate tumor type from the normal pancreas, but also for the differential diagnosis of chronic pancreatitis. Using miRNA microarrays on surgically removed PDAC, normal pancreas and chronic pancreatitis tissues, Bloomston et al. (2007) reported that 15 over-expressed and 8 under-expressed miRNAs can differentiate pancreatic cancer from chronic pancreatitis with an accuracy of 93%. In another study, Szafranska et al. (2007) showed a clear discrimination between PDAC, chronic pancreatitis and normal pancreas can be achieved by use of a subset of 20 miRNAs. These studies showed that neoplastic and non-neoplastic tissues can be secernated. As mentioned earlier, *miR-21* expression is strongly increased in PDAC. It was shown that 79% of the PDAC samples demonstrated strong staining by ISH, compared to 8% of begin pancreas and to 27% of chronic pancreatitis (Dillhoff et al. 2008). Interestingly, none of the non-cancerous tissues demonstrated strong *miR-21* staining. Endoscopic ultrasound (EUS) has become an essential modality in diagnosis and management of pancreatic malignancies. Gene expression analysis in EUS-guided fine needle aspirations (FNA) material is now possible on a routine basis. Szafranska (2008) reported that miRNA altered expression can be measured in PDAC FNA samples and that the combination of *miR-196a* and *miR-217* expression can segregate PDAC samples from other pancreatic disease samples. We recently demonstrated that *let-7* expression is repressed in the same type of PDAC samples. We also showed that the measurement of hypermethylation of *miR-148a* encoding DNA region from FNA samples can serve as a useful diagnostic tool to differentiate PDAC and pseudo-tumor forms of chronic pancreatitis (Hanoun et al. 2010) (Fig. 13.2).

Recently, the discovery of miRNAs in serum establishes the possibility of using miRNAs as biomarkers of disease. Wang et al. (2009) were the first to report the detection of miRNA in the blood of PDAC patients. Indeed, miRNA profiling in plasma can differentiate PDAC patients from healthy controls. Of importance, analyses of plasma levels revealed a 64% sensitivity and 89% specificity for a panel of four miRNAs (*miR-21*, *miR-210*, *miR-196a*, and *miR-155*). In a recent study, miRNA expression was measured by qRT-PCR in plasma of 2 different cohorts of patients with PDAC and compared to an age-matched control cohort. The authors demonstrated that circulating *miR-210* level is significantly more elevated in PDAC patients and may potentially serve as a useful biomarker for PDAC diagnosis (Ho et al.). Lastly, Li et al. (2010a) demonstrated that both *miR-200a* and *miR-200b* were significantly elevated in the sera of PDAC and chronic pancreatitis patients compared with healthy controls.

13.6.2 MiRNAs as Prognostic Markers

One main reason for the poor survival rate associated to PDAC is that only a small fraction of patients (~15%) are diagnosed while they have small, surgically resectable cancers. Thus, there is considerable interest in identifying reliable prognostic markers, especially for the vast majority (85%) of patients that can undergo surgery due to a locally advanced or metastatic disease state.

In studies conducted from resected tissues, Bloomston et al. (2007) reported that the SAM application identified 6 miRNAs that were differentially overexpressed in patients with a longer survival rates (greater than 2 years) (*miR-452*, *miR-105*, *miR-127*, *miR-518a-2*, *miR-187*, and *miR-30a-3p*). Moreover, 2 miRNAs were of considerable interest. First, tumors with high expression of *miR-196a-2* have a median survival of 14.3 months compared with 26.5 months for those with low expression. Second, the median survival in patients with high expression of *miR-219* was 13.6 months compared with 23.8 for those with low expression whereas no correlation was found with increased expression of *TP53* (31/54), loss of *p16* expression (49/56) and lost of *SMAD4/DPC4* expression (39/56). In another study, Greither et al. (2010) measured the expression of *miR-155*, *miR-200*, *miR-203*, and *miR-205* in 56 microdissected PDAC samples and showed that elevated expression of these miRNAs is associated with poorer survival rates. Additionally, this study demonstrated that patients with tumors presenting with elevated levels of all 4 miRNAs retained a 6.2-fold increased risk of tumor-related death compared to patients whose tumors expressed a lower expression of these miRNAs. As mentioned earlier, strong expression of *miR-21* in PDAC was reported in a plethora of studies. Although its strong expression in cancer does not correlate with tumor size, differentiation, nodal status or T stage, it predicts limited survival in patients with node-negative (Dillhoff et al. 2008). It is of interest that these results differ from those from Tanaka's group, which observed no correlation between *miR-21* expression and prognosis or other clinicopathologic features (Moriyama et al. 2009). More recently, by use of microdissected tissues, Giovannetti et al. (2010) found that patients with high *miR-21* expression have a significantly shorter overall survival both in metastatic and adjuvant settings. Multivariate analysis confirmed the prognostic significance of *miR-21* expression. The same group recently published that low *miR-21* expression is associated with benefiting from adjuvant treatment in two independent cohorts of PDAC cases (Hwang et al. 2010). Lastly, Yu et al. (2010a) investigated the relationship between *E-cadherin* and *miR-200c* expression in a panel of 14 PDAC cell lines and in macrodissected formalin-fixed paraffin-embedded (FFPE) tissue samples obtained from 99 patients who underwent pancreatectomy. The authors reported that patients with high levels of *miR-200c* expression manifest with a significantly better survival rate as compared to those with low levels of *miR-200c* expression.

13.7 Potential Interests in MiRNAs as Therapeutic Tool for PDAC

It is now well established that miRNAs play pivotal roles in diverse processes involved in cancer, such as differentiation, control of cell proliferation and death, stress response and metabolism. As such, the activation of tumor-suppressive miRNAs and/or the inhibition of oncogenic miRNAs by small molecules or gene transfer may potentially provide a fundamental approach for the development of cancer therapeutics.

To date, most of the approaches used to control miRNA expression are based on the use of synthetic mimics or small interfering RNAs. Tsuda et al. (2006) reported that synthetic miRNA designed to target *glioma-associated antigen 1* transcription factor inhibits division and induces late apoptosis in pancreatic tumor cells. In another study, *miR-21* and *miR-221* antisense oligonucleotides were demonstrated to inhibit PDAC cell proliferation (cell cycle arrest in G1) and increases their sensitivity to gemcitabine treatment (Moriyama et al. 2009; Park et al. 2009). Down-regulation of proliferation-inducing miRNA can also be triggered by external stimuli. Weiss et al. (2009) reported that *miR-10a* expression promotes metastatic behavior of PDAC-derived cells and that repression of *miR-10a* is sufficient to inhibit invasion and metastasis formation. Interestingly, the authors found that *miR-10a* is targeted by retinoid acid and that retinoic acid receptor antagonists effectively repress *miR-10a* expression and fully block metastasis. On the other hand, *miR-34* enforced expression by mimics or lentiviral transduction inhibits invasion, promotes apoptosis and cell cycle arrest. As a result, treated cells are more sensitive to chemotherapy and radiation. Following a similar strategy, Li et al. (2010b) demonstrated that increasing cellular *miR-146a* content using mimics or the natural products 3,3′-diinodolylmethane (DIM) or isoflavone, inhibited the invasive capacity of PDAC-derived cells with concomitant down-regulation of *EGFR* and the *NF-kappaB* regulatory *kinase interleukin 1 receptor-associated kinase 1* (*IRAK-1*). We recently demonstrated that restoring *let-7* expression strongly affects the proliferation of a PDAC-derived cell line (Torrisani et al. 2009). Unfortunately, we failed to antagonize in vivo pancreatic tumors growth using the same approach by delivering *let-7*. Akin to our findings, Watanabe et al. (2009) suggested that ectopically expressed *let-7* miRNAs may not evidently have a therapeutic role in PDAC cells that originally express *let-7*.

13.8 Perspectives and Challenges

MiRNAs have proven effective for PDAC classification, prognostic stratification and drug-response prediction. These small but potent molecular markers can be detected and quantified not only in frozen tissues, but also in FFPE tissues, as well as serum/plasma samples. Whether this will translate into clinical application is still highly debated. Nevertheless, circulating miRNAs are expected to be proven as specific and sensitive for PDAC as compared to current biomarkers and therefore raise promising perspectives for PDAC detection. Concerning the use of miRNA as anti-cancer targets or molecules for PDAC in vivo, the efficiency of these approaches is still to be demonstrated. In vivo use of so called antagomirs may be an opportunity to achieve significant down regulation of oncomirs in solid tumors. However, cell targeting and distribution of miRs by this approach remains challenging. This issue may be resolved by the intratumoral transduction of tumors using targeted viral vectors encoding for pro- or anti-miRNA.

Acknowledgments We thank Dr Dina Arvanitis (Centre de Biologie du Développement, Toulouse) for critical reading of the manuscript. J.T. was funded by the Ligue Nationale contre le Cancer.

References

Bird AP. The relationship of DNA methylation to cancer. Cancer Surv. 1996;28:87–101.

Bloomston M, Frankel WL, Petrocca F, et al. MicroRNA expression patterns to differentiate pancreatic adenocarcinoma from normal pancreas and chronic pancreatitis. JAMA. 2007;297:1901–8.

Bournet B, Souque A, Senesse P, et al. Endoscopic ultrasound-guided fine-needle aspiration biopsy coupled with KRAS mutation assay to distinguish pancreatic cancer from pseudotumoral chronic pancreatitis. Endoscopy. 2009;41:552–7.

Burk U, Schubert J, Wellner U, et al. A reciprocal repression between ZEB1 and members of the miR-200 family promotes EMT and invasion in cancer cells. EMBO Rep. 2008;9:582–9.

Calin GA, Croce CM. MicroRNA signatures in human cancers. Nat Rev Cancer. 2006;6:857–66.

Calin GA, Dumitru CD, Shimizu M, et al. Frequent deletions and down-regulation of micro-RNA genes miR15 and miR16 at 13q14 in chronic lymphocytic leukemia. Proc Natl Acad Sci USA. 2002;99:15524–9.

Chang TC, Wentzel EA, Kent OA, et al. Transactivation of miR-34a by P53 broadly influences gene expression and promotes apoptosis. Mol Cell. 2007;26:745–52.

Chang TC, Yu D, Lee YS, et al. Widespread microRNA repression by myc contributes to tumorigenesis. Nat Genet. 2008;40:43–50.

Crosby ME, Devlin CM, Glazer PM, et al. Emerging roles of microRNAs in the molecular responses to hypoxia. Curr Pharm Des. 2009;15:3861–6.

Dillhoff M, Liu J, Frankel W, et al. MicroRNA-21 is overexpressed in pancreatic cancer and a potential predictor of survival. J Gastrointest Surg. 2008;12:2171–6.

du Rieu MC, Torrisani J, Selves J, et al. MicroRNA-21 is induced early in pancreatic ductal adenocarcinoma precursor lesions. Clin Chem. 2010;56:603–12.

Feldmann G, Beaty R, Hruban RH, et al. Molecular genetics of pancreatic intraepithelial neoplasia. J Hepatobiliary Pancreat Surg. 2007;14:224–32.

Giovannetti E, Funel N, Peters GJ, et al. MicroRNA-21 in pancreatic cancer: correlation with clinical outcome and pharmacologic aspects underlying its role in the modulation of gemcitabine activity. Cancer Res. 2010;70:4528–38.

Gironella M, Seux M, Xie MJ, et al. Tumor protein 53-induced nuclear protein 1 expression is repressed by miR-155, and its restoration inhibits pancreatic tumor development. Proc Natl Acad Sci USA. 2007;104:16170–5.

Greither T, Grochola LF, Udelnow A, et al. Elevated expression of microRNAs 155, 203, 210 and 222 in pancreatic tumors is associated with poorer survival. Int J Cancer. 2010;126:73–80.

Habbe N, Koorstra JB, Mendell JT, et al. MicroRNA miR-155 is a biomarker of early pancreatic neoplasia. Cancer Biol Ther. 2009;8:4.

Hanoun N, Delpu Y, Suriawinata AA, et al. The silencing of microRNA 148a production by DNA hypermethylation is an early event in pancreatic carcinogenesis. Clin Chem. 2010;56:1107–18.
Hingorani SR, Petricoin EF, Maitra A, et al. Preinvasive and invasive ductal pancreatic cancer and its early detection in the mouse. Cancer Cell. 2003;4:437–50.
Ho AS, Huang X, Cao H, et al. Circulating miR-210 as a novel hypoxia marker in pancreatic cancer. Transl Oncol. 2010;3:109–13.
Hruban RH, Adsay NV. Molecular classification of neoplasms of the pancreas. Hum Pathol. 2009;40:612–23.
Hruban RH, Adsay NV, Albores-Saavedra J, et al. Pancreatic intraepithelial neoplasia: a new nomenclature and classification system for pancreatic duct lesions. Am J Surg Pathol. 2001;25:579–86.
Hruban RH, Goggins M, Parsons J, et al. Progression model for pancreatic cancer. Clin Cancer Res. 2000;6:2969–72.
Huang X, Ding L, Bennewith KL, et al. Hypoxia-inducible miR-210 regulates normoxic gene expression involved in tumor initiation. Mol Cell. 2009;35:856–67.
Hwang JH, Voortman J, Giovannetti E, et al. Identification of microRNA-21 as a biomarker for chemoresistance and clinical outcome following adjuvant therapy in resectable pancreatic cancer. PLoS One. 2010;5:e10630.
Jemal A, Siegel R, Xu J, et al. Cancer statistics. CA Cancer J Clin. 2010; 60:277–300.
Ji Q, Hao X, Zhang M, et al. MicroRNA miR-34 inhibits human pancreatic cancer tumor-initiating cells. PLoS One. 2009;4:e6816.
Jung M, Schaefer A, Steiner I, et al. Robust microRNA stability in degraded RNA preparations from human tissue and cell samples. Clin Chem. 2010;56:998–1006.
Kent OA, Mullendore M, Wentzel EA, et al. A resource for analysis of microRNA expression and function in pancreatic ductal adenocarcinoma cells. Cancer Biol Ther. 2009;8:21.
Lee EJ, Gusev Y, Jiang J, et al. Expression profiling identifies microRNA signature in pancreatic cancer. Int J Cancer. 2007;120:1046–54.
Lee KH, Lotterman C, Karikari C, et al. Epigenetic silencing of microRNA miR-107 regulates cyclin-dependent kinase 6 expression in pancreatic cancer. Pancreatology. 2009;9:293–301.
Li A, Omura N, Hong SM, et al. Pancreatic cancers epigenetically silence SIP1 and hypomethylate and overexpress miR-200a/200b in association with elevated circulating miR-200a and miR-200b levels. Cancer Res. 2010a;70:5226–37.
Li Y, VandenBoom TG 2nd, Kong D, et al. Up-regulation of miR-200 and let-7 by natural agents leads to the reversal of epithelial-to-mesenchymal transition in gemcitabine-resistant pancreatic cancer cells. Cancer Res. 2009;69:6704–12.
Li Y, VandenBoom TG 2nd, Wang Z, et al. MiR-146a suppresses invasion of pancreatic cancer cells. Cancer Res. 2010b;70:1486–95.
Lodygin D, Tarasov V, Epanchintsev A, et al. Inactivation of miR-34a by aberrant cpg methylation in multiple types of cancer. Cell Cycle. 2008;7:2591–600.
Lu J, Getz G, Miska EA, et al. MicroRNA expression profiles classify human cancers. Nature. 2005;435:834–8.
Maitra A, Hruban RH. Pancreatic cancer. Annu Rev Pathol. 2008;3:157–88.
Mathew LK, Simon MC. MiR-210: a sensor for hypoxic stress during tumorigenesis. Mol Cell. 2009;35:737–8.
Mees ST, Mardin WA, Sielker S, et al. Involvement of CD40 targeting miR-224 and miR-486 on the progression of pancreatic ductal adenocarcinomas. Ann Surg Oncol. 2009;16:2339–50.
Mees ST, Mardin WA, Wendel C, et al. EP300 – a miRNA-regulated metastasis suppressor gene in ductal adenocarcinomas of the pancreas. Int J Cancer. 2010;126:114–24.
Moriyama T, Ohuchida K, Mizumoto K, et al. MicroRNA-21 modulates biological functions of pancreatic cancer cells including their proliferation, invasion, and chemoresistance. Mol Cancer Ther. 2009;12:12.

Morris JP 4th, McManus MT. Slowing down the Ras lane: miRNAs as tumor suppressors? Sci STKE. 2005;2005:pe41.

Muniyappa MK, Dowling P, Henry M, et al. MiRNA-29a regulates the expression of numerous proteins and reduces the invasiveness and proliferation of human carcinoma cell lines. Eur J Cancer. 2009;45:3104–18.

Park JK, Lee EJ, Esau C, et al. Antisense inhibition of microRNA-21 or -221 arrests cell cycle, induces apoptosis, and sensitizes the effects of gemcitabine in pancreatic adenocarcinoma. Pancreas. 2009;38:e190–9.

Ryu JK, Hong SM, Karikari CA, et al. Aberrant microRNA-155 expression is an early event in the multistep progression of pancreatic adenocarcinoma. Pancreatology. 2010;10: 66–73.

Safioleas MC, Moulakakis KG. Pancreatic cancer today. Hepatogastroenterology. 2004;51: 862–8.

Saito Y, Liang G, Egger G, et al. Specific activation of microRNA-127 with downregulation of the proto-oncogene BCL6 by chromatin-modifying drugs in human cancer cells. Cancer Cell. 2006;9:435–43.

Sato N, Maitra A, Fukushima N, et al. Frequent hypomethylation of multiple genes overexpressed in pancreatic ductal adenocarcinoma. Cancer Res. 2003;63:4158–66.

Scott GK, Mattie MD, Berger CE, et al. Rapid alteration of microRNA levels by histone deacetylase inhibition. Cancer Res. 2006;66:1277–81.

Seux M, Iovanna J, Dagorn JC, et al. MicroRNAs in pancreatic ductal adenocarcinoma: new diagnostic and therapeutic clues. Pancreatology. 2008;9:66–72.

Sun M, Estrov Z, Ji Y, et al. Curcumin (diferuloylmethane) alters the expression profiles of microRNAs in human pancreatic cancer cells. Mol Cancer Ther. 2008;7:464–73.

Szafranska AE, Davison TS, John J, et al. MicroRNA expression alterations are linked to tumorigenesis and non-neoplastic processes in pancreatic ductal adenocarcinoma. Oncogene. 2007;26:4442–52.

Szafranska AE, Doleshal M, Edmunds HS, et al. Analysis of microRNAs in pancreatic fine-needle aspirates can classify benign and malignant tissues. Clin Chem. 2008;54:1716–24.

Torrisani J, Bournet B, du Rieu MC, et al. Let-7 microRNA transfer in pancreatic cancer-derived cells inhibits in vitro cell proliferation but fails to alter tumor progression. Hum Gene Ther. 2009;20:831–41.

Torrisani J, Buscail L. Molecular pathways of pancreatic carcinogenesis. Ann Pathol. 2002;22:349–55.

Tsuda N, Ishiyama S, Li Y, et al. Synthetic microRNA designed to target glioma-associated antigen 1 transcription factor inhibits division and induces late apoptosis in pancreatic tumor cells. Clin Cancer Res. 2006;12:6557–64.

Volinia S, Calin GA, Liu CG, et al. A microRNA expression signature of human solid tumors defines cancer gene targets. Proc Natl Acad Sci USA. 2006;103:2257–61.

Wang J, Chen J, Chang P, et al. MicroRNAs in plasma of pancreatic ductal adenocarcinoma patients as novel blood-based biomarkers of disease. Cancer Prev Res (Phila). 2009;2:807–13.

Watanabe S, Ueda Y, Akaboshi S, et al. HMGA2 maintains oncogenic RAS-induced epithelial-mesenchymal transition in human pancreatic cancer cells. Am J Pathol. 2009;174: 854–68.

Weiss FU, Marques IJ, Woltering JM, et al. Retinoic acid receptor antagonists inhibit miR-10a expression and block metastatic behavior of pancreatic cancer. Gastroenterology. 2009;137:2136–45.

Wellner U, Schubert J, Burk UC, et al. The EMT-activator ZEB1 promotes tumorigenicity by repressing stemness-inhibiting microRNAs. Nat Cell Biol. 2009;11:1487–95.

Yan HJ, Wu JX, Liu WS, et al. MicroRNA-20a overexpression inhibited proliferation and metastasis of pancreatic carcinoma cells. Hum Gene Ther. 2010;21:1723–34.

Yu S, Lu Z, Liu C, et al. MiRNA-96 suppresses KRAS and functions as a tumor suppressor gene in pancreatic cancer. Cancer Res. 2010b;70:6015–25.

Yu J, Ohuchida K, Mizumoto K, et al. MicroRNA, hsa-miR-200c, is an independent prognostic factor in pancreatic cancer and its upregulation inhibits pancreatic cancer invasion but increases cell proliferation. Mol Cancer. 2010a;9:169.

Zhang S, Cai X, Huang F, et al. Effect of trichostatin a on viability and microRNA expression in human pancreatic cancer cell line BxPC-3. Exp Oncol. 2008;30:265–8.

Zhang Y, Li M, Wang H, et al. Profiling of 95 microRNAs in pancreatic cancer cell lines and surgical specimens by real-time PCR analysis. World J Surg. 2009;33:698–709.

Zhu Z, Gao W, Qian Z, et al. Genetic variation of miRNA sequence in pancreatic cancer. Acta Biochim Biophys Sin (Shanghai). 2009;41:407–13.

Chapter 14
MicroRNAs in Epithelial Ovarian Cancer

Chunsheng Li, Yi Feng, George Coukos, and Lin Zhang

Abstract Epithelial ovarian cancer is one of the most common gynecological cancers worldwide. Despite decades of researches, it remains a major cause of morbidity and mortality among women in the United States. However, progress has been made on understanding the cellular and molecular mechanisms for ovarian cancer. Frequent mutations in the p53/Rb tumor suppressor pathways and activation of oncogenic signaling pathways such as c-Myc, K-ras, and Akt greatly contribute to disease progression. Recently, microRNAs (miRNAs) have been demonstrated to be a novel class of regulators in cell proliferation, differentiation, and apoptosis. Functional studies revealed that miRNAs play an important role during tumorigenesis, cancer progression, and therapeutic response. Gene expression profiling studies found aberrant miRNA expressions in essentially all types of tumors studied. In this chapter, we will review recent researches on the role of miRNAs in ovarian cancer and discuss how these studies can improve our understanding of ovarian cancer pathogenesis. We also discuss the perspectives of miRNAs as diagnostic and prognostic tools in clinical practice, and as new avenues for therapeutic development.

14.1 Introduction

The ovaries are a pair of multipurpose female reproductive organs in the pelvis connected by the fallopian tubes. First of all, it harbors, nurtures and guides the development of the egg (ova) so that when a mature egg is extruded from the ovary (i.e. ovulation) it is ready for the journey through the fallopian tube, penetration by a sperm, and its eventual implantation in the uterus. Second, the ovary is a sophisticated endocrine organ which secretes many hormones essential for pregnancy as well as the onset and cyclical perpetuation of menstruation. These hormones are also in involved in the development of the female body physique that transforms a prepubertal girl into a mature woman.

L. Zhang (✉)
Department of Obstetrics and Gynecology, Center for Research on Early Detection and Cure of Ovarian Cancer, University of Pennsylvania School of Medicine, Philadelphia, PA 19104, USA
e-mail: linzhang@mail.med.upenn.edu

W.C.S. Cho (ed.), *MicroRNAs in Cancer Translational Research*,
DOI 10.1007/978-94-007-0298-1_14, © Springer Science+Business Media B.V. 2011

14.2 Overview of Epithelial Ovarian Cancer

According to the American Cancer Society, there are 21,880 new cases of ovarian cancer and 13,850 deaths from ovarian cancer estimated in 2010 (Jemal et al. 2010). In the United States, a woman dies of ovarian cancer every 45 min. Despite decades of research, ovarian cancer, one of the most common gynecological cancers, remains the fifth leading cause of morbidity and mortality among women in the United States.

The poor survival rate is mainly due to the fact that most ovarian cancer patients are diagnosed in late stages when the tumors respond poorly to therapy and have high metastasis potential. A huge hurdle to improve ovarian cancer patient care is the lack of reliable means of early detection except for genetic screening for a small percentage of high-risk individuals (Herbst 1994). The primary treatment for ovarian cancer is cyto-reductive surgery followed by chemotherapy. The majority of ovarian cancers, although initially responsive to platinum-based first-line chemotherapy, will ultimately progress and often become resistant to second-line chemotherapy treatments.

14.2.1 Pathological Complexity of Ovarian Cancers

Ovarian cancer is not a single disease. Tumors can arise from the ovarian sex-cord stroma, germ cells or epithelium, which classifies ovarian cancer into multiple types with distinct biological and clinical behaviors. Approximately 90% of invasive ovarian neoplasms are considered to originate from epithelium, referred to as epithelial ovarian carcinoma (EOC), whereas the rest originate from granulose cells, or more rarely from the germ cells or the stroma. This book chapter will focus on EOC since it represents most cases of the disease.

One of the most striking aspects of EOC is the unusual cellular differentiation during neoplastic progression. Unlike most other carcinomas, in which epithelial cells become less differentiated during neoplastic progression, in EOC normal simple primitive ovarian surface epithelial (OSE) cells acquired more complex characteristics of Mullerian duct-derived epithelia after progression to carcinomas. The unusual differentiation serves as the basis for classification of EOC into serous (fallopian tube-like), endometrioid (endometrium-like), mucinous (endocervical-like) and clear cell carcinomas. These histotype groups have distinct clinical behaviors, treatment responses and biomarker profiles.

14.2.2 The Origin of EOC

The etiology and early events in the progression of EOC are among the least understood of all major human malignancies. There are no identifiable premalignant lesions that most pathologists can agree on and the histological region for the neoplasm origin remains enigmatic. These two uncertainties contribute directly to the high mortality rate of EOC since most of the patients are diagnosed at an advanced stage. Indeed, early detection would greatly lessen the mortality because patients with cancer confined in ovaries have good prognosis (Herbst 1994).

For decades it was widely accepted that most, if not all, EOC arises in the OSE, which is the pelvic mesothelium overlying the ovaries and linking to the ovarian epithelium cysts. Several theories have been proposed to understand the epidemiologic data for ovarian carcinoma susceptibility. These epidemiological studies implicated possible racial, geographical, social and hormonal factors. Nulliparity and hyperovulation increases the risk whereas pregnancies and oral contraceptives are protective, supporting the incessant ovulation theory proposed by Fathalla (1971). This hypothesis postulates that repetitive ruptures of the follicles and post-ovulatory repair processes stimulate OSE cell proliferation and the chances of subsequent stepwise accumulation of genetic abnormalities. Inflammation is postulated to be a risk factor in EOC development since hysterectomies and tubal ligation seems to reduce the risk by preventing the spread of environmental pathogens to initiate inflammation (Ness and Cottreau 1999). Another risk factor is germline mutations, such as *BRCA1* and *BRCA2* mutations, which account for ~5–10% of cases (Xu and Solomon 1996). However (a) *BRCA* mutations increase cancer incidence of mostly in the breast, ovarian and prostate, and importantly (b) not all the carriers with the predisposing *BRCA* mutations develop ovarian cancer, suggesting the presence of cross-talk between *BRCA* and other unknown genetic and/or epigenetic factors, such as hormonal influences.

The histological origin of ovarian cancer has been challenged recently. Historically, EOC was considered to arise from epithelial components, either in the single cell layer of surface epithelium or in the morphologic altered architecture thereof, such as surface epithelium lined clefts and cortical inclusion cysts (Auersperg et al. 1998; Radisavljevic 1977). However, Dubeau argued for an alternative origin of EOC in the components of the Mullerian tracts, such as rete ovarii, inclusion cysts, fallopian tube fimbriae, endosalpingiosis, endometriosis and endocervicosis (Dubeau 1999, 2008). The Mullerian systems are conveniently located adjacent to or within the ovaries. From a developmental biology point of view, the ovarian surface epithelia connect with the peritoneal mesothelium, which has a similar morphology. Paradoxically, typical EOCs resemble Mullerian-derived tissues such as the fallopian tube, endometrium and endocervix rather than the ovarian mesothelioma. These observations suggest that if the ovarian coelomic epithelium is indeed the origin of EOC, it need metaplasia and change to Mullerian-like epithelium (Seidman et al. 2002). The Mullerian theory is also supported by the fact that some clear cell and endometrioid EOC arise in endometriotic implants in the ovary (Dubeau 1999). The tumors from Mullerian-derived tissues may spread into the ovary and are later diagnosed as primary ovarian cancer.

A subset of low grade EOC, namely type I tumors (including low-grade serous carcinomas, endometrioid, clear cell and mucinous adenocarcinomas, and malignant Brenner tumors) follow the adenoma-carcinoma scheme where an epithelial cell undergoes genetic alterations of increasing complexity and evolves phenotypically toward a tumor cell (Smith Sehdev et al. 2003). In most epithelial tumors, the earliest genetic changes in premalignant lesions are the activation of DNA damage repair and abnormal *p53* (Halazonetis et al. 2008). However, another subset namely type II EOCs, consist of high grade serous ovarian carcinomas (HG-SOC) and carcinosarcomas, lack an identifiable pre-existing pathways such as DNA damage repair and adenoma-carcinoma pathway. Furthermore, distinct genetic alterations are suggested to separate type I and type II EOCs. However, the premalignant lesion of type II has been enigmatic. Existing paradigms are further transformed by recent molecular genetics studies of type II EOC. The "fallopian tube model" has originated from Piek et al. (2001) who observed a high incidence of dysplastic and neoplastic lesions in fallopian tubes that were prophylactically removed from women predisposed to EOC with *BRCA*-mutations. A "p53 signature" shows strong p53 immunostaining among benign tubal mucosa from *BRCA*-mutation carriers (Lee et al. 2007). The "p53 signature" identified in the tubal fimbriae were associated with *TP53* mutations in the regions of tubal intraepithelial carcinoma and its associated EOC has identical *TP53* mutations (Folkins et al. 2008). From risk-reducing salpingectomy cases, the multiple foci of tubal intraepithelial carcinoma and focus of invasive carcinoma showed similar gene copy number changes within each case, suggesting a monoclonal origin (Salvador et al. 2008). Crum and collaborators studied the "p53 signature" in ovarian cortical inclusion cysts and found such signatures were often observed in the salpingeal epithelium but not in the corresponding ovaries (Crum et al. 2007; Folkins et al. 2009). Since the data supporting the fallopian tube model are all from patients with *BRCA* mutations, it is possible that this model is more relevant in this particular population and additional data from women without *BRCA* mutations are needed. Moreover, direct evidence demonstrating the progression of these lesions to carcinoma will be needed to prove the fallopian tube model. The new diagnostic and therapeutic opportunities and unmet needs unmasked by such a paradigm shift warrant further study.

14.2.3 Biomarkers of EOC

Taking EOC as a whole group, about 80–90% of all cases express elevated levels of the cancer antigen 125 glycoprotein (CA125), which can be shed into circulation and use as serum marker for monitoring chemotherapy response and tumor recurrence. Changes in CA125 serum levels correlated with the course of the disease and response to treatment (Bast et al. 1983; Hawkins et al. 1989; Rustin et al. 2004). Although it is detected in almost all serous adenocarcinomas, it is undetectable in about 50% of mucinous carcinoma. In addition, CA125 failed to detect almost one third of the EOC recurrence. Therefore, novel biomarkers are needed, alone or in combination with CA125. Carcinoembryonic antigen (CEA) and inhibin are useful biomarkers to monitor the mucinous carcinomas but are not useful for other EOC (Phocas et al. 1996). However, CA125 levels vary among different subtypes. In patients where CA125 was not expressed, human epididymis factor 4 (HE4) can be a sensitive marker for detection of EOC (Moore et al. 2008). Other tumor markers, such as carbohydrate antigen 19-9 (CA19-9), tissue polypeptide antigen and immunosuppressive acidic protein, have also been suggested (Kamiya et al. 1990).

Although protein biomarkers already contribute to the management of EOC, emergence of additional markers, such as microRNAs (miRNAs), will add significant diagnostic and prognostic power to detect and manage this deadly gynecological malignancy.

14.3 MiRNA Biology

MiRNAs are ~22 nucleotides long, non-coding, endogenous RNA molecules which regulate protein expression by degrading or repressing the translation of the corresponding messenger RNAs (mRNAs). This class of RNA moiety was first identified in *C. elegans* by Lee and Wightman, who found that a small non-coding RNA, *lin-4*, can serve as a regulator for the timing of the nematode development through incomplete base pairing to the 3′ untranslated region (3′UTR) of target mRNA, *lin-14*, to repress its translation (Lee et al. 1993; Wightman et al. 1993). Since then, experimental and computational analyses have identified hundreds of miRNAs in animals, plants and even viruses. To date, 1,048 *Homo sapiens*, 672 *Mus musculus*, 176 *Drosophila melanogaster*, 175 *C. elegans*, 186 *Xenopus tropicalis*, and 358 *Danio rerio* miRNAs are reported in miR-Base (http://microrna.sanger.ac.uk) (Griffiths-Jones 2010). It is estimated that up to 3% of human genes encode for miRNAs, and more than 30% of human protein-coding genes may be regulated by miRNAs (Bartel 2004; Lewis et al. 2005).

14.3.1 MiRNA Biogenesis

One interesting feature of miRNA genomic organization is that many miRNA genes are in clusters. These miRNA clusters are often transcribed as long polycistronic transcripts and the mature products of these miRNAs often function in concert. However, some miRNA genes are located in regions with no other annotated genes and have their own promoters. A third type of miRNA gene resides in intronic regions of previously annotated protein-coding genes. They are transcribed from the same promoters as the host genes and often regulate the expression of their host genes. It is noteworthy that many miRNAs have multiple genomic copies that generate identical mature products. It is reasoned that these miRNAs play such essential roles that multiple copies of genes evolved to ensure their normal functions in organisms (Li et al. 2009; Olena and Patton 2010).

The transcriptions of most miRNA genes are performed by RNA polymerase II (Pol II), and can be subsequently regulated by various Pol II-associated transcription factors. In some cases, pri-miRNAs can be spliced out from the intronic sequence of a protein-coding transcript (Bartel 2004; Olena and Patton 2010). MiRNAs that have upstream *Alu*, transfer RNA- or mammalian-wide interspersed repeat (MWIR)-based promoter elements are transcribed by RNA polymerase III (Borchert et al. 2006). Once a pri-miRNA is synthesized in the nucleus, it undergoes two cleavage steps to reach its mature form. The first processing event releases a ~70 nt pre-miRNA from pri-miRNA by Drosha, a type III nuclear RNase with an essential cofactor, the DiGeorge syndrome critical region gene 8 protein (DGCR8) (Kim et al. 2009). After Drosha cleavage, pre-miRNAs are exported to the cytoplasm and further trimmed by Dicer, a highly conserved RNase III type endoribonuclease. The miRNA duplexes generated by Dicer then incorporate into the RNA-induced silencing complex (RISC), where one strand (the passenger strand, miRNA*) is degraded while the other strand (the guide strand or miRNA) guides the RISC to bind to target transcripts for repressing their expressions (Kim et al. 2009).

14.3.2 MiRNA Directed Gene Silencing

The core component of the RISC is the Argonaute protein which is composed of four domains: (i) the PAZ domain, which binds to the 3′ terminus of the guide strand; (ii) the PIWI domain, which cleaves target transcripts; (iii) MID-domain, which binds to the 5′ terminus of the guide strand, and (iv) the N-domain (Kim et al. 2009). The targeting specificity mainly depends on the sequence complementarity between the mRNA target site and the seed sequence (nucleotide sequence from position 2 to 8) at the 5′ end of miRNA. MiRNA sequences at 3′ end can also contribute to target selection, especially when the base-paring interaction between the seed sequence and mRNA target is weak. Target sequences in mRNA often reside in the 3′UTRs, though recent studies suggest that miRNA can also target 5′UTR and coding sequences to repress mRNA expression (Tay et al. 2008).

Once engaged with a target transcript, the miRNA incorporated RISC complex can lead to either translational repression or mRNA decay, which depends on the degree of miRNA-mRNA complementarity. Perfect base-pairing, a rare event in animals, triggers Ago-catalyzed cleavage of messenger transcripts, while mismatches promote translational repression. Two models on the mechanism of translational repression have been proposed. One model proposes that RISC complex stalk translation at its initiation step; while the other model suggests that miRNA mediated translational repression is a post-initiation event. It is possible that miRNA-bound RISC can inhibit translation at multiple steps during translation (Carthew and Sontheimer 2009).

14.3.3 Regulating MiRNA Function

MiRNA function can be modulated at four levels: genomic alteration, transcriptional regulation, post-transcriptional modification and regulation of miRNA processing. Chromosomal deletion and amplification generated miRNA expression alteration have been reported by many studies on miRNA expression in cancer (Spizzo et al. 2009). In addition, genomic mutations and polymorphisms can also alter miRNA processing or targeting specificity. One example of such a modulation is that a common G/C polymorphism in *miR-146a* can modulate the level of pre- and mature *miR-146a* (Jazdzewski et al. 2008). It is also worth mentioning that the single nucleotide polymorphism (SNP) in mRNA sequences can be another mechanism of modulating miRNA function by disrupting the miRNA:mRNA interaction (Sethupathy and Collins 2008).

As the majority of miRNA genes are transcribed by Pol II, epigenetic modifications, especially DNA methylation, is found to be an important regulator for miRNA expression. Aberrant hypermethylation, which is often found during cancer development can affect miRNA expression significantly and is considered a major contributor to aberrant miRNA expression in cancer (Han et al. 2007). Moreover, DNA methylation can alter the level of transcription factors, which in turn control the expression of miRNAs.

The third miRNA regulatory mechanism is post-transcriptional modification, such as miRNA editing and miRNA uridylation. It is estimated that about 16% of human pri-miRNA can be regulated by adenosine-to-inosine (A-to-I) RNA editing, a RNA modification process converting adenosine (A) residues to inosine (I), therefore changing an A-U base pair into an I-G base pair (Cai et al. 2009; Gott and Emeson 2000). Such a conversion can affect miRNA biogenesis, diversify mature miRNA products and alter miRNA targeting specificity (Kawahara and Nishikura 2009; Li et al. 2009). In addition to editing, It is reported recently that Lin28, a repressor of *let-7* miRNA, acts primarily by uridylating *let-7* pre-miRNA, resulting in blockade of Dicer cleavage and pre-miRNA degradation (Heo et al. 2008).

Finally, multiple proteins involved in miRNA processing have been found to be tightly controlled. Deregulations of *Dicer*, *Drosha* and *Ago* have been reported in various types of cancer (Carmell et al. 2002; Karube et al. 2005; Merritt et al. 2008). Drosha's cofactor, *DGCR8*, is regulated by tumor suppressor ING1 (Gomez-Cabello et al. 2010). Mutations of *TRBP2*, the protein complexes with Dicer, were found to impair miRNA processing in cancer (Melo et al. 2009).

14.4 MiRNA in Human Cancer

Since miRNA functions are involved in regulating proliferation, differentiation and apoptosis, it is not surprising that miRNA expression is often altered in cancer. The first example was reported by Calin et al. (2002), who found that the *13q14* locus, a site often deleted in chronic lymphocytic leukemia (CLL), contains two miRNA genes, *miR-15a* and *miR-16-1*. Later, it was found that *miR-15a* and *miR-16-1* can target *Bcl-2*, a gene often up-regulated in CLL. This finding provided a molecular mechanism by which losing *miR-15a* and *miR-16-1* can cause CLL (Cimmino et al. 2005). Since the report of miRNA alteration in CLL, aberrant miRNA expression has been reported in almost all the tumors that have been studied, strongly supporting a functional involvement of miRNA in tumorigenesis and progression (Calin and Croce 2006; Spizzo et al. 2009).

14.4.1 MiRNA as a Tumor Suppressor and Oncogene

As protein-coding genes, miRNA can act as an oncogene or a tumor suppressor. miRNAs that possess oncogenic or tumor-suppressive activities are termed "oncomir" (Esquela-Kerscher and Slack 2006). It is noteworthy that the oncogenic or tumor suppressive function of a given miRNA is cell-dependent. Therefore, one should be cautious to define a miRNA as an oncogene or a tumor suppressor.

Many miRNAs have been found to function as tumor suppressor. In addition to *miR-15a* and *miR-16-1*, which are often deleted in CLL, one of the best studied tumor suppressive miRNA gene is *let-7*. The role of *let-7* in cancer was first documented by the Slack group when they found that the *let-7* negatively regulates *Let-60/Ras* in *C. elegans* (Johnson et al. 2005). Moreover, *let-7* family members are down-regulated in many tumors, including lung and breast cancer. Interestingly, many of the *let-7* family members are located in fragile sites associated with lung, breast and cervical cancer (Calin et al. 2004). In addition to *Ras*, *let-7* targets include *HMGA2* and *c-Myc*. Over-expressing *let-7* can induce apoptosis and cell cycle arrest in different cancer cell lines (Kumar et al. 2008). *let-7* probably influences these cellular processes by targeting various genes that regulate cell proliferation, survival, cell cycle as well as differentiation (Roush and Slack 2008).

A third example of a miRNA tumor suppressor is *miR-29*, whose targets include anti-apoptotic gene *MCL-1* and oncogene *TCL-1*. Over-expressing *miR-29* can induce apoptosis in cancer cell lines (Fabbri et al. 2007; Mott et al. 2007; Pekarsky et al. 2006). *MiR-29* genes are located in a chromosome region that is frequently deleted in myelo-dysplasia and therapy-related acute myeloid leukemia (AML) (Pedersen-Bjergaard et al. 1995). *MiR-29* down-regulation has been reported in CLL, lung cancer, invasive breast cancer, AML, and cholangiocarcinoma (Calin et al. 2005; Garzon et al. 2008; Iorio et al. 2005; Mott et al. 2007; Yanaihara et al. 2006).

However, evidence demonstrating tumor suppressive function of miRNA in animals is lacking. It is possible that since tumor suppressive miRNA genes often have multiple genomic locations, knocking out any single gene is not sufficient to induce tumorigenesis.

The only example of alteration in a single miRNA that is sufficient to induce tumorigenesis in an animal model was reported by the Croce group in 2006. Transgenic mice with *miR-155* over-expression have enlarged spleens, highly proliferative preleukemic premature B cells and develop B cell leukemia and high grade lymphoma at 6–7 months of age (Costinean et al. 2006). Consistent with its oncogenic activity, high expression of *miR-155* has been reported in several types of cancer and was found to indicate poor prognosis in lung cancer, diffuse large B cell lymphoma, and aggressive CLL (Costinean et al. 2006; Eis et al. 2005).

A second example of oncogenic miRNAs is the *miR-17-92* cluster, which comprises six miRNAs (*miR-17*, *miR-18a*, *miR-19a*, *miR-20a*, *miR-19b-1*, and *miR-92-1*). The *miR-17-92* cluster is located within a 800 base pair region that is frequently amplified in follicular lymphoma and diffuse large B cell lymphoma (Ota et al. 2004). Elevated expression of the *miR-17-92* cluster miRNAs are reported in a variety of solid tumors and hematological malignancies (Mendell 2008). Mice bearing deletions within this cluster died prematurely and have a deficiency in B cell development (Ventura et al. 2008). Accordingly, once over-expressed in cooperation with c-Myc, this miRNA cluster induced lymphoma in mice (He et al. 2005). Interestingly, as a collaborator with c-Myc to induce lymphoma in vivo, the *miR-17-92* cluster is also a downstream transcriptional target of c-Myc (O'Donnell et al. 2005). In addition, mice with a higher expression of *miR-17-92* in lymphocytes developed lymphoproliferative disease and autoimmunity and died prematurely (Xiao et al. 2008). These observations can be explained in part by the finding that *miR-17-92* miRNA target *PTEN*, *Bim* and *E2F* transcription factors. Conversely, both E2F1 and E2F3 can activate the *miR-17-92* cluster, suggesting a regulatory loop between miRNA and their targets (Woods et al. 2007). In aggregate, the evidence mentioned above strongly suggests that the *miR-17-92* cluster is oncogenic. However, down-regulation of *miR-17-5p* has been reported in breast cancer cell lines and restoring its expression inhibits cell proliferation by inhibiting *AIB1* (amplified in breast cancer I) (Hossain et al. 2006). Consistent with this observation, deletion of the *miR-17-92* genomic locus has been described in 16.5% of ovarian cancers, 21.9% of breast cancers, and 20% of melanomas (Mendell 2008). Therefore, depending on the cellular context, *miR-17-92* cluster may function as a tumor suppressor or an oncogene.

MiR-21, whose targets include *PTEN*, *PDCD4* and *TPM1*, represents another example of oncogenic miRNA (Selcuklu et al. 2009). Reducing *miR-21* level inhibited cell growth and induced apoptosis while over-expressing *miR-21* in glioblastoma cells blocked apoptosis (Chan et al. 2005). Accordingly, *miR-21* is up-regulated in a wide variety of tumors (Garzon et al. 2009). However, expression profile studies in ovarian cancer reported *miR-21* expression may either be up or down-regulated (Dahiya et al. 2008; Iorio et al. 2007; Lou et al. 2010; Nam et al. 2008; Pan et al. 2008; Taylor and Gercel-Taylor 2008). It is possible that *miR-21* functions differentially in distinct cell compartments within ovarian tumors and the difference in *miR-21* expression alteration in various samples reflects the inconsistency of tumor cellular content.

14.4.2 Mechanism of MiRNA Deregulation in Cancer

As mentioned in Section 14.3.3, miRNA can be regulated at the genomic, transcriptional, post-transcriptional levels and at the steps of miRNA maturation. Deregulation of miRNA in cancer can result from alteration at any of these four steps. First, genomic abnormality causes aberrant miRNA expression in cancer. Approximately 50% of annotated human miRNAs are located in fragile sites of cancer. Indeed, genomic alterations in miRNA loci were revealed in human melanomas, ovarian and breast cancers (Calin et al. 2004; Zhang et al. 2006, 2008a). Mutations and polymorphisms of miRNAs are also reported in tumor samples. For example, mutations in *miR-15a* and *miR-16-1* pri-miRNAs resulted in lower expression of corresponding miRNAs in CLL and breast cancer patients (Calin et al. 2005;

Raveche et al. 2007). In addition to genomic alteration, epigenetic regulations of miRNAs are also reported in cancer. It is reported that TP53 can directly regulate the transcription of *miR-34* family members (Chang et al. 2007; He et al. 2007; Raver-Shapira et al. 2007) and c-Myc can regulate the transcription of multiple miRNAs, including the *miR-17-92* cluster and *miR-29* (Chang et al. 2008). Since these proteins themselves are regulators of proliferation, differentiation and apoptosis, it is likely that TP53 and c-Myc coordinate the expression of both mRNA and miRNA transcripts to elicit its oncogenic/tumor suppressive activity. In addition, modulating promoter methylation and histone deacetylase activity can also affect miRNA expression (Saito et al. 2006; Scott et al. 2006; Zhang et al. 2008a). More recently, miRNA editing was found to regulate miRNA biogenesis and targeting. The expression of ADARs, the enzymes catalyzing RNA editing, were found to be altered in many types of tumors, suggesting a potential functional impact of miRNA editing in tumor development and progression (Paz et al. 2007). Lastly, Dicer and Drosha, the two ribonucleases that process miRNA are differentially expressed in tumor tissues in comparison to their normal counterparts (Karube et al. 2005; Merritt et al. 2008), and the *Ago* locus in chromosome 1 is often found deleted in Wilms tumors and neuroectodermal tumors (Carmell et al. 2002). In summary, miRNA deregulations are frequent in cancer samples and at least four different mechanisms can result in such alterations.

14.4.3 MiRNA as Prognostic Marker and Therapeutic Agent in Cancer

In recent years, miRNAs have been reported to serve as markers for predicting prognosis, metastasis, treatment response as well as survival (Cho 2010a). Lu et al. (2005) demonstrated that miRNA profiling can help identify developmental lineage and differentiation state of the tumors and performed better for classifying poorly differentiated tumors than mRNA profiles. Associations between miRNA expression signatures and prognostic factors, disease progression, survival rate or/and chemoresponses in various types of cancers have been reported (Cho 2010a). As specific and sensitive assays that can capture and detect mature miRNAs in isolated tumor cells, xenografts, sera from patients as well as from formalin-fixed tissues are being developed and optimized, using miRNAs as reliable and non-invasive biomarkers for cancer diagnosis in clinic is feasible (Cho 2010a).

The therapeutic potential of miRNAs has also been explored. Several reagents aiming to modulate miRNA functions have been tested in vitro and in vivo for their specificity and efficacy (Li et al. 2009). LNA modified RNA molecules exhibit some very promising characteristics for inhibiting the activity of specific miRNAs as therapeutic approaches (Cho 2010b). For the replacement therapy, retroviral expression of tumor suppressive miRNAs has been reported to be effective in several mouse models. Recently, Intratumoral and intranasal delivery of *let-7* have been shown to effectively reduce tumor size in a lung cancer xenograft model (Trang et al. 2010). The results from these studies are promising. However, given the complexity of miRNA affect, altering one miRNA may affect the expression of unexpected genes, therefore causing unwanted side effect; on the other hand, since each mRNA has target sites of multiple miRNAs, single miRNA alteration may not be efficient enough (Cho 2010b). In addition, there are many technical challenges regarding improving the efficacy, specificity and stability of miRNA based therapeutic reagent. Further investigations are much needed for successful application of miRNA therapy in cancer care.

14.5 MiRNA Deregulation in Epithelial Ovarian Cancer

Incomplete understanding of the underlying mechanism of ovarian cancer tumorigenesis and progression is the main impediment to improve the ovarian cancer care. Since miRNA has been identified as a novel class of regulators in tumorigenesis, the involvement of miRNA in ovarian cancer has also been recently explored. Multiple studies have compared the miRNA expression patterns between malignant tissues and normal controls or between ovarian cancer cell lines and primary ovarian epithelia (Chung et al. 2009; Creighton et al. 2010; Dahiya et al. 2008; Eitan et al. 2009; Giannakakis et al. 2008; Hu et al. 2009; Iorio et al. 2007; Kuo et al. 2009; Lee et al. 2009; Lodes et al. 2009; Nam et al. 2008; Resnick et al. 2009; Taylor and Gercel-Taylor 2008; Wu et al. 2009; Yang et al. 2008a; Zhang et al. 2006, 2008a). Multiple lines of evidence suggest that aberrant miRNA expression is associated with ovarian cancer prognosis, response to treatments and rate of survival. Genomic abnormality, epigenetic modification as well as changes in miRNA biogenesis have all been reported as mechanisms responsible for the changes of miRNA expression in ovarian cancer.

14.5.1 Genomic Alteration of MiRNA in Ovarian Cancer

Genomic alteration of miRNA genes in ovarian cancer was first reported by Zhang et al. (2006). In their study, 93 primary ovarian cancer tissues and 16 cell lines were subjected to array comparative genomic hybridization (aCGH) analysis to identify a gain or loss in miRNA gene copy numbers. About 37.1% (105/283) of the miRNA loci in ovarian cancer tissues was significantly altered with 52% (55/105) being copy number gain and 48% (50/105) being copy number loss. Interestingly, a number of miRNA gene copy number alterations in ovarian cancer were shared by breast cancer and melanoma. For example, the *miR-17-92* locus is lost in all three tumor types; whereas the *miR-15a/16-1* locus is lost in 24 and 25% of ovarian cancer and breast cancer, respectively (Zhang et al. 2006). These observations suggest that some miRNA genomic alterations may be important to tumorigenesis in general. This work was complemented by a later analysis on serous ovarian cancer tissues by Kuo et al., who observed higher number of genomic alterations in high grade serous ovarian cancer in comparison to low grade serous cancer by aCGH. In addition, hemizygous 1q36, which contains *miR-34a*, was present in most low grade serous carcinomas but not in serous borderline tumors (Kuo et al. 2009). In a more comprehensive analysis conducted by Zhang et al. (2008a), it was estimated that genomic deletion accounts for ~15% miRNA down-regulation, supporting the notion that genomic alteration is an important mechanism of miRNA deregulation in ovarian cancer.

14.5.2 Epigenetic Alteration of MiRNA in Ovarian Cancer

DNA methylation and histone modification have significant impacts on the expression of protein-coding genes and are often associated with molecular, clinical and pathological features of ovarian carcinomas (Asadollahi et al. 2010). Recent studies on miRNA expression revealed that these epigenetic modifications also contribute to alterations in miRNA expression. Iorio et al. treated OVCAR3 cells with 5-aza-2-deoxycytidine, a demethylating agent, and found that the expression of *miR-21, miR-203* and *miR-205* were significantly increased. As *miR-21, miR-203* and *miR-205* are often up-regulated in cancer, this observation suggests that DNA hypomethylation can be a mechanism responsible for their over-expression (Iorio et al. 2007). In the study published by Zhang et al. (2008a), 44 miRNAs were found to be down-regulated in late stage ovarian cancer in comparison to early stage tumors and 16 of the 44 miRNAs can be restored by demethylation or histone deacetylase (HDAC) inhibition, suggesting that epigenetic silencing accounts for at least ~36% of miRNA down-regulation (Zhang et al. 2008a). In the same study, corresponding mRNA expression profiles from the tumors were also analyzed by affymetrix microarray, and the result clearly demonstrated that miRNA down-regulation contributes to a global deregulation of mRNAs in ovarian cancer (Zhang et al. 2008a).

14.5.3 Alteration of MiRNA Biogenesis in Ovarian Cancer

Alterations of proteins involved in miRNA biogenesis pathways in ovarian cancer have also been under investigation. While no difference in mRNA and protein expressions of *Dicer* and *Drosha* are observed between early and late stage ovarian cancer, or between EOC cell lines and immortalized ovarian surface epithelial cells by Zhang et al. (2008a), alteration of *Dicer* and *Drosha* in ovarian cancer has been reported by other groups. For example, Faggad et al. (2010) observed down-regulation of *Dicer* in serous subtype of ovarian carcinomas and found that decreased *Dicer* expression is associated with nodule status, disease stage and survival rate. Pampalakis et al. (2010) analyzed *Dicer* expression in 34 samples (10 normal ovaries, 8 benign, 16 ovarian cancer) by quantitative reverse transcription-polymerase chain reaction (qRT-PCR) and found down-regulation of *Dicer* in tumor tissues when compared to normal ovary. Merritt et al. (2008) studied *Dicer* and *Drosha* expression in 111 patients by qRT-PCR and found that decreased *Dicer* and *Drosha* expression were present in 60 and 51% of cancer specimens, respectively. Moreover, they observed an association between low *Dicer* expression with advanced stage, low *Drosha* expression with suboptimal surgical cyto-reduction and a correlation between high *Dicer* and *Drosha* expressions with increased survival.

Flavin et al. (2008) quantified Dicer protein expression in a tissue array containing 66 serous carcinoma specimen and *Dicer* mRNA expression in an independent set of 50 patients with serous carcinomas. Compared to normal ovary, *Dicer* is up-regulated in a significant proportion of serous carcinomas and high *Dicer* expression significantly correlated with absence of lymph node metastases. In addition, they discovered that *eIF6*, a component in silencing machinery, correlated with lymph node metastasis and that a low *eIF6* expression is associated with decreased disease free survival in ovarian cancer patients. However, the same group did not observe any correlation between *Dicer* expression and disease free or overall survival.

Recently, multiple lines of evidence suggest that p53 exerts its tumor suppressive roles at least partially by regulating miRNA biogenesis. For example, p53 directly regulate *miR-34* transcription. Moreover, it was reported that p53 can directly interact with microprocessor complex and enhance miRNA maturation in colon and ovarian cancer cell lines. This interaction requires the DNA binding domain of p53 whose mutations were thought to result in decrease in miRNA maturation by weakening the interaction between pri-miRNA and Drosha or interfere with the formation of microprocessing complex (Suzuki et al. 2009). These findings can be used to explain the dominant miRNA down-regulation observed in late stage EOC, since tumors at late stages often have mutant or over-expression of *p53* (Mezzanzanica et al. 2010). In aggregate, modulation of miRNA biogenesis is an important mechanism for miRNA deregulation in ovarian cancer.

14.5.4 Subtype Specific MiRNA Deregulation

Ovarian cancer is a heterogeneous disease whose pathological origin is still under debate. Currently, it is widely accepted the majority of ovarian cancers arise from the surface epithelium. Epithelial ovarian cancer can be categorized into four subtypes including serous, mucinous, endometrioid, and clear cell carcinomas. Different subtypes have distinct pathological characteristics as well as specific miRNA expression patterns (van Jaarsveld et al. 2010). Iorio et al. (2007) compared miRNA expression profiles in 31 serous, 8 endometrioid and 4 clear cell ovarian cancer samples. Common miRNAs alterations in all three subtypes include up-regulation of *miR-200a* and *miR-200c* and down-regulation of *miR-125b1*, *miR-199a* and *miR-140*. In addition, *miR-200b* and *miR-141* are up-regulated in endometrioid and serous subtypes but not in clear cell samples. *miR-21, miR-203*, and *miR-205* up-regulations are specific to the endometrioid subtype. *MiR-145* is down-regulated in serous and clear cell carcinomas, *miR-222* is down-regulated in endometrioid and clear cell subtypes. More recently, using next generation sequencing, Wyman et al. (2009) also found subtype-specific miRNAs in ovarian cancer: 8 miRNAs (*miR-9, miR-196a, miR-196b, miR-96, miR-182, miR-205, miR-183*, and *miR-375*) were specifically up-regulated in endometrioid subtype, 4 miRNAs (*miR-7, miR-34c-5p, miR-146b-5p*, and *miR-449*) were up-regulated in serous cancer and 3 miRNAs (*miR-30a, miR-30a**, and *miR-486-5p*) were up-regulated in clear cell cancer. These data suggest that the miRNA signature can serve as a new biomarker for specific ovarian cancer subtypes.

14.5.5 Factors Contributing to the Inter-study Conflicts

Significant discrepancies exist between the results from different studies. Three major factors contribute to these inconsistencies. First, different groups used different platforms, including qRT-PCR, microarray and next generation sequencing, to profile miRNA expression and the analytical tools used by different studies are not the same (van Jaarsveld et al. 2010). Second, the control tissues used between different studies vary. While some used immortalized human ovarian surface cell cultures (Dahiya et al. 2008; Wyman et al. 2009; Yang et al. 2008a; Zhang et al. 2006, 2008a), others used normal ovaries (Iorio et al. 2007; Nam et al. 2008). Both controls have their drawbacks. Lacking a standardized protocol for generating immortalized ovarian surface epithelium may introduce unwanted differences in their miRNA expression. However, as the epithelium represents less than 1% of the cellular content of the ovary, the normal ovarian tissue does not serve as the most appropriate control either. Recently, fallopian tube tissues have also been used as controls in miRNA expression profiling studies (Lee et al. 2009), adding another variable when comparing results from different studies. Standardized controls will be much appreciated in future studies on miRNA expression profiles in ovarian cancer. Lastly, as EOC is a heterogeneous disease, it is very difficult to get miRNAs exclusively from EOC cells but not adjacent normal and stromal cells, which would cause variations in the expression of miRNAs. To minimize this variation, laser-capture microdissection to enrich tumor cells from clinical samples and separating the inputs from different cellular compartments *via* a bioinformatics approach will be of great help.

Despite all the inconsistencies, down-regulation appeared to be dominant in ovarian cancer, especially in late-stage and high-grade samples. The fact that some miRNA alterations were independently identified by different studies strongly suggests their involvement in ovarian tumorigenesis and progression.

14.6 MiRNA Function in Epithelial Ovarian Cancer

As mentioned above, some miRNA alterations in ovarian cancer were consensus among different studies. It is suggested that these miRNAs may be functionally involved in ovarian cancer tumorigenesis. In the following section, we will summarize these miRNAs and their involvement in ovarian cancer development.

14.6.1 Tumor Suppressive MiRNA in Ovarian Cancer

The best studied tumor-suppressive miRNA to date is *let-7*. Reduced *let-7* expression has been documented in various types of cancer (Bussing et al. 2008; Roush and Slack 2008; Takamizawa et al. 2004; Yang et al. 2008b). *Let-7* functions to suppress cell growth and induce apoptosis in vitro and in vivo (Esquela-Kerscher et al. 2008; Johnson et al. 2005; Kumar et al. 2008). Validated targets of *let-7* include *Ras,*

HMGA2, CDK6, IMP-1, CDC25A, and *Kallikrein 10*, most of which are deregulated in EOC (Cho 2010a, 2010b; Li et al. 2010; White et al. 2010). Down-regulation of *let-7* in EOC was documented by almost all the expression profile studies published (Lee et al. 2009; Nam et al. 2008; Wyman et al. 2009; Yang et al. 2008a; Zhang et al. 2006, 2008b). It is speculated that reduction of *let-7* initiates the de-differentiation process which often takes place at the early phase of ovarian carcinogenesis. This is supported by the observation that primary EOC tissues and metastatic tumors both have high expression of a *let-7* target gene, *HMGA2* (Mahajan et al. 2010).

MiR-15a and *miR-16-1* were first identified as tumor suppressive miRNAs by Calin et al. (2002) in CLL. Recent studies by Bhattacharya et al. (2009) found that *miR-15a* and *miR-16-1* were under-expressed in ovarian cell lines and in primary ovarian tissues. They also demonstrated that oncogene Bmi-1 is a direct target of *miR-15a* and *miR-16-1*. In ovarian cancer cell lines, *miR-15a* and *miR-16-1* expression reduced Bmi-1 protein expression as well as cell proliferation and clonal growth. In patients, an inverse correlation of Bmi-1 protein levels and *miR-15a/16-1* expression was also observed. These findings suggest that *miR-15a* and *miR-16-1* function as tumor suppressors in ovarian cancer and indicate the possibility of restoring *miR-15a* and *miR-16-1* expression in ovarian cancer as an anti-tumor treatment (Bhattacharya et al. 2009).

The *miR-34* family is composed of three miRNAs (*miR-34a, miR-34b*, and *miR-34c*), which are induced by p53 and are functionally involved in p53 mediated cellular response to DNA damage or oncogenic stress (He et al. 2007). *MiR-34a* expression is decreased in more than 90% of EOC, regardless of their p53 status, while *miR-34 b/c* is also down-regulated in 72% of tumors with *p53* mutations. Furthermore, expression of *miR-34b*/c* is significantly lower in advanced stage tumors (Corney et al. 2010). In addition to p53 mediated regulation, expression of *miR-34* can result from genomic deletion, since the *miR-34a* locus, *1p36* is frequently deleted in low grade serous ovarian carcinomas (Kuo et al. 2009). Moreover, promoter methylation was also observed in *miR-34* genes in ovarian cancer (Corney et al. 2010). Known targets of *miR-34* include *Bcl-2, Cdk4, CCDN1* and *Met* (Corney et al. 2007, 2010). In *p53* mutant EOC cells, forced expression of *miR-34* reduces cell proliferation, motility, and invasion. As *Met* is a *miR-34* target, *miR-34* down-regulation may serve as a mechanism for *Met* over-expression, which is often present in advanced stage EOC (Corney et al. 2010).

MiR-9 may also serve as a tumor suppressor in ovarian cancer. Supporting evidence include decreased expression of *miR-9* in ovarian cancer relative to normal ovary and suppression of cell growth by *miR-9* over-expression in vitro (Guo et al. 2009; Laios et al. 2008). *NF-κB1* is an important target of *miR-9* in ovarian cancer cells, therefore it is hypothesized that regulating *NF-κB1* by *miR-9* is an important mechanism for *miR-9* to inhibit ovarian cancer proliferation (Guo et al. 2009). Intriguingly, as NF-κB also plays important roles in the inflammatory response and inflammation accompanying ovulation can induce genetic lesions in OSE cells. Therefore, it has been suggested that reduced *miR-9* expression may result in up-regulated NF-κB expression in the stromal compartment, therefore promote the inflammatory response, which in turn leads to ovarian cancer initiation. It is an interesting theory since *miR-199a* which targets *IKKβ* is also down-regulated in ovarian cancer and ovarian cancer cells with low *miR-199a* expression have the capacity to constitutively secrete pro-inflammatory/pro-tumor cytokines and therefore promote tumor progression and chemoresistance (Chen et al. 2008). It is possible that *miR-9* function in both tumor and stromal cells and it cooperates with *miR-199* to promote a pro-inflammatory and pro-tumorigenic environment to support ovarian cancer tumorigenesis.

14.6.2 Oncogenic MiRNA in Ovarian Cancer

Evidence demonstrating miRNAs as oncogenes in ovarian cancer is limited. *MiR-21* over-expression has been reported in various types of cancer. Supporting its oncogenic activity, known tumor-suppressors that include *TPM1, PTEN, SPRY2, PDCD,* and *maspin* were found to be targeted by *miR-21*. In addition, an indirect regulation of *Bcl-2* by *miR-21* has also been shown in breast cancer (Selcuklu et al. 2009). However, its role in ovarian cancer is controversial; some groups report that *miR-21* is up-regulated (Iorio et al. 2007; Nam et al. 2008) while another group found it down-regulated (Dahiya et al. 2008) in ovarian cancer. The disagreement may be due to the aformended technical differences. Further experiments are needed to evaluate the role of *miR-21* in ovarian cancer.

14.6.3 Epithelial-mesenchymal Transition (EMT) Associated MiRNA in Ovarian Cancer

EMT, which describes the transition from epithelial phenotype to the mesenchymal phenotype, plays an important role during embryonic development, wound-healing as well as cancer progression Tumorigenesis is often accompanied by a dedifferentiation process, in which more differentiated epithelial cells progress into less differentiated, mesenchymal cells (Park et al. 2008). Consistent with this process, the *let-7* family, whose major function include promoting cell differentiation, is down-regulated during cancer progression in various human cancers, including EOC (Mezzanzanica et al. 2010). However, it is a little complicated in the case of *miR-200* family, another EMT regulator. *MiR-200* miRNAs are composed of *miR-200a, b, c, miR-141*, and *miR-429*. It has been demonstrated that this miRNA family is a general marker for E-cadherin positive and Vimentin-negative cells (Park et al. 2008). *MiR-200* miRNAs target *ZEB1/2* transcription factors, which can inhibit the transcription of *miR-200* and suppress E-cadherin expression, therefore promoting EMT. *MiR-200* family members are expressed at low level in normal ovarian surface epithelial cells, whereas the expression of ZEB1 and ZEB2 are high (Bendoraite et al. 2010). In EOC, while multiple studies reported an up-regulation of *miR-200* miRNA family (Iorio et al. 2007; Mezzanzanica et al. 2010; Nam et al. 2008; Wyman et al. 2009; Yang et al. 2008a), some studies observed down-regulations (Dahiya et al. 2008; Zhang et al. 2008a). It may reflect the technical differences in different studies, such as the choice of controls and the procedure for generating normal OES cell lines. For example, the increased expression of the *miR-200* family miRNAs in ovarian tumors can be the result of comparing tumors that contain massive amounts of epithelial-like tumor cells (high *miR-200*, low ZEB, high E-cadherin) to the normal ovary whose major cellular components are stromal cells (low *miR-200*, high ZEB, low E-cadherin). Nevertheless, the increased *miR-200* expression supports a mesothelial to epithelial model for ovarian cancer development (Li et al. 2010). However, it is possible that mesenchymal EOC cells invade into distal sites and then re-epithelialize, therefore the stages of tumor progression can affect *miR-200* expression alterations (Mezzanzanica et al. 2010).

14.7 MiRNAs as Biomarkers and Prognostic Tools in Ovarian Cancer

The lack of symptoms at the early stages of ovarian cancer represents a major obstacle to accurate diagnosis and effective treatment of the disease. Many patients are not identified until late stages when the disease is characterized by aggressive tumors that are therapy resistant and highly metastatic. Unfortunately, neither mRNA nor protein profiling identified sensitive and specific biomarkers for ovarian cancer early detection. With the discovery of miRNA and the increasing amount of evidence demonstrating the importance of this class of molecules as modulators of ovarian cancer development, it is reasonable that profiling miRNAs may serve as a better diagnostic tool.

14.7.1 Serum MiRNA as Biomarker in Ovarian Cancer

Serum biomarkers can be very useful in screening for early-stage ovarian cancer because of its non-invasive nature. Recent studies showed that tumor-derived miRNAs can be detected in serum and are relatively stable (Gilad et al. 2008). Circulating miRNAs have been evaluated as biomarkers in B cell lymphoma, lung and colorectal cancers (Lawrie et al. 2008; Ng et al. 2009; Rabinowits et al. 2009; Rosell et al. 2009). In ovarian cancer, eight miRNAs were found to be differentially expressed between normal and patient sera, suggesting the possibility of using serum miRNA profiling to screen for ovarian cancer patients at early stages (Mitchell et al. 2008). In addition, a recent study compared miRNA expression in tumor cell secreted exosomes in blood to that in primary tumors and found a high correlation between the two groups. Since exosomal miRNAs were not detected in normal controls and have a distinct expression pattern from patients with benign disease, these findings strongly suggest the feasibility of using sera exosomal miRNA profiling as biomarkers for diagnosis (Taylor and Gercel-Taylor 2008).

14.7.2 MiRNA as Tools to Classify Ovarian Cancer Subtypes

Cancer is a heterogeneous disease and different subtypes often have different pathological features and differential therapeutic responses. Therefore, accurate tumor classification is important to help clinicians choose the optimal treatments for individual patients. microRNAs can classify breast tumors with high accuracy (Iorio et al. 2005). In ovarian cancer, there are four different subtypes with distinct pathological features. Although some miRNAs alterations are common across subtypes, specific miRNA expression alterations associated with different subtypes have been reported by several groups (Iorio et al. 2007; Wyman et al. 2009). These miRNAs may be used as markers for accurate classification of ovarian cancer subtypes.

14.7.3 MiRNA as Prognostic Marker

MiRNA profiles have been found to correlate with disease progression and patient survival in various types of tumors. In ovarian cancer, expression of several miRNAs has been correlated with disease progression (Table 14.1). These results suggest that miRNAs can be used to predict disease outcome in ovarian cancer. However, inconsistent results from different group represent a major challenge in defining the prognostic value for a given miRNA. Future analysis with a standardized control tissue, platform and analytical tools will help reduce the inconsistencies and promote the identification of reliable biomarkers for ovarian cancer prognosis. In addition, non-linear regression-based algorithms will help to generate more robust miRNA signatures to serve as a prognostic indicator for ovarian cancer.

14.7.4 MiRNA as Predictor for Therapy Response

Platinum and taxane based chemotherapy are the first line treatments for ovarian cancer patients. Even though the initial response rate can be as high as 90%, chemoresistance is almost inevitable and poses as a major threat to patient survival. Biomarkers that can predict the response of a patient to a specific therapy would greatly help to optimize treatment. Several miRNAs have been found to correlate with chemoresistance, either in patients or in ovarian cancer cell lines (Eitan et al. 2009; Sorrentino et al. 2008; van Jaarsveld et al. 2010; Yang et al. 2008a, 2008b). In the cases of *let-7i* and *miR-214*, down-regulation of these miRNAs reduced sensitivity to cisplatin treatment, suggesting that they not only serve as a biomarker but also function as a modulator for chemosensitivity (Yang et al. 2008a, 2008b). Accordingly, some miRNAs associated with chemoresistance are differentially expressed in late-stage tumors in comparison to early-stage tumors. Since late stage cancer is often refractory to chemotherapy, these miRNAs may be used as indicators of therapy response as well as potential therapeutic targets.

14.8 MiRNA as Therapeutic Agents in Ovarian Cancer

Given all the evidence implicating miRNAs in tumorigenesis and cancer progression, it is not surprising that investigations using miRNAs as anti-tumor agents or novel therapeutic targets are underway. Strategies to inhibit oncogenic miRNA activities include anti-miRNA oligonucleotides, microRNA sponges, miRNA masking, and small molecule inhibitors (Li et al. 2009). For tumor

Table 14.1 Correlation of microRNA expression and disease progression in ovarian cancer

MicroRNAs	Clinical correlation in ovarian cancer	References	Validated targets
let-7a	Lower expression in ovarian cancer vs. normal control	Iorio et al. (2007); Yang et al. (2008b)	*HMGA2, KRAS, NRAS, E2F2, KLK10*
	Let-7a hypermethylation associated with favorable survival	Lu et al. (2007)	
let-7b	Lower expression correlate with poor prognosis	Nam et al. (2008)	
	Lower expression in ovarian cancer vs. normal control	Nam et al. (2008); Yang et al. (2008b)	
	Lower expression in high vs. low grade tumors	Lee et al. (2009)	
let-7c	Lower expression in ovarian cancer vs. normal control	Iorio et al. (2007); Yang et al. (2008b)	
let-7d	Lower expression in ovarian cancer vs. normal control	Dahiya et al. (2008); Iorio et al. (2007); Zhang et al. (2008a)	
let-7e	Higher expression in paclitaxel resistant cells and paclitaxel/cyclosporine sensitive cells	Sorrentino et al. (2008)	
	Higher expression in doxorubicin resistance cells	Boren et al. (2009)	
let-7g	Higher expression in platinum resistant tumors	Eitan et al. (2009)	
let-7i	Lower expression in chemoresistant tumors and cells	Yang et al. (2008a)	
	Higher expression in recurrent vs. primary tumors	Laios et al. (2008)	
miR-1	Higher expression in late vs. early stage tumors	Zhang et al. (2008a)	*HAND2, TMSB4, HDAC4, GJA1, KCNJ2*
miR-1b-1	Higher expression in chemoresistant tumor and cells	Yang et al. (2008a)	–
miR-9	Lower expression in ovarian cancer vs normal control	Guo et al. (2009)	*NF-κB, TrkC*
	Lower expression in recurrent vs. primary tumors	Laios et al. (2008)	
miR-10a	Lower expression in high vs. low grade tumors	Lee et al. (2009)	*HOXA1*
miR-10b	Higher expression in ovarian cancer vs. normal control	Dahiya et al. (2008); Wyman et al. (2009)	*RhoC*
	Lower expression in ovarian cancer vs. normal control	Nam et al. (2008)	
	Higher expression in recurrent vs. primary tumors	Laios et al. (2008)	
miR-16	Higher expression in ovarian cancer vs. normal control	Nam et al. (2008); Wyman et al. (2009)	*Bmi-1, Bcl2, COX2, CGI-38, BRCA1, AKT3, MAP3KIP3, RPS6*
miR-18a	Higher expression correlate with poor prognosis	Nam et al. (2008)	*ER, ESR1*

Table 14.1 (continued)

MicroRNAs	Clinical correlation in ovarian cancer	References	Validated targets
miR-20a	Higher expression in ovarian cancer vs. normal control	Nam et al. (2008); Wyman et al. (2009)	*E2F1, RUNX1, E2F2, E2F3, STAT3, CDKN1A, p21, CCND1, HIF1A*
miR-20b	Lower expression in paclitaxel-resistant cells	Sorrentino et al. (2008)	–
miR-21	Higher expression in serous ovarian cancer vs. normal control	Iorio et al. (2007); Nam et al. (2008)	*PTEN, TPM1, EGFR, HER2/Neu,PDCD4, Bcl-2, BTG2*
	Higher expression in serum from patients vs. control serum	Resnick et al. (2009)	
	Higher expression in paclitaxel-resistant cells	Sorrentino et al. (2008)	
	Higher expression in doxorubicin resistance cells	Boren et al. (2009)	
	Lower expression in ovarian cancer vs normal control	Dahiya et al. (2008)	
miR-23a	Higher expression in platinum resistant tumors and correlate with poor survival	Eitan et al. (2009)	*CXCL12, FLJ13158, c-Myc*
	Higher expression in recurrent vs. primary tumors	Laios et al. (2008)	
miR-23b	Higher expression in late vs. early stage tumors	Eitan et al. (2009)	*Notch-1*
	Higher expression in platinum resistant tumors	Boren et al. (2009)	
miR-24-2	Higher expression correlates with poor survival	Eitan et al. (2009)	*Notch-1, MAPK14*
miR-26a	Higher expression in ovarian cancer vs normal control	Wyman et al. (2009); Zhang et al. (2008a)	*PLAG1, Cyclin D2, Cyclin E2*
	Lower expression in ovarian cancer vs. normal control	Nam et al. (2008)	
miR-26b	Higher expression in ovarian cancer vs. normal control	Wyman et al. (2009); Zhang et al. (2008a)	–
miR-27a	Higher expression in platinum resistant tumors and correlate with poor survival	Eitan et al. (2009)	*FOXO1*
miR-29a	Higher expression in serum from patients vs. control serum	Resnick et al. (2009)	*DNMT3A, PIK3R1, CDC42, COL1A1, COL1A2, COL3A1, COL4A1, LAMC1, TCL1, B7-H3, TTP*

Table 14.1 (continued)

MicroRNAs	Clinical correlation in ovarian cancer	References	Validated targets
	Higher expression in ovarian cancer vs. normal control	Dahiya et al. (2008); Yang et al. (2008b)	
	Lower expression in ovarian cancer vs. normal control	Iorio et al. (2007); Nam et al. (2008)	
	Lower expression in serum from patients vs. control serum	Resnick et al. (2009)	
	Lower expression in high vs. low grade tumors	Lee et al. (2009)	
miR-29b	Higher expression correlate with poor survival	Flavin et al. (2009)	
miR-29c	Higher expression in ovarian cancer vs. normal control	Dahiya et al. (2008); Wyman et al. (2009)	
	Lower expression in ovarian cancer vs. normal control	Iorio et al. (2007)	
	Lower expression in high vs. low grade tumor	Lee et al. (2009)	
	Lower expression in paclitaxel resistant tumors	Boren et al. (2009)	
*miR-30a**	Higher expression in recurrent vs. primary tumors	Laios et al. (2008)	*CTGF, KRT7, Ubc9*
miR-30c	Higher expression in platinum resistant tumors	Eitan et al. (2009)	
	Lower expression in chemoresistant cells	Sorrentino et al. (2008)	
miR-30e	Higher expression in ovarian cancer vs. normal control	Dahiya et al. (2008); Wyman et al. (2009)	
miR-31	Lower expression in recurrent vs. primary tumors	Laios et al. (2008)	*RhoA*
miR-34a	Lower expression in late vs. early stage tumors	Zhang et al. (2008a)	*E2F3, DLL, Notch-1, c-Met, Cyclin E, p53, CREB*
miR-34b/c	Lower expression in high vs. low grade tumors	Lee et al. (2009)	
	Lower expression in high vs. low grade tumors	Lee et al. (2009); Zhang et al. (2008a),	
miR-92	Higher expression in serum from patients vs. control serum	Resnick et al. (2009)	*P63, pVHL*
	Lower expression in high vs. low grade tumor	Lee et al. (2009)	
miR-93	Higher expression correlate with poor prognosis	Nam et al. (2008)	*P21*
	Higher expression in serum from patients vs. control serum	Resnick et al. (2009)	
	Higher expression in recurrent vs. primary tumors	Laios et al. (2008)	
miR-99a	Higher expression in ovarian cancer vs. normal control	Dahiya et al. (2008); Wyman et al. (2009)	–
	Lower expression in ovarian cancer vs. normal control	Iorio et al. (2007); Nam et al. (2008); Zhang et al. (2008a)	

Table 14.1 (continued)

MicroRNAs	Clinical correlation in ovarian cancer	References	Validated targets
	Lower expression in recurrent vs. primary tumors	Laios et al. (2008)	
miR-99b	Lower expression in serum from patients vs. control serum	Resnick et al. (2009)	–
	Higher expression in chemoresistant tumors	Boren et al. (2009)	
miR-100	Lower expression in clear cell ovarian cancer cell lines vs. normal ovarian surface epithelial culture	Nagaraja et al. (2010)	*Plk1*
	Lower expression in ovarian cancer vs. normal control	Iorio et al. (2007); Nam et al. (2008); Wyman et al. (2009); Yang et al. (2008b); Zhang et al. (2008a)	
	Lower expression in recurrent vs. primary tumors	Laios et al. (2008)	
miR-106a	Lower expression in platinum resistant tumors and cell lines	Boren et al. (2009); Sorrentino et al. (2008)	*RB1, p21, p73*
miR-124a	Lower expression in ovarian cancer vs. normal control	Iorio et al. (2007); Zhang et al. (2008a)	–
miR-125a-3/5p	Higher expression in late vs. early stage tumors	Eitan et al. (2009)	–
miR-125a	Lower expression in ovarian tumors vs. normal control	Iorio et al. (2007); Nam et al. (2008)	*ERBB2, ERBB3, Lin28, TrkC, CYP24*
miR-125b	Lower expression in ovarian cancer vs. normal control	Laios et al. (2008)	
	Lower expression in ovarian cancer vs. normal control	Iorio et al. (2007); Nam et al. (2008); Yang et al. (2008b),	
miR-126	Lower expression in serum from patients vs. normal serum	Resnick et al. (2009)	–
	Higher expression in chemosensitive cells	Boren et al. (2009); Sorrentino et al. (2008)	
	Higher expression in platinum resistant tumors	Yang et al. (2008a)	
miR-127	Lower expression in ovarian cancer vs. normal control	Iorio et al. (2007); Zhang et al. (2008a)	*BCL6*
	Higher expression in serum from patients vs. normal serum	Resnick et al. (2009)	
miR-129-2	Hypermethylated *miR-129-2* correlate with poor survival	Huang et al. (2009)	*GALNT1, SOX4*

Table 14.1 (continued)

MicroRNAs	Clinical correlation in ovarian cancer	References	Validated targets
miR-130a	Higher expression in late vs. early stage tumors	Eitan et al. (2009)	*m-CSF*
	Lower expression in chemoresistant cells	Sorrentino et al. (2008)	
miR-134	Lower expression in ovarian cancer vs. normal control	Dahiya et al. (2008); Zhang et al. (2008a)	*Limk-1*
miR-137	Lower expression in recurrent vs. primary tumors	Laios et al. (2008)	–
miR-140	Lower expression in ovarian cancer vs. normal control	Iorio et al. (2007); Zhang et al. (2008a)	–
miR-141	Higher expression in ovarian cancer vs. normal control	Iorio et al. (2007); Nam et al. (2008); Wyman et al. (2009)	*Clock, TGFb, SIP-1*
	Higher expression correlate with poor prognosis	Nam et al. (2008)	
	Lower expression in ovarian cancer vs. normal control	Dahiya et al. (2008)	
miR-142	Higher expression in ovarian cancer vs. normal control	Dahiya et al. (2008); Wyman et al. (2009)	–
miR-143	Lower expression in ovarian cancer vs. normal control	Iorio et al. (2007); Nam et al. (2008)	*ERK5, KRAS, DNMT3A, FNDC3B*
	Higher expression in ovarian cancer vs. normal control	Wyman et al. (2009)	
miR-145	Lower expression in ovarian cancer vs. normal control	Iorio et al. (2007); Nam et al. (2008); Zhang et al. (2008a)	*FLJ21308, EGFR, IGF-1R*
miR-146a	G to C SNP correlate with early diagnosis	Shen et al. (2008)	*BRMS1*
miR-146b-5p	Higher expression in late vs. early stage tumors	Eitan et al. (2009)	
miR-152	Lower expression in ovarian cancer vs. normal control	Wyman et al. (2009); Zhang et al. (2008a)	–
miR-154	Lower expression in late vs. early stage tumors	Zhang et al. (2008a)	–
miR-155	Lower expression in ovarian cancer vs. normal control	Dahiya et al. (2008); Zhang et al. (2008a)	*AGTR-1*
	Higher expression in patients serum vs. normal serum	Resnick et al. (2009)	
miR-181- a/b	Higher expression in chemosensitive cells	Boren et al. (2009)	*Tcl-1, CDX2, GATA6, NLK*
miR-182	Higher expression in ovarian cancer vs. normal control	Iorio et al. (2007); Zhang et al. (2008a),	*FOXO1, FOXO3*
miR-184	Lower expression in late vs. early stage tumors	Zhang et al. (2008a)	–
miR- 185	Lower expression in platinum resistant cells	Boren et al. (2009)	–
miR-193a	Higher expression in late vs. early stage tumors	Eitan et al. (2009)	–
miR-195	Lower expression in ovarian cancer vs. normal control	Dahiya et al. (2008); Iorio et al. (2007)	*Cyclin D1, E2F3*
	Higher expression in ovarian cancer vs. normal control	Wyman et al. (2009)	
miR-198	Higher expression in platinum resistant tumors	Yang et al. (2008a)	–

Table 14.1 (continued)

MicroRNAs	Clinical correlation in ovarian cancer	References	Validated targets
miR-199a	Lower expression correlate with poor prognosis	Nam et al. (2008)	*KRT7*
	Lower expression in ovarian cancer vs. normal control	Iorio et al. (2007)	
miR-200a	Higher expression in ovarian cancer vs. normal control	Iorio et al. (2007); Nam et al. (2008); Wyman et al. (2009); Yang et al. (2008b)	*ZeB1* and *2FN1, NTRK2* and *QKI, TUBB3*
	Lower expression in ovarian cancer vs. normal control	Dahiya et al. (2008)	
	Lower expression in late vs. early stage tumors	Eitan et al. (2009)	
	Higher expression correlate with poor prognosis	Nam et al. (2008)	
miR-200b	Higher expression in ovarian cancer vs. normal control	Iorio et al. (2007); Nam et al. (2008); Wyman et al. (2009); Yang et al. (2008b)	
	Lower expression in late vs. early stage tumors	Eitan et al. (2009)	
	Higher expression correlate with poor prognosis	Nam et al. (2008)	
miR-200c	Higher expression in ovarian cancer vs. normal control	Iorio et al. (2007); Nam et al. (2008); Wyman et al. (2009); Yang et al. (2008b)	
miR-203	Higher expression in ovarian cancer vs. normal control	Iorio et al. (2007); Wyman et al. (2009)	–
miR-204	Lower expression in ovarian cancer vs. normal control	Iorio et al. (2007)	*Bcl-2, Mcl-1*
miR-206	Higher in recurrent vs. primary tumors	Laios et al. (2008)	*Fstl1, Utrn, GHA1, ESR1, c-Met*
miR-210	High frequency deletion in ovarian cancer patients	Giannakakis et al. (2008)	*EFNA3, E2F3, NPTX1, RAD52, MNT*
miR 213	Higher expression in chemoresistant cells	Boren et al. (2009)	–
miR-214	Lower expression in ovarian cancer vs. normal control	Iorio et al. (2007); Nam et al. (2008)	*PTEN, TPM1*
	Higher expression in ovarian cancer vs. normal control	Yang et al. (2008b)	
	Higher expression in platinum resistant tumors	Yang et al. (2008a)	
miR-216	Higher expression in platinum resistant tumors	Yang et al. (2008a)	–
miR-221	Higher expression in ovarian cancer vs. normal control	Dahiya et al. (2008)	*KIT, p27, p57*
	Low *miR-221/miR-222* ratio correlates with worst survival	Wurz et al. (2010)	
miR-223	Higher expression in recurrent vs. primary tumors	Laios et al. (2008)	*NFLA*
miR-321	Lower expression in platinum resistant tumors	Yang et al. (2008a)	–

Table 14.1 (continued)

MicroRNAs	Clinical correlation in ovarian cancer	References	Validated targets
miR-331	Lower expression in platinum resistant cells	Boren et al. (2009)	*HER2/Neu*
miR-335	Higher expression in chemosensitive cells	Sorrentino et al. (2008)	–
miR-337	Lower expression in high vs. low grade tumors	Zhang et al. (2008a)	–
miR-338	Higher expression in recurrent vs. primary tumors	Laios et al. (2008)	–
miR-339	Lower expression in gemcitabine resistant cells	Boren et al. (2009)	–
miR-340	Higher expression in platinum resistant cells	Boren et al. (2009)	–
miR-348	Lower expression in high vs. low grade tumors	Zhang et al. (2008a)	–
	Higher expression in recurrent vs. primary tumors	Laios et al. (2008)	
miR-370	Higher expression in platinum resistant tumors	Yang et al. (2008a)	–
miR-371	Higher expression in docetaxel resistant cells	Boren et al. (2009)	–
miR-373	Higher expression in ovarian cancer vs. normal control	Iorio et al. (2007)	–
miR-376	Lower expression in high vs. low grade tumors	Zhang et al. (2008a)	–
miR-376a	Lower expression in late vs. early stage tumors	Zhang et al. (2008a)	–
miR-378	Lower expression in platinum resistant cells	Eitan et al. (2009)	–
miR-381	Higher expression in platinum resistant cells	Boren et al. (2009)	–
miR-422b	Lower expression associate with poor survival	Lee et al. (2009)	–
miR-423-3p	Higher expression in late vs. early stage tumors	Eitan et al. (2009)	–
miR-424	Higher expression in ovarian cancer vs. normal control	Dahiya et al. (2008); Zhang et al. (2008a)	–
	Higher expression in late vs. early stage tumors	Zhang et al. (2008a)	
miR-429	Higher expression correlate with poor prognosis	Nam et al. (2008)	–
	Higher expression in ovarian cancer vs. normal control	Wyman et al. (2009)	
miR-449	Lower expression in late vs. early stage tumors	Eitan et al. (2009)	*HDAC1*
	Lower expression correlate with poor prognosis	Eitan et al. (2009)	
miR-451	Higher expression in late vs. early stage tumors	Eitan et al. (2009)	*MIF*
miR-491-5p	Higher expression in late vs. early stage tumors	Eitan et al. (2009)	–
miR-495	Lower expression in late vs. early stage tumors	Eitan et al. (2009)	–
miR-503	Lower expression in high vs. low grade tumors	Zhang et al. (2008a)	–
miR-508	Lower expression in high vs. low grade tumors	Lee et al. (2009); Zhang et al. (2008a)	–

Table 14.1 (continued)

MicroRNAs	Clinical correlation in ovarian cancer	References	Validated targets
miR-509	Lower expression in high vs. low grade tumors	Lee et al. (2009); Zhang et al. (2008a)	–
	Lower expression in late vs. early stage tumors	Zhang et al. (2008a)	
	Lower expression in platinum resistant tumors	Yang et al. (2008a)	
miR-509-3/5p	Lower expression in late vs. early stage tumors	Eitan et al. (2009)	–
miR-513	Lower expression in late vs. early stage tumors	Zhang et al. (2008a)	–
	Lower in high vs. low grade tumors		
miR-514	Lower expression in late vs. early stage tumors	Lee et al. (2009)	–
	Lower in high vs. low grade tumors	Zhang et al. (2008a)	
	Higher expression in chemoresistant cells	Boren et al. (2009)	
miR-515-5p	Higher expression in docetaxel resistant cells	Boren et al. (2009)	–
miR-519d	Lower expression in late vs. early stage tumors	Zhang et al. (2008a)	–
miR-519e	Higher expression in platinum resistant tumors	Yang et al. (2008a)	–
miR-520e	Higher expression in platinum resistant tumors	Yang et al. (2008a)	–
miR-520f	Higher expression in platinum resistant cells	Boren et al. (2009)	–
miR-521	Higher expression in platinum resistant tumors	Yang et al. (2008a)	–
miR-574-5p	Lower expression in late vs. early stage tumors	Eitan et al. (2009)	–
miR-625	Lower expression in platinum resistant cells	Nam et al. (2008)	–

Validated targets for specific microRNAs are from Penn center for bioinformatics (Tarbase – http://www.diana.pcbi.upenn.edu/cgi-bin/search.cgi), Cho (2010a, 2010b) and Li et al. (2010).

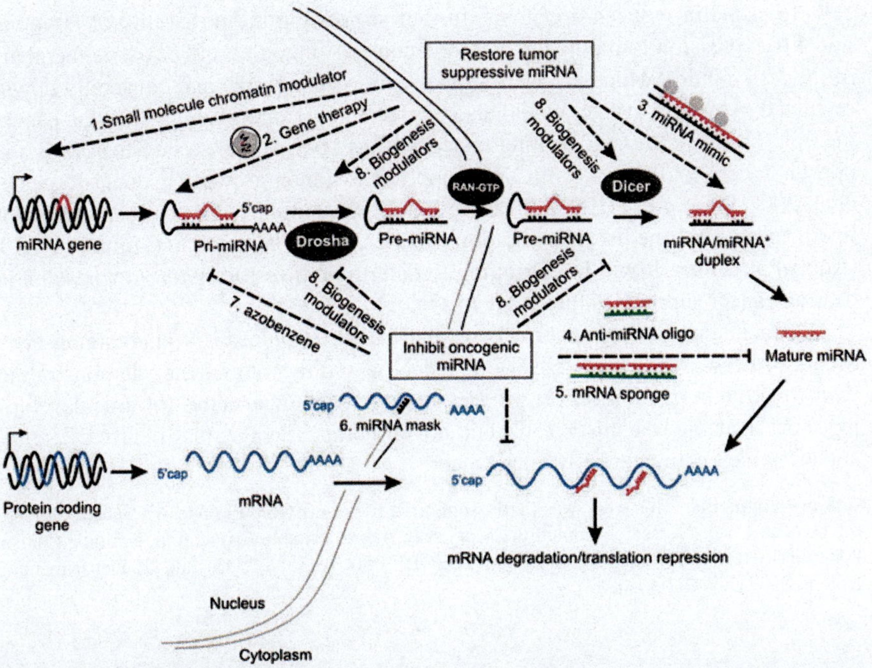

Fig. 14.1 Schematic diagram of miRNA-based therapeutic strategies. Strategy *1–3* aim to restore tumor suppressive miRNA activity: *1* Chromatin modulator, such as demethylation agents and HDAC inhibitors; *2* gene therapy using virus delivery system; *3* molecular mimics of tumor suppressive miRNA delivered via nanoparticles. Strategy *4–7* aim to inhibit oncogenic miRNA activity: *4* Anti-miRNA oligonucleotides; *5* mRNA sponges with multiple binding sites for a specific miRNA; *6* miRNA mask, a complementary DNA molecule containing miRNA binding site which in turn block the binding of a specific miRNA to its endogenous targets; *7* azobenzene, a small molecule inhibitor to *miR-21*. *8* targeting Drosha and Dicer mediated biogenesis have been proposed, yet further evaluation is needed

suppressive miRNAs, forced expression of miRNAs with a viral based system has been tested to restore their expression. Since a global decrease of miRNA expression has been observed in several tumor types, enhancing miRNA biogenesis is considered as a way to treat cancer with low levels of miRNA biogenesis. The aforementioned therapeutic strategies of modulating miRNA are illustrated in Fig. 14.1.

MiRNA-based ovarian cancer therapy is still in its infancy. However, recent studies on miRNA function for ovarian cancer cell lines suggested that some miRNAs can serve as potential targets in ovarian cancer therapy. For example, forced expression of *miR-9* in ovarian cancer cell line can suppress cell growth (Guo et al. 2009); and over-expression of *miR-34b* and *miR-34c* can suppress cell proliferation and anchorage-independent growth (Corney et al. 2007). In addition, expressing *miR-200c* reduced the expression of class III β-tubulin, therefore sensitizing the cells to paclitaxel treatment (Cochrane et al. 2009). On the other hand, blocking *miR-214* expression, which up-regulates PTEN activity, sensitized ovarian cancer cells to cisplatin-induced apoptosis, further suggesting the possibility of targeting miRNA expression to modulate chemoresponse (Yang et al. 2008a). The members of the *let-7* family, which are well-known tumor suppressors and constantly down-regulated in ovarian cancer, would also be interesting targets for anti-ovarian cancer therapy. In mouse models of lung cancer, it has been demonstrated that viral and non-viral delivery of *let-7* miRNAs suppressed tumor growth (Esquela-Kerscher et al. 2008; Kumar et al. 2008; Trang et al. 2010). Similarly, in a mouse model for liver cancer, viral-mediated restoration of *miR-26* demonstrated anti-tumor activity (Kota et al. 2009). It will be intriguing to determine if restoring *let-7* expression in ovarian cancer suppresses tumor growth in vivo.

In conclusion, curing epithelial ovarian cancer remains very challenging after decades of research. The discovery of miRNAs and recent evidence demonstrating their function in ovarian cancer progression opened new avenues towards developing more accurate diagnostic tools and novel therapeutic agents for early detection and effective treatments to ovarian cancer.

Acknowledgments This work was partly supported by grants from the Breast Cancer Alliance, the Ovarian Cancer Research Found, the Mary Kay Ash Charitable Foundation, National Cancer Institute (R01CA142776 and Ovarian Cancer SPORE P50CA83638-7951), and US Department of Defense (W81XWH-10-1-0082).

References

Asadollahi R, Hyde CA, Zhong XY. Epigenetics of ovarian cancer: From the lab to the clinic. Gynecol Oncol. 2010;18:81–7.

Auersperg N, Edelson MI, Mok SC, et al. The biology of ovarian cancer. Semin Oncol. 1998;25:281–304.

Bartel DP. MicroRNAs: genomics, biogenesis, mechanism, and function. Cell. 2004;116: 281–97.

Bast RC Jr, Klug TL, St John E, et al. A radioimmunoassay using a monoclonal antibody to monitor the course of epithelial ovarian cancer. N Engl J Med. 1983;309:883–7.

Bendoraite A, Knouf EC, Garg KS, et al. Regulation of miR-200 family microRNAs and ZEB transcription factors in ovarian cancer: evidence supporting a mesothelial-to-epithelial transition. Gynecol Oncol. 2010;116:117–25.

Bhattacharya R, Nicoloso M, Arvizo R, et al. MiR-15a and miR-16 control Bmi-1 expression in ovarian cancer. Cancer Res. 2009;69:9090–5.

Borchert GM, Lanier W, Davidson BL. RNA polymerase III transcribes human microRNAs. Nat Struct Mol Biol. 2006;13:1097–101.

Boren T, Xiong Y, Hakam A, et al. MicroRNAs and their target messenger RNAs associated with ovarian cancer response to chemotherapy. Gynecol Oncol. 2009;113:249–55.

Bussing I, Slack FJ, Grosshans H. Let-7 microRNAs in development, stem cells and cancer. Trends Mol Med. 2008;14:400–9.

Cai Y, Yu X, Hu S, et al. A brief review on the mechanisms of miRNA regulation. Genomics Proteomics Bioinformatics. 2009;7:147–54.

Calin GA, Croce CM. MicroRNA-cancer connection: the beginning of a new tale. Cancer Res. 2006;66:7390–4.

Calin GA, Dumitru CD, Shimizu M, et al. Frequent deletions and down-regulation of micro-RNA genes miR15 and miR16 at 13q14 in chronic lymphocytic leukemia. Proc Natl Acad Sci USA. 2002;99:15524–9.

Calin GA, Ferracin M, Cimmino A, et al. A microRNA signature associated with prognosis and progression in chronic lymphocytic leukemia. N Engl J Med. 2005;353: 1793–801.

Calin GA, Sevignani C, Dumitru CD, et al. Human microRNA genes are frequently located at fragile sites and genomic regions involved in cancers. Proc Natl Acad Sci USA. 2004;101:2999–3004.

Carmell MA, Xuan Z, Zhang MQ, et al. The argonaute family: tentacles that reach into RNAi, developmental control, stem cell maintenance, and tumorigenesis. Genes Dev. 2002;16: 2733–42.

Carthew RW, Sontheimer EJ. Origins and mechanisms of miRNAs and siRNAs. Cell. 2009;136:642–55.

Chan JA, Krichevsky AM, Kosik KS. MicroRNA-21 is an antiapoptotic factor in human glioblastoma cells. Cancer Res. 2005;65:6029–33.

Chang TC, Wentzel EA, Kent OA, et al. Transactivation of miR-34a by p53 broadly influences gene expression and promotes apoptosis. Mol Cell. 2007;26:745–52.

Chang TC, Yu D, Lee YS, et al. Widespread microRNA repression by Myc contributes to tumorigenesis. Nat Genet. 2008;40:43–50.

Chen R, Alvero AB, Silasi DA, et al. Regulation of IKKbeta by miR-199a affects NF-kappaB activity in ovarian cancer cells. Oncogene. 2008;27:4712–23.

Cho WC. MicroRNAs: potential biomarkers for cancer diagnosis, prognosis and targets for therapy. Int J Biochem Cell Biol. 2010a;42:1273–81.

Cho WC. MicroRNAs in cancer – from research to therapy. Biochim Biophys Acta. 2010b;1805:209–17.

Chung TK, Cheung TH, Huen NY, et al. Dysregulated microRNAs and their predicted targets associated with endometrioid endometrial adenocarcinoma in Hong Kong women. Int J Cancer. 2009;124:1358–65.

Cimmino A, Calin GA, Fabbri M, et al. MiR-15 and miR-16 induce apoptosis by targeting BCL2. Proc Natl Acad Sci USA. 2005;102:13944–9.

Cochrane DR, Spoelstra NS, Howe EN, et al. MicroRNA-200c mitigates invasiveness and restores sensitivity to microtubule-targeting chemotherapeutic agents. Mol Cancer Ther. 2009;8:1055

Corney DC, Flesken-Nikitin A, Godwin AK, et al. MicroRNA-34b and microRNA-34c are targets of p53 and cooperate in control of cell proliferation and adhesion-independent growth. Cancer Res. 2007;67:8433–8.

Corney DC, Hwang CI, Matoso A, et al. Frequent down-regulation of miR-34 family in human ovarian cancers. Clin Cancer Res. 2010;16:1119–28.

Costinean S, Zanesi N, Pekarsky Y, et al. Pre-B cell proliferation and lymphoblastic leukemia/high-grade lymphoma in E(mu)-miR155 transgenic mice. Proc Natl Acad Sci USA. 2006;103: 7024–9.

Creighton CJ, Fountain MD, Yu Z, et al. Molecular profiling uncovers a p53-associated role for microRNA-31 in inhibiting the proliferation of serous ovarian carcinomas and other cancers. Cancer Res. 2010;70:1906–15.

Crum CP, Drapkin R, Miron A, et al. The distal fallopian tube: a new model for pelvic serous carcinogenesis. Curr Opin Obstet Gynecol. 2007;19:3–9.

Dahiya N, Sherman-Baust CA, Wang TL, et al. MicroRNA expression and identification of putative miRNA targets in ovarian cancer. PLoS One. 2008;3:e2436.

Dubeau L. The cell of origin of ovarian epithelial tumors and the ovarian surface epithelium dogma: does the emperor have no clothes? Gynecol Oncol. 1999;72:437–42.

Dubeau L. The cell of origin of ovarian epithelial tumours. Lancet Oncol. 2008;9:1191–7.

Eis PS, Tam W, Sun L, et al. Accumulation of miR-155 and BIC RNA in human B cell lymphomas. Proc Natl Acad Sci USA. 2005;102:3627–32.

Eitan R, Kushnir M, Lithwick-Yanai G, et al. Tumor microRNA expression patterns associated with resistance to platinum based chemotherapy and survival in ovarian cancer patients. Gynecol Oncol. 2009;114:253–9.

Esquela-Kerscher A, Slack FJ. Oncomirs – microRNAs with a role in cancer. Nat Rev Cancer. 2006;6:259–69.

Esquela-Kerscher A, Trang P, Wiggins JF, et al. The let-7 microRNA reduces tumor growth in mouse models of lung cancer. Cell Cycle. 2008;7:759–64.

Fabbri M, Garzon R, Cimmino A, et al. MicroRNA-29 family reverts aberrant methylation in lung cancer by targeting DNA methyltransferases 3A and 3B. Proc Natl Acad Sci USA. 2007;104:15805–10.

Faggad A, Budczies J, Tchernitsa O, et al. Prognostic significance of Dicer expression in ovarian cancer-link to global microRNA changes and oestrogen receptor expression. J Pathol. 2010;220:382–91.

Fathalla MF. Incessant ovulation – a factor in ovarian neoplasia? Lancet. 1971;2:163.

Flavin R, Smyth P, Barrett C, et al. MiR-29b expression is associated with disease-free survival in patients with ovarian serous carcinoma. Int J Gynecol Cancer. 2009;19:641–7.

Flavin RJ, Smyth PC, Finn SP, et al. Altered eIF6 and Dicer expression is associated with clinicopathological features in ovarian serous carcinoma patients. Mod Pathol. 2008;21:676–84.

Folkins AK, Jarboe EA, Saleemuddin A, et al. A candidate precursor to pelvic serous cancer (p53 signature) and its prevalence in ovaries and fallopian tubes from women with BRCA mutations. Gynecol Oncol. 2008;109:168–73.

Folkins AK, Saleemuddin A, Garrett LA, et al. Epidemiologic correlates of ovarian cortical inclusion cysts (CICs) support a dual precursor pathway to pelvic epithelial cancer. Gynecol Oncol. 2009;115:108–11.

Garzon R, Calin GA, Croce CM. MicroRNAs in cancer. Annu Rev Med. 2009;60:167–79.

Garzon R, Volinia S, Liu CG, et al. MicroRNA signatures associated with cytogenetics and prognosis in acute myeloid leukemia. Blood. 2008;111:3183–9.

Giannakakis A, Sandaltzopoulos R, Greshock J, et al. MiR-210 links hypoxia with cell cycle regulation and is deleted in human epithelial ovarian cancer. Cancer Biol Ther. 2008;7: 255–64.

Gilad S, Meiri E, Yogev Y, et al. Serum microRNAs are promising novel biomarkers. PLoS One. 2008;3:e3148.

Gomez-Cabello D, Callejas S, Benguria A, et al. Regulation of the microRNA processor DGCR8 by the tumor suppressor ING1. Cancer Res. 2010;70:1866–74.

Gott JM, Emeson RB. Functions and mechanisms of RNA editing. Annu Rev Genet. 2000;34: 499–531.

Griffiths-Jones S. MiRBase: microRNA sequences and annotation. Curr Protoc Bioinformatics. 2010;Chapter 12:Unit 12.9.1–12.9.10.

Guo LM, Pu Y, Han Z, et al. MicroRNA-9 inhibits ovarian cancer cell growth through regulation of NF-kappaB1. FEBS J. 2009;276:5537–46.

Halazonetis TD, Gorgoulis VG, Bartek J. An oncogene-induced DNA damage model for cancer development. Science. 2008;319:1352–5.

Han L, Witmer PD, Casey E, et al. DNA methylation regulates microRNA expression. Cancer Biol Ther. 2007;6:1284–8.

Hawkins RE, Roberts K, Wiltshaw E, et al. The clinical correlates of serum CA125 in 169 patients with epithelial ovarian carcinoma. Br J Cancer. 1989;60:634–7.

He L, He X, Lim LP, et al. A microRNA component of the p53 tumour suppressor network. Nature. 2007;447:1130–4.

He L, Thomson JM, Hemann MT, et al. A microRNA polycistron as a potential human oncogene. Nature. 2005;435:828–33.

Heo I, Joo C, Cho J, et al. Lin28 mediates the terminal uridylation of let-7 precursor microRNA. Mol Cell. 2008;32:276–84.

Herbst AL. The epidemiology of ovarian carcinoma and the current status of tumor markers to detect disease. Am J Obstet Gynecol. 1994;170:1099–107.

Hossain A, Kuo MT, Saunders GF. Mir-17-5p regulates breast cancer cell proliferation by inhibiting translation of AIB1 mRNA. Mol Cell Biol. 2006;26:8191–201.

Hu X, Macdonald DM, Huettner PC, et al. A miR-200 microRNA cluster as prognostic marker in advanced ovarian cancer. Gynecol Oncol. 2009;114:457–64.

Huang YW, Liu JC, Deatherage DE, et al. Epigenetic repression of microRNA-129-2 leads to over-expression of SOX4 oncogene in endometrial cancer. Cancer Res. 2009;69: 9038–46.

Iorio MV, Ferracin M, Liu CG, et al. MicroRNA gene expression deregulation in human breast cancer. Cancer Res. 2005;65:7065–70.

Iorio MV, Visone R, Di Leva G, et al. MicroRNA signatures in human ovarian cancer. Cancer Res. 2007;67:8699–707.

Jazdzewski K, Murray EL, Franssila K, et al. Common SNP in pre-miR-146a decreases mature miR expression and predisposes to papillary thyroid carcinoma. Proc Natl Acad Sci USA. 2008;105:7269–74.

Jemal A, Siegel R, Xu J, et al. Cancer statistics, 2010. CA Cancer J Clin. 2010;60:277–300.

Johnson SM, Grosshans H, Shingara J, et al. RAS is regulated by the let-7 microRNA family. Cell. 2005;120:635–47.

Kamiya N, Mizuno K, Kawai M, et al. Simultaneous measurement of CA 125, CA 19-9, tissue polypeptide antigen, and immunosuppressive acidic protein to predict recurrence of ovarian cancer. Obstet Gynecol. 1990;76:417–21.

Karube Y, Tanaka H, Osada H, et al. Reduced expression of Dicer associated with poor prognosis in lung cancer patients. Cancer Sci. 2005;96:111–5.

Kawahara Y, Nishikura K. Regulation of the miRNA function by RNA editing. Tanpakushitsu Kakusan Koso. 2009;54:1133–40.

Kim VN, Han J, Siomi MC. Biogenesis of small RNAs in animals. Nat Rev Mol Cell Biol. 2009;10:126–39.

Kota J, Chivukula RR, O'Donnell KA, et al. Therapeutic microRNA delivery suppresses tumorigenesis in a murine liver cancer model. Cell. 2009;137:1005–17.

Kumar MS, Erkeland SJ, Pester RE, et al. Suppression of non-small cell lung tumor development by the let-7 microRNA family. Proc Natl Acad Sci USA. 2008;105:3903–8.

Kuo KT, Guan B, Feng Y, et al. Analysis of DNA copy number alterations in ovarian serous tumors identifies new molecular genetic changes in low-grade and high-grade carcinomas. Cancer Res. 2009;69:4036–42.

Laios A, O'Toole S, Flavin R, et al. Potential role of miR-9 and miR-223 in recurrent ovarian cancer. Mol Cancer. 2008;7:35.

Lawrie CH, Gal S, Dunlop HM, et al. Detection of elevated levels of tumour-associated microRNAs in serum of patients with diffuse large B-cell lymphoma. Br J Haematol. 2008;141:672–5.

Lee RC, Feinbaum RL, Ambros V. The *C. elegans* heterochronic gene lin-4 encodes small RNAs with antisense complementarity to lin-14. Cell. 1993;75:843–54.

Lee Y, Miron A, Drapkin R, et al. A candidate precursor to serous carcinoma that originates in the distal fallopian tube. J Pathol. 2007;211:26–35.

Lee CH, Subramanian S, Beck AH, et al. MicroRNA profiling of BRCA1/2 mutation-carrying and non-mutation-carrying high-grade serous carcinomas of ovary. PLoS One. 2009;4:e7314.

Lewis BP, Burge CB, Bartel DP. Conserved seed pairing, often flanked by adenosines, indicates that thousands of human genes are microRNA targets. Cell. 2005;120:15–20.

Li C, Feng Y, Coukos G, et al. Therapeutic microRNA strategies in human cancer. AAPS J. 2009;11:747–57.

Li SD, Zhang JR, Wang YQ, et al. The role of microRNAs in ovarian cancer initiation and progression. J Cell Mol Med. 2010;14:2240–9.

Lodes MJ, Caraballo M, Suciu D, et al. Detection of cancer with serum miRNAs on an oligonucleotide microarray. PLoS One. 2009;4:e6229.

Lou YH, Yang XS, Wang FL, et al. Expression of microRNA-21 in ovarian epithelial carcinoma and its clinical significance. Nan Fang Yi Ke Da Xue Xue Bao. 2010;30:608–10.

Lu J, Getz G, Miska EA, et al. MicroRNA expression profiles classify human cancers. Nature. 2005;435:834–8.

Lu L, Katsaros D, de la Longrais IA, et al. Hypermethylation of let-7a-3 in epithelial ovarian cancer is associated with low insulin-like growth factor-II expression and favorable prognosis. Cancer Res. 2007;67:10117–22.

Mahajan A, Liu Z, Gellert L, et al. HMGA2: a biomarker significantly over-expressed in high-grade ovarian serous carcinoma. Mod Pathol. 2010;23:673–81.

Melo SA, Ropero S, Moutinho C, et al. A TARBP2 mutation in human cancer impairs microRNA processing and DICER1 function. Nat Genet. 2009;41:365–70.

Mendell JT. Miriad roles for the miR-17-92 cluster in development and disease. Cell. 2008;133:217–22.

Merritt WM, Lin YG, Han LY, et al. Dicer, drosha, and outcomes in patients with ovarian cancer. N Engl J Med. 2008;359:2641–50.

Mezzanzanica D, Bagnoli M, De Cecco L, et al. Role of microRNAs in ovarian cancer pathogenesis and potential clinical implications. Int J Biochem Cell Biol. 2010;42:1262–72.

Mitchell PS, Parkin RK, Kroh EM, et al. Circulating microRNAs as stable blood-based markers for cancer detection. Proc Natl Acad Sci USA. 2008;105:10513–8.

Moore RG, Brown AK, Miller MC, et al. The use of multiple novel tumor biomarkers for the detection of ovarian carcinoma in patients with a pelvic mass. Gynecol Oncol. 2008;108:402–8.

Mott JL, Kobayashi S, Bronk SF, et al. Mir-29 regulates Mcl-1 protein expression and apoptosis. Oncogene. 2007;26:6133–40.

Nagaraja AK, Creighton CJ, Yu Z, et al. A link between mir-100 and FRAP1/mTOR in clear cell ovarian cancer. Mol Endocrinol. 2010;24:447–63.

Nam EJ, Yoon H, Kim SW, et al. MicroRNA expression profiles in serous ovarian carcinoma. Clin Cancer Res. 2008;14:2690–5.

Ness RB., Cottreau C. Possible role of ovarian epithelial inflammation in ovarian cancer. J Natl Cancer Inst. 1999;91:1459–67.

Ng EK, Chong WW, Jin H, et al. Differential expression of microRNAs in plasma of patients with colorectal cancer: a potential marker for colorectal cancer screening. Gut. 2009;58:1375–81.

O'Donnell KA, Wentzel EA, Zeller KI, et al. c-Myc-regulated microRNAs modulate E2F1 expression. Nature. 2005;435:839–43.

Olena AF, Patton JG. Genomic organization of microRNAs. J Cell Physiol. 2010;222:540–5.

Ota A, Tagawa H, Karnan S, et al. Identification and characterization of a novel gene, C13orf25, as a target for 13q31-q32 amplification in malignant lymphoma. Cancer Res. 2004;64: 3087–95.

Pampalakis G, Diamandis EP, Katsaros D, et al. Down-regulation of dicer expression in ovarian cancer tissues. Clin Biochem. 2010;43:324–7.

Pan Q, Luo X, Chegini N. Differential expression of microRNAs in myometrium and leiomyomas and regulation by ovarian steroids. J Cell Mol Med. 2008;12:227–40.

Park SM, Gaur AB, Lengyel E, et al. The miR-200 family determines the epithelial phenotype of cancer cells by targeting the E-cadherin repressors ZEB1 and ZEB2. Genes Dev. 2008;22: 894–907.

Paz N, Levanon EY, Amariglio N, et al. Altered adenosine-to-inosine RNA editing in human cancer. Genome Res. 2007;17:1586–95.

Pedersen-Bjergaard J, Pedersen M, Roulston D, et al. Different genetic pathways in leukemogenesis for patients presenting with therapy-related myelodysplasia and therapy-related acute myeloid leukemia. Blood. 1995;86:3542–52.

Pekarsky Y, Santanam U, Cimmino A, et al. Tcl1 expression in chronic lymphocytic leukemia is regulated by miR-29 and miR-181. Cancer Res. 2006;66:11590–3.

Phocas I, Sarandakou A, Sikiotis K, et al. A comparative study of serum alpha-beta A immunoreactive inhibin and tumor-associated antigens CA125 and CEA in ovarian cancer. Anticancer Res. 1996;16:3827–31.

Piek JM, van Diest PJ, Zweemer RP, et al. Dysplastic changes in prophylactically removed fallopian tubes of women predisposed to developing ovarian cancer. J Pathol. 2001;195:451–6.

Rabinowits G, Gercel-Taylor C, Day JM, et al. Exosomal microRNA: a diagnostic marker for lung cancer. Clin Lung Cancer. 2009;10:42–6.

Radisavljevic SV. The pathogenesis of ovarian inclusion cysts and cystomas. Obstet Gynecol. 1977;49:424–9.

Raveche ES, Salerno E, Scaglione BJ, et al. Abnormal microRNA-16 locus with synteny to human 13q14 linked to CLL in NZB mice. Blood. 2007;109:5079–86.

Raver-Shapira N, Marciano E, Meiri E, et al. Transcriptional activation of miR-34a contributes to p53-mediated apoptosis. Mol Cell. 2007;26:731–43.

Resnick KE, Alder H, Hagan JP, et al. The detection of differentially expressed microRNAs from the serum of ovarian cancer patients using a novel real-time PCR platform. Gynecol Oncol. 2009;112:55–9.

Rosell R, Wei J, Taron M. Circulating microRNA signatures of tumor-derived exosomes for early diagnosis of non-small-cell lung cancer. Clin Lung Cancer. 2009;10:8–9.

Roush S, Slack FJ. The let-7 family of microRNAs. Trends Cell Biol. 2008;18:505–16.

Rustin GJ, Bast RC Jr, Kelloff GJ, et al. Use of CA-125 in clinical trial evaluation of new therapeutic drugs for ovarian cancer. Clin Cancer Res. 2004;10:3919–26.

Saito Y, Liang G, Egger G, et al. Specific activation of microRNA-127 with down-regulation of the proto-oncogene BCL6 by chromatin-modifying drugs in human cancer cells. Cancer Cell. 2006;9:435–43.

Salvador S, Rempel A, Soslow RA, et al. Chromosomal instability in fallopian tube precursor lesions of serous carcinoma and frequent monoclonality of synchronous ovarian and fallopian tube mucosal serous carcinoma. Gynecol Oncol. 2008;110:408–17.

Scott GK, Mattie MD, Berger CE, et al. Rapid alteration of microRNA levels by histone deacetylase inhibition. Cancer Res. 2006;66:1277–81.

Seidman JD, Russell P, Kurman RJ. Surface epithelial tumor of the ovary. In: Kurman RJ, editor. Blaustein's pathology of the female genital tract. 5th ed. New York, NY: Springer; 2002. pp. 791–904.

Selcuklu SD, Donoghue MT, Spillane C. MiR-21 as a key regulator of oncogenic processes. Biochem Soc Trans. 2009;37:918–25.

Sethupathy P, Collins FS. MicroRNA target site polymorphisms and human disease. Trends Genet. 2008;24:489–97.

Shen J, Ambrosone CB, DiCioccio RA, et al. A functional polymorphism in the miR-146a gene and age of familial breast/ovarian cancer diagnosis. Carcinogenesis. 2008;29:1963–6.

Smith Sehdev AE, Sehdev PS, Kurman RJ. Noninvasive and invasive micropapillary (low-grade) serous carcinoma of the ovary: a clinicopathologic analysis of 135 cases. Am J Surg Pathol. 2003;27:725–36.

Sorrentino A, Liu CG, Addario A, et al. Role of microRNAs in drug-resistant ovarian cancer cells. Gynecol Oncol. 2008;111:478–86.

Spizzo R, Nicoloso MS, Croce CM, et al. Snapshot: microRNAs in cancer. Cell. 2009; 137:586.

Suzuki HI, Yamagata K, Sugimoto K, et al. Modulation of microRNA processing by p53. Nature. 2009;460:529–33.

Takamizawa J, Konishi H, Yanagisawa K, et al. Reduced expression of the let-7 microRNAs in human lung cancers in association with shortened postoperative survival. Cancer Res. 2004;64:3753–6.

Tay Y, Zhang J, Thomson AM, et al. MicroRNAs to Nanog, Oct4 and Sox2 coding regions modulate embryonic stem cell differentiation. Nature. 2008;455:1124–8.

Taylor DD, Gercel-Taylor C. MicroRNA signatures of tumor-derived exosomes as diagnostic biomarkers of ovarian cancer. Gynecol Oncol. 2008;110:13–21.

Trang P, Medina PP, Wiggins JF, et al. Regression of murine lung tumors by the let-7 microRNA. Oncogene. 2010;29:1580–7.

van Jaarsveld MT, Helleman J, Berns EM, et al. MicroRNAs in ovarian cancer biology and therapy resistance. Int J Biochem Cell Biol. 2010;42:1282–90.

Ventura A, Young AG, Winslow MM, et al. Targeted deletion reveals essential and overlapping functions of the miR-17 through 92 family of miRNA clusters. Cell. 2008;132:875–86.

White NM, Chow TF, Mejia-Guerrero S, et al. Three dysregulated miRNAs control kallikrein 10 expression and cell proliferation in ovarian cancer. Br J Cancer. 2010;102:1244–53.
Wightman B, Ha I, Ruvkun G. Post-transcriptional regulation of the heterochronic gene lin-14 by lin-4 mediates temporal pattern formation in *C. elegans*. Cell. 1993;75:855–62.
Woods K, Thomson JM, Hammond SM. Direct regulation of an oncogenic micro-RNA cluster by E2F transcription factors. J Biol Chem. 2007;282:2130–4.
Wu W, Lin Z, Zhuang Z, et al. Expression profile of mammalian microRNAs in endometrioid adenocarcinoma. Eur J Cancer Prev. 2009;18:50–5.
Wurz K, Garcia RL, Goff BA, et al. MiR-221 and miR-222 alterations in sporadic ovarian carcinoma: relationship to CDKN1B, CDKNIC and overall survival. Genes Chromosomes Cancer. 2010;49:577–84.
Wyman SK, Parkin RK, Mitchell PS, et al. Repertoire of microRNAs in epithelial ovarian cancer as determined by next generation sequencing of small RNA cDNA libraries. PLoS One. 2009;4:e5311.
Xiao C, Srinivasan L, Calado DP, et al. Lymphoproliferative disease and autoimmunity in mice with increased miR-17-92 expression in lymphocytes. Nat Immunol. 2008;9:405–14.
Xu CF, Solomon E. Mutations of the BRCA1 gene in human cancer. Semin Cancer Biol. 1996;7: 33–40.
Yanaihara N, Caplen N, Bowman E, et al. Unique microRNA molecular profiles in lung cancer diagnosis and prognosis. Cancer Cell. 2006;9:189–98.
Yang N, Kaur S, Volinia S, et al. MicroRNA microarray identifies let-7i as a novel biomarker and therapeutic target in human epithelial ovarian cancer. Cancer Res. 2008b;68:10307–14.
Yang H, Kong W, He L, et al. MicroRNA expression profiling in human ovarian cancer: miR-214 induces cell survival and cisplatin resistance by targeting PTEN. Cancer Res. 2008a;68:425–33.
Zhang L, Huang J, Yang N, et al. MicroRNAs exhibit high frequency genomic alterations in human cancer. Proc Natl Acad Sci USA. 2006;103:9136–41.
Zhang L, Volinia S, Bonome T, et al. Genomic and epigenetic alterations deregulate microRNA expression in human epithelial ovarian cancer. Proc Natl Acad Sci USA. 2008a;105:7004–9.
Zhang L, Yang N, Coukos G. MicroRNA in human cancer: one step forward in diagnosis and treatment. Adv Exp Med Biol. 2008b;622:69–78.

Chapter 15
MicroRNAs in Brain Tumors

Tara Spence, Johnny Nguyen, Eric Bouffet, and Annie Huang

Abstract Malignant brain tumors are one of the leading causes of cancer-related deaths or disability in adults and children. Although current brain tumor therapy is primarily based on diagnostic histopathology, it is increasingly clear that tumors with similar or indistinguishable histologies can exhibit substantial molecular heterogeneity that dictate clinical phenotypes. Recent whole genome studies indicate molecular diversity of brain tumors not only result from diverse mechanisms of tumor initiation/progression but may also reflect distinct cellular derivation. Indeed, molecular defects that drive brain tumor development/progression overlap substantially with key pathways that regulate normal central nervous system development. Increasing evidence implicates microRNAs (miRNAs) in critical signaling pathways that regulate cycling, differentiation, death and ultimately cell fate/identity of normal neural progenitor/precursor cells. Not surprisingly, dysregulated miRNA expression occurs in a spectrum of brain tumors; many miRNA loci map up- or downstream of canonical brain tumor signaling pathways, and can modulate oncogenic phenotypes in brain tumor cells. Importantly, presence of recurrent genetic alterations which target specific miRNAs (e.g. C19MC, *miR-17-92*) in primary brain tumors further highlight causative roles for miRNAs in brain tumor pathogenesis. These collective discoveries not only offer clinicians exciting prospects for novel therapeutic strategies but also provide important new tools to refine and deliver greater precision in prognostication and treatment planning/stratification at the bedside. This chapter reviews the role of specific miRNAs in malignant gliomas and childhood embryonal tumors – medulloblastoma and primitive neuroectodermal brain tumors, in which miRNAs have been most extensively investigated.

A. Huang (✉)
Pediatric Brain Tumour Program, Program in Cell Biology, Division of Hematology Oncology, Department of Pediatrics, Sonia and Arthur Labatt Brain Tumour Research Centre, Hospital for Sick Children, Toronto, ON, Canada
e-mail: annie.huang@sickkids.ca

W.C.S. Cho (ed.), *MicroRNAs in Cancer Translational Research*,
DOI 10.1007/978-94-007-0298-1_15,

15.1 Introduction

Intrinsic brain/CNS tumors remain one of the most devastating and challenging neoplasms to treat. It is estimated that 600,000 adults and 30,000 children living in the United States will be diagnosed with a primary brain tumor and ~140,000 and 19,000 respectively of adult and pediatric brain tumors will be malignant (Porter et al. 2010). Surgery and radiotherapy remains an important mainstay of brain tumor treatments, but are associated with significant neuro-cognitive sequelae in survivors of all ages. New therapeutic opportunities are beginning to emerge with advances in understanding of cellular origins of brain tumors and discovery of non-coding loci, including miRNAs, as critical contributors to brain tumor growth.

Brain tumors encompass a heterogeneous spectrum of diseases. Pathologic diagnosis is most often established according to the World Health Organization (WHO) criteria which is based on tumor location, morphologic resemblance to the three major (astrocytic/glial, oligodendroglial or neuronal) CNS cell lineages, and histologic grading (I–IV) (Louis et al. 2007). Malignant gliomas (grades III–IV) represent the most aggressive and frequent group of tumors diagnosed in adults, and encompasses several subtypes with varying degrees of astrocytic or oligodendroglial differentiation. Grade IV malignant gliomas or glioblastoma (GBM) are highly infiltrative and invasive tumors, that represents the most difficult subtype to cure. Malignant gliomas are generally treated with extensive surgery followed by radiation with or without chemotherapy. Despite advances in neurosurgery, radiotherapy and trials with a spectrum of novel therapeutics, median survival for GBM has remained largely static with OS of 12–15 months (Clarke et al. 2010; DeAngelis 2001). GBMs diagnosed in younger individuals (median age of 45 years) arise from transformation of a lower grade glioma and have better outcomes than primary GBM which characteristically arise de novo as grade IV tumors in the 6th decade of life. Primary and secondary GBM are known to differ in frequency of specific genetic alterations (*TP53* mutations, *EGFR* amplification) characteristic of gliomas (Louis et al. 2002; Ohgaki and Kleihues 2007). Significantly, recent global genomic investigations indicate far greater complexity and genetic heterogeneity in tumors with a histopathological diagnosis of GBM.

Intrinsic brain tumors are the most common solid neoplasm in children with embryonal brain tumors. Presently, embryonal tumors are classified based on histology and location into three broad categories – medulloblastoma (MB), central nervous system primitive neuroectodermal brain tumors (CNS-PNET) and rhabdoid tumors. With the exception of rhabdoid tumors, which are identified by *INI1* gene alterations, no specific diagnostic markers are known for other embryonal brain tumors, including MB, therefore tumor diagnosis/classification, and treatment approaches remain imprecise. As embryonal tumors frequently exhibit cranio-spinal metastasis, whole brain-spine radiation and multi-agent chemotherapy represent important post-operative treatment. Multi-modality treatment have substantially improved survival in non-metastatic medulloblastoma (> 75–80% 5 year OS), however survival for metastatic MB and other embryonal tumor subtypes remains poor (range of 20–50% OS) (Gajjar and Pizer 2010). Pervasive neurocognitive and growth deficits, especially in younger survivors (Mulhern et al. 2001) represent a significant limitation of current curative treatments (Crawford et al. 2007; Lafay-Cousin et al. 2009; Packer and Vezina 2008). Recent discovery of molecular alterations associated with clinically relevant subgroups of MB and CNS-PNET promises significant advances in treatment stratification and new insights into biological therapeutics for embryonal brain tumors.

15.1.1 MiRNAs in Normal and Neoplastic Brain Growth

Major advances in understanding brain tumor pathogenesis have come from developmental neurobiology as significant parallels exist between cellular and molecular processes that underlie normal and neoplastic brain growth. Indeed genes and signaling pathways which regulate neural stem/progenitor cell renewal, differentiation, cycling and death are frequently implicated as drivers in brain tumor formation and progression (Wechsler-Reya and Scott 2001). These include miRNA loci which are now increasingly recognized as important in normal and neoplastic brain growth. Studies in mouse and zebra fish show that disruption of miRNA synthesis machinery or specific miRNAs profoundly alters brain morphogenesis. Specifically, Nestin-Cre or Emx-Cre (De Pietri Tonelli et al. 2008; Kawase-Koga et al. 2009) mediated knockout of the central miRNAs processing enzyme, Dicer, in neural precursors or progenitors leads to defective neural differentiation and cortical hypoplasia. Genetic studies demonstrate specific rescue of brain morphogenesis defects in Dicer$^{-/-}$ zebra fish by members of the large *miR-430* miRNA cluster (Giraldez et al. 2005) and a requirement for *miR-9*, a highly brain enriched miRNA, for murine mid-hind brain patterning (Leucht et al. 2008), thus highlighting direct roles for specific miRNAs in neural development. Furthermore, studies in cellular models correlate manipulation of specific miRNAs, including *miR-9*, *miR-124*, *miR-125*, with changes in lineage specific gene expression and neuronal-glial phenotypes. Of note, *miR-9* (Packer et al. 2008) and *miR-124* (Conaco et al. 2006) are repressed by transcription factor REST, a potent transcriptional repressor and inhibitor of neuronal differentiation which synergizes with Myc to inhibit neural differentiation and induce MB-like tumors in cerebellar granule neural progenitors (Lawinger et al. 2000; Su et al. 2006), the presumed cell of origin for a subset of human MB.

As has been demonstrated in various other human neoplasms, altered miRNA expression has now been described in a spectrum of brain tumors including GBM, oligodendroglioma and childhood embryonal brain tumors – MB, CNS-PNET, rhabdoid tumors (summarized in Table 15.1). Detection of genomic and epigenetic alterations targeting specific miRNAs in primary tumors (C19MC, *miR-17-92* clusters and *miR-26a*), support direct etiologic roles for these miRNAs in brain tumor pathogenesis. Functional roles for a subset of miRNAs with altered expression in tumors, including *miR-21* and *miR-124*, have been validated – however, the significance of many other miRNA loci with altered expression in different primary brain tumors remains to be elucidated. Given the developmental specific roles demonstrated for a significant proportion of the "brain tumor miRNAs", altered expression

Table 15.1 MiRNAs implicated in malignant brain tumors

MicroRNA	Tumor type	Chromosome location	Expression in tumors	Validated or putative targets	References
miR-21	Glioblastoma	17q23.1	High	EGFR, PTEN, HNRPK, TaP63, Tropomyosin 1, PDCD4, Maspin, RhoB, MARCKS, RECK, TIMP3	Connolly et al. (2010); Li et al. (2009a); Papagiannakopoulos et al. (2008); Ren et al. (2010)
miR-221/222	Glioblastoma, CNS Rhabdoid Tumors	Xp11.3	High	STAT3, Akt, p27^{Kip1}, c-kit	Gillies and Lorimer (2007); Zhang et al. (2010)
miR-128-1	Glioblastoma	2q22.1	Low	Bmi-1	Godlewski et al. (2008)
miR-7	Glioblastoma	7q26.1	Low	IRS1/2, Raf1	Kefas et al. (2008); Webster et al. (2009)
miR-181a/181b	Glioblastoma	1q32.1	Low		
miR-9	Oligodendro-glioma	*miR-9-1*: 1q22 *miR-9-1*: 5q14.3 *miR-9-1*: 15q26.1	High	REST, BAF34a	Ferretti et al. (2009); Silber et al. (2008); Yoo et al. (2009b)
miR-30b/d	Medulloblastoma	8q24.22-q24.23	High		
miR-124	Medulloblastoma	*miR-124-1*: 8p23.1 *miR-124-2*: 8q13.1 *miR-124-3*: 20q13.33	Low	REST, BAF34a, CDK6, RB1, TrkC receptor, CDK6, SLC16A1	Conaco et al. (2006); Ferretti et al. (2009); Li et al. (2009a); Pierson et al. (2008); Silber et al. (2008); Yoo et al. (2009b)
miR-17-92 cluster	Medulloblastoma	13q31.3	High	TSP-1, Bmi-1, PTEN, PP2A, SHH signaling	Dews et al. (2006); Mavrakis et al. (2010)
miR-326	Medulloblastoma	11q13.4	Low	PKM2, SMO, NOTCH 1/2	Ferretti et al. (2008); Kefas et al. (2010)
miR-125b, *miR-324-5*	Medulloblastoma	*miR-125b-1*: 11q24.1 *miR-125b-2*: 21q21.1 *miR324-5p*: 17p13.1	Low	SMO, SHH signaling	Ferretti et al. (2009)
C19MC	CNS-PNET	19q13.42: *miR-520g*; *miR-517c*; *miR-520h*	High	WNT/JNK signaling; WNT/JNK signalling; ABCG2	Li et al. (2009b); Wang et al. (2010b)

of some miRNAs likely reflects specific neural lineages and cellular derivation of tumors and may serve not only as powerful tools for tumor classification but also for improved delineation of the biology of tumor subgroups. In contrast to other tumor types, relatively few studies (total of 82 to date) have investigated the role of miRNAs in brain tumors with most studies focused on gliomas and embryonal brain tumors.

15.2 Malignant Gliomas

15.2.1 Genetic and Molecular Features of Malignant Gliomas

Malignant gliomas, which includes grade III–IV tumors, exhibit recurrent alterations of a spectrum of oncogenes and tumor suppressor genes with *TP53*, receptor tyrosine kinase – EGFR, PDGFR, RB1, and PI3K/PTEN/Akt pathway genes representing the most frequently targeted loci. Etiologic and functional contributions of the majority of these loci to various aspects of glioma biology have been extensively studied and experimentally validated (Bleau et al. 2009; Furnari et al. 2007; Holland et al. 2000; Uhrbom et al. 2002; Zheng et al. 2008a). However, clinical correlates have been established for only a subset of the genomic alterations seen in gliomas. These include 1p/19q loss which is associated with chemo-sensitivity in oligodendroglioma and *TP53* mutation and *EGFR* amplification which are seen differentially in primary versus secondary GBM. The significance of the large spectra of gene alterations, including those recently reported for the *NF1* and *IDH1* genes, to clinical phenotypes in glioblastoma has just started to emerge from systematic, global integrated genomic and proteomic analysis performed by several groups (Colman et al. 2010; Parsons et al. 2008; Verhaak et al. 2010). These studies in aggregate indicate that diverse genetic alterations in almost all GBMs converge on several core signaling pathways regulated by receptor tyrosine kinases (EGFR/PDGFR), RB1, TP53, and PI3K/PTEN, thus underscoring the importance of these pathways in glioma therapy. In addition to these studies, Phillips et al. (2006) revealed at least four molecular GBM subtypes (Proneural, Neural, Classical, Mesenchymal) that mirror neurogenesis and are distinguishable on the basis of gene expression profiles and specific genomic and signaling alterations. Importantly, molecular subtypes of GBM appear to correlate with clinical phenotypes, tumor characteristics and treatment outcomes (Verhaak et al. 2010). These promising studies highlight the genetic diversity of GBM and underscore the need for rigorous molecular classification in order to design appropriate therapeutics and to accurately evaluate therapeutic efficacy in well-defined molecular subsets of GBM.

15.2.2 MiRNAs in GBM Pathogenesis

GBM multiforme histopathology is characterized by hallmark features of high cellular and microvascular proliferation, intra-tumoral necrosis and diffuse brain invasion/infiltration. Defects in cell cycle, cell death and migration, and enhanced neoangiogenesis which underlie these cellular phenotypes are critical to the highly malignant GBM phenotype. Disruption of the p53/pRB signaling pathways are known to contribute to the cell cycle/proliferative/apoptotic defects in GBM cells, while alterations in the EGFR/PTEN/Akt pathway have been linked to enhanced survival of GBM cells. Changes in tumor metabolism and response to hypoxia, which result in part from activation of the hypoxia-inducible factor-1 (HIF-1) alpha pathways, are also now increasingly recognized as key contributors to GBM pathogenesis (Ohgaki and Kleihues 2007). Direct genetic alterations of key signaling pathway components are seen in a large proportion, but not all GBMs, thus alternative mechanisms which sustain these key oncogenic pathways must exist (Oliver et al. 2009). In addition to post-translational mechanisms, there is increasing evidence that post-transcriptional or epigenetic modulators also target key oncogenic signaling pathways and contribute to the hallmark cellular defects of GBM cells. Indeed a subset of miRNA differentially expressed in malignant gliomas act directly to modulate cell cycle and survival, invasiveness, self-renewal and differentiation properties of glioma cells.

15.2.2.1 Cell Cycle/Proliferative/Survival Defects

MiRNAs that have been consistently and most significantly correlated with cell cycle/proliferation/survival defects, and implicated as oncogenic or tumor suppressor miRNAs in gliomas include *miR-21*, *miR-26a*, *miR-221/222*, *miR-7*, and *miR-34a*, all of which target one or more components of EGFR/PTEN/Akt and the p53/pRB signaling pathways. *MiR-21* represents one of the most extensively studied miRNA in GBM. Initially identified as one of the most significantly overexpressed miRNA in GBM (up to 100X higher expression in tumor versus normal brain) (Chan et al. 2005), substantial cellular and molecular evidence now show *miR-21* regulates several aspects of GBM biology including cell survival and migration (Gabriely et al. 2008), and implicate *miR-21* as an important oncogenic miRNA in GBM. Specifically, inhibition of *miR-21* has been shown to confer cell survival and also enhance sensitivity of GBM cells to a number of chemo-therapeutic agents (Chan et al. 2005; Li et al. 2009c; Ren et al. 2010) and TNF-related ligands (Corsten et al. 2007). The anti-apoptotic effect of *miR-21* appears to be effected in part via regulation of cell cycle as *miR-21* knockdown induces G0-G1 cell cycle arrest. At the molecular level, *miR-21* has been shown to activate the EGFR/Akt cell survival pathways through direct targeting of PTEN and may also function via PTEN-independent pathway (Ren et al. 2010; Zhou et al. 2010). In addition, *miR-21* also confers survival and cell cycle arrest via the p53-mediated and mitochondrial apoptotic pathways in part via direct regulation of HNRPK and Tap63, a p53 homologue (Papagiannakopoulos et al. 2008).

MiR-26a was identified as a target of gene amplification together with RB1 and PI3K/Akt pathway oncogenes, CDK4 and CENTG, in ~15% of 176 primary GBMs characterized in the TCGA database (Kim et al. 2010). Interestingly, upregulation of *miR-26a*, which has been shown to directly target PTEN, occurs at greater frequency in primary tumors with PTEN LOH and monoallelic expression.

Huse et al. (2009) demonstrate *miR-26a* expression limits loss of the remaining PTEN allele in a PDGF driven PTEN$^{-/+}$ model of glioma, thus suggesting that *miR-26a* over-expression functionally substitutes for PTEN loss in vivo (Brennan et al. 2009). In addition to PTEN, *miR-26a* directly regulates RB1 and MEKK2, and *miR-26a* expression augments CDK4 oncogenic effects in glioma cells in vitro and in vivo. These findings therefore suggest that the *miR-26a*/CDK4/CENTG oncogenic cluster seen in primary glioma cooperatively regulate multiple targets to modulate the cell survival/proliferative and apoptotic functions mediated by the Akt, RB1 as well as the JNK pathways in malignant gliomas (Kim et al. 2010). Other oncogenic miRNAs which are over-expressed and target the cell cycle and cell survival pathways in GBM include *miR-221/222* which has been reported to promote oncogenesis in vitro and in vivo by regulating the STAT3/Akt pathway (Zhang et al. 2010) as well as by direct post-transcriptional regulation of tumor suppressor p27^{Kip1} (Gillies and Lorimer 2007; Zhang et al. 2010) in glioma cells.

MiRNAs with tumor suppressor activity in GBM include *miR-7* and *miR-34a* which also have inhibitory effects on cell cycle and proliferation. Specifically, ectopic *miR-7* expression has been shown to increase cell death in established, primary and tumor derived GBM stem cell lines in part by targeting EGFR and controlling Akt activity via translational inhibition of upstream Akt activators, IRS1 and 2 (Kefas et al. 2008). Using gene expression analysis of miRNA transfected cells, Webster et al. (2009) mapped *miR-7* target genes to several different signaling pathways downstream of EGFR, and demonstrated that Raf1, an effector of EGFR signaling in the oncogenic Raf-MEK-ERK cascade, is also directly regulated by *miR-7* at the transcriptional level (Kefas et al. 2008; Webster et al. 2009). Thus diminished *miR-7* expression in primary GBM is predicted to impact on multiple EGFR mediated signaling pathways.

The *miR-34a* locus, which is frequently epigenetically silenced in a spectrum of tumors, was first identified as a direct p53 transcriptional target and is an important component of the p53 tumor suppressor network (He et al. 2007). The highly conserved *miR-34a* locus, which maps to 1p36, a region frequently lost in gliomas, is expressed at relatively low levels in primary GBM compared to normal brain (Guessous et al. 2010), and has been shown to directly target multiple oncogenes including c-MET, NOTCH 1 and 2, and CDK6 in glioma cells. Consistent with a tumor suppressor function, ectopic *miR-34a* in GBM cells suppresses xenograft formation, diminishes G1/S cell cycle progression and promotes cell death (Li et al. 2009d). Thus loss of *miR-34a* expression may promote gliomagenesis by inhibiting p53-mediated apoptosis in primary GBMs cells. Figure 15.1 summarizes the signaling pathways implicated in glioblastoma pathogenesis that are targeted by dysregulated miRNAs, including the EGFR/PTEN/Akt and RAS/Raf/ERK signalling pathways, as well as the Rb G1-S cell cycle checkpoint.

15.2.2.2 Cell Migration and Invasion

The role of miRNAs in glioma cell invasion and migration have not been extensively studied. To date *miR-21* and *miR-10b* have been implicated as positive

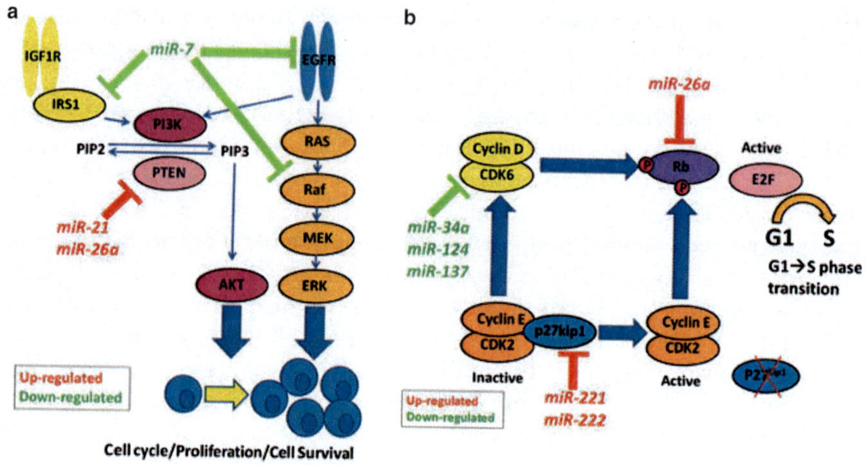

Fig. 15.1 Signaling pathways implicated in glioblastoma pathogenesis are targeted by dysregulated microRNAs. (**a**) The EGFR/PTEN/Akt and RAS/Raf/ERK signalling pathways; (**b**) Rb G1→S cell cycle checkpoint have previously been shown to be genetically and epigenetically altered in glioblastoma. MicroRNAs that are dysregulated in glioblastoma and target these pathways may exacerbate this aberrant signalling and further promote glioblastoma pathogenesis. Up-regulated microRNAs are shown in *red* and down-regulated microRNAs are shown in *green*

mediators while *miR-29b*, *miR-125a*, and *miR-146b* have been demonstrated to act as suppressors of GBM invasion and migration. In addition to its function in cell proliferation/cycling, high *miR-21* expression also promotes cell migration and invasion in variety of cancer cell types including GBM, breast and prostate cancers. Gene expression analysis of glioma cells with *miR-21* knockdown shows *miR-21* targets are enriched for functions in extracellular matrix modeling and vascular development. Specifically, RECK and TIMP3, which are inhibitors of matrix metalloproteases (MMPs), have been shown to be directly targeted by *miR-21* to promote GBM cell migration and invasion (Connolly et al. 2010; Gabriely et al. 2008). The Tropomyosin 1, PDCD4, Maspin, RhoB, and MARCKS loci, which have been shown to be important *miR-21* targets in breast and/or prostate cancer cell migration (Connolly et al. 2010; Li et al. 2009c) are also likely to play important roles in glioma cell migration.

MiR-10b was first reported to promote breast cancer cell migration and invasion, by up-regulating expression of pro-metastatic gene *RhoC*, indirectly via inhibition of the homeobox gene *HOXD* (Ma et al. 2007). High *miR-10b* expression has been shown to correlate with increased grade and invasive phenotypes in gliomas (Sasayama et al. 2009). Importantly, *miR-10b* represents one of the most highly expressed miRNA (up to 208 fold higher than normal brain) reported by two other studies of primary GBM (Huse et al. 2009; Silber et al. 2008). Although direct targets of *miR-10b* in gliomas remains to be identified, Sasayama et al. (2009) observed a significant correlation of *miR-10b* expression in gliomas with up-regulation of

RHOC as well as the urokinase receptor uPAR which has been shown independently to directly promote glioma cell invasion and migration (Veeravalli et al. 2010).

MiR-29b, *miR-125a*, and *miR-146b* are all expressed at low levels in primary GBM. Ectopic expression of *miR-29b* and *miR-125a*, inhibits invasive and migratory phenotypes in glioma cell lines in part via regulation of podoplanin (Cortez et al. 2010) while *miR-146b* has been shown to target MMP16 expression (Xia et al. 2009). Many of the miRNAs implicated in glioma cell cycle/survival/death, including *miR-34a* and *miR-7* also modulate migratory and invasive phenotypes of GBM cells. Similarly, *miR-222*, which is up-regulated and targets the Akt pathway in GBM, has been shown to mediate squamous cell carcinoma metastasis by targeting MMP1 (Liu et al. 2009a). Thus we can anticipate that further insights into how specific miRNA mediate glioma cell migration and invasion will emerge from studies in other neoplasms.

15.2.2.3 Angiogenesis and Tumor Metabolism

Microvascular proliferation and neo-angiogenesis is one of the most intensely studied and prominent histo-pathological hallmarks of GBM and has been a major area of focus for novel therapeutics. It is increasingly recognized that tumor angiogenesis is driven by a complex balance of pro-angiogenic (VEGF/VEGFR, PDGF/PDGFR) and anti-angiogenic (TSP-1/2) factors elaborated by the tumor cells and the surrounding niche, as well as a network of genes that regulate tissue remodeling/degradation (MMPs, TIMPs), cellular motility and maintain or promote tissue hypoxia (HIFs, MYC) (Yoo et al. 2009a). Thus many of the oncogenes and tumor suppressor genes as well as miRNAs implicated in cell migration and invasion are also likely to contribute to angiogenesis and tumor metabolism in GBM.

The importance of the tumor:niche relationship in GBM angiogenesis was most recently highlighted in a study by Wurdinger et al. (2008) which discovered *miR-296* as a pro-angiogenic downstream effector of VEGF produced by human microvascular endothelial cells after co-culture with U87 glioma cells (Wurdinger et al. 2008). Using loss and gain of function experiments they demonstrated that *miR-296* promoted tubulogenesis in in vitro models of angiogenesis. Notably, endothelial cells isolated from primary human GBM tumor vessels expressed higher levels of *miR-296* than normal endothelial cells and administration of *miR-296* antagomirs significantly inhibited angiogenesis in a glioma xenograft model. Human hepatocyte growth factor substrate (HGS), which normally regulates degradation and levels of the PDGFR-B and VEGFR2 in endothelial cells, was identified as a direct *miR-296* target, thus high levels of *miR-296* promotes angiogenesis indirectly via up-regulation of receptors for the pro-angiogenic ligands PDGF and VEGF in tumor associated endothelial cells.

Although the spectrum of miRNAs involved in glioma angiogenesis remains to be defined, it is anticipated that many miRNAs with functions in normal vascular development and angiogenesis will have overlapping roles in tumor-driven angiogenesis. Indeed, *miR-34a*, which has tumor suppressor activity in glioma cells has been recently shown to inhibit physiologic angiogenesis by inducing senescence in endothelial progenitor cells (Zhao et al. 2010), and *miR-221/222* which has oncogenic effects in GBM, modulates the activity of the angiogenic factor SCF on endothelial cells by targeting its receptor c-kit (Poliseno et al. 2006).

Hypoxia has been shown to be a potent driver and determinant of GBM biology, including tumor grade, invasiveness and vascularity (Bao et al. 2006; Evans et al. 2004; Vordermark 2005). Functional synergy between the MYC oncoprotein and the HIF family of transcription factors (HIF-1/2), which are frequently dysregulated and respectively shown to have oncogenic and tumor suppressor roles in GBM (Huse et al. 2009; Li et al. 2009f; Mendez et al. 2010) have now been demonstrated to play central roles in the adaptive response of tumor cells to hypoxic environments by promoting anaerobic metabolism and angiogenesis. Both c-Myc and HIF-1 have been shown to directly activate a number of miRNAs whose target genes have functions in metabolism and angiogenesis. Specifically, HIF and Myc synergistically activate VEGFA and glycolytic enzymes HK2 (hexokinase 2), LDHA, TFRC, and PDK1 (pyruvate dehydrogenase kinase 1) (Kim et al. 2007). The Myc activated *miR-17-92* cluster of miRNAs inhibit translation of TSP-1 a potent anti-angiogenic tumor suppressor (Dews et al. 2006) while *miR-23a/b* promotes mitochondrial glutamine metabolism (Chang et al. 2008; Gao et al. 2009). Similarly, HIF-1 activated *miR-210* targets ISCU (iron-sulfur cluster protein) an essential cofactor for oxidative metabolism (Favaro et al. 2010). Recent demonstration of HIF-1a as a target of the *miR-17-92* cluster in lung cancer cell lines indicates a complex feedback loop (Taguchi et al. 2008), further highlights the role of miRNAs in modulating the role of the Myc-HIF axis in tumor hypoxia and metabolism. Other miRNAs implicated in glioma metabolism include *miR-451* and *miR-326* which respectively target the glycolytic enzymes LKB1/AMPK (Godlewski et al. 2010) and PKM2 (Kefas et al. 2010). Significantly, miRNAs of the *miR-17-92* cluster, *miR-210* and other hypoxia associated miRNAs (Kulshreshtha et al. 2007) are among the most highly up-regulated miRNAs detected in primary GBM. Expression of many of these miRNAs also parallel acquisition of the hallmark traits of GBM during astrocytic tumor progression and have pro-survival functions thus indicating that they may be of therapeutic significance.

15.2.2.4 Differentiation and Self-renewal

Substantial evidence now support important physiological roles of miRNAs in regulating renewal and differentiation in stem cells derived from a spectrum of tissues and cell types including neural stem/progenitors cells (NSCs) (Gangaraju and Lin 2009), the presumed cell of origin for malignant brain tumors. Recently a small fraction of tumor cells named "cancer stem cells (CSCs)", which exhibit the hallmark stem cell property of self-renewal and multipotent differentiation potential, and potent tumor initiation and maintenance properties, have been isolated from a variety of solid tumors including GBM (Galli et al. 2004; Singh et al. 2004). GBM derived CSCs are distinguished and isolated experimentally through their specific expression of the marker CD133 (Beier et al. 2007; Liu et al. 2006). Phenotypically, these cancer stem cells have a higher potential for proliferation, invasion, and resistance to chemotherapy and radiotherapy (Bao et al. 2006; Liu et al. 2006). Notch signaling has been implicated in mediating the growth and tumorigenic potential of GBM-CSCs (Xing et al. 2010).

RNA and miRNA expression profiles of primary GBM indicate substantial overlap with the transcriptomes of multipotent NSCs. Specifically miRNAs that are highly enriched in pluri-potent ESCs including the *miR-17* family, *miR-183-182*, *miR-302-367*, and *miR-371-373* clusters are also upregulated in GBMs and NSCs. In addition miRNAs which are important for neural differentiation (*miR-7*, *miR-9*, *miR-124*, *miR-137*) are underrepresented in GBMs and undifferentiated NSCs (Lavon et al. 2010). Whether GBM and GBM-CSCs derive directly from transformation of normal neural precursor cells (NPC) or whether CSCs represent variably committed/differentiated NPCs that have been reprogrammed by specific cancer-specific genetic alterations to maintain "stem cell properties" of self-renewal and multi-potent differentiation capacity (Zheng et al. 2008a) remains unclear. Importantly, manipulation of specific miRNAs, with normal functions in stem cell renewal/differentiation, can alter the tumorigenic potential of GBM cells and GBM-CSCs in vitro and in vivo. Specifically, ectopic expression of *miR-124* and *miR-137* potently induced differentiation and cell cycle arrest in GBM and GBM-CSCs, by targeting CDK6 and activating the RB1 pathway (Silber et al. 2008), while *miR-128* expression inhibited proliferation and self renewal in GBM cells by targeting *Bmi-1*, an oncogenic neural stem cell renewal factor (Godlewski et al. 2008).

Activation of the SHH and NOTCH developmental signaling pathways which normally regulates NSC fate and differentiation also promotes growth and radioresistance of GBM-CSCs (Bar et al. 2010; Clement et al. 2007; Wang et al. 2010a; Xu et al. 2008). Notably, expression of *miR-34a*, which is found at low levels in GBM, inhibits GBM-CSCs differentiation in part via regulation of Notch 1 and 2 (Guessous et al. 2010).

Studies to date on the role of miRNAs in GBM or GBM-CSCs have been limited by sample numbers and microarray platforms. Of note, consistent with the emerging theme of molecular heterogeneity in GBM, mRNA expression profiling studies indicate GBM-CSCs may also be segregated on the basis of specific signaling pathway signatures (Beier et al. 2007; Lottaz et al. 2010). Thus more extensive investigations of the complete miRNA spectra expressed in CSCs isolated from large number of GBMs will help to identify miRNAs that sustain core signaling pathways required for maintenance of the potent tumorigenic potential of GBM-CSCs.

15.3 Embryonal Brain Tumors

15.3.1 Medulloblastoma

Embryonal brain tumors comprise a genetically heterogeneous group (Pomeroy et al. 2002); among these, MB are the most common and have also been most extensively studied. MBs affect ~1/200,000 school age children per year in the USA (McNeil et al. 2002) and are diagnosed on the basis of primitive neuroectodermal histology and cerebellar location. Up to 30–40% of MB exhibit loss of 17p with or without gain of 17q; 9q, 6q and 10q loss are also seen at lower frequency (Lamont et al. 2004; Pfister et al. 2009a). Unlike in GBM, focal recurrent genomic alterations are relatively infrequent and generally found in only 10–20% of primary MB targeted loci including the *MYCC/MYCN*, *CDK/CYCLIN D*, *GLI*, and *OTX2* oncogenes (Northcott et al. 2010). Recent studies reveal deletions of multiple genes which control histone methylation and suggest epigenetic alterations may contribute substantially to MB pathogenesis (Northcott et al. 2009b). Alterations in several signaling pathways that regulate normal cerebellar and neural stem development, including Sonic Hedgehog (SHH), Wingless (WNT) and NOTCH have been shown to be important in MB development (Gilbertson and Ellison 2008).

In addition to animal models, causative roles for SHH, WNT, and NOTCH pathways have been established through studies of Gorlin and Turcot syndromes (Goodrich et al. 1997) and demonstration of direct alterations of SHH (PTCH, SUFU, SMO, GLI) or WNT pathway (APC, β-catenin) genes in 10–20% of sporadic medulloblastoma (Ellison et al. 2005; Northcott et al. 2010; Taylor et al. 2002; Yokota et al. 2002) and *NOTCH* gene amplification (Fan et al. 2004) in a proportion of MB. In addition, the p53/RB1 pathway is implicated in MB based on predisposition to brain tumors in Li Fraumeni syndrome patients, animal studies (Shakhova et al. 2006; Wetmore et al. 2001), and amplification/up-regulation of *CDK/CYCLIN D* genes in primary tumors (Mendrzyk et al. 2005).

Medulloblastoma can be grouped on the basis of histology into three major subgroups – classic, desmoplastic, and large cell anaplastic (LCA), which have been correlated by various studies with clinical phenotypes (Lamont et al. 2004). Most consistently, favorable outcomes have been reported for desmoplastic medulloblastoma arising in young children (Rutkowski et al. 2005), while large cell anaplastic tumors are reported to have more aggressive clinical phenotypes and are enriched for *MYCC/MYCN* amplification (Eberhart and Burger 2003; Gajjar et al. 2004; Gilbertson and Gajjar 2005).

Recent large gene expression profiling studies by several groups have resulted in the classification of MB into 4 or 5 molecular subgroups with distinct gene expression, genomic as well as clinical characteristics (Kool et al. 2008; Northcott et al. 2009a; Thompson et al. 2006). These included two well defined groups distinguished by WNT or SHH signaling, which correlated respectively with classic and desmoplastic histology; both groups also exhibited NOTCH/PDGF signaling. The other 2 or 3 groups were closely related and less well defined with signatures enriched for neuronal or photoreceptor differentiation genes, increased c-Myc expression, metastatic phenotypes and overall worse survival. Substantial evidence indicate SHH-associated MB arise from cerebellar granule neural precursors (CGNPs) while initiating cells for the other subtypes remain unknown but are likely to be more primitive neural progenitors residing in subventricular germinal centres in the brain.

15.3.2 MiRNAs in Medulloblastoma

15.3.2.1 MB MiRNAs, MYC, and SHH Signaling

SHH serves as a vital mitogen during post-natal cerebellar development to trigger cell cycle progression in CGNPs by binding PTCH and releasing SMO, from PTCH-mediated inhibition. SMO acts via the GLI family of proteins, through yet unclear mechanisms, to modulate expression of critical cell cycle mediators such N-Myc and the D-cyclins. In primary MB, mutations of PTCH, SMO, and amplification of *GLI*, *MYC*, and *CDK/CYCLIN* genes, contribute to up-regulation of SHH signaling (Oliver and Wechsler-Reya 2004). Not surprisingly several recent studies have implicated miRNAs in regulation of SHH signaling in CGNPs and MB cells. Initial studies by Ferretti et al. (2008) utilized miRNA profiling of 32 MB and high Gli expression to identify SHH signaling miRNAs and showed interestingly that many miRNAs were down-regulated with SHH signaling. Significantly, they demonstrated that a subset of three miRNAs with low expression in high Gli tumors, *miR-125b*, *miR-324-5p*, and *miR-326* commonly targeted SMO activity and inhibited growth and transforming activity in several MB cell lines. Furthermore they observed that depletion of all three miRNAs inhibited differentiation, and promoted SHH-induced proliferation of normal murine CGNPs, consistent with functional importance in SHH-driven MB transformation (Ferretti et al. 2008).

Furthermore, several recent global profiling studies demonstrate that *miR-17-92*, a miRNA cluster with proven oncogenic activity in several human neoplasms, is up-regulated in murine and human MB. Uziel et al. (2009) discovered using deep sequencing that *miR-17-92* miRNAs were over-expressed in MB from Ptc$^{-/-}$ mice (Uziel et al. 2009) while genomic amplification of *miR-17-92* was reported in 6% of 201 primary MB characterized using genomic arrays high resolution copy number and FISH analysis (Northcott et al. 2009a). Notably, both groups demonstrated that *miR-17-92* up-regulation significantly correlated with SHH as well as *MYCC/MYCN* activation in MB. Indeed, *miR-17-92* expression was increased by Shh-treatment or ectopic Myc expression in primary CGNPs, consistent with prior reports that *miR-17-92* is a transcriptional target of c-Myc in B cell lymphoma (O'Donnell et al. 2005). Importantly, ectopic *miR-17-92* accelerated MB development in Ptc heterozygote mice (Uziel et al. 2009), confirming a functional relationship between *miR-17-92* and SHH in both normal and neoplastic CGNP growth. The specific mechanism by which *miR-17-92* augments oncogenic signaling by the SHH pathway is not known, but is probably related to the transforming functions of MYC in MB development. Of note, Northcott et al. (2009a) also observed high *miR-17-92* levels in non-SHH tumors with high Myc levels, thus indicating a tight functional relationship between the 2 loci (Northcott et al. 2009a). It is possible that as in lymphoma (He et al. 2005), *miR-17-92* activation is needed for full manifestation of Myc oncogenic function in MB. Although specific targets of *miR-17-92* in MB remain to be identified, several *miR-17-92* predicted targets, including *Bmi-1*, *PTEN*, and *PP2A* identified recently in leukemia (Mavrakis et al. 2010), are known to contribute to MYC transformation (Arnold and Sears 2008; Haupt et al. 1993; Zheng et al. 2008b), and represent attractive candidate targets.

15.3.2.2 MB MiRNAs and NOTCH Signaling

NOTCH signaling plays an important role during normal CNS development to inhibit neuronal differentiation and maintain neural stem/progenitor cell pools. In addition to gliomas, NOTCH signaling has also been implicated in MB based on up-regulation and/or gene amplification of various NOTCH pathway components including NOTCH1/2 and HES1/5 in primary MB and MB models (Fan et al. 2004; Hallahan et al. 2004; Northcott et al. 2009a). Notably, pharmacologic inhibition of NOTCH signaling promotes cell death and differentiation and inhibits transformation and xenograft growth of MB cells. Furthermore, NOTCH inhibition diminishes a proportion of $CD133^+$ MB cells in culture indicating that NOTCH activation is important for maintenance of CSCs pools (Fan et al. 2006).

Two miRNAs, *miR-199b* and *miR-326*, which are differentially expressed and correlated with more aggressive MB phenotypes, have been implicated in NOTCH signaling (Ferretti et al. 2009; Garzia et al. 2009). Garzia et al. (2009) demonstrated that expression of *miR-199b* targets the NOTCH effector – HES 1, and inhibits MB cell proliferation. Notably, *miR-199b* expression inhibited xenograft formation and depleted stem-like cell pools, indicating that loss of *miR-199b* expression may contribute to aberrant NOTCH signaling that maintains and promotes the survival of MB stem-like cells. A recent study identified NOTCH 1 and 2 as direct targets of *miR-326* in malignant glioma (Kefas et al. 2010), thus *miR-326* may similarly function to regulate NOTCH signaling in MB.

15.3.2.3 MB-associated MiRNAs and Neuronal Differentiation

Several miRNAs including *miR-7*, *miR-9*, and *miR-124* have been consistently reported to be underexpressed in MB (Ferretti et al. 2009; Northcott et al. 2009a; Uziel et al. 2009). Ectopic expression of *miR-9* and *miR-124* suppress MB proliferation and survival by targeting the TrkC receptor (Ferretti et al. 2009), CDK6 (Pierson et al. 2008) and SLC16A1, a lactic acid efflux carrier in aerobic glycolysis (Li et al. 2009a). Interestingly, both *miR-9* and *miR-124a* have pro-neurogenic functions and act in a feedback loop to antagonize REST – a critical regulator of normal neurogenesis which normally functions to represses neurogenesis by acting together with chromatin redmodelling complex proteins to inhibit expression of neuronal genes in non-neuronal tissues (Conaco et al. 2006; Packer et al. 2008). Notably BAF34a, a chromatin complex protein and an important REST co-factor has been identified as a direct target of *miR-9* and *miR-124* (Yoo et al. 2009b). Therefore, downregulation of pro-neural miRNAs in MB may reflect a general process during transformation of NPCs to favor the maintenance of multi- or pluripotent state of tumor cells.

Other miRNAs implicated in MB include *miR-30b* and *miR-30d*, which were reported as part of an amplicon containing the *KHDRBS3* gene on 8q24.22 in

MB cell lines (Lu et al. 2009). *MiR-30b/d* and *KHDRBS3* expression correlated with copy number and was up-regulated in a subset of primary medulloblastoma. However, the function of these putative oncogenic genes and their significance to MB biology remain to be investigated. Interestingly, *miR-30b/d* was identified as a Myc repressed target in a model of B cell lymphoma, though no significant phenotype was observed with ectopic expression *miR-30b/d* (Chang et al. 2008).

15.3.3 CNS-PNETs

CNS-PNETs are embryonal tumors that arise at about 1/10 the frequency of MB. Although frequently histologically indistinguishable from MB, CNS-PNET may arise in different intracranial locations, most commonly in the cerebrum of much younger children (most < 5 years of age), and have substantially inferior outcomes than MB. Diagnosis and classification of these tumors is highly challenging as they may share resemblance with high grade hemispheric gliomas; furthermore several subtypes are described based on location (supra-tentorial PNET) or variant histological features (Gessi et al. 2009; Louis et al. 2007). In addition, due to the relative rarity of CNS-PNET, very few studies have sought to characterize the molecular basis for these tumors (Li et al. 2005). Cumulative data, however, indicate that they are distinguishable from MB in gene expression (Pomeroy et al. 2002) and global patterns of genomic alterations (Li et al. 2009b). Notably, they rarely exhibit an isochromosome 17 alteration or *MYC* gene amplification that characterizes a proportion of MB. CNS-PNET may arise in the context of the Li-Fraumeni, DNA mismatch repair or RB cancer predisposition syndrome (Li et al. 2005). Signaling pathways implicated in CNS-PNET mirror those implicated in MB and include the RB1/p53, NOTCH, and WNT pathways. In contrast to MB, specific animal models of CNS-PNET are limited. Transgenic mice with SV40-targeted inactivation of RB1 develop midbrain tumors with features of PNET (Brinster et al. 1984), while CNS-PNET in various locations (brain stem, cerebellum, cerebrum, pineal) develop in $p53^{-/-}$ mice with INKc/d allelic deletions (Saab et al. 2009; Zindy et al. 2003). Furthermore our studies of primary CNS-PNET reveal genomic alterations targeting one or more of the *CDK/CYCLIN D* genes (Li et al. 2009b), thus providing further support that alterations of the p53/RB1/INK4 pathway is important in a subgroup of CNS-PNET. NOTCH pathway activation associated with amplification of NOTCH1/2 ligand genes have been reported in a proportion of CNS-PNET (Fan et al. 2004). Evidence supporting a role for WNT activation in CNS-PNET include amplification or up-regulation of WNT pathway genes in CNS-PNET (Li et al. 2009b; Rogers et al. 2009), and induction of PNET like tumors in $p53^{-/-}$ mice by expression of c-Myc and activated β-catenin (Momota et al. 2008).

It is presently unclear whether the spectrum of non-MB CNS-PNET, have common cellular origins. It is presumed that they derive from subventricular zone NPCs, although interestingly Momota et al. (2008) reported successful generation of tumors with CNS-PNET phenotypes by targeting c-Myc and activated β-catenin to GFAP but not Nestin expressing cells.

15.3.3.1 MiRNAs in CNS-PNET

We recently reported a recurrent large amplicon targeting two miRNA clusters – the C19MC and the *miR-371-373* cluster on 19q13.41, using copy number analysis of CNS-PNET arising in the cerebrum (also called supra-tentorial PNET) (Li et al. 2009b). Notably, only a subset of miRNAs mapping to the C19MC cluster were over-expressed in tumors with genomic amplification. We also demonstrated that C19MC amplification or up-regulation of specific C19MC miRNAs correlated with very poor survival indicating that activation of C19MC may identify a clinically distinct group of CNS-PNET. Indeed, in more recent studies we (Huang and Hawkins, unpublished) and others (Pfister et al. 2009b) detected the C19MC amplicon in CNS-PNET arising from various locations indicating that these tumors potentially share common lineages. Consistent with this hypothesis, we demonstrated common gene expression signatures in C19MC amplified tumors characterized by enrichment of early cell lineage genes, WNT pathway components and cell survival genes which were suggestive of C19MC miRNA functions in neural differentiation and cell survival. Indeed, ectopic expression of *miR-520g* and *miR-517c*, two of most highly expressed C19MC miRNAs, inhibited differentiation of human NSCs, promoted survival of various PNET cell lines and importantly enhanced tumor xenograft growth.

The C19MC locus which encodes 54 predicted miRNAs represents the largest known miRNA cluster in the human genome and was only recently discovered likely due to its lack of strict conservation in non-primates (Bentwich et al. 2005). Recently up-regulation of C19MC secondary to 19q13.41 translocations was reported in a proportion of thyroid adenomas and up-regulation of *miR-520g*, *miR-516-3p*, and *miR-520c* have been linked to more aggressive breast cancer phenotypes (Foekens et al. 2008; Lowery et al. 2009) and ectopic *miR-520c* has been shown to promote breast cancer cell migration and invasion by targeting CD44 (Huang et al. 2008). The ABCG2 transporter has also been reported as a target of *miR-520h*, which shares nearly identical seed sequence with *miR-520g* (Wang et al. 2010b). However, the biological and molecular function of the majority of miRNAs in this large cluster and their role in oncogenesis remains unclear. Of note, a significant number of C19MC miRNAs are highly expressed in ESCs and down-regulated upon differentiation (Bar et al. 2008). Our experimental data indicates expression of *miR-520g* and *miR-517c* in hNSCs modulates and favors non-canonical WNT/JNK signaling which has been implicated in stem cell maintenance (Rattis et al. 2004). Intriguingly, C19MC miRNAs share seed sequence similarities with members of the Zebra fish *miR-430* family which are also encoded by a large, poorly conserved miRNA cluster and have functions in brain morphogenesis (Giraldez et al. 2005). Specific targets of these miRNAs remain to be established, and is hindered by the lack of robust conservation of C19MC miRNA target seed sequence to other miRNA families in humans and other organisms and will likely be advanced by studies of functional orthologos such as the *miR-430* cluster.

15.4 Prospects for Translation into the Clinic

15.4.1 MiRNAs as Brain Tumor Diagnostic and Predictive Tools

Current treatment decisions for most brain tumors primarily rest on classification of likely cell lineage based on tumor morphology and semi-quantitative immune-histochemical assays. Although gliomas represent the most common group of tumors diagnosed in adults and children, criteria for tumor grading which is critical for treatment stratification and prognostication, remains poorly defined as evidenced by up to 30% discordance in central pathology reviews in therapeutics trials (Pollack et al. 2003). Similarly, diagnoses of embryonal brain tumors which characteristically exhibit varying neuro/glial or epithelial differentiation, can be frequently contested even amongst experts (Burger 2006), as reflected in new descriptions and reclassification of histologic subtypes (Gessi et al. 2009; Judkins and Ellison 2008; Louis et al. 2007).

Various studies have now shown that miRNA expression differs markedly between tumor and normal tissue, and between tumor subtypes and thus has the potential to be exploited for improved cancer diagnosis (Calin et al. 2005; Cho 2009, 2010a; Rosenfeld et al. 2008; Volinia et al. 2006). Recently, a 48 miRNA expression pattern was shown to identify tissue of origin in up to 80% of 336 primary and 22 metastatic tumors. Notably, *miR-92b* and *miR-9/9** expression specifically identifies and distinguishes intrinsic from metastatic tumors in the brain (Nass et al. 2009). Lu et al. (2005) performed miRNA profiling in a large cohort of normal and neoplastic tissues and demonstrated that expression signatures of ~200 miRNAs provided more robust classification than expression profiles of > 16,000 mRNAs and importantly identified tumor types, lineage and differentiation state (Lu et al. 2005).

Recent mRNA expression studies demonstrating several subgroups of tumors within the morphologic umbrella of GBM underscores the molecular diversity of gliomas and challenges in developing more specific treatments. Importantly, studies of gliomas, ependymomas and medulloblastoma indicate tumor cell of origin as predicted from gene expression signatures of primary tumors, may determine tumor phenotypes and therapeutic response. As a substantial proportion of brain-enriched miRNAs are highly lineage or cell type restricted (Sempere et al. 2004), miRNA expression may represent more powerful diagnostic and predictive tools in clinical management of brain tumors.

To date, most published studies on brain tumor miRNA expression have been performed on a small number of tumors and with limited miRNA array platforms. Nonetheless these studies confirm that miRNA profiles in normal and neoplastic brain tissues are distinguishable and can also identify specific miRNAs which correlate with brain tumor progression and patient prognosis (Ferretti et al. 2009; Silber et al. 2008). Specifically, increased levels of *miR-21*, which is one of the most highly expressed miRNAs in GBM, high *miR-221* and decreased *miR-125*, *miR-137* (Silber et al. 2008), and *miR-181b* expression has been reported to parallel progression of grade II to grade III/IV gliomas (Conti et al. 2009; Malzkorn et al. 2009). These studies highlight the potential for miRNAs to be used for glioma classification and prognostication. Global miRNA profiles have been established as part of the TCAG project on GBM. However, comprehensive, integrated analysis of large data sets of miRNA profiles with mRNA and genomic profiles, and how they relate to the different molecular subgroups of GBM described recently remains to be undertaken. It is anticipated, given the data from studies of other tumors and the importance of miRNAs in determination of neural lineages, that miRNA analysis will further delineate and refine the clinical relevance of the proneural, neural and mesenchymal subgroups of GBM.

Studies of the *miR-17-92* locus in MB and C19MC amplicon in CNS-PNET, indicate that miRNA expression patterns can also define subgroups of embryonal tumors. We first reported an oncogenic role for the C19MC miRNAs in CNS-PNET; importantly, our studies indicate C19MC identifies a highly aggressive subgroup of CNS-PNET with common gene expression signatures and very likely, also common cell of origin. To date, we have examined over 500 malignant childhood brain tumors including malignant glioma, ependymomas, choroid plexus carcinomas, MB and CNS-PNET arising in various locations, and detected C19MC amplification only in a subset of CNS-PNET distinguished by ependymal differentiation, very poor survival and younger age (< 4 years) (Li et al. 2009b). Similarly, Pfister et al. (2009b) reported that C19MC amplification was found almost exclusively in CNS-PNET with features of ependymoblastic differentiation (Pfister et al. 2009b), thus C19MC amplification may be an important diagnostic and treatment stratification tool, similar to INI1 gene alteration, for childhood CNS tumors.

Ferretti et al. (2009) showed using miRNA profiles of 34 MB that expression levels of specific miRNAs can segregate histologic subtypes of MB and clinical risk groups (Ferretti et al. 2009). Of note, *miR-17-92* has been linked to Myc activation – which is generally correlated with negative outcome and more aggressive anaplastic MB histology, but also with a SHH signature which is characteristic of more favorable desmoplastic MB subtype (Lamont et al. 2004). Clearly, future studies which correlate large miRNA/mRNA/genomic data sets with histo-pathological and patient information will be needed to develop a more robust model of subtype MB for diagnostic and prognostic application.

15.4.2 MiRNAs as Biomarkers

There is now increasing recognition with development of pathway specific cancer therapeutics that appropriate biomarkers are needed not only for tailoring treatment but also to evaluate therapeutic efficacy. For example, PTEN loss has been shown to modulate and predict response of GBM with EGFR amplification/over-expression to EGFR inhibitors (Wang et al. 2006). Similarly miRNAs expression may also predict or modify glioma response to radiation or specific chemotherapy. Specifically, *miR-21* expression has been correlated with radiation resistance (Chaudhry et al. 2010), and sensitivity to taxol (Ren et al. 2010) and VM26 (Li et al. 2009c). To date, use of blood or serum to detect conventional signaling molecules as biomarkers for CNS tumors has met with limited success. With reports of surprising miRNA stability in body fluid (Mitchell et al. 2008), miRNAs have thus emerged as attractive and easily accessible blood-based biomarkers for CNS tumors. Furthermore, the detection of miRNAs in acellular serum microvesicles (Skog et al. 2008) suggest miRNAs may also be successfully detected in cell-free CSF. The ability to detect tumor specific miRNAs in CSF will have significant therapeutic implications particularly in treatment of embryonal brain tumors, such as MB and CNS-PNET, where CSF cytology, a relatively crude analysis, remains the gold standard method for staging and treatment risk stratification.

15.4.3 MiRNAs as Targets for Brain Tumor Treatment

The implication of miRNAs in hundreds of cancers has brought obvious attention to targeting miRNAs in cancer treatment. A single miRNA often targets multiple mRNAs that comprises one or several pathways that specify multiple cellular functions such as differentiation, proliferation, apoptosis or cell growth. Therefore, inhibition or correction of a critical tumor oncogenic or tumor suppressor miRNA at a critical regulatory hub potentially represents a potent and efficient cancer therapeutic.

15.4.3.1 Antisense MiRNA Treatment

MiRNA based therapy rely on reintroducing significantly down-regulated miRNAs or delivering anti-miRNA oligonucleotides to competitively bind or enhance degradation of miRNAs up-regulated in cancer. Ideal properties of an anti-miRNA include in vitro and in vivo stability, and high specificity and binding affinity for miRNA targets. Unmodified antisense DNA-based oligonucleotides, first used in *Drosophila* embryos (Boutla et al. 2003), have poor efficacy in vivo due to low target affinity (Davis et al. 2006) and susceptibility to cellular nucleases. Since initial reports that 2′-*O*-methyl modification enhances anti-miRNA stability, various modifications including phosphorothioated linkage substitutions in the DNA backbone to increase RNAse resistance, lipophilicity and tissue uptake (Davis et al. 2006), have been incorporated into “anti-miRNAs”. LNA modification, which introduces a rigid bicycle bridge between the 2′-oxygen and the 4′-position of the anti-miRNA through a methylene linker, provides excellent specificity and mismatch discrimination (You et al. 2006), and high in vivo stability, low toxicity, and high solubility (Vester and Wengel 2004), and have shown promising effects in mice (Elmen et al. 2008b; Fluiter et al. 2003), and non-human primates (Elmen et al. 2008a). Triply modified anti-miRNAs with a 2′-*O*-methyl/LNA chimeric oligonucleotide and phosphorothioated backbone exhibit highly potent and specific miRNA knockdown in vitro even at nanomolar concentrations, and thus may represent anti-miRNA pharmacologics with greater in vivo therapeutic index. In recent miRNA therapeutic models, lentiviruses or adenoviruses expressing one or multiple miRNAs or antagomirs have been successfully used to silence or restore expression of single or multiple miRNAs, respectively (Kota et al. 2009; Scherr et al. 2007). Similarly adenovirus co-expressing p53 and p21-targeting miRNAs has been used successfully to enhance tumor cell chemo-sensitivity in vitro and in vivo (Idogawa et al. 2009). These studies highlight the scope of anti-miRNA applications that remain to be realized for cancer therapeutics.

15.4.3.2 Selecting Candidate MiRNAs to Target for Treatment

MiRNA expression is highly cell type and tissue specific, therefore a therapeutic target that is most differentially expressed between tumor and normal tissue, and is a significant driver of tumor growth is ideal for cancer therapy (Cho 2010b). Several miRNAs, including *miR-21*, *miR-17-92*, and C19MC, implicated in brain tumors exhibit high level genomic amplification or expression reminiscent of traditional oncogenes, such as MYC. Similar to the MYC oncogenes in childhood MB and neuroblastoma, the C19MC amplicon has been reported to be selected for and represented in almost all tumor cells in recurrent CNS-PNET (Pfister et al. 2009b), underscoring the importance of the C19MC oncogenic miRNAs as critical drivers in CNS-PNET. Similarly the *miR-17-92* locus, which has been shown to potently enhance c-MYC driven lymphomagenesis, is an attractive target for medulloblastoma (He et al. 2005). A recent study used an inducible *miR-21* to trigger lymphomagenesis, thus demonstrating that *miR-21* is a robust oncogene in vivo (Medina et al. 2010) and an attractive candidate for glioma therapy. Significantly, this study also demonstrated that expression of *miR-21* alone is sufficient to induce lymphomagenesis and that tumor regression accompanied withdrawal of *miR-21* expression, indicating that tumor cells may exhibit "addiction to miRNA oncogenes" in the same manner as seen with conventional oncogenes such as MYC (Jain et al. 2002). This study highlights oncomirs such as C19MC, *miR-17-92* and *miR-21* as highly attractive anti-cancer targets.

As monotherapy is exceptional in clinical oncology, it is likely that specific anti-miRNAs will be most useful in combination with other novel therapies or conventional chemo- or radio-therapeutics. For example, anti-*miR-21* has been shown to synergize with secreted TRAIL (Corsten et al. 2007) and conventional chemotherapeutics Taxol (Ren et al. 2010) and VM26 (Li et al. 2009) to induce apoptosis in glioma cell lines in vitro and in vivo. Furthermore, as many miRNAs have functions in differentiation, manipulation of specific miRNAs, such as those of the C19MC oncogenic cluster, may be effective as differentiation therapies alone or in combination with conventional agents, particularly for embryonal CNS tumors.

15.4.3.3 Challenges to Brain Tumor Treatment Using Anti-miRNA Oligonucleotides

As in conventional brain tumor therapy, the relative impermeability of the blood brain barrier will represent one of the most significant obstacles in applying anti-miRNAs therapy to brain tumors. Although systemic administration of anti-miRNAs can effectively target miRNA levels in various tumors (Elmen et al. 2008a, 2008b; Fluiter et al. 2003; Krutzfeldt et al. 2007), and anti-miRNAs effectively silence target miRNAs when directly injected into mouse cortex, systemically delivered anti-miRs do not effectively cross the blood-brain barrier to alter target miRNA expression (Krutzfeldt et al. 2007). However, recent studies raise exciting possibilities of effective delivery of miRNA-based pharmacologics across the blood-brain barrier via nanoparticles (Liu et al. 2010; Zhang et al. 2004) and immunoliposomes (Zhang et al. 2004).

Additional challenges in realizing miRNA-based therapies in brain tumors may be the relative abundance of CNS-enriched miRNAs, indicating critical regulatory roles for miRNAs in neurogenesis (Kapsimali et al. 2007). Furthermore many of the large miRNA clusters, such as C19MC, which encode highly related miRNAs that differ by only a single nucleotide, may be particularly difficult to target with specific anti-miRNAs. Thus studies of the normal function of miRNAs in brain development will be critical for designing the best miRNA-based therapeutics.

15.5 Conclusion

This chapter has outlined the enormous potential for use of miRNAs in brain tumor diagnosis, prognosis and therapy. We can anticipate development of robust algorithms for brain tumor classification/diagnosis/prognosis and increased number of potential miRNAs to be targeted in brain tumors with emerging data from the rapidly advancing deep sequencing technologies. The field however remains at its infancy, with relative paucity of data regarding the physiological function of specific miRNAs in normal brain development. A significant limitation is the lack of genetically engineered brain tumors models driven by specific miRNAs, such as *miR-17-92*, C19MC, or *miR-21*, in neural progenitor/precursor cells. Such models are essential not only for elucidating the etiological importance and mechanisms of action of specific miRNAs in brain tumor pathogenesis but are pre-requisites for translating miRNA-based pharmacologics into clinical use for brain tumor treatment.

References

Arnold HK, Sears RC. A tumor suppressor role for PP2A-B56alpha through negative regulation of c-Myc and other key oncoproteins. Cancer Metastasis Rev. 2008;27:147–58.

Bao S, Wu Q, McLendon RE, et al. Glioma stem cells promote radioresistance by preferential activation of the DNA damage response. Nature. 2006;444:756–60.

Bar EE, Lin A, Mahairaki V, et al. Hypoxia increases the expression of stem-cell markers and promotes clonogenicity in glioblastoma neurospheres. Am J Pathol. 2010;177:1491–502.

Bar M, Wyman SK, Fritz BR, et al. MicroRNA discovery and profiling in human embryonic stem cells by deep sequencing of small RNA libraries. Stem Cells. 2008;26:2496–505.

Beier D, Hau P, Proescholdt M, et al. $CD133^{(+)}$ and $CD133^{(-)}$ glioblastoma-derived cancer stem cells show differential growth characteristics and molecular profiles. Cancer Res. 2007;67:4010–5.

Bentwich I, Avniel A, Karov Y, et al. Identification of hundreds of conserved and nonconserved human microRNAs. Nat Genet. 2005;37:766–70.

Bleau AM, Hambardzumyan D, Ozawa T, et al. PTEN/PI3K/AKT pathway regulates the side population phenotype and ABCG2 activity in glioma tumor stem-like cells. Cell Stem Cell. 2009;4:226–35.

Boutla A, Delidakis C, Tabler M. Developmental defects by antisense-mediated inactivation of micro-RNAs 2 and 13 in drosophila and the identification of putative target genes. Nucleic Acids Res. 2003;31:4973–80.

Brennan C, Momota H, Hambardzumyan D, et al. Glioblastoma subclasses can be defined by activity among signal transduction pathways and associated genomic alterations. PLoS One. 2009;4:e7752.

Brinster RL, Chen HY, Messing A, et al. Transgenic mice harboring SV40 T-antigen genes develop characteristic brain tumors. Cell. 1984;37:367–79.

Burger PC. Supratentorial primitive neuroectodermal tumor (sPNET). Brain Pathol. 2006;16:86.

Calin GA, Ferracin M, Cimmino A, et al. A microRNA signature associated with prognosis and progression in chronic lymphocytic leukemia. N Engl J Med. 2005;353:1793–801.

Chan JA, Krichevsky AM, Kosik KS. MicroRNA-21 is an antiapoptotic factor in human glioblastoma cells. Cancer Res. 2005;65:6029–33.

Chang TC, Yu D, Lee YS, et al. Widespread microRNA repression by myc contributes to tumorigenesis. Nat Genet. 2008;40:43–50.

Chaudhry MA, Sachdeva H, Omaruddin RA. Radiation-induced micro-RNA modulation in glioblastoma cells differing in DNA-repair pathways. DNA Cell Biol. 2010;29:553–61.

Cho WC. Role of miRNAs in lung cancer. Expert Rev Mol Diagn. 2009;9:773–6.

Cho WC. MicroRNAs: potential biomarkers for cancer diagnosis, prognosis and targets for therapy. Int J Biochem Cell Biol. 2010a;42:1273–81.

Cho WC. MicroRNAs in cancer – from research to therapy. Biochim Biophys Acta. 2010b;1805:209–17.

Clarke J, Butowski N, Chang S. Recent advances in therapy for glioblastoma. Arch Neurol. 2010;67:279–83.

Clement V, Sanchez P, de Tribolet N, et al. HEDGEHOG-GLI1 signaling regulates human glioma growth, cancer stem cell self-renewal, and tumorigenicity. Curr Biol. 2007;17:165–72.

Colman H, Zhang L, Sulman EP, et al. A multigene predictor of outcome in glioblastoma. Neuro Oncol. 2010;12:49–57.

Conaco C, Otto S, Han JJ, et al. Reciprocal actions of rest and a microRNA promote neuronal identity. Proc Natl Acad Sci USA. 2006;103:2422–7.

Connolly EC, Van Doorslaer K, Rogler LE, et al. Overexpression of miR-21 promotes an in vitro metastatic phenotype by targeting the tumor suppressor RHOB. Mol Cancer Res. 2010;8: 691–700.

Conti A, Aguennouz M, La Torre D, et al. MiR-21 and 221 upregulation and miR-181b downregulation in human grade II-IV astrocytic tumors. J Neurooncol. 2009;93: 325–32.

Corsten MF, Miranda R, Kasmieh R, et al. MicroRNA-21 knockdown disrupts glioma growth in vivo and displays synergistic cytotoxicity with neural precursor cell delivered s-trail in human gliomas. Cancer Res. 2007;67:8994–9000.

Cortez MA, Nicoloso MS, Shimizu M, et al. MiR-29b and miR-125a regulate podoplanin and suppress invasion in glioblastoma. Genes Chromosomes Cancer. 2010;49:981–90.

Crawford JR, MacDonald TJ, Packer RJ. Medulloblastoma in childhood: new biological advances. Lancet Neurol. 2007;6:1073–85.

Davis S, Lollo B, Freier S, et al. Improved targeting of miRNA with antisense oligonucleotides. Nucleic Acids Res. 2006;34:2294–304.

De Pietri Tonelli D, Pulvers JN, Haffner C, et al. MiRNAs are essential for survival and differentiation of newborn neurons but not for expansion of neural progenitors during early neurogenesis in the mouse embryonic neocortex. Development. 2008;135:3911–21.

DeAngelis LM. Brain tumors. N Engl J Med. 2001;344:114–23.

Dews M, Homayouni A, Yu D, et al. Augmentation of tumor angiogenesis by a myc-activated microRNA cluster. Nat Genet. 2006;38:1060–5.

Eberhart CG, Burger PC. Anaplasia and grading in medulloblastomas. Brain Pathol. 2003;13: 376–85.

Ellison DW, Onilude OE, Lindsey JC, et al. Beta-catenin status predicts a favorable outcome in childhood medulloblastoma: The United Kingdom children's cancer study group brain tumour committee. J Clin Oncol. 2005;23:7951–7.

Elmen J, Lindow M, Schutz S, et al. LNA-mediated microRNA silencing in non-human primates. Nature. 2008a;452:896–9.

Elmen J, Lindow M, Silahtaroglu A, et al. Antagonism of microRNA-122 in mice by systemically administered LNA-antimiR leads to up-regulation of a large set of predicted target mRNAs in the liver. Nucleic Acids Res. 2008b;36:1153–62.

Evans SM, Judy KD, Dunphy I, et al. Hypoxia is important in the biology and aggression of human glial brain tumors. Clin Cancer Res. 2004;10:8177–84.

Fan X, Matsui W, Khaki L, et al. Notch pathway inhibition depletes stem-like cells and blocks engraftment in embryonal brain tumors. Cancer Res. 2006;66:7445–52.

Fan X, Mikolaenko I, Elhassan I, et al. Notch1 and Notch2 have opposite effects on embryonal brain tumor growth. Cancer Res. 2004;64:7787–93.

Favaro E, Ramachandran A, McCormick R, et al. MicroRNA-210 regulates mitochondrial free radical response to hypoxia and krebs cycle in cancer cells by targeting iron sulfur cluster protein iscu. PLoS One. 2010;5:e10345.

Ferretti E, De Smaele E, Miele E, et al. Concerted microRNA control of Hedgehog signalling in cerebellar neuronal progenitor and tumour cells. EMBO J. 2008;27:2616–27.

Ferretti E, De Smaele E, Po A, et al. MicroRNA profiling in human medulloblastoma. Int J Cancer. 2009;124:568–77.

Fluiter K, ten Asbroek AL, de Wissel MB, et al. In vivo tumor growth inhibition and biodistribution studies of locked nucleic acid (LNA) antisense oligonucleotides. Nucleic Acids Res. 2003;31:953–62.

Foekens JA, Sieuwerts AM, Smid M, et al. Four miRNAs associated with aggressiveness of lymph node-negative, estrogen receptor-positive human breast cancer. Proc Natl Acad Sci USA. 2008;105:13021–6.

FuRNAri FB, Fenton T, Bachoo RM, et al. Malignant astrocytic glioma: genetics, biology, and paths to treatment. Genes Dev. 2007;21:2683–710.

Gabriely G, Wurdinger T, Kesari S, et al. MicroRNA 21 promotes glioma invasion by targeting matrix metalloproteinase regulators. Mol Cell Biol. 2008;28:5369–80.

Gajjar A, Hernan R, Kocak M, et al. Clinical, histopathologic, and molecular markers of prognosis: toward a new disease risk stratification system for medulloblastoma. J Clin Oncol. 2004;22:984–93.

Gajjar A, Pizer B. Role of high-dose chemotherapy for recurrent medulloblastoma and other cns primitive neuroectodermal tumors. Pediatr Blood Cancer. 2010;54:649–51.

Galli R, Binda E, Orfanelli U, et al. Isolation and characterization of tumorigenic, stem-like neural precursors from human glioblastoma. Cancer Res. 2004;64:7011–21.

Gangaraju VK, Lin H. MicroRNAs: key regulators of stem cells. Nat Rev Mol Cell Biol. 2009;10:116–25.

Gao P, Tchernyshyov I, Chang TC, et al. C-myc suppression of miR-23a/b enhances mitochondrial glutaminase expression and glutamine metabolism. Nature. 2009;458:762–5.

Garzia L, Andolfo I, Cusanelli E, et al. MicroRNA-199b-5p impairs cancer stem cells through negative regulation of hes1 in medulloblastoma. PLoS One. 2009;4:e4998.

Gessi M, Giangaspero F, Lauriola L, et al. Embryonal tumors with abundant neuropil and true rosettes: a distinctive CNA primitive neuroectodermal tumor. Am J Surg Pathol. 2009;33:211–7.

Gilbertson RJ, Ellison DW. The origins of medulloblastoma subtypes. Annu Rev Pathol. 2008;3:341–65.

Gilbertson RJ, Gajjar A. Molecular biology of medulloblastoma: will it ever make a difference to clinical management? J Neurooncol. 2005;75:273–8.

Gillies JK, Lorimer IA. Regulation of $p27^{kip1}$ by *MiRNA* 221/222 in glioblastoma. Cell Cycle. 2007;6:2005–9.

Giraldez AJ, Cinalli RM, Glasner ME, et al. MicroRNAs regulate brain morphogenesis in zebrafish. Science. 2005;308:833–8.

Godlewski J, Nowicki MO, Bronisz A, et al. Targeting of the Bmi-1 oncogene/stem cell renewal factor by microRNA-128 inhibits glioma proliferation and self-renewal. Cancer Res. 2008;68:9125–30.

Godlewski J, Nowicki MO, Bronisz A, et al. MicroRNA-451 regulates LKB1/AMPK signaling and allows adaptation to metabolic stress in glioma cells. Mol Cell. 2010;37:620–32.

Goodrich LV, Milenkovic L, Higgins KM, et al. Altered neural cell fates and medulloblastoma in mouse patched mutants. Science. 1997;277:1109–13.

Guessous F, Zhang Y, Kofman A, et al. MicroRNA-34a is tumor suppressive in brain tumors and glioma stem cells. Cell Cycle. 2010;9:1031–6.

Hallahan AR, Pritchard JI, Hansen S, et al. The SmoA1 mouse model reveals that notch signaling is critical for the growth and survival of sonic hedgehog-induced medulloblastomas. Cancer Res. 2004;64:7794–800.

Haupt Y, Bath ML, Harris AW, et al. Bmi-1 transgene induces lymphomas and collaborates with myc in tumorigenesis. Oncogene. 1993;8:3161–4.

He L, He X, Lim LP, et al. A microRNA component of the p53 tumour suppressor network. Nature. 2007;447:1130–4.

He L, Thomson JM, Hemann MT, et al. A microRNA polycistron as a potential human oncogene. Nature. 2005;435:828–33.

Holland EC, Celestino J, Dai C, et al. Combined activation of Ras and Akt in neural progenitors induces glioblastoma formation in mice. Nat Genet. 2000;25:55–7.

Huang Q, Gumireddy K, Schrier M, et al. The microRNAs miR-373 and miR-520c promote tumour invasion and metastasis. Nat Cell Biol. 2008;10:202–10.

Huse JT, Brennan C, Hambardzumyan D, et al. The PTEN-regulating microRNA miR-26a is amplified in high-grade glioma and facilitates gliomagenesis in vivo. Genes Dev. 2009;23: 1327–37.

Idogawa M, Sasaki Y, Suzuki H, et al. A single recombinant adenovirus expressing p53 and p21-targeting artificial microRNAs efficiently induces apoptosis in human cancer cells. Clin Cancer Res. 2009;15:3725–32.

Jain M, Arvanitis C, Chu K, et al. Sustained loss of a neoplastic phenotype by brief inactivation of myc. Science. 2002;297:102–4.

Judkins AR, Ellison DW. Ependymoblastoma: dear, damned, distracting diagnosis, farewell!* Brain Pathol. 2008;20:133–9.

Kapsimali M, Kloosterman WP, de Bruijn E, et al. MicroRNAs show a wide diversity of expression profiles in the developing and mature central nervous system. Genome Biol. 2007;8:R173.

Kawase-Koga Y, Otaegi G, Sun T. Different timings of dicer deletion affect neurogenesis and gliogenesis in the developing mouse central nervous system. Dev Dyn. 2009;238: 2800–12.

Kefas B, Comeau L, Erdle N, et al. Pyruvate kinase M2 is a target of the tumor-suppressive microRNA-326 and regulates the survival of glioma cells. Neuro Oncol. 2010;12: 1102–12.

Kefas B, Godlewski J, Comeau L, et al. MicroRNA-7 inhibits the epidermal growth factor receptor and the Akt pathway and is down-regulated in glioblastoma. Cancer Res. 2008;68: 3566–72.

Kim JW, Gao P, Liu YC, et al. Hypoxia-inducible factor 1 and dysregulated c-myc cooperatively induce vascular endothelial growth factor and metabolic switches hexokinase 2 and pyruvate dehydrogenase kinase 1. Mol Cell Biol. 2007;27:7381–93.

Kim H, Huang W, Jiang X, et al. Integrative genome analysis reveals an oncomir/oncogene cluster regulating glioblastoma survivorship. Proc Natl Acad Sci USA. 2010;107:2183–8.

Kool M, Koster J, Bunt J, et al. Integrated genomics identifies five medulloblastoma subtypes with distinct genetic profiles, pathway signatures and clinicopathological features. PLoS One. 2008;3:e3088.

Kota J, Chivukula RR, O'Donnell KA, et al. Therapeutic microRNA delivery suppresses tumorigenesis in a murine liver cancer model. Cell. 2009;137:1005–17.

Krutzfeldt J, Kuwajima S, Braich R, et al. Specificity, duplex degradation and subcellular localization of antagomirs. Nucleic Acids Res. 2007;35:2885–92.

Kulshreshtha R, Ferracin M, Wojcik SE, et al. A microRNA signature of hypoxia. Mol Cell Biol. 2007;27:1859–67.

Lafay-Cousin L, Bouffet E, Hawkins C, et al. Impact of radiation avoidance on survival and neurocognitive outcome in infant medulloblastoma. Curr Oncol. 2009;16:21–8.

Lamont JM, McManamy CS, Pearson AD, et al. Combined histopathological and molecular cytogenetic stratification of medulloblastoma patients. Clin Cancer Res. 2004;10:5482–93.

Lavon I, Zrihan D, Granit A, et al. Gliomas display a microRNA expression profile reminiscent of neural precursor cells. Neuro Oncol. 2010;12:422–33.

Lawinger P, Venugopal R, Guo ZS, et al. The neuronal repressor REST/NRSF is an essential regulator in medulloblastoma cells. Nat Med. 2000;6:826–31.

Leucht C, Stigloher C, Wizenmann A, et al. MicroRNA-9 directs late organizer activity of the midbrain-hindbrain boundary. Nat Neurosci. 2008;11:641–8.

Li Z, Bao S, Wu Q, et al. Hypoxia-inducible factors regulate tumorigenic capacity of glioma stem cells. Cancer Cell. 2009f;15:501–13.

Li MH, Bouffet E, Hawkins CE, et al. Molecular genetics of supratentorial primitive neuroectodermal tumors and pineoblastoma. Neurosurg Focus. 2005;19:E3.

Li Y, Guessous F, Zhang Y, et al. MicroRNA-34a inhibits glioblastoma growth by targeting multiple oncogenes. Cancer Res. 2009d;69:7569–76.

Li M, Lee KF, Lu Y, et al. Frequent amplification of a chr19q13.41 microRNA polycistron in aggressive primitive neuroectodermal brain tumors. Cancer Cell. 2009b;16:533–46.

Li T, Li D, Sha J, et al. MicroRNA-21 directly targets marcks and promotes apoptosis resistance and invasion in prostate cancer cells. Biochem Biophys Res Commun. 2009c;383:280–5.

Li Y, Li W, Yang Y, et al. MicroRNA-21 targets LRRFIP1 and contributes to VM-26 resistance in glioblastoma multiforme. Brain Res. 2009e;1286:13–8.

Li KK, Pang JC, Ching AK, et al. MiR-124 is frequently down-regulated in medulloblastoma and is a negative regulator of SLC16A1. Hum Pathol. 2009a;40:1234–43.

Liu HL, Hua MY, Yang HW, et al. Magnetic resonance monitoring of focused ultrasound/magnetic nanoparticle targeting delivery of therapeutic agents to the brain. Proc Natl Acad Sci USA. 2010;107:15205–10.

Liu X, Yu J, Jiang L, et al. MicroRNA-222 regulates cell invasion by targeting matrix metalloproteinase 1 (MMP1) and manganese superoxide dismutase 2 (SOD2) in tongue squamous cell carcinoma cell lines. Cancer Genomics Proteomics. 2009;6:131–9.

Liu G, Yuan X, Zeng Z, et al. Analysis of gene expression and chemoresistance of CD133$^+$ cancer stem cells in glioblastoma. Mol Cancer. 2006;5:67.

Lottaz C, Beier D, Meyer K, et al. Transcriptional profiles of CD133$^+$ and CD133$^-$ glioblastoma-derived cancer stem cell lines suggest different cells of origin. Cancer Res. 2010;70:2030–40.

Louis DN, Ohgaki H, Wiestler OD, et al. The 2007 who classification of tumours of the central nervous system. Acta Neuropathol. 2007;114:97–109.

Louis DN, Pomeroy SL, Cairncross JG. Focus on central nervous system neoplasia. Cancer Cell. 2002;1:125–8.

Lowery AJ, Miller N, Devaney A, et al. MicroRNA signatures predict oestrogen receptor, progesterone receptor and HER2/neu receptor status in breast cancer. Breast Cancer Res. 2009;11:R27.

Lu J, Getz G, Miska EA, et al. MicroRNA expression profiles classify human cancers. Nature. 2005;435:834–8.

Lu Y, Ryan SL, Elliott DJ, et al. Amplification and overexpression of hsa-miR-30b, hsa-miR-30d and khdrbs3 at 8q24.22-q24.23 in medulloblastoma. PLoS One. 2009;4:e6159.

Ma L, Teruya-Feldstein J, Weinberg RA. Tumour invasion and metastasis initiated by microRNA-10b in breast cancer. Nature. 2007;449:682–8.

Malzkorn B, Wolter M, Liesenberg F, et al. Identification and functional characterization of microRNAs involved in the malignant progression of gliomas. Brain Pathol. 2009;20:539–50.

Mavrakis KJ, Wolfe AL, Oricchio E, et al. Genome-wide RNA-mediated interference screen identifies miR-19 targets in notch-induced T-cell acute lymphoblastic leukaemia. Nat Cell Biol. 2010;12:372–9.

McNeil DE, Cote TR, Clegg L, et al. Incidence and trends in pediatric malignancies medulloblastoma/primitive neuroectodermal tumor: a seer update. Surveillance epidemiology and end results. Med Pediatr Oncol. 2002;39:190–4.

Medina PP, Nolde M, Slack FJ. Oncomir addiction in an in vivo model of microRNA-21-induced pre-B-cell lymphoma. Nature. 2010;467:86–90.

Mendez O, Zavadil J, Esencay M, et al. Knock down of hif-1alpha in glioma cells reduces migration in vitro and invasion in vivo and impairs their ability to form tumor spheres. Mol Cancer. 2010;9:133.

Mendrzyk F, Radlwimmer B, Joos S, et al. Genomic and protein expression profiling identifies cdk6 as novel independent prognostic marker in medulloblastoma. J Clin Oncol. 2005;23: 8853–62.

Mitchell PS, Parkin RK, Kroh EM, et al. Circulating microRNAs as stable blood-based markers for cancer detection. Proc Natl Acad Sci USA. 2008;105:10513–8.

Momota H, Shih AH, Edgar MA, et al. C-myc and beta-catenin cooperate with loss of p53 to generate multiple members of the primitive neuroectodermal tumor family in mice. Oncogene. 2008;27:4392–401.

Mulhern RK, Palmer SL, Reddick WE, et al. Risks of young age for selected neurocognitive deficits in medulloblastoma are associated with white matter loss. J Clin Oncol. 2001;19:472–9.

Nass D, Rosenwald S, Meiri E, et al. MiR-92b and miR-9/9* are specifically expressed in brain primary tumors and can be used to differentiate primary from metastatic brain tumors. Brain Pathol. 2009;19:375–83.

Northcott PA, FeRNAndez LA, Hagan JP, et al. The miR-17/92 polycistron is up-regulated in sonic hedgehog-driven medulloblastomas and induced by N-myc in sonic hedgehog-treated cerebellar neural precursors. Cancer Res. 2009a;69:3249–55.

Northcott PA, Nakahara Y, Wu X, et al. Multiple recurrent genetic events converge on control of histone lysine methylation in medulloblastoma. Nat Genet. 2009b;41:465–72.

Northcott PA, Rutka JT, Taylor MD. Genomics of medulloblastoma: from giemsa-banding to next-generation sequencing in 20 years. Neurosurg Focus. 2010;28:E6.

O'Donnell KA, Wentzel EA, Zeller KI, et al. C-myc-regulated microRNAs modulate e2f1 expression. Nature. 2005;435:839–43.

Ohgaki H, Kleihues P. Genetic pathways to primary and secondary glioblastoma. Am J Pathol. 2007;170:1445–53.

Oliver L, Olivier C, Marhuenda FB, et al. Hypoxia and the malignant glioma microenvironment: regulation and implications for therapy. Curr Mol Pharmacol. 2009;2:263–84.

Oliver TG, Wechsler-Reya RJ. Getting at the root and stem of brain tumors. Neuron. 2004;42: 885–8.

Packer RJ, Vezina G. Management of and prognosis with medulloblastoma: therapy at a crossroads. Arch Neurol. 2008;65:1419–24.

Packer AN, Xing Y, Harper SQ, et al. The bifunctional microRNA miR-9/miR-9* regulates rest and corest and is downregulated in huntington's disease. J Neurosci. 2008;28:14341–6.

Papagiannakopoulos T, Shapiro A, Kosik KS. MicroRNA-21 targets a network of key tumor-suppressive pathways in glioblastoma cells. Cancer Res. 2008;68:8164–72.

Parsons DW, Jones S, Zhang X, et al. An integrated genomic analysis of human glioblastoma multiforme. Science. 2008;321:1807–12.

Pfister S, Remke M, Benner A, et al. Outcome prediction in pediatric medulloblastoma based on DNA copy-number aberrations of chromosomes 6q and 17q and the myc and mycn loci. J Clin Oncol. 2009a;27:1627–36.

Pfister S, Remke M, Castoldi M, et al. Novel genomic amplification targeting the microRNA cluster at 19q13.42 in a pediatric embryonal tumor with abundant neuropil and true rosettes. Acta Neuropathol. 2009b;117:457–64.

Phillips HS, Kharbanda S, Chen R, et al. Molecular subclasses of high-grade glioma predict prognosis, delineate a pattern of disease progression, and resemble stages in neurogenesis. Cancer Cell. 2006;9:157–73.

Pierson J, Hostager B, Fan R, et al. Regulation of cyclin dependent kinase 6 by microRNA 124 in medulloblastoma. J Neurooncol. 2008;90:1–7.

Poliseno L, Tuccoli A, Mariani L, et al. MicroRNAs modulate the angiogenic properties of huvecs. Blood. 2006;108:3068–71.

Pollack IF, Boyett JM, Yates AJ, et al. The influence of central review on outcome associations in childhood malignant gliomas: results from the CCG-945 experience. Neuro Oncol. 2003;5: 197–207.

Pomeroy SL, Tamayo P, Gaasenbeek M, et al. Prediction of central nervous system embryonal tumour outcome based on gene expression. Nature. 2002;415:436–42.

Porter KR, McCarthy BJ, Freels S, et al. Prevalence estimates for primary brain tumors in the united states by age, gender, behavior, and histology. Neuro Oncol. 2010;12:520–7.

Rattis FM, Voermans C, Reya T. Wnt signaling in the stem cell niche. Curr Opin Hematol. 2004;11:88–94.

Ren Y, Zhou X, Mei M, et al. MicroRNA-21 inhibitor sensitizes human glioblastoma cells u251 (PTEN-mutant) and ln229 (PTEN-wild type) to taxol. BMC Cancer. 2010;10:27.

Rogers HA, Miller S, Lowe J, et al. An investigation of WNT pathway activation and association with survival in central nervous system primitive neuroectodermal tumours (CNS PNET). Br J Cancer. 2009;100:1292–302.

Rosenfeld N, Aharonov R, Meiri E, et al. MicroRNAs accurately identify cancer tissue origin. Nat Biotechnol. 2008;26:462–9.

Rutkowski S, Bode U, Deinlein F, et al. Treatment of early childhood medulloblastoma by postoperative chemotherapy alone. N Engl J Med. 2005;352:978–86.

Saab R, Rodriguez-Galindo C, Matmati K, et al. P18ink4c and p53 act as tumor suppressors in cyclin d1-driven primitive neuroectodermal tumor. Cancer Res. 2009;69:440–8.

Sasayama T, Nishihara M, Kondoh T, et al. MicroRNA-10b is overexpressed in malignant glioma and associated with tumor invasive factors, uPAR and RhoC. Int J Cancer. 2009;125: 1407–13.

Scherr M, Venturini L, Battmer K, et al. Lentivirus-mediated antagomir expression for specific inhibition of miRNA function. Nucleic Acids Res. 2007;35:e149.

Sempere LF, Freemantle S, Pitha-Rowe I, et al. Expression profiling of mammalian microRNAs uncovers a subset of brain-expressed microRNAs with possible roles in murine and human neuronal differentiation. Genome Biol. 2004;5:R13.

Shakhova O, Leung C, van Montfort E, et al. Lack of RB and p53 delays cerebellar development and predisposes to large cell anaplastic medulloblastoma through amplification of N-Myc and Ptch2. Cancer Res. 2006;66:5190–200.

Silber J, Lim DA, Petritsch C, et al. MiR-124 and miR-137 inhibit proliferation of glioblastoma multiforme cells and induce differentiation of brain tumor stem cells. BMC Med. 2008;6:14.

Singh SK, Hawkins C, Clarke ID, et al. Identification of human brain tumour initiating cells. Nature. 2004;432:396–401.

Skog J, Wurdinger T, van Rijn S, et al. Glioblastoma microvesicles transport RNA and proteins that promote tumour growth and provide diagnostic biomarkers. Nat Cell Biol. 2008;10:1470–6.

Su X, Gopalakrishnan V, Stearns D, et al. Abnormal expression of REST/NRSF and Myc in neural stem/progenitor cells causes cerebellar tumors by blocking neuronal differentiation. Mol Cell Biol. 2006;26:1666–78.

Taguchi A, Yanagisawa K, Tanaka M, et al. Identification of hypoxia-inducible factor-1 alpha as a novel target for miR-17-92 microRNA cluster. Cancer Res. 2008;68:5540–5.

Taylor MD, Liu L, Raffel C, et al. Mutations in SUFU predispose to medulloblastoma. Nat Genet. 2002;31:306–10.

Thompson MC, Fuller C, Hogg TL, et al. Genomics identifies medulloblastoma subgroups that are enriched for specific genetic alterations. J Clin Oncol. 2006;24:1924–31.

Uhrbom L, Dai C, Celestino JC, et al. Ink4a-Arf loss cooperates with KRas activation in astrocytes and neural progenitors to generate glioblastomas of various morphologies depending on activated Akt. Cancer Res. 2002;62:5551–8.

Uziel T, Karginov FV, Xie S, et al. The miR-17~92 cluster collaborates with the Sonic Hedgehog pathway in medulloblastoma. Proc Natl Acad Sci USA. 2009;106:2812–7.
Veeravalli KK, Chetty C, Ponnala S, et al. MMP-9, uPAR and cathepsin B silencing downregulate integrins in human glioma xenograft cells in vitro and in vivo in nude mice. PLoS One. 2010;5:e11583.
Verhaak RG, Hoadley KA, Purdom E, et al. Integrated genomic analysis identifies clinically relevant subtypes of glioblastoma characterized by abnormalities in PDGFRA, IDH1, EGFR, and NF1. Cancer Cell. 2010;17:98–110.
Vester B, Wengel J. LNA (locked nucleic acid): high-affinity targeting of complementary RNA and DNA. Biochemistry. 2004;43:13233–41.
Volinia S, Calin GA, Liu CG, et al. A microRNA expression signature of human solid tumors defines cancer gene targets. Proc Natl Acad Sci USA. 2006;103:2257–61.
Vordermark D. Significance of hypoxia in malignant glioma. Re: Evans, et al. Hypoxia is important in the biology and aggression of human glial brain tumors. Clin Cancer Res. 2005;11:3966–7. author reply 7–8. Clin cancer res 2004;10:8177–84.
Wang MY, Lu KV, Zhu S, et al. Mammalian target of rapamycin inhibition promotes response to epidermal growth factor receptor kinase inhibitors in PTEN-deficient and PTEN-intact glioblastoma cells. Cancer Res. 2006;66:7864–9.
Wang J, Wakeman TP, Lathia JD, et al. Notch promotes radioresistance of glioma stem cells. Stem Cells. 2010a;28:17–28.
Wang F, Xue X, Wei J, et al. Hsa-miR-520h downregulates ABCG2 in pancreatic cancer cells to inhibit migration, invasion, and side populations. Br J Cancer. 2010b;103:567–74.
Webster RJ, Giles KM, Price KJ, et al. Regulation of epidermal growth factor receptor signaling in human cancer cells by microRNA-7. J Biol Chem. 2009;284:5731–41.
Wechsler-Reya R, Scott MP. The developmental biology of brain tumors. Annu Rev Neurosci. 2001;24:385–428.
Wetmore C, Eberhart DE, Curran T. Loss of p53 but not arf accelerates medulloblastoma in mice heterozygous for patched. Cancer Res. 2001;61:513–6.
Wurdinger T, Tannous BA, Saydam O, et al. MiR-296 regulates growth factor receptor overexpression in angiogenic endothelial cells. Cancer Cell. 2008;14:382–93.
Xia H, Qi Y, Ng SS, et al. MicroRNA-146b inhibits glioma cell migration and invasion by targeting mmps. Brain Res. 2009;1269:158–65.
Xing Q, Ye Q, Fan M, et al. Porphyromonas gingivalis lipopolysaccharide inhibits the osteoblastic differentiation of preosteoblasts by activating notch1 signaling. J Cell Physiol. 2010;225: 106–14.
Xu Q, Yuan X, Liu G, et al. Hedgehog signaling regulates brain tumor-initiating cell proliferation and portends shorter survival for patients with PTEN-coexpressing glioblastomas. Stem Cells. 2008;26:3018–26.
Yokota N, Nishizawa S, Ohta S, et al. Role of wnt pathway in medulloblastoma oncogenesis. Int J Cancer. 2002;101:198–201.
Yoo YG, Hayashi M, Christensen J, et al. An essential role of the HIF-1alpha-C-myc axis in malignant progression. Ann NY Acad Sci. 2009a;1177:198–204.
Yoo AS, Staahl BT, Chen L, et al. MicroRNA-mediated switching of chromatin-remodelling complexes in neural development. Nature. 2009b;460:642–6.
You Y, Moreira BG, Behlke MA, et al. Design of LNA probes that improve mismatch discrimination. Nucleic Acids Res. 2006;34:e60.
Zhang C, Han L, Zhang A, et al. Global changes of mRNA expression reveals an increased activity of the interferon-induced signal transducer and activator of transcription (STAT) pathway by repression of miR-221/222 in glioblastoma U251 cells. Int J Oncol. 2010;36: 1503–12.
Zhang Y, Zhang YF, Bryant J, et al. Intravenous RNA interference gene therapy targeting the human epidermal growth factor receptor prolongs survival in intracranial brain cancer. Clin Cancer Res. 2004;10:3667–77.

Zhao T, Li J, Chen AF. MicroRNA-34a induces endothelial progenitor cell senescence and impedes its angiogenesis via suppressing silent information regulator 1. Am J Physiol Endocrinol Metab. 2010;299:E110–6.

Zheng H, Ying H, Yan H, et al. P53 and PTEN control neural and glioma stem/progenitor cell renewal and differentiation. Nature. 2008a;455:1129–33.

Zheng H, Ying H, Yan H, et al. PTEN and p53 converge on C-myc to control differentiation, self-renewal, and transformation of normal and neoplastic stem cells in glioblastoma. Cold Spring Harb Symp Quant Biol. 2008b;73:427–37.

Zhou X, Ren Y, Moore L, et al. Downregulation of miR-21 inhibits EGFR pathway and suppresses the growth of human glioblastoma cells independent of PTEN status. Lab Invest. 2010;90: 144–55.

Zindy F, Nilsson LM, Nguyen L, et al. Hemangiosarcomas, medulloblastomas, and other tumors in Ink4c/p53-null mice. Cancer Res. 2003;63:5420–7.

Chapter 16
MicroRNAs and Cancer Stem Cells

Zuoren Yu and Richard G. Pestell

Abstract MicroRNAs (miRNAs) are a class of non-coding RNAs that are believed to play important roles during tumorigenesis and cancer metastasis. Growing evidence has shown altered regulation of miRNAs in cancer stem cell populations. In this chapter, the expression profiles of miRNA in embryonic stem cells and cancer stem cells are summarized. The individual miRNAs which may regulate cancer stem cells and their target genes are described. Several miRNAs, including *miR-302* and *miR-181*, function to promote the cancer stem cell phenotype. Conversely, other miRNAs including *let-7, miR-145, miR-200* family, *miR-203, miR-128, miR-34*, and *miR-199b*, suppress stemness and promote differentiation of cancer stem cells. The recent evidence for a role of miRNA in regulating cancer stem cells, epithelial-mesenchymal transition, and cancer metastasis are described. We introduce the potential of miRNA for cancer diagnostics and therapeutics based on current tests and studies of miRNA treatment on cancer. The current challenges to apply miRNA-based cancer therapeutics are also discussed with an emphasis on recent evidence for miRNA-mediated heterotypic signals. The miRNA regulation of factors that are secreted into the blood stream creates an attractive new approach to managing miRNA-driven disease processes.

16.1 Introduction

16.1.1 Definition of the Stem Cell and Cancer Stem Cell

Stem cells are cells that are capable of self renewal, capable of differentiation and thereby giving rise to specialized cell types. This concept has been extended to cancer and a growing body of evidence has indicated a subpopulation of stem-like

R.G. Pestell (✉)
Departments of Cancer Biology and Medical Oncology, Kimmel Cancer Center, Thomas Jefferson University, Philadelphia, PA 19107, USA
e-mail: Richard.pestell@jefferson.edu

W.C.S. Cho (ed.), *MicroRNAs in Cancer Translational Research*,
DOI 10.1007/978-94-007-0298-1_16, © Springer Science+Business Media B.V. 2011

cells within tumors, known as cancer stem cells (CSCs). A definition of CSCs has been based on a self-renewal capacity, an ability to differentiate into populations of non-tumorigenic cell progeny, and their ability to seed tumors when transplanted into animal hosts. As recently reviewed, cancer stem cells have specific alterations in their cell cycle compared with normal cells, which may be important in targeting cancer stem cells (Rosen and Jordan 2009; Velasco-Velázquez et al. 2009).

16.1.2 Discovery of the Cancer Stem Cell

The evidence of SCs in cancer originated from a research on human acute myeloid leukaemia (AML) in 1994 by Dick (Lapidot et al. 1994), who identified an AML-initiating cell population by transplantation into severe combined immune-deficient (SCID) mice. These cells homed to the bone marrow and showed a pattern of dissemination and leukaemic cell morphology similar to that seen in the original patients. The leukaemic stem-like cells, named SCID leukemia-initiating cell (SL-IC), possessed differentiative and proliferative capacities and a potential for self-renewal. They were fractioned on the basis of cell surface marker expression ($CD34^+$ $CD38^-$). CSCs have been demonstrated in several solid tumors including human breast cancer and brain cancer (Al-Hajj et al. 2003; Singh et al. 2003), and later in colon, pancreas, lung, prostate, melanoma and glioblastoma cancers. Clarke et al. identified and isolated mammary tumorigenic stem-like cells as $CD44^+CD^{24-/low}Lineage^-$ in eight of nine breast cancer patients. As few as 100 cells with this phenotype were able to form tumors in nonobese diabetic/severe combined immunodeficient (NOD/SCID) mice indicating the CSCs properties of this cell fraction (Al-Hajj et al. 2003).

The initial description of CSCs claimed that CSCs represent only a small fraction of the total cancer cell population in a solid tumor, which had the ability to self-renew and maintain the tumor. However recent findings suggest that as many as 25% of cancer cells may have the properties of CSCs, and co-exist with non-CSCs (Kelly et al. 2007; Quintana et al. 2008). The proportion of CSCs population in tumors requires further experimental clarification and may relate to the original methods of isolation. CSC-specific cell surface markers such as CD44, CD24 and/or CD133 enrich CSCs from cancer cells (Al-Hajj et al. 2003; Singh et al. 2003), but these markers may depend upon the type of tumor. Thus, breast CSCs are characterized as $CD44^+CD24^{-/low}$; colon CSC are isolated by cell sorting with $CD133^+$ (Ricci-Vitiani et al. 2007). Recent reports have extended the CSCs-related cell surface markers to include markers such as epithelial-specific antigen (ESA) and aldehyde dehydrogenase-1 (ALDH-1) (Ginestier et al. 2007; Li et al. 2007). The variety of CSC markers in different tumors or tissue types and even within the same tissue type induced by distinct oncogenes suggests significant complexity of CSCs (Table 16.1).

Table 16.1 Cancer stem cell markers in human cancer

Cancer type	Cancer stem cell markers	References
Breast	$CD44^{+}/CD24^{-/low}/Lin^{-}ALDH1^{+}$	Al-Hajj et al. (2003); Ginestier et al. (2007)
Brain	$CD133^{+}$	Singh et al. (2003)
Leukemia	$CD34^{+}/CD38^{-}$	Lapidot et al. (1994)
Head and neck	$CD44^{+}$	Prince et al. (2007)
Lung	$CD133^{+}$	Eramo et al. (2008)
Liver	$CD90^{+}$	Yang et al. (2008)
Pancreas	$CD44^{+}/CD24^{+}/ESA^{+}$	Li et al. (2007)
Colon	$CD133^{+}$	Ricci-Vitiani et al. (2007)

16.2 MicroRNA

MicroRNAs (miRNAs) are a class of non-coding RNAs that regulate the stability or translational efficiency of targeted messenger RNAs. Over 1,000 miRNAs have been already identified from human (miRBase Sequence Database Release 16.0 in September 2010). As each mRNA contains binding sites for different miRNAs, and each miRNA is predicted to target as many as hundreds of genes, it is predicted that about one-third of human mRNAs could potentially be regulated by miRNAs. MiRNAs are believed to play important roles in a broad range of biological processes including embryonic development, differentiation, apoptosis, cell division, cell proliferation, tumorigenesis and cancer metastasis (Sotiropoulou et al. 2009; Yu et al. 2010a). Aberrant expression of miRNAs in many types of tumors (below) indicates the involvement of miRNAs in the regulatory net of tumorigenesis.

16.2.1 MiRNA and the Stem Cell

MiRNAs have been linked to stem cells from the original discovery of miRNA as the first two miRNAs, *lin-4* and *let-7*, were involved in regulation of developmental timing in *C. elegans* (Reinhart et al. 2000). A subset of miRNAs was cloned and sequenced from mouse embryonic stem (mES) cells in 2003 (Houbaviy et al. 2003), and later from human embryonic stem (hES) cells (Suh et al. 2004). Subsequent studies indicated the regulation of miRNAs at distinct stages of the onset and progression of tumorigenesis and cancer stem cells. *Let-7* was reported to act as a regulatory switch for stem cell self-renewal and differentiation in breast tumor-initiating cells (Yu et al. 2007); *miR-145* expression was low in self-renewing hES cells and upregulated during differentiation. *miR-145* functioned via stem cell reprogramming factors (*OCT4, SOX2, KLF4*) to regulate stem cell populations (Xu et al. 2009). Given the aberrant expression of miRNAs in tumors and the crucial regulatory function of miRNA in tumorigenesis and cancer stem cells, it is likely that miRNAs will be found to play important roles in diagnosis and therapy of cancers.

16.2.2 MiRNA Expression Profile in ES Cells

A global search of miRNA expression in embryonic stem cells was performed by Dr Sharp and Dr Kim using a small RNA cloning and sequencing method (Houbaviy et al. 2003; Suh et al. 2004). From mES cells 53 small RNAs were identified as miRNAs including 6 embryonic stem cell-specific miRNAs (*miR-290, 291, 292, 293, 294,* and *295*) which were either silenced or down-regulated upon differentiation. From hES cells, 36 small RNAs met the criteria of miRNA including the *miR-302* cluster and the *miR-371-373* cluster. The *miR-371-373* cluster on chromosome 19 is the human homologues of the *miR-290-295* cluster in mES cells. The *miR-302* cluster on human chromosome 4 is a close homologue of the *miR-302* cluster on mouse chromosome 3. The *miR-302* cluster and the *miR-371-373* cluster are expressed in hES, mES and embryonic carcinoma (EC) cells suggesting conserved roles in mammalian pluripotent stem cells. Expression of several mouse and human miRNA clusters (*miR-15/16, let-7* family, *miR-34* family, and *miR-17* cluster) are observed in both mES and hES cells suggesting their function in stem cells may be conserved across species.

MiRNA microarray analysis and miRNA quantitative reverse transcription-polymerase chain reaction (qRT-PCR) are more sensitive than RNA cloning method and have been used to detect the expression profiles of miRNAs. Li et al. (2009) examined 250 human miRNAs expression in a hES cell line hES-T3 using TaqMan real-time PCR assays. They confirmed 10 hES cell-specific miRNAs including *miR-302a**, *302b**, *302c**, *302d, 367, 368, 371, 372, 373**, *200c*, and predicted their target genes as well. Differences in miRNA expression between undifferentiated hES cells and their corresponding differentiated cells using both microarray analysis and qRT-PCR methods confirmed the identity of a miRNA signature profile in pluripotent cells, which comprised a small subset of differentially expressed miRNAs in hES cells (Lakshmipathy et al. 2007).

16.2.3 MiRNA Expression in Cancer Stem Cells

MiRNAs regulate self-renewal and differentiation of ES cells and adult tissue stem cells, and they also regulate cancer stem cells. The miRNA expression profiles of cancer stem cells are a key starting point. Zhu et al. reported altered expression of miRNAs (*miR-142-3p, miR-451, miR-106a, miR-142-5p, miR-15b, miR-20a, miR-106b, miR-25*, and *miR-486*) in bronchioalveolar "stem-like cells" which were proposed to be lung cancer progenitor cells (Qian et al. 2008). Clarke et al. recently compared miRNA expression profiles between human breast cancer stem cells ($CD44^{+}CD24^{-/low}lineage^{-}$) and the remaining $lineage^{-}$ nontumorigenic breast cancer cells (Shimono et al. 2009). 460 miRNAs were measured. 37 miRNAs were found to be deregulated in breast cancer stem cells. MiRNAs that were upregulated included *miR-199, miR-125b*, and *miR-34b*, whereas down-regulated miRNAs included the *miRNA-200c-141* cluster, the *miR-200b-200a-429* cluster, and the *miR-183-96-182* cluster. Some of these miRNAs regulate cancer stem cells and tumorigenesis (below).

16.3 Individual MiRNAs

16.3.1 MiR-302

The *miR-302* cluster was initially cloned from mES and hES cells (see above). It is highly expressed in the ES cell population but expression decreases rapidly after cell differentiation, and is not expressed in somatic cells (Barroso-del Jesus et al. 2009). The promoter of the *miR-302* cluster is regulated by *Oct4* and *Sox2* (Card et al. 2008), two transcription factors required for pluripotency during early embryogenesis and for maintenance of "stemness". Recent elegant study by Archer identified the key G_1 cell cycle regulator *cyclin D1* as a target of the *miR-302* cluster in ES cells (Card et al. 2008). It has been proposed therefore that *Oct4/Sox2-miR-302-cyclin D1* network is crucial for ES cell pluripotency and self-renewal properties.

MiR-302 over-expression converts cancer cells into ES-like pluripotent stem cells with high expression of several ES cell markers (*Oct3/4, SSEA-3, SSEA-4, Sox2*, and *Nanog*). These *miR-302* induced pluripotent stem cells differentiate into distinct tissue cell types in vitro including neurons, chondrocytes, fibroblasts, and spermatogonia-like primordial cells. The authors concluded that the *mir-302* cluster reprograms cancer cells into an ES-like pluripotent state and maintains this state under feeder-free culture conditions (Lin et al. 2008).

16.3.2 MiR-17 Cluster

The *miR-17* cluster, which encodes six mature miRNAs within a 1 kb genomic region, inhibits tumor growth in breast cancer (Yu et al. 2008, 2010b) and promotes cell growth in both lung cancer and lymphomas (Hayashita et al. 2005; He et al. 2005). These findings suggest that the *miR-17* cluster functions in a cell type-dependent manner. The *miR-17* cluster coordinates the timing of cell cycle and tumorigenesis by targeting multiple transcription factors such as *E2F, c-myc, Rb*, and *cyclin D1* (Yu et al. 2008). Within the *miR-17* cluster, *miR-19* may be the key oncogenic component rather than *miR-17-20* (Mu et al. 2009; Olive et al. 2009). *MiR-19* is required and largely sufficient to recapitulate the oncogenic properties of the entire cluster. It is possible that in a given tumor type, distinct components of the cluster drive the distinct oncogene/tumor suppressor phenotypes. Emerging evidence indicates that the *miR-17* cluster is involved in regulating the stem cell phenotype. During lung development, expression of the *miR-17* cluster is high at early stages, but declines as development proceeds. Transgenic mouse expressing the *miR-17* cluster develop proliferative lung epithelial cells with high levels of *Sox9*, suggesting that the *miR-17* cluster may promote the proliferation of lung epithelial progenitor cells (Lu et al. 2007). The *miR-17* cluster is differentially expressed in developing mouse embryos and function to control differentiation of stem cells perhaps via its target *STAT3* which is a known ES cell regulator (Foshay and Gallicano 2009).

16.3.3 MiR-181

MiR-181 is upregulated during mammalian skeletal-muscle differentiation to establish the muscle phenotype by down-regulating the homeobox protein *Hox-A11* (a repressor of the differentiation process) (Naguibneva et al. 2006). In B-lymphocytes, *miR-181a* was upregulated in differentiated cells compared with undifferentiated progenitor cells, suggesting it is a positive regulator for B cell differentiation (Chen et al. 2004). In contrast, *miR-181* family members were up-regulated in hepatocellular carcinoma (HCC), particularly within EpCAM-positive hepatic cancer stem/progenitor cells isolated from alpha-fetoprotein $(AFP)^+$ liver tumors (Ji et al. 2009b). Ji et al. (2009b) found that *miR-181* family members were highly expressed in embryonic livers and isolated hepatic stem cells. *MiR-181* may contribute to "stemness" through regulating HCC cell differentiation (via suppressing *GATA6* and *CDX2*) and activating the *Wnt/β-catenin* pathway (by down-regulating the *NLK* gene).

16.3.4 Let-7

Let-7 regulation has been studied during development of *C. elegans* (Reinhart et al. 2000). *Let-7* was undetectable at embryonic, L1 or L2 stages, expressed at low level at the early L3 stage, and highly expressed at the early L4 and adult stages. Loss of *let-7* causes larval cell fates during the adult stage, whereas increased *let-7* gene dosage causes precocious expression of adult fates during larval stages. Several heterochronic genes such as *lin-41* and *daf-12* have been confirmed as targets of *let-7* (Reinhart et al. 2000).

In mouse and human, the *let-7* family is broadly expressed in adult tissues. Expression of mature *let-7* family members is absent in human and mouse embryonic stem cells and pluripotent cell populations, and show increased expression upon differentiation (Büssing et al. 2008; Wulczyn et al. 2007). Collectively these observations suggested a conserved function of *let-7* in regulating self-renewal and cell differentiation. The abundance of *let-7* family members is reduced in several types of cancer including lung and breast cancer (Takamizawa et al. 2004; Yu et al. 2007). *Let-7* over-expression suppressed cell cycle progression and tumorigenesis by inhibiting oncogenes including *Ras* and *HMGA* (Johnson et al. 2005; 2007; Mayr et al. 2007; Tsang and Kwok 2008). Multiple important cell cycle control genes are repressed by *let-7* including *cyclin D1, cyclin D3, cyclin A, CDK 4* (Schultz et al. 2008) and *CCNA2, CDC25A, CDK6, CDK8* (Johnson et al. 2007). Yu et al. (2007) compared miRNA expression profiles in self-renewing and differentiated cells from breast cancer lines and in breast tumor-initiating cells (BT-IC)

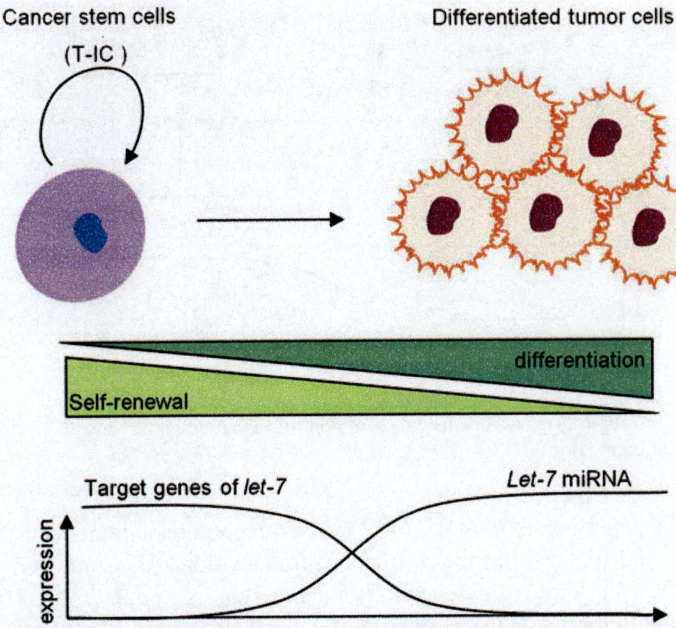

Fig. 16.1 *Let-7* miRNA regulation of cancer stem cell differentiation

and non BT-IC. *Let-7* expression was low in BT-IC and increased with differentiation. *Let-7* over-expression in BT-IC reduced cell proliferation and mammosphere formation. Enforced *let-7* expression in BT-IC interfered with tumor initiation and self-renewing capacity in vivo (Yu et al. 2007). Reduced *let-7* maintains the undifferentiated state and proliferative potential of mammospheres and BT-IC. Two known targets of *let-7*, *H-RAS*, and *HMGA2* were examined for their contribution to the BT-IC characteristics. Silencing *H-RAS* in a BT-IC-enriched cell line reduced self renewal but had no effect on differentiation, while silencing *HMGA2* enhanced differentiation but did not affect self renewal (Yu et al. 2007). Thus *let-7* acts to promote stem cell differentiation (Fig. 16.1).

16.3.5 MiR-145

MiR-145 was reported as a tumor suppressor (Cho et al. 2009; Kent and Mendell 2006). *MiR-145* inhibits human cancer cell growth in vitro via the 3′UTR of the type 1 insulin-like growth factor receptor (*IGF-IR*) and its docking protein, the insulin receptor substrate-1 (*IRS-1*) (Shi et al. 2007). In undifferentiated mES cells *IRS-1* expression is high, but decreases when cells differentiate (Rubin et al. 2007). *IRS-1* over-expression inhibits mES cell differentiation. Since *miR-145* regulates ES cell, *miR-145* regulation of stem cell differentiation may be mediated in part via *IRS-1*. *MiR-145* expression is low in self-renewing hES cells, and upregulated during differentiation (Xu et al. 2009). The pluripotency genes *OCT4*, *SOX2*, and *KLF4*

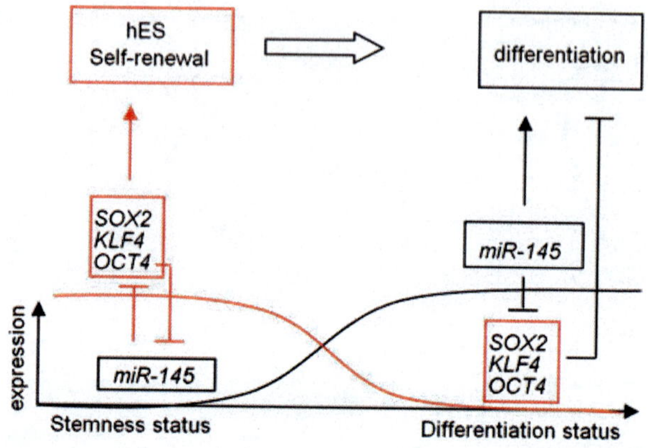

Fig. 16.2 *OCT4-miR-145* regulatory feedback in control of human embryonic stem (hES) cell stemness and differentiation

are the direct targets of *miR-145*. *MiR-145* over-expression inhibits self-renewal of hES cells and induces lineage-restricted differentiation. As a regulatory feedback loop, *OCT4* in turn represses *miR-145* expression (Xu et al. 2009). *MiR-145* regulates cell differentiation in smooth muscle cells (Cordes et al. 2009) where *miR-145* was necessary for myocardin-induced reprogramming of adult fibroblasts into smooth muscle cells. In addition to *KLF4*, *myocardin*, and *Elk-1* are targets of *miR-145*. Like *miR-145*, *miR-143* promotes differentiation, and represses smooth muscle cell proliferation (Cordes et al. 2009) (Fig. 16.2).

16.3.6 MiR-203

MiR-203 functions as a switch between proliferation and differentiation during skin development (Yi et al. 2008). *MiR-203* is expressed poorly in epidermis from embryonic day 13.5 but emerges as one of the most abundant epidermal miRNAs from embryonic day 15.5 onwards. *MiR-203* promotes epidermal differentiation by inhibiting the proliferative potential of epidermal stem cells and inducing cell-cycle exit through suppressing *p63*. *p63* is a transcription factor member of *p53* family which plays an essential role for stem-cell maintenance in stratified epithelial tissues.

In pancreatic cancer stem cells *miR-203* inhibits "stemness" (Wellner et al. 2009). *MiR-203* itself is suppressed by *ZEB1* which is an epithelial-mesenchymal transition (EMT)-activator and necessary for the self-renewing capacity of pancreatic and colorectal cancer stem cells. Knockdown of *ZEB1* led to a reduction of the pancreatic cancer stem cell ($CD24^+/CD44^+$) population, reducing sphere formation in undifferentiated pancreatic cancer cell line. *ZEB1* may thus link EMT-activation and maintenance of "stemness" by suppressing miRNAs (Fig. 16.3) (Wellner et al. 2009).

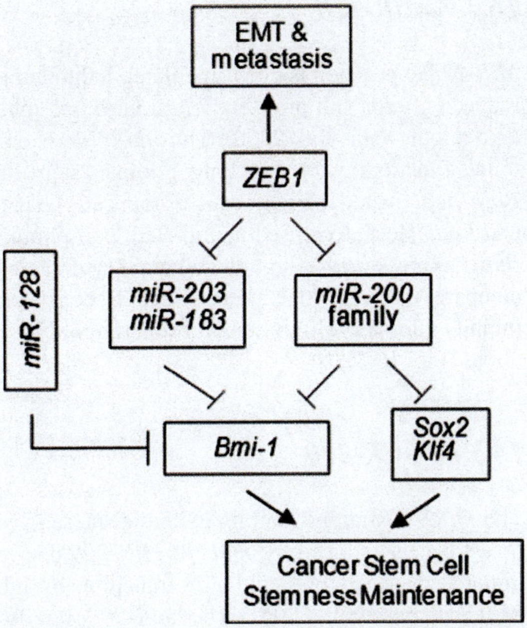

Fig. 16.3 The epithelial-mesenchymal transition (EMT)-activator *ZEB1* regulates cancer stem cell population mediated by microRNAs

16.3.7 MiR-34

The *miR-34* family was initially cloned from hES and mES cells, and is abundant in male germ cells (Yu et al. 2005). *MiR-34* expression induces cell cycle arrest, induces apoptosis and inhibits cell proliferation and colony formation in part via *p53* (Corney et al. 2007; He et al. 2007; Tarasov et al. 2007). *MiR-34* over-expression also reduced pancreatic cancer stem cells, inhibiting tumor-sphere growth in vitro and tumor formation in vivo. Two stem cell regulators, *notch* and *bcl-2*, were identified as targets of *miR-34* (Ji et al. 2009a).

16.3.8 MiR-199b

As a tumor suppressor miRNA, *miR-199b* expression is lost in metastatic medulloblastoma tumors. *MiR-199b* over-expression negatively regulated cell proliferation and impaired the oncogenic potential of brain tumor cell lines (Garzia et al. 2009). Enforced *miR-199b* expression suppressed cancer stem-cell gene expression and decreased the medulloblastoma cancer stem-cell-like ($CD133^{+}$) subpopulation, which resulted in impairment of medulloblastoma tumor development in the cerebellum xenograft mouse model in vivo. *MiR-199b* was shown to target *HES1*, a transcription factor of the notch pathway, thereby regulating cancer stem cell self-renewal (Garzia et al. 2009).

16.3.9 MiR-128

MiR-128 expression is down-regulated in human glioblastoma. *MiR-128* expression reduced glioma cell proliferation in vitro and inhibited glioma xenograft growth in vivo (Godlewski et al. 2008). Thus *miR-128* is considered a suppressor of glioblastoma, functioning by blocking glioma "stem-like" cell self-renewal via *Bmi-1*. *Bmi-1* is activated in breast cancer stem cells $CD44^{+}CD24^{-/low}Lin^{-}$ (Liu et al. 2006), and *Bmi-1* loss decreased neural stem cell population. In glioblastoma specimens, *Bmi-1* expression was significantly up-regulated while *miR-128* was down-regulated compared with normal brain, which is consistent with the finding that *miR-128* inhibits glioma proliferation and self-renewal by targeting *Bmi-1* (Godlewski et al. 2008) (Fig. 16.3).

16.3.10 MiR-200

The *miR-200* family inhibits epithelial-mesenchymal transition (EMT) in human breast cancer induced by *TGF-β*. *MiR-200* over-expression in mesenchymal cells initiated mesenchymal-epithelial transition by inhibiting expression of *ZEB1* and *ZEB2* (Eger et al. 2005). The *miR-200* family member *miR-200a, miR-200b,* and *miR-200c* were down-regulated in human breast cancer stem cells, normal human and murine mammary stem/progenitor cells, and embryonal carcinoma cells (Shimono et al. 2009; Wellner et al. 2009). *MiR-200c* inhibits stemness of normal mammary stem cell differentiation into mammary duct and tumor formation driven by human breast cancer stem cells in vivo (Shimono et al. 2009). *MiR-200c* inhibits stemness of pancreatic cancer stem cell of (Wellner et al. 2009) and breast cancer stem cell (Shimono et al. 2009). *Bmi-1* is a common target of stem cell-related miRNAs including *miR-200* family, *miR-203* and *miR-183* (Fig. 16.3).

16.4 MiRNA, Cancer Stem Cell, EMT, and Cancer Metastasis

Metastasis is a complex process by which primary solid tumor cells invade adjacent and distant tissues and grow into secondary tumors. EMT is believed to be an essential early step in tumor metastasis. MiRNA regulates EMT and cancer metastasis (Ma and Weinberg 2008; Valastyan and Weinberg 2009; Yu et al. 2010a). Recent reports have linked miRNA on one hand to the regulation of cancer EMT and metastasis and on the other hand to the regulation of the cancer stem cell phenotype in human breast cancer (Shimono et al. 2009; Tavazoie et al. 2008). *MiR-335* suppresses human breast cancer metastasis and inhibits *SOX4* expression (Tavazoie et al. 2008). *SOX4* is known to promote progenitor cell development and migration. *MiR-335* suppression of *SOX4* may thereby decrease breast cancer stem cell expansion, and thereby decrease breast cancer metastasis.

The *miR-200* family (*miR-200a*, *miR-200b*, *miR-200c*, *miR-141*, *miR-429*) and *miR-205* are down-regulated in cells that have undergone EMT (Gregory et al. 2008;

Peter 2010). Enforced expression of the *miR-200* family prevents *TGF-β* induced EMT. *ZEB1* and *ZEB2* are targeted by *miR-200* family, and *ZEB1* represses *miR-200* family. *MiR-200* is down-regulated in breast cancer stem cells and in normal mammary stem cells. *Bmi-1*, a promoter of stem cell self-renewal as noted previously, is targeted by *miR-200c*. *ZEB1* represses a stemness-inhibitor, *miR-203,* in pancreatic and colorectal cancer (Wellner et al. 2009). *Sox2* and *Klf4* are also repressed by *miR-200c*. Collectively, miRNA provide a link between EMT (*ZEB1*) and stem cells (*Sox2/Klf4*) via a *ZEB1/ miR-200/ Sox2/Klf4* loop (Fig. 16.3).

16.5 MiRNA, Cancer Stem Cell, and Cancer Therapy

Cancer stem cells are being targeted by a number of investigators and are discussed elsewhere (Gupta et al. 2009). MiRNAs also represent an excellent target for therapeutic and prognostic action (Cho 2010).

Oncogenic miRNAs are targets for inactivation and for therapy. The tumor suppressor miRNAs may be ideal candidates for the development of new drugs to inhibit tumorigenesis by enhancing their biogenesis, abundance and function. So far synthetic miRNA mimics, miRNA expression vectors and chemically modified anti-miRNAs (antisense oligonucleotides as miRNA inhibitors) have been successfully applied in bench work for the purpose of modifying miRNA abundance and activity in cells and experimental animals. The tumor suppressor function of *let-7* was demonstrated in non-small cell lung cancer by applying miRNA knock-in and knock-down techniques in a mouse model in vivo (Trang et al. 2010). By combining a successful protocol for delivery of small RNAs by intranasal passage with established mRNA antisense technology, they delivered anti-*let-7* into mouse lung increasing tumor burden. Intranasal delivery of *let-7* reduced *K-ras*-dependent lung tumors. Intratumoral delivery of *let-7* reduced lung cancer size in a Xenograft. *MiR-26a*, which is highly expressed in normal tissues and lost in hepatocellular carcinoma cells, was delivered into a mouse model of hepatocellular carcinoma, and protected from disease progression (Kota et al. 2009).

16.6 MiRNA-mediated Heterotypic Signals in Tumorigenesis

Until quite recently a prevailing model of tumorigenesis focused on the epithelial cell rather than the tissue microenvironment. It is now known that the local tissue environment contributes to the onset and progression of tumorigenesis. Similarly it is now known that tumors secrete key factors that promote tumor progression and metastasis. In this regard tumors secrete interleukin-8 (IL-8), and antibodies directed to IL-8 prevent breast tumor metastasis in vivo (Wu et al. 2008).

Recent studies have demonstrated that miRNA also regulates the secretion of key factors that are essential for the migration phenotype in breast cancer cells. In this regard *miR-17/20*, which inhibits breast cancer cell proliferation, suppresses breast tumor invasion and migration through regulating secreted factors, including IL-8

(Yu et al. 2010b). *MiR-17/20* abundance is reduced in highly invasive breast cancer cells and in node positive breast cancer (Yu et al. 2010b). *MiR-17/20* inhibited secreted plasminogen activators and reduced IL-8 production by interaction with the *IL-8* 3′UTR. Of interest, *miR-17/20* inhibited the secreted cytokeratin 8 (CK8) through a mechanism that required cyclin D1 (Yu et al. 2010b), the G1 cell-cycle regulator.

As noted above, miRNAs can regulate cell migration by secreted factors. Ultimately therapies directed to miRNA mediated disease can be directed to the cell making the miRNA, or to the mediator induced by the miRNA such as secreted growth factors or cytokines.

16.7 The Challenges of MiRNA-based Therapeutic

A miRNA-based therapeutic targeting intracellular miRNA faces three technical challenges.

16.7.1 How to Alter the MiRNA Level in Tumors?

An oncogenic miRNA can be reduced via chemically modified antisense oligos (2′-*O*-methyl RNA oligonucleotides, locked nucleic acid oligonucleotides and peptide nucleic acid (PNA)) (Fabani and Gait 2008). PNA oligonucleotides show high affinity and sequence specificity for complementary RNA, and are stable. These three approaches have been successfully used to knock down miRNA in vitro and in vivo (Fabani et al. 2010; Krützfeldt et al. 2005). Two other approaches, named "miR-sponge" and "miR-mask", were developed to block miRNA-mRNA interaction (Ebert et al. 2007; Choi et al. 2007). Tumor suppressor miRNA can be increased via synthetic miRNA mimics or miRNA expression vectors carrying either a pre-miRNA sequence or an artificial miRNA hairpin sequence.

16.7.2 How to Deliver MiRNA to Patients?

Several small interfering RNA delivery systems for cancer treatment have been developed and applied on humans (Oh and Park 2009). Cationic liposomes are attractive vehicles for small RNA because of their high transfection efficiency, low toxicity and protection of miRNAs from enzymatic degradation. Polymer-based delivery systems include synthetic polymers which have high transfection efficiency but induce cell death, and natural cationic polymers which are non-toxic, biocompatible, and biodegradable such as chitosan which is widely used. Physical delivery via electroporation is used for small RNA delivery in vivo.

Viral vector mediated delivery has high miRNA transfer efficiency with sustained and consistent gene modulation. However the residual viral elements are immunogenic in humans although modification to virus structure and delivery method has minimized the risk. A modified recombinant adeno-associated virus (rAAV) delivered *miR-26a* to mouse live tumors (Kota et al. 2009). Intranasal administration is ideal for respiratory therapy with no major adverse immune reactions or nasal-mediated RNA detected in other organs (Bitko and Barik 2008). This technique was used to deliver *let-7* miRNA to mouse lung and block lung cancer in vivo (Trang et al. 2010).

16.7.3 How to Locate MiRNA to the Tumor Tissue?

Targeting miRNA delivery to avoid off-target effects is essential for miRNA cancer therapy. Tissue-specific miRNA expressing or tumor-targeting miRNA delivery systems are being established. A tumor-specific, ligand-targeting system for gene therapy of cancer has been developed and patented (DeSano and Xu 2009; Xu et al. 2002) employing transferrin (Tf) or the scFv against the transferrin receptor (TfR) as a tumor-targeting ligand. The transferrin receptor is over-expressed in most human cancers, warranting FDA approval for a clinical trial (DeSano and Xu 2009).

An alternative strategy to restrict gene expression is the use of temporally and spatially controlled gene delivery systems. These systems include tetracycline and ecdysone based delivery systems (Albanese et al. 2002). With the advent of photouncaging, delivery can be directed at a single cell level, improving therapeutic efficiency to a theoretical maximum (Lin et al. 2002).

Acknowledgments This work was supported in part by awards from National Institutes of Health [R01CA70896, R01CA75503, and R01CA86072 to R.G.P.]. Work conducted at the Kimmel Cancer Center was supported by the NIH Cancer Center Core grant [P30CA56036 to R.G.P.]. This project is supported by a generous grant from the Dr. Ralph and Marian C. Falk Medical Research Trust and was funded and supported in part by a grant from the Pennsylvania Department of Health. The Department specifically disclaims responsibility for any analyses, interpretations or conclusions.

References

Albanese C, Hulit J, Sakamaki T, et al. Recent advances in inducible expression in transgenic mice. Semin Cell Dev Biol. 2002;13:129–41.

Al-Hajj M, Wicha MS, Benito-Hernandez A, et al. Prospective identification of tumorigenic breast cancer cells. Proc Natl Acad Sci USA. 2003;100:3983–8.

Barroso-del Jesus A, Lucena-Aguilar G, Menendez P. The miR-302-367 cluster as a potential stemness regulator in ESCs. Cell Cycle. 2009;8:394–8.

Bitko V, Barik S. Nasal delivery of siRNA. Methods Mol Biol. 2008;442:75–82.

Büssing I, Slack FJ, Grosshans H. Let-7 microRNAs in development, stem cells and cancer. Trends Mol Med. 2008;14:400–9.

Card DA, Hebbar PB, Li L, et al. Oct4/Sox2-regulated miR-302 targets cyclin D1 in human embryonic stem cells. Mol Cell Biol. 2008;28:6426–38.

Chen CZ, Li L, Lodish HF, et al. MicroRNAs modulate hematopoietic lineage differentiation. Science. 2004;303:83–6.

Cho WC. MicroRNAs: potential biomarkers for cancer diagnosis, prognosis and targets for therapy. Int J Biochem Cell Biol. 2010;42:1273–81.

Cho WC, Chow AS, Au JS. Restoration of tumour suppressor hsa-miR-145 inhibits cancer cell growth in lung adenocarcinoma patients with epidermal growth factor receptor mutation. Eur J Cancer. 2009;45:2197–206.

Choi WY, Giraldez AJ, Schier AF. Target protectors reveal dampening and balancing of Nodal agonist and antagonist by miR-430. Science. 2007;318:271–4.

Cordes KR, Sheehy NT, White MP, et al. MiR-145 and miR-143 regulate smooth muscle cell fate and plasticity. Nature. 2009;460:705–10.

Corney DC, Flesken-Nikitin A, Godwin AK, et al. MicroRNA-34b and MicroRNA-34c are targets of p53 and cooperate in control of cell proliferation and adhesion-independent growth. Cancer Res. 2007;67:8433–8.

DeSano JT, Xu L. MicroRNA regulation of cancer stem cells and therapeutic implications. AAPS J. 2009;11:682–92.

Ebert MS, Neilson JR, Sharp PA. MicroRNA sponge: competitive inhibitors of small RNAs in mammalian cells. Nat Methods. 2007;4:721–6.

Eger A, Aigner K, Sonderegger S, et al. DeltaEF1 is a transcriptional repressor of E-cadherin and regulates epithelial plasticity in breast cancer cells. Oncogene. 2005;24:2375–85.

Eramo A, Lotti F, Sette G, et al. Identification and expansion of the tumorigenic lung cancer stem cell population. Cell Death Differ. 2008;15:504–14.

Fabani MM, Abreu-Goodger C, Williams D, et al. Efficient inhibition of miR-155 function in vivo by peptide nucleic acids. Nucleic Acids Res. 2010;38:4466–75.

Fabani MM, Gait MJ. MiR-122 targeting with LNA/2′-O-methyl oligonucleotide mixmers, peptide nucleic acids (PNA), and PNA-peptide conjugates. RNA. 2008;14:336–46.

Foshay KM, Gallicano GI. MiR-17 family miRNAs are expressed during early mammalian development and regulate stem cell differentiation. Dev Biol. 2009;326:431–43.

Garzia L, Andolfo I, Cusanelli E, et al. MicroRNA-199b-5p impairs cancer stem cells through negative regulation of HES1 in medulloblastoma. PLoS One. 2009;4(3):e4998.

Ginestier C, Hur MH, Charafe-Jauffret E, et al. ALDH1 is a marker of normal and malignant human mammary stem cells and a predictor of poor clinical outcome. Cell Stem Cell. 2007;1: 555–67.

Godlewski J, Nowicki MO, Bronisz A, et al. Targeting of the Bmi-1 oncogene/stem cell renewal factor by microRNA-128 inhibits glioma proliferation and self-renewal. Cancer Res. 2008;68:9125–30.

Gregory PA, Bert AG, Paterson EL, et al. The miR-200 family and miR-205 regulate epithelial to mesenchymal transition by targeting ZEB1 and SIP1. Nat Cell Biol. 2008;10:593–601.

Gupta PB, Onder TT, Jiang G, et al. Identification of selective inhibitors of cancer stem cells by high-throughput screening. Cell. 2009;138:645–59.

Hayashita Y, Osada H, Tatematsu Y, et al. A polycistronic microRNA cluster, miR-17-92, is overexpressed in human lung cancers and enhances cell proliferation. Cancer Res. 2005;65:9628–32.

He L, He X, Lim LP, et al. A microRNA component of the p53 tumour suppressor network. Nature. 2007;447:1130–4.

He L, Thomson JM, Hemann MT, et al. A microRNA polycistron as a potential human oncogene. Nature. 2005;435:828–33.

Houbaviy HB, Murray MF, Sharp PA. Embryonic stem cell-specific MicroRNAs. Dev Cell. 2003;5:351–8.

Ji Q, Hao X, Zhang M, et al. MicroRNA miR-34 inhibits human pancreatic cancer tumor-initiating cells. PLoS One. 2009a;4:e6816.

Ji J, Yamashita T, Budhu A, et al. Identification of microRNA-181 by genome-wide screening as a critical player in EpCAM-positive hepatic cancer stem cells. Hepatology. 2009b;50:472–80.

Johnson CD, Esquela-Kerscher A, Stefani G, et al. The let-7 microRNA represses cell proliferation pathways in human cells. Cancer Res. 2007;67:7713–22.

Johnson SM, Grosshans H, Shingara J, et al. RAS is regulated by the let-7 microRNA family. Cell. 2005;120:635–47.

Kelly PN, Dakic A, Adams JM, et al. Tumor growth need not be driven by rare cancer stem cells. Science. 2007;317:337.

Kent OA, Mendell JT. Small piece in the cancer puzzle: microRNAs as tumor suppressors and oncogenes. Oncogene. 2006;25:6188–96.

Kota J, Chivukula RR, O'Donnell KA, et al. Therapeutic microRNA delivery suppresses tumorigenesis in a murine liver cancer model. Cell. 2009;137:1005–17.

Krützfeldt J, Rajewsky N, Braich R, et al. Silencing of microRNAs in vivo with 'antagomirs'. Nature. 2005;438:685–9.

Lakshmipathy U, Love B, Goff LA, et al. MicroRNA expression pattern of undifferentiated and differentiated human embryonic stem cells. Stem Cells Dev. 2007;16:1003–16.

Lapidot T, Sirard C, Vormoor J, et al. A cell initiating human acute myeloid leukaemia after transplantation into SCID mice. Nature. 1994;367:645–8.

Li C, Heidt DG, Dalerba P, et al. Identification of pancreatic cancer stem cells. Cancer Res. 2007;67:1030–7.

Li SS, Yu SL, Kao LP, et al. Target identification of microRNAs expressed highly in human embryonic stem cells. J Cell Biochem. 2009;106:1020–30.

Lin W, Albanese C, Pestell RG, et al. Spatially discrete, light-driven protein expression. Chem Biol. 2002;9:1347–53.

Lin SL, Chang DC, Chang-Lin S, et al. MiR-302 reprograms human skin cancer cells into a pluripotent ES-cell-like state. RNA. 2008;14:2115–24.

Liu S, Dontu G, Mantle ID, et al. Hedgehog signaling and Bmi-1 regulate self-renewal of normal and malignant human mammary stem cells. Cancer Res. 2006;66:6063–71.

Lu Y, Thomson JM, Wong HY, et al. Transgenic over-expression of the microRNA miR-17-92 cluster promotes proliferation and inhibits differentiation of lung epithelial progenitor cells. Dev Biol. 2007;310:442–53.

Ma L, Weinberg RA. Micromanagers of malignancy: role of microRNAs in regulating metastasis. Trends Genet. 2008;4:448–56.

Mayr C, Hemann MT, Bartel DP. Disrupting the pairing between let-7 and Hmga2 enhances oncogenic transformation. Science. 2007;315:1576–9.

Mu P, Han YC, Betel D, et al. Genetic dissection of the miR-17~92 cluster of microRNAs in Myc-induced B-cell lymphomas. Genes Dev. 2009;23:2806–11.

Naguibneva I, Ameyar-Zazoua M, Polesskaya A, et al. The microRNA miR-181 targets the homeobox protein Hox-A11 during mammalian myoblast differentiation. Nat Cell Biol. 2006;8(3):278–84.

Oh YK, Park TG. siRNA delivery systems for cancer treatment. Adv Drug Deliv Rev. 2009;61:850–62.

Olive V, Bennett MJ, Walker JC, et al. MiR-19 is a key oncogenic component of miR-17-92. Genes Dev. 2009;23:2839–49.

Peter ME. Regulating cancer stem cells the miR way. Cell Stem Cell. 2010;6:4–6.

Prince ME, Sivanandan R, Kaczorowski A, et al. Identification of a subpopulation of cells with cancer stem cell properties in head and neck squamous cell carcinoma. Proc Natl Acad Sci USA. 2007;104:973–8.

Qian S, Ding JY, Xie R, et al. MicroRNA expression profile of bronchioalveolar stem cells from mouse lung. Biochem Biophys Res Commun. 2008;377:668–73.

Quintana E, Shackleton M, Sabel MS, et al. Efficient tumour formation by single human melanoma cells. Nature. 2008;456:593–8.

Reinhart BJ, Slack FJ, Basson M, et al. The 21-nucleotide let-7 RNA regulates developmental timing in *Caenorhabditis elegans*. Nature. 2000;403:901–6.

Ricci-Vitiani L, Lombardi DG, Pilozzi E, et al. Identification and expansion of human colon-cancer-initiating cells. Nature. 2007;445:111–5.

Rosen JM, Jordan CT. The increasing complexity of the cancer stem cell paradigm. Science. 2009;4:1670–3.

Rubin R, Arzumanyan A, Soliera AR, et al. Insulin receptor substrate (IRS)-1 regulates murine embryonic stem (mES) cells self-renewal. J Cell Physiol. 2007;213:445–53.

Schultz J, Lorenz P, Gross G, et al. MicroRNA let-7b targets important cell cycle molecules in malignant melanoma cells and interferes with anchorage-independent growth. Cell Res. 2008;18:549–57.

Shi B, Sepp-Lorenzino L, Prisco M, et al. MicroRNA 145 targets the insulin receptor substrate-1 and inhibits the growth of colon cancer cells. J Biol Chem. 2007;282:32582–90.

Shimono Y, Zabala M, Cho RW, et al. Downregulation of miRNA-200c links breast cancer stem cells with normal stem cells. Cell. 2009;138:592–603.

Singh SK, Clarke ID, Terasaki M, et al. Identification of a cancer stem cell in human brain tumors. Cancer Res. 2003;63:5821–8.

Sotiropoulou G, Pampalakis G, Lianidou E, et al. Emerging roles of microRNAs as molecular switches in the integrated circuit of the cancer cell. RNA. 2009;15:1443–61.

Suh MR, Lee Y, Kim JY, et al. Human embryonic stem cells express a unique set of microRNAs. Dev Biol. 2004;270:488–98.

Takamizawa J, Konishi H, Yanagisawa K, et al. Reduced expression of the let-7 microRNAs in human lung cancers in association with shortened postoperative survival. Cancer Res. 2004;64:3753–6.

Tarasov V, Jung P, Verdoodt B, et al. Differential regulation of microRNAs by p53 revealed by massively parallel sequencing: miR-34a is a p53 target that induces apoptosis and G1-arrest. Cell Cycle. 2007;6:1586–93.

Tavazoie SF, Alarcon C, Oskarsson T, et al. Endogenous human microRNAs that suppress breast cancer metastasis. Nature. 2008;451:147–52.

Trang P, Medina PP, Wiggins JF, et al. Regression of murine lung tumors by the let-7 microRNA. Oncogene. 2010;29:1580–7.

Tsang WP, Kwok TT. Let-7a microRNA suppresses therapeutics-induced cancer cell death by targeting caspase-3. Apoptosis. 2008;13:1215–22.

Valastyan S, Weinberg RA. MicroRNAs: crucial multi-tasking components in the complex circuitry of tumor metastasis. Cell Cycle. 2009;8:3506–12.

Velasco-Velázquez MA, Yu Z, Jiao X, et al. Cancer stem cells and the cell cycle: targeting the drive behind breast cancer. Expert Rev Anticancer Ther. 2009;9:275–9.

Wellner U, Schubert J, Burk UC, et al. The EMT-activator ZEB1 promotes tumorigenicity by repressing stemness-inhibiting microRNAs. Nat Cell Biol. 2009;11:1487–95.

Wu K, Katiyar S, Li A, et al. Dachshund inhibits oncogene-induced breast cancer cellular migration and invasion through suppression of interleukin-8. Proc Natl Acad Sci USA. 2008;105:6924–9.

Wulczyn FG, Smirnova L, Rybak A, et al. Post-transcriptional regulation of the let-7 microRNA during neural cell specification. FASEB J. 2007;21:415–26.

Xu L, Frederik P, Pirollo KF, et al. Self-assembly of a virus-mimicking nanostructure system for efficient tumor-targeted gene delivery. Hum Gene Ther. 2002;13:469–81.

Xu N, Papagiannakopoulos T, Pan G, et al. MicroRNA-145 regulates OCT4, SOX2, and KLF4 and represses pluripotency in human embryonic stem cells. Cell. 2009;137:647–58.

Yang ZF, Ho DW, Ng MN, et al. Significance of $CD90^+$ cancer stem cells in human liver cancer. Cancer Cell. 2008;13:153–66.

Yi R, Poy MN, Stoffel M, et al. A skin microRNA promotes differentiation by repressing 'stemness'. Nature. 2008;452:225–9.

Yu Z, Baserga R, Chen L, et al. MicroRNA, cell cycle, and human breast cancer. Am J Pathol. 2010a;176:1058–64.

Yu Z, Raabe T, Hecht NB. MicroRNA MiR-122a reduces expression of the posttranscriptionally regulated germ cell transition protein 2 (Tnp2) messenger RNA (mRNA) by mRNA cleavage. Biol Reprod. 2005;73:427–33.

Yu Z, Wang C, Wang M, et al. A cyclin D1/microRNA 17/20 regulatory feedback loop in control of breast cancer cell proliferation. J Cell Biol. 2008;182:509–17.

Yu Z, Willmarth NE, Zhou J, et al. MicroRNA 17/20 inhibits cellular invasion and tumor metastasis in breast cancer by heterotypic signaling. Proc Natl Acad Sci USA. 2010b;107:8231–6.

Yu F, Yao H, Zhu P, et al. Let-7 regulates self renewal and tumorigenicity of breast cancer cells. Cell. 2007;131:1109–23.

Chapter 17
MicroRNAs in Cancer Invasion and Metastasis

**Ashhar S. Ali, Shadan Ali, Aamir Ahmad,
Philip A. Philip, and Fazlul H. Sarkar**

Abstract The field of cancer research has received invaluable gifts over the last few decades through novel innovations in molecular understanding and drug development. One area that is currently receiving much attention is that of microRNAs (miRNAs). The miRNAs are small, non-coding molecules that inhibit gene expression post-transcriptionally, and emerging evidence suggests that miRNAs are involved in cell growth, differentiation, and apoptosis. These developments could serve as the catalyst for further research focusing on finding a possible molecular link between miRNAs and cancer. This revolutionary research in the field of cancer has shown great promise in understanding the regulatory role of miRNAs in the development and progression of cancer, emphasizing its biochemical and pathological implications, and in particular, its significant role in cancer invasion and metastasis. For example, it has now been widely accepted that certain miRNAs are oncogenic while others act as tumor suppressors. Additionally, studies have shown that miRNAs can be used to alter sensitivity of drug-resistant tumor cells in order to improve the effects of conventional therapeutics. Furthermore, natural agents have been shown to alter miRNA expression, leading to possible inhibition of cancer cell growth and induction of apoptosis, which may contribute to the inhibition of tumor cell migration, invasion, and metastases. Therefore, selective up- and down-regulation of miRNAs holds a great promise for cancer therapy especially for those patients with invasive and metastatic disease. In this chapter, we will summarize the state of our knowledge regarding the role of miRNAs in cancer invasion and metastasis, and will also provide some information on how miRNAs could be regulated for therapeutic interventions.

F.H. Sarkar (✉)
Department of Pathology, School of Medicine, Karmanos Cancer Institute, Wayne State University, 740 Hudson Webber Cancer Research Center, Detroit, MI 48201, USA
e-mail: fsarkar@med.wayne.edu

W.C.S. Cho (ed.), *MicroRNAs in Cancer Translational Research*,
DOI 10.1007/978-94-007-0298-1_17,

17.1 Introduction

Over the last several years, research in elucidating the role of microRNAs (miRNAs) has significantly increased. Many areas of research have benefited as a result of this innovation, including the role of these non-coding molecules in cancer. The miRNAs have been shown to be involved in the development and progression of cancers, including its biochemical and pathological implications (Iorio et al. 2005; Lee and Dutta 2009; Nicoloso et al. 2009; Vandenboom et al. 2008). Specifically, we are now beginning to learn more about their significant role in the invasion and spread of cancers, both of which appears to be mediated through the oncogenic and tumor-suppressing functions of miRNAs as documented by our recent review articles (Li et al. 2010b; Sarkar et al. 2010; Wang et al. 2010) and summarized in Table 17.1.

Primary tumor growth does not always result in the death of patients. More often, the cause of cancer-related death is due to metastasis of the primary tumor to other regions of the body. Metastatic cancer has taken the lives of many, and its resistant nature to therapy has made it all the more important, suggesting that newer avenues must be discovered and new areas of research must be molecularly investigated. As we learn about the many areas of miRNAs research, we realize that these promising molecules may hold great promise in the field of cancer diagnosis, prognosis and treatment, and the knowledge gained from the emerging studies will have significant impact on the lives of patients diagnosed with malignancies. One area of miRNA research that has been receiving significant attention is in the areas of cancer invasion and metastasis (De and Naldini 2010), and thus we are summarizing the state of our knowledge only focusing on this area.

The connection between cancer development and miRNAs was made approximately a decade ago. Prior to this discovery, it was widely accepted that normal cell processes such as differentiation, proliferation, and apoptosis, are dysregulated during the pathogenesis of cancer where miRNAs are found to be involved, and thus research has really bloomed in this field (Iorio et al. 2005; Lee and Dutta 2009; Nicoloso et al. 2009; Vandenboom et al. 2008). Over the years, the attention has become more focused on specific roles of miRNA up-regulation or down-regulation and the impact of such regulation on their target genes in various types of cancer. Emerging evidences are accumulating showing the roles of miRNAs in the development and progression of cancer including the processes of cancer cell migration, invasion, and metastasis (Fig. 17.1).

Here, we provide the results of our recent survey on the role of miRNAs in various cancers that have been reported. We also surveyed the critical advancements in the control of miRNAs and its downstream effects on individual cancers and in cancer as a whole. It must be noted that while some miRNAs are involved specifically in metastasis but not in tumor formation, and others may function both in primary tumor formation and metastasis (Iorio et al. 2005; Lee and Dutta 2009; Nicoloso et al. 2009; Vandenboom et al. 2008); thus, overlap between different biological processes inevitably exists; however, the focus of this chapter is restricted by documenting the role of miRNAs in cancer invasion and metastasis. This chapter is also subdivided by specific biological functions of miRNAs wherever relevant; however,

Table 17.1 Oncogenic or tumor suppressive microRNAs in various tumors

Tumor type	MicroRNAs up-regulated	MicroRNAs down-regulated	References
Breast cancer	*21, 155, 373, 9, 10b, 21, 520c*	*125b, 145, 205, 101, 335, 126, 206, 31, 146a, 17/20, let-7*	Baranwal and Alahari (2010); Barh et al. (2010); Gregory et al. (2008a); Huang et al. (2008); Iorio et al. (2005); Lee and Dutta (2009); Ma and Weinberg (2008); Ma et al. (2007, 2010); Negrini and Calin (2008); Qian et al. (2009); Shi and Guo (2009); Tavazoie et al. (2008); Valastyan et al. (2009); Volinia et al. (2006); Wu et al. (2009); Yu et al. (2010)
Colorectal cancer	*25, 92, 31, 96, 135b, 183, 21, 34, 181b*	*133b, 145, 101*	Asangani et al. (2008); Iwatsuki et al. (2010); Lee and Dutta (2009)
Esophageal carcinoma	*21, 10b*	*205*	Gregory et al. (2008b); Hiyoshi et al. (2009); Tian et al. (2010)
Glioblastoma	*21, 10b, 221*	*146b, 128, 181a, b*	Lee and Dutta (2009); Ma and Weinberg (2008); Xia et al. (2009)
Head and neck carcinoma	*31, 566-pre*	*138, let-7*	Jiang et al. (2010); Liu et al. (2009b)
Hepatocellular carcinoma	*21, 221, 224, 18, 9*	*let-7, 122, 199a, 200a, 34a, 15a*	Hao-Xiang et al. (2009); Lan et al. (2011); Lee and Dutta (2009); Li et al. (2009a); Ma and Weinberg (2008); Tsai et al. (2009)
Leukemia	*128a, 128b, 21, 150, 155*	*let-7b, 223, 92, 222, 29b, 181b, 16-1, 15a*	Lee and Dutta (2009)
Lung cancer	*21, 205*	*1, 126, 34, 145, let-7*	Baranwal and Alahari (2010); Barh et al. (2010); He et al. (2010); Lee and Dutta (2009); Sachdeva and Mo (2010)

Table 17.1 (continued)

Tumor type	MicroRNAs up-regulated	MicroRNAs down-regulated	References
Nasopharyngeal carcinoma	*151, 192*	*29c, 34b, 34c, 212, 216, 217*	Sengupta et al. (2008)
Ovarian cancer	*200a, 141, 21, 205*	*199a, 140, 145, 125b, 9*	Gregory et al. (2008b); Laios et al. (2008); Lee and Dutta (2009)
Pancreatic cancer	*21, 221, 222, 181a, b, d, 155, 196a, 125b, 212, 301, 376a, 10a, 10b*	*375, 155, 34, 217, 146a, 200b, 200c, let-7*	Ali et al. (2010); Lee and Dutta (2009); Li et al. (2009c, 2010b); Ma and Weinberg (2008); Weiss et al. (2009)
Prostate cancer	*21*	*let-7c,125b, 145, 205, 101, 146a*	Baranwal and Alahari (2010); Huang et al. (2008); Lee and Dutta (2009); Li et al. (2009b); Lin et al. (2008); Varambally et al. (2008)
Thyroid cancer	*197, 346, 221, 222, 181b, 146b*	*30d, 125b, 26a*	Lee and Dutta (2009)

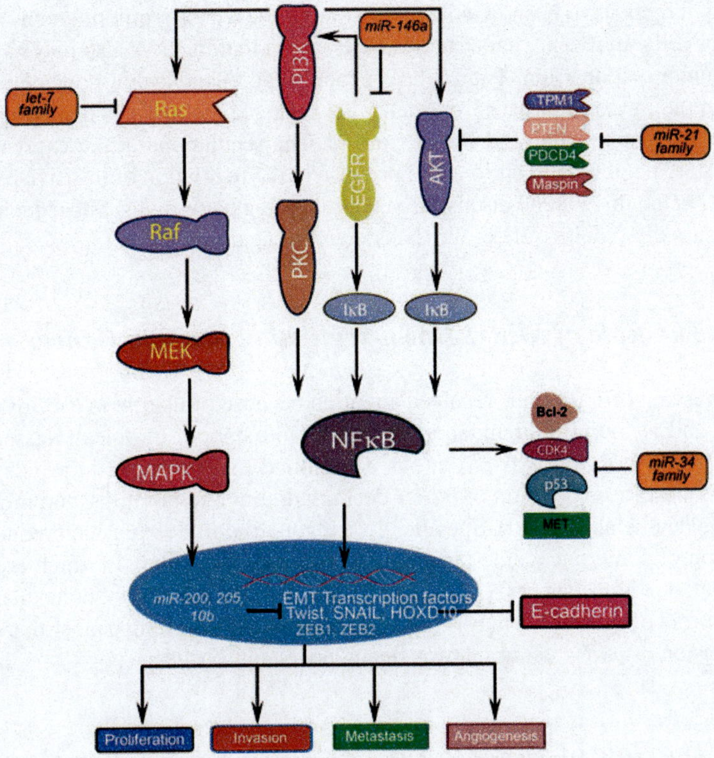

Fig. 17.1 Role of microRNAs in development and progression of cancer

this succinct chapter cannot provide all that is available in the literature, and thus for those who are interested in additional reading could consult recently published review articles, although none have focused on tumor cell invasion and metastasis (Iorio et al. 2005; Li et al. 2010a; Sarkar et al. 2010; Vandenboom et al. 2008; Wang et al. 2010).

17.2 The Role of MiRNAs in the Regulation of Tumor Cell Invasion

17.2.1 Cell Invasion as the 1st Step of Cancer Progression

Cancerous tumors differ from benign tumors in their ability to invade and metastasize. Due to their ability to invade neighboring tissues, first and foremost, they are able to establish themselves such that metastatic progression can occur (Lee and Dutta 2009). Tumor invasion is thus the first step of cancer progression, and altering its ability to do so provides for a means to limit its progression and subsequent downward spiral leading to deaths.

Many targets have been discovered that are involved in promoting cell invasion by tumor cells. In recent years, studies have shown that miRNA also plays a crucial role in tumor cell invasion. Thus, it has become a target and is now considered a key player in the invasion of cancerous cells. For instance, its role has been pinpointed in various hepatocellular and breast cancers. The regulations in the expression of miRNAs are beginning to show some promise in the regulation of invasive behavior of cancer cells; this will be discussed in the following paragraphs. Subsequently, we will discuss the specific role of miRNAs in tumor aggressiveness.

17.2.2 The Role of MiR-1 Down-regulation in Lung Cancer

The expression of *miR-1* has received attention because of its role in cardiovascular disease, and is found in abundance in cardiac and smooth muscles (Nasser et al. 2008). However, *miR-1* expression has been found to be reduced in primary lung cancer tissues. Re-expression of *miR-1* decreased tumor cell motility and migration (Baranwal and Alahari 2010). Specifically, exogenous *miR-1* expression reduced the expression of some oncogenic targets, including MET and Pim-1, which are often up-regulated in lung cancer (Nasser et al. 2008). As more studies are done on finding the exact role of *miR-1*, we may witness its potential role in lung cancer therapy by re-expression of *miR-1* using innovative and novel approaches.

17.2.3 The Role of MiR-21: A Major MiRNA in Tumor Cell Invasion

One miRNA that has received much attention is *miR-21*, which has been found to be up- regulated in many solid tumors, and functions as an oncogene and metastasis promoter (Connolly et al. 2010; Ma and Weinberg 2008). It is one of the most frequently up-regulated miRNAs in cancers (Lee and Dutta 2009). Moreover, it is considered as an oncomir due to its ability to suppress tumor suppressor genes and promote tumor growth, invasion, and metastasis (Qian et al. 2009). More specifically, it suppresses the expression of many tumor suppressor proteins, including TIMP1, TIMP3, MARCK5, BMD6, and RACK. Additionally, it up-regulates the expression of HER2, MMP2, and MMP9 (Baranwal and Alahari 2010). It may also promote invasion by direct targeting of and down-regulation of metalloprotease inhibitors, such as TIMP3 and RECK (Lawler and Chiocca 2009).

In addition, *miR-21* promotes cell motility and invasion through its target PTEN, a well-known tumor suppressor that is responsible for cell invasion. The *miR-21* acts on PTEN, which favors the Akt survival pathway and the Akt-dependent modulation of focal adhesion kinase phosphorylation and matrix metalloprotease expression level (Ali et al. 2010; Santarpia et al. 2010). Interestingly, we have shown that *miR-21* was significantly up-regulated in drug-resistant pancreatic cancer cells, and the inactivation of *miR-21* by a natural agent led to the reversal of drug-resistance and induction of cell death (Ali et al. 2010). In colorectal cancers, *miR-21* assists in invasion and intravasation through its posttranscriptional down-regulation of Pdcd4 in tumor tissues (Asangani et al. 2008). The *miR-21* has received much attention due in part to its wider function in primary tumor growth and progression compared with other microRNAs. For instance, it is increased and acts as an anti-apoptotic factor in human glioblastoma cells, and is also globally expressed in pancreatic tumors, which leads to increased cells proliferation and metastasis to the liver (Ma and Weinberg 2008). In addition, it is associated with a poor disease-free survival in early stage breast cancer patients. Elevated *miR-21* is associated with features of aggressive disease and may facilitate tumor progression, possibly through up-regulation by TGF-β (Qian et al. 2009). Suppression of *miR-21* in metastatic breast cancer cells shows considerably reduced cell invasion, likely through regulation of Pdcd4 and Maspin (Shi and Guo 2009).

Interestingly, *miR-21* over-expression has been linked with a higher tumor grade and lower survival in early stage patients (Shi and Guo 2009). Furthermore, while not definitive, *miR-21* promotes motility and invasion in prostate cancer cells, perhaps due to its regulation of Pdcd4, TPM1, and MARCKS (Li et al. 2009b), suggesting that the inhibition of *miR-21* may provide a novel and useful tool in prostate cancer therapy. Further studies suggest that *miR-21* could also play an important role in esophageal squamous cell carcinoma, which could be mediated through the regulation of Pdcd4 at the posttranscriptional level, resulting in the inhibition of cell proliferation and invasion, which suggest that the down-regulation on *miR-21* could be a therapeutic approach for esophageal cancer (Hiyoshi et al. 2009).

17.2.4 The Role of MiR-29 Family in Cancer

The *miR-29* family members (which include *miR-29a*, *miR-29b*, *miR-29c*) activate the tumor suppressor gene, *p53* (Park et al. 2009), and thus it has been linked to several cancers. A recent study has shown the re-expression of *miR-29c* in mesothelioma cell lines resulted in decreased invasion, migration, and colony formation (Pass et al. 2010), suggesting that the evaluation of the expression of *miR-29* may serve as a prognostic biomarker for malignant mesothelioma. In addition, *miR-29c* which targets ECM proteins, including collagens and laminin-γ, has been found to be significantly reduced in nasopharyngeal carcinoma invasion and metastasis (Sengupta et al. 2008).

The evidence presented above clearly shows that there are several miRNA which play important role in multiple malignancies. In the subsequent paragraphs we will discuss specific roles of selected miRNA in cancer cell invasion and metastasis.

17.2.5 Down-regulation of MiR-205 in Several Cancers

The *miR-205* plays a significant role in cell invasion. It was found to be highly expressed in both normal breast tissue and nonmalignant breast tissues whereas its expression was found to be lost, and re-expression of *miR-205* showed reduced cell proliferation, clonogenic survival, and anchorage-independent growth (Wu et al. 2009). Additionally, it was found to be significantly down- regulated in both androgen-dependent and androgen-independent prostate cancer cells whereas its expression was found to be high in normal prostatic tissues (Baranwal and Alahari 2010), suggesting that the re-expression of *miR-205* may lead to decreased prostate cancer cell invasion. However, further in-depth studies are needed to find agents or some novel avenues by which *miR-205* could be modulated, which is likely to contribute for the design of novel therapeutic approaches.

17.3 The Functional Role of MiRNA in the Regulation of Cancer Cell Migration, Invasion, and Metastasis

Once tumor cells have become invasive, they have gained the ability to migrate and spread to neighboring and distant areas of the body. This migration often makes the difference between a patient's ability to survive because migration and subsequent metastasis is an unfavorable prognostic factor. Targeting causes that lead to migration and metastasis has, thus, become a key area of cancer research. Multiple factors are responsible in the decision making of whether a cancer will metastasize or not. To that end areas of genetics, pathology, molecular biology and biochemistry have collaboratively shed light into our understanding. Biologically-speaking, the process of tumor spread consist of multiple steps. First, cancer cells must escape from the primary tumor site and invade the local tissue component. Then, it enters the blood or lymphatic circulation, where most cells will die due to unfavorable conditions. Once the cell survives, the cells will leave circulation through the endothelial vessel lining and progress into surrounding tissues. Eventually, the cancer cell will attempt to colonize into new sites through invasion and grow as a metastatic lesion (Iwatsuki et al. 2010).

In recent years, research has shown that miRNA could function as an important molecule playing a significant role in the migration and metastases of various cancers (Cho 2010a; Di and Croce 2010). The role of miRNAs to alter the signaling pathways has been carefully investigated in many cancers; however, the progress has been more visible in breast and pancreatic cancer. Cataloguing the role of miRNAs role in tumor cell migration and metastasis, we must first subdivide our topic into metastatic miRNAs and anti-metastatic miRNAs. Metastatic miRNAs tend to be over-expressed in metastatic tumors and many human malignancies. On the other hand, anti-metastatic miRNAs are molecules that can be over-expressed in metastatic tumors or under-expressed, depending on the specific molecule. This can also vary by the type of cancer cells. A comprehensive list of miRNAs, their cellular targets as well as effect on invasion and metastasis is provided in Table 17.2. The aggressiveness of cancer is often associated with cellular changes, and thus we will discuss the aggressive behavior of cancer cells and will catalogue the role of specific miRNA as detailed in the following paragraphs.

Table 17.2 Oncogenic or tumor suppressive microRNAs in various tumors as mediators of invasion/metastasis

MicroRNAs	MicroRNAs deregulated	Targets	Tumor type	Stimulates/suppress, invasion, and metastasis
Let-7	Down	*Ras, c-myc, HMGA2*	Pancreatic, prostate, hepatocellular, leukemia, oral cancers	Suppress cancer growth
miR-1	Down	*MET, Pim-1,*	Lung cancer	Suppress invasion and metastasis
miR-101	Down	*EZH2* (a histone methyl transferase)	Gastric, colorectal cancers, breast, prostate cancers	Suppress progression and metastasis
miR-10a	Up	*HOXB1, HOXB3*	Pancreatic cancer	Stimulate metastasis
miR-10b	Up	*Twist, SNAIL, ZEB, HOXD10, KLF4*	Breast, glioblastomas, pancreatic carcinoma, hepatocellular, prostate, esophageal cancers	Stimulate invasion and metastasis
miR-122	Down	*Cyclin G1*	Hepatocellular carcinoma	Suppress invasion and metastasis
miR-126	Down	Unknown	Breast, lung cancers	Suppresses metastasis
miR-138	Down	*Rhoc, ROCK2*	Head and neck carcinoma	Suppresses metastasis
miR-145	Down	*mucin 1*	Lung cancer	Suppresses metastasis and cell invasion
miR-146a	Down	*EGFR, MTA-2, IRAK-1, NF-κB*	Pancreatic cancer	Suppress invasion and metastasis
miR-146b	Down	*MMP16*	Glioblastoma, thyroid carcinoma	Suppress invasion and migration
miR-155	Up	*TGF, Smad4*	Breast cancer	Stimulate invasion and metastasis
miR-182	Up	*FOXO-3*	Melanoma	Promotes migration and survival
miR-200	Down	*Twist, SNAIl, ZEB1, ZEB2, TGFβ2*	Hepatocellular, pancreatic, bladder cancers	Suppress tumor
miR-205	Down	*ErbB3, VEGF, PKC-ϵ*	Breast, prostate, esophageal tumors	Suppress invasion and metastasis
miR-21	Up	*PTEN, PDCD4, TPM1, RECK, SERPINB5*	Pancreatic, Breast, hepatocellular, prostate, lung cancers	Stimulate invasion and metastasis
miR-211	Up	*LacZ*	Oral carcinoma	Stimulate metastasis

Table 17.2 (continued)

MicroRNAs	MicroRNAs deregulated	Targets	Tumor type	Stimulates/suppress, invasion, and metastasis
miR-29	Down	*p53, collagens, laminin-γ*	Mesothelioma, nasopharyngeal carcinoma	Suppresses invasion and metastasis
miR-31	Down	*RhoA, Fzd3, RDX*	Breast cancer	Suppresses metastasis
miR-335	Down	*SOX4, TNC*	Breast, lung cancers	Suppresses metastasis
miR-34	Down	*Bcl2, CDK4, p53, c-Met, Notch*	Lung, pancreatic, hepatocellular, cervical carcinomas, and choriocarcinoma cells	Suppress invasion and migration
miR-373	Up	*CD44*	Breast cancer	Stimulates invasion and migration
miR-520c	Up	*CD44*	Breast cancer	Stimulates invasion and migration

17.3.1 *Epithelial-mesenchymal Transition (EMT) as a Prerequisite for Cancer Metastasis*

Recently, it has been shown that one of the first steps in tumor progression may be mediated through the acquisition of EMT phenotype of cancer cells. The processes of EMT is defined as a process by which epithelial cells lose their cell to cell contacts such as E-cadherin and γ-catenin and subsequently attain characteristics of mesenchymal cells gaining the expression of protein markers such as vimentin and fibronectin (Gregory et al. 2008b; Li et al. 2009c). Once it acquires these properties, the EMT-type cells acquire the ability to migrate, which is increased significantly. In fact, cells that undergo EMT are able to detach from the primary tumor site and enter into circulation (Gregory et al. 2008b), which is believed to be responsible for tumor cell metastasis.

The process of EMT is a normal process in embryonic cell development. In tumor progression, it is a migratory escape strategy, where adherence junctions and desmosomes are partially dissociated (i.e. lose their cell to cell contact). This changes the normal behavior of cells, leading to acquisition of invasive characteristics including cell detachment and migration (Friedl 2004). Recent studies have also shown that EMT is linked to drug resistance against conventional therapeutics, and these cells are typically responsible for cancer cell metastasis (Li et al. 2009c). We will now provide evidence in support of the role of specific miRNAs associated with the processes of EMT.

17.3.2 *Role of MiR-101 in Modulating E-cadherin Expression*

E-cadherin is a transmembrane glycoprotein that is expressed in epithelial cells and tissues and is responsible for maintaining tight junctions that link adjacent cells to one another. The most common change occurring in EMT is, in fact, down-regulation of surface E-cadherin expression, which consequently results in the loss of homotypic adhesion (McConkey et al. 2009). E-cadherin down-regulation and under-expression, thus, may result in tumor progression and metastasis, and as such has been linked to gastric and colorectal cancers (Iwatsuki et al. 2010). In addition, it has been found to be down-regulated in several breast cancers (Baranwal and Alahari 2010).

The *miR-101* plays a significant role in EMT through Enhancer of Zeste homolog (*EZH2*). *EZH2* is a histone methyl transferase, which contributes to the silencing of E-cadherin as well as other genes, resulting in the regulation of cell survival and metastasis (Cao et al. 2008). *EZH2* has been found to be elevated in some localized prostate cancers; however, essentially all metastatic prostate cancers showed increased expression of *EZH2*, which appears to be due to decreased expression of *miR-101*, and consequently silencing the expression of E-cadherin, resulting in tumor progression (Varambally et al. 2008). Thus, activation of *miR-101* by novel approaches could become a promising target for the inhibition of cancer cell invasion and metastasis.

17.3.3 Role of MiR-520h in EMT

The biological function of adenovirus type 5 E1A (*E1A*) gene has been linked with the process of tumorigenicity and it has been suggested that the expression of *E1A* gene causes induction of apoptosis. One of the mechanism by which *E1A* induces tumor-suppressor effects is by reversal of EMT; however, the molecular mechanisms of such activity are not clearly understood (Su et al. 2010). Recently, it has been reported that *E1A* down-regulates *miR-520h*, which is a critical step in the *E1A*-mediated inhibition of invasion and metastasis (Su et al. 2010). Through a cascade of signaling molecules, down-regulation of *miR-520h* leads to the inhibition of NF-κB pathway and Twist which might explain the observed effects of *E1A* on EMT, tumor cell invasion and metastasis of cancer cells.

17.3.4 Elevation of MiR-155 Correlates with Invasive Breast Cancer

The *miR-155* is an important molecule that contributes to the TGF-induced epithelial cell plasticity. This leads to migration and metastasis, as it has been noted as a downstream effector of TGF and Smad4. Furthermore, it is associated with TGF-induced *RhoA* suppression, leading to dissolution of cell-cell tight junctions (Kong et al. 2008). Additionally, *miR-155* has been found to be over-expressed in invasive breast cancer, indicating its role in the spread of breast cancer (Iorio et al. 2005; Volinia et al. 2006). We will now present some evidence on the role of metastasis specific miRNAs in human malignancies.

17.3.5 The Role of MiR-200, MiR-205, and Modulation of ZEB Expression

Because EMT is now considered a prerequisite for cancer metastasis, it seems that targeting this transitional process will provide for a promising strategy in limiting the spread of various cancers. Increasing evidence suggest the role of miRNAs in the processes of EMT. Hence we will now discuss the role of miRNAs in inducing EMT with respect to cancer invasion and metastasis. The entire *miR-200* family as well as *miR-205* has been found to be significantly down-regulated during the processes of EMT (Gregory et al. 2008a), therefore mechanistic role of these miRNAs deserve further discussion.

The *miR-200* family appears to engage in a signaling cascade with the E-cadherin transcriptional repressors ZEB1, ZEB2, and TGFβ2. This maintains stable epithelial or mesenchymal states, yet allows for switching between the two, especially when influenced by EMT effectors (Gregory et al. 2008b). The *miR-200c* strongly activates epithelial differentiation in breast cancer cells by enhancing the expression of

E-cadherin and inhibition of the expression of ZEB1 (Shi and Guo 2009). However, TGF-β is up-regulated and is responsible for up-regulation of ZEB1 and ZEB2 (Gregory et al. 2008b).

Nevertheless, it must be noted that the *miR-200* family and *miR-205* have shown differential expression. It is known that *miR-200* family of miRNAs regulates EMT in bladder cancer cells (McConkey et al. 2009) while *miR-205* appears to be down-regulated in breast cancer and esophageal tumors, yet up-regulated in ovarian and bladder tumors (Gregory et al. 2008b), suggesting that the regulation and function of miRNAs could be context-dependent. Moreover, the expression of *miR-200* appears to suppress *WAVE3,* thereby inhibiting cancer cell invasion in certain cell lines, and the re-introducing *miR-200*-resistant *WAVE3* gene reverses this inhibition (Sossey-Alaoui et al. 2009). Of importance is determining the role of EMT in determining responses to various cancer therapies. Recent studies have shown that EMT-type cells are resistant to both chemotherapy and radiation therapies (McConkey et al. 2009), and thus, novel ways to deregulate miRNAs could be useful for targeted therapy by reversing resistance phenotype of EMT-like cells. Thus, further research into the regulatory roles of *miR-200* family and *miR-205* must be done in order to gain further insight on the role of these miRNAs in cancer metastases. Interestingly, we have shown that *miR-200* was down-regulated in EMT-type cells and the up-regulation of *miR-200* could be easily achieved by natural agents, which led to the reversal of EMT phenotype and cancer cell death (Li et al. 2009b). These results further suggest that therapeutic approaches could be found targeted towards the up-regulation of critical miRNAs that are lost in invasive cancers especially with EMT phenotype.

17.4 Metastatic MiRNAs

17.4.1 Over-expression of MiR-10b in Metastatic Tumors

The *miR-10b* is a key metastasis promoter in many cell types, and it has been found to be up-regulated in the context of the expression of Twist, and EMT inducing transcription factor (Iwatsuki et al. 2010). The *miR-10b* has been found to be over-expressed in breast cancers, glioblastomas, and pancreatic carcinoma amongst others (Ma and Weinberg 2008). In breast cancer, *miR-10b* was found to be over-expressed in metastatic cancer cells specifically. It plays a role in regulating cell migration and invasion in vitro, and can also initiate tumor invasion in vivo (Ma and Weinberg 2008). In fact, knock-down approaches have indicated that *miR-10b* is required for in vitro invasiveness, but not required for cell viability (Baranwal and Alahari 2010). Additionally, ectopic expression of *miR-10b* in breast cancer cells showed increased cell motility and invasiveness without affecting cell proliferation (Ma et al. 2007). Moreover, the level of *miR-10b* expression in primary breast carcinomas was found to be correlated with clinical progression

(Ma et al. 2007). In addition, *miR-10b* levels also correlated with cell motility and invasiveness in squamous cell carcinoma of the esophagus by targeting *KLF4*, a tumor suppressor gene which suppresses esophageal cancer cell migration (Tian et al. 2010).

In addition, *miR-10b* indirectly activates *RhoC*, a pro-metastatic gene, by suppressing *HOXD10*, leading to tumor invasion and metastasis (Negrini and Calin 2008). Of note, over-expression of HOXD10 or inhibition of RhoC showed nearly complete reversal of *miR-10b* induced cell migration and invasion (Ma and Weinberg 2008). In contrast, over-expression of the *miR-10b* indirectly increases RhoC protein levels by repression of its negative transcriptional regulator, HOXD10 (Ma et al. 2007). This leads to an increase in cellular motility and metastasis in breast tumors (Ma et al. 2007). It has thus been suggested that *miR-10b* plays a specific role in the metastasis processes, but not in primary tumor formation, suggesting that *miR-10b* could be a pivotal target for the development of miRNA-associated metastasis-specific drugs (Ma et al. 2007).

17.4.2 The Role of MiR-9 and MiR-10a in Pancreatic and Other Cancers

The *miR-10a* has been found to be a vital mediator of metastatic behavior of infiltrating ductal adenocarcinomas of the pancreas, arguably amongst the most aggressive metastatic cancer types. Through suppression of HOXB1 and HOXB3, it promotes metastasis formation, suggesting that the inactivation of *miR-10a* could be a useful approach for the treatment of metastatic pancreatic and other cancers (Weiss et al. 2009). The *miR-9* has been initially reported as a tumor suppressor miRNA in recurrent ovarian cancers (Laios et al. 2008), and the expression of *miR-9* has been proposed to function as a reliable marker to distinguish brain primary tumors from tumors derived from other (non-brain) tissues (Nass et al. 2009). Most significantly, *miR-9* was found to be a good prognostic marker for brain metastases with 88% sensitivity and 100% sensitivity. While these studies identified a suppressive effect of *miR-9*, a few more recent reports have found a positive correlation between *miR-9* expression and cancer metastasis and invasion (Hao-Xiang et al. 2009; Ma et al. 2010). In hepatocellular carcinoma, *miR-9* was found to be significantly up-regulated in metastatic cells and its suppression led to reduced invasion. E-cadherin, the marker of epithelial phenotype, was reported as a putative target of *miR-9* based on its up-regulation by *miR-9* inhibitor (Hao-Xiang et al. 2009). Similar activity of *miR-9* has been reported in breast cancer as well (Ma et al. 2010) where it has been shown to induce EMT in human breast epithelial cell line by down-regulation of E-cadherin and activation of beta-catenin signaling. Thus, emerging evidence connects *miR-9* with EMT, invasion and metastasis of multiple cancers although *miR-9* could still play tumor suppressive role in other malignancies.

17.4.3 The Role of MiR-211 in Oral Carcinoma

A significant association has been found with high *miR-211* expression and oral carcinoma. Elevated *miR-211* levels indicate advanced nodal metastasis and vascular invasion. Thus, the expression of *miR-211* may be associated with the progression of oral carcinoma and metastasis, with significant clinical implications (Chang et al. 2008).

17.4.4 The Role of MiR-373 and MiR-520c Expressions in Promoting Metastasis Through Suppression of CD44

The *miR-373* and *miR-520* were initially discovered due to their ability to transform normal epithelial cells into a tumorigenic cells (Ma and Weinberg 2008), and was found to be associated with significantly increased cell migration and invasion. Further studies were focused on the identification of a specific target of these miRNAs, and it was found that the expression of CD44, which is an adhesion molecule that promotes cell adhesiveness, motility, and proliferation, was suppressed by *miR-373* and *miR-520c*. These results suggest that these miRNAs could induce migration and invasion through direct suppression of CD44 (Huang et al. 2008). Additional evidence suggest that CD44 could also suppress metastasis in colon and prostate cancer (Choi et al. 2000; Gao et al. 1997) and thus, the role of CD44 and its regulation by specific miRNAs require further in-depth investigations.

17.5 Anti-metastatic MiRNA

17.5.1 The Emerging Role of Let-7 Family Members in Malignancies

The *let-7* (lethal-7) family of miRNAs has been studied extensively over the past few years, with a strong focus on its role in tumorigenesis and tumor growth. It is one of the earliest discovered miRNAs, and although nearly absent during embryonic stages, high expression of *let-7* has been observed in most differentiated tissues (Lee and Dutta 2009). It is well recognized because it plays a pivotal role in tumorigenesis through its function as a potential growth suppressor (Lan et al. 2011). The reduction of *let-7* expression has been found in advanced stages of cancer cells that are poorly differentiated (Lee and Dutta 2009). For example, down-regulation of *let-7b* was found in oral cancer cells (possibly through elevated levels of Dicer targeting the miRNA), which was linked with increased cell proliferation (Jakymiw et al. 2010). *Let-7* is under expressed in many cancers, and restoration of its normal levels has been linked with inhibition of cancer cell growth through its target of oncogenes, as well as inhibition of key regulators of mitogenic pathways (Barh et al. 2010). In hepatocellular carcinoma, its tumor-suppressor activities seem to be at work. One study found that *let-7* acts as a tumor suppressor in hepatocellular cell proliferation through its down-regulation of the oncogene *c-Myc* (Lan et al. 2011). Another noted a link between over-expression of *let-7a* and lung cancer growth inhibition (He et al. 2010).

While much of the focus of *let-7* has been on cancer growth and differentiation, its role in tumor invasion and metastasis are being investigated. A recent study indicated that down-regulation of *let-7*, with a subsequent over-expression of HMGA1 and HMGA2, could be associated with tumor proliferation and invasion in retinoblastoma (Mu et al. 2010). Moreover, others have found that *let-7* could be involved in the regulation of "stemness", perhaps contributing to metastases that can arise from cancer stem cells (Li et al. 2007).

The exact function and role of *let-7* is thus, still unclear. Existing evidence suggests that *let-7* could play a significant role in suppressing cancer cell growth, and emerging evidence suggest that re-expression of *let-7* in cancer cells could be a useful therapeutic approach (Boyerinas et al. 2010). It will be interesting to see its role in tumor metastasis for which in-depth molecular mechanistic studies are required. Interestingly, we have shown that *let-7* was down-regulated in EMT-type cells and the up-regulation of *let-7* could be easily achieved by treating cancer cells with natural agents, which led to the reversal of EMT phenotype and cancer cell death (Li et al. 2009c). These results clearly suggest that therapeutic approaches could be found by targeted up-regulation of critical miRNAs that are lost in invasive cancers, which would likely revolutionize the treatment of human malignancies.

17.5.2 The Role of MiR-17/20 Cluster in Human Malignancies

The expression of *miR-17-5p/miR-20* cluster leads to suppression of cell proliferation and tumor colony formation (Hossain et al. 2006; Yu et al. 2008) as well as invasion of breast cancer cells (Yu et al. 2010). Conditioned media from non-invasive cells, rich in *miR-17/20* cluster, inhibits the invasiveness of cancer cells by down-regulation of cytokines and inhibition of factors that are involved in degradation of extracellular matrix. In contrast, there are reports supporting the oncogenic role of *miR-17-5p* (Chow et al. 2010; Cloonan et al. 2008; Takakura et al. 2008; Yang et al. 2010). Thus, there appears to be different effects of *miR-17-5p* which might be cancer cell-specific or may be context-dependent in the tumor microenvironment. However, it is clear that *miR-17/20* cluster may be biologically very important, and thus further in-depth investigation is required for ascertaining the role of this cluster of miRNA in cancer cell invasion and metastases. Once further knowledge is gained then attempt could be made for finding ways to modulate this miRNA cluster toward cancer therapy in the future.

17.5.3 The Role of MiR-31 as an Anti-metastatic Human MiRNA

The expression of *miR-31* plays a pivotal role in the inhibition of breast cancer metastasis. Several steps and multiple mechanisms are regulated, including invasion, extravasation, and metastatic colonization (Valastyan et al. 2009). Repression of several genes could be due to the expression of *miR-31*, which is very attractive because over-expression of one miRNA could elicit pleiotropic effects by regulating multiple downstream target genes (Valastyan et al. 2009).

One recent discovery illustrates the role of *miR-31* as a metastasis inhibiting molecule. In breast cancer cells and tissues, *miR-31* was increased in cells that were less metastatic, and consequently, decreased levels of *miR-31* resulted in increased invasion and metastasis (Valastyan et al. 2009). The expression status of *miR-31* was evaluated in breast cancer tissues and found that increased expression of *miR-31* had prolonged overall survival (Schmittgen 2010), suggesting that the loss of *miR-31* is a negative prognostic factor.

17.5.4 The Regulatory Role of MiR-34a Expression

It has been reported that *miR-34a* suppresses the expression of many genes, causing G1 arrest and apoptosis, and reducing cancer invasion in a wide variety of cancers. Its transcription is activated by *p53*, and thus it has been linked with *p53* inactivation mediated loss of expression (Subramanian et al. 2010). In human hepatocellular carcinoma, a suppressive role has been observed whereby the expression of *miR-34a* could limit tumor cell migration and invasion through modulation of *c-Met* (Li et al. 2009a). Additionally, there is evidence that *miR-34a* inhibits the invasiveness of cervical carcinoma and choriocarcinoma cells, through its down-regulation of the Notch-pathway and its downstream matrix degrading enzyme (Pang et al. 2010). It also plays a potential role in the malignant transformation processes in peripheral nerve sheath tumors (Subramanian et al. 2010). Furthermore, *miR-34a* acts as a tumor suppressor in melanoma cell migration, through down-regulation of *c-Met* (Yan et al. 2009).

17.5.5 The Role of MiR-122 in Liver Cancer

The expression of *miR-122* appears to play a major role in the repression of liver cancer metastasis. It is a marker of hepatocyte-specific differentiation (Coulouarn et al. 2009). This tumor suppressor microRNA prevents intrahepatic metastasis by angiogenesis suppression (Tsai et al. 2009). Loss of *miR-122* results in an increase cell migration and invasion of hepatocellular carcinoma, and thus the expression of *miR-122* could serve as a prognostic and/or diagnostic marker of tumor progression (Tsai et al. 2009) especially because *miR-122* expression in tumor cells dictates specific gene expression profiles that are linked with invasion and tumor progression (Santarpia et al. 2010).

17.5.6 The Role of MiR-138 in Head and Neck Squamous Cell Carcinoma

Reduced expression of *miR-138* has been found in head and neck squamous cell carcinoma (HNSCC). More specifically, expression of *miR-138* reduces the expression of RhoC and ROCK2 in tongue squamous cell carcinoma, leading to less motile cells, suppressing cell migration and invasion (Jiang et al. 2010). Conversely, knockdown of *miR-138* enhances cell invasion (Jiang et al. 2010). Thus, approaches by which *miR-138* could be up-regulated will become potential targeted therapeutic strategy for HNSCC patients who are at risk of metastatic disease (Liu et al. 2009b).

17.5.7 The Role of MiR-145 and MiR-146a/146b

The inhibitory role of *miR-145* in tumor cell growth has been known for some time, and it is also involved in tumor cell invasion through suppressing the expression of *Mucin 1* gene, which appears to be down-regulated in tumor metastasis especially in lung cancer (Sachdeva and Mo 2010). The role of *miR-146* has been investigated in breast cancer, and it was found to be over-expressed in BRMS1 (breast cancer metastasis suppressor 1) expressing cells (Hurst et al. 2009). Additionally, re-expression of *miR-146a* or *miR-146b* into MDA-MB231 cells inhibited invasion and migration (Hurst et al. 2009). Recently a study was published showing decreased expression of *miR-146a* in pancreatic cancer cells compared to normal human pancreatic duct epithelial cells, and the re-expression of this miRNA inhibited invasion of pancreatic cancer cells (Li et al. 2010b).

In addition, the expression of *miR-146a* was found to be lost in metastatic breast cancer. This is believed to be as a result of direct inhibition of the expression of ROCK1, a Rho target that affects cell movement (Bhaumik et al. 2008). ROCK1 is the target of *miR-146a* in androgen-independent prostate cancer as well, and over-expression of this miRNA suppresses cell invasion and metastatic behavior (Lin et al. 2008) although the precise role of *miR-146b* is still unclear. The expression of *miR-146b* was significantly dysregulated in human glioblastoma tissues. It was found to be involved in glioma cell migration and invasion through matrix metalloproteinase genes, and the expression of *miR-146b* could be associated with the inhibition of metastasis in glioma (Xia et al. 2009). More specifically, inhibition of *miR-146b* does not appear to have any effect on its growth, yet over-expression of *miR-146b* reduced invasion and cellular migration, implicating independent cell invasion effects (Baranwal and Alahari 2010). The expression of MMP16 has emerged as a target of *miR-146b*, which is important because MMP16 appears to play an important role in the regulation of cell migration, and thus the regulation in the expression of MMP16 by *miR-146b* could be important in glioblastoma (Xia et al. 2009). Interestingly, we have shown the mechanistic role of *miR-146a* in the regulation of epidermal growth factor-related signaling associated with cancer cell invasion (Li et al. 2010b), suggesting that attenuation of *miR-146a* could be therapeutically useful.

17.5.8 *Over-expression of MiR-335, MiR-206, and MiR-126 Inhibits Metastasis*

The expression of *miR-335*, *miR-206* and *miR-126* has been identified as breast cancer metastasis suppressors. More specifically, loss of *miR-335* led to SOX4 activation, which promote differentiation, and also caused activation of TNC, resulting in the modulation of ECM composition (Tavazoie et al. 2008). These changes lead to the acquisition of invasive and metastatic properties (Negrini and Calin 2008). Furthermore, the expression of *miR-335* and *miR-126* was found to be lost in the majority of primary breast cancer tumors from patients who relapse, and the loss of expression has been associated with poor distal metastasis-free survival (Tavazoie et al. 2008). In addition, the loss of expression was also predictive of metastasis and tumor recurrence (Ma and Weinberg 2008), and further evidence suggested that the ectopic expression of either of these miRNAs could reduce both lung and bone metastases (Ma and Weinberg 2008). These results suggest that *miR-335*, *miR-206*, and *miR-126* could function as metastatic suppressors (Tavazoie et al. 2008).

17.6 The Role of MiRNAs in Angiogenesis

The process of angiogenesis refers to the generation of new blood vessels, which is a crucial process for the sustenance of cancer cell growth in vivo. It is believed that targeted inhibition of angiogenesis would therefore be a desired property of any anti-cancer therapy. In addition, emerging evidence suggests the role of miRNAs in various processes/stages of cancer progression as described above. It appears that miRNAs may also play important roles in the process of angiogenesis (Buysschaert et al. 2008; Cho 2009; Fish and Srivastava 2009; Kuehbacher et al. 2008; Suarez and Sessa 2009; Urbich et al. 2008; Wang and Olson 2009; Wurdinger and Tannous 2009). Specifically, the *miR-126* has been reported to inhibit VEGF, a classical marker of angiogenesis, thus leading to anti-angiogenic effects (Fish et al. 2008; Liu et al. 2009a; Wang et al. 2008). The *miR-34a* also inhibits angiogenesis by targeting of silent information regulator 1 (Sirt1) (Zhao et al. 2010) whereas *miR-519c* and *miR-107* inhibits angiogenesis by targeting hypoxia-inducible factor-1 alpha (HIF-1α) (Cha et al. 2010) and HIF-1β (Yamakuchi et al. 2010) respectively. The Involvement of various miRNAs of the *miR-17-92* cluster has also been suggested to play important roles in the processes of angiogenesis but the mechanism is still unclear, and it is still under active investigation (Bonauer et al. 2009; Doebele et al. 2010; Suarez et al. 2008). On the other hand, *miR-296* has been suggested to positively regulate angiogenesis (Wurdinger et al. 2008). Similarly, *miR-130a* has been shown to down-regulate anti-angiogenic homeobox genes *GAX* and *HOXA5* (Chen and Gorski 2008). These evidences clearly suggest that miRNAs are beginning to be recognized as important players in the regulation of angiogenesis, which is mediated by a complex regulation of specific genes targeted by specific miRNAs.

17.7 Conclusions and Perspectives

17.7.1 *Critical Role of MiRNA in Tumor Cell Invasion, Migration, and Metastases*

It is now widely accepted that miRNAs play a vital role in all steps of cancer progression. Studies discussed above as well as many other studies that are currently being conducted continue to suggest that miRNAs have distinct regulatory functions in the early steps in tumor progression, tumor cell invasion and metastasis. Furthermore, many miRNAs are emerging that are deregulated in cancers but have not been studied in-depth and perhaps many others are yet to be discovered. As we learn more about the implications behind the dysregulations, we are bound to learn more about their specific functions in the regulation of critical genes that are important in the regulation of tumor invasion and metastasis. However, the field is still in its infancy and further rapid growth will certainly add to our understanding of the complexities of the role of miRNAs in the processes of tumor cell migration, invasion, and distant metastasis. Such understanding will lead us to find novel avenues by which the level of expression of specific miRNA could be modulated toward cancer therapy.

17.7.2 *The Role of MiRNAs as Diagnostic/Prognostic Markers*

Up-regulation of miRNAs found in certain cancers have been found down-regulated in others. Often this varies by specific miRNAs, the types of cancer, and the stages of the disease. However, as illustrated in this chapter, measuring levels of miRNAs is beginning to assist us in identifying markers of cancer progression. The role of miRNAs as diagnostic and prognostic markers is currently being explored (Cho 2010b), with early indications that the expression of miRNAs would become an important player not only for diagnostic and prognostic purposes but it could also be useful in monitoring therapeutic response. The role of miRNAs would not be limited to the earlier steps in cancer progression nor would it be limited to invasive and metastatic carcinomas but it could also be useful as diagnostic and prognostic markers of human chronic diseases other than cancer.

17.7.3 *Development of MiRNA Targeted Therapies for Cancers*

It is becoming clear that the most debilitating step in cancer progression is that of cancer cell migration, invasion and metastasis, which is responsible for most deaths due to cancer. Therefore more efforts must be invested toward understanding the molecular complexities of tumor cell migration, invasion and metastasis focusing on miRNA, which will lead to the development of novel approaches by which one could either down-regulate or up-regulate specific miRNA for cancer therapy. Several studies have exploited targeting miRNA for therapeutic purposes, especially by targeting those miRNAs that are believed to be responsible for cancer cell migration, invasion and metastasis. While it is too early to tell how much therapeutic promise could be realized by targeting miRNAs, early evidence clearly suggests that the field is moving in the right direction as evidenced by our recent publications (Ali et al. 2010; Li et al. 2009c, 2010b). Therefore, we are hopeful that novel approaches for the suppression of metastatic miRNAs and up-regulation of anti-metastatic miRNAs would be found in the immediate future which will certainly be useful either alone or in combination with conventional therapeutics for improving the long-term survival of patients diagnosed with cancers.

References

Ali S, Ahmad A, Banerjee S, et al. Gemcitabine sensitivity can be induced in pancreatic cancer cells through modulation of miR-200 and miR-21 expression by curcumin or its analogue CDF. Cancer Res. 2010;70:3606–17.

Asangani IA, Rasheed SA, Nikolova DA, et al. MicroRNA-21 (miR-21) post-transcriptionally downregulates tumor suppressor Pdcd4 and stimulates invasion, intravasation and metastasis in colorectal cancer. Oncogene. 2008;27:2128–36.

Baranwal S, Alahari SK. MiRNA control of tumor cell invasion and metastasis. Int J Cancer. 2010;126:1283–90.

Barh D, Malhotra R, Ravi B, et al. MicroRNA let-7: an emerging next-generation cancer therapeutic. Curr Oncol. 2010;17:70–80.

Bhaumik D, Scott GK, Schokrpur S, et al. Expression of microRNA-146 suppresses NF-kappaB activity with reduction of metastatic potential in breast cancer cells. Oncogene. 2008;27: 5643–7.

Bonauer A, Carmona G, Iwasaki M, et al. MicroRNA-92a controls angiogenesis and functional recovery of ischemic tissues in mice. Science. 2009;324:1710–3.

Boyerinas B, Park SM, Hau A, et al. The role of let-7 in cell differentiation and cancer. Endocr Relat Cancer. 2010;17:F19–F36.

Buysschaert I, Schmidt T, Roncal C, et al. Genetics, epigenetics and pharmaco-(epi)genomics in angiogenesis. J Cell Mol Med. 2008;12:2533–51.

Cao Q, Yu J, Dhanasekaran SM, et al. Repression of E-cadherin by the polycomb group protein EZH2 in cancer. Oncogene. 2008;27:7274–84.

Cha ST, Chen PS, Johansson G, et al. MicroRNA-519c suppresses hypoxia-inducible factor-1alpha expression and tumor angiogenesis. Cancer Res. 2010;70:2675–85.

Chang KW, Liu CJ, Chu TH, et al. Association between high miR-211 microRNA expression and the poor prognosis of oral carcinoma. J Dent Res. 2008;87:1063–8.

Chen Y, Gorski DH. Regulation of angiogenesis through a microRNA (miR-130a) that down-regulates antiangiogenic homeobox genes GAX and HOXA5. Blood. 2008;111: 1217–26.

Cho WC. Updates in cancer research: insights from the AACR 100th Annual Meeting. Expert Rev Mol Diagn. 2009;9:411–6.

Cho WC. MicroRNAs in cancer – from research to therapy. Biochim Biophys Acta. 2010a;1805:209–17.

Cho WC. MicroRNAs: potential biomarkers for cancer diagnosis, prognosis and targets for therapy. Int J Biochem Cell Biol. 2010b;42:1273–81.

Choi SH, Takahashi K, Eto H, et al. CD44 s expression in human colon carcinomas influences growth of liver metastases. Int J Cancer. 2000;85:523–6.

Chow TF, Mankaruos M, Scorilas A, et al. The miR-17-92 cluster is over expressed in and has an oncogenic effect on renal cell carcinoma. J Urol. 2010;183:743–51.

Cloonan N, Brown MK, Steptoe AL, et al. The miR-17-5p microRNA is a key regulator of the G1/S phase cell cycle transition. Genome Biol. 2008;9:R127.

Connolly EC, Van DK, Rogler LE, et al. Overexpression of miR-21 promotes an in vitro metastatic phenotype by targeting the tumor suppressor RHOB. Mol Cancer Res. 2010;8:691–700.

Coulouarn C, Factor VM, Andersen JB, et al. Loss of miR-122 expression in liver cancer correlates with suppression of the hepatic phenotype and gain of metastatic properties. Oncogene. 2009;28:3526–36.

De PM, Naldini L. Antagonizing metastasis. Nat Biotechnol. 2010;28:331–2.

Di LG, Croce CM. Roles of small RNAs in tumor formation. Trends Mol Med. 2010;16:257–67.

Doebele C, Bonauer A, Fischer A, et al. Members of the microRNA-17-92 cluster exhibit a cell intrinsic anti-angiogenic function in endothelial cells. Blood. 2010;115:4944–50.

Fish JE, Srivastava D. MicroRNAs: opening a new vein in angiogenesis research. Sci Signal. 2009;2:e1.

Fish JE, Santoro MM, Morton SU, et al. MiR-126 regulates angiogenic signaling and vascular integrity. Dev Cell. 2008;15:272–84.

Friedl P. Prespecification and plasticity: shifting mechanisms of cell migration. Curr Opin Cell Biol. 2004;16:14–23.

Gao AC, Lou W, Dong JT, et al. CD44 is a metastasis suppressor gene for prostatic cancer located on human chromosome 11p13. Cancer Res. 1997;57:846–9.

Gregory PA, Bert AG, Paterson EL, et al. The miR-200 family and miR-205 regulate epithelial to mesenchymal transition by targeting ZEB1 and SIP1. Nat Cell Biol. 2008a;10:593–601.

Gregory PA, Bracken CP, Bert AG, et al. MicroRNAs as regulators of epithelial-mesenchymal transition. Cell Cycle. 2008b;7:3112–8.

Hao-Xiang T, Qian W, Lian-Zhou C, et al. MicroRNA-9 reduces cell invasion and E-cadherin secretion in SK-Hep-1 cell. Med Oncol. 2010;27:654–60.

He XY, Chen JX, Zhang Z, et al. The let-7a microRNA protects from growth of lung carcinoma by suppression of k-Ras and c-Myc in nude mice. J Cancer Res Clin Oncol. 2010;136: 1023–8.

Hiyoshi Y, Kamohara H, Karashima R, et al. MicroRNA-21 regulates the proliferation and invasion in esophageal squamous cell carcinoma. Clin Cancer Res. 2009;15:1915–22.

Hossain A, Kuo MT, Saunders GF, Mi R-. 17-5p regulates breast cancer cell proliferation by inhibiting translation of AIB1 mRNA. Mol Cell Biol. 2006;26:8191–201.

Huang Q, Gumireddy K, Schrier M, et al. The microRNAs miR-373 and miR-520c promote tumour invasion and metastasis. Nat Cell Biol. 2008;10:202–10.

Hurst DR, Edmonds MD, Scott GK, et al. Breast cancer metastasis suppressor 1 up-regulates miR-146, which suppresses breast cancer metastasis. Cancer Res. 2009;69:1279–83.

Iorio MV, Ferracin M, Liu CG, et al. MicroRNA gene expression deregulation in human breast cancer. Cancer Res. 2005;65:7065–70.

Iwatsuki M, Mimori K, Yokobori T, et al. Epithelial-mesenchymal transition in cancer development and its clinical significance. Cancer Sci. 2010;101:293–9.

Jakymiw A, Patel RS, Deming N, et al. Overexpression of dicer as a result of reduced let-7 MicroRNA levels contributes to increased cell proliferation of oral cancer cells. Genes Chromosomes Cancer. 2010;49:549–59.

Jiang L, Liu X, Kolokythas A, et al. Downregulation of the Rho GTPase signaling pathway is involved in the microRNA-138 mediated inhibition of cell migration and invasion in tongue squamous cell carcinoma. Int J Cancer. 2010;127:505–12.

Kong W, Yang H, He L, et al. MicroRNA-155 is regulated by the transforming growth factor beta/Smad pathway and contributes to epithelial cell plasticity by targeting RhoA. Mol Cell Biol. 2008;28:6773–84.

Kuehbacher A, Urbich C, Dimmeler S. Targeting microRNA expression to regulate angiogenesis. Trends Pharmacol Sci. 2008;29:12–5.

Laios A, O'Toole S, Flavin R, et al. Potential role of miR-9 and miR-223 in recurrent ovarian cancer. Mol Cancer. 2008;7:35.

Lan FF, Wang H, Chen YC, et al. Hsa-let-7g inhibits proliferation of hepatocellular carcinoma cells by down-regulation of c-Myc and up-regulation of p16(INK4A). Int J Cancer. 2011;128: 319–31.

Lawler S, Chiocca EA. Emerging functions of microRNAs in glioblastoma. J Neurooncol. 2009;92:297–306.

Lee YS, Dutta A. MicroRNAs in cancer. Annu Rev Pathol. 2009;4:199–227.

Li N, Fu H, Tie Y, et al. MiR-34a inhibits migration and invasion by down-regulation of c-Met expression in human hepatocellular carcinoma cells. Cancer Lett. 2009a;275:44–53.

Li Y, Kong D, Wang Z, et al. Regulation of microRNAs by natural agents: an emerging field in chemoprevention and chemotherapy research. Pharm Res. 2010a;27:1027–41.

Li T, Li D, Sha J, et al. MicroRNA-21 directly targets MARCKS and promotes apoptosis resistance and invasion in prostate cancer cells. Biochem Biophys Res Commun. 2009b;383: 280–5.

Li F, Tiede B, Massague J, et al. Beyond tumorigenesis: cancer stem cells in metastasis. Cell Res. 2007;17:3–14.

Li Y, Vandenboom TG, Kong D, et al. Up-regulation of miR-200 and let-7 by natural agents leads to the reversal of epithelial-to-mesenchymal transition in gemcitabine-resistant pancreatic cancer cells. Cancer Res. 2009c;69:6704–12.

Li Y, Vandenboom TG, Wang Z, et al. MiR-146a suppresses invasion of pancreatic cancer cells. Cancer Res. 2010b;70:1486–95.

Lin SL, Chiang A, Chang D, et al. Loss of miR-146a function in hormone-refractory prostate cancer. RNA. 2008;14:417–24.

Liu X, Jiang L, Wang A, et al. MicroRNA-138 suppresses invasion and promotes apoptosis in head and neck squamous cell carcinoma cell lines. Cancer Lett. 2009b;286:217–22.

Liu B, Peng XC, Zheng XL, et al. MiR-126 restoration down-regulate VEGF and inhibit the growth of lung cancer cell lines in vitro and in vivo. Lung Cancer. 2009a;66:169–75.

Ma L, Teruya-Feldstein J, Weinberg RA. Tumour invasion and metastasis initiated by microRNA-10b in breast cancer. Nature. 2007;449:682–8.

Ma L, Weinberg RA. Micromanagers of malignancy: role of microRNAs in regulating metastasis. Trends Genet. 2008;24:448–56.

Ma L, Young J, Prabhala H, et al. MiR-9, a MYC/MYCN-activated microRNA, regulates E-cadherin and cancer metastasis. Nat Cell Biol. 2010;12:247–56.

McConkey DJ, Choi W, Marquis L, et al. Role of epithelial-to-mesenchymal transition (EMT) in drug sensitivity and metastasis in bladder cancer. Cancer Metastasis Rev. 2009;28: 335–44.

Mu G, Liu H, Zhou F, et al. Correlation of overexpression of HMGA1 and HMGA2 with poor tumor differentiation, invasion, and proliferation associated with let-7 down-regulation in retinoblastomas. Hum Pathol. 2010;41:493–502.

Nass D, Rosenwald S, Meiri E, et al. MiR-92b and miR-9/9* are specifically expressed in brain primary tumors and can be used to differentiate primary from metastatic brain tumors. Brain Pathol. 2009;19:375–83.

Nasser MW, Datta J, Nuovo G, et al. Down-regulation of micro-RNA-1 (miR-1) in lung cancer. Suppression of tumorigenic property of lung cancer cells and their sensitization to doxorubicin-induced apoptosis by miR-1. J Biol Chem. 2008;283:33394–405.

Negrini M, Calin GA. Breast cancer metastasis: a microRNA story. Breast Cancer Res. 2008;10:203.

Nicoloso MS, Spizzo R, Shimizu M, et al. MicroRNAs–the micro steering wheel of tumour metastases. Nat Rev Cancer. 2009;9:293–302.

Pang RT, Leung CO, Ye TM, et al. MicroRNA-34a suppresses invasion through down-regulation of Notch1 and Jagged1 in cervical carcinoma and choriocarcinoma cells. Carcinogenesis. 2010;31:1037–44.

Park SY, Lee JH, Ha M, et al. MiR-29 miRNAs activate p53 by targeting p85 alpha and CDC42. Nat Struct Mol Biol. 2009;16:23–9.

Pass HI, Goparaju C, Ivanov S, et al. Hsa-miR-29c* is linked to the prognosis of malignant pleural mesothelioma. Cancer Res. 2010;70:1916–24.
Qian B, Katsaros D, Lu L, et al. High miR-21 expression in breast cancer associated with poor disease-free survival in early stage disease and high TGF-beta1. Breast Cancer Res Treat. 2009;117:131–40.
Sachdeva M, Mo YY. MicroRNA-145 suppresses cell invasion and metastasis by directly targeting mucin 1. Cancer Res. 2010;70:378–87.
Santarpia L, Nicoloso M, Calin GA. MicroRNAs: a complex regulatory network drives the acquisition of malignant cell phenotype. Endocr Relat Cancer. 2010;17:F51–F75.
Sarkar FH, Li Y, Wang Z, et al. Implication of microRNAs in drug resistance for designing novel cancer therapy. Drug Resist Updat. 2010;13:57–66.
Schmittgen TD. MiR-31: a master regulator of metastasis? Future Oncol. 2010;6: 17–20.
Sengupta S, den Boon JA, Chen IH, et al. MicroRNA 29c is down-regulated in nasopharyngeal carcinomas, up-regulating mRNAs encoding extracellular matrix proteins. Proc Natl Acad Sci USA. 2008;105:5874–8.
Shi M, Guo N. MicroRNA expression and its implications for the diagnosis and therapeutic strategies of breast cancer. Cancer Treat Rev. 2009;35:328–34.
Sossey-Alaoui K, Bialkowska K, Plow EF. The miR200 family of microRNAs regulates WAVE3-dependent cancer cell invasion. J Biol Chem. 2009;284:33019–29.
Su JL, Chen PB, Chen YH, et al. Downregulation of MicroRNA miR-520h by E1A Contributes to Anticancer Activity. Cancer Res. 2010;70:5096–108.
Suarez Y, Fernandez-Hernando C, Yu J, et al. Dicer-dependent endothelial microRNAs are necessary for postnatal angiogenesis. Proc Natl Acad Sci USA. 2008;105:14082–7.
Suarez Y, Sessa WC. MicroRNAs as novel regulators of angiogenesis. Circ Res. 2009;104: 442–54.
Subramanian S, Thayanithy V, West RB, et al. Genome-wide transcriptome analyses reveal p53 inactivation mediated loss of miR-34a expression in malignant peripheral nerve sheath tumours. J Pathol. 2010;220:58–70.
Takakura S, Mitsutake N, Nakashima M, et al. Oncogenic role of miR-17-92 cluster in anaplastic thyroid cancer cells. Cancer Sci. 2008;99:1147–54.
Tavazoie SF, Alarcon C, Oskarsson T, et al. Endogenous human microRNAs that suppress breast cancer metastasis. Nature. 2008;451:147–52.
Tian Y, Luo A, Cai Y, et al. MicroRNA-10b promotes migration and invasion through KLF4 in human esophageal cancer cell lines. J Biol Chem. 2010;285:7986–94.
Tsai WC, Hsu PW, Lai TC, et al. MicroRNA-122, a tumor suppressor microRNA that regulates intrahepatic metastasis of hepatocellular carcinoma. Hepatology. 2009;49:1571–82.
Urbich C, Kuehbacher A, Dimmeler S. Role of microRNAs in vascular diseases, inflammation, and angiogenesis. Cardiovasc Res. 2008;79:581–8.
Valastyan S, Reinhardt F, Benaich N, et al. A pleiotropically acting microRNA, miR-31, inhibits breast cancer metastasis. Cell. 2009;137:1032–46.
Vandenboom TG, Li Y, Philip PA, et al. MicroRNA and cancer: tiny molecules with major implications. Curr Genomics. 2008;9:97–109.
Varambally S, Cao Q, Mani RS, et al. Genomic loss of microRNA-101 leads to overexpression of histone methyltransferase EZH2 in cancer. Science. 2008;322:1695–9.
Volinia S, Calin GA, Liu CG, et al. A microRNA expression signature of human solid tumors defines cancer gene targets. Proc Natl Acad Sci USA. 2006;103:2257–61.
Wang S, Aurora AB, Johnson BA, et al. The endothelial-specific microRNA miR-126 governs vascular integrity and angiogenesis. Dev Cell. 2008;15:261–71.
Wang Z, Li Y, Kong D, et al. Cross-talk between miRNA and Notch signaling pathways in tumor development and progression. Cancer Lett. 2010;292:141–8.
Wang S, Olson EN. AngiomiRs–key regulators of angiogenesis. Curr Opin Genet Dev. 2009;19:205–11.

Weiss FU, Marques IJ, Woltering JM, et al. Retinoic acid receptor antagonists inhibit miR-10a expression and block metastatic behavior of pancreatic cancer. Gastroenterology. 2009;137:2136–45.

Wu H, Zhu S, Mo YY. Suppression of cell growth and invasion by miR-205 in breast cancer. Cell Res. 2009;19:439–48.

Wurdinger T, Tannous BA. Glioma angiogenesis: towards novel RNA therapeutics. Cell Adh Migr. 2009;3:230–5.

Wurdinger T, Tannous BA, Saydam O, et al. MiR-296 regulates growth factor receptor overexpression in angiogenic endothelial cells. Cancer Cell. 2008;14:382–93.

Xia H, Qi Y, Ng SS, et al. MicroRNA-146b inhibits glioma cell migration and invasion by targeting MMPs. Brain Res. 2009;1269:158–65.

Yamakuchi M, Lotterman CD, Bao C, et al. P53-induced microRNA-107 inhibits HIF-1 and tumor angiogenesis. Proc Natl Acad Sci USA. 2010;107:6334–9.

Yan D, Zhou X, Chen X, et al. MicroRNA-34a inhibits uveal melanoma cell proliferation and migration through downregulation of c-Met. Invest Ophthalmol Vis Sci. 2009;50:1559–65.

Yang F, Yin Y, Wang F, et al. MiR-17-5p Promotes migration of human hepatocellular carcinoma cells through the p38 mitogen-activated protein kinase-heat shock protein 27 pathway. Hepatology. 2010;51:1614–23.

Yu Z, Wang C, Wang M, et al. A cyclin D1/microRNA 17/20 regulatory feedback loop in control of breast cancer cell proliferation. J Cell Biol. 2008;182:509–17.

Yu Z, Willmarth NE, Zhou J, et al. MicroRNA 17/20 inhibits cellular invasion and tumor metastasis in breast cancer by heterotypic signaling. Proc Natl Acad Sci USA. 2010;107:8231–6.

Zhao T, Li J, Chen AF. MicroRNA-34a induces endothelial progenitor cell senescence and impedes its angiogenesis via suppressing silent information regulator 1. Am J Physiol Endocrinol Metab. 2010;299:E110–6.

Chapter 18
MicroRNAs in Predicting Radiotherapy and Chemotherapy Response

Emily J. Noonan, Robert F. Place, and Long-Cheng Li

Abstract Currently, many of the standard first line treatments for cancer consist of a combination of surgical removal, chemotherapy (CT), and radiotherapy (RT) of diseased tissue. During disease progression, tumor cells evolve and adapt to physiological states of resistance. Tumor biomarkers are proven to be useful in predicting response to CT, RT, and risk of recurrence. MicroRNAs (miRNAs) are small endogenous regulatory RNAs that are frequently dysregulated in cancer. A growing number of miRNAs are being identified in a variety of cancers with tumor suppressive and oncogenic functions. MiRNAs are also developing as a class of biomarkers that have been shown in both in vitro and in vivo studies to be useful in identifying malignant disease, classifying tumor subtypes, and as prognostic indicators. Additionally, circulating miRNAs are highly stable and detectable in tissue, urine, saliva, stool, sputum, and serum/plasma making them ideal candidates as cancer biomarkers. With these, a growing number of miRNAs have been identified as altering sensitivity to RT and CT. There is a number of previously established tumor suppressive and oncogenic miRNAs that function by regulating genes involved in cell cycle, apoptosis, multidrug resistance, and epithelial-mesenchymal transition. Use of miRNAs in predicting response to chemotherapeutics may give clinicians more accurate and/or sensitive methods to determine appropriate treatment choices.

18.1 Introduction

18.1.1 Current Classification of Malignancies

Cancer is a multistage heterogeneous disease involving successive changes in a number of molecular pathways to provide growth and survival advantage to rapidly evolving cells. Staging of malignancies involves a number of different criteria and

E.J. Noonan (✉)
Department of Medicine and Department of Hematology, Center for Molecular Biology in Medicine, Stanford University School of Medicine, Veterans Affairs Palo Alto Health Care System, Palo Alto, CA, USA
e-mail: emily.noonan@gmail.com

W.C.S. Cho (ed.), *MicroRNAs in Cancer Translational Research*,
DOI 10.1007/978-94-007-0298-1_18, © Springer Science+Business Media B.V. 2011

provides clinicians with information that is used to determine diagnosis, prognosis, and treatment options. The systems of cancer staging are specific to type and continue to change as we learn more about each disease. Most solid tumors use a system called TNM staging (Tumor, Nodes, Metastasis), which describes size and spread of primary tumor to nearby tissue (T), involvement of regional lymph nodes (N), and incidence of distant metastasis (M). Blood cancers use different classification systems that include normal and abnormal cell counts, region, lymph node involvement, and spread to extra-lymphatic organs (i.e. liver, lungs, etc). Histological grading is also used to determine cancer progression at the tissue level. Pathologists examine tissue samples for signs of irregular tissue structure, loss of differentiation, and abnormal cellular growth or appearance. Cytogenetic analysis, detection of tumor biomarkers, and immuno-phenotyping provide clinicians with molecular information about changes in cell physiology. For instance, tumor biomarkers found in tissue, blood, urine, or other body fluids aid in disease detection and/or treatment. Identifying methods that refine cancer staging allow physicians to accurately estimate the course of the disease, design optimal treatment strategies, and determine the risk of recurrence.

18.1.2 Biomarkers in Sensitization and Resistance to Radiotherapy (RT) and Chemotherapy (CT)

Cancer treatment varies depending on diagnosis. Surgical removal, RT, CT, or combinations thereof are common in cancer intervention. RT and CT can be toxic and/or ineffective depending on a patient's individual genetic background or through acquired resistance. Currently, there are a number of biomarkers that have been identified and correlate with sensitivity or resistance to certain therapies. Identifying targets that can help predict treatment response is critical for developing effective therapies for each patient.

18.1.2.1 RT

RT is used as a localized treatment for primary tumors or as adjuvant therapy to disseminated disease. It is well-established that proliferating cells are more sensitive to RT during the G2 and mitotic (M) phases of the cell cycle (Pawlik and Keyomarsi 2004). Cyclooxygenase-2 (COX-2), an enzyme involved in synthesizing pro-inflammatory prostanoids, is up-regulated in many cancers. Inhibition of COX-2 activity promotes G2/M arrest and has been shown to increase radiosensitivity in colon cancer cell lines (Jun et al. 2009). Phase II clinical trials show that pre-treating patients with COX-2 inhibitor celexobib sensitizes rectal cancer to RT (Debucquoy et al. 2009). As such, synchronizing cells in G2/M phases of the cell cycle is a potential strategy for improving effectiveness of RT.

One of the most well studied predictors to RT is tumor hypoxia. The idea is that hypoxia promotes accumulation of reactive oxygen species causing DNA damage, point mutations, and chromosomal instability. This leads to altered gene expression and mediates a selective advantage for tumor cells to grow in a nutrient- and oxygen-deprived environment (Giaccia 1996; Semenza 2000). As such, hypoxia in the tumor microenvironment is associated with decreased radiosensitivity (Evans et al. 1997; Li and Jackson 2002). Hypoxia promotes expression/activation of HIF1-α, HIF1-β, and NF-κB transcription factors (Koong et al. 1994; Royds et al. 1998; Wang et al. 1995). HIF1-α and/or NF-κB promote *VEGF* expression, in association with a number of other oncogenes, to promote angiogenesis and cell survival (Bhat-Nakshatri et al. 1998; Carmeliet et al. 1996; Zhang et al. 1995). Interestingly, inhibitors of angiogenesis that act to normalize vascularization and improve oxygenation have been shown to increase sensitivity to RT (Mauceri et al. 1998; Siemann and Rojiani 2002). Regions of hypoxia can be heterogeneous within tumors because of fluctuations in blood flow. Currently, polarographic electrodes are used to determine the partial pressure of oxygen, and expression of *HIF1-α*, carbonic anhydrase, pimonidazole, and plasma osteopontin are used as biomarkers for hypoxia within the tumor microenvironment. More recently, the use of positron emission tomography allows physicians to directly image tumor hypoxia and plan treatments accordingly (Bentzen 2005). However, more sensitive or less invasive techniques would be useful in determining hypoxia in cancerous tissue and sensitivity to RT.

Epidermal growth factor receptor tyrosine kinases (EGFRs) are a family of proteins involved in cell proliferation, growth, and differentiation. Among these EGFRs, endothelial growth factor receptor (*EGFR/HER1/ERBB1*) is involved in malignant transformation. Tumors that over-express EGFRs are generally more resistant to RT (Milas et al. 2004). As such, inhibition of EGFR has become one strategy for sensitizing tumor cells to RT (Krause et al. 2009; Milas et al. 2003; Thariat et al. 2007). *HER2/ERBB2* is another EGFR family member aberrantly expressed in breast and bladder cancer. Treatment with anti-HER2 antibody, trastuzumab, in combination with RT has had favorable results in vitro (Liang et al. 2003; Pietras et al. 1999). Additionally, a Phase II clinical trial has shown that trastuzumab promotes radiosensitivity in breast cancer (Horton et al. 2010).

18.1.2.2 CT

Chemotherapeutics can be used alone or in combination with surgery, RT, or other drugs to treat cancer. Many types of cancer display chemotherapeutic resistance because of their heterogeneous nature. In the past few decades, a lot of work has focused on determining biomarkers that associate with resistance or sensitivity to specific chemotherapeutics. Currently, therapies that counteract resistance are designed to target factors that sensitize cancer cells.

Apart from their role in RT resistance, growth factors are also over-expressed in certain tumors leading to CT resistance (Song et al. 2000). Inhibition of growth factors has been shown to sensitize cancer cells to other chemotherapeutics (Pardo et al. 2009). For instance, *EGFR/HER1/ERBB1* status is used to assess prognosis and resistance in colorectal cancer (CRC) (Jackson and Cunningham 2007). Inhibition of EGFRs can overcome resistance in CRC, lung, and head and neck cancer (Vokes and Chu 2006).

BCL2 is an anti-apoptotic oncogene that is over-expressed in a number of tumor types and is especially common in B cell malignancies. Chemotherapeutic resistance has been associated to Bcl-2 over-expression in cancer (Gascoyne et al. 1997; Hermine et al. 1996). As such, Bcl-2 antisense and Bcl-2 mimetics are being used to increase sensitivity to other chemotherapies (Stolz et al. 2008; Paoluzzi et al. 2008a, b; Liu and Gazitt 2003).

Drug transport proteins of the ABC super family are another group of biomarkers that confer resistance to chemotherapeutics (Haimeur et al. 2004). Aberrant expression of drug transport proteins (i.e. P-glycoprotein, MRPs, MDRs, and ABCBs) is found in leukemia, breast, hepatocellular, and ovarian cancers (Auner et al. 2010; Coley 2008; Kang et al. 2010; Noguchi et al. 2009; Sun et al. 2010). Inhibitors of P-glycoprotein are currently in clinical trials to overcome resistance (Coley 2010).

Accumulating research in therapeutic resistance indicates that many types of cancers harbor populations of cancer stem cells (CSCs). CSCs are believed to play a role in resistance in sarcomas (Honoki 2010; Jiang et al. 2010), osteosarcomas (Di Fiore et al. 2009), colorectal cancer (Cammareri et al. 2010; Dallas et al. 2009; Todaro et al. 2010), breast cancer (Nakanishi et al. 2010; To et al. 2010), and ovarian cancer (Ahmed et al. 2010). In pancreatic cancer, resistance is believed to stem from populations of CSCs and additionally cells that have an enhanced capacity to enter epithelial-mesenchymal transition (EMT) and become metastatic (Arumugam et al. 2009; Sarkar et al. 2009).

Along with this, a number of cancer types have shown resistance to chemotherapeutics through expression of cell cycle genes (Akli et al. 2010; Duncan et al. 2010; Ferrandiz et al. 2010; Johansson and Persson 2008; Shah and Schwartz 2003), histone deacetylase genes (Banwell et al. 2003; Inoue et al. 2006; Xu et al. 2007), and DNA repair genes (Kinsella 2009; Viktorsson et al. 2005). Cancer cells that survive initial CT can remain in a drug tolerant state due to chromatin modifications, but secondary treatments with HDAC inhibitors re-sensitize cells to CT (Sharma et al. 2010).

18.2 MiRNAs in Cancer: Tumor Suppressors and Oncogenes

MiRNAs have been implicated in a wide range of cellular processes including cellular growth, differentiation, and development. As such, aberrant miRNA expression has been linked to a number of human cancers. Depending on downstream targets, miRNAs can function as tumor suppressors and/or oncogenes to influence cancer development.

18.2.1 Tumor Suppressive MiRNAs

Let-7, a family of tumor suppressive miRNAs universally down-regulated in cancers, is lost early in cancer progression (Akao et al. 2006; Johnson et al. 2005). *Let-7* family members have been shown to negatively regulate the proto-oncogenes

RAS and *MYC* (Akao et al. 2006; Johnson et al. 2005). The *let-7* miRNAs also deplete the expression of genes involved in promoting EMT (i.e. HMGA2) (Park et al. 2007; Shell et al. 2007), cell cycle progression (i.e. Cyclin D2, E2F transcription factors, and NIRF) (Dong et al. 2010; He et al. 2009), and apoptosis (i.e. Bcl-xL and caspase-3) (Shimizu et al. 2010; Tsang and Kwok 2008).

MiR-15 and *miR-16* are miRNAs with potent tumor suppressive activity that are well-known for promoting apoptosis by inhibiting the expression of anti-apoptotic factor Bcl-2. These miRNAs are commonly deleted in CLL (Cimmino et al. 2005; Faderl et al. 2002) and down-regulated in gastric cancer (Xia et al. 2008), pituitary cancers (Bottoni et al. 2005), prostate cancer (Takeshita et al. 2010), squamous cancer, and lung cancer (Bandi et al. 2009). In prostate cancer, *miR-15* and *miR-16* have been shown to target Cyclin D1 and *WNT3A* to promote proliferation and survival (Bonci et al. 2008). Furthermore, injection of *miR-16* into human tumor xenografts inhibits tumor growth and down-regulates a number of genes involved in cell cycle progression (Takeshita et al. 2010). In lung cancer cell lines, *miR-15* and *miR-16* down-regulate Cyclin D1, Cyclin D2, and Cyclin E1 to inhibit growth in a manner dependent on functional retinoblastoma (Rb) protein (Bandi et al. 2009).

The *miR-34* family (i.e. *miR-34a/b/c*) of tumor suppressor miRNAs is transcriptionally regulated by the gatekeeper protein, p53. Activation of *miR-34a* by p53 promotes cell cycle arrest through modulation of E2F proteins (He et al. 2007; Tarasov et al. 2007). *miR-34a* also targets Bcl-2, Cyclin D1, CDK6 (Sun et al. 2008), SIRT1 (Yamakuchi et al. 2008), Notch-1 (Pang et al. 2010), MYC (Christoffersen et al. 2010), and c-Met (Li et al. 2009b) emphasizing its role as an inhibitor of tumor initiation, progression, and metastasis. In support, introduction of *miR-34a* into cancer cells suppresses in vivo tumor growth (Tazawa et al. 2007). Other *miR-34* family members target oncogenes involved in tumor development and progression. For instance, *miR-34b* targets CREB (Pigazzi et al. 2009) and *miR-34b/c* target MET (Migliore et al. 2008). Interestingly, a close cousin to *miR-34a/b/c*, *miR-449a*, is also down-regulated in prostate cancer tissue and causes cell cycle arrest, apoptosis, and senescence, in part, by depleting HDAC-1 (histone deacetylase 1) expression (Noonan et al. 2009). Both *miR-449a* and *miR-449b* regulate Rb activity by targeting CDK6 and CDC25a to interfere with cell proliferation (Yang et al. 2009b).

18.2.2 Oncogenic MiRNAs

MiR-21 is an oncogenic miRNA that depletes the expression of several tumor suppressor genes involved in promoting cell migration (i.e. RHOB) (Connolly et al. 2010), cell survival (i.e. FAS-ligand and PCD4) (Sayed et al. 2010), cell proliferation (i.e. PTEN) (Meng et al. 2007), and invasion (i.e. RECK, TIMP3, and TPM1) (Zhu et al. 2007).

MiR-221/222 are oncogenic miRNAs aberrantly co-expressed in a number of cancers. In hepatocellular cancer, *miR-221/222* target DDIT4, a component of the mTOR pathway involved in cell growth (Pineau et al. 2010), and cyclin-dependent kinase inhibitors p27/*CDKN1B* and p57/*CDKN1C* to promote cell proliferation

(Fornari et al. 2008). Both *miR-221* and *miR-222* target *p27* in gliomas (Zhang et al. 2009), thyroid (Visone et al. 2007b), melanoma (Felicetti et al. 2008), breast (Miller et al. 2008), and prostate cancer (Galardi et al. 2007; Mercatelli et al. 2008). *MiR-221/222* have also been shown to deplete estrogen receptor alpha (ERα) in breast cancer cells (Zhao et al. 2008) and pro-apoptotic protein BIM in neuronal cells (Di Leva et al. 2010). Furthermore, *miR-221* and *miR-222* target *FOXO3*, an upstream transcriptional regulator of both *BIM* and p27/*CDKNIB* (Di Leva et al. 2010). *PTEN* and *TIMP3* are also targeted by *miR-221* and *miR-222* in hepatocellular carcinoma and lung cancer cell lines promoting a more metastatic and resistant phenotype (Garofalo et al. 2009).

18.2.3 MiRNAs Involved in Cell Cycle and Apoptosis

Precise control of cell cycle is important for normal proliferation. In cancer, aberrant proliferation is caused by dysregulation of a number of key genes including miRNAs. In addition to the miRNAs mentioned earlier, *miR-339* regulates cell cycle progression by targeting E2F1 and suppressing Akt phosphorylation in prostate cancer cells (Lee et al. 2009a).

Apoptosis is another critical cellular process that is dysregulated during cancer leading to cell survival. By targeting pro-survival or pro-apoptotic genes, miRNA dysregulation can also have a profound effect on apoptotic pathways. For instance, *miR-29a/b* has been shown to deplete Mcl-1 and Tcl-1 promoting apoptosis in cholangiocarcinoma and CLL (Mott et al. 2007; Pekarsky et al. 2006), whereas oncogenic *miR-155* targets pro-apoptotic gene *TP53INP1* (TP53-induced nuclear protein 1) in pancreatic ductal carcinoma to promote tumor development (Gironella et al. 2007).

The *miR-17-92* cluster plays a role in both cell cycle and apoptosis. Family member *miR-20a* has been shown to target transcription factors E2F2 and E2F3 in prostate cancer cell lines affecting cell growth and survival. Interestingly, E2F2 and E2F3 also activate the transcription of the *miR-17-92* cluster forming an autoregulatory loop (Sylvestre et al. 2007). Furthermore, *miR-92*, a member of the *miR-17-92* cluster, down-regulates cell cycle gene p63/*TP63* and increases proliferation in myeloid cells (Manni et al. 2009). While in gastric cancer, the *miR-17-92* cluster promotes apoptosis and cell cycle arrest by interfering with TGF-β signaling (Petrocca et al. 2008).

18.2.4 MiRNAs in Metastasis and EMT

The *miR-200* family and *miR-205* target transcriptional repressors and EMT regulators ZEB1 and ZEB2 (Gregory et al. 2008; Peter 2009). Loss of expression of these miRNAs promotes EMT and a more invasive phenotype in bladder (Adam et al. 2009; Wiklund et al. 2010), breast (Tryndyak et al. 2010), pancreatic (Kent et al. 2009), lung (Gibbons et al. 2009), and prostate (Kong et al. 2009) cancer cells. In breast cancer, *miR-335* and *miR-31* have been shown to function as tumor suppressor miRNAs by inhibiting the expression of a number of factors (i.e. SOX4, FZD3, MMP16, etc) involved in promoting invasion and metastasis (Tavazoie et al. 2008). *MiR-9* has recently been determined to play a key role in EMT by directly targeting E-cadherin and promoting β-catenin activity (Ma et al. 2010). *MiR-373* and *miR-520c* promote metastasis in vitro and in vivo in breast cancer (Huang et al. 2008) and inhibits translation of surface receptor CD44 to promote invasion of prostate cancer cells (Yang et al. 2009a). *MiR-10b* is another miRNA that was first identified for promoting metastasis in breast cancer. The transcription factor Twist activates *miR-10b* expression, which targets HOXD10 promoting expression of *RHOC*, a gene involved in metastasis (Ma et al. 2007). In esophageal cancer cell lines, *miR-10b* also blocks the expression of KLF4 tumor suppressor, which inhibits migration (Tian et al. 2010).

Hypoxia is another process also involved in tumor progression regulated, in part, by miRNAs. *MiR-210*, *miR-26*, and *miR-107* are induced in tumor cells grown under hypoxic conditions and correlate with inhibition of apoptosis (Kulshreshtha et al. 2007). Moreover, *miR-210* has been associated with hypoxia in head and neck, ovarian, and breast cancers (Camps et al. 2008; Gee et al. 2010; Giannakakis et al. 2008). Key hypoxic transcriptional regulator HIF-1α has also been shown to be a direct target of *miR-17-92* (Taguchi et al. 2008) and *miR-20b* (Lei et al. 2009). *MiR-20b* additionally targets VEGF and STAT*3* (Cascio et al. 2010; Lei et al. 2009). In head and neck cancer, HIF inhibitory factor, FIH, is targeted by *miR-31* to enhance HIF-1α signaling (Chen et al. 2010). Tumor suppressor *miR-449a* has also been identified as being down-regulated in hypoxic fibroblasts to possibly play a role in the hypoxic response (Muth et al. 2010).

18.3 MiRNAs as Biomarkers of Cancer

Because miRNAs play such an important functional role in cancer development and progression, it should come as no surprise they are being explored as potential cancer biomarkers. A number of studies have been initiated using miRNA expression profiling to determine markers of disease. One of the earliest studies comparing normal and tumorigenic tissue examined 217 different miRNAs within several cancer types and showed that the miRNA expression signature was able to classify tumors based on developmental lineage and differentiation state. The authors also found that miRNAs were more reliable predictors of poorly differentiated tumors underscoring the significance of using miRNAs as biomarkers (Lu et al. 2005). Following this study, miRNA expression profiles have been used to define a number of normal and cancerous tissues from thyroid, kidney, bladder, esophageal, liver, etc (Calin et al. 2004; Cohn et al. 2010; Cosmopoulos et al. 2009; Gottardo et al. 2007; Guo et al. 2008; Katada et al. 2009; Langer et al. 2008; Li et al. 2009a; Patnaik et al. 2010; Porkka et al. 2007; Schepeler et al. 2008; Varnholt et al. 2008; Visone et al. 2007a; Yan et al. 2008). Furthermore, miRNA profiling has also been used to classify tumor subtypes in breast cancer, mesothelioma, kidney cancer, lung cancer, and neuroblastomas (Buckley et al. 2010; Guled et al. 2009; Petillo et al. 2009; Rossi et al. 2009).

18.3.1 MiRNAs in Determining Cancer Prognosis

MiRNAs are also valuable tools for determining cancer prognosis (Cho 2009). In lung cancer, a signature of miRNAs has been shown to independently predict for survival and relapse (Yu et al. 2008). This is just one of the hundreds of studies examining the correlation of miRNA signatures with survival, recurrence of disease, or metastasis (Calin and Croce 2006; Jay et al. 2007; Nana-Sinkam and Croce 2010; Yu et al. 2007). In these studies, a number of the up-regulated miRNAs associated with poor survival, relapse or advanced stage, and maintaining oncogenic function; whereas, down-regulated miRNAs associated with tumor suppressor-like function.

18.3.2 Circulating MiRNAs

A good reason miRNAs make reliable biomarkers is their inherent stability in a number of different biological tissues and fluids. Use of blood and other bodily fluids provides a less invasive way to detect cancer. MiRNAs are generally resistant to degradation by RNases and can be isolated from the serum, saliva, semen, vaginal secretions, urine, stool, and sputum (Ahmed et al. 2009; Melkonyan et al. 2008; Michael et al. 2010; Mitchell et al. 2008; Zubakov et al. 2010). One of earliest studies in serum indicated *miR-141*, a miRNA over-expressed in prostate cancer, as a biomarker that could distinguish between cancer patients and normal controls (Lawrie et al. 2008). Studies performed in B cell lymphoma, lung, colorectal, ovarian, and breast cancers have also shown similar prognostic value for miRNAs detected in serum (Chen et al. 2008; Hu et al. 2010b; Lawrie et al. 2008; Lodes et al. 2009; Resnick et al. 2009; Zhu et al. 2009). Two studies have shown the usefulness of detecting miRNAs in saliva of oral cancer patients. Elevated *miR-31* was indicated as a biomarker for oral squamous cell carcinoma in both saliva and serum, while *miR-125a* and *miR-200a* were found at lower levels in the saliva of oral squamous cell carcinoma patients (Liu et al. 2010; Park et al. 2009b). Sputum is a fluid often used for cytological detection of lung cancer. Researchers have extended the idea to detect aberrant miRNA expression in the sputum of lung cancer patients. As such, *miR-21* has been identified as a more sensitive marker for lung cancer than standard sputum cytology (Xie et al. 2010). An additional study identified four miRNAs in the sputum of lung cancer patients that could be biomarkers for early detection (Yu et al. 2010). The urine is also another readily available source for screening miRNAs. Preliminary research shows that miRNA signatures found in the urine of bladder cancer patients have potential use as biomarkers. The ratio of *miR-126:miR-152* was found to be a highly-sensitive and specific indicator for bladder cancer (Hanke et al. 2010). MiRNAs have also been detected in the stool and shown to correlate with stages of colorectal cancer (Ahmed et al. 2009).

18.3.3 MiRNA Biomarkers in Clinical Trials

Clinical trials are currently underway examining miRNAs as biomarkers for cancer. Renal cell carcinoma (RCC) is a type of cancer that is difficult to detect early. There is a strong interest in identifying early markers for this disease in order to improve patient survival. An observational study is underway to identify miRNAs involved in RCC development, progression, pathological subtypes, and prognosis (NCT00743054: www.clinicaltrials.gov). A similar observational study is underway examining miRNA expression profiles of children in Taiwan with acute lymphoblastic leukemia (ALL) (NCT00526084) to more accurately classify subgroups. The prognostic significance of miRNA expression profiling is also being examined in a Phase II clinical trial for treatment of non-small cell lung cancer (NSCLC) with erlotinib and docetaxel (NCT00840125). Additionally, another study is underway comparing blood from NSCLC patients and healthy controls to determine diagnostic significance of miRNA expression profiling (NCT00897234). The National Cancer Institute is also evaluating miRNA profiles to predict clinical outcome in acute myeloid leukemia (AML), as well as studying the functional role of *miR-34a* and *miR-194* in differentiation and proliferation (NCT00898092, NCT01057199). MiRNA tissue profiling is also being examined in hard-to-cure solid tumors (i.e. tissue sarcomas, osteosarcomas, and neuroblastomas) in children to obtain a better understanding of development and progression of these specific types of cancer (NCT01050296). Multiple myeloma (MM), another hard-to-treat cancer, is the subject of an observational study that will attempt to identify prognostic miRNA indicators during relapse (NCT00639054). A number of other studies are underway examining miRNAs as diagnostic biomarkers in lymphoblastic leukemia (NCT00896766), neuroblastomas (NCT00958659), subtypes of breast cancer (NCT00581750), and endometrial cancer (NCT01119573). Since detecting and differentiating between different cancer subtypes can be a challenge, miRNA signatures are a promising new source of biomarkers to improve diagnosis.

18.4 MiRNAs in Altering Sensitivity to RT and CT

Just as miRNAs can control tumorigenesis, they can also regulate sensitivity and resistance to RT and CT. Of these are a number of previously established tumor suppressive and oncogenic miRNAs involved in cell cycle, apoptosis, and EMT. One of the first comprehensive miRNA profiling studies examined the NCI-60, a set of 59 human cancer cell lines derived from diverse tissues including the brain, blood and bone marrow, breast, colon, kidney, lung, ovary, prostate and skin, and showed that miRNA expression correlated with resistance and sensitivity to certain compounds (Blower et al. 2007). Over 100,000 different potential anticancer compounds were evaluated that identified hundreds of thousands of correlations. Of interest, oncogenic *miR-221* and *miR-222* strongly correlated with resistance to several compounds, while *miR-214* and *miR-342* correlated with chemosensitivity. *MiR-214* is down-regulated in certain tumor types and has been shown to play a role in both development and survival (Flynt et al. 2007; Lee et al. 2009b; Yang et al. 2008a; 2009c). *MiR-342* is differentially expressed in breast cancer subtypes with high expression in luminal B type tumors and decreased expression in the therapeutically difficult, estrogen receptor/progesterone receptor/HER-2 negative tumors (Lowery et al. 2009). This indicates that, based on expression of select miRNAs, various types of tumors and tumor subtypes may be more sensitive or resistant to particular treatments.

18.4.1 MiRNAs in Predicting Response to RT

Radiation therapy is an integral part of most cancer treatment regimens; however, RT can have toxic consequences in patients. As such, there is much interest in identifying markers that may predict a patient's response to RT. MiRNAs are one such novel group of gene regulators recently recognized to correlate with RT resistance. For a summary of miRNAs involved in RT response, please see Table 18.1.

18.4.1.1 Radiosensitivity

One of the first papers studying miRNAs in RT identified the *let-7* family for their role in modulating sensitivity in lung cancer (Weidhaas et al. 2007). This study showed that the expression of *let-7* family members decreases following RT to correlate with acquired resistance. It demonstrated that over-expression of select *let-7* members promotes radiosensitivity, while knockdown increases resistance both in vitro and in vivo. Subsequent studies in lung and pancreatic cell lines have indicated that *let-7* modulates radiosensitivity by targeting the *KRAS* oncogene; *let-7* over-expression suppresses *KRAS* levels to improve response to RT (Oh et al. 2010). *MiR-181a* is another miRNA that has been identified as playing a role in radiosensitivity of malignant glioma cells. Expression profiling has revealed that *miR-181a* is down-regulated in response to radiation treatment, whereas transient over-expression of *miR-181a* augments radiosensitivity. Reintroduction of *miR-181a* correlates with decreased expression of anti-apoptotic protein Bcl-2 and radiosensitivity (Chen et al. 2010). MiRNA samples taken from gastric cancer patients following surgical resection and radiation therapy have correlated down-regulation of *miR-451* with poor prognosis. Over-expression of *miR-451* in colorectal cancer cell lines decreases proliferation and renders gastric cancer cells sensitive to RT (Bandres et al. 2009). Another strategy for exploiting radiosensitivity is synchronizing cells into radiosensitive phases of the cell cycle. Reintroduction or activation of miRNAs promoting arrest in the G2/M phase of the cell cycle (i.e. *let-7*, *miR-34a*, etc) may render cells more sensitive to subsequent RT.

18.4.1.2 Radioresistance

Expression of certain miRNAs can also correlate with resistance to RT. In one study, miRNA expression was assessed before and after combination of CT and RT of

Table 18.1 MicroRNAs involved in radiotherapy and chemotherapy response

Tissue	MicroRNA	Target	Therapeutic agent	MicroRNA therapeutic effect	References
Breast					
–	*miR-221/ 222*	ERα, p-27	Tamoxifen	Promotes resistance	Zhao et al. (2008)
–	*miR-21*	PDCD4, MDR1[a]	Taxol, doxorubicin	Promotes resistance	Bourguignon et al. (2009); Mei et al. (2010)
–	*miR-30c*	–	Tamoxifen	Sensitizes	Rodriguez-Gonzalez et al. (2010)
–	*miR-326*	MRP-1	Etoposide, doxorubicin	Sensitizes	Liang et al. (2010)
–	*miR-451*	MDR1	Doxorubicin	Sensitizes	Kovalchuk et al. (2008)
–	*miR-328*	ABCG2	Mitoxantrone	Sensitizes	Pan et al. (2009)
–	*miR-205*	HER-3	Gefitinib, lapatinib	Sensitizes	Iorio et al. (2009)
–	*miR-125b*	BAK1	Taxol	Sensitizes	Zhou et al. (2010)
Lung					
NSCLC	*miR-21*	–	AG148	Promotes resistance	Seike et al. (2009)
–	*miR-630*	–	Cisplatin	Promotes resistance	Chen et al. (2010)
–	*miR-128b*	EGFR	Gefitinib	Promotes resistance	Weiss et al. (2008)
–	*Let-7a*	p-Akt[b]	Gefitinib	Sensitizes	Zhong et al. (2010)
–	*miR-126*	p-Akt[b]	Gefitinib	Sensitizes	Zhong et al. (2010)
–	*miR-145*	p-Akt[b]	Gefitinib	Sensitizes	Zhong et al. (2010)
–	*miR-181*	–	Cisplatin	Sensitizes	Chen et al. (2010)
–	*let-7* family	KRAS	RT	Sensitizes	Oh et al. (2010)
–	*miR-1*	MET, Pim-1, FoxP1, HDAC4	Doxorubicin	Sensitizes	Nasser et al. (2008)
SCLC	*miR-134*	MRP-1[b]	Cisplatin, etoposide, doxorubicin	Sensitizes	Guo et al. (2010)
Blood					
CLL	*miR-34a*	–	γ-irradiation	Sensitizes	Zenz et al. (2009)
–	*miR-15/16*	–	Nutlin	Sensitizes	Salerno et al. (2009)
CML	*miR-21*	–	Arsenic	Promotes resistance	Li et al. (2010)

Table 18.1 (continued)

Tissue	MicroRNA	Target	Therapeutic agent	MicroRNA therapeutic effect	References
–	*miR-199*	SIRT1[c], ABCC5[c]	Imatinib mesylate	Sensitizes	San Jose-Eneriz et al. (2009)
–	*miR-520a*	RAB11A[c]	Imatinib mesylate	Sensitizes	San Jose-Eneriz et al. (2009)
–	*miR-183*	ABCA1[b]	Imatinib mesylate	Sensitizes	San Jose-Eneriz et al. (2009)
–	*miR-29c*	ABCB6[c]	Imatinib mesylate	Sensitizes	San Jose-Eneriz et al. (2009)
ALL	*miR-128b*	MLL, AF4, MLL-AF4 (fusion gene)	Glucocorticoids	Sensitizes	Kotani et al. (2009)
–	*miR-221*	p27	Glucocorticoids	Sensitizes	Kotani et al. (2009)
–	*miR-15/ 16*	–	Glucocorticoids	Sensitizes	Rainer et al. (2009)
Pancreas					
–	*miR-21*	–	Gemcitabine, fluorouracil, RT	Promotes resistance	Giovannetti et al. (2010); Park et al. (2009a); Hwang et al. (2010)
–	*miR-221*	–	Gemcitabine	Promotes resistance	Park et al. (2009a)
–	*miR-200b/c*	E-cadherin[a], vimentin[b], ZEB1[b]	Gemcitabine	Sensitizes	Li et al. (2009d)
–	*let-7b/c/d/e*	–	Gemcitabine	Sensitizes	Li et al. (2009d)
–	*miR-34a*	Bcl-2, Notch-1,2[b]	Docetaxel, cisplatin, gemcitabine, RT	Sensitizes	Ji et al. (2009)
Ovarian					
–	*let-7i*	–	Cisplatin	Sensitizes	Yang et al. (2008b)
–	*miR-200c*	ZEB1, ZEB2, β-tubulin	Paclitaxel, epothilone-B, vincristine	Sensitizes	Cochrane et al. (2009)
CNS					
Glioma	*miR-21*	–	Fluorouracil, etoposide, S-TRAIL	Promotes resistance	Corsten et al. (2007); Li et al. (2009c); Ren et al. (2010)
–	*miR-451*	–	Imatinib	Sensitizes	Gal et al. (2008)
–	*miR-181a*	Bcl-2[b]	RT	Sensitizes	Chen et al. (2010)

Table 18.1 (continued)

Tissue	MicroRNA	Target	Therapeutic agent	MicroRNA therapeutic effect	References
Liver					
–	*Let-7a*	caspase-3	Interferon-γ, doxorubicin, paclitaxel	Promotes resistance	Tsang and Kwok (2008)
–	*miR-122*	IGF1R, Cyclin G1	Sorafenib, doxorubicin	Sensitizes	Bai et al. (2009); Fornari et al. (2009)
–	*miR-193*	Mcl1	Sorafenib	Sensitizes	Braconi et al. (2010)
Colon					
–	*miR-125*	–	Capecitabine	Promotes resistance	Svoboda et al. (2008)
–	*miR-137*	–	Capecitabine	Promotes resistance	Svoboda et al. (2008)
–	*miR-181b*	–	Nucleoside analog S-1	Sensitizes	Nakajima et al. (2006)
–	*let-7g*	–	Nucleoside analog S-1	Sensitizes	Nakajima et al. (2006)
–	*miR-143*	NF-κB[b], Bcl-2[b], ERK5[b]	Fluorouracil	Sensitizes	Borralho et al. (2009)
–	*miR-519c*	ABCG2	Mitoxantrone	Sensitizes	To et al. (2009)
–	*miR-125b*	–	RT	Promotes resistance	Svoboda et al. (2008)
–	*miR-137*	–	RT	Promotes resistance	Svoboda et al. (2008)
Gastric and colon					
–	*miR-451*	–	RT	Sensitizes	Bandres et al. (2009)
Bile duct					
–	*miR-21*	–	Gemcitabine	Promotes resistance	Meng et al. (2006)
–	*miR-200b*	–	Gemcitabine	Promotes resistance	Meng et al. (2006)
Bladder					
–	*miR-200b*	ERRFI-1	Cetuximab, erlotinib	Sensitizes	Adam et al. (2009)
Prostate					
–	*miR-221/222*	–	Androgens	Promotes resistance	Sun et al. (2009)
–	*miR-34a*	–	Camptothecin	Sensitizes	Fujita et al. (2008)

RT, Radiotherapy.
[a]Indicates expression positively correlates or,
[b]Inversely correlates with miRNA (not a direct target).
[c]Putative targets (no expression data).

rectal cancer patients. Although the researchers discovered inter-tumoral variability, they did identify elevated expression levels of *miR-125b* and *miR-137* as predictors of poor response (Svoboda et al. 2008). Another study evaluating miRNA expression in normal human endothelial cells identified a number of miRNAs that became dysregulated following radiation treatment. *Let-7g*, *miR-16*, *miR-20a*, *miR-21*, and *miR-29c*, were found to be up-regulated, while *miR-18a*, *miR-125a*, *miR-127*, *miR-148b*, *miR-189*, and *miR-503* were all down-regulated. Of these, *miR-125a* and *miR-189* conferred radioresistance, whereas *let-7g* and *miR-127* promoted radiosensitivity (Wagner-Ecker et al. 2010).

Because hypoxia plays a key role in tumor resistance to RT, identifying miRNA signatures of hypoxia is also important for overcoming radioresistance. As such, interfering with miRNAs (i.e. *miR-9*, *miR-20b*, *miR-205*, etc) involved in regulating hypoxia and other radioresistance pathways (i.e. VEGF, HER-2, etc) may have potential for increasing tumor cell sensitivity to RT.

18.4.2 MiRNAs in Predicting Response to CT

Resistance to CT is a major obstacle to successful treatment. Identifying components of resistance, including miRNAs, will enable researchers and physicians to find better regimens of therapy for each patient. For a summary of miRNAs involved in chemotherapeutic response, please see Table 18.1.

18.4.2.1 Breast Cancer

Breast cancer is one of leading causes of cancer deaths in women and resistance to CT is a common issue. Expression of ERα selects for patients that may respond well to hormonal therapy. However, a subset of these tumors will lose ERα expression during therapy and become resistant. A study examining survival and outcome of patients with ER-positive breast cancer found *miR-30c* to positively correlate with tamoxifen sensitivity; an ERα antagonist. *MiR-30c* associated with altered EGFR and RAC1 signaling, which affect cell growth and metastasis, respectively (Rodriguez-Gonzalez et al. 2010). Loss of ERα may also be regulated, in part, by miRNAs. MiRNA expression profiling of ER-positive and ER-negative breast cancer cell lines and tumors identified elevation of *miR-221* and *miR-222* in ER-negative breast cancer cells. *MiR-221* and *miR-222* were shown to directly suppress ERα expression and promote resistance to tamoxifen. Furthermore, knockdown of *miR-221* and *miR-222* renders resistant ER-negative cells sensitive to tamoxifen (Zhao et al. 2008). In another study, *miR-221* and *miR-222* were also associated with HER2/neu positive breast cancer tissue samples; an aggressive breast cancer subtype generally resistant to tamoxifen therapy. Ectopic expression of *miR-221* and *miR-222* was found to promote tamoxifen resistance in otherwise sensitive cells by targeting and suppressing cell cycle inhibitory protein p27 (Miller et al. 2008).

As previously mentioned, multidrug resistance and drug transport proteins (e.g. MDR1, MRP-1, and P-glycoprotein) play an important role in CT resistance. *MRP-1* is highly expressed in the topoisomerase II inhibitor-resistant breast cancer cell line MCF-7/Vp. MRP-1 has been found to be targeted and suppressed by *miR-326*. As such, restoration of *miR-326* expression sensitizes MCF-7/Vp cells to the topoisomerase II inhibitor, etoposide, and a DNA intercalating agent, doxorubicin (Liang et al. 2010). Elevated levels of *miR-21* have also been shown in the doxorubicin and verapamil (P-glycoprotein inhibitor) resistant breast cancer cell line MCF-7/AdrVp (Mei et al. 2010). The stimulation of surface receptor CD44 in breast cancer cells up-regulates *miR-21* and *MDR1* (multidrug resistance protein 1); an essential protein involved in doxorubicin resistance. *MDR1* expression is dependent on *miR-21* as *miR-21* inhibition interferes with CD44-mediated activation of *MDR1* (Bourguignon et al. 2009). Similarly, *miR-21* antagonism promotes breast cancer cell line sensitivity to the mitotic inhibitor taxol (Mei et al. 2010). Expression profiling also identified depletion of *miR-451* in the doxorubicin-resistant breast cancer cell line MCF-7/Dox. Loss of *miR-451* was found to promote *MDR1* levels as *miR-451* inversely regulates *MDR1* expression (Kovalchuk et al. 2008). Drug transport protein ABCG2 is also an essential determinant of chemoresistance. In mitoxantrone (topoisomerase inhibitor) resistant breast cancer cells, *ABCG2* inversely correlates with *miR-328* levels. Re-expression of *miR-328* increases breast cancer cells sensitivity to mitoxantrone (Pan et al. 2009).

MiRNA regulation of growth factor expression is another common form of CT resistance. Elevated expression of EGFR family member *HER2/ERBB2* is associated with an aggressive subtype of breast cancer resistant to multiple chemotherapeutic agents. HER2 is functionally dependent, in part, on its binding partner *HER3/ERBB3* for signal transduction. *MiR-205* has been shown to directly target HER3 in breast cancer and down-regulate its expression in tumor tissue compared to controls. Restoration of *miR-205* in breast cancer cells decreases clonogenicity and renders cells sensitive to EGFR inhibitors gefitinib and lapatinib (Iorio et al. 2009).

MiRNAs that alter apoptotic signaling can also function to promote resistance in breast cancer. A study examining miRNA signatures found that *miR-125b*, *miR-221*, *miR-222*, and *miR-923* are up-regulated in breast cancer cell lines resistant to taxol. As previously mentioned, *miR-125b* targets and inhibits the expression of pro-apoptotic gene BAK1. Reintroduction of BAK1 or *miR-125b* into breast cancer cell lines rescues taxol sensitivity further indicating the significance of miRNAs in chemotherapeutic resistance (Zhou et al. 2010).

18.4.2.2 Lung Cancer

Lung cancer has the highest mortality rate among all cancers and is the second most common cancer in both men and women. Although resistance to standard therapeutics remains a problem, recent data suggests miRNAs play a key role in regulating sensitivity. *EGFR/HER1/ERBB1* growth factor receptor expression is one of the major avenues of resistance in lung cancer. Lung tumor tissue samples exhibit loss of heterozygosity of *miR-128b* relative to normal tissue, which inversely correlates to EGFR expression. In this case, loss of *miR-128b* has been described as a marker for EGFR dependence and susceptibility to anti-EGFR drugs. As such, *miR-128b* depletion associates with increased disease control in response to EGFR inhibitor gefitinib and improved survival (Weiss et al. 2008). Correlation has also been found between phosphorylated-EGFR (p-EGFR) and *miR-21* in lung carcinoma cell lines. Suppression of p-EGFR by tyrosine kinase inhibitor, AG1478, down-regulates *miR-21* indicating that EGFR signaling may positively regulate *miR-21* expression to correlate with resistance. As such, inhibition of *miR-21* in lung cancer cell lines promotes apoptosis and enhances sensitivity to AG1478 (Seike et al. 2009). In NSCLC cell lines, EGFR-independent activation of Akt and ERK confers resistance to gefitinib. Transient transfection with tumor suppressor miRNAs *let-7a*, *miR-126*, and *miR-145* promote growth arrest both in vitro and in vivo and enhance sensitivity to gefitinib treatment, in part, by interfering with Akt and ERK activity (Zhong et al. 2010).

A549 human lung cancer cells were also examined for alterations in miRNA profile in response to cisplatin; a platinum compound that induces apoptosis by DNA cross-linking. In this study, *miR-181a* was identified to sensitize cells to cisplatin, while *miR-630* promoted resistance. As only proliferating cells are sensitive to cisplatin, *miR-630* activates p53/DNA repair and cell cycle arrest to evade cisplatin sensitivity. Interestingly, *miR-181a* has been shown to possess tumor suppressor function in human glioma cells by targeting Bcl-2 (Chen et al. 2010). This activity may also contribute to sensitizing lung cancer cells to cisplatin-induced apoptosis.

MiR-1 is lost in primary lung cancer tissue and lung cancer cell lines. *MiR-1* expression is reactivated in response to HDAC inhibitor trichostatin A (TSA). Reintroduction of *miR-1* has been shown to suppress the expression of a number of oncogenes including MET, Pim-1, FoxP1, and HDAC-4, as well as increase sensitivity to doxorubicin. Interestingly, activation of *miR-1* by HDAC inhibitors is one potential treatment strategy for sensitizing NSCLC cells to doxorubicin (Nasser et al. 2008).

Resistance to anti-mitotic drug docetaxel is another therapeutic issue in NSCLC. Preliminary studies evaluating miRNA signatures in docetaxel-resistant NSCLC cells have identified down-regulation of *miR-200b*, *miR-194*, *miR-212* and up-regulation of *miR-192*, *miR-424*, *miR-98* (Rui et al. 2010). Although interesting, more work is needed to elucidate their specific roles in docetaxel resistance.

Evaluating miRNA expression in small cell lung cancer (SCLC) has also revealed that inhibition of *miR-134* promotes sensitivity to cisplatin, etoposide, and doxorubicin in resistant SCLC cells. Although the exact molecular function of *miR-134* has not been determined, MRP-1 was found to negatively correlate with *miR-134* levels (Guo et al. 2010).

18.4.2.3 Hematological Malignancies

Cancers of the blood are no exception to miRNA dysregulation and resistance. In chronic lymphocytic leukemia (CLL), p53/*TP53* deletion identifies patients who are more likely to be resistant to CT. The *miR-34* family is well-established as downstream transcriptional targets of p53 and directly linked to its tumor suppressor function (Tarasov et al. 2007). Interestingly, CLL cells that minimally express *miR-34a* show resistance to DNA damaging agents regardless of p53 status. This implies that the *miR-34* family may play a significant role in CT resistance of CLL (Zenz et al. 2009). *MiR-15* and *miR-16* have also been well characterized for their role in hematological malignancies including CLL. In a mouse model for CLL, restoration of *miR-15* and *miR-16* sensitizes cancer cells to MDM2 antagonist, nutlin, and natural tyrosine kinase inhibitor, genistein (Salerno et al. 2009).

Chronic myeloid leukemia (CML) is a blood cancer that usually responds favorably to imatinib mesylate (IM); an inhibitor of the oncogene *BCR-ABL1* that results from a chromosomal translocation known as the Philadelphia chromosome. However, approximately 25% of patients treated with IM are non-responders or become resistant to treatment. Expression profiling performed on CML patients has identified miRNAs that are able to predict resistance to IM. Two miRNAs downregulated in CML, *miR-199a* and *miR-520a*, potentially target and suppress the function of IM resistance-promoting proteins SIRT1 and RAB11A, respectively. Other miRNAs depleted in CML (i.e. *miR-199a, miR-183*, and *miR-29c*) potentially target and inhibit such drug trafficking proteins as ABCC5, ABCA1, and ABCB6, respectively; each of which has been implicated in IM resistance (San Jose-Eneriz et al. 2009). *MiR-21* has also been shown to have oncogenic function in CML by playing a role in apoptosis and cell migration (Hu et al. 2010a). Furthermore, inhibition of *miR-21* enhances CML cells to arsenic-induced apoptosis. Although arsenic is highly toxic and possesses unfavorable side effects, combination therapies that sensitize CML cells to minimal doses of arsenic are clinically relevant (Li et al. 2010).

MLL-AF4 acute lymphocytic leukemia (ALL) is another type of blood cancer associated with poor prognosis and steroid resistance that results from a balanced translocation between *MLL* and *AF4* genes at t(4;11). Interestingly, *miR-128b* and *miR-221* expression is depleted in MLL-AF4 ALL and co-expression of these two miRNAs increases sensitivity to glucocorticoid therapy. It has been shown that *miR-128b* directly targets and suppresses MLL, AF4, and the fusion gene, while *miR-221* promotes cellular retention in S-phase of the cell cycle by targeting p27 sensitizing MLL-AF4 ALL cells (Kotani et al. 2009). Tumor suppressor *miR-15* and *miR-16* have also been implicated in ALL steroid resistance. Restoration of *miR-15* and *miR-16* increases ALL cell sensitivity to glucocorticoid therapy, as well (Rainer et al. 2009).

18.4.2.4 Pancreatic Cancer

Pancreatic cancer is one of the most aggressive and difficult cancers to cure. Expression profiling of pancreatic tumor cells has identified several miRNAs that play important roles in drug resistance. Both *miR-21* and *miR-221* are aberrantly expressed in pancreatic cancer. Resistance to nucleoside analog, gemcitabine, and shortened patient survival have been correlated to high-levels of *miR-21* in pancreatic tumor samples (Giovannetti et al. 2010). Inhibition of *miR-21* and *miR-221* with antisense antagonists decreases cell viability and synergistically improves sensitivity to gemcitabine (Park et al. 2009a). In pancreatic ductal adenocarcinoma (PDAC) cell lines, *miR-21* expression enhances resistance to pyrimidine analog, fluorouracil (5-FU), and combination treatment of RT plus gemcitabine (Hwang et al. 2010).

Down-regulation of *miR-200b*, *miR-200c*, *let-7b*, *let-7c*, *let-7d*, and *let-7e* has also been linked to gemcitabine resistance in pancreatic cancer cell lines. Gemcitabine resistance associates with EMT and elevated levels of ZEB1 and vimentin; key regulators of EMT directly targeted by several of these dysregulated miRNAs. Reintroduction of *miR-200b* results in normalized expression of EMT markers and increases sensitivity of otherwise resistant cells to gemcitabine (Li et al. 2009d).

Resistance to therapy has also been linked to *miR-34a* depletion in pancreatic cancer. Pancreatic cancer cells with p53 loss-of-function mutations express low-levels of *miR-34a*. Reintroduction of *miR-34a* increases their sensitivity to CT (i.e. docetaxel, cisplatin, and gemcitabine) and RT. Furthermore, *miR-34a* introduction causes an 87% decrease in the population of cancer stem cells (tumor initiating cells) and inhibits tumor-sphere growth in vitro and tumor formation in vivo (Ji et al. 2009).

18.4.2.5 Ovarian Cancer

Ovarian cancer has a high mortality rate, which is attributed, in part, to therapy resistance. Reduced *let-7i* miRNA levels have been shown to associate with shorter survival in late stage ovarian cancer, as well as to CT resistance. Reintroduction of *let-7i* improves sensitivity to platinum-based CT cisplatin, while knockdown of *let-7i* promotes resistance (Yang et al. 2008b). *MiR-200c*, another miRNA identified in regulating resistance, is depleted in poorly-differentiated ovarian cells. As such, loss of *miR-200c* correlates with EMT and over-expression of downstream targets ZEB1 and ZEB2; transcription factors that directly facilitate EMT. Class III β-tubulin (TUBB3) is also an additional target of *miR-200c* and reintroduction of *miR-200c* significantly improves sensitivity of ovarian cancer cells to microtubule-targeting chemotherapeutic agents (Cochrane et al. 2009). Another study has also identified *let-7e*, *miR-30c*, *miR-125b*, *miR-130a*, and *miR-335* as dysregulated miRNAs in cisplatin and/or paclitaxel resistant ovarian cancer cells. *MiR-30c*, *miR-130a*, and *miR-335* were down-regulated in all resistant cell types, while *let-7e* and *miR-30c* were selectively dysregulated (Sorrentino et al. 2008). Associations between miRNA and resistance are a good way for identifying candidate miRNAs involved in improving ovarian cancer therapeutics.

18.4.2.6 Other Cancers

The role of miRNAs in CT resistance has also been cited in a number of other malignancies. In cancers of the central nervous system, *miR-21* has been identified as an oncogenic miRNA that promotes resistance, cell growth, proliferation, and survival of gliomas. Inhibition of *miR-21* sensitizes U251 glioma cells to fluorouracil, glioblastoma multiforme cells to etoposide, and gliomas in mice to tumor necrosis factor-related apoptosis inducing ligand S-TRAIL (Corsten et al. 2007; Li et al. 2009c; Ren et al. 2010). Glioblastoma stem cells also play an important role in chemoresistance of glioblastoma tumors. *MiR-451* has been identified as being down-regulated in glioblastoma stem cell populations. Reintroduction of *miR-451* in combination with frequently used glioblastoma treatment, imatinib, inhibits the growth of glioblastoma stem cells and the formation of neurospheres (Gal et al. 2008).

In hepatocellular carcinoma (HCC), one of the most extensively studied miRNAs is liver-specific *miR-122*. *MiR-122* is down-regulated in HCC, which promotes resistance, growth, proliferation, EMT, and metastasis. Insulin growth factor 1 tyrosine kinase receptor (Igf1R) is targeted and suppressed by *miR-122* in normal liver cells; however, depletion of *miR-122* in HCC increases Igf1R levels. As such, reintroduction of *miR-122* inhibits Igf1R in HCC and promotes sensitivity to tyrosine kinase inhibitor sorafenib (Bai et al. 2009). *MiR-122* also targets and suppresses the expression of cyclin G1, which functions to negatively regulate p53 by promoting Mdm2 activity. Inhibition of cyclin G1 by *miR-122* stabilizes p53 protein and sensitizes HCC to doxorubicin (Fornari et al. 2009). Hepatitis C virus (HCV) modulation of miRNA expression has also been shown to play a role in the chemosensitivity of HCC. Introduction of HCV into malignant hepatocytes promotes *miR-193b* over-expression, which targets and suppresses the expression of anti-apoptotic factor Mcl1. Because Mcl1 also modulates sensitivity to sorafenib, introduction of HCV or *miR-193b* alone enhanced apoptosis and HCC cell sensitivity to sorafenib (Braconi et al. 2010). *Let-7a* expression in HepG2 hepatocellular carcinoma and A431 epidermal carcinoma cell lines increases resistance to a number of chemotherapeutic drugs including interferon-γ, doxorubicin, and paclitaxel. Inhibition of *let-7a* improves response to CT, in part, through up-regulation of downstream target caspase-3. *Let-7a*-mediated resistance is dependent on caspase-3 as tumor cells that do not express caspase-3 retain resistance regardless of *let-7a* status (Tsang and Kwok 2008).

In colorectal cancers, a number of miRNAs have been associated with predicting response to nucleoside analogs. *MiR-181b* and *let-7g*, while over-expressed in colon cancer, are indicative of a favorable prognosis for treatment with nucleoside analog S-1 (Nakajima et al. 2006). *MiR-143* is down-regulated in colon cancer, targets *NF-κB*, *Bcl-2*, *ERK5*, and has been shown to increase sensitivity to fluorouracil in HCT-166 colon cancer cell lines (Borralho et al. 2009). In rectal cancer patients, miRNA expression analysis performed before and after CT treatment identified *miR-125b* and *miR-137* associated with poor response to capecitabine; a pro-drug that is enzymatically converted to fluorouracil (Svoboda et al. 2008). Dysregulation of miRNA-regulated repression of drug transport proteins has also been shown to play a role in resistance of colon cancer cells. *MiR-519c* targets and suppresses ABCG2 in cell lines sensitive to mitoxantrone; whereas, *miR-519c* inhibition increases ABCG2 levels and chemoresistance. In resistant cell lines, ABCG2 mRNA possesses a shortened 3′UTR, which results in the loss of a *miR-519c* target site and high-levels of ABCG2 protein (To et al. 2009). Cholangiocarcinoma is a rare type of cancer formed in bile ducts that has poor prognosis without surgical removal. In malignant cholangiocytes, inhibition of *miR-21* and *miR-200b* improves sensitivity to gemcitabine (Meng et al. 2006).

Differential expression of miRNAs in urological cancers has also been linked to sensitivity and resistance to therapy. In bladder cancer, miRNAs involved in EMT play a role in sensitizing cells to EGFR inhibitor therapy. Reintroduction of *miR-200* family members into mesenchymal-like cells inhibits EMT regulators ZEB1 and ZEB2 and improves sensitivity to EGFR inhibitors (Adam et al. 2009). In prostate cancer, *miR-221* and *miR-222* are over-expressed in the castration-resistant cell line LNCaP-Abl. Over-expression of each miRNA in androgen-dependent cell lines promotes androgen-independent growth. Alternatively, inhibition of *miR-221* and *miR-222* promotes androgen dependence in LNCaP-Abl cells (Sun et al. 2009). *MiR-34a* has also been linked to resistance to topoisomerase I inhibitor, camptothecin, in prostate cancer cells. PC-3 prostate cancer cells, which lack *miR-34a* expression due to their p53-null status, are generally resistant to camptothecin; however, restoration of *miR-34a* enhances chemosensitivity (Fujita et al. 2008).

18.4.2.7 MiRNA Predictors in Clinical Trials

Use of miRNAs in predicting response to chemotherapeutics may give clinicians a more accurate or sensitive method to determine appropriate treatment choices. MiRNA expression signatures are being examined in clinical trials typically as secondary outcomes or in addition to other molecular markers such as single nucleotide polymorphisms (SNPs), mRNA, or protein expression patterns. This resulting data may be useful in determining novel markers for therapeutic resistance or sensitivity.

Currently, a Phase II clinical trial is underway to evaluate the effect of VEGF inhibitor, Avastin, in combination with CT or endocrine therapy as a preoperative treatment in patients with HER2/ERBB2 negative breast cancer. MiRNA expression changes will be examined following treatment as a secondary outcome (NCT00773695). An observational study in metastatic prostate cancer is also evaluating miRNA signatures as potential markers of resistance or sensitivity to treatment (NCT01050504). In an interventional study for NSCLC, miRNA and SNP signatures are being characterized as means to predict outcome for response to CT treatment (NCT00864266). Additionally, two Phase II clinical trials in NSCLC are underway in which miRNAs will be used to determine prognostic value in response to specific drug regimens. One study is designed to examine miRNA expression in stage IIIA N2+ NSCLC tumors treated with taxol and carboplatin (alkylating agent) followed by RT plus cetuximab (monoclonal antibody) (NCT00979212). The other study is examining miRNA profiles in patients with stage IIIB-IV NSCLC following treatment with erlotinib and docetaxel (NCT00840125).

In hematological malignancies, two observational studies are underway to identify miRNA biomarkers that can be used to predict response to treatment. The first will examine miRNAs in response to standard chemotherapeutic treatment in ALL of Taiwanese children (NCT00526084), while the second will evaluate miRNA signatures at diagnosis and relapse in MM patients (NCT00639054). Additionally, a Phase I clinical trial in AML is underway in which miRNA expression will be correlated with response to a regimen of Azacytidine (nucleoside analog) and Bortezomib (proteasome inhibitor) (NCT00624936). Clinical trials investigating miRNAs that correlate with response to treatment may uncover novel markers for disease sensitivity or resistance. Furthermore, understanding the specific function of these miRNAs in subsequent studies could prove useful in designing more targeted therapies to overcome persistent disease.

18.4.3 Targeting MiRNAs for More Effective Treatment

Because miRNAs have pleiotrophic effects in tumorigenesis, targeting miRNAs as part of a therapy regimen is a promising new treatment strategy. Based on miRNA function, potential therapeutic approaches include miRNA inhibition by antisense technology or replacement therapy utilizing synthetic miRNA mimics. One of the first examples of use of antisense oligonucleotides on miRNAs in vivo was performed in Drosophila to determine the function of specific miRNAs during development (Leaman et al. 2005). A number of studies have employed antisense technology to knock down miRNAs for both in vitro and in vivo studies. For instance, targeted knockdown of *miR-21* by medicinally-modified antisense oligos in gliomas significantly improves the therapeutic benefits of S-TRAIL (Corsten et al. 2007). By this method, antisense technology offers a novel approach at inhibiting oncogenic miRNAs or miRNAs involved in resistance to improve chemotherapeutic strategies. Alternatively, miRNA replacement therapies offer means at restoring miRNA function that may benefit disease control. In themselves, miRNAs have desirable therapeutic characteristics in that they are small in size and generally low in toxicity. In a recent study, viral delivery of tumor suppressor miRNA *let-7* into NSCLC cells reduced in vivo tumor growth in mice (Trang et al. 2010). This approach offers restoration of tumor suppressor miRNAs or miRNAs involved in sensitizing cells to CT or RT.

18.4.4 MiRNA Diagnostics

Predictors of disease response are valuable in developing treatment strategies and have become part of standard clinical evaluation. As gene expression profiling can provide insight into dysregulated pathways of individualized disease, clinical tools have become available to examine panels of genes commonly associated with certain diseases. In breast cancer, a diagnostic tool called Oncotype DX (Genomic Health Inc.) examines 16 genes involved in breast cancer tumorigenesis to predict the likelihood of recurrence and response to standard CT (Cobleigh et al. 2005; Paik et al. 2004). Agendia Inc. produces another clinical test called MammaPrint, which examines a panel of 70 genes to determine recurrence and treatment strategy in breast cancer (Cardoso et al. 2008; Knauer et al. 2010). An Israeli-based company called Rosetta Genomics is currently developing miRNA profiling platforms called miRview^{TM} to more clearly diagnose cancers of unknown origin, lung cancer subtypes, and invasiveness of bladder cancer (Lebanony et al. 2009; Meiri et al. 2010; Rosenwald et al. 2010). Characterizing such features will provide useful information in regards for treatment strategies. Considering the role miRNAs also play in predicting CT and RT sensitivity or resistance, developing a panel of miRNAs to predict response would overcome some uncertainty in treatment selection.

18.5 Conclusions

A growing number of miRNAs have been identified as altering sensitivity to RT and CT. Establishment of miRNAs as markers and the ability to assay them in the clinic will arm clinicians with the knowledge they need to design and implement better treatment strategies. Currently, miRNA expression profiling has been the primary method for identifying relationships between miRNA and resistance. However, physical location of miRNAs at chromosomal breakpoints or frequent sites of insertion, deletion, or amplification in certain cancers may also correlate with RT and CT resistance. SNPs in miRNAs or their target sites are another promising area of research that has only been minimally explored. For instance, one study has shown that a polymorphism found at the target site for *miR-24* in the 3′UTR of dihydrofolate reductase (*DHFR*) associates with resistance to methotrexate (Mishra et al. 2007). Characterizing polymorphisms and miRNA location will provide us with important information that may further help identify patients who are resistant or sensitive to specific treatments. Using predictive markers, including miRNAs, in combination with our current cancer staging systems is essential for developing more effective and more personalized medicine.

References

Adam L, Zhong M, Choi W, et al. MiR-200 expression regulates epithelial-to-mesenchymal transition in bladder cancer cells and reverses resistance to epidermal growth factor receptor therapy. Clin Cancer Res. 2009;15:5060–72.

Ahmed N, Abubaker K, Findlay J, et al. Epithelial mesenchymal transition and cancer stem cell-like phenotypes facilitate chemoresistance in recurrent ovarian cancer. Curr Cancer Drug Targets. 2010;10:268–78.

Ahmed FE, Jeffries CD, Vos PW, et al. Diagnostic microRNA markers for screening sporadic human colon cancer and active ulcerative colitis in stool and tissue. Cancer Genomics Proteomics. 2009;6:281–95.

Akao Y, Nakagawa Y, Naoe T. Let-7 microRNA functions as a potential growth suppressor in human colon cancer cells. Biol Pharm Bull. 2006;29:903–6.

Akli S, Bui T, Wingate H, et al. Low-molecular-weight cyclin E can bypass letrozole-induced G1 arrest in human breast cancer cells and tumors. Clin Cancer Res. 2010;16:1179–90.

Arumugam T, Ramachandran V, Fournier KF, et al. Epithelial to mesenchymal transition contributes to drug resistance in pancreatic cancer. Cancer Res. 2009;69:5820–8.

Auner V, Sehouli J, Oskay-Oezcelik G, et al. Abc transporter gene expression in benign and malignant ovarian tissue. Gynecol Oncol. 2010;117:198–201.

Bai S, Nasser MW, Wang B, et al. MicroRNA-122 inhibits tumorigenic properties of hepatocellular carcinoma cells and sensitizes these cells to sorafenib. J Biol Chem. 2009;284:32015–27.

Bandi N, Zbinden S, Gugger M, et al. MiR-15a and miR-16 are implicated in cell cycle regulation in a RB-dependent manner and are frequently deleted or down-regulated in non-small cell lung cancer. Cancer Res. 2009;69:5553–9.

Bandres E, Bitarte N, Arias F, et al. MicroRNA-451 regulates macrophage migration inhibitory factor production and proliferation of gastrointestinal cancer cells. Clin Cancer Res. 2009;15:2281–90.

Banwell CM, Singh R, Stewart PM, et al. Proliferative signalling by 1,25$(OH)_2D_3$ in prostate and breast cancer is suppressed by a mechanism involving histone deacetylation. Recent Results Cancer Res. 2003;164:83–98.

Bentzen SM. Theragnostic imaging for radiation oncology: dose-painting by numbers. Lancet Oncol. 2005;6:112–7.

Bhat-Nakshatri P, Newton TR, Goulet R Jr, et al. NF-kappaB activation and interleukin 6 production in fibroblasts by estrogen receptor-negative breast cancer cell-derived interleukin 1alpha. Proc Natl Acad Sci USA. 1998;95:6971–6.

Blower PE, Verducci JS, Lin S, et al. MicroRNA expression profiles for the NCI-60 cancer cell panel. Mol Cancer Ther. 2007;6:1483–91.

Bonci D, Coppola V, Musumeci M, et al. The miR-15a-miR-16-1 cluster controls prostate cancer by targeting multiple oncogenic activities. Nat Med. 2008;14:1271–7.

Borralho PM, Kren BT, Castro RE, et al. MicroRNA-143 reduces viability and increases sensitivity to 5-fluorouracil in hct116 human colorectal cancer cells. FEBS J. 2009;276:6689–700.

Bottoni A, Piccin D, Tagliati F, et al. MiR-15a and miR-16-1 down-regulation in pituitary adenomas. J Cell Physiol. 2005;204:280–5.

Bourguignon LY, Spevak CC, Wong G, et al. Hyaluronan-CD44 interaction with protein kinase c(epsilon) promotes oncogenic signaling by the stem cell marker nanog and the production of microRNA-21, leading to down-regulation of the tumor suppressor protein pdcd4, anti-apoptosis, and chemotherapy resistance in breast tumor cells. J Biol Chem. 2009;284: 26533–46.

Braconi C, Valeri N, Gasparini P, et al. Hepatitis C virus proteins modulate microRNA expression and chemosensitivity in malignant hepatocytes. Clin Cancer Res. 2010;16:957–66.

Buckley PG, Alcock L, Bryan K, et al. Chromosomal and miRNA expression patterns reveal biologically distinct subgroups of 11q- neuroblastoma. Clin Cancer Res. 2010;16:2971–8.

Calin GA, Croce CM. MicroRNA signatures in human cancers. Nat Rev Cancer. 2006;6: 857–66.

Calin GA, Liu CG, Sevignani C, et al. MicroRNA profiling reveals distinct signatures in b cell chronic lymphocytic leukemias. Proc Natl Acad Sci USA. 2004;101:11755–60.

Cammareri P, Scopelliti A, Todaro M, et al. Aurora-a is essential for the tumorigenic capacity and chemoresistance of colorectal cancer stem cells. Cancer Res. 2010;70:4655–65.

Camps C, Buffa FM, Colella S, et al. Hsa-miR-210 is induced by hypoxia and is an independent prognostic factor in breast cancer. Clin Cancer Res. 2008;14:1340–8.

Cardoso F, Van't Veer L, Rutgers E, et al. Clinical application of the 70-gene profile: the mindact trial. J Clin Oncol. 2008;26:729–35.

Carmeliet P, Ferreira V, Breier G, et al. Abnormal blood vessel development and lethality in embryos lacking a single vegf allele. Nature. 1996;380:435–9.

Cascio S, D'Andrea A, Ferla R, et al. MiR-20b modulates vegf expression by targeting HIF-1 alpha and STAT3 in MCF-7 breast cancer cells. J Cell Physiol. 2010;224:242–9.

Chen X, Ba Y, Ma L, et al. Characterization of microRNAs in serum: a novel class of biomarkers for diagnosis of cancer and other diseases. Cell Res. 2008;18:997–1006.

Chen G, Zhu W, Shi D, et al. MicroRNA-181a sensitizes human malignant glioma U87MG cells to radiation by targeting bcl-2. Oncol Rep. 2010;23:997–1003.

Cho WC. MicroRNAs: potential biomarkers for cancer diagnosis, prognosis and targets for therapy. Int J Biochem Cell Biol. 2009;42:1273–81.

Christoffersen NR, Shalgi R, Frankel LB, et al. P53-independent up-regulation of miR-34a during oncogene-induced senescence represses myc. Cell Death Differ. 2010;17:236–45.

Cimmino A, Calin GA, Fabbri M, et al. MiR-15 and miR-16 induce apoptosis by targeting bcl2. Proc Natl Acad Sci USA. 2005;102:13944–9.

Cobleigh MA, Tabesh B, Bitterman P, et al. Tumor gene expression and prognosis in breast cancer patients with 10 or more positive lymph nodes. Clin Cancer Res. 2005;11(24 Pt 1):8623–31.

Cochrane DR, Spoelstra NS, Howe EN, et al. MicroRNA-200c mitigates invasiveness and restores sensitivity to microtubule-targeting chemotherapeutic agents. Mol Cancer Ther. 2009;8: 1055–66.

Cohn DE, Fabbri M, Valeri N, et al. Comprehensive miRNA profiling of surgically staged endometrial cancer. Am J Obstet Gynecol. 2010;22:656.

Coley HM. Mechanisms and strategies to overcome chemotherapy resistance in metastatic breast cancer. Cancer Treat Rev. 2008;34:378–90.

Coley HM. Overcoming multidrug resistance in cancer: clinical studies of p-glycoprotein inhibitors. Methods Mol Biol. 2010;596:341–58.

Connolly EC, Van Doorslaer K, Rogler LE, et al. Over-expression of miR-21 promotes an in vitro metastatic phenotype by targeting the tumor suppressor rhob. Mol Cancer Res. 2010;8: 691–700.

Corsten MF, Miranda R, Kasmieh R, et al. MicroRNA-21 knockdown disrupts glioma growth in vivo and displays synergistic cytotoxicity with neural precursor cell delivered s-trail in human gliomas. Cancer Res. 2007;67:8994–9000.

Cosmopoulos K, Pegtel M, Hawkins J, et al. Comprehensive profiling of Epstein-Barr virus microRNAs in nasopharyngeal carcinoma. J Virol. 2009;83:2357–67.

Dallas NA, Xia L, Fan F, et al. Chemoresistant colorectal cancer cells, the cancer stem cell phenotype, and increased sensitivity to insulin-like growth factor-I receptor inhibition. Cancer Res. 2009;69:1951–7.

Debucquoy A, Roels S, Goethals L, et al. Double blind randomized Phase II study with radiation+5-fluorouracil+/-celecoxib for resectable rectal cancer. Radiother Oncol. 2009;93:273–8.

Di Fiore R, Santulli A, Ferrante RD, et al. Identification and expansion of human osteosarcoma-cancer-stem cells by long-term 3-aminobenzamide treatment. J Cell Physiol. 2009;219:301–13.

Di Leva G, Gasparini P, Piovan C, et al. MicroRNA cluster 221-222 and estrogen receptor {alpha} interactions in breast cancer. J Natl Cancer Inst. 2010;102:706–21.

Dong Q, Meng P, Wang T, et al. MicroRNA let-7a inhibits proliferation of human prostate cancer cells in vitro and in vivo by targeting E2F2 and CCND2. PLoS One. 2010;5:e10147.

Duncan TJ, Al-Attar A, Rolland P, et al. Cytoplasmic p27 expression is an independent prognostic factor in ovarian cancer. Int J Gynecol Pathol. 2010;29:8–18.

Evans SM, Jenkins WT, Shapiro M, et al. Evaluation of the concept of "Hypoxic fraction" As a descriptor of tumor oxygenation status. Adv Exp Med Biol. 1997;411:215–25.

Faderl S, Keating MJ, Do KA, et al. Expression profile of 11 proteins and their prognostic significance in patients with chronic lymphocytic leukemia (CLL). Leukemia. 2002;16: 1045–52.

Felicetti F, Errico MC, Bottero L, et al. The promyelocytic leukemia zinc finger-microRNA-221/-222 pathway controls melanoma progression through multiple oncogenic mechanisms. Cancer Res. 2008;68:2745–54.

Ferrandiz N, Caraballo JM, Albajar M, et al. P21(Cip1) confers resistance to imatinib in human chronic myeloid leukemia cells. Cancer Lett. 2010;292:133–9.

Flynt AS, Li N, Thatcher EJ, et al. Zebrafish miR-214 modulates Hedgehog signaling to specify muscle cell fate. Nat Genet. 2007;39:259–63.

Fornari F, Grameri L, Ferracin M, et al. MiR-221 controls CDKN1C/p57 and CDKN1B/p27 expression in human hepatocellular carcinoma. Oncogene. 2008;27:5651–61.

Fornari F, Grameri L, Giovannini C, et al. MiR-122/cyclin G1 interaction modulates p53 activity and affects doxorubicin sensitivity of human hepatocarcinoma cells. Cancer Res. 2009;69:5761–7.

Fujita Y, Kojima K, Hamada N, et al. Effects of miR-34a on cell growth and chemoresistance in prostate cancer PC3 cells. Biochem Biophys Res Commun. 2008;377:114–9.

Gal H, Pandi G, Kanner AA, et al. MiR-451 and imatinib mesylate inhibit tumor growth of glioblastoma stem cells. Biochem Biophys Res Commun. 2008;376:86–90.

Galardi S, Mercatelli N, Giorda E, et al. MiR-221 and miR-222 expression affects the proliferation potential of human prostate carcinoma cell lines by targeting p27Kip1. J Biol Chem. 2007;282:23716–24.

Garofalo M, Di Leva G, Romano G, et al. MiR-221 & 222 regulate trail resistance and enhance tumorigenicity through PTEN and TIMP3 down-regulation. Cancer Cell. 2009;16:498–509.

Gascoyne RD, Adomat SA, Krajewski S, et al. Prognostic significance of bcl-2 protein expression and bcl-2 gene rearrangement in diffuse aggressive non-Hodgkin's lymphoma. Blood. 1997;90:244–51.

Gee HE, Camps C, Buffa FM, et al. Hsa-miR-210 is a marker of tumor hypoxia and a prognostic factor in head and neck cancer. Cancer. 2010;116:2148–58.

Giaccia AJ. Hypoxic stress proteins: survival of the fittest. Semin Radiat Oncol. 1996;6:46–58.

Giannakakis A, Sandaltzopoulos R, Greshock J, et al. MiR-210 links hypoxia with cell cycle regulation and is deleted in human epithelial ovarian cancer. Cancer Biol Ther. 2008;7:255–64.

Gibbons DL, Lin W, Creighton CJ, et al. Contextual extracellular cues promote tumor cell emt and metastasis by regulating miR-200 family expression. Genes Dev. 2009;23:2140–51.

Giovannetti E, Funel N, Peters GJ, et al. MicroRNA-21 in pancreatic cancer: correlation with clinical outcome and pharmacologic aspects underlying its role in the modulation of gemcitabine activity. Cancer Res. 2010;70:4528–38.

Gironella M, Seux M, Xie MJ, et al. Tumor protein 53-induced nuclear protein 1 expression is repressed by miR-155, and its restoration inhibits pancreatic tumor development. Proc Natl Acad Sci USA. 2007;104:16170–5.

Gottardo F, Liu CG, Ferracin M, et al. Micro-RNA profiling in kidney and bladder cancers. Urol Oncol. 2007;25:387–92.

Gregory PA, Bert AG, Paterson EL, et al. The miR-200 family and miR-205 regulate epithelial to mesenchymal transition by targeting ZEB1 and SIP1. Nat Cell Biol. 2008;10:593–601.

Guled M, Lahti L, Lindholm PM, et al. CDKN2A, NF2, and JUN are dysregulated among other genes by miRNAs in malignant mesothelioma – a miRNA microarray analysis. Genes Chromosomes Cancer. 2009;48:615–23.

Guo Y, Chen Z, Zhang L, et al. Distinctive microRNA profiles relating to patient survival in esophageal squamous cell carcinoma. Cancer Res. 2008;68:26–33.

Guo L, Liu Y, Bai Y, et al. Gene expression profiling of drug-resistant small cell lung cancer cells by combining microRNA and cdna expression analysis. Eur J Cancer. 2010;46:1692–702.

Haimeur A, Conseil G, Deeley RG, et al. The MRP-related and BCRP/ABCG2 multidrug resistance proteins: biology, substrate specificity and regulation. Curr Drug Metab. 2004;5:21–53.

Hanke M, Hoefig K, Merz H, et al. A robust methodology to study urine microRNA as tumor marker: microRNA-126 and microRNA-182 are related to urinary bladder cancer. Urol Oncol. 2010;28:655–61.

He X, Duan C, Chen J, et al. Let-7a elevates p21(WAF1) levels by targeting of NIRF and suppresses the growth of A549 lung cancer cells. FEBS Lett. 2009;583:3501–7.

He L, He X, Lim LP, et al. A microRNA component of the p53 tumour suppressor network. Nature. 2007;447:1130–4.

Hermine O, Haioun C, Lepage E, et al. Prognostic significance of bcl-2 protein expression in aggressive non-Hodgkin's lymphoma. Groupe d'etude des lymphomes de l'adulte (gela). Blood. 1996;87:265–72.

Honoki K. Do stem-like cells play a role in drug resistance of sarcomas? Expert Rev cancer Ther. 2010;10:261–70.

Horton JK, Halle J, Ferraro M, et al. Radiosensitization of chemotherapy-refractory, locally advanced or locally recurrent breast cancer with trastuzumab: a Phase II trial. Int J Radiat Oncol Biol Phys. 2010;76:998–1004.

Hu Z, Chen X, Zhao Y, et al. Serum microRNA signatures identified in a genome-wide serum microRNA expression profiling predict survival of non-small-cell lung cancer. J Clin Oncol. 2010b;28:1721–6.

Hu H, Li Y, Gu J, et al. Sense oligonucleotide against miR-21 inhibits migration and induces apoptosis in leukemic K562 cells. Leuk Lymphoma. 2010a;51:694–701.

Huang Q, Gumireddy K, Schrier M, et al. The microRNAs miR-373 and miR-520c promote tumour invasion and metastasis. Nat Cell Biol. 2008;10:202–10.

Hwang JH, Voortman J, Giovannetti E, et al. Identification of microRNA-21 as a biomarker for chemoresistance and clinical outcome following adjuvant therapy in resectable pancreatic cancer. PLoS One. 2010;5:e10630.

Inoue S, Mai A, Dyer MJ, et al. Inhibition of histone deacetylase class I but not class II is critical for the sensitization of leukemic cells to tumor necrosis factor-related apoptosis-inducing ligand-induced apoptosis. Cancer Res. 2006;66:6785–92.

Iorio MV, Casalini P, Piovan C, et al. MicroRNA-205 regulates HER3 in human breast cancer. Cancer Res. 2009;69:2195–200.

Jackson C, Cunningham D. A retrospective on the inhibition of epidermal growth factor receptor as a therapeutic strategy for patients with relapsed metastatic colorectal cancer: impact on treatment of today's patients. Clin Colorectal Cancer. 2007;7(Suppl 1):S8–S15.

Jay C, Nemunaitis J, Chen P, et al. MiRNA profiling for diagnosis and prognosis of human cancer. DNA Cell Biol. 2007;26:293–300.

Ji Q, Hao X, Zhang M, et al. MicroRNA miR-34 inhibits human pancreatic cancer tumor-initiating cells. PLoS One. 2009;4:e6816.

Jiang X, Gwye Y, Russell D, et al. CD133 expression in chemo-resistant ewing sarcoma cells. BMC Cancer. 2010;10:116.

Johansson M, Persson JL. Cancer therapy: targeting cell cycle regulators. Anticancer Agents Med Chem. 2008;8:723–31.

Johnson SM, Grosshans H, Shingara J, et al. Ras is regulated by the let-7 microRNA family. Cell. 2005;120:635–47.

Jun HJ, Kim YM, Park SY, et al. Modulation of ionizing radiation-induced G2 arrest by cyclooxygenase-2 and its inhibitor celecoxib. Int J Radiat Oncol Biol Phys. 2009;75: 225–34.

Kang MR, Lee K, Kang JS, et al. KBH-A42, a histone deacetylase inhibitor, inhibits the growth of doxorubicin-resistant leukemia cells expressing p-glycoprotein. Oncol Rep. 2010;23:801–9.

Katada T, Ishiguro H, Kuwabara Y, et al. MicroRNA expression profile in undifferentiated gastric cancer. Int J Oncol. 2009;34:537–42.

Kent OA, Mullendore M, Wentzel EA, et al. A resource for analysis of microRNA expression and function in pancreatic ductal adenocarcinoma cells. Cancer Biol Ther. 2009;8:2013–24.

Kinsella TJ. Coordination of DNA mismatch repair and base excision repair processing of chemotherapy and radiation damage for targeting resistant cancers. Clin Cancer Res. 2009;15:1853–9.

Knauer M, Mook S, Rutgers EJ, et al. The predictive value of the 70-gene signature for adjuvant chemotherapy in early breast cancer. Breast Cancer Res Treat. 2010;120:655–61.

Kong D, Li Y, Wang Z, et al. MiR-200 regulates PDGF-D-mediated epithelial-mesenchymal transition, adhesion, and invasion of prostate cancer cells. Stem Cells. 2009;27: 1712–21.

Koong AC, Chen EY, Giaccia AJ. Hypoxia causes the activation of nuclear factor kappa B through the phosphorylation of I kappa B alpha on tyrosine residues. Cancer Res. 1994;54:1425–30.

Kotani A, Ha D, Hsieh J, et al. MiR-128b is a potent glucocorticoid sensitizer in MLL-AF4 acute lymphocytic leukemia cells and exerts cooperative effects with miR-221. Blood. 2009;114:4169–78.

Kovalchuk O, Filkowski J, Meservy J, et al. Involvement of microRNA-451 in resistance of the MCF-7 breast cancer cells to chemotherapeutic drug doxorubicin. Mol Cancer Ther. 2008;7:2152–9.

Krause M, Gurtner K, Deuse Y, et al. Heterogeneity of tumour response to combined radiotherapy and egfr inhibitors: differences between antibodies and tk inhibitors. Int J Radiat Biol. 2009;85:943–54.

Kulshreshtha R, Ferracin M, Wojcik SE, et al. A microRNA signature of hypoxia. Mol Cell Biol. 2007;27:1859–67.

Langer C, Radmacher MD, Ruppert AS, et al. High baalc expression associates with other molecular prognostic markers, poor outcome, and a distinct gene-expression signature in cytogenetically normal patients younger than 60 years with acute myeloid leukemia: a cancer and leukemia group B (CALGB) sstudy. Blood. 2008;111:5371–9.

Lawrie CH, Gal S, Dunlop HM, et al. Detection of elevated levels of tumour-associated microRNAs in serum of patients with diffuse large B-cell lymphoma. Br J Haematol. 2008;141: 672–5.

Leaman D, Chen PY, Fak J, et al. Sense-mediated depletion reveals essential and specific functions of microRNAs in drosophila development. Cell. 2005;121:1097–108.

Lebanony D, Benjamin H, Gilad S, et al. Diagnostic assay based on hsa-miR-205 expression distinguishes squamous from nonsquamous non-small-cell lung carcinoma. J Clin Oncol. 2009;27:2030–7.

Lee YB, Bantounas I, Lee DY, et al. Twist-1 regulates the miR-199a/214 cluster during development. Nucleic Acids Res. 2009b;37:123–8.

Lee KH, Chen YL, Yeh SD, et al. MicroRNA-330 acts as tumor suppressor and induces apoptosis of prostate cancer cells through E2F1-mediated suppression of AKT phosphorylation. Oncogene. 2009a;28:3360–70.

Lei Z, Li B, Yang Z, et al. Regulation of HIF-1alpha and VEGF by miR-20b tunes tumor cells to adapt to the alteration of oxygen concentration. PLoS One. 2009;4:e7629.

Li N, Fu H, Tie Y, et al. MiR-34a inhibits migration and invasion by down-regulation of c-MET expression in human hepatocellular carcinoma cells. Cancer Lett. 2009b;275:44–53.

Li C, Jackson RM. Reactive species mechanisms of cellular hypoxia-reoxygenation injury. Am J Physiol Cell Physiol. 2002;282:C227–C41.

Li C, Kim SW, Rai D, et al. Copy number abnormalities, myc activity, and the genetic fingerprint of normal B cells mechanistically define the microRNA profile of diffuse large B-cell lymphoma. Blood. 2009a;113:6681–90.

Li Y, Li W, Yang Y, Lu Y, et al. MicroRNA-21 targets LRRFIP1 and contributes to VM-26 resistance in glioblastoma multiforme. Brain Res. 2009c;1286:13–8.

Li Y, VandenBoom TG 2nd, Kong D, et al. Up-regulation of miR-200 and let-7 by natural agents leads to the reversal of epithelial-to-mesenchymal transition in gemcitabine-resistant pancreatic cancer cells. Cancer Res. 2009d;69:6704–12.

Li Y, Zhu X, Gu J, et al. Anti-miR-21 oligonucleotide sensitizes leukemic K562 cells to arsenic trioxide by inducing apoptosis. Cancer Sci. 2010;101:948–54.

Liang K, Lu Y, Jin W, et al. Sensitization of breast cancer cells to radiation by trastuzumab. Mol Cancer Ther. 2003;2:1113–20.

Liang Z, Wu H, Xia J, et al. Involvement of miR-326 in chemotherapy resistance of breast cancer through modulating expression of multidrug resistance-associated protein 1. Biochem Pharmacol. 2010;79:817–24.

Liu Q, Gazitt Y. Potentiation of dexamethasone-, paclitaxel-, and Ad-p53-induced apoptosis by Bcl-2 sense oligodeoxynucleotides in drug-resistant multiple myeloma cells. Blood. 2003;101:4105–14.

Liu CJ, Kao SY, Tu HF, et al. Increase of microRNA miR-31 level in plasma could be a potential marker of oral cancer. Oral Dis. 2010;16:360–4.

Lodes MJ, Caraballo M, Suciu D, et al. Detection of cancer with serum miRNAs on an oligonucleotide microarray. PLoS One. 2009;4:e6229.

Lowery AJ, Miller N, Devaney A, et al. MicroRNA signatures predict oestrogen receptor, progesterone receptor and HER2/NEU receptor status in breast cancer. Breast Cancer Res. 2009;11:R27.

Lu J, Getz G, Miska EA, et al. MicroRNA expression profiles classify human cancers. Nature. 2005;435:834–8.

Ma L, Teruya-Feldstein J, Weinberg RA. Tumour invasion and metastasis initiated by microRNA-10b in breast cancer. Nature. 2007;449:682–8.

Ma L, Young J, Prabhala H, et al. MiR-9, a MYC/MYCN-activated microRNA, regulates E-cadherin and cancer metastasis. Nat Cell Biol. 2010;12:247–56.

Manni I, Artuso S, Careccia S, et al. The microRNA miR-92 increases proliferation of myeloid cells and by targeting p63 modulates the abundance of its isoforms. FASEB J. 2009;23: 3957–66.

Mauceri HJ, Hanna NN, Beckett MA, et al. Combined effects of angiostatin and ionizing radiation in tumour therapy. Nature. 1998;394:287–91.

Mei M, Ren Y, Zhou X, et al. Down-regulation of miR-21 enhances chemotherapeutic effect of taxol in breast carcinoma cells. Technol Cancer Res Treat. 2010;9:77–86.

Meiri E, Levy A, Benjamin H, et al. Discovery of microRNAs and other small RNAs in solid tumors. Nucleic Acids Res. 2010;38:6234–46.

Melkonyan HS, Feaver WJ, Meyer E, et al. Transrenal nucleic acids: from proof of principle to clinical tests. Ann NY Acad Sci. 2008;1137:73–81.

Meng F, Henson R, Lang M, et al. Involvement of human micro-RNA in growth and response to chemotherapy in human cholangiocarcinoma cell lines. Gastroenterology. 2006;130:2113–29.

Meng F, Henson R, Wehbe-Janek H, et al. MicroRNA-21 regulates expression of the PTEN tumor suppressor gene in human hepatocellular cancer. Gastroenterology. 2007;133: 647–58.

Mercatelli N, Coppola V, Bonci D, et al. The inhibition of the highly expressed miR-221 and miR-222 impairs the growth of prostate carcinoma xenografts in mice. PLoS One. 2008;3:e4029.

Michael A, Bajracharya SD, Yuen PS, et al. Exosomes from human saliva as a source of microRNA biomarkers. Oral Dis. 2010;16:34–8.

Migliore C, Petrelli A, Ghiso E, et al. MicroRNAs impair MET-mediated invasive growth. Cancer Res. 2008;68:10128–36.

Milas L, Fan Z, Andratschke NH, et al. Epidermal growth factor receptor and tumor response to radiation: in vivo preclinical studies. Int J Radiat Oncol Biol Phys. 2004;58:966–71.

Milas L, Mason KA, Ang KK. Epidermal growth factor receptor and its inhibition in radiotherapy: in vivo findings. Int J Radiat Biol. 2003;79:539–45.

Miller TE, Ghoshal K, Ramaswamy B, et al. MicroRNA-221/222 confers tamoxifen resistance in breast cancer by targeting p27Kip1. J Biol Chem. 2008;283:29897–903.

Mishra PJ, Humeniuk R, Longo-Sorbello GS, et al. A miR-24 microRNA binding-site polymorphism in dihydrofolate reductase gene leads to methotrexate resistance. Proc Natl Acad Sci USA. 2007;104:13513–8.

Mitchell PS, Parkin RK, Kroh EM, et al. Circulating microRNAs as stable blood-based markers for cancer detection. Proc Natl Acad Sci USA. 2008;105:10513–8.

Mott JL, Kobayashi S, Bronk SF, et al. MiR-29 regulates MCL-1 protein expression and apoptosis. Oncogene. 2007;26:6133–40.

Muth M, Theophile K, Hussein K, et al. Hypoxia-induced down-regulation of microRNA-449a/b impairs control over targeted serpine1 (PAI-1) mRNA – a mechanism involved in serpine1 (PAI-1) over-expression. J Transl Med. 2010;8:33.

Nakajima G, Hayashi K, Xi Y, et al. Non-coding microRNAs hsa-let-7g and hsa-miR-181b are associated with chemoresponse to S-1 in colon cancer. Cancer Genomics Proteomics. 2006;3:317–24.

Nakanishi T, Chumsri S, Khakpour N, et al. Side-population cells in luminal-type breast cancer have tumour-initiating cell properties, and are regulated by HER2 expression and signalling. Br J Cancer. 2010;102:815–26.

Nana-Sinkam P, Croce CM. MicroRNAs in diagnosis and prognosis in cancer: what does the future hold? Pharmacogenomics. 2010;11:667–9.

Nasser MW, Datta J, Nuovo G, et al. Down-regulation of micro-RNA-1 (miR-1) in lung cancer. Suppression of tumorigenic property of lung cancer cells and their sensitization to doxorubicin-induced apoptosis by miR-1. J Biol Chem. 2008;283:33394–405.

Noguchi K, Katayama K, Mitsuhashi J, et al. Functions of the breast cancer resistance protein (BCRP/ABCG2) in chemotherapy. Adv Drug Deliv Rev. 2009;61:26–33.

Noonan EJ, Place RF, Pookot D, et al. MiR-449a targets HDAC-1 and induces growth arrest in prostate cancer. Oncogene. 2009;28:1714–24.

Oh JS, Kim JJ, Byun JY, et al. Lin28-let7 modulates radiosensitivity of human cancer cells with activation of k-Ras. Int J Radiat Oncol Biol Phys. 2010;76:5–8.

Paik S, Shak S, Tang G, et al. A multigene assay to predict recurrence of tamoxifen-treated, node-negative breast cancer. N Engl J Med. 2004;351:2817–26.

Pan YZ, Morris ME, Yu AM. MicroRNA-328 negatively regulates the expression of breast cancer resistance protein (BCRP/ABCG2) in human cancer cells. Mol Pharmacol. 2009;75:1374–9.

Pang RT, Leung CO, Ye TM, et al. MicroRNA-34a suppresses invasion through down-regulation of Notch1 and Jagged1 in cervical carcinoma and choriocarcinoma cells. Carcinogenesis. 2010;31:1037–44.

Paoluzzi L, Gonen M, Bhagat G, et al. The BH3-only mimetic ABT-737 synergizes the neoplastic activity of proteasome inhibitors in lymphoid malignancies. Blood. 2008a;112: 2906–16.

Paoluzzi L, Gonen M, Gardner JR, et al. Targeting bcl-2 family members with the BH3 mimetic at-101 markedly enhances the therapeutic effects of chemotherapeutic agents in in vitro and in vivo models of B-cell lymphoma. Blood. 2008b;111:5350–8.

Pardo OE, Latigo J, Jeffery RE, et al. The fibroblast growth factor receptor inhibitor PD173074 blocks small cell lung cancer growth in vitro and in vivo. Cancer Res. 2009;69:8645–51.

Park JK, Lee EJ, Esau C, et al. sense inhibition of microRNA-21 or -221 arrests cell cycle, induces apoptosis, and sensitizes the effects of gemcitabine in pancreatic adenocarcinoma. Pancreas. 2009a;38:e190–9.

Park SM, Shell S, Radjabi AR, et al. Let-7 prevents early cancer progression by suppressing expression of the embryonic gene HMGA2. Cell Cycle. 2007;6:2585–90.

Park NJ, Zhou H, Elashoff D, et al. Salivary microRNA: discovery, characterization, and clinical utility for oral cancer detection. Clin Cancer Res. 2009b;15:5473–7.

Patnaik SK, Kannisto E, Knudsen S, et al. Evaluation of microRNA expression profiles that may predict recurrence of localized stage I non-small cell lung cancer after surgical resection. Cancer Res. 2010;70:36–45.

Pawlik TM, Keyomarsi K. Role of cell cycle in mediating sensitivity to radiotherapy. Int J Radiat Oncol Biol Phys. 2004;59:928–42.

Pekarsky Y, Santanam U, Cimmino A, et al. TCL1 expression in chronic lymphocytic leukemia is regulated by miR-29 and miR-181. Cancer Res. 2006;66(24):11590–3.

Peter ME. Let-7 and miR-200 microRNAs: guardians against pluripotency and cancer progression. Cell Cycle. 2009;8:843–52.

Petillo D, Kort EJ, Anema J, et al. MicroRNA profiling of human kidney cancer subtypes. Int J Oncol. 2009;35:109–14.

Petrocca F, Visone R, Onelli MR, et al. E2F1-regulated microRNAs impair TGFbeta-dependent cell-cycle arrest and apoptosis in gastric cancer. Cancer Cell. 2008;13:272–86.

Pietras RJ, Poen JC, Gallardo D, et al. Monoclonal antibody to HER-2/NEU receptor modulates repair of radiation-induced DNA damage and enhances radiosensitivity of human breast cancer cells over-expressing this oncogene. Cancer Res. 1999;59:1347–55.

Pigazzi M, Manara E, Baron E, et al. MiR-34b targets cyclic AMP-responsive element binding protein in acute myeloid leukemia. Cancer Res. 2009;69:2471–8.

Pineau P, Volinia S, McJunkin K, et al. MiR-221 over-expression contributes to liver tumorigenesis. Proc Natl Acad Sci USA. 2010;107:264–9.

Porkka KP, Pfeiffer MJ, Waltering KK, et al. MicroRNA expression profiling in prostate cancer. Cancer Res. 2007;67:6130–5.
Rainer J, Ploner C, Jesacher S, et al. Glucocorticoid-regulated microRNAs and mirtrons in acute lymphoblastic leukemia. Leukemia. 2009;23:746–52.
Ren Y, Kang CS, Yuan XB, et al. Co-delivery of as-miR-21 and 5-FU by poly(amidoamine) dendrimer attenuates human glioma cell growth in vitro. J Biomater Sci Polym Ed. 2010;21: 303–14.
Resnick KE, Alder H, Hagan JP, et al. The detection of differentially expressed microRNAs from the serum of ovarian cancer patients using a novel real-time pcr platform. Gynecol Oncol. 2009;112:55–9.
Rodriguez-Gonzalez FG, Sieuwerts AM, Smid M, et al. MicroRNA-34a suppresses invasion through down-regulation of Notch1 and Jagged1 in cervical carcinoma and choriocarcinoma cells. Breast Cancer Res Treat. 2010;31:1037–44.
Rosenwald S, Gilad S, Benjamin S, et al. Validation of a microRNA-based qRT-PCR test for accurate identification of tumor tissue origin. Mod Pathol. 2010;23:814–23.
Rossi G, Papotti M, Barbareschi M, et al. Morphology and a limited number of immuno-histochemical markers may efficiently subtype non-small-cell lung cancer. J Clin Oncol. 2009;27:e.
Royds JA, Dower SK, Qwarnstrom EE, et al. Response of tumour cells to hypoxia: role of p53 and NFKB. Mol Pathol. 1998;51:55–61.
Rui W, Bing F, Hai-Zhu S, et al. Identification of microRNA profiles in docetaxel-resistant human non-small cell lung carcinoma cells (SPC-A1). J Cell Mol Med. 2010;14:206–14.
Salerno E, Scaglione BJ, Coffman FD, et al. Correcting miR-15a/16 genetic defect in new zealand black mouse model of CLL enhances drug sensitivity. Mol Cancer Ther. 2009;8:2684–92.
San Jose-Eneriz E, Roman-Gomez J, Jimenez-Velasco A, et al. MicroRNA expression profiling in imatinib-resistant chronic myeloid leukemia patients without clinically significant ABL1-mutations. Mol Cancer. 2009;8:69.
Sarkar FH, Li Y, Wang Z, et al. Pancreatic cancer stem cells and EMT in drug resistance and metastasis. Minerva Chir. 2009;64:489–500.
Sayed D, He M, Hong C, et al. MicroRNA-21 is a downstream effector of AKT that mediates its apoptotic effects via suppression of Fas ligand. J Biol Chem. 2010;285:20281–90.
Schepeler T, Reinert JT, Ostenfeld MS, et al. Diagnostic and prognostic microRNAs in stage II colon cancer. Cancer Res. 2008;68:6416–24.
Seike M, Goto A, Okano T, et al. MiR-21 is an EGFR-regulated anti-apoptotic factor in lung cancer in never-smokers. Proc Natl Acad Sci USA. 2009;106:12085–90.
Semenza GL. Hypoxia, clonal selection, and the role of HIF-1 in tumor progression. Crit Rev Biochem Mol Biol. 2000;35:71–103.
Shah MA, Schwartz GK. Cyclin-dependent kinases as targets for cancer therapy. Cancer Chemother Biol Response Modif. 2003;21:145–70.
Sharma SV, Lee DY, Li B, et al. A chromatin-mediated reversible drug-tolerant state in cancer cell subpopulations. Cell. 2010;141:69–80.
Shell S, Park SM, Radjabi AR, et al. Let-7 expression defines two differentiation stages of cancer. Proc Natl Acad Sci USA. 2007;104:11400–5.
Shimizu S, Takehara T, Hikita H, et al. The let-7 family of microRNAs inhibits bcl-xl expression and potentiates sorafenib-induced apoptosis in human hepatocellular carcinoma. J Hepatol. 2010;52:698–704.
Siemann DW, Rojiani AM. Enhancement of radiation therapy by the novel vascular targeting agent ZD6126. Int J Radiat Oncol Biol Phys. 2002;53:164–71.
Song S, Wientjes MG, Gan Y, et al. Fibroblast growth factors: an epigenetic mechanism of broad spectrum resistance to cancer drugs. Proc Natl Acad Sci USA. 2000;2000(97): 8658–63.
Sorrentino A, Liu CG, Addario A, et al. Role of microRNAs in drug-resistant ovarian cancer cells. Gynecol Oncol. 2008;111:478–86.

Stolz C, Hess G, Hahnel PS, et al. Targeting bcl-2 family proteins modulates the sensitivity of B-cell lymphoma to rituximab-induced apoptosis. Blood. 2008;112:3312–21.
Sun F, Fu H, Liu Q, et al. Down-regulation of CCND1 and CDK6 by miR-34a induces cell cycle arrest. FEBS Lett. 2008;582:1564–8.
Sun T, Wang Q, Balk S, et al. The role of microRNA-221 and microRNA-222 in androgen-independent prostate cancer cell lines. Cancer Res. 2009;69:3356–63.
Sun Z, Zhao Z, Li G, et al. Relevance of two genes in the multidrug resistance of hepatocellular carcinoma: in vivo and clinical studies. Tumori. 2010;96:90–6.
Svoboda M, Izakovicova Holla L, Sefr R, et al. Micro-RNAs miR125b and miR137 are frequently up-regulated in response to capecitabine chemoradiotherapy of rectal cancer. Int J Oncol. 2008;33:541–7.
Sylvestre Y, De Guire V, Querido E, et al. An E2F/miR-20a autoregulatory feedback loop. J Biol Chem. 2007;282:2135–43.
Taguchi A, Yanagisawa K, Tanaka M, et al. Identification of hypoxia-inducible factor-1 alpha as a novel target for miR-17-92 microRNA cluster. Cancer Res. 2008;68:5540–5.
Takeshita F, Patrawala L, Osaki M, et al. Systemic delivery of synthetic microRNA-16 inhibits the growth of metastatic prostate tumors via down-regulation of multiple cell-cycle genes. Mol Ther. 2010;18:181–7.
Tarasov V, Jung P, Verdoodt B, et al. Differential regulation of microRNAs by p53 revealed by massively parallel sequencing: miR-34a is a p53 target that induces apoptosis and G1-arrest. Cell Cycle. 2007;6:1586–93.
Tavazoie SF, Alarcon C, Oskarsson T, et al. Endogenous human microRNAs that suppress breast cancer metastasis. Nature. 2008;451:147–52.
Tazawa H, Tsuchiya N, Izumiya M, et al. Tumor-suppressive miR-34a induces senescence-like growth arrest through modulation of the E2F pathway in human colon cancer cells. Proc Natl Acad Sci USA. 2007;104:15472–7.
Thariat J, Yildirim G, Mason KA, et al. Combination of radiotherapy with EGFR antagonists for head and neck carcinoma. Int J Clin Oncol. 2007;12:99–110.
Tian Y, Luo A, Cai Y, et al. MicroRNA-10b promotes migration and invasion through KLF4 in human esophageal cancer cell lines. J Biol Chem. 2010;285:7986–94.
To K, Fotovati A, Reipas KM, et al. Y-box binding protein-1 induces the expression of CD44 and CD49f leading to enhanced self-renewal, mammosphere growth, and drug resistance. Cancer Res. 2010;70:2840–51.
To KK, Robey RW, Knutsen T, et al. Escape from hsa-miR-519c enables drug-resistant cells to maintain high expression of ABCG2. Mol Cancer Ther. 2009;8:2959–68.
Todaro M, Francipane MG, Medema JP, et al. Colon cancer stem cells: promise of targeted therapy. Gastroenterology. 2010;138:2151–62.
Trang P, Medina PP, Wiggins JF, et al. Regression of murine lung tumors by the let-7 microRNA. Oncogene. 2010;29:1580–7.
Tryndyak VP, Beland FA, Pogribny IP. E-cadherin transcriptional down-regulation by epigenetic and microRNA-200 family alterations is related to mesenchymal and drug-resistant phenotypes in human breast cancer cells. Int J Cancer. 2010;126:2575–83.
Tsang WP, Kwok TT. Let-7a microRNA suppresses therapeutics-induced cancer cell death by targeting caspase-3. Apoptosis. 2008;13:1215–22.
Varnholt H, Drebber U, Schulze F, et al. MicroRNA gene expression profile of hepatitis C virus-associated hepatocellular carcinoma. Hepatology. 2008;47:1223–32.
Viktorsson K, De Petris L, Lewensohn R. The role of p53 in treatment responses of lung cancer. Biochem Biophys Res Commun. 2005;331:868–80.
Visone R, Pallante P, Vecchione A, et al. Specific microRNAs are down-regulated in human thyroid anaplastic carcinomas. Oncogene. 2007a;26:7590–5.
Visone R, Russo L, Pallante P, et al. MicroRNAs (miR)-221 and miR-222, both over-expressed in human thyroid papillary carcinomas, regulate p27/Kip1 protein levels and cell cycle. Endocr Relat Cancer. 2007b;14:791–8.

Vokes EE, Chu E. Anti-EGFR therapies: clinical experience in colorectal, lung, and head and neck cancers. Oncology (Williston Park). 2006;20(5 Suppl 2):15–25.

Wagner-Ecker M, Schwager C, Wirkner U, et al. MicroRNA expression after ionizing radiation in human endothelial cells. Radiat Oncol. 2010;5:25.

Wang GL, Jiang BH, Rue EA, et al. Hypoxia-inducible factor 1 is a basic-helix-loop-helix-pas heterodimer regulated by cellular O_2 tension. Proc Natl Acad Sci USA. 1995;92:5510–4.

Weidhaas JB, Babar I, Nallur SM, et al. MicroRNAs as potential agents to alter resistance to cytotoxic cancer therapy. Cancer Res. 2007;67:11111–6.

Weiss GJ, Bemis LT, Nakajima E, et al. EGFR regulation by microRNA in lung cancer: correlation with clinical response and survival to gefitinib and EGFR expression in cell lines. Ann Oncol. 2008;19:1053–9.

Wiklund ED, Bramsen JB, Hulf T, et al. Coordinated epigenetic repression of the miR-200 family and miR-205 in invasive bladder cancer. Int J Cancer. 2010. doi:10.1002/ijc.25461.

Xia L, Zhang D, Du R, et al. MiR-15b and miR-16 modulate multidrug resistance by targeting bcl2 in human gastric cancer cells. Int J Cancer. 2008;123:372–9.

Xie Y, Todd NW, Liu Z, et al. Altered miRNA expression in sputum for diagnosis of non-small cell lung cancer. Lung Cancer. 2010;67:170–6.

Xu WS, Parmigiani RB, Marks PA. Histone deacetylase inhibitors: molecular mechanisms of action. Oncogene. 2007;26:5541–52.

Yamakuchi M, Ferlito M, Lowenstein CJ. MiR-34a repression of SIRT1 regulates apoptosis. Proc Natl Acad Sci USA. 2008;105:13421–6.

Yan LX, Huang XF, Shao Q, et al. MicroRNA miR-21 over-expression in human breast cancer is associated with advanced clinical stage, lymph node metastasis and patient poor prognosis. RNA. 2008;14:2348–60.

Yang Z, Chen S, Luan X, et al. MicroRNA-214 is aberrantly expressed in cervical cancers and inhibits the growth of HeLa cells. IUBMB Life. 2009c;61:1075–82.

Yang X, Feng M, Jiang X, et al. MiR-449a and miR-449b are direct transcriptional targets of E2F1 and negatively regulate PRB-E2F1 activity through a feedback loop by targeting CDK6 and CDC25A. Genes Dev. 2009b;23:2388–93.

Yang K, Handorean AM, Iczkowski KA. MicroRNAs 373 and 520c are down-regulated in prostate cancer, suppress CD44 translation and enhance invasion of prostate cancer cells in vitro. Int J Clin Exp Pathol. 2009a;2:361–9.

Yang N, Kaur S, Volinia S, et al. MicroRNA microarray identifies let-7i as a novel biomarker and therapeutic target in human epithelial ovarian cancer. Cancer Res. 2008b;68:10307–14.

Yang H, Kong W, He L, et al. MicroRNA expression profiling in human ovarian cancer: miR-214 induces cell survival and cisplatin resistance by targeting pten. Cancer Res. 2008a;68:425–33.

Yu SL, Chen HY, Chang GC, et al. MicroRNA signature predicts survival and relapse in lung cancer. Cancer Cell. 2008;13:48–57.

Yu SL, Chen HY, Yang PC, et al. Unique microRNA signature and clinical outcome of cancers. DNA Cell Biol. 2007;26:283–92.

Yu L, Todd NW, Xing L, et al. Early detection of lung adenocarcinoma in sputum by a panel of microRNA markers. Int J Cancer. 2010;127:2870–8.

Zenz T, Mohr J, Eldering E, et al. MiR-34a as part of the resistance network in chronic lymphocytic leukemia. Blood. 2009;113:3801–8.

Zhang HT, Craft P, Scott PA, et al. Enhancement of tumor growth and vascular density by transfection of vascular endothelial cell growth factor into MCF-7 human breast carcinoma cells. J Natl Cancer Inst. 1995;87:213–9.

Zhang C, Kang C, You Y, et al. Co-suppression of miR-221/222 cluster suppresses human glioma cell growth by targeting p27Kip1 in vitro and in vivo. Int J Oncol. 2009;34:1653–60.

Zhao JJ, Lin J, Yang H, et al. MicroRNA-221/222 negatively regulates estrogen receptor alpha and is associated with tamoxifen resistance in breast cancer. J Biol Chem. 2008;283:31079–86.

Zhong M, Ma X, Sun C, et al. MicroRNAs reduce tumor growth and contribute to enhance cytotoxicity induced by gefitinib in non-small cell lung cancer. Chem Biol Interact. 2010;184:431–8.
Zhou M, Liu Z, Zhao Y, et al. MicroRNA-125b confers the resistance of breast cancer cells to paclitaxel through suppression of pro-apoptotic bcl-2 antagonist killer 1 (BAK1). J Biol Chem. 2010;285:21496–507.
Zhu W, Qin W, Atasoy U, et al. Circulating microRNAs in breast cancer and healthy subjects. BMC Res Notes. 2009;2:89.
Zhu S, Si ML, Wu H, et al. MicroRNA-21 targets the tumor suppressor gene tropomyosin 1 (TPM1). J Biol Chem. 2007;282:14328–36.
Zubakov D, Boersma AW, Choi Y, et al. MicroRNA markers for forensic body fluid identification obtained from microarray screening and quantitative RT-PCR confirmation. Int J Legal Med. 2010;124:217–26.

Chapter 19
Role of MicroRNAs in Anti-cancer Drug Resistance

Erik A.C. Wiemer

Abstract MicroRNAs (miRNAs) are key regulators of various cellular and developmental processes, and are closely linked to the pathogenesis of a number of diseases including cancer. In recent years, several miRNAs have been identified that are capable of modulating sensitivity/resistance to conventional or targeted anti-cancer drugs. These miRNAs may function in the development and maintenance of drug resistance as it appears in the clinic. This chapter introduces the phenomenon of (multi)drug resistance occurring in cancer cells and its underlying causes. Further, it outlines the role of miRNAs in biology and cancer, and describes anti-cancer drugs that affect miRNA expression. MiRNAs associated with drug resistance are listed, some miRNAs are experimentally shown to regulate drug sensitivity. These miRNAs operate by repressing the expression of novel and known resistance-related genes and/or signaling pathways. However, the relevant target genes are often not known. Special attention is given to *miR-519c* involved in regulating *ABCG2*, and to *miR-21* and *miR-214* that both target *PTEN* thereby regulating the PI3K/Akt pathway. *MiR-221* and *miR-222* are described, they play a role in tumor necrosis factor related apoptosis inducing ligand (TRAIL) and tamoxifen resistance through the repression of multiple genes including $p27^{Kip1}$. Similarly, the *miR-200* family, essential for the regulation of epithelial-mesenchymal transition associated with cancer progression and drug resistance, is reviewed. Finally, the possibilities for diagnostic and therapeutic applications of miRNAs and how they may help to overcome drug resistance by re-sensitizing resistant cancer cells are discussed.

19.1 Drug Resistance

Chemotherapy is an important treatment modality for many types of (metastatic) cancer. It is often used in combination with surgery and/or radiotherapy, either in a neo-adjuvant setting i.e. before surgical resection of the tumor or radiotherapy,

---

E.A.C. Wiemer (✉)
Department of Medical Oncology, Josephine Nefkens Institute, Erasmus Medical Center, 3015 GE Rotterdam, The Netherlands
e-mail: e.wiemer@erasmusmc.nl

W.C.S. Cho (ed.), *MicroRNAs in Cancer Translational Research*,
DOI 10.1007/978-94-007-0298-1_19,

or in an adjuvant setting after surgery or radiotherapy to prevent tumor metastasis. Despite its success in prolonging survival and quality of life, the efficacy of chemotherapy is frequently hampered due to the occurrence of drug resistance (DR). DR is encountered in cancer patients that do not, or only transiently, respond to individual anti-cancer drugs or drug combinations. Insensitivity to anti-cancer drugs can be an intrinsic characteristic of the tumor or can be acquired over time after exposure to chemotherapy. The drug resistant cancer cells make the tumor less prone to the cytotoxic or cytostatic action of the anti-cancer drugs thereby enabling the tumor to progress. A confounding factor is that cancers often display a so-called multi-drug resistant (MDR) phenotype implying insensitivity to chemically and mechanistically unrelated classes of anti-cancer agents, even ones to which the tumor was never exposed.

In the last 3 decades the phenomenon of MDR and its underlying molecular causes, has been the subject of many studies investigating drug resistance in cell lines, in various animal models including xenografts as well as in cancer patients. Multiple molecular, cellular and systemic mechanisms have been identified that are capable of conferring drug resistance up to a certain level (Broxterman et al. 2009). The mechanisms of resistance may involve genetic (gene rearrangements, mutations and gene amplifications) or epigenetic changes affecting the expression and function of key genes (Fojo 2007). In general two main categories of resistance mechanisms are recognized, one characterized by an impaired delivery, uptake or retention of drugs in the cancer cell, the other by molecular changes in the cancer cells affecting drug sensitivity (Gottesman et al. 2002). The expression of drug efflux pumps of the ABC transporter family is an example from the first category (Gottesman et al. 2002) whereas faulty damage effector pathways like apoptosis and senescence (Kahlem et al. 2004; Rodriguez-Nieto and Zhivotovsky 2006) and activated DNA repair (Martin et al. 2008) fall in the second category.

Much of the work on MDR relates to the classic cytotoxic drugs which still form the mainstay of chemotherapy as it is applied in the clinic. These conventional anti-cancer treatments are usually composed of a combination of different anti-cancer agents with the purpose of simultaneously attacking the tumor at multiple levels thereby preventing the development of drug resistance. Frequently used, are drugs that target the DNA of the cancer cell either by binding covalently (e.g. cisplatin, carboplatin, oxaliplatin) or by binding non-covalently like the anthracyclines (e.g. doxorubicin). Anti-metabolites like 5-fluorouracil, fludarabine and cytarabine affect the synthesis of nucleic acids whereas other drugs target the cytoskeleton notably the microtubules, interfering with their essential role in cell division. Taxanes (paclitaxel, docetaxel) stabilize microtubules whereas vinblastine and vincristine depolymerise microtubules. Topoisomerase I or II inhibitors such as camptothecin, topotecan, etoposide or teniposide affect DNA replication and transcription. The advent of small molecules (e.g. imatinib, gefitinib, erlotinib, sunitinib, sorafenib, lapatinib, and everolimus) targeting specific signalling intermediates and/or oncogenic pathways does not solve the drug resistance problem as also treatment with these novel compounds eventually give rise to resistance and tumor progression (Broxterman et al. 2009; Fojo 2007; Stegmeier et al. 2010). The current notion is that drug resistance is multifactorial, with many factors contributing at the same time to the resistant phenotype.

19.2 MicroRNAs: Biogenesis, Targets, and Function

MicroRNAs (miRNAs) are small, evolutionarily conserved, 16–29 nucleotide RNA molecules that repress gene expression in a post-transcriptional fashion (Bartel 2004, 2009). MiRNAs are derived from RNA polymerase II transcripts, the primary miRNAs (pri-miRNA) can be several kb in length and may contain several miRNAs. The pri-miRNA assumes a characteristic stem-loop structure which is recognized and cleaved by a nuclear RNA III endonuclease called Drosha thereby generating a 60–70 nucleotide precursor miRNA (pre-miRNA) which is exported to the cytoplasm where it is further processed by another endonuclease named Dicer. The final outcome is a ~22 nucleotide RNA duplex that associates with the proteins of the RNA-induced silencing complex (RISC). Usually, but not always, one of the strands (the passenger strand) is degraded leaving only the mature single stranded miRNA or guide strand, attached to the RISC. MiRNAs carry out their silencing function by binding to target messenger RNAs in the context of RISC. Most miRNAs base-pair to specific sequences in the 3′ untranslated region (3′UTR) of specific mRNAs causing translational repression and/or mRNA degradation thereby leading to reduced levels of the proteins encoded by the targeted mRNAs. In mammals, including humans, miRNAs display an imperfect complementarity to their target mRNA making the identification of potential miRNA regulated genes rather difficult. Computer algorithms have been developed that take into account that only 6–7 nucleotides of the miRNA usually nucleotides 2–8, the seed-sequence, is critical for target recognition. Target prediction programs like TargetScan (http://www.targetscan.org), PicTar (http://pictar.mdc-berlin.de) or microrna.org (http://www.microrna.org) reveal that each miRNA has multiple mRNA targets and that most mRNAs have binding sites for multiple miRNAs. The 1,048 human miRNAs currently listed in the miRNA registry (release 16, http://microrna.sanger.ac.uk) are believed to target about 60% of all human genes (Friedman et al. 2009). It is therefore not surprising that miRNAs have been shown to be involved in a variety of cellular processes mainly dealing with developmental and metabolic processes e.g. cell proliferation, cell differentiation, developmental timing, fat metabolism, apoptosis, insulin secretion, stem cell maintenance, neuronal patterning, hematopoietic differentiation, immune cell development and the control of signal transduction (Williams 2008). Aberrant miRNA expression has been linked to various diseases including cardiovascular diseases (Chen et al. 2008; Thum et al. 2008), neurological disorders (Maes et al. 2009; Schratt 2009) as well as cancer (Esquela-Kerscher and Slack 2006; Garzon et al. 2009; Wiemer 2007).

19.3 MiRNAs and Cancer

In recent years it has become clear that miRNAs are intimately involved in the initiation and progression of numerous cancers and display oncogenic or tumor suppressor activities (Di Leva and Croce 2010; Iorio and Croce 2009). Both Drosha and Dicer, two essential components for miRNA biogenesis, are frequently aberrantly expressed in cancers (Chiosea et al. 2006, 2007; Merritt et al. 2008; Sand et al. 2010; Wu et al. 2010). Furthermore multiple miRNAs were found to be regulators of cancer related pathways like apoptosis and cell proliferation (Chen et al. 2010a; Ernst et al. 2010; Qin et al. 2010; Tsukamoto et al. 2010; Veronese et al. 2010; Yu et al. 2010a). In addition it was recognized that miRNA genes are frequently found in cancer associated regions and fragile sites in the genome (Calin et al. 2004). Numerous miRNA profiling studies have reported altered miRNA expression patterns in cancer (Di Leva and Croce 2010; Iorio and Croce 2009; Wiemer 2007). The molecular basis of the changed miRNA expression is not always known but may involve genomic deletions or amplifications, mutated miRNA loci, epigenetic silencing or the dysregulation of transcription factors that target specific miRNA genes (Croce 2009). Proof that miRNAs fulfil pivotal roles in cancer as oncogene or tumor suppressor, or sometimes both depending on the precise conditions, comes from studies in which the levels of specific miRNAs or miRNA clusters are modulated in established in vivo cancer models (Costinean et al. 2006; He et al. 2005; Kumar et al. 2008; Yu et al. 2007).

As miRNAs function in all kinds of cellular processes in (cancer) cells, the question arises whether they can also play a determining role in the sensitivity to anti-cancer compounds. A clear indication that this is indeed the case comes from a study by Bu et al. who showed that a knockdown of *Dicer* in MCF-7 breast cancer cells made them more sensitive to the anti-cancer agent cisplatin (Bu et al. 2009). Apparently miRNAs are necessary for the cells to properly deal with the DNA damage caused by cisplatin.

19.4 Anti-cancer Drugs Affect MiRNA Expression

Exposure of mammalian cells to sub-lethal concentrations of anti-cancer drugs elicits an evident miRNA response. Table 19.1 lists some of the compounds that have been tested and the cells or cell lines that were exposed. Certain miRNAs were found to be significantly up-regulated whereas the expression of others was reduced (Table 19.1). However, as all studies have a different set-up the findings are difficult to compare. Obviously the specific cellular reactions, of which the differential miRNA expression forms an integral part, will vary and depend on many factors like drug concentration, drug half-life, exposure time, and the cells and/or tissues involved. Carefully executed studies are needed to map the miRNA responses in time and to link them to specific cellular changes e.g. cell cycle arrest, apoptosis or DNA repair. This information is necessary if one wants to elucidate the precise role of the miRNAs that are differentially expressed. An interesting paper in this respect describes the miRNA responses to UV-induced DNA damage (8 J/m^2 of UVC) in primary human fibroblasts (Pothof et al. 2009). 31 miRNAs were significantly up or down-regulated and it was noted that the miRNA response was for a large part transient, as the expression levels of most miRNAs returned to normal at 24 h after UV exposure. Some miRNAs displayed an early response i.e. were found to be regulated during the first hours after UV exposure whereas others showed an early regulation lasting for more than 24 h or were up-regulated late after UV. At least part of the miRNA response upon UV damage was also observed to occur in HeLa cells (cervix

Table 19.1 Anti-cancer drugs that affect microRNA expression

Drug	Concentration/ time	Cell type/cell line (cancer type)	References
Curcumin[a] and gemcitabine combination	1 μmol/L (Cur) 10 nmol/L (Gem) 72 h	BxPC-3; MiAPaCa-E; MiAPaCa-M (pancreatic cancer)	Ali et al. (2010)
Curcumin[a]	10 μmol/L 72 h	BxPC-3 (pancreatic cancer)	Sun et al. (2008)
Capecitabine[b]	825 mg/m^2 twice a day 2 weeks	Tumor biopsies (rectal cancer)	Svoboda et al. (2008)
Doxorubicin	10 μmol/L 16 h	Neonatal rat cardiac myocytes	Horie et al. (2010)
Doxorubicin	100 ng/mL 16 h	HCT-116 (colon cancer)	Tazawa et al. (2007)
Epigallocatechin gallate[c]	100 μmol/L 24 h	HepG2 (hepatocellular carcinoma)	Tsang and Kwok (2010)
5- Fluorouracil	10 μmol/L 6 days	HT-29, HCT-116 (colon cancer)	Rossi et al. (2007)
5-Fluorouracil	1 μmol/L 4 days	KM12C (colon cancer)	Zhao et al. (2008b)
5-Fluorouracil	10 μmol/L 24 h	HCT-8 (colon cancer)	Zhou et al. (2010b)
5-Fluorouracil	10 μmol/L 24 h	HCT-116 (colon cancer)	Zhou et al. (2010b)
Imatinib	400 mg/day 2 weeks	Peripheral blood cells (chronic myeloid leukaemia)	Flamant et al. (2010)
Oxaliplatin	4.25 μmol/L 24 h	HCT-8 (colon cancer)	Zhou et al. (2010b)
Oxaliplatin	3 μmol/L 24 h	HCT-116 (colon cancer)	Zhou et al. (2010b)

[a]Curcumin is an anti-oxidant and food supplement with anti-cancer activity.
[b]Capecitabine is combined with radiotherapy.
[c]Epigallocathechin-3-gallate is a bioflavonoid and anti-oxidant with putative anti-cancer activity.

carcinoma cells) in particular the short term fast response whereas the late responses are defective in this cancer cell line. The exact role of most of these differentially regulated miRNAs is not yet clear. At least one miRNA *miR-16* was found to fulfil a central role in the regulation of cell proliferation and the G1-S checkpoint; an important mRNA target in this respect was *CDC25a*.

19.5 MiRNAs Capable of Modulating Sensitivity to Anti-cancer Drugs

In order to identify miRNAs that determine sensitivity to anti-cancer drugs, a frequently taken approach is to compare steady-state miRNA expression profiles from drug sensitive and isogenic drug resistant derivatives (Table 19.2). Many miRNAs

Table 19.2 MicroRNA expression profiles differ between sensitive and drug resistant cells

Drug	Cancer type/cell lines	Up-regulation in resistant cells	Down-regulation in resistant cells	MicroRNA platform/remarks	References
Cisplatin	Breast cancer; MCF-7; MCF-7/CDDP	*miR-10a* *miR-10b* *miR-29a* *miR-29b* *miR-132* *miR-146a* *miR-194* *miR-206* *miR-214* *miR-221* *miR-222*	*miR-7* *miR-126* *miR-127* *miR-141* *miR-196* *miR-200b* *miR-200c* *miR-205* *miR-342* *miR-345* *miR-429* *miR-489*	Microarray, RT-PCR/In total 46 up-regulated and 57 down-regulated miRNAs were found	Pogribny et al. (2010)
	Ovarian cancer; frozen epithelial ovarian cancer samples, complete response ($n = 42$) vs. non-complete response ($n = 27$)	*miR-1b-1* *miR-22* *miR-126* *miR-152* *miR-196a* *miR-198* *miR-214* *miR-216* *miR-223* *miR-370* *miR-509* *miR-519e* *miR-520e* *miR-521*	*Let-7i* *miR-321*	Microarray, RT-PCR/In total 24 miRNAs were up-regulated in the non-complete response group and 10 were up-regulated in the complete response group	Yang et al. (2008a)

Table 19.2 (continued)

Drug	Cancer type/cell lines	Up-regulation in resistant cells	Down-regulation in resistant cells	MicroRNA platform/remarks	References
	Ovarian cancer; A2780; A2780CIS		*Let-7e* *miR-20b* *miR-30c* *miR-106a* *miR-126* *miR-130b* *miR-335*	Microarray	Sorrentino et al. (2008)
	Ovarian cancer; FFPE epithelial ovarian cancer samples, resistant ($n = 12$) vs. sensitive ($n = 25$)	*Let-7g* *miR-23a* *miR-27a* *miR-30c* *miR-199a-3p*	*miR-378* *miR-625*	Microarray	Eitan et al. (2009)
	Tongue squamous cell carcinoma; Tca8113; Tca/cisplatin	*Let-7c* *Let-7d* *Let-7e* *Let-7g* *miR-20b* *miR-23a* *miR-30d* *miR-181d* *miR-188* *miR-214* *miR-373** *miR-432* *miR-498* *miR-518c**	*miR-21* *miR-342*	Microarray, RT-PCR	Yu et al. (2010b)

Table 19.2 (continued)

Drug	Cancer type/cell lines	Up-regulation in resistant cells	Down-regulation in resistant cells	MicroRNA platform/remarks	References
		miR-584 *miR-608* *miR-628*			
	Ovarian cancer (OV19, TOV112D, TOV21G, FUOV1, C13, OV2008, A2780Cp, A2780S, IGROV1, T8, OVCAR5, IMCC3, A2008, OVCAR3, SKOV3, IOSER)	*miR-23b* *miR-381* *miR-340* *miR-520f*	*miR-106a* *miR-185* *miR-331*	Microarray	Boren et al. (2009)
Docetaxel[a]	NSCLC; SPC-A1; SPC-A1/docetaxel	*miR-98* *miR-192* *miR-424*	*miR-194* *miR-200b* *miR-212*	Microarray, RT-PCR/in total 52 dysregulated miRNAs	Rui et al. (2010)
	Ovarian cancer (OV19, TOV112D, TOV21G, FUOV1, C13, OV2008, A2780Cp, A2780S, IGROV1, T8, OVCAR5, IMCC3, A2008, OVCAR3, SKOV3, IOSER)	*miR-99b* *miR-371* *miR-502* *miR-514* *miR-515-5p* *miR-518c-AS*		Microarray	Boren et al. (2009)

Table 19.2 (continued)

Drug	Cancer type/cell lines	Up-regulation in resistant cells	Down-regulation in resistant cells	MicroRNA platform/remarks	References
Doxorubicin	NSCLC; NCI-H69; NCI-H69AR	*miR-29a* *miR-99a* *miR-99b* *miR-100* *miR-125a* *miR-143* *miR-145* *miR-195* *miR-214* *miR-224* *miR-324* *miR-432*	*miR-127* *miR-134* *miR-200b* *miR-216a* *miR-217* *miR-299* *miR-329* *miR-335* *miR-375* *miR-376a* *miR-376c* *miR-379* *miR-382* *miR-409* *miR-432* *miR-485* *miR-487b* *miR-494* *miR-495* *miR-654*	Microarray, RT-PCR/in total 24 up-regulated and 37 down-regulated miRNAs were found	Guo et al. (2010)
	Breast cancer; MCF-7; MCF-7/DOX	*miR-21* *miR-28* *miR-106a* *miR-206*	*Let-7* *miR-27b* *miR-34a* *miR-127* *miR-200c*	Microarray, RT-PCR/in total 63 up-regulated and 75 down-regulated miRNAs were found	Kovalchuk et al. (2008)

Table 19.2 (continued)

Drug	Cancer type/cell lines	Up-regulation in resistant cells	Down-regulation in resistant cells	MicroRNA platform/remarks	References
	Ovarian cancer; A2780; A2780DX5	*miR-27a* *miR-99a* *miR-100* *miR-125b1* *miR-451*		Microarray	Zhu et al. (2008)
	Ovarian cancer (OV19, TOV112D, TOV21G, FUOV1, C13, OV2008, A2780Cp, A2780S, IGROV1, T8, OVCAR5, IMCC3, A2008, OVCAR3, SKOV3, IOSER)	*Let-7e* *miR-21* *miR-181a* *miR-181b* *miR-213* *miR-520f*	*miR-518c*	Microarray	Boren et al. (2009)
Etoposide	Breast cancer; MCF-7; MCF-7/VP	*miR-9* *miR-17* *miR-19b* *miR-23b* *miR-26b* *miR-29a* *miR-29b* *miR-29c* *miR-30d* *miR-91* *miR-197* *miR-320*	*miR-7* *miR-92-2* *miR-187* *miR-326* *miR-429*	Microarray	Liang et al. (2010)

Table 19.2 (continued)

Drug	Cancer type/cell lines	Up-regulation in resistant cells	Down-regulation in resistant cells	MicroRNA platform/remarks	References
Fulvestrant	Breast cancer; MCF-7; MCF-7-FR	*Let-7i* *miR-181a* *miR-191* *miR-199b* *miR-204* *miR-211* *miR-212* *miR-216* *miR-328* *miR-346* *miR-373** *miR-424* *miR-638* *miR-768-3p*	*miR-221* *miR-222*	Microarray	Xin et al. (2009)
Gemcitabine	Pancreatic cancer; L3.6pl; Colo357; MiaPaCa-2; Panc-1		*Let-7b* *Let-7c* *Let-7d* *Let-7e* *miR-200b* *miR-200c*	Microarray, RT-PCR	Li et al. (2009b)

Table 19.2 (continued)

Drug	Cancer type/cell lines	Up-regulation in resistant cells	Down-regulation in resistant cells	MicroRNA platform/remarks	References
	Ovarian cancer (OV19, TOV112D, TOV21G, FUOV1, C13, OV2008, A2780Cp, A2780S, IGROV1, T8, OVCAR5, IMCC3, A2008, OVCAR3, SKOV3, IOSER)	*miR-181a* *miR-181b* *miR-213*	*miR-132* *miR-330* *miR-339*	Microarray	Boren et al. (2009)
	Pancreatic cancer; BxPC-3; MiAPaCa-E; MiAPaCa-M	*miR-21*	*miR-200b* *miR-200c*	RT-PCR	Ali et al. (2010)
Imatinib	CML; bone marrow mononuclear cells; resistant vs. sensitive pts	*miR-191*	*miR-7* *miR-23a* *miR-26a* *miR-29a* *miR-29c* *miR-30b* *miR-30c* *miR-100* *miR-126** *miR-134* *miR-141* *miR-183* *miR-196b* *miR-199a* *miR-224*	RT-PCR	San Jose-Eneriz et al. (2009)

Table 19.2 (continued)

Drug	Cancer type/cell lines	Up-regulation in resistant cells	Down-regulation in resistant cells	MicroRNA platform/remarks	References
			miR-326 *miR-422b* *miR-520a*		
Paclitaxel/taxol	Ovarian carcinoma; A2780; A2780TAX; A2780TC1; A2780TC3	*Let-7e* *miR-21* *miR-100* *miR-321**	*Let-7e* *miR-20b* *miR-30c* *miR-106a* *mir-126* *miR-130b* *mir-321** *miR-335*	Microarray, *let-7e* was both found up- and down-regulated depending on the cell line	Sorrentino et al. (2008)
	Ovarian cancer (OV19, TOV112D, TOV21G, FUOV1, C13, OV2008, A2780Cp, A2780S, IGROV1, T8, OVCAR5, IMCC3, A2008, OVCAR3, SKOV3, IOSER)	*miR-99b* *miR-126* *miR-514*	*miR-29c*	Microarray	Boren et al. (2009)
	Breast cancer; MDA-435; MDA-435TR1; MDA-435TRP	*miR-100* *miR-106b* *miR-125b* *miR-140-3p* *miR-221* *miR-222* *miR-620*	*miR-15b* *miR-16* *miR-17* *miR-20b* *miR-27a* *miR-29a* *miR-29b*	Microarray, RT-PCR	Zhou et al. (2010a)

Table 19.2 (continued)

Drug	Cancer type/cell lines	Up-regulation in resistant cells	Down-regulation in resistant cells	MicroRNA platform/remarks	References
		miR-923	*miR-30a* *miR-30b* *miR-146a* *miR-146b-5p* *miR-363* *miR-801*		
Temozolomide	Glioblastoma; U251; U251R	*miR-10a** *miR-132* *miR-193b** *miR-193a-5p* *miR-195* *miR-452* *miR-455-3p* *miR-500** *miR-502-3p* *miR-503* *miR-584*	*miR-106b** *miR-210*	Microarray, RT-PCR	Ujifuku et al. (2010)
Tamoxifen	Breast cancer; MCF-7; MCF-7/OHT^{R}	*miR-32* *miR-171* *miR-181* *miR-203* *miR-213* *miR-221* *miR-222* *miR-375*	*miR-21* *miR-23* *miR-24* *miR-27* *miR-200* *miR-342* *miR-489*	Microarray, RT-PCR	Miller et al. (2008)

Table 19.2 (continued)

Drug	Cancer type/cell lines	Up-regulation in resistant cells	Down-regulation in resistant cells	MicroRNA platform/remarks	References
Topotecan	Ovarian cancer; (OV19, TOV112D, TOV21G, FUOV1, C13, OV2008, A2780Cp, A2780S, IGROV1, T8, OVCAR5, IMCC3, A2008, OVCAR3, SKOV3, IOSER)	*miR-34b* *miR-431* *miR-518c-AS*	*miR-142-5p*	Microarray	Boren et al. (2009)
TRAIL	NSCLC; A459; A549; CALU-1	*miR-15b* *miR-100* *miR-125b* *miR-221* *miR-222*	*miR-9* *miR-96*	Microarray, RT-PCR	Garofalo et al. (2008)
Vinblastine	Epidermoid carcinoma; KB-3-1; KB-V1	*miR-27a* *miR-99a* *miR-100* *miR-125b1* *miR-451*		Microarray	Zhu et al. (2008)

Table 19.2 (continued)

Drug	Cancer type/cell lines	Up-regulation in resistant cells	Down-regulation in resistant cells	MicroRNA platform/remarks	References
Vincristine	Gastric cancer; SGC7901; SGC7901/VCR	*miR-302b* *miR-492*	*Let-7a* *miR-15b* *miR-16* *miR-17-5p* *miR-20a* *miR-23b* *miR-106a* *miR-106b* *miR-196a* *miR-320*	Microarray, RT-PCR	Xia et al. (2008)

[a]Docetaxel (Taxotere) is a semi-synthetic analog of Paclitaxel (Taxol).

appear to be differentially expressed in cancer cell lines made resistant to various anti-cancer drugs; the most significant miRNAs are summarized in Table 19.2. A straightforward comparison between the different studies referred to in Table 19.2 should be looked at with care as the various profiling platforms used were not cross-validated. Most authors, however, validated their findings with an alternative technique usually a RT-PCR based method. Moreover, simple miRNA expression comparisons do not distinguish between direct and indirect (i.e. bystander) effects. Another factor to keep in mind is that the number of human miRNAs in the miRNA registry almost doubled over the last 3 years meaning that in recent studies far more miRNAs were monitored. Nevertheless some miRNAs, like members of the *miR-200* family (*miR-200a*, *miR-200b*, *miR-200c*, *miR-141* and *miR-429*), *miR-21*, *miR-214*, *miR-221*, and *miR-222* were independently identified in multiple studies and are described in more detail below.

To prove a direct influence on drug sensitivity selected miRNAs can be ectopically up and down-regulated after which the cellular effects to the drug exposure (level of resistance) can be quantitatively determined. Table 19.3 enumerates all miRNAs that are able to modulate resistance to different sorts of anti-cancer drugs. In more than half the cases one or more target genes are postulated through which the miRNA affects drug sensitivity. Perhaps not surprisingly a number of publications focus on miRNAs that regulate genes involved in apoptosis like the pro-apoptotic *Bcl2* and *Bcl-xL* and the anti-apoptotic *Bak1* (Table 19.3). Also miRNAs are described that regulate drug transporters of the ATP binding cassette (ABC) family such as ABCB1 (MDR1 or P-gp), ABCC1 (MRP1) and ABCG2 (BCRP). These drug efflux transporters play an important role in clinical pharmacology and are often found highly expressed in in vitro selected drug resistant cancer cell lines in which the exact mechanisms governing their over-expression are not always clear. To et al. investigated the molecular mechanisms regulating *ABCG2* expression in S1 colon cancer cells and its drug resistant derivative S1M180 (To et al. 2008, 2009). It was discovered that the 3′UTR of the *ABCG2* mRNA was more than 1,500 base pairs in length in the drug sensitive cell line, considerable longer than the 3′UTR of 300–400 base pairs detected in the drug resistant cells. In addition to size differences the *ABCG2* mRNA turned out to have a variable half life ranging from 6 h in S1 cells to 16 h in the resistant cells. *MiR-519c* was shown to exclusively bind to the longer transcript repressing *ABCG2* expression both by mRNA degradation and translational inhibition. As the binding site for *miR-519c* is missing in the truncated *ABCG2* 3′UTR of the resistant cells its repressive effect on *ABCG2* is released, causing the ABCG2 over-expression observed in the resistant cells. This novel way of regulating ABCG2 levels was also present in other drug resistant cell lines implying it is a common mechanism (To et al. 2009). At least two additional miRNAs i.e. *miR-520h* (Liao et al. 2008) and *miR-328* (Pan et al. 2009) have been reported to target the *ABCG2* mRNA. As *miR-520h* basepairs close to the stop codon in the 3′UTR it is still capable of regulating the expression of the truncated transcripts observed in resistant cells. However as the resistant cells express relatively low quantities of *miR-520h* it is postulated to be all sequestered on the over-expressed *ABCG2* mRNA and unable to significantly limit ABCG2 expression (To et al. 2009).

Table 19.3 MicroRNAs capable of modulating drug resistance

MicroRNA	Ectopic up-/down-regulation	Resistance	Drug	Cell/cancer type	mRNA targets	References
Let-7a	+	–	Gefitinib	NSCLC	Unknown	Zhong et al. (2010)
	+/–	+/–	Doxorubicin Interferon γ Paclitaxel	Epidermoid carcinoma Hepatocellular carcinoma	*CASP 3*	Tsang and Kwok (2008)
Let-7c	+	–	Sorafenib	Hepatocellular carcinoma	*BCL-xL*	Shimizu et al. (2010)
Let-7i	+/–	–/+	Cisplatin	Breast cancer Ovarian cancer	Unknown	Yang et al. (2008a)
	+/–	+/–	Cytotoxic compounds	NSCLC	Unknown	Blower et al. (2008)
miR-1	+	–	Doxorubicin	NSCLC	*MET*	Nasser et al. (2008)
miR-7	+ Assayed together with *miR-345*	–	Cisplatin	Breast cancer	*ABCC1*	Pogribny et al. (2010)
*miR-10a**	+	+	Temozolomide	Glioblastoma	Unknown	Ujifuku et al. (2010)
miR-15b	+/–	+/–	Cisplatin Doxorubicin Etoposide Vincristine	Gastric cancer	*BCL2*	Xia et al. (2008)
miR-16	+/–	+/–	Cisplatin Doxorubicin Etoposide Vincristine	Gastric cancer	*BCL2*	Xia et al. (2008)
	+/–	–/+	Cytotoxic compounds	NSLC	Unknown	Blower et al. (2008)
miR-21	–	–	Arsenic trioxide	CML	*PDCD4*	Li et al. (2010)
	–	–	Teniposide	Glioblastoma	*LRRFIP1*	Li et al. (2009a)
	+	+	Gemcitabine	Pancreatic ductal carcinoma	*PTEN*	Giovannetti et al. (2010)
	+/–	+/–	Gemcitabine	Cholangio-carcinoma	*PTEN*	Meng et al. (2006)

Table 19.3 (continued)

MicroRNA	Ectopic up-/down-regulation	Resistance	Drug	Cell/cancer type	mRNA targets	References
	–	–	Fludarabine	CLL	Unknown	Ferracin et al. (2010)
	+/–	–/+	Cytarabine Various cytotoxic compounds	Glioblastoma NSCLC Ovarian cancer	Unknown	Blower et al. (2008)
	+/–	+/–	Geldanamycin Cytotoxic compounds	NSCLC Ovarian cancer	Unknown	Blower et al. (2008)
	+	–	Cisplatin	Tongue squamous cell carcinoma	Unknown	Yu et al. (2010b)
	–	–	AG1478 (EGFR inhibitor)	Lung adeno-carcinoma	Unknown	Seike et al. (2009)
miR-23a	–	–	Cisplatin	Tongue squamous cell carcinoma	Unknown	Yu et al. (2010b)
miR-24	+	+	Methotrexate	Colon cancer	*DHFR*	Mishra et al. (2007, 2009)
	+/–	–/+	Bleomycin Cisplatin	CML Hepatocellular carcinoma	*H2AX*	Lal et al. (2009)
miR-27a	–	–	Vinblastine	Ovarian cancer	*ABCB1*	Zhu et al. (2008)
miR-34a	+	–	Camptothecin	Prostate cancer	*SIRT1*	Fujita et al. (2008)
	+	–	Cisplatin Docetaxel Doxorubicin Gemcitabine	Gastric cancer	*BCL2*	Ji et al. (2008)
	+	–	Topotecan	Retinoblastoma	Unknown	Dalgard et al. (2009)
	+/–	–/+		Prostate cancer	Unknown	Rokhlin et al. (2008)
	Assayed together with *miR-34c* –	–	Cisplatin	Proximal tubular kidney cells	Unknown	Bhatt et al. (2010)

Table 19.3 (continued)

MicroRNA	Ectopic up-/ down-regulation	Resistance	Drug	Cell/cancer type	mRNA targets	References
miR-34c	+/– Assayed together with *miR-34a*	–/+		Prostate cancer	Unknown	Rokhlin et al. (2008)
miR-93	+/– Cooperates with *miR-25*	+/–	TGFβ	Gastric cancer	*E2F1* *CDKN1A*	Petrocca et al. (2008)
miR-98	+	+	Cisplatin Doxorubicin	Head and neck squamous cell carcinoma	*HMGA2*	Hebert et al. (2007)
miR-100	+	–	Everolimus	Ovarian cancer	*mTOR*	Nagaraja et al. (2010)
miR-101	+	–	Curcumin Doxorubicin Etoposide	Hepatocellular carcinoma	*MCL1*	Su et al. (2009)
miR-106	–	–	Cisplatin Demethyl-cantharidin	NSCLC	Unknown	Wang et al. (2010)
miR-106b	+/– Cooperates with *miR-25*	+/–	TGFβ	Gastric cancer	*E2F1* *CDKN1A*	Petrocca et al. (2008)
	+	+	Doxorubicin	Mammary epithelial cells	*CDKN1A*	Ivanovska et al. (2008)
miR-122	+	–	Sorafenib	Hepatocellular carcinoma	Unknown	Bai et al. (2009)
	+	–	Doxorubicin	Hepatocellular carcinoma	*CCNG1*	Fornari et al. (2009)
miR-125b	+	+	Paclitaxel	Breast cancer	*BAK1*	Zhou et al. (2010a)
miR-125b-2	+/–	+/–	Doxorubicin Staurosporin	ALL (ETV6/ RUNX1)	Unknown	Gefen et al. (2010)
miR-126	+	–	Gefitinib	NSCLC	Unknown	Zhong et al. (2010)
miR-128a	–	–	TGFβ1	Breast cancer	*TGFβRI*	Masri et al. (2010)

Table 19.3 (continued)

MicroRNA	Ectopic up-/down-regulation	Resistance	Drug	Cell/cancer type	mRNA targets	References
miR-128b	+ Cooperates with *miR-221*	–	Dexamethasone	ALL (MLL rearranged)	*MLL* *AF4* *MLL-AF4* *AF4-MLL*	Kotani et al. (2009, 2010)
	–	–	Gefitinib	NSCLC	*EGFR*	Weiss et al. (2008)
miR-134	+	–	Cisplatin Doxorubicin Etoposide	SCLC	*ABCC1*	Guo et al. (2010)
miR-140	+	+	5-Fluorouracil Methotrexate	Colon cancer Osteosarcoma	*HDAC4*	Song et al. (2009)
miR-143	+	–	5-Fluorouracil	Colon cancer	*ERK5*	Borralho et al. (2009)
miR-145	+	–	Gefitinib	NSCLC	Unknown	Zhong et al. (2010)
miR-148a	+	–	Paclitaxel	Prostate cancer (hormone refractory)	*MSK1*	Fujita et al. (2010)
miR-150	–	–	Cisplatin Demethyl-cantharidin	NSCLC	Unknown	Wang et al. (2010)
miR-155	+/–	+/–	Doxorubicin Etoposide Paclitaxel	Breast cancer	*FOXO3a*	Kong et al. (2010)
miR-181a	+	–	Cisplatin	NSCLC	Unknown	Galluzzi et al. (2010)
miR-181b	+	–	Cisplatin Vincristine	Gastric cancer NSCLC	*BCL2*	Zhu et al. (2010)
miR-195	+	+	Temozolomide	Glioblastoma	Unknown	Ujifuku et al. (2010)
miR-199a	–	–	Cisplatin	Cervical cancer	Unknown	Lee et al. (2008)
miR-199a-3p	+	–	Doxorubicin	Hepatocellular carcinoma	*mTOR* *c-MET*	Fornari et al. (2010)
miR-200b	+	–	Gemcitabine	Pancreatic cancer	Unknown	Li et al. (2009b)

Table 19.3 (continued)

MicroRNA	Ectopic up-/ down-regulation	Resistance	Drug	Cell/cancer type	mRNA targets	References
miR-200c	+/–	+/–	Gemcitabine	Cholangio-carcinoma	*PTPN12*	Meng et al. (2006)
	+	–	Doxorubicin	Breast cancer	Unknown	Tryndyak et al. (2010)
	+	–	Erlotinib Gefitinib	Bladder cancer	*ERRF1*	Adam et al. (2009)
miR-205	+	–	Doxorubicin	Breast cancer	Unknown	Tryndyak et al. (2010)
	+	–	Gefitinib Lapatinib	Breast cancer	*HER3*	Iorio et al. (2009)
miR-212	+	–	TRAIL	NSCLC	*PED/PEA15*	Incoronato et al. (2010)
miR-214	+/–	+/–	Cisplatin	Ovarian cancer	*PTEN*	Yang et al. (2008b)
	–	–	Cisplatin	Tongue squamous cell carcinoma	Unknown	Yu et al. (2010b)
miR-215	+	+	Methotrexate Raltitrexed	Colon cancer Osteosarcoma	*DLT*	Song et al. (2010)
miR-221	+/–	+/–	TRAIL	NSCLC	$P27^{Kip1}$ *PTEN* *TIMP3*	Garofalo et al. (2008, 2009)
	+ Cooperates with *miR-128b*	–	Dexamethasone	ALL (MLL rearranged)	*MLL* *AF4* *MLL-AF4* *AF4-MLL*	Kotani et al. (2009, 2010)
	+/–	+/–	Tamoxifen	Breast cancer	*ERα* $P27^{Kip1}$	Miller et al. (2008); Zhao et al. (2008a)
miR-222	+/–	+/–	TRAIL	NSCLC	$P27^{Kip1}$ *PTEN* *TIMP3*	Garofalo et al. (2008, 2009)
	+	+	Tamoxifen	Breast cancer	*ERα* $P27^{Kip1}$	Miller et al. (2008); Zhao et al. (2008a)
	–	–	Fludarabine	CLL	Unknown	Ferracin et al. (2010)

Table 19.3 (continued)

MicroRNA	Ectopic up-/ down-regulation	Resistance	Drug	Cell/cancer type	mRNA targets	References
miR-296	–	–	Cisplatin Doxorubicin 5-Fluorouracil Vincristine	Esophageal cancer	*ABCB1*	Hong et al. (2010)
miR-326	+	–	Doxorubicin Etoposide	Breast cancer	*ABCC1*	Liang et al. (2010)
miR-328	+	–	Mitoxantrone	Breast cancer	*ABCG2*	Pan et al. (2009)
miR-345	+ Assayed together with *miR-7*	–	Cisplatin	Breast cancer	*ABCC1*	Pogribny et al. (2010)
miR-379	+	–	Cisplatin Doxorubicin Etoposide	SCLC	Unknown	Guo et al. (2010)
miR-451	+	–	Doxorubicin Vinblastine	Breast cancer Ovarian cancer	*ABCB1*	Kovalchuk et al. (2008); Zhu et al. (2008)
	+	–	Imatinib	Glioblastoma	Unknown	Gal et al. (2008)
miR-455-3p	+	+	Temozolomide	Glioblastoma	Unknown	Ujifuku et al. (2010)
miR-504	+	+	Etoposide	Large cell lung cancer Osteosarcoma	*P53*	Hu et al. (2010)
miR-512-3p	+	–	Taxol	Hepatocellular carcinoma	*c-FLIP*	Chen et al. (2010b)
miR-519c	–	+	Mitoxantrone	Colon cancer	*ABCG2*	To et al. (2008, 2009)
miR-520h	–	+	Mitoxantrone	Colon cancer	*ABCG2*	To et al. (2008, 2009)
miR-630	+	+	Cisplatin	NSCLC	Unknown	Galluzzi et al. (2010)

ALL, Acute lymphoblastic leukaemia; CLL, Chronic lymphocytic leukaemia; CML, Chronic myeloid leukaemia; NSCLC, Non-small cell lung cancer; SSLC, Small cell lung cancer; TNFβ, Tumor necrosis factor beta; TRAIL, Tumor necrosis factor related apoptosis inducing ligand.

19.5.1 MiR-21

MiR-21 is highly expressed in many cancers (Volinia et al. 2006) in which it is considered to function as oncogene. Over-expression of *miR-21* also frequently correlates with tumor resistance. For example Meng et al. (2006) showed that *miR-21* knockdown in malignant cholangiocarcinoma cell lines sensitized the cells for gemcitabine. In this context *miR-21* targets the tumor suppressor phosphatase and tensin homolog deleted on chromosome 10 (*PTEN*), a phosphatase that antagonizes the survival promoting activity of the PI3K/Akt signalling pathway. In fact PTEN counteracts the activity of phosphatidylinositol 3-kinase (PI3K) by dephosphorylating phosphatidylinositol-3,4,5-triphosphate (PIP3) into phosphatidylinositol-4,5-diphosphate (PIP2) thereby making the cells more susceptible to cell cycle arrest and apoptosis. The high *miR-21* levels down-regulate PTEN resulting in a more active survival signalling rendering the cells relatively resistant to anti-cancer drugs. Blower and co-workers reported that variable *miR-21* levels can change sensitivity to the cytotoxic action of drugs in opposite directions suggesting the existence of different mechanisms determining toxicity in cancer cell lines (Blower et al. 2008). In addition to *PTEN* more *miR-21* targets have been described that may explain its relation to drug resistance. The tumor suppressor programmed cell death 4 (*PDCD4*) involved in growth, apoptosis, invasion and the cell cycle, also appears to be regulated by *miR-21* (Li et al. 2010). Down-regulation of *miR-21* sensitized the leukemic K562 cells for arsenic trioxide by inducing growth inhibition and apoptosis partially due to the up-regulation of PDCD4. When a small interfering RNA (siRNA) directed against *PDCD4* was used in combination with an anti-*miR-21* growth inhibition and apoptosis were prevented, providing conclusive evidence for a role of PDCD4.

19.5.2 MiR-214

In cisplatin resistant breast cancer and tongue squamous carcinoma cell lines as well as in platinum resistant ovarian cancer samples *miR-214* was found to be over-expressed (Table 19.2). Also in non-small cell lung cancer (NSCLC) cell lines resistant to doxorubicin high *miR-214* levels were detected (Table 19.2). Just like *miR-21*, *miR-214* targets *PTEN* (Jindra et al. 2010; Yang et al. 2008b) modulating resistance through the PI3 kinase/AKT pathway. The genes for *miR-199a-2* and *miR-214* are located about 6 kb apart on chromosome 1q24.2 in an intron of the Dynamin-3 (*DNM3*) gene. *MiR-214* and *miR-199a-2* form a single transcriptional unit that is regulated by Twist1, a nuclear basic helix-loop-helix transcription factor (Lee et al. 2009; Yin et al. 2010). Interestingly in various cancers high Twist1 expression has been associated with resistance to microtubule-targeting anti-cancer drugs (taxol, vincristine) (Wang et al. 2004) and cisplatin (Li et al. 2007). Moreover Twist1 expression is correlated with poor disease outcome and an increased risk of cancer recurrence (Hosono et al. 2007; Shibata et al. 2008). Twist1 may influence these processes in part through *miR-214* and/or *miR-199a-2*. Whereas *miR-214* operates through reducing PTEN levels, it is not yet clear if and how increased *miR-199a-2* levels contribute to the resistant phenotype. Note that *miR-199a-2* is identical to *miR-199a-5p* of which the gene is located on chromosome 19. Silencing of *miR-199a*, most likely *miR-199a-5p*, in cervical cancer cell lines resulted in a growth suppression which was potentiated by cisplatin (Lee et al. 2008) implying an increased sensitivity to this drug. These data suggest that *miR-199a* up-regulation could confer resistance, at least to cisplatin. IKK-β (encoded by the *IKBKB* gene), which is a protein subunit of IκB kinase, was shown to be regulated by *miR-199a-2* (Yin et al. 2010). Down-regulation of the expression of this gene, by high *miR-199a-2* levels, would lead to a reduction of phosphorylated IκB preventing the activation of NF-κB. Paradoxically, however, activation of the transcription factor NF-κB is generally associated with drug resistance (Bentires-Alj et al. 2003) suggesting that if *miR-199a-2* is involved in drug resistance it will be through the suppression of another target gene.

19.5.3 MiR-200 Family

The epithelial-mesenchymal transition (EMT) is a reversible cellular process in which differentiated epithelial cells detach from each other, acquire mesenchymal properties and become motile. EMT plays a critical role in normal embryonic development and wound healing. EMT as it occurs in cancer is closely linked to cancer progression into more dedifferentiated and malignant states (Adam et al. 2009; Iwatsuki et al. 2010; Sabbah et al. 2008; Thiery 2003; Thiery and Sleeman 2006). Moreover there seems to be a link between EMT and the gain of epithelial (cancer) stem cell properties (Mani et al. 2008; Wellner et al. 2009) which may be a prerequisite for the formation of metastases from disseminated tumor cells. Cancer cells that have gone through EMT display an increased metastatic and invasive potential. Biochemically EMT is characterized by the loss and redistribution of epithelial markers like E-cadherin, β-catenin, cytokeratin and the gain of mesenchymal markers such as N-cadherin, vimentin, and fibronectin. Crucial for EMT are a group of transcriptional repressors (SNAIL, SLUG, Twist1, ZEB1, and ZEB2) of which the levels go up upon EMT repressing the expression of epithelial identity genes, e.g. E-cadherin.

A screen to identify miRNAs of which the expression changes during EMT revealed the *miR-200* family (*miR-141*, *miR-200a*, *miR-200b*, *miR-200c*, *miR-429*) and *miR-205* as being down-regulated in cells undergoing EMT (Gregory et al. 2008a,b). It was demonstrated that both the *miR-200* family and *miR-205* are key regulators of EMT through repression of the EMT inducers ZEB 1 and ZEB2. Both the *ZEB1* and *ZEB2* mRNAs harbor multiple binding sites in their 3′UTR for *miR-200* family members and *miR-205*. Each of the miRNAs does function separately in repressing *ZEB1* and *ZEB2* but together they operate in a synergistic manner (Burk et al. 2008; Gregory et al. 2008a; Korpal et al. 2008; Park et al. 2008). Conversely, ZEB1 suppresses the expression of the *miR-200* family indicating the *miR-200* members and ZEB1 reciprocally control each other (Burk et al. 2008). The regulatory importance of the *miR-200* family is emphasized by Gibbons et al. (2009) who showed that metastatic lung adenocarcinoma could revert between epithelial and mesenchymal states, a property which is governed by these miRNAs. EMT is closely associated with drug resistance (Iwatsuki et al. 2010; Voulgari and Pintzas 2009). Table 19.2 clearly depicts that *miR-200* family members are exclusively down-regulated in cancer cell lines that exhibit resistance to different drugs. In agreement are the studies referred to in Table 19.3 in which ectopic over-expression of *miR-200b* and *miR-200c* almost always correlates with a reduced level of resistance. It is not clear how exactly the *miR-200* family members influence drug sensitivity. In addition to *ZEB1* and *ZEB2* at least two other target genes *PTPN12* (Meng et al. 2006) and *ERRF1* (Adam et al. 2009) have been described.

19.5.4 MiR-221 and 222

Another set of miRNAs known to affect the sensitivity to anti-cancer drugs is the *miR-221/222* cluster. The genes for these miRNAs are found close to each other on the short arm of chromosome X at p11.3 and are probably co-expressed. A miRNA profiling study aimed at identifying miRNAs that determine tumor necrosis factor related apoptosis inducing ligand (TRAIL) sensitivity in NSCLC cell lines demonstrated at least four miRNAs to be up-regulated namely *miR-221* and *miR-222* but also *miR-100* and *miR-125b* (Garofalo et al. 2008). Up- and down-modulation of the *miR-221*, *miR-222*, and *miR-100* levels, but not *miR-125b*, correlated with TRAIL sensitivity. High *miR* levels associated with an increased resistance to TRAIL induced apoptosis. The functional role of *miR-221* and *miR-222* was further investigated, predicted targets for these miRNAs are the proto-oncogene *c-KIT* and the tumor suppressor $p27^{Kip1}$ (Felli et al. 2005; le Sage et al. 2007). Subsequent experiments in which both *c-KIT* and $p27^{Kip1}$ were silenced, using specific siRNAs, followed by an assessment of the sensitivity to TRAIL indicated that the resistance modulation was solely mediated through $\mathrm{p27^{Kip1}}$ (Garofalo et al. 2008). In a follow-up study Garofolo et al. also marked *PTEN* and *TIMP3* as direct targets of *miR-221* and *miR-222* in NSCLC and hepatocellular carcinoma cells (Garofalo et al. 2009). TIMP3, a tissue inhibitor of metalloproteinases, was previously shown to inhibit invasion and promote apoptotic cell death when over-expressed in vascular smooth muscle cells and melanoma cell lines (Ahonen et al. 1998; Baker et al. 1998). Down modulation of *PTEN* and *TIMP3* by *miR-221* and *miR-222* conferred resistance to TRAIL in NSCLC and hepatocellular carcinoma cell lines (Garofalo et al. 2009). It was concluded that the silencing of multiple targets by the action of *miR-221* and *miR-222* constitutes the phenomenon of TRAIL resistance. *MiR-221* and *miR-222* have also been linked to tamoxifen resistance. Tamoxifen is a drug which is used to treat women with estrogen receptor positive (ER^+) breast cancer (Miller et al. 2008; Zhao et al. 2008a). Breast cancer cell lines made resistant to 4-hydroxytamoxifen were characterized by a high expression of *miR-221* and *miR-222* compared to the levels observed in the tamoxifen sensitive parental cells. Also in this case, tamoxifen resistance seems to be mediated through the repression of $p27^{Kip1}$ by *miR-221/222* (Miller et al. 2008). A major additional factor contributing to tamoxifen resistance may be the fact that the estrogen receptor α – being the protein through which tamoxifen exerts its anti-cancer action – is also being targeted by *miR-221* and *miR-222* (Zhao et al. 2008a).

19.6 Future Prospects

MiRNAs display a cell and tissue specific distribution making them useful as diagnostic biomarkers. MiRNA expression signatures can be used to classify tumors according to their differentiation state and developmental origin (Lu et al. 2005). Furthermore miRNAs have been successfully used to pinpoint the tissue of origin of metastatic cancers of unknown primary origin with overall high-confidence accuracy (Rosenfeld et al. 2008) or to distinguish cancer subtypes (Lebanony et al. 2009). In addition prognostic miRNA expression signatures have been identified within specific tumor groups that predict high progression risk, poor survival, risk of recurrence or metastatic capability (Calin et al. 2005; Childs et al. 2009; Foekens et al. 2008; Roldo et al. 2006; Schetter et al. 2008; Yanaihara et al. 2006).

If one would be able to modulate the levels of disease-associated miRNAs, one might be able to correct a critical molecular defect and potentially cure the disease. Increased miRNA levels can be reduced by the introduction of antisense oligonucleotides (anti-miRs) interfering with miRNA activity. Alternatively miRNAs mimics may be over-expressed in specific target tissues or tumors. That systemic administration of anti-miRs is feasible was demonstrated in a seminal paper by Krutzfeldt et al. (2005) who reported that *miR-122* in adult mice was effectively and specifically silenced in all tissues investigated with the exception of the brain. Apparently the anti-miRs cannot cross the blood-brain barrier. Also in non-human primates *miR-122* could be effectively antagonized by simple unconjugated locked nucleic acid-modified oligonucleotides (LNA-anti-miRs) with just three dosages of 10 mg/kg injected into the bloodstream (Elmen et al. 2008). Recently it was reported that therapeutic silencing of *miR-122* in hepatitis C virus infected chimpanzees led to a long-lasting suppression of HCV viremia (Lanford et al. 2010). Kota et al. (2009) investigated the efficacy of an miRNA-based therapy for hepatocellular carcinoma (HCC). HCC displays a reduced expression of *miR-26a*. Systemic delivery of *miR-26a* to a HCC mouse model resulted in inhibition of cancer cell proliferation, induction of tumor-specific apoptosis and reduced disease progression identifying miRNAs as a valid target for therapeutic intervention.

The fact that a systemic knockdown of *miR-122* gave no or little adverse effects is encouraging. However, this may not be the case for every miRNA as it depends on the precise functions the miRNA fulfils in the different cells and tissues of the body. If unwanted side effects do occur in non-target tissues alternative, more directed, approaches, other than systemic use, need to be devised particularly if long term use is intended. One can envisage that therapeutic miRNA modulation may also be able to overcome resistance and re-sensitize tumors to the action of regular anti-cancer drugs. Appealing in this respect is that miRNAs are able to hit multiple targets at the same time; if more targets play part in the resistant phenotype this is a definite advantage. A systemic administration – most likely in combination with other anti-cancer agents – will be favored because in this way miRNA-based therapeutics is able to reach possible metastases throughout the whole body. An essential question is which miRNAs should be targeted. Our current knowledge regarding the function of most miRNAs, particularly in relation to drug resistance, is still fragmentary and limited and more preclinical research is needed to make a balanced choice. Eventually miRNA based therapies will appear in the clinic and may help to solve the drug resistance problem.

References

Adam L, Zhong M, Choi W, et al. MiR-200 expression regulates epithelial-to-mesenchymal transition in bladder cancer cells and reverses resistance to epidermal growth factor receptor therapy. Clin Cancer Res. 2009;15:5060–72.

Ahonen M, Baker AH, Kahari VM. Adenovirus-mediated gene delivery of tissue inhibitor of metalloproteinases-3 inhibits invasion and induces apoptosis in melanoma cells. Cancer Res. 1998;58:2310–5.

Ali S, Ahmad A, Banerjee S, et al. Gemcitabine sensitivity can be induced in pancreatic cancer cells through modulation of miR-200 and miR-21 expression by curcumin or its analogue CDF. Cancer Res. 2010;70:3606–17.

Bai S, Nasser MW, Wang B, et al. MicroRNA-122 inhibits tumorigenic properties of hepatocellular carcinoma cells and sensitizes these cells to sorafenib. J Biol Chem. 2009;284:32015–27.

Baker AH, Zaltsman AB, George SJ, et al. Divergent effects of tissue inhibitor of metalloproteinase-1, -2, or -3 overexpression on rat vascular smooth muscle cell invasion, proliferation, and death in vitro. TIMP-3 promotes apoptosis. J Clin Invest. 1998;101:1478–87.

Bartel DP. MicroRNAs: genomics, biogenesis, mechanism, and function. Cell. 2004;116:281–97.

Bartel DP. MicroRNAs: target recognition and regulatory functions. Cell. 2009;136:215–33.

Bentires-Alj M, Barbu V, Fillet M, et al. NF-kappaB transcription factor induces drug resistance through MDR1 expression in cancer cells. Oncogene. 2003;22:90–7.

Bhatt K, Zhou L, Mi QS, et al. MicroRNA-34a is induced via p53 during cisplatin nephrotoxicity and contributes to cell survival. Mol Med. 2010;16:409–19.

Blower PE, Chung JH, Verducci JS, et al. MicroRNAs modulate the chemosensitivity of tumor cells. Mol Cancer Ther. 2008;7:1–9.

Boren T, Xiong Y, Hakam A, et al. MicroRNAs and their target messenger RNAs associated with ovarian cancer response to chemotherapy. Gynecol Oncol. 2009;113:249–55.

Borralho PM, Kren BT, Castro RE, et al. MicroRNA-143 reduces viability and increases sensitivity to 5-fluorouracil in HCT116 human colorectal cancer cells. Febs J. 2009;276:6689–700.

Broxterman HJ, Gotink KJ, Verheul HM. Understanding the causes of multidrug resistance in cancer: a comparison of doxorubicin and sunitinib. Drug Resist Updat. 2009;12:114–26.

Bu Y, Lu C, Bian C, et al. Knockdown of Dicer in MCF-7 human breast carcinoma cells results in G1 arrest and increased sensitivity to cisplatin. Oncol Rep. 2009;21:13–7.

Burk U, Schubert J, Wellner U, et al. A reciprocal repression between ZEB1 and members of the miR-200 family promotes EMT and invasion in cancer cells. EMBO Rep. 2008;9:582–9.

Calin GA, Ferracin M, Cimmino A, et al. A MicroRNA signature associated with prognosis and progression in chronic lymphocytic leukemia. N Engl J Med. 2005;353:1793–801.

Calin GA, Sevignani C, Dumitru CD, et al. Human microRNA genes are frequently located at fragile sites and genomic regions involved in cancers. Proc Natl Acad Sci USA. 2004;101: 2999–3004.

Chen X, Gong J, Zeng H, et al. MicroRNA145 targets BNIP3 and suppresses prostate cancer progression. Cancer Res. 2010a;70:2728–38.

Chen JF, Murchison EP, Tang R, et al. Targeted deletion of Dicer in the heart leads to dilated cardiomyopathy and heart failure. Proc Natl Acad Sci USA. 2008;105:2111–6.

Chen F, Zhu HH, Zhou LF, et al. Inhibition of c-FLIP expression by miR-512-3p contributes to Taxol-induced apoptosis in hepatocellular carcinoma cells. Oncol Rep. 2010b;23:1457–62.

Childs G, Fazzari M, Kung G, et al. Low-level expression of microRNAs let-7d and miR-205 are prognostic markers of head and neck squamous cell carcinoma. Am J Pathol. 2009;174:736–45.

Chiosea S, Jelezcova E, Chandran U, et al. Up-regulation of dicer, a component of the MicroRNA machinery, in prostate adenocarcinoma. Am J Pathol. 2006;169:1812–20.

Chiosea S, Jelezcova E, Chandran U, et al. Overexpression of Dicer in precursor lesions of lung adenocarcinoma. Cancer Res. 2007;67:2345–50.

Costinean S, Zanesi N, Pekarsky Y, et al. Pre-B cell proliferation and lymphoblastic leukemia/high-grade lymphoma in E(mu)-miR155 transgenic mice. Proc Natl Acad Sci USA. 2006;103: 7024–9.

Croce CM. Causes and consequences of microRNA dysregulation in cancer. Nat Rev Genet. 2009;10:704–14.

Dalgard CL, Gonzalez M, deNiro JE, et al. Differential microRNA-34a expression and tumor suppressor function in retinoblastoma cells. Invest Ophthalmol Vis Sci. 2009;50:4542–51.

Di Leva G, Croce CM. Roles of small RNAs in tumor formation. Trends Mol Med. 2010;16: 257–67.

Eitan R, Kushnir M, Lithwick-Yanai G, et al. Tumor microRNA expression patterns associated with resistance to platinum based chemotherapy and survival in ovarian cancer patients. Gynecol Oncol. 2009;114:253–9.

Elmen J, Lindow M, Schutz S, et al. LNA-mediated microRNA silencing in non-human primates. Nature. 2008;452:896–9.

Ernst A, Campos B, Meier J, et al. De-repression of CTGF via the miR-17-92 cluster upon differentiation of human glioblastoma spheroid cultures. Oncogene. 2010;29:3411–22.

Esquela-Kerscher A, Slack FJ. Oncomirs – microRNAs with a role in cancer. Nat Rev Cancer. 2006;6:259–69.

Felli N, Fontana L, Pelosi E, et al. MicroRNAs 221 and 222 inhibit normal erythropoiesis and erythroleukemic cell growth via kit receptor down-modulation. Proc Natl Acad Sci USA. 2005;102:18081–6.

Ferracin M, Zagatti B, Rizzotto L, et al. MicroRNAs involvement in fludarabine refractory chronic lymphocytic leukemia. Mol Cancer. 2010;9:123.

Flamant S, Ritchie W, Guilhot J, et al. Micro-RNA response to imatinib mesylate in patients with chronic myeloid leukemia. Haematologica. 2010;95:1325–33.

Foekens JA, Sieuwerts AM, Smid M, et al. Four miRNAs associated with aggressiveness of lymph node-negative, estrogen receptor-positive human breast cancer. Proc Natl Acad Sci USA. 2008;105:13021–6.

Fojo T. Multiple paths to a drug resistance phenotype: mutations, translocations, deletions and amplification of coding genes or promoter regions, epigenetic changes and microRNAs. Drug Resist Updat. 2007;10:59–67.

Fornari F, Gramantieri L, Giovannini C, et al. MiR-122/cyclin G1 interaction modulates p53 activity and affects doxorubicin sensitivity of human hepatocarcinoma cells. Cancer Res. 2009;69:5761–7.

Fornari F, Milazzo M, Chieco P, et al. MiR-199a-3p regulates mTOR and c-Met to influence the doxorubicin sensitivity of human hepatocarcinoma cells. Cancer Res. 2010;70:5184–93.

Friedman RC, Farh KK, Burge CB, et al. Most mammalian mRNAs are conserved targets of microRNAs. Genome Res. 2009;19:92–105.

Fujita Y, Kojima K, Hamada N, et al. Effects of miR-34a on cell growth and chemoresistance in prostate cancer PC3 cells. Biochem Biophys Res Commun. 2008;377:114–9.

Fujita Y, Kojima K, Ohhashi R, et al. MiR-148a attenuates paclitaxel-resistance of hormone-refractory, drug-resistant prostate cancer PC3 cells by regulating MSK1 expression. J Biol Chem. 2010;285:19076–84.

Gal H, Pandi G, Kanner AA, et al. MIR-451 and Imatinib mesylate inhibit tumor growth of Glioblastoma stem cells. Biochem Biophys Res Commun. 2008;376:86–90.
Galluzzi L, Morselli E, Vitale I, et al. MiR-181a and miR-630 regulate cisplatin-induced cancer cell death. Cancer Res. 2010;70:1793–803.
Garofalo M, Di Leva G, Romano G, et al. MiR-221&222 regulate TRAIL resistance and enhance tumorigenicity through PTEN and TIMP3 downregulation. Cancer Cell. 2009;16:498–509.
Garofalo M, Quintavalle C, Di Leva G, et al. MicroRNA signatures of TRAIL resistance in human non-small cell lung cancer. Oncogene. 2008;27:3845–55.
Garzon R, Calin GA, Croce CM. MicroRNAs in cancer. Annu Rev Med. 2009;60:167–79.
Gefen N, Binder V, Zaliova M, et al. Hsa-miR-125b-2 is highly expressed in childhood ETV6/RUNX1 (TEL/AML1) leukemias and confers survival advantage to growth inhibitory signals independent of p53. Leukemia. 2010;24:89–96.
Gibbons DL, Lin W, Creighton CJ, et al. Contextual extracellular cues promote tumor cell EMT and metastasis by regulating miR-200 family expression. Genes Dev. 2009;23:2140–51.
Giovannetti E, Funel N, Peters GJ, et al. MicroRNA-21 in pancreatic cancer: correlation with clinical outcome and pharmacologic aspects underlying its role in the modulation of gemcitabine activity. Cancer Res. 2010;70:4528–38.
Gottesman MM, Fojo T, Bates SE. Multidrug resistance in cancer: role of ATP-dependent transporters. Nat Rev Cancer. 2002;2:48–58.
Gregory PA, Bert AG, Paterson EL, et al. The miR-200 family and miR-205 regulate epithelial to mesenchymal transition by targeting ZEB1 and SIP1. Nat Cell Biol. 2008a;10:593–601.
Gregory PA, Bracken CP, Bert AG, et al. MicroRNAs as regulators of epithelial-mesenchymal transition. Cell Cycle. 2008b;7:3112–8.
Guo L, Liu Y, Bai Y, et al. Gene expression profiling of drug-resistant small cell lung cancer cells by combining microRNA and cDNA expression analysis. Eur J Cancer. 2010;46:1692–702.
He L, Thomson JM, Hemann MT, et al. A microRNA polycistron as a potential human oncogene. Nature. 2005;435:828–33.
Hebert C, Norris K, Scheper MA, et al. High mobility group A2 is a target for miRNA-98 in head and neck squamous cell carcinoma. Mol Cancer. 2007;6:5.
Hong L, Han Y, Zhang H, et al. The prognostic and chemotherapeutic value of miR-296 in esophageal squamous cell carcinoma. Ann Surg. 2010;251:1056–63.
Horie T, Ono K, Nishi H, et al. Acute doxorubicin cardiotoxicity is associated with miR-146a-induced inhibition of the neuregulin-ErbB pathway. Cardiovasc Res. 2010;87:656–64.
Hosono S, Kajiyama H, Terauchi M, et al. Expression of Twist increases the risk for recurrence and for poor survival in epithelial ovarian carcinoma patients. Br J Cancer. 2007;96:314–20.
Hu W, Chan CS, Wu R, et al. Negative regulation of tumor suppressor p53 by microRNA miR-504. Mol Cell. 2010;38:689–99.
Incoronato M, Garofalo M, Urso L, et al. MiR-212 increases tumor necrosis factor-related apoptosis-inducing ligand sensitivity in non-small cell lung cancer by targeting the antiapoptotic protein PED. Cancer Res. 2010;70:3638–46.
Iorio MV, Casalini P, Piovan C, et al. MicroRNA-205 regulates HER3 in human breast cancer. Cancer Res. 2009;69:2195–200.
Iorio MV, Croce CM. MicroRNAs in cancer: small molecules with a huge impact. J Clin Oncol. 2009;27:5848–56.
Ivanovska I, Ball AS, Diaz RL, et al. MicroRNAs in the miR-106b family regulate p21/CDKN1A and promote cell cycle progression. Mol Cell Biol. 2008;28:2167–74.
Iwatsuki M, Mimori K, Yokobori T, et al. Epithelial-mesenchymal transition in cancer development and its clinical significance. Cancer Sci. 2010;101:293–9.
Ji Q, Hao X, Meng Y, et al. Restoration of tumor suppressor miR-34 inhibits human p53-mutant gastric cancer tumorspheres. BMC Cancer. 2008;8:266.
Jindra PT, Bagley J, Godwin JG, et al. Costimulation-dependent expression of microRNA-214 increases the ability of T cells to proliferate by targeting PTEN. J Immunol. 2010;185: 990–7.

Kahlem P, Dorken B, Schmitt CA. Cellular senescence in cancer treatment: friend or foe? J Clin Invest. 2004;113:169–74.

Kong W, He L, Coppola M, et al. MicroRNA-155 regulates cell survival, growth, and chemosensitivity by targeting FOXO3a in breast cancer. J Biol Chem. 2010;285:17869–79.

Korpal M, Lee ES, Hu G, et al. The miR-200 family inhibits epithelial-mesenchymal transition and cancer cell migration by direct targeting of E-cadherin transcriptional repressors ZEB1 and ZEB2. J Biol Chem. 2008;283:14910–14.

Kota J, Chivukula RR, O'Donnell KA, et al. Therapeutic microRNA delivery suppresses tumorigenesis in a murine liver cancer model. Cell. 2009;137:1005–17.

Kotani A, Ha D, Hsieh J, et al. MiR-128b is a potent glucocorticoid sensitizer in MLL-AF4 acute lymphocytic leukemia cells and exerts cooperative effects with miR-221. Blood. 2009;114:4169–78.

Kotani A, Ha D, Schotte D, et al. A novel mutation in the miR-128b gene reduces miRNA processing and leads to glucocorticoid resistance of MLL-AF4 acute lymphocytic leukemia cells. Cell Cycle. 2010;9:1037–42.

Kovalchuk O, Filkowski J, Meservy J, et al. Involvement of microRNA-451 in resistance of the MCF-7 breast cancer cells to chemotherapeutic drug doxorubicin. Mol Cancer Ther. 2008;7:2152–9.

Krutzfeldt J, Rajewsky N, Braich R, et al. Silencing of microRNAs in vivo with 'antagomirs'. Nature. 2005;438:685–9.

Kumar MS, Erkeland SJ, Pester RE, et al. Suppression of non-small cell lung tumor development by the let-7 microRNA family. Proc Natl Acad Sci USA. 2008;105:3903–8.

Lal A, Pan Y, Navarro F, et al. MiR-24-mediated downregulation of H2AX suppresses DNA repair in terminally differentiated blood cells. Nat Struct Mol Biol. 2009;16:492–8.

Lanford RE, Hildebrandt-Eriksen ES, Petri A, et al. Therapeutic silencing of microRNA-122 in primates with chronic hepatitis C virus infection. Science. 2010;327:198–201.

le Sage C, Nagel R, Egan DA, et al. Regulation of the p27(Kip1) tumor suppressor by miR-221 and miR-222 promotes cancer cell proliferation. EMBO J. 2007;26:3699–708.

Lebanony D, Benjamin H, Gilad S, et al. Diagnostic assay based on hsa-miR-205 expression distinguishes squamous from nonsquamous non-small-cell lung carcinoma. J Clin Oncol. 2009;27:2030–7.

Lee YB, Bantounas I, Lee DY, et al. Twist-1 regulates the miR-199a/214 cluster during development. Nucleic Acids Res. 2009;37:123–8.

Lee JW, Choi CH, Choi JJ, et al. Altered MicroRNA expression in cervical carcinomas. Clin Cancer Res. 2008;14:2535–42.

Li Y, Li W, Yang Y, et al. MicroRNA-21 targets LRRFIP1 and contributes to VM-26 resistance in glioblastoma multiforme. Brain Res. 2009a;1286:13–8.

Li Y, VandenBoom TG 2nd, Kong D, et al. Up-regulation of miR-200 and let-7 by natural agents leads to the reversal of epithelial-to-mesenchymal transition in gemcitabine-resistant pancreatic cancer cells. Cancer Res. 2009b;69:6704–12.

Li J, Wood WH 3rd, Becker KG, et al. Gene expression response to cisplatin treatment in drug-sensitive and drug-resistant ovarian cancer cells. Oncogene. 2007;26:2860–72.

Li Y, Zhu X, Gu J, et al. Anti-miR-21 oligonucleotide sensitizes leukemic K562 cells to arsenic trioxide by inducing apoptosis. Cancer Sci. 2010;101:948–54.

Liang Z, Wu H, Xia J, et al. Involvement of miR-326 in chemotherapy resistance of breast cancer through modulating expression of multidrug resistance-associated protein 1. Biochem Pharmacol. 2010;79:817–24.

Liao R, Sun J, Zhang L, et al. MicroRNAs play a role in the development of human hematopoietic stem cells. J Cell Biochem. 2008;104:805–17.

Lu J, Getz G, Miska EA, et al. MicroRNA expression profiles classify human cancers. Nature. 2005;435:834–8.

Maes OC, Chertkow HM, Wang E, et al. MicroRNA: implications for alzheimer disease and other human CNS disorders. Curr Genomics. 2009;10:154–68.

Mani SA, Guo W, Liao MJ, et al. The epithelial-mesenchymal transition generates cells with properties of stem cells. Cell. 2008;133:704–15.

Martin SA, Lord CJ, Ashworth A. DNA repair deficiency as a therapeutic target in cancer. Curr Opin Genet Dev. 2008;18:80–86.

Masri S, Liu Z, Phung S, et al. The role of microRNA-128a in regulating TGFbeta signaling in letrozole-resistant breast cancer cells. Breast Cancer Res Treat. 2010;124:89–99.

Meng F, Henson R, Lang M, et al. Involvement of human micro-RNA in growth and response to chemotherapy in human cholangiocarcinoma cell lines. Gastroenterology. 2006;130: 2113–29.

Merritt WM, Lin YG, Han LY, et al. Dicer, Drosha, and outcomes in patients with ovarian cancer. N Engl J Med. 2008;359:2641–50.

Miller TE, Ghoshal K, Ramaswamy B, et al. MicroRNA-221/222 confers tamoxifen resistance in breast cancer by targeting p27Kip1. J Biol Chem. 2008;283:29897–903.

Mishra PJ, Humeniuk R, Mishra PJ, et al. A miR-24 microRNA binding-site polymorphism in dihydrofolate reductase gene leads to methotrexate resistance. Proc Natl Acad Sci USA. 2007;104:13513–18.

Mishra PJ, Song B, Mishra PJ, et al. MiR-24 tumor suppressor activity is regulated independent of p53 and through a target site polymorphism. PLoS One. 2009;4:e8445.

Nagaraja AK, Creighton CJ, Yu Z, et al. A link between miR-100 and FRAP1/mTOR in clear cell ovarian cancer. Mol Endocrinol. 2010;24:447–63.

Nasser MW, Datta J, Nuovo G, et al. Down-regulation of micro-RNA-1 (miR-1) in lung cancer. Suppression of tumorigenic property of lung cancer cells and their sensitization to doxorubicin-induced apoptosis by miR-1. J Biol Chem. 2008;283:33394–405.

Pan YZ, Morris ME, Yu AM, MicroRN A-. 328 negatively regulates the expression of breast cancer resistance protein (BCRP/ABCG2) in human cancer cells. Mol Pharmacol. 2009;75:1374–9.

Park SM, Gaur AB, Lengyel E, et al. The miR-200 family determines the epithelial phenotype of cancer cells by targeting the E-cadherin repressors ZEB1 and ZEB2. Genes Dev. 2008;22: 894–907.

Petrocca F, Visone R, Onelli MR, et al. E2F1-regulated microRNAs impair TGFbeta-dependent cell-cycle arrest and apoptosis in gastric cancer. Cancer Cell. 2008;13:272–86.

Pogribny IP, Filkowski JN, Tryndyak VP, et al. Alterations of microRNAs and their targets are associated with acquired resistance of MCF-7 breast cancer cells to cisplatin. Int J Cancer. 2010;127:1785–94.

Pothof J, Verkaik NS, van IW, et al. MicroRNA-mediated gene silencing modulates the UV-induced DNA-damage response. EMBO J. 2009;28:2090–9.

Qin W, Shi Y, Zhao B, et al. MiR-24 regulates apoptosis by targeting the open reading frame (ORF) region of FAF1 in cancer cells. PLoS One. 2010;5:e9429.

Rodriguez-Nieto S, Zhivotovsky B. Role of alterations in the apoptotic machinery in sensitivity of cancer cells to treatment. Curr Pharm Des. 2006;12:4411–25.

Rokhlin OW, Scheinker VS, Taghiyev AF, et al. MicroRNA-34 mediates AR-dependent p53-induced apoptosis in prostate cancer. Cancer Biol Ther. 2008;7:1288–96.

Roldo C, Missiaglia E, Hagan JP, et al. MicroRNA expression abnormalities in pancreatic endocrine and acinar tumors are associated with distinctive pathologic features and clinical behavior. J Clin Oncol. 2006;24:4677–84.

Rosenfeld N, Aharonov R, Meiri E, et al. MicroRNAs accurately identify cancer tissue origin. Nat Biotechnol. 2008;26:462–9.

Rossi L, Bonmassar E, Faraoni I. Modification of miR gene expression pattern in human colon cancer cells following exposure to 5-fluorouracil in vitro. Pharmacol Res. 2007;56:248–53.

Rui W, Bing F, Hai-Zhu S, et al. Identification of microRNA profiles in docetaxel-resistant human non-small cell lung carcinoma cells (SPC-A1). J Cell Mol Med. 2010;14:206–14.

Sabbah M, Emami S, Redeuilh G, et al. Molecular signature and therapeutic perspective of the epithelial-to-mesenchymal transitions in epithelial cancers. Drug Resist Updat. 2008;11: 123–51.

San Jose-Eneriz E, Roman-Gomez J, Jimenez-Velasco A, et al. MicroRNA expression profiling in imatinib-resistant chronic myeloid leukemia patients without clinically significant ABL1-mutations. Mol Cancer. 2009;8:69.

Sand M, Gambichler T, Skrygan M, et al. Expression levels of the microRNA processing enzymes Drosha and dicer in epithelial skin cancer. Cancer Invest. 2010;28:649–53.

Schetter AJ, Leung SY, Sohn JJ, et al. MicroRNA expression profiles associated with prognosis and therapeutic outcome in colon adenocarcinoma. JAMA. 2008;299:425–36.

Schratt G. MicroRNAs at the synapse. Nat Rev Neurosci. 2009;10:842–9.

Seike M, Goto A, Okano T, et al. MiR-21 is an EGFR-regulated anti-apoptotic factor in lung cancer in never-smokers. Proc Natl Acad Sci USA. 2009;106:12085–90.

Shibata K, Kajiyama H, Ino K, et al. Twist expression in patients with cervical cancer is associated with poor disease outcome. Ann Oncol. 2008;19:81–5.

Shimizu S, Takehara T, Hikita H, et al. The let-7 family of microRNAs inhibits Bcl-xL expression and potentiates sorafenib-induced apoptosis in human hepatocellular carcinoma. J Hepatol. 2010;52:698–704.

Song B, Wang Y, Titmus MA, et al. Molecular mechanism of chemoresistance by miR-215 in osteosarcoma and colon cancer cells. Mol Cancer. 2010;9:96.

Song B, Wang Y, Xi Y, et al. Mechanism of chemoresistance mediated by miR-140 in human osteosarcoma and colon cancer cells. Oncogene. 2009;28:4065–74.

Sorrentino A, Liu CG, Addario A, et al. Role of microRNAs in drug-resistant ovarian cancer cells. Gynecol Oncol. 2008;111:478–86.

Stegmeier F, Warmuth M, Sellers WR, et al. Targeted cancer therapies in the twenty-first century: lessons from imatinib. Clin Pharmacol Ther. 2010;87:543–52.

Su H, Yang JR, Xu T, et al. MicroRNA-101, down-regulated in hepatocellular carcinoma, promotes apoptosis and suppresses tumorigenicity. Cancer Res. 2009;69:1135–42.

Sun M, Estrov Z, Ji Y, et al. Curcumin (diferuloylmethane) alters the expression profiles of microRNAs in human pancreatic cancer cells. Mol Cancer Ther. 2008;7:464–73.

Svoboda M, Izakovicova Holla L, Sefr R, et al. Micro-RNAs miR125b and miR137 are frequently upregulated in response to capecitabine chemoradiotherapy of rectal cancer. Int J Oncol. 2008;33:541–7.

Tazawa H, Tsuchiya N, Izumiya M, et al. Tumor-suppressive miR-34a induces senescence-like growth arrest through modulation of the E2F pathway in human colon cancer cells. Proc Natl Acad Sci USA. 2007;104:15472–7.

Thiery JP. Epithelial-mesenchymal transitions in development and pathologies. Curr Opin Cell Biol. 2003;15:740–6.

Thiery JP, Sleeman JP. Complex networks orchestrate epithelial-mesenchymal transitions. Nat Rev Mol Cell Biol. 2006;7:131–42.

Thum T, Catalucci D, Bauersachs J. MicroRNAs: novel regulators in cardiac development and disease. Cardiovasc Res. 2008;79:562–70.

To KK, Robey RW, Knutsen T, et al. Escape from hsa-miR-519c enables drug-resistant cells to maintain high expression of ABCG2. Mol Cancer Ther. 2009;8:2959–68.

To KK, Zhan Z, Litman T, et al. Regulation of ABCG2 expression at the 3′ untranslated region of its mRNA through modulation of transcript stability and protein translation by a putative microRNA in the S1 colon cancer cell line. Mol Cell Biol. 2008;28:5147–61.

Tryndyak VP, Beland FA, Pogribny IP. E-cadherin transcriptional down-regulation by epigenetic and microRNA-200 family alterations is related to mesenchymal and drug-resistant phenotypes in human breast cancer cells. Int J Cancer. 2010;126:2575–83.

Tsang WP, Kwok TT. Let-7a microRNA suppresses therapeutics-induced cancer cell death by targeting caspase-3. Apoptosis. 2008;13:1215–22.

Tsang WP, Kwok TT. Epigallocatechin gallate up-regulation of miR-16 and induction of apoptosis in human cancer cells. J Nutr Biochem. 2010;21:140–6.

Tsukamoto Y, Nakada C, Noguchi T, et al. MicroRNA-375 is downregulated in gastric carcinomas and regulates cell survival by targeting PDK1 and 14-3-3zeta. Cancer Res. 2010;70:2339–49.

Ujifuku K, Mitsutake N, Takakura S, et al. MiR-195, miR-455-3p and miR-10a(*) are implicated in acquired temozolomide resistance in glioblastoma multiforme cells. Cancer Lett. 2010;296:241–8.

Veronese A, Lupini L, Consiglio J, et al. Oncogenic role of miR-483-3p at the IGF2/483 locus. Cancer Res. 2010;70:3140–9.

Volinia S, Calin GA, Liu CG, et al. A microRNA expression signature of human solid tumors defines cancer gene targets. Proc Natl Acad Sci USA. 2006;103:2257–61.

Voulgari A, Pintzas A. Epithelial-mesenchymal transition in cancer metastasis: mechanisms, markers and strategies to overcome drug resistance in the clinic. Biochim Biophys Acta. 2009;1796:75–90.

Wang PY, Li YJ, Zhang S, et al. Regulating A549 cells growth by ASO inhibiting miRNA expression. Mol Cell Biochem. 2010;339:163–71.

Wang X, Ling MT, Guan XY, et al. Identification of a novel function of TWIST, a bHLH protein, in the development of acquired taxol resistance in human cancer cells. Oncogene. 2004;23: 474–82.

Weiss GJ, Bemis LT, Nakajima E, et al. EGFR regulation by microRNA in lung cancer: correlation with clinical response and survival to gefitinib and EGFR expression in cell lines. Ann Oncol. 2008;19:1053–9.

Wellner U, Schubert J, Burk UC, et al. The EMT-activator ZEB1 promotes tumorigenicity by repressing stemness-inhibiting microRNAs. Nat Cell Biol. 2009;11:1487–95.

Wiemer EA. The role of microRNAs in cancer: no small matter. Eur J Cancer. 2007;43:1529–44.

Williams AE. Functional aspects of animal microRNAs. Cell Mol Life Sci. 2008;65:545–62.

Wu JF, Shen W, Liu NZ, et al. Down-regulation of Dicer in hepatocellular carcinoma. Med Oncol. 2010. doi:10.1007/s12032-010-9520–5.

Xia L, Zhang D, Du R, et al. MiR-15b and miR-16 modulate multidrug resistance by targeting BCL2 in human gastric cancer cells. Int J Cancer. 2008;123:372–9.

Xin F, Li M, Balch C, et al. Computational analysis of microRNA profiles and their target genes suggests significant involvement in breast cancer antiestrogen resistance. Bioinformatics. 2009;25:430–4.

Yanaihara N, Caplen N, Bowman E, et al. Unique microRNA molecular profiles in lung cancer diagnosis and prognosis. Cancer Cell. 2006;9:189–98.

Yang N, Kaur S, Volinia S, et al. MicroRNA microarray identifies Let-7i as a novel biomarker and therapeutic target in human epithelial ovarian cancer. Cancer Res. 2008a;68:10307–14.

Yang H, Kong W, He L, et al. MicroRNA expression profiling in human ovarian cancer: miR-214 induces cell survival and cisplatin resistance by targeting PTEN. Cancer Res. 2008b;68:425–33.

Yin G, Chen R, Alvero AB, et al. TWISTing stemness, inflammation and proliferation of epithelial ovarian cancer cells through MIR199A2/214. Oncogene. 2010;29:3545–53.

Yu F, Deng H, Yao H, et al. MiR-30 reduction maintains self-renewal and inhibits apoptosis in breast tumor-initiating cells. Oncogene. 2010a;29:4194–204.

Yu F, Yao H, Zhu P, et al. let-7 regulates self renewal and tumorigenicity of breast cancer cells. Cell. 2007;131:1109–23.

Yu ZW, Zhong LP, Ji T, et al. MicroRNAs contribute to the chemoresistance of cisplatin in tongue squamous cell carcinoma lines. Oral Oncol. 2010b;46:317–22.

Zhao JJ, Lin J, Yang H, et al. MicroRNA-221/222 negatively regulates estrogen receptor alpha and is associated with tamoxifen resistance in breast cancer. J Biol Chem. 2008a;283:31079–86.

Zhao HY, Ooyama A, Yamamoto M, et al. Down regulation of c-Myc and induction of an angiogenesis inhibitor, thrombospondin-1, by 5-FU in human colon cancer KM12C cells. Cancer Lett. 2008b;270:156–63.

Zhong M, Ma X, Sun C, et al. MicroRNAs reduce tumor growth and contribute to enhance cytotoxicity induced by gefitinib in non-small cell lung cancer. Chem Biol Interact. 2010;184:431–8.

Zhou M, Liu Z, Zhao Y, et al. MicroRNA-125b confers the resistance of breast cancer cells to paclitaxel through suppression of pro-apoptotic Bcl-2 antagonist killer 1 (Bak1). J Biol Chem. 2010a;285:21496–507.

Zhou J, Zhou Y, Yin B, et al. 5-fluorouracil and oxaliplatin modify the expression profiles of microRNAs in human colon cancer cells in vitro. Oncol Rep. 2010b;23:121–8.

Zhu W, Shan X, Wang T, et al. MiR-181b modulates multidrug resistance by targeting BCL2 in human cancer cell lines. Int J Cancer. 2010;127:2520–9.

Zhu H, Wu H, Liu X, et al. Role of microRNA miR-27a and miR-451 in the regulation of MDR1/P-glycoprotein expression in human cancer cells. Biochem Pharmacol. 2008;76:582–8.

Chapter 20
MicroRNAs and Cancer Metabolism

Ping Gao

Abstract Over 80 years ago, Otto Warburg discovered the phenomena of aerobic glycolysis, known now as the Warburg effect, whereby cancer cells relies on glycolysis, rather than the more efficient process of oxidative phosphorylation, for ATP production, even when oxygen is available. Recent progress in the field of cancer metabolism documents that alterations of many oncogenes, such as *MYC*, or tumor suppressor genes are involved in the metabolic switch of cancer cells to aerobic glycolysis. Myc was the first oncogenic transcription factor reported to regulate microRNAs (miRNAs), which are non-coding RNAs of ~22-nucleotides that act as post-transcriptional regulators of gene expression. Hundreds of miRNAs have been identified in the human genome thus far that are estimated to regulate about 30% of human genes involved in diverse processes including development, cell differentiation, cell proliferation, apoptosis, and metabolism. In this chapter, I will review the characteristics of cancer metabolism and its molecular basis, with an emphasis on the interplay of miRNAs and oncogenes/tumor suppressors in the field. It is hoped that understanding of molecular basis of cancer metabolism, including the involvement of miRNAs, will provide new and powerful strategies to target this deadly disease.

20.1 Introduction

Cancer metabolism is the intersection of two fields of research, cancer and metabolism. In the past decade, molecular elucidation of cancer genes and their pathways that contribute to the characteristics of cancer metabolism offers new understandings of fundamental mechanisms behind carcinogenesis, representing a new, powerful strategy to target this deadly disease. It is hoped that scientists may

P. Gao (✉)
Division of Hematology, Department of Medicine, Johns Hopkins University School of Medicine, Baltimore, MD, USA; School of Life Sciences, University of Science and Technology of China, Hefei, Anhui 230027, PR China
e-mail: pgao2@jhmi.edu; pgao2@ustc.edu.cn

W.C.S. Cho (ed.), *MicroRNAs in Cancer Translational Research*,
DOI 10.1007/978-94-007-0298-1_20,

find Achilles' heel for cancer therapy by targeting cancer metabolism (Dang et al. 2008; Yuneva 2008).

MicroRNAs (miRNAs) act as post-transcriptional regulators of gene expression by inhibiting translation, increasing mRNA degradation, or in rare instances increasing translation under stressed conditions. More than 900 miRNAs have been annotated in the human genome thus far and it is estimated that about 30% of human genes are regulated by miRNAs. By targeting multiple transcripts, miRNAs are involved in diverse processes including development, cell differentiation, cell proliferation, apoptosis and metabolism (Chang and Mendell 2009; Rana 2009). In this chapter, I will review the characteristics of cancer metabolism and its molecular basis, with an emphasis on the interplay of microRNAs and oncogenes/tumor suppressors in the field.

20.2 Aerobic Glycolysis, or Warburg Effect, Is the Central Hallmark of Cancer Metabolism

Normal tissues convert glucose to pyruvate and generate ATP and NADH via the glycolytic pathway. Pyruvate is taken up and then further metabolized by mitochondria through the Krebs or tricarboxylic acid (TCA) cycle. From a single pyruvate molecule, the TCA cycle donates eight high-energy electrons to the mitochondrial transport chain, which creates a proton gradient across the inner mitochondrial membrane, with oxygen as the terminal electron acceptor. Dissipation of the proton gradient produces ATP through oxidative phosphorylation. In the absence of oxygen, pyruvate is instead converted to lactate through a process termed as glycolysis. Only 2 ATP molecules are generated from one molecule of glucose via glycolysis, compared with 36 from oxidative phosphorylation.

Otto Warburg first discovered in 1924 the phenomena of aerobic glycolysis, whereby cancer cells, in contrast to normal cells, rely on glycolysis, rather than the more efficient process of oxidative phosphorylation, for ATP production, even when oxygen is available. Warburg (1956) believed that the metabolic disturbance he observed in cancer cells was the origin of tumorigenesis. He hypothesized that cancer developed in a two step process; first the injury or defect in mitochondrial respiration, which is then followed by an increase in fermentation, the glycolysis. Over 80 years after this seminal finding, the phenomena of aerobic glycolysis in most, but not all, cancer cells are widely accepted nowadays. A clinical correlation to the increase glucose flux to lactate is found in cancers that are visualized using 18-fluoro-2-deoxyglucose with positron emission tomography (FDG-PET). However, the mechanisms responsible for the metabolic switch are ardently debated for decades as the molecular basis of aerobic glycolysis has remained largely unexplained until recent years when landmark discoveries have been made to offer molecular insight into the mechanisms of Warburg effect (Kim and Dang 2006; Vander Heiden et al. 2009). In particular, it is being revealed that alteration in oncogenes and tumor suppressors has been critical nodes in facilitating the metabolic switch to glycolysis in cancer cells.

20.3 Genetic Alteration of Oncogenes and Tumor Suppressors and Its Interplay with MiRNAs in Cancer Metabolism

The first evidence that established the role of an oncogene in regulating cancer metabolism was documented by Shim et al. when they discovered that proto-oncogene *c-Myc* regulated the expression of lactate dehydrogenase-A (*LDHA*) gene, whose product participates in normal anaerobic glycolysis and is frequently increased in human cancers (Shim et al. 1997). Now it is clear that the alteration of many oncogenes or tumor suppressor genes is involved in the metabolic switch of cancer cells to glycolysis. Herein, this chapter will focus on several important oncogenes or tumor suppressors, whose alterations have contributed extensively to the metabolic phenotype of cancer cells, with an emphasis on the interplay of miRNAs and those oncogenes/tumor suppressors in cancer metabolism.

20.3.1 c-Myc

The *c-Myc* proto-oncogene was discovered three decades ago as the cellular homolog of the retroviral v-*Myc* gene that is sufficient to cause a variety of chicken tumors (Sheiness and Bishop 1979). *c-Myc* belongs to the Myc family of genes including *N-Myc*, and *L-Myc*. The *MYC* gene encodes a protein c-Myc, which is a transcription factor and master regulator that directs and integrates cell proliferation with cell metabolism. Many studies document that *c-Myc* is frequently altered in human cancers. Deregulated expression of the *c-Myc* oncogene occurs in ~30% of human cancers, including many commonly occurring cancers, such as colon, prostate and breast carcinomas (Cole 1986; Jamerson et al. 2004). Subsequent in vitro studies also demonstrated that forced expression of *c-Myc* in lymphoid tissues resulted in lymphoid hyperplasia and lymphomas. Moreover virtually all other studies of constitutive or inducible *c-Myc* in tissues from skin to liver also resulted in neoplastic transformation of the targeted tissue (Cory et al. 1987; Langdon et al. 1986; Lau et al.1987, Leder et al. 1986).

Early studies established *Myc* as a transcription factor through the identification of the transactivation domain and specific DNA binding domain. Many genes have been identified as direct and indirect *c-Myc* target genes thanks to high throughput approaches such as gene expression microarray studies and chromatin immunoprecipitation, which maps direct *c-Myc* genomic binding sites (Zeller et al. 2006). It is well established in the in vitro and in vivo systems that those genes have been implicated in a variety of cellular processes such as cell cycle regulation, cellular size control, proliferation, death and intriguingly cellular metabolism. Recently, reports from our laboratory and others document a role for *c-Myc* in regulating miRNAs and cell metabolism that play key roles in cancer cell biology.

Since the discovery by Shim et al. that proto-oncogene *c-Myc* up-regulated LDHA in cancer cells in 1997, *Myc* is now known to regulate the majority of the enzymes involved in glycolysis pathways (Kim et al. 2004). More recently, *Myc* is reported to regulate miRNAs that have impact in wide range of cellular events including cell metabolism. O'Donnell et al. (2005) reported that c-*Myc* directly up-regulates a pro-tumorigenic group of miRNAs known as the *miR-17-92* cluster. They further found that two miRNAs in this cluster, *miR-17-5p* and *miR-20a*, negatively regulates another transcription factor *E2F1*. These findings expand the known classes of transcripts within the *c-Myc* target gene network, and reveal a mechanism through which c-Myc simultaneously activates *E2F1* transcription and limits its translation, allowing a tightly controlled proliferative signal via miRNAs. It is interesting to note that two groups recently reported that over-expression of the *miR-17-92* cluster miRNAs led to the inhibition of tumor suppressor *PTEN*, an antagonist of PI3K/Akt/mTOR pathway which plays important roles in facilitating aerobic glycolysis in cancer cells (Olive et al. 2009; Xiao et al. 2008). Myc's effect on miRNAs is not limited to the *miR-17-92* cluster. Chang et al. subsequently demonstrated that Myc's role on miRNAs was much wider than expected and that widespread miRNA repression by Myc contributes to tumorigenesis in a B lymphoma system (Chang et al. 2008). Gao et al. further discovered that c-Myc suppression of *miR-23a/b* enhances mitochondrial glutaminase (GLS) expression and glutamine metabolism. Through proteomic analysis of mitochondria from high Myc-expressing human B lymphocytes as compared with control lymphocytes, Gao et al. identified GLS, among seven other mitochondrial proteins, as a target with > 10 fold induction by Myc. GLS converts glutamine to glutamate, which is further catabolized through the TCA cycle for the production of ATP or serves as substrate for glutathione synthesis. Further analysis revealed that unlike glutamine transporters (ASCT2 and SLC7A25), which are direct *Myc* target genes, GLS protein level was induced by Myc through direct suppression of miRNAs *miR-23a* and *miR-23b*. Northern blot confirmed that *miR-23a* and *miR-23b* are indeed suppressed by Myc and profoundly diminished by antisense *miR-23a* and *miR-23b* LNAs. Reporter assay demonstrated that *miR-23a/b* target the *GLS* mRNA 3′UTR. Importantly, the suppression of GLS expression under low-Myc conditions in P493 lymphoma and PC3 prostate cancer cells was rescued by the treatment with antisense *miR-23a* and *miR-23b* LNAs, indicating that the real mediators of Myc suppression of GLS protein expression are *miR-23a* and *miR-23b*. Since GLS is essential for Myc-mediated cell proliferation and its expression is frequently increased in human cancers, the unique means by which Myc regulates GLS via suppression of *miR-23a/b* uncovers a previously unsuspected link between oncogenes and miRNAs in modulating cancer metabolism (Gao et al. 2009). Independently, Wise et al. (2008) also documented a role for Myc in regulation glutamine metabolism.

It is a common feature for a miRNA to have multiple targets. In this regard, it is interesting that Phang et al. (2010) described that *miR-23b**, the less predominant form of the expressed precursor of the *miR-23b*, regulates tumor suppressor *POX/PRODH* in renal carcinomas. *POX/PRODH*, proline oxidase, a.k.a. proline dehydrogenase, is a mitochondrial inner-membrane enzyme catalyzing the transfer of electrons from proline, producing P5C. Phang and colleagues (2010) found that *miR-23b** potently suppresses the expression of *POX/PRODH* primarily at the level of protein translation. Not only does mimic RNA of *miR-23b* suppress *POX/PRODH* expression, its antagomir also increases *POX/PRODH* expression in tumor cells with low *POX/PRODH* expression. The binding of *miR-23b** to the 3′UTR of the *POX/PRODH* gene was shown using a luciferase assay. To make the critical correlation of this *miR-23b** to *POX/PRODH* for human cancer, they obtained frozen tumors and paired normal tissues from 16 histologically defined clear cell renal carcinomas and found that 13 of the 16 samples had markedly decreased *POX/PRODH* expression as monitored by Western analysis. Furthermore, the tumors had increased levels of *miR-23b**. Importantly, the inverse relationship between the levels of the *miR-23b** and *POX/PRODH* was statistically significant. Using in situ hybridization, they also confirmed markedly increased expression of *miR-23b** in the tumor tissues compared to normal kidney tissues. Thus, not only is *POX/PRODH* a tumor suppressor, but also the mechanism of its down-regulation has been established as due to *miR-23b**. This relationship was further examined by using antagomir and mimics for *miR-23b** to monitor *miR-23b**-mediated regulation of *POX/PRODH* and its functional effects. Using the antagomir for *miR-23b**, they observed an increased *POX/PRODH* expression in a number of colorectal cancer cells and a decrease in cell growth and augmentation of the blockade at the S/G2 checkpoint, whereas *miR-23b** mimics not only decreased the induction of *POX/PRODH* but also modulated the decrease in hypoxia-inducible factor (HIF)-1α, a recapitulation of the effect of *POX/PRODH* on levels of HIF-1α reported earlier. This was especially relevant because the HIF-1α signaling system plays a special role in renal carcinogenesis. Thus, the suppression of *POX/PRODH* tumor suppressor by *miR-23b** and its connection to HIF-1α may be of special clinical significance. Intriguingly, *miR-23* has now been linked to suppression of *PTEN* in addition to GLS and *POX/PRODH*, suggesting its diverse effect on cancer cell metabolism (Kim et al. 2010).

20.3.2 HIF-1

Adaptation of cancer cells to their microenvironment is an important driving force for the tumorigenic progression. When a tumor grows, areas with low O_2 availability (hypoxia) develops, and a major mechanism mediating adaptive responses to hypoxia is the regulation of transcription by a transcription factor HIF-1. The HIF family comprises three members, HIF-1, HIF-2, and HIF-3, with HIF-1 being ubiquitously expressed, HIF-2 only expressed in endothelial cells and in the kidney, heart, lungs, and small intestine, whereas the function of HIF-3 being poorly understood. HIF-1 is a heterodimeric protein that is composed of a constitutively expressed HIF-1β subunit and an O_2-regulated HIF-1α subunit. HIF-1α is subjected to O_2-dependent hydroxylation on proline residue 402 and/or 564 by prolyl hydroxylase domain protein 2 (PHD2) and this modification creates an interface for interaction with the von Hippel–Lindau tumor suppressor protein (VHL), which recruits an E3 ubiquitin-protein ligase that catalyzes polyubiquitination of HIF-1α, thereby targeting it for proteasomal degradation. Under hypoxic conditions, hydroxylation is inhibited and HIF-1α rapidly accumulates, dimerizes with HIF-1β, binds to the core DNA binding sequence in target genes, recruits co-activators and activates transcription (Kaelin 2005; Semenza 2003; Semenza et al. 1994).

The stabilized HIF heterodimers bind and transactivate downstream target genes. There are hundreds of genes identified as HIF-1 targets that are primarily involved in pH regulation, glucose transport, glycolysis and vasculogenesis, allowing for cellular adaptation to hypoxia. HIF-1 not only activates several genes involved in glycolysis, it also actively inhibits mitochondrial respiration. Two recent studies have demonstrated that HIF-1 induces pyruvate dehydrogenase kinase 1 (PDK1), which phosphorylates and inactivates pyruvate dehydrogenase (PDH), a mitochondrial enzyme that catalyses the conversion of pyruvate to acetyl-CoA, which is a key substrate for the Krebs cycle and the ensuing oxidative phosphorylation. Thus, the activation of PDK1 by HIF1 results in the depletion of acetyl-CoA and diminishes mitochondrial respiration, facilitating the metabolic switch to glycolysis (Kim et al. 2006; Papandreou et al. 2006). Presumably, a benefit of metabolic switch from mitochondrial respiration to glycolysis in hypoxia is the reduction of the amount of toxic reactive oxygen species (ROS) generated by inefficient respiration. In this regard, Fukuda et al. (2007) reported that HIF-1 also reduces levels of ROS by optimizing respiration efficiency through inducing cytochrome *c* oxidase subunit IV, isoform 2 (COX4I2) to replace the less efficient non-hypoxic COX4I1. These isoforms are components of cytochrome *c* oxidase, which is located in the mitochondrial inner membrane and transfers electrons to oxygen. More recently, Zhang et al. (2008) reported that mitochondrial autophagy is an HIF-1-dependent adaptive metabolic response to hypoxia. Taken together, HIF-1 is extensively involved in mitochondrial function and metabolic adaptation in cancer cells.

It has been reported by several groups that many miRNAs are induced by HIF-1 during hypoxia, in particular, *miR-210* is specifically induced by HIF-1α (Kulshreshtha et al. 2007). The cell cycle regulator E2F3, the receptor tyrosine kinase ligand ephrin A3, and the DNA repair protein RAD52 have been documented as targets for repression by *miR-210* (Crosby et al. 2009; Fasanaro et al. 2008; Giannakakis et al. 2008). Recently, two groups identified *miR-210* as a factor that may dynamically modulate mitochondrial metabolism. By analysis of multiple bioinformatic algorithms, a highly conserved binding site for *miR-210* is predicted in the 3′UTR of the transcripts of the iron-sulfur cluster assembly proteins, ISCU1 and ISCU2, two splice isoforms of *ISCU* existing in mammalian cells. ISCU1/2 facilitate the assembly of [4Fe-4S] and [2Fe-2S] iron-sulfur clusters, prosthetic groups that promote electron transport and oxidation-reduction reactions integral to numerous cellular processes, such as ribosome biogenesis, purine catabolism, heme biosynthesis, DNA repair, and iron metabolism. In particular, iron-sulfur clusters are incorporated into enzymes that are responsible for mitochondrial respiration and energy production, including aconitase, which are integral to the TCA cycle, and the mitochondrial respiratory complexes (Complexes I, II, and III), which facilitate electron transport. Depending upon the level of ambient oxygen exposure, alteration of these and other iron-sulfur dependent mitochondrial metabolism can lead to distinct downstream consequences on ROS production and cellular survival.

Using human pulmonary arterial endothelial cells (HPAEC), in which robust up-regulation of *miR-210* in response to hypoxia was reported, Chan et al. (2009) identified ISCU1/2 as direct targets for repression by *miR-210*. By down-regulating expression of ISCU1/2 during hypoxia, *miR-210* decreases the specific activity of prototypical iron-sulfur enzymes regulating mitochondrial function, mitochondrial Complex I and aconitase, leading to downstream phenotypic consequences integral to the Pasteur effect. Consequently, *miR-210* represses mitochondrial respiration and associated downstream functions. Another group independently reported similar findings in human cancer cell lines and tumors. Favaro et al. (2010) found that hypoxia-induced *miR-210* targets the mitochondrial iron sulfur scaffold protein ISCU, required for assembly of iron-sulfur clusters, cofactors for key enzymes involved in the Krebs cycle, electron transport, iron metabolism, and ROS production. ISCU suppression reduced mitochondrial complex 1 activity and aconitase activity, caused a shift to glycolysis in normoxia and enhanced cell survival. Importantly, analysis of over 900 patients with different tumor types showed that suppression of ISCU is strongly correlated with a worse prognosis. This study thus reveals a new pathway activated in hypoxic tumors, mediated by *miR-210* affecting mitochondrial metabolic activity and free radical generation and highlights the importance of mitochondrial metabolism in hypoxia biology. Taken together, induction of these major hallmarks of cancer shows that a single microRNA, *miR-210*, mediates a new mechanism of adaptation to hypoxia, by regulating mitochondrial function via iron-sulfur cluster metabolism and free radical generation.

20.3.3 LKB1/AMPK

During metabolic stress with an increase in AMP:ATP ratio, AMPK (AMP kinase) is activated to promote energy conservation and glucose uptake, allowing cells to survive when energy availability is not sufficient. LKB1 is a kinase that phosphorylates and activates AMPK in the presence of AMP when intracellular levels of ATP are low. AMPK, in turn, increases ATP production through stimulation of pathways including fatty acid oxidation (Sanders et al. 2007). LKB1 is active as a heterotrimeric complex with Ste-20-related adaptor (STRAD) and calcium-binding protein 39 (CAB39), a scaffold protein that increases markedly the activity of LKB1 (Baas et al. 2004; Boudeau et al. 2003; Hawley et al. 2003). Mutations of the *LKB1* tumor suppressor result in the rare autosomal dominant form of Peutz-Jeghers syndrome that is characterized by early age onset of gastrointestinal polyposis. AMPK also activates two other tumor suppressors, *TSC1* and *TSC2*, which are mutated in the familial syndrome tuberous sclerosis characterized by benign tumors commonly affecting the brain as well as the kidneys, heart, eyes, lungs, and skin. *TSC1* and *TSC2* form heterodimers that inhibit mTOR, the ubiquitous kinase which stimulates cell growth and protein synthesis in response to growth factors. Hence, loss of *LKB1*, *TSC1* or *TSC2*, activates the mTOR pathway. Hyperactivation of the mTOR pathway through loss of tumor suppressors also appears to increase translation of

HIF-1α mRNA and thereby stimulates the metabolic switch to glycolysis in cancer cells (Easton and Houghton 2006; Inoki et al. 2002).

Godlewski et al. (2010) showed recently that a single miRNA *miR-451* regulates LKB1/AMPK signaling and allows adaptation to metabolic stress in glioma cells. They identified *miR-451* that controls glioma cell proliferation, migration, and responsiveness to glucose deprivation. Abundant glucose allows relatively high *miR-451* expression, promoting cell growth. In low glucose, *miR-451* levels decrease, slowing proliferation but enhancing migration and survival. They suggest that this allows cells to survive metabolic stress and seek out favorable growth conditions. In glioblastoma patients, elevated *miR-451* is associated with shorter survival. The effects of *miR-451* are mediated by LKB1, which it represses through targeting its binding partner, CAB39. Over-expression of *miR-451* sensitized cells to glucose deprivation, suggesting that its down-regulation is necessary for robust activation of LKB1 in response to metabolic stress. Thus, *miR-451* is a regulator of the LKB1/AMPK pathway, and this may represent a fundamental mechanism that contributes to cellular adaptation in response to altered energy availability during glioma development. It will be interesting to know if this unique regulatory pathway mediated by *miR-451* also exists in other major human cancers.

20.3.4 P53

The *p53* tumor suppressor gene, one of the most frequently mutated genes in human cancers, encodes a transcription factor p53 that has been shown to be involved in the induction of apoptosis, cell-cycle arrest, senescence, and differentiation, responses that prevent further proliferation of stressed or damaged cells and so protect from the outgrowth of cells harboring malignant alterations. A role for p53 in the repair of DNA damage has also been described and the ability of p53 to induce reversible cell-cycle arrest may contribute to the ability of cells to repair and recover from damage before reentering a normal proliferative state. Stress-induced activation of p53 leads to the induction of expression of a large number of p53 target genes, several of which have been shown to play an important role in mediating the various responses to p53. For example, activation of expression of the cyclin-dependent kinase inhibitor p21 by p53 plays an important role in the induction of G1 cell-cycle arrest and the expression of the BH3-domain protein PUMA is an essential component of p53-activated apoptotic pathways. Several lines of evidence suggest that one mechanism by which the response to p53 can be regulated is through differential activation of different groups of p53 target genes (Oren 2003; Vogelstein et al. 2000; Vousden and Lane 2007). In this regard, it is very interesting to note that, besides its well established role in the induction of apoptosis, cell-cycle arrest, senescence, and differentiation, P53 was recently shown to directly activate genes that enhance cellular respiration, such that loss of P53 favors the conversion of glucose to lactate and cell survival (Bensaad et al. 2006; Matoba et al. 2006).

Matoba et al. (2006) recently demonstrated the direct regulation of aerobic respiration by p53. They showed that p53 modulates the balance between the utilization of respiratory and glycolytic pathways. They identify synthesis of cytochrome c oxidase 2 (SCO2) as the downstream mediator of this effect in mice and human cancer cell lines. SCO2 is critical for regulating the cytochrome c oxidase complex, the major site of oxygen utilization in the eukaryotic cell. Disruption of the *SCO2* gene in human cancer cells with wild-type p53 recapitulated the metabolic switch toward glycolysis that is exhibited by p53-deficient cells. Thus, SCO2 coupling p53 to mitochondrial respiration provides a possible explanation for the Warburg effect and offers new clues as to how p53 might affect aging and metabolism. Additionally, Bensaad et al. (2006) have identified a p53-inducible gene named *TIGAR* (*T*P53-*i*nduced glycolysis and *a*poptosis *r*egulator), whose expression lowered fructose-2,6-bisphosphate levels in cells, resulting in an inhibition of glycolysis and an overall decrease in intracellular ROS levels. These functions of TIGAR correlated with an ability to protect cells from ROS-associated apoptosis, and consequently, knockdown of endogenous TIGAR expression sensitized cells to p53-induced death. Expression of TIGAR may therefore modulate the apoptotic response to p53, allowing survival in the face of mild or transient stress signals that may be reversed or repaired. Hence, loss of the tumor suppressor *p53* would be expected to enhance glycolysis, diminish respiration, and decrease ROS-associated apoptosis of cancer cells.

Several groups documented the interplay of p53 and miRNAs in cancer cells (Hermeking 2007). While a lot of reports describing the role of p53 on miRNAs are focused on *miR34* and apoptosis pathways, evidence suggests that miRNAs might modulate a much wider range of p53-mediated effects.

When studying miRNAs influenced by p53 and radiation in HCT116 human colon carcinoma cells, Shin et al. (2009) discovered that, while many miRNAs are down-regulated in the p53-null HCT116 cells, three miRNAs, *miR-210*, *miR-500*, and *miR-516-5p*, were up-regulated. The up-regulation of *miR-210* in P53-null cells is particularly interesting since *miR-210* is also known as being up-regulated by HIF-1. As discussed above, *miR-210* has been reported by several groups to regulate cellular respiration of mammalian cells via *ISCU1/2*. As a consequence, p53 mutation might render cancer cells a more glycolytic metabolism phenotype via a potential *p53-miR210-ISCU1/2* pathway, reminiscent of early observations when oncogenic HIF-1 is activated.

p53 is also reported to regulate cMyc, an important proto-oncogene that is known to play extensive roles in cancer metabolism, by a miRNA *miR-145*. Sachdeva et al. (2009) reported that a putative tumor suppressor, *miR-145*, is expressed through the phosphoinositide 3-kinase (PI3K)/Akt and p53 pathways. They reported that p53 transcriptionally induces the expression of *miR-145* by interacting with a potential p53 response element (p53RE) in the *miR-145* promoter. They further showed that c-Myc is a direct target for *miR-145* and, while *miR-145* mimic silences the expression of c-Myc, anti-*miR-145* enhances its expression. This specific silencing of c-Myc by *miR-145* may account at least in part for the *miR-145*-mediated inhibition of tumor cell growth both in vitro and in vivo. Finally, the blockade of *miR-145* by anti-*miR-145* is able to reverse the p53-mediated c-Myc repression. Together, these results define the role of *miR-145* in the post-transcriptional regulation of c-Myc by p53 and suggest that *miR-145* provides a direct link between p53 and c-Myc in the regulation of cellular events such as proliferation and metabolism of cancer cells.

The impact of p53 on tumor suppression via miRNAs is probably much more widespread than previously expected. Suzuki et al. (2009) reported that p53 enhances the post-transcriptional maturation of several miRNAs with growth-suppressive function, including *miR-16-1, miR-143*, and *miR-145*, in response to DNA damage. In HCT116 cells and human diploid fibroblasts, p53 interacts with the Drosha processing complex through the association with DEAD-box RNA helicase p68 (also known as DDX5) and facilitates the processing of primary miRNAs to precursor miRNAs. They also found that transcriptionally inactive p53 mutants interfere with a functional assembly between Drosha complex and p68, leading to attenuation of miRNA processing activity. These findings suggest that transcription-independent modulation of miRNA biogenesis is intrinsically embedded in a tumor suppressive program governed by p53. This study reveals a previously unrecognized function of p53 in miRNA processing, which may underlie broad aspects of cancer biology including cancer metabolism.

Not only that miRNAs are regulated by P53, reports also indicated that P53 is regulated by miRNAs. In a screen for miRNAs that modulate p53 activity, Park et al. (2009) found that *miR-29* family members (*miR-29a, miR-29b*, and *miR-29c*) up-regulate p53 levels and induce apoptosis in a p53-dependent manner. They further found that *miR-29* family members directly suppress p85*a* (the regulatory subunit of PI3 kinase) and CDC42 (a Rho family GTPase), both of which negatively regulate p53. Their findings provide new insights into the role of miRNAs in the p53 pathway. Given that p53 plays essential roles in modulating cellular respiration and mitochondrial functions in cancer cells as indicated by induction of downstream targets such as SCO2 and TIGAR, one could presume that *miR-29* members or other miRNAs might modulate cancer metabolism by regulating P53.

20.4 Conclusion and Future Outlook

The involvement of miRNAs in carcinogenesis has been known for almost a decade (Cho 2010) and scientists have roughly categorized those miRNAs to oncogenic miRNAs and tumor suppressor miRNAs. As with oncogenes and tumor suppressors, which have now been linked to regulation of cell metabolism, so too are expected of the miRNAs. It stands to reason then, that there would be an extensive interplay between those miRNAs and well-known oncogenes and tumor suppressors, such as *cMyc, HIF-1, p53*, in cancer metabolism (Fig. 20.1). For years, cancer researchers have been seeking molecular mechanisms underlying the Warburg effect with the hope to exploit the characteristic cancer metabolism for therapy. In this regard, it is particularly significant to realize that small miRNAs play big roles in mediating the critical effects of those oncogenes and tumor suppressors. In the next 5–10 years, we will witness the identification of more nodes in the ever expanding network of cancer metabolism, oncogenes and tumor suppressors, and hopefully will benefit from these findings.

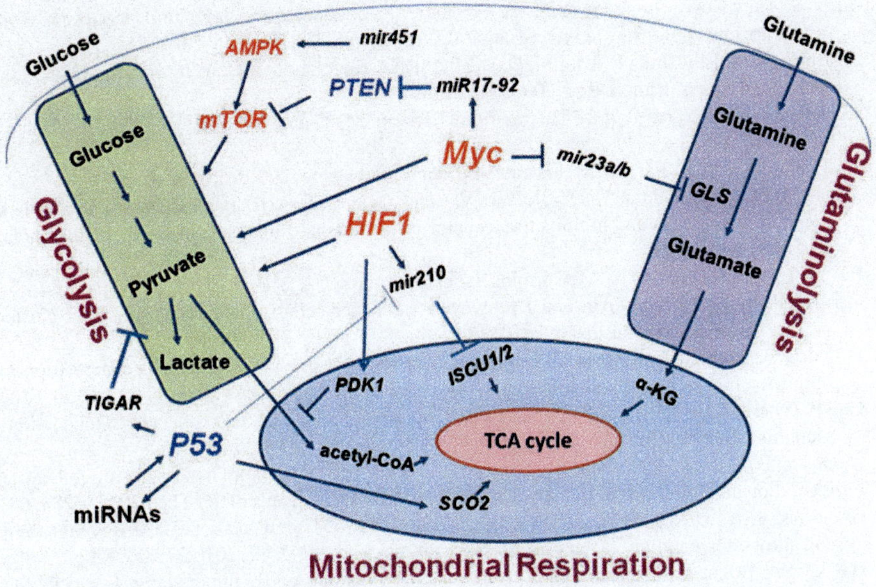

Fig. 20.1 The interplay between microRNAs and oncogenes/tumor suppressors in cancer metabolism. The links of microRNAs and key oncogenes (*cMyc, HIF-1, mTOR, AMPK*, etc, indicated in *red*) and tumor suppressors (*P53* and *PTEN*, indicated in *blue*) to glycolysis, mitochondrial respiration and glutaminolysis pathways of cancer cells are illustrated

Acknowledgments I thank Dr. Chi Dang for his critical suggestions. Our original work is supported by NIH, Leukemia and Lymphoma Society, and Maryland Stem Cell Research Foundation grants.

References

Baas AF, Kuipers J, van der Wel NN, et al. Complete polarization of single intestinal epithelial cells upon activation of LKB1 by STRAD. Cell. 2004;116:457–66.

Bensaad K, Tsuruta A, Selak MA, et al. TIGAR, a p53-inducible regulator of glycolysis and apoptosis. Cell. 2006;126:107–20.

Boudeau J, Baas AF, Deak M, et al. MO25alpha/beta interact with STRADalpha/beta enhancing their ability to bind, activate and localize LKB1 in the cytoplasm. EMBO J. 2003;22:5102–14.

Chan SY, Zhang YY, Hemann C, et al. MicroRNA-210 controls mitochondrial metabolism during hypoxia by repressing the iron-sulfur cluster assembly proteins ISCU1/2. Cell Metab. 2009;10:273–84.

Chang TC, Mendell JT. MicroRNAs in vertebrate physiology and human disease. Annu Rev Genomics Hum Genet. 2009;8:215–39.

Chang TC, Yu D, Lee YS, et al. Widespread microRNA repression by Myc contributes to tumorigenesis. Nat Genet. 2008;40:43–50.

Cho WC. MicroRNAs in cancer – from research to therapy. Biochim Biophys Acta. 2010;1805:209–17.

Cole MD. The myc oncogene: its role in transformation and differentiation. Annu Rev Genet. 1986;20:361–84.

Cory S, Harris AW, Langdon WY, et al. The myc oncogene and lymphoid neoplasia: from translocations to transgenic mice. Hamatol Bluttransfus. 1987;31:248–51.

Crosby ME, Kulshreshtha R, Ivan M, et al. MicroRNA Regulation of DNA repair gene expression in hypoxic stress. Cancer Res. 2009;69:1221–9.

Dang CV, Kim JW, Gao P, et al. The interplay between MYC and HIF in cancer. Nat Rev Cancer. 2008;8:51–6.

Easton JB, Houghton PJ. mTOR and cancer therapy. Oncogene. 2006;25:6436–46.

Fasanaro P, D'Alessandra Y, Di Stefano V, et al. MicroRNA-210 modulates endothelial cell response to hypoxia and inhibits the receptor tyrosine kinase ligand Ephrin-A3. J Biol Chem. 2008;283:15878–83.

Favaro E, Ramachandran A, McCormick R, et al. MicroRNA-210 regulates mitochondrial free radical response to hypoxia and Krebs cycle in cancer cells by targeting iron sulfur cluster protein ISCU. PLoS One. 2010;5:e10345.

Fukuda R, Zhang H, Jim JW, et al. HIF-1 regulates cytochrome oxidase subunit composition to optimize efficiency of respiration in hypoxic cells. Cell. 2007;129:111–22.

Gao P, Tchernyshyov I, Chang TC, et al. c-Myc suppression of miR-23a/b enhances mitochondrial glutaminase expression and glutamine metabolism. Nature. 2009;458:762–5.

Giannakakis A, Sandaltzopoulos R, Greshock J, et al. MiR-210 links hypoxia with cell cycle regulation and is deleted in human epithelial ovarian cancer. Cancer Biol Ther. 2008;7:255–64.

Godlewski J, Nowicki MO, Bronisz A, et al. MicroRNA-451 regulates LKB1/AMPK signaling and allows adaptation to metabolic stress in glioma cells. Mol Cell. 2010;37:620–32.

Hawley SA, Boudeau J, Reid JL, et al. Complexes between the LKB1 tumor suppressor, STRAD alpha/beta and MO25 alpha/beta are upstream kinases in the AMP-activated protein kinase cascade. J Biol. 2003;2:28.

Hermeking H. p53 enters the MicroRNA world. Cancer Cell. 2007;12:414–8.

Inoki K, Li Y, Zhu T, et al. TSC2 is phosphorylated and inhibited by Akt and suppresses mTOR signalling. Nat Cell Biol. 2002;4:648–57.

Jamerson MH, Johnson MD, Dickson RB. Of mice and Myc: c-Myc and mammary tumorigenesis. J Mammary Gland Biol Neoplasia. 2004;9:27–37.

Kaelin WG Jr.. ROS: really involved in oxygen sensing. Cell Metab. 2005;1:357–8.

Kim JW, Dang CV. Cancer's molecular sweet tooth and the Warburg effect. Cancer Res. 2006;66:8927–30.

Kim JW, Mori S, Nevins JR. Myc-induced microRNAs integrate Myc-mediated cell proliferation and cell fate. Cancer Res. 2010;70:4820–8.

Kim JW, Tchernyshyov I, Semenza GL, et al. HIF-1-mediated expression of pyruvate dehydrogenase kinase: a metabolic switch required for cellular adaptation to hypoxia. Cell Metab. 2006;3:177–85.

Kim JW, Zeller KI, Wang Y, et al. Evaluation of Myc E-box phylogenetic footprints in glycolytic genes by chromatin immunoprecipitation assays. Mol Cell Biol. 2004;24:5923–36.

Kulshreshtha R, Ferracin M, Wojcik S, et al. A microRNA signature of hypoxia. Mol Cell Biol. 2007;27:1859–67.

Langdon WY, Harris AW, Cory S, et al. The c-myc oncogene perturbs B lymphocyte development in E-mu-myc transgenic mice. Cell. 1986;47:11–18.

Lau LF, Nathans D. Expression of a set of growth-related immediate early genes in BALB/c 3T3 cells: coordinate regulation with c-fos or c-myc. Proc Natl Acad Sci USA. 1987;84:1182–6.

Leder A, Pattengale PK, Kuo A, et al. Consequences of widespread deregulation of the c-myc gene in transgenic mice: multiple neoplasms and normal development. Cell. 1986;45:485–95.

Matoba S, Kang JG, Patino WD, et al. P53 regulates mitochondrial respiration. Science. 2006;312:1650–3.

O'Donnell KA, Wentzel EA, Zeller KI, et al. c-Myc-regulated microRNAs modulate E2F1 expression. Nature. 2005;435:839–43.

Olive V, Bennett MJ, Walker JC, et al. MiR-19 is a key oncogenic component of miR-17-92. Genes Dev. 2009;23:2839–49.

Oren M. Decision making by p53: life, death and cancer. Cell Death Differ. 2003;10:431–42.
Papandreou I, Cairns RA, Fontana L, et al. HIF-1 mediates adaptation to hypoxia by actively downregulating mitochondrial oxygen consumption. Cell Metab. 2006;3:187–97.
Park SY, Lee JH, Ha M, et al. MiR-29 miRNAs activate p53 by targeting p85 alpha and CDC42. Nat Struct Mol Biol. 2009;16:23–9.
Phang JM, Liu W, Zabirnyk O. Proline metabolism and microenvironmental stress. Annu Rev Nutr. 2010;30:441–63.
Rana TM. Illuminating the silence: understanding the structure and function of small RNAs. Nat Rev Mol Cell Biol. 2009;8:23–36.
Sachdeva M, Zhu S, Wu F, et al. P53 represses c-Myc through induction of the tumor suppressor miR-145. Proc Natl Acad Sci USA. 2009;106:3207–12.
Sanders MJ, Ali ZS, Hegarty BD, et al. Defining the mechanism of activation of AMP-activated protein kinase by the small molecule A-769662, a member of the thienopyridone family. J Biol Chem. 2007;282:32539–48.
Semenza GL. Targeting HIF-1 for cancer therapy. Nature Rev Cancer. 2003;3:721–32.
Semenza GL, Roth PH, Fang HM, et al. Transcriptional regulation of genes encoding glycolytic enzymes by hypoxia-inducible factor 1. J Biol Chem. 1994;269:23757–63.
Sheiness D, Bishop JM. DNA and RNA from uninfected vertebrate cells contain nucleotide sequences related to the putative transforming gene of avian myelocytomatosis virus. J Virol. 1979;31:514–21.
Shim H, Dolde C, Lewis BC, et al. C-Myc transactivation of LDH-A: implications for tumor metabolism and growth. Proc Natl Acad Sci USA. 1997;94:6658–63.
Shin S, Cha HJ, Lee EM, et al. MicroRNAs are significantly influenced by p53 and radiation in HCT116 human colon carcinoma. Int J Oncol. 2009;34:1645–52.
Suzuki HI, Yamagata K, Sugimoto K, et al. Modulation of microRNA processing by p53. Nature. 2009;460:529–33.
Vander Heiden MG, Cantley LC, Thompson CB. Understanding the Warburg effct: the metabolic requirements of cell proliferation. Science. 2009;324:1029–33.
Vogelstein B, Lane D, Levine AJ. Surfing the p53 network. Nature. 2000;408:307–10.
Vousden KH, Lane DP. p53 in health and disease. Nat Rev Mol Cell Biol. 2007;8:275–83.
Warburg O. On respiratory impairment in cancer cells. Science. 1956;124:269–70.
Wise DR, DeBerardinis RJ, Mancuso A, et al. Myc regulates a transcriptional program that stimulates mitochondrial glutaminolysis and leads to glutamine addiction. Proc Natl Acad Sci USA. 2008;105:18782–7.
Xiao C, Srinivasan L, Calado DP, et al. Lymphoproliferative disease and autoimmunity in mice with increased miR-17-92 expression in lymphocytes. Nat Immunol. 2008;9:405–14.
Yuneva M. Finding an "Achilles' heel" of cancer: the role of glucose and glutamine metabolism in the survival of transformed cells. Cell Cycle. 2008;7:2083–9.
Zeller KI, Zhao X, Lee CW, et al. Global mapping of c-Myc binding sites and target gene networks in human B cells. Proc Natl Acad Sci USA. 2006;103:17834–9.
Zhang H, Bosch-Marce M, Shimoda LA, et al. Mitochondrial autophagy is an HIF-1-dependent adaptive metabolic response to hypoxia. J Biol Chem. 2008;283:10892–903.

Chapter 21
MicroRNAs as Blood-based Biomarkers of Cancer

Xi Chen, Junfeng Zhang, Ke Zen, and Chen-Yu Zhang

Abstract The detection of human cancer at an early stage presents a daunting challenge due to the lack of a specific non-invasive marker. The discovery of microRNAs (miRNAs), a class of non-coding RNAs that regulate target gene expression at the post-transcriptional level, has opened a new avenue for tumor diagnosis. Once thought to be unstable RNA molecules, miRNAs are now shown to be stably expressed in serum, plasma, urine, saliva, milk, and other body fluids. Results of the last 2 years establish the quantification of circulating miRNAs in serum/plasma as an extremely promising approach for cancer diagnosis, especially at an early stage. The aim of this chapter is to review the recently reported studies on the circulating miRNAs in cancer patients and to estimate their great potential as a class of highly specific and sensitive blood-based biomarkers for tumor classification and prognostication. Meanwhile, this chapter also addresses certain critical issues that hinder the wide application of the new approach. Some potential challenges for the transition of circulating miRNAs from a research setting to a clinical application are also highlighted. As a blood-based biomarker of cancer, circulating miRNAs have the potential to uncover cancer at an early stage and greatly reduce the worldwide health burden of cancer.

21.1 Introduction

Identifying the sensitive, specific, and non-invasive biomarkers for detecting tumors at an early stage has been regarded as a key point in tumor prevention and therapy. In general, human cancers, like other diseases, are easier to treat and control when detected at an early stage of disease. For example, 65% of non-small cell lung cancer (NSCLC) patients have advanced disease at the time of diagnosis (Brambilla et al. 2003; Patz et al. 2000). Most NSCLC cases, particularly at stages I and II,

C.-Y. Zhang (✉)
State Key Laboratory of Pharmaceutical Biotechnology, School of Life Sciences, Jiangsu Diabetes Center, Nanjing University, Nanjing, Jiangsu 210093, PR China
e-mail: cyzhang@nju.edu.cn

W.C.S. Cho (ed.), *MicroRNAs in Cancer Translational Research*,
DOI 10.1007/978-94-007-0298-1_21, © Springer Science+Business Media B.V. 2011

rarely show symptoms and are difficult to be detected (Brambilla et al. 2003; Patz et al. 2000; Rossi et al. 2005). The 5-year survival rate following surgical resection is approximately 70% for patients with stage I NSCLC, but that rate drops to only 30% for patients with stage III NSCLC (Dominioni et al. 2000). Thus, the earlier detection of NSCLC will greatly facilitate more effective management of the disease. Likewise, early detection of cancer is highly demanded for patients with other cancers, including colon cancer, cervical cancer, ovarian cancer, breast cancer, etc. The development of non-invasive or minimally invasive tests capable to uncover cancer at an early stage can greatly reduce the worldwide health burden of cancers.

Unfortunately, few effective tests are available so far. To date, the reference gold standard in diagnosing human cancers, such as NSCLC, is pathologic evidence of malignant cells, which typically requires an invasive strategy, such as surgery, biopsy, needle aspiration, bronchoscopy, or thoracotomy. Chest X-ray and computed tomography screening can detect some cancers at an early stage (Henschke et al. 2006; Oken et al. 2005), but the diagnostic procedures and the hazards of the associated radiation may outweigh the potential benefits (Mascalchi et al. 2006; Taplin et al. 2008). For breast cancer, mammography is currently the gold standard diagnostic tool; however, it also has limitations, including ionizing radiation and a false positive rate of 8–10% (Taplin et al. 2008). These concerns have led researchers to seek novel biomarkers and new diagnostic assays to non-invasively assess tumors.

Blood-based biomarkers are appealing because they can be used in screening asymptomatic patients. It is generally believed that disease-related biomarkers, including antigens, enzymes, and lipid components, could shed from the tissues or cells with ongoing diseases and be present in blood. Thus, a variety of protein markers, genetic markers, and epigenetic markers have been extensively investigated in blood in an attempt to seek the markers for early stage tumors. For example, several currently available serum/plasma biomarkers, such as carcinoembryonic antigen (CEA), neuron-specific enolase (NSE), chromogranin A (CgA), carbohydrate antigen 15-3 (CA 15-3), carbohydrate antigen 19-9 (CA19-9), alpha-fetoprotein (AFP), and human chorionic gonadotropin (hCG), offer some promise of analyzing tumors comprehensively without the need to carry out a biopsy or a surgical procedure (Dbouk et al. 2007; Huber et al. 1992; Schneider et al. 2000; Sorio et al. 2006; Tarro et al. 2005). However, up to date, only limited success has been achieved in identifying markers those are clinically effective in diagnostic tests. The major concern of these markers is their low sensitivities and specificities (Harris et al. 2007). Even for prostate-specific antigen (PSA), which has been shown to be sensitive for advanced prostate cancer, the specificity is still limited, as evidenced by a high frequency of falsely elevated values in men with benign prostatic hyperplasia (Stenman et al. 1999). Conventional methodologies for the detection and quantification of antigens have also been limited in clinical use due to their low-throughput nature. Recent studies in proteomics have led to the development and broad use of mass spectrometry, a technology that shows the potential to identify unknown proteins in plasma (Gerszten et al. 2010; Gygi et al. 2002; Makawita and Diamandis 2010; Ward et al. 2006; Zhou et al. 2002). With the new technologies and better sample preparation protocols, the sensitivity of mass spectrometry is extremely high, with a detection limit of 1 pg/mL. However, such type of assay is expensive and labor-intensive, which makes it difficult for widespread application in clinical diagnosis. Although the enzyme-linked immunosorbent assay (ELISA) can be performed in a high-throughput manner and is suitable for the detection of diagnostic antigens, its detection sensitivity is greatly dependent on antibody affinities. Therefore, it is important to develop new methods and novel diagnostic biomarkers for the detection of early events of cancer.

Recently, a new class of RNA regulatory genes known as microRNAs (miRNAs) has been found to introduce a whole new layer of gene regulation in eukaryotes (Bartel 2004; He and Hannon 2004). MiRNAs are endogenous non-coding RNAs, consisting of 19–24 nucleotides in length (Bartel 2004; He and Hannon 2004). They play an important role in regulating gene expression by base-pairing to the complementary sites on the target mRNAs, thus blocking the translation or triggering the degradation of the target mRNAs (Bartel 2004; He and Hannon 2004). Since the discovery of the first miRNA *lin-4* in *C. elegans*, thousands of miRNAs have been identified experimentally or computationally from a variety of species (Bartel 2004; He and Hannon 2004). MiRNAs are currently estimated to comprise 1–5% of animal genes and collectively regulate up to 30% of genes, making them one of the most abundant classes of regulators (Bartel 2004; He and Hannon 2004). Over the past several years, it has been well demonstrated that alteration of the miRNA expression contributes to the pathogenesis of many human malignancies (Calin and Croce 2006; Cho 2010; Esquela-Kerscher and Slack 2006). A number of miRNAs, including *miR-15a/miR-16-1* (Calin et al. 2002), *miR-21* (Meng et al. 2007; Ribas et al. 2009; Selaru et al. 2009), *miR-155* (Eis et al. 2005; Gironella et al. 2007), *miR-200* family (Gregory et al. 2008), *miR-210* (Camps et al. 2008; Mathew and Simon 2009), etc, have been found to be differentially expressed in various tumor tissues and cells. Based on their differential roles in carcinogenesis, miRNAs are categorized as either oncogenes or tumor suppressors. These miRNAs not only serve as specific indicators of various aspects of carcinogenesis and tumor development, but also provide important sources for the development of future miRNA-based cancer treatment and therapies (Calin and Croce 2006; Cho 2010; Esquela-Kerscher and Slack 2006).

Recent studies by our group and others have showed that miRNAs are stably present in many body fluids, including serum (Chen et al. 2008; Gilad et al. 2008; Mitchell et al. 2008), plasma (Chen et al. 2008; Mitchell et al. 2008), saliva (Park et al. 2009), urine (Hanke et al. 2010), milk (Kosaka et al. 2010), and cell culture supernatants (Skog et al. 2008; Valadi et al. 2007). Once thought as unstable RNA molecules, circulating miRNAs are in fact highly stable and readily detected in serum and plasma. Biochemical analyses indicate that they are resistant to RNase digestion, as well as extreme pH and temperature (Chen et al. 2008; Mitchell et al. 2008). More importantly, the expression patterns of circulating miRNAs are tightly correlated with various diseases including cancer, diabetes, and tissue injury. These results firmly establish the quantification of the tumor-derived miRNAs in serum/plasma as an extremely promising approach for cancer diagnostics.

In this chapter, we provide an update and overview of the new findings that reveal the potential use of circulating miRNAs as clinically diagnostic biomarkers of various cancers and other diseases. Additionally, we summarize the approaches used to detect circulating miRNAs and address some critical issues in the quantification of these molecules. Finally, we estimate their impact on making the ongoing research closer to clinical application.

21.2 Detection and Characterization of Circulating MiRNAs in Serum and Plasma

21.2.1 MiRNAs and Human Cancer

MiRNAs have been studied intensively in the field of oncology research, and emerging evidence suggests that miRNAs are involved in the pathogenesis of cancers, mainly by regulating the translation of oncogenes and tumor suppressors (Calin and Croce 2006; Cho 2010; Esquela-Kerscher and Slack 2006). By employing the techniques including miRNA cloning, quantitative reverse transcription-polymerase chain reaction (qRT-PCR), microarray, and bead-based profiling method, changes in the expression profiles of miRNAs have been detected in a broad spectrum of hematological malignancies and solid tumors (Cummins et al. 2006; Iorio et al. 2005; Lu et al. 2005; Volinia et al. 2006; Yanaihara et al. 2006). Some miRNAs are frequently found to be up-regulated or down-regulated in tumors compared with normal tissues, in agreement with their dual role in carcinogenesis as either "oncogenes" or "tumor suppressors" (Calin and Croce 2006; Cho 2010; Esquela-Kerscher and Slack 2006). Although expression differences may not necessarily reflect causal events of carcinogenesis, such changes may, nevertheless, be useful for classifying tumors and predicting their outcomes (Lu et al. 2005; Rosenfeld et al. 2008; Yanaihara et al. 2006). Other reports also showed that miRNAs are remarkably stable in formalin-fixed, paraffin-embedded tissues, as well as fresh snap-frozen specimens (Li et al. 2007; Xi et al. 2007). Such information is particularly useful for applying tumor miRNA expression profiles as biomarkers to diminish the diagnostic uncertainty. Since the biopsy for pathologic analysis is a common routine procedure to diagnose solid tumors, the expression profile of tissue miRNAs from biopsy samples become a novel biomarker in early tumor detection. Many studies have demonstrated the differential expression of miRNAs between cancer and the related normal tissue and indicated the potential use of these molecules as biomarkers. However, for developing a non-invasive or minimally invasive human cancer diagnostic tool, more intriguing question is whether the tumor-derived miRNAs are presented and detectable in serum, plasma, and other human body fluids.

21.2.2 Existence of MiRNAs in Serum and Plasma

Dysregulated expression of miRNAs in various tissues has been associated with a variety of human cancers. More recently, miRNAs' occurrence in the serum and plasma of humans has been repeatedly observed. The first description of the presence of circulating miRNAs in serum and their potential as cancer markers was reported by Lawrie et al. (2008) (Table 21.1). By comparing the levels of several

Table 21.1 Circulating microRNAs as a biomarker for cancer diagnosis, classification, and prognostication

Studies	Diseases	Body fluids	Design	Samples	MicroRNAs examined	Techniques	MicroRNAs differentially expressed
Lawrie et al. (2008)	DLBCL	Serum	Tumor vs. normal; correlation with survival	60 patients vs. 43 controls	3	qRT-PCR	*miR-155*, *miR-210*, *miR-21*: higher in patients; *miR-21*: high level is associated with improved relapse-free survival
Mitchell et al. (2008)	Prostate cancer	Plasma	Tumor vs. normal	25 patients vs. 25 controls	6	qRT-PCR	*miR-141*: higher in patients
Chen et al. (2008)	NSCLC	Serum	Tumor vs. normal	11 patients vs. 21 controls in screening phase; 152 patients vs. 75 controls in validation phase	All in screening phase; 2 in validation phase	Solexa sequencing in screening phase, qRT-PCR in validation phase	*miR-25*, *miR-223*: higher in patients

Table 21.1 (continued)

Studies	Diseases	Body fluids	Design	Samples	MicroRNAs examined	Techniques	MicroRNAs differentially expressed
Ng et al. (2009)	CRC	Plasma	Tumor vs. normal; correlation with tumor resection	5 patients vs. 5 controls in screening phase; 25 patients vs. 20 controls in the first validation phase; 90 CRC patients, 50 controls, and 40 patients with other disease in the second validation phase; 10 patients collected before and after the tumor resection	95 in screening phase; 5 in the first validation phase; 2 in the second validation phase	qRT-PCR	*miR-92*, *miR-17-3p*: higher in patients; decrease after tumor resection
Resnick et al. (2009)	Ovarian cancer	Serum	Tumor vs. normal	9 patients vs. 4 controls in screening phase; 19 patients vs. 11 controls in validation phase	365 in screening phase; 21 in validation phase	TaqMan miRNA array in screening phase; qRT-PCR in validation phase	*miR-21*, *miR-92*, *miR-93*, *miR-126*, *miR-29a*: higher in patients; *miR-155*, *miR-127*, *miR-99b*: lower in patients

Table 21.1 (continued)

Studies	Diseases	Body fluids	Design	Samples	MicroRNAs examined	Techniques	MicroRNAs differentially expressed
Huang et al. (2010)	CRC	Plasma	Tumor vs. normal; correlation with tumor resection; correlation with TNM stage	20 carcinomas vs. 20 controls in screening phase; 100 carcinomas, 37 advanced adenomas, and 59 controls in validation phase; 20 carcinomas collected before and after the tumor resection	12 in screening phase; 2 in validation phase	qRT-PCR	*miR-29a, miR-92a*: higher in carcinomas; decrease after tumor resection; higher in advanced TNM stage
Wang et al. (2009a)	Pancreatic ductal adeno-carcinoma	Serum	Tumor vs. normal	49 patients vs. 36 controls	4	qRT-PCR	*miR-21, miR-210, miR-155, miR-196a*: higher in patients
Tanaka et al. (2009)	Acute leukemia	Plasma	Tumor vs. normal	2 patients vs. 7 controls in screening phase; 54 patients vs. 16 controls in validation phase	All in screening phase; 1 in validation phase	Microarray in screening phase, qRT-PCR in validation phase	*miR-92a:miR-638* ratio: lower in patients
Zhu et al. (2009)	Breast cancer	Serum	Subtype classification of tumors	9 progesterone receptor positive vs. 3 negative	3	qRT-PCR	*miR-155*: higher in progesterone receptor positive patients

Table 21.1 (continued)

Studies	Diseases	Body fluids	Design	Samples	MicroRNAs examined	Techniques	MicroRNAs differentially expressed
Tsujiura et al. (2010)	Gastric cancer	Plasma	Tumor vs. normal; correlation with tumor resection	69 patients vs. 30 controls; 10 patients collected before and after the tumor resection	5	qRT-PCR	*miR-17-5p, miR-21, miR-106a, miR-106b*: higher in patients; *let-7a*: lower in patients; *miR-21, miR-106a*: decrease after tumor resection
Liu et al. (2010)	Oral squamous cell carcinoma	Plasma, saliva	Tumor vs. normal; correlation with tumor resection	43 patients vs. 21 controls, 43 patients collected before and after the tumor resection; saliva: 9 patients vs. 8 controls, 9 patients collected before and after the tumor resection	1	qRT-PCR	*miR-31*: higher in plasma and saliva of patients; decrease in plasma and saliva after tumor resection

Table 21.1 (continued)

Studies	Diseases	Body fluids	Design	Samples	MicroRNAs examined	Techniques	MicroRNAs differentially expressed
Brase et al. (2011)	Prostate cancer	Serum	Correlation with tumor progression	7 metastatic and 14 localized prostate cancer in screening phase; 45 patients in the first validation phase; 71 patients in the second validation phase	667 in screening phase; 5 in the first validation phase; 2 in the second validation phase	TaqMan miRNA array in screening phase; qRT-PCR in validation phase	*miR-375, miR-141*: higher in patients with high-risk prostate cancer
Hu et al. (2010)	NSCLC	Serum	Correlation with survival	30 patients with longer survival vs. 30 patients with shorter survival in screening phase; 120 patients in training set; 123 patients in testing set	All in screening phase; 4 in training and testing set	Solexa sequencing and qRT-PCR in screening phase, qRT-PCR in training and testing set	*miR-486, miR-30d, miR-1, miR-499*: associate with overall survival

CRC, Colorectal cancer; DLBCL, Diffuse large B cell lymphoma; MiRNA, MicroRNA; NSCLC, Non-small cell lung cancer; qRT-PCR, Quantitative reverse transcription-polymerase chain reaction; TNM, Tumor-node-metastasis.

tumor-associated miRNAs, such as *miR-155*, *miR-210*, and *miR-21*, in the serum of diffuse large B cell lymphoma (DLBCL) patients with those of healthy controls, Lawrie et al. (2008) determined that patients with DLBCL had high serum levels of *miR-21*. More interestingly, *miR-21* levels were found to be associated with the relapse-free survival of these patients (Lawrie et al. 2008).

Around the same time, Mitchell et al. (2008) showed that miRNAs were present in human plasma in a remarkably stable form and were protected from endogenous RNase activity (Table 21.1). Using a mouse xenograft model where a human prostate cancer cell line was implanted into mice, they showed that there were tumor-derived miRNAs circulating in blood (Mitchell et al. 2008). By filtering plasma samples through a 0.22 μm filter or subjecting these samples to a series of centrifugations followed by RNA extraction and miRNA detection, they showed that those plasma miRNAs were not associated with cells or the broken cellular parts (Mitchell et al. 2008). Finally, the expression of certain miRNAs in cancer patient plasma, particularly the prostate cancer-associated *miR-141*, was found to be significantly different from that in control plasma (Mitchell et al. 2008). These data indicate that tumor-derived miRNAs in human plasma could serve as important biomarkers for the blood-based detection of human cancer.

In our study, we give the first comprehensive tally of miRNAs in blood serum and identify patterns of miRNAs that distinguish patients with two kinds of cancer and diabetes from healthy subjects (Chen et al. 2008) (Table 21.1). Firstly, we detected miRNAs in the sera of various animal species including human, mouse, rat, bovine, and horse by employing a stem-loop qRT-PCR assay (Chen et al. 2008). Secondly, we showed that serum miRNAs were highly stable even in the harsh conditions such as extreme temperature and pH, or RNase digestion, and serum miRNAs can be almost fully recovered from boiling, freeze-thaw cycles or long-time storage in frozen condition (Chen et al. 2008). Moreover, we showed that the levels of miRNAs in serum were reproducible and consistent among individuals of the same species (Chen et al. 2008). In order to obtain the complete expression profile of miRNAs in serum, we further performed high-throughput Solexa sequencing using various serum samples pooled from healthy controls or cancer patients. As a high-throughput sequencing technology producing highly accurate, reproducible, and quantitative readouts of small RNAs, Solexa can detect and define all small RNA molecules including miRNA with various expression levels, and can unambiguously distinguish miRNAs from other small RNAs (Chen et al. 2008, 2009). By this method, we investigated the types and levels of miRNAs in the serum samples pooled from healthy volunteers (male and female) or patients with various diseases including NSCLC, colorectal cancer (CRC), and diabetes. For each disease, we identified a unique expression pattern of miRNAs that differed from that of healthy people (Chen et al. 2008). Our study presented the first genome-wide expression profile of miRNAs in human serum. Interestingly, we also observed striking differences between the miRNAs in the serum and blood cells of cancer patients, while there were no differences between the serum and blood cells of healthy individuals (Chen et al. 2008). The data suggest that some serum miRNAs of cancer patients cannot be derived from circulating blood cells but likely from tumor cells or tissues that are associated with tumorigenesis. It is of interest to note that NSCLC patients shared a large number of serum miRNAs with CRC patients, and Pearson correlation further indicated that the levels of miRNAs in serum from NSCLC patients and CRC patients were consistent (Chen et al. 2008), suggesting that there are some "common" tumor-related miRNAs in serum from cancer patients. More specifically, our results determined that *miR-25* and *miR-223*, two miRNAs previously shown to be involved in tumorigenesis, were highly expressed in NSCLC serum compared to those in normal serum (Chen et al. 2008). Therefore, these two serum-miRNAs could serve as novel blood-based biomarkers of NSCLC.

Taken together, these early reports by us and other groups clearly showed that miRNAs circulate in a stable, cell-free form in the bloodstream and that the abundance of specific miRNAs in plasma or serum can serve as biomarkers of cancers.

21.2.3 Circulating MiRNAs as a Biomarker for Cancer Diagnosis

More recently, miRNAs' occurrence in the serum and plasma of humans has been repeatedly observed (Table 21.1). By comparing the plasma samples of patients with CRC to those of healthy controls, Ng et al. (2009) found that both *miR-17-3p* and *miR-92* are significantly elevated in the plasma of CRC patients. Interestingly, the levels of these miRNAs in plasma were significantly reduced after surgery in 10 patients with CRC (Ng et al. 2009), suggesting that plasma miRNAs may serve as indicators for monitoring cancer patients. Further validation with an independent set of plasma samples indicated that *miR-92* differentiated CRC not only from normal subjects but also from gastric cancer and inflammatory bowel disease (Ng et al. 2009). In discriminating CRC from control subjects, this marker yielded an ROC curve area of 88.5% (the sensitivity was 89% and the specificity was 70%) (Ng et al. 2009). Thus, *miR-92* has reasonable sensitivity for CRC detection and can be employed as a potential non-invasive blood-based biomarker for CRC screening. In accordance with this observation, Huang et al. (2010) also measured the levels of 12 miRNAs in plasma samples from patients with advanced colorectal neoplasia and healthy controls using qRT-PCR and found that plasma *miR-29a* and *miR-92a* have significant diagnostic value for advanced neoplasia.

Detection of the expression of individual or a panel of circulating miRNAs in serum and plasma from cancer patients has also been widely reported by other investigators (Table 21.1). To determine the utility of serum miRNAs as biomarkers for epithelial ovarian cancer, Resnick et al. (2009) compared the serum miRNA levels in cancer specimens to normal specimens with qRT-PCR assay. They identified eight miRNAs that were significantly differentially expressed between cancer and normal specimens, of which *miR-21*, *miR-92*, *miR-93*, *miR-126*, and *miR-29a* were over-expressed while *miR-155*, *miR-127*, and *miR-99b* were under-expressed (Resnick et al. 2009). Employing a pan-human miRNA, high density microarray, Lodes et al. (2009) evaluated miRNA expression patterns in human serum for five types of human cancer, and found that miRNAs present in 1 mL of serum were sufficient to detect miRNA expression patterns, without the need for amplification techniques. Their findings provide a clue to use these expression patterns in correctly discriminating between normal and cancer patient samples. Moreover, in the process of identifying differentially-expressed circulating miRNAs in pancreatic cancer, Wang et al. (2009a) revealed that four pancreatic cancer-specific miRNAs, including *miR-21*, *miR-210*, *miR-155*, and *miR-196a*, had elevated expression levels in the serum of pancreatic cancer patients compared to that of normal controls. By both microarray and qRT-PCR analysis, Tanaka et al. (2009) revealed that *miR-638* is stably present in human plasma while *miR-92a* dramatically decreased in the plasma of acute leukemia patients, and the ratio of *miR-92a*/*miR-638* in plasma was useful for distinguishing leukemia patients from healthy controls. In sum, the miRNA level in serum and plasma has strong potential for clinical application as a novel biomarker for detection of various human cancers.

21.2.4 Circulating MiRNAs as a Biomarker for Tumor Classification and Prognostication

MiRNAs in serum and plasma may also be used as a minimally invasive method for tumor classification and prognostication (Table 21.1). After confirming that RNA species can be detected in archived serum samples, Zhu et al. (2009) reported that the level of *miR-155* was elevated in the serum of progesterone receptor positive patients compared to those with progesterone receptor negative breast cancer. Tsujiura et al. (2010) showed that the plasma concentrations of four miRNAs including *miR-17-5p*, *miR-21*, *miR-106a*, and *miR-106b* were significantly higher in gastric cancer patients than controls, whereas *let-7a* was lower in gastric cancer patients. Interestingly, they found that the levels of *miR-21* and *miR-106a* were significantly reduced in post-operative samples (Tsujiura et al. 2010). Liu et al. (2010) showed that *miR-31* in plasma was significantly elevated in oral squamous cell carcinoma patients relative to age and sex-matched control individuals. They also showed that the plasma *miR-31* in patients was remarkably reduced after tumor resection (Liu et al. 2010). In order to investigate the potential relevance of circulating miRNAs with tumor progression in prostate cancer, Brase et al. (2011) screened miRNA expression in serum samples from patients with metastatic and localized prostate cancer and found that various miRNAs were highly abundant in the sera of patients with metastatic disease. After selecting and validating five up-regulated miRNAs in two independent sample sets, they demonstrated that *miR-375* and *miR-141* were pronounced markers for high-risk tumors (Brase et al. 2011). In sum, their study indicates that circulating miRNAs offer promise as non-invasive biomarkers for tumor progression. In a recent study, Hu et al. (2010) investigate the role of serum miRNA in predicting the prognosis of NSCLC by a genome-wide serum miRNA expression analysis. Specifically, they used Solexa sequencing followed by individual qRT-PCR assays to test the difference in levels of serum miRNAs between patients with longer survival and patients with shorter survival. Levels of four miRNAs including *miR-486*, *miR-30d*, *miR-1*, and *miR-499* were found to be significantly associated with overall survival. The results revealed a great potential of serum miRNA signature as disease fingerprints to predict survival.

21.2.5 Circulating MiRNAs as an Indicator to Reflect Other Diseases

The profile of circulating miRNAs could also be used as an indicator to reflect other diseases or even subtle physiological condition changes (Table 21.2). For instance, some findings suggest the potential of tissue-specific circulating miRNAs as sensitive and informative biomarkers of tissue injury (Ji et al. 2009; Laterza et al. 2009; Wang et al. 2009b). To identify reliable and predictive markers to detect the early signs of drug-induced injury to the liver, one of the most vulnerable organs in the body, Wang et al. (2009b) used acetaminophen overdose-induced liver injury in the mouse as a model system and observed significant differences in the spectrum and levels of miRNAs in both liver tissues and in plasma between control and overdosed mice. A number of elevated circulating miRNAs in plasma collected from acetaminophen-overdosed animals are highly expressed in the liver (Wang et al. 2009b). In particular, *miR-122* and *miR-192*, which are enriched in the liver tissue, exhibit dose- and exposure duration-dependent changes in the plasma that parallel serum aminotransferase levels and the histopathology of liver degeneration, but their changes can be detected significantly earlier (Wang et al. 2009b). The results suggest the potential of using specific circulating miRNAs as sensitive and informative biomarkers for drug-induced liver injury. Another study by Laterza et al. (2009) investigated the use of liver-, muscle-, and brain-specific miRNAs as circulating biomarkers of tissue injury. Tissue-specific miRNAs, *miR-122*, *miR-133a*, and *miR-124*, in plasma samples from rats treated with liver or muscle toxicants and from a rat surgical model of stroke, were determined by a highly sensitive qRT-PCR assay (Laterza et al. 2009). Increases in plasma concentrations of *miR-122*, *miR-133a*, and *miR-124* corresponding to injuries in liver, muscle, and brain, respectively, were observed. Using isoproterenol-induced myocardial injury in rats as a model, Ji et al. (2009) showed that plasma concentrations of *miR-208* increased significantly after myocardial injury and showed a similar time course to the concentration of plasma cardiac troponin I (*cTnI*), a classic biomarker of myocardial injury. Interestingly, Wang et al. (2010b) also found that elevated cardiac-specific *miR-208a* in plasma could be a novel biomarker for early detection of acute myocardial infarction (AMI) in humans. Likewise, Ai et al. (2010) also showed that muscle-specific *miR-1* was significantly higher in plasma from AMI patients compared with non-AMI subjects and its level dropped to normal on discharge following medication. Moreover, Tijsen et al. (2010) identify 6 miRNAs that are elevated in patients with heart failure (HF), among which *miR-423-5p* is most strongly related to the clinical diagnosis of heart failure. Wang et al. (2010a) showed that serum *miR-146a* and *miR-223* were significantly reduced in septic patients compared with systemic inflammatory response syndrome (SIRS) patients and healthy controls, suggesting that serum *miR-146a* and *miR-223* might serve as new biomarkers for sepsis.

Table 21.2 Circulating microRNAs as an indicator to reflect other diseases

Studies	Diseases	Body fluids	Design	Samples	MicroRNAs examined	Techniques	MicroRNAs differentially expressed
Wang et al. (2009b)	Liver injury	Plasma	Disease vs. normal	Acetaminophen overdose-induced liver injury in mouse model	All	Microarray, qRT-PCR	*miR-122*, *miR-192*: elevate in early stage of liver injury
Laterza et al. (2009)	Liver, muscle, and brain injury	Plasma	Disease vs. normal	Rats treated with liver or muscle toxicants and a rat surgical model of stroke	3	qRT-PCR	*miR-122*, *miR-133a*, *miR-124*: increase in plasma corresponding to injuries in liver, muscle, and brain, respectively
Ji et al. (2009)	Myocardial injury	Plasma	Disease vs. normal	Isoproterenol-induced myocardial injury in rat model	1	qRT-PCR	*miR-208*: increase in plasma of rats with myocardial injury

Table 21.2 (continued)

Studies	Diseases	Body fluids	Design	Samples	MicroRNAs examined	Techniques	MicroRNAs differentially expressed
Wang et al. (2010b)	AMI	Plasma	Disease vs. normal, correlation with therapeutic efficacy	AMI rat model, 33 AMI patients, 30 controls, 16 patients with non-AMI coronary heart disease, 17 patients with other cardiovascular diseases, and 5 AMI patients received medical treatment	4	qRT-PCR, microarray	*miR-208, miR-1, miR-133a, miR-499*: increase in AMI rats; higher in AMI patients than in controls and non-AMI patients; lower after medical treatment
Ai et al. (2010)	AMI	Plasma	Disease vs. normal, correlation with therapeutic efficacy, correlation with clinicopathological factors	93 patients vs. 66 controls 83 patients discharged following medication	2	qRT-PCR	*miR-1:* higher in patients, drop to normal on discharge following medication, correlate with QRS

Table 21.2 (continued)

Studies	Diseases	Body fluids	Design	Samples	MicroRNAs examined	Techniques	MicroRNAs differentially expressed
Tijsen et al. (2010)	HF	Plasma	Disease vs. normal	12 HF patients vs. 12 controls in screening phase; 30 HF patients, 20 patients with non-HF forms of dyspnea, and 39 controls in validation phase	All in screening phase16 in validation phase	MiRNA array (Illumina beadchip, human v2 miRNA panel), qRT-PCR	*miR-423-5p*: elevate in HF patients
Wang et al. (2010a)	Sepsis	Serum	Disease vs. normal	50 patients with sepsis, 30 patients with SIRS, and 20 controls	7	qRT-PCR	*miR-146a, miR-223*: decrease in septic patients compared with SIRS patients and healthy controls
Gilad et al. (2008)	Pregnancy	Serum	Pregnant vs. non-pregnant	10 non-pregnant women, 10 women in the first trimester, and 10 women in the third trimester	28	qRT-PCR	*miR-526a, miR-527, miR-520d-5p*: higher in the sera of pregnant women

Table 21.2 (continued)

Studies	Diseases	Body fluids	Design	Samples	MicroRNAs examined	Techniques	MicroRNAs differentially expressed
Chim et al. (2008)	Pregnancy	Maternal plasma	Pre-delivery vs. post-delivery, correlation with gestational age	5 third-trimester placentas, 5 samples of maternal blood cells, and 5 samples of post-delivery maternal plasma in screening phase; 10 pre-delivery and 10 post-delivery plasma samples in validation phase; 10 plasma samples each obtained at the first, second, and third trimesters in validation phase	157 in screening phase 4 in validation phase	TaqMan miRNA array in screening phase, qRT-PCR in validation phase	*miR-141*, *miR-149*, *miR-299-5p*, *miR-135b*: are detectable during pregnancy and show reduced detection rates in post-delivery plasma *miR-141*: increase with gestational age

AMI, Acute myocardial infarction; HF, Heart failure; MiRNA, MicroRNA; qRT-PCR, Quantitative reverse transcription-polymerase chain reaction; SIRS, Systemic inflammatory response syndrome.

Furthermore, Gilad et al. (2008) reported that healthy individuals shared a similar serum miRNA profile, while human placenta-associated miRNAs such as *miR-527* and *miR-526a* were significantly elevated in pregnant women compared to those in non-pregnant women. Chim and co-workers (2008) also reported that placental miRNAs can be detected in serum and plasma, and the levels of these miRNAs correlate with the stages of pregnancy.

21.3 Detection and Characterization of Circulating MiRNAs in Other Body Fluids

MiRNAs have also been shown to be stably expressed in body fluids including urine (Hanke et al. 2010), saliva (Park et al. 2009), milk (Kosaka et al. 2010), and cell culture supernatants (Skog et al. 2008; Valadi et al. 2007) (Table 21.3). In a recent study, Hanke et al. (2010) develop a clinically applicable and sensitive protocol for the preparation and molecular analysis of miRNAs from urine samples obtained from bladder cancer (BCa) patients or healthy volunteers. They showed that many miRNAs had higher abundance in the urine samples from BCa patients compared to those from healthy volunteers (Hanke et al. 2010). Among 157 miRNA species being analyzed, the differential expression of *miR-126* and *miR-182* in urine was found to be able to identify BCa (Hanke et al. 2010). Several studies have shown the presence of miRNA in saliva and their potential as an additional set of oral cancer biomarkers. Park and co-workers (2009) detected approximately 50 miRNAs in both the whole and supernatant saliva, and showed that endogenous saliva miRNA degraded much slower compared with exogenous miRNA. By comparing the expressions of miRNAs in saliva of 50 oral squamous cell carcinoma patients and 50 healthy matched control subjects by qRT-PCR, they identified two miRNAs, *miR-125a* and *miR-200a*, whose expression levels were significantly lower in the saliva of oral squamous cell carcinoma patients than in control subjects (Park et al. 2009). In an attempt to identify putative body fluid-specific miRNAs, Hanson et al. (2009) reported the first comprehensive evaluation of miRNA expression in dried, forensically relevant biological fluids including blood, semen, saliva, vaginal secretions, and menstrual blood. They identified a panel of nine miRNAs (*miR-451*, *miR-16*, *miR-135b*, *miR-10b*, *miR-658*, *miR-205*, *miR-124a*, *miR-372*, and *miR-412*) that can be used as sensitive biomarkers to identify the body fluid origin of forensic biological stain. The results of this study provide an initial indication that miRNA profiling may provide a promising alternative approach for identification of body fluid in forensic casework.

Based on the findings discussed above, we conclude that: (i) miRNAs are stably expressed in human serum, plasma, and other body fluids and can be readily detected by various techniques including Solexa sequencing, miRNA microarray, and qRT-PCR; and (ii) the expression levels and patterns of circulating miRNAs are correlated with various human dysfunctional conditions. Given their size, abundance, tissue specificity, and relative stability, circulating miRNAs in serum, plasma, and other body fluids hold a great promise as non-invasive or minimally invasive biomarkers of various diseases including cancers.

Table 21.3 Circulating microRNAs in other body fluids

Studies	Diseases	Body fluids	Design	Samples	MicroRNAs examined	Techniques	MicroRNAs differentially expressed
Hanke et al. (2010)	BCa	Urine	Tumor vs. normal	9 patients with low-grade BCa, 9 patients with high-grade BCa, 9 patients with urinary tract infections, 9 healthy controls in screening phase; 11 patients with low-grade BCa, 18 patients with high-grade BCa, 7 patients with urinary tract infections, 11 healthy controls in validation phase	157 in screening phase 4 in validation phase	TaqMan microRNA array in screening phase; qRT-PCR in validation phase	*miR-126:miR-152* and *miR-182:miR-152* ratio: higher in patients
Park et al. (2009)	Oral squamous cell carcinoma	Saliva	Tumor vs. normal	12 controls in screening phase; 50 patients vs. 50 controls in validation phase	314 in screening phase 4 in validation phase	RT-pre-amplification-qPCR	*miR-125a*, *miR-200a*: lower in patients

Table 21.3 (continued)

Studies	Diseases	Body fluids	Design	Samples	MicroRNAs examined	Techniques	MicroRNAs differentially expressed
Hanson et al. (2009)	–	Blood, semen, saliva, vaginal secretions, menstrual blood	Provide the evaluation of miRNAs in dried, forensically relevant biological fluids	19 blood samples, 11 semen samples, 18 saliva samples, 11 vaginal secretion samples, and 11 menstrual blood samples	452 in screening phase 9 in validation phase	qRT-PCR	*miR-451, miR-16, miR-135b, miR-10b, miR-658, miR-205, miR-124a, miR-372, miR-412*: a panel of nine miRNAs permits the identification of the body fluid origin of forensic biological stains
Kosaka et al. (2010)	–	Human breast milk	Provide the evaluation of miRNAs in human breast milk	8 human breast milk samples	All	Microarray, qRT-PCR	Several immune-related miRNAs: abundant in the milk

BCa, Bladder cancer; MiRNA, MicroRNA; qRT-PCR, Quantitative reverse transcription-polymerase chain reaction.

21.4 Experimental Techniques and Issues in Analysis of Circulating MiRNAs

Quantification of circulating miRNAs requires effective tools. The explosion of interest in miRNAs over the recent years results in the development of various technique platforms. High-throughput profiling techniques such as Solexa sequencing, miRNA microarray, and bead-based miRNA profiling are effective tools for obtaining expression profiles of circulating miRNAs. These high-throughput techniques greatly facilitate the process of circulating miRNA expression profiling; in this sense, they are superior to the existing low-throughput techniques such as northern blotting and cloning. However, since high-throughput techniques generally require a relative large volume of serum/plasma to enrich total RNAs, they are usually used in initial screening processes. qRT-PCR methodologies have been widely applied in miRNA research, especially in assessing the low level of certain serum miRNAs in individual samples. To date, the most widely-used and successful approach in terms of specificity and sensitivity is a two-step approach using looped miRNA-specific reverse transcription primers and TaqMan probes (Chen et al. 2005; Tang et al. 2006). So far, both high and low-throughput techniques are extremely useful in miRNA study and are responsible for the majority of studies regarding to miRNA research. To make the detection of circulating miRNAs more simple, some new techniques have recently been developed (Lusi et al. 2009). Using an electrochemical genosensor, Lusi et al. (2009) were able to directly detect miRNAs without the need of PCR and a labeling reaction, and the test is simple, very fast and ultrasensitive, with a detection limit of 0.1 pmol. This approach, as well as other revolutionary technologies, may greatly facilitate the application of circulating miRNAs as blood-based biomarkers for the diagnosis of various human cancers.

The lack of a house-keeping gene for normalization is a major technique issue in the quantification of circulating miRNAs. In contrast to tissue or cellular miRNAs, of which the expression levels of certain miRNAs can be normalized against U6 snRNA or house-keeping miRNAs (Peltier and Latham 2008), circulating miRNAs have no such internal control for miRNA normalization. Although miRNAs are stable in serum and plasma, their levels may alter under various conditions. Moreover, there is considerable sample-to-sample variability in both protein and lipid content of plasma and serum samples, which could affect efficiency of RNA extraction. Therefore, how to normalize the level of circulating miRNA is an important issue the investigators concerning from the beginning. Since there is no common circulating miRNA that can be used as a standard, normalizing the level of circulating miRNAs by the volume of serum or plasma samples might be currently the most feasible way to solve the problem. Furthermore, since the yield of RNA from small volume of plasma or serum samples (i.e. 100 μL) is obviously below the limit of accurate quantification by spectrophotometry, it is reasonable to use a fixed volume of serum/plasma sample as input for qRT-PCR rather than to use a fixed mass of input RNA. However, this approach cannot rule out the contamination by technical variations. A more ideal approach is to normalize circulating miRNA levels using exogenous non-human (e.g. *C. elegans* or plants) miRNAs, which were usually spiked-in after the initial denaturation of plasma or serum samples (Mitchell et al. 2008; Hu et al. 2010; Kroh et al. 2010). The inclusion of the spiked-in miRNAs is important for adjusting for technical variations in RNA extraction. An effective normalization strategy for biological variability, however, is currently not well-developed. The use of so-called "invariant" miRNAs as endogenous controls, such as *miR-16* (Lawrie et al. 2008) and *miR-638* (Tanaka et al. 2009), has been proposed by some investigators. However, future studies should be performed to confirm that these miRNAs are indeed expressed at high levels in plasma and serum and are relatively invariant across large numbers of samples.

Furthermore, the clinical effectiveness of circulating miRNAs as biomarkers is likely to be affected by pre-analytic factors. For example, when generating plasma, there is a risk of contamination of the plasma supernatant by cells from the cellular pellet when aspirating (Kroh et al. 2010). Moreover, the type of anti-coagulant used in plasma collection tubes is also important to consider. Whereas EDTA and citrate are both acceptable anti-coagulants for downstream qRT-PCR, the use of heparin as the anti-coagulant potently inhibits subsequent PCR (Kroh et al. 2010). Furthermore, there are other pre-analytic variables that have yet to be studied carefully. Although miRNAs appear to be stable to extended room temperature incubation of plasma (Chen et al. 2008; Mitchell et al. 2008), it is not yet known whether the duration of time taken between collecting and processing of plasma or serum affects miRNA levels (Kroh et al. 2010). In the absence of data on this pre-analytic variable, it is prudent in designing studies to standardize conditions as much as possible with respect to time elapsed between collecting blood and processing for plasma or serum. These are some other factors, such as the diurnal variation in miRNA levels, fasting vs. non-fasting state at blood collection, etc, that may plausibly affect either the amount of miRNA present in a given plasma or serum sample (Kroh et al. 2010). It is prudent to try to match as many variables as possible in the collection and/or selection of case and control samples for research studies.

21.5 Circulating MiRNAs Serving as Novel Potential Biomarkers for Early Tumor Detection, Diagnosis, and Prognosis

21.5.1 A Panel of Circulating MiRNAs Instead of Individual Circulating MiRNA as a Biomarker for Tumor Detection

Ideal blood-based biomarkers of tumors must fit two criteria: (a) specificity, i.e. their presence is associated with the occurrence of only a specific type of tumor; and (b) proportionality, i.e., their levels in serum and plasma should be correlated with the extent of tumor development. Early studies clearly demonstrated that circulating miRNAs perfectly satisfied these standards. Tremendous efforts have been devoted to identify circulating miRNA-based novel non-invasive biomarkers for early tumor detection, diagnosis, and prognosis.

Early reports on circulating miRNAs as disease fingerprints mainly focused on individual or several tumor-related miRNAs. This approach has its advantage because detecting a single miRNA molecule or a few target miRNAs in circulation is much simpler and more straightforward than comprehensively detecting all circulating miRNAs. Successful examples of using this approach include the first two studies in this field, conducted by Lawrie et al. (2008) and Mitchell et al. (2008), respectively. However, in general, the specificity of biomarkers based on a single miRNA or several individual miRNAs is relatively poor. For instance, elevated liver-specific *miR-122* in plasma or serum could result from not only hepatocellular carcinoma but also hepatitis B virus infection, cirrhosis, and general liver injury (Wang et al. 2009b). The molecular basis for the limitation of individual miRNA as a tumor biomarker is the diverse, complex nature of malignancy. For a normal cell transforming into a tumor cell, many different genes would be involved. Accordingly, there should be many miRNAs that target these genes contributing to this complex process. In other word, there are many miRNAs that are dysregulated in each type of cancer. For hepatocellular carcinoma, the altered miRNAs can be derived not only from liver but also immune system. To identify circulating miRNA-based biomarkers for various types of cancer, it is necessary to screen the serum/plasma miRNAs in a genome-wide manner and find all differentially-expressed miRNAs in serum/plasma of patients with that specific type of cancer. From this regard, employing a miRNA expression panel instead of an individual miRNA as a biomarker represents a rational option to circumvent the limitation in miRNAs utilization in tumor diagnosis.

21.5.2 A Working Model for Identifying Circulating MiRNA-based Biomarkers for Diseases

In the process of searching circulating miRNA-based cancer biomarker, the scientists have developed a working model to identify and refine differentially expressed miRNAs in cancer serum/plasma samples compared to control samples (Fig. 21.1). The analysis was separated into three steps: (i) initial screening by high-throughput techniques such as Solexa sequencing, microarray, or miRNA cloning using pooled or several serum samples; (ii) qRT-PCR validation in a large number of individual serum samples arranged in multiple training and testing sets; and (iii) statistical evaluation of the diagnostic or prognostic value of the circulating miRNA profiling system by algorithms such as ROC curve, Risk score analysis, and cluster analysis. Briefly, after the initial screening, a panel of differentially-expressed miRNAs could be derived by comparing the serum/plasma miRNA profile in healthy volunteers and cancer patients. Since the sample sizes used in the initial screening stage were small, the miRNAs selected had to be further validated by qRT-PCR assay at an individual level. This model has been proven to be greatly successful in identifying serum miRNA-based biomarkers for lung cancer classification and predicting the survival of lung cancer patients (Hu et al. 2010).

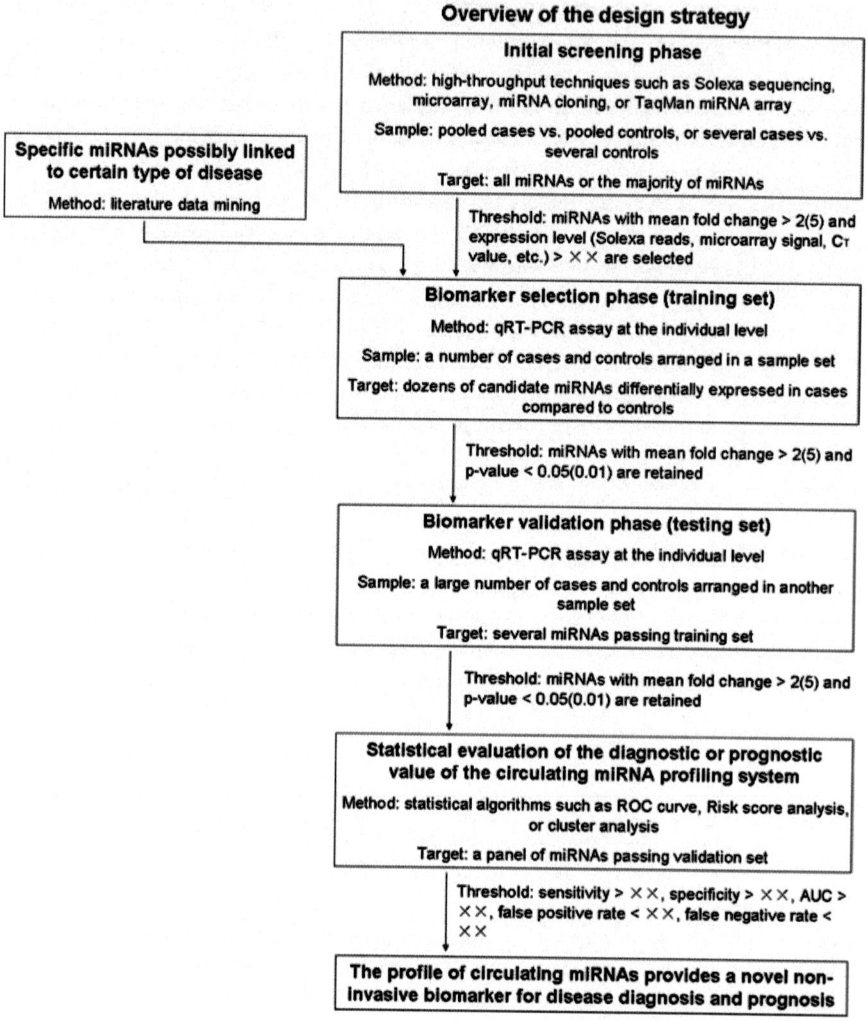

Fig. 21.1 Schematic description of the working model for identifying circulating miRNA-based biomarkers for diseases. AUC, Area under the ROC curve; MiRNA, MicroRNA; qRT-PCR, Quantitative reverse transcription-polymerase chain reaction; ROC curve, Receiver operating characteristic curve

In general, initial screening by high-throughput techniques, such as Solexa and microarray, is to identify circulating miRNAs that are differentially expressed in patients' serum, plasma, and other body fluids. However, if we know the specific miRNAs that are linked to certain type of diseases, we can directly determine their levels using miRNA qRT-PCR assay. One of alternative approaches is to use multiplex qRT-PCR assay to identify a panel of specific circulating miRNAs, which will be faster and simpler than sequencing or microarray studies.

21.6 Some Key Issues Regarding the Sources and Biological Functions of Circulating MiRNAs

21.6.1 Sources of Circulating MiRNAs

To establish circulating miRNAs as novel disease biomarkers, a key issue is to clarify their sources. Theoretically, there are two major pathways that have been proposed for miRNAs to enter the circulation: (i) passive leakage from broken cells and (ii) active secretion via cell-derived microvesicles (MVs) (Fig. 21.2).

Passive leakage from broken cells: Although direct leakage of cellular miRNAs into the circulating is not common under normal circumstance, it still occurs at the time of tissue damage or cell apoptosis. Under the circumstances of tumor metastasis or chronic inflammation, tissues or cells at the primary site may break up; therefore, miRNAs may leak into the circulating blood from these cells. This type

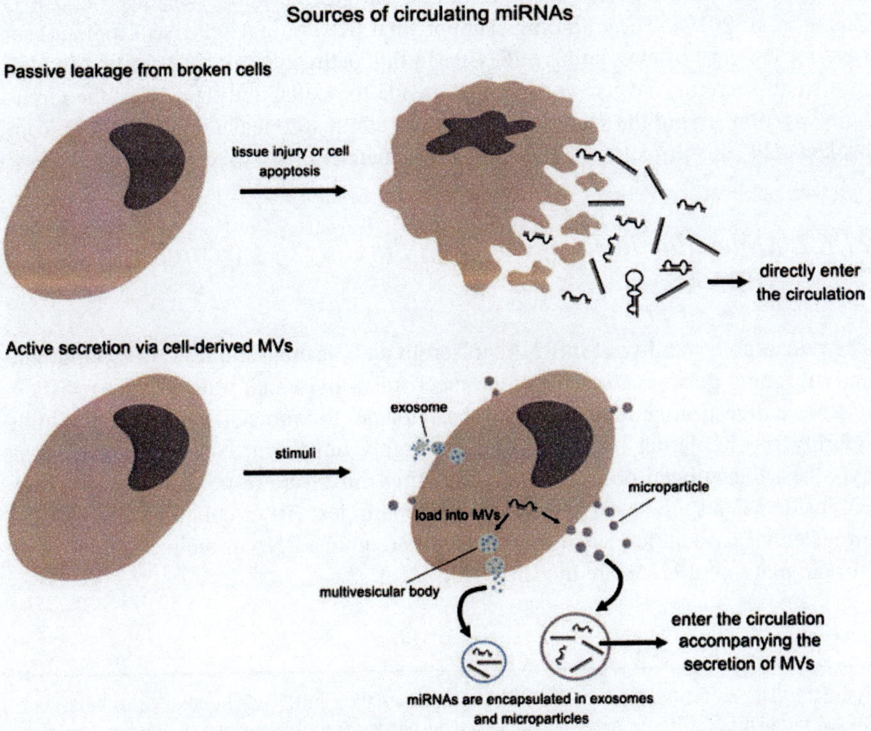

Fig. 21.2 Schematic description of the sources of circulating miRNAs. There are two major pathways that have been proposed for miRNAs to enter the circulation: (i) passive leakage from broken cells and (ii) active secretion via MVs. In the first pathway, tissues or cells at the primary site may break up under the conditions of tissue damage or cell apoptosis, and the miRNAs may leak into the circulating blood from these cells. In the second pathway, miRNAs are first loaded into MVs (exosomes and microparticles) inside the cells, and then enter the circulation accompanying the secretion of MVs. MiRNA, MicroRNA; MVs, Microvesicles

of miRNA release may also happen in certain blood cells, such as platelets and monocytes, because of their relatively short half-lives. In general, passive leakage from broken cells and tissues is a process that does not require energy.

Active secretion via cell-derived MVs: MVs are small vesicles that are shed from almost all cell types under both normal and pathological conditions (Cocucci et al. 2009; Thery et al. 2002). They generally include exosomes and microparticles, whose size and vesicular structures are quite different from each other. These secreted MVs bear surface receptors/ligands of the original cells and have the potential to selectively interact with specific target cells and mediate intercellular communication by transporting bioactive lipids, mRNA, or proteins between cells (Thery et al. 2002). MVs have been identified from many cellular sources including monocytes, macrophages, endothelial cells, leukocytes, and tumor cells and are now thought to have pivotal roles in tumor invasion and metastases, inflammation, coagulation, and stem-cell renewal and expansion (Cocucci et al. 2009; Thery et al. 2002). Recent studies clearly showed that MVs from cultured cells and human peripheral blood contain miRNAs (Hunter et al. 2008; Skog et al. 2008; Valadi et al. 2007; Zhang et al. 2010). These findings support that MVs might serve as efficient carriers for delivery of circulating miRNAs. In this pathway, miRNAs are first loaded into small secretory vesicles or granules inside the cells, and then enter the circulation accompanying the secretion of MVs. In contrast to leakage of miRNAs from broken cells, this process is active and ATP-dependent.

21.6.2 Molecular Basis of the High Stability of Circulating MiRNAs

The remarkable stability of miRNAs in serum and plasma samples raises important and intriguing questions regarding the mechanism by which miRNAs are resistant to RNase digestion and harsh conditions. To date, the molecular basis of the high stability of circulating miRNAs remains largely unknown. Nevertheless, several hypotheses have been proposed: (i) circulating miRNAs are protected by packaging inside MVs such as exosomes and microparticles; (ii) circulating miRNAs are protected via association with other molecules (e.g. in a RNA-protein complex); and (iii) circulating miRNAs are modified (Fig. 21.3).

Fig. 21.3 The molecular basis of the stability of circulating miRNAs. In contrast to naked miRNAs, circulating miRNAs are well preserved in harsh conditions including RNase digestion, extreme temperature and pH, extended storage, freeze-thaw cycles, etc. Several hypotheses regarding the molecular basis of the high stability of circulating miRNAs have been proposed: (i) circulating miRNAs are encapsulated in MVs such as exosomes and microparticles, and the membrane structures of MVs protect miRNAs from degradation; (ii) circulating miRNAs are embedded in RISC and are therefore stabilized via association with these miRNA-binding proteins; and (iii) circulating miRNAs are protected by modifications including methylation, adenylation, and uridylation. MiRNA, MicroRNA; MVs, Microvesicles; RISC, RNA-induced silencing complex

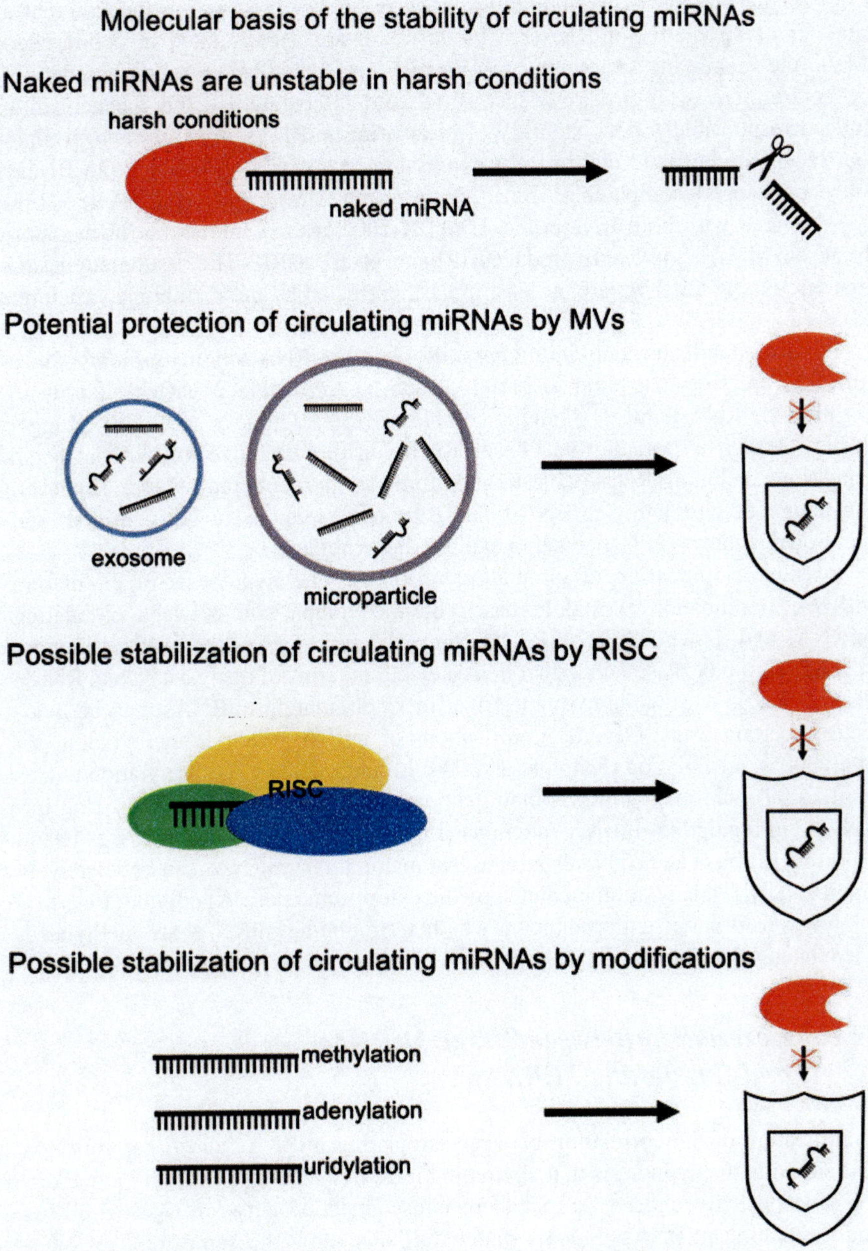

Fig. 21.3 (continued)

Potential protection of circulating miRNAs by MVs: Based on the fact that a number of circulating miRNAs in bloodstream are encapsulated in cell-derived MVs, the membrane structures of MVs may protect miRNAs from degradation. Although there is no direct evidence so far to demonstrate that this mechanism is indeed responsible for the stability of circulating miRNAs in serum and plasma, our recent study found that miRNA expression levels in MVs treated with RNase were unchanged compared to those in untreated MVs, while miRNA levels in MVs treated with both RNase and Triton X-100 (destroy the double layer membrane of MVs) significantly reduced (Zhang et al. 2010). The results suggested that circulating miRNAs are, at least in part, protected by the membrane structures of MVs.

Possible stabilization of circulating miRNAs by miRNA-binding proteins: Since miRNAs are shown to bind to proteins such as Argonaute 2 (AGO2) family in forming the RNA-induced silencing complex (RISC) (Gregory et al. 2005; Lingel et al. 2003), it is possible that these proteins in turn stabilize miRNAs in harsh conditions including RNase digestion, extreme temperature and pH, etc. However, the nature of circulating miRNA-binding proteins remains to be identified, and additional studies will be needed to explore the hypothesis.

Possible modification of circulating miRNAs: The hypothesis of circulating miRNA modification is mainly based on the comparison between circulating miRNAs and tissue/cellular miRNAs. Since the same individual miRNA is more stable in serum or plasma than that in tissue/cell, in terms of the resistance to RNase digestion, it is reasonable to suggest that these circulating miRNAs may be modified in certain ways. General modifications of miRNAs include methylation (Yu et al. 2005), adenylation (Katoh et al. 2009; Lu et al. 2009), and uridylation (Jones et al. 2009), and plant miRNA modifications play critical roles in stabilizing miRNAs and regulating miRNA functions (Lu et al. 2005; Yu et al. 2005). A recent report by Katoh et al. (2009) also found that mammalian miRNAs can be selectively stabilized by 3′ adenylation mediated by the cytoplasmic poly(A) polymerase *GLD-2*. However, it is currently unknown whether circulating miRNAs are methylated, adenylated, or uridylated.

21.6.3 Correlation Between Tissue MiRNAs and Circulating MiRNAs

Identification of the correlation between circulating miRNAs and tissue miRNAs also supports the hypothesis that circulating miRNAs can serve as ideal biomarkers for cancers. It is conceivable that there is a connection between tissue miRNAs and circulating miRNAs. Indeed, many miRNAs show the same trend of alteration, either increase or decrease, in the plasma/serum and tumor tissues of patients with various types of cancer. For examples, the expression of *miR-25* and *miR-223* was found increased in both the serum of lung cancer patients (Chen et al. 2008) and their lung tumor tissues (Volinia et al. 2006). The level of *miR-155* was also found to be elevated in the tumor tissues/cells (Eis et al. 2005) and plasma of lymphoma patients (Lawrie et al. 2008). The correlation between tissue and circulating miRNAs provides evidence for the hypothesis that circulating miRNAs can reflect various aspects of human physiological status and serve as fingerprints for disease diagnosis.

21.6.4 Potential Biological Functions of Circulating MiRNAs

Although the detection of circulating miRNAs in serum, plasma, and other body fluids has been widely reported, the biological and physiological functions of circulating miRNAs are largely unknown. Recent studies have shown that miRNAs can be secreted and delivered into target cells via MVs, and that these exogenous miRNAs can regulate the expression of target genes and cellular functions in recipient cells (Pegtel et al. 2010; Skog et al. 2008; Valadi et al. 2007; Yuan et al. 2009). These findings provide a mechanism for transport and exchange of miRNAs among non-adjacent cells, and open up the possibility that miRNAs may serve as signaling molecules allowing for coordinated intercellular regulation of gene expression.

The ability of MVs to transfer miRNA raises very exciting possibilities for therapeutic uses. Cells engineered to express miRNA may be capable of delivering these small molecules to local cellular environments via MVs. These engineered cells can be encapsulated to provide sustained local delivery. Since current techniques for gene transfer use viral or synthetic agents as delivery carriers, their replacement by MVs released from autologous transplants of engineered cells will offer the advantage of a virus-free approach and make the prospects of gene therapy safer.

From this point of view, secreted miRNAs may represent a class of signaling molecules that play an important role in mediating intercellular communication. Moreover, the secretion and targeting of miRNAs among the different cells establish a highly regulated complex network under various physiological and pathological conditions. The research in secreted miRNAs will not only provide valuable information for developing biomarkers for disease diagnosis, but also shed light on our understanding of the highly complex cell communication network under various physiological and pathological conditions.

21.7 Concluding Remarks and Future Directions

Since the association of miRNAs with tumorigenesis was discovered several years ago, many attempts have been made to develop the sensitive and robust miRNA-based tests for early tumor diagnostics. Both miRNAs derived from tumor cells and affected tissues have been extensively evaluated as a diagnostic and prognostic tool to monitor cancers (Cummins et al. 2006; Iorio et al. 2005; Lu et al. 2005; Volinia et al. 2006; Yanaihara et al. 2006). Recently, we and other groups have found that human serum/plasma contained a large amount of stable miRNAs, and that the expression pattern of serum/plasma miRNA altered in reflection of various disease conditions, including cancers. The enormous potential of circulating miRNAs as a class of ideal cancer biomarkers is based on the following facts: (i) they are remarkably stable molecules, well preserved in harsh conditions, and resistant to RNase activity; (ii) their expression profiles are specifically correlated with certain type of cancer; and (iii) they are easily accessible and can be detected in a relatively non-invasive manner by various techniques.

Although circulating miRNAs have triggered much excitement in clinical and scientific communities, this field is only now emerging. Much of the work on circulating miRNAs is still in its infancy and requires further exploration. Since the first identification of serum/plasma miRNAs in 2008 (Chen et al. 2008; Gilad et al. 2008; Lawrie et al. 2008; Mitchell et al. 2008), numerous studies have shown the presence of miRNAs in circulation and their potential use as novel biomarkers for diseases and pathophysiological status, including malignancy, diabetes mellitus, pregnancy, and acute tissue injuries. However, these early studies have been limited by their small sample sizes, their lack of constant standards in quantification, and inconsistencies in methodologies (Chin and Slack 2008). Like many other novel biomarkers at their early stages of research, circulating miRNAs require extensive investigation to validate their great potential. Several areas may need to be focused on in future studies: (i) establish a simple standard assay for the quantification of circulating miRNAs in various body fluids; (ii) test the specificity and sensitivity of circulating miRNA profile-based biomarkers in a large number of samples; in particular, compare the expression of serum/plasma miRNA in different types of cancer to identify the specific biomarkers for specific cancer; and (iii) perform the perspective studies such as cohort studies to determine whether circulating miRNAs can serve as a diagnostic tool to detect cancer at its early stage.

As the functional roles of miRNAs in cancer biology are further uncovered and the methods of circulating miRNA detection and analysis are improved, circulating miRNAs will serve as novel minimally invasive or non-invasive biomarkers for various types of cancer. Their wide applicability and potential importance will probably initiate a revolution in clinical management, including detecting the early stage of tumors, estimating prognosis, predicting therapeutic efficacy, maintaining surveillance, and forecasting disease recurrence.

References

Ai J, Zhang R, Li Y, et al. Circulating microRNA-1 as a potential novel biomarker for acute myocardial infarction. Biochem Biophys Res Commun. 2010;391:73–7.

Bartel DP. MicroRNAs: genomics, biogenesis, mechanism, and function. Cell. 2004;116:281–97.

Brambilla C, Fievet F, Jeanmart M, et al. Early detection of lung cancer: role of biomarkers. Eur Respir J Suppl. 2003;39:36s–44s.

Brase JC, Johannes M, Schlomm T, et al. Circulating miRNAs are correlated with tumor progression in prostate cancer. Int J Cancer. 2011;128:608–16.

Calin GA, Croce CM. MicroRNA signatures in human cancers. Nat Rev Cancer. 2006;6:857–66.

Calin GA, Dumitru CD, Shimizu M, et al. Frequent deletions and down-regulation of micro-RNA genes miR15 and miR16 at 13q14 in chronic lymphocytic leukemia. Proc Natl Acad Sci USA. 2002;99:15524–9.

Camps C, Buffa FM, Colella S, et al. Hsa-miR-210 Is induced by hypoxia and is an independent prognostic factor in breast cancer. Clin Cancer Res. 2008;14:1340–8.

Chen X, Ba Y, Ma L, et al. Characterization of microRNAs in serum: a novel class of biomarkers for diagnosis of cancer and other diseases. Cell Res. 2008;18:997–1006.

Chen X, Li Q, Wang J, et al. Identification and characterization of novel amphioxus microRNAs by Solexa sequencing. Genome Biol. 2009;10:R78.

Chen C, Ridzon DA, Broomer AJ, et al. Real-time quantification of microRNAs by stem-loop RT-PCR. Nucleic Acids Res. 2005;33:e179.

Chim SS, Shing TK, Hung EC, et al. Detection and characterization of placental microRNAs in maternal plasma. Clin Chem. 2008;54:482–90.

Chin LJ, Slack FJ. A truth serum for cancer-microRNAs have major potential as cancer biomarkers. Cell Res. 2008;18:983–4.

Cho WC. MicroRNAs: potential biomarkers for cancer diagnosis, prognosis and targets for therapy. Int J Biochem Cell Biol. 2010;42:1273–81.

Cocucci E, Racchetti G, Meldolesi J. Shedding microvesicles: artefacts no more. Trends Cell Biol. 2009;19:43–51.

Cummins JM, He Y, Leary RJ, et al. The colorectal microRNAome. Proc Natl Acad Sci USA. 2006;103:3687–92.

Dbouk HA, Tawil A, Nasr F, et al. Significance of CEA and VEGF as diagnostic markers of colorectal cancer in Lebanese patients. Open Clin Cancer J. 2007;1:1–5.

Dominioni L, Imperatori A, Rovera F, et al. Stage I nonsmall cell lung carcinoma: analysis of survival and implications for screening. Cancer. 2000;89:2334–44.

Eis PS, Tam W, Sun L, et al. Accumulation of miR-155 and BIC RNA in human B cell lymphomas. Proc Natl Acad Sci USA. 2005;102:3627–32.

Esquela-Kerscher A, Slack FJ. Oncomirs – microRNAs with a role in cancer. Nature Reviews Cancer. 2006;6:259–69.

Gerszten RE, Carr SA, Sabatine M. Integration of proteomic-based tools for improved biomarkers of myocardial injury. Clin Chem. 2010;56:194–201.

Gilad S, Meiri E, Yogev Y, et al. Serum microRNAs are promising novel biomarkers. PLoS One. 2008;3:e3148.

Gironella M, Seux M, Xie MJ, et al. Tumor protein 53-induced nuclear protein 1 expression is repressed by miR-155, and its restoration inhibits pancreatic tumor development. Proc Natl Acad Sci USA. 2007;104:16170–5.

Gregory PA, Bert AG, Paterson EL, et al. The miR-200 family and miR-205 regulate epithelial to mesenchymal transition by targeting ZEB1 and SIP1. Nat Cell Biol. 2008;10:593–601.

Gregory RI, Chendrimada TP, Cooch N, et al. Human RISC couples microRNA biogenesis and posttranscriptional gene silencing. Cell. 2005;123:631–40.

Gygi SP, Rist B, Griffin TJ, et al. Proteome analysis of low-abundance proteins using multidimensional chromatography and isotope-coded affinity tags. J Proteome Res. 2002;1:47–54.

Hanke M, Hoefig K, Merz H, et al. A robust methodology to study urine microRNA as tumor marker: microRNA-126 and microRNA-182 are related to urinary bladder cancer. Urol Oncol. 2010;28:655–61.

Hanson EK, Lubenow H, Ballantyne J. Identification of forensically relevant body fluids using a panel of differentially expressed microRNAs. Anal Biochem. 2009;387:303–14.

Harris L, Fritsche H, Mennel R, et al. American society of clinical oncology 2007 update of recommendations for the use of tumor markers in breast cancer. J Clin Oncol. 2007;25:5287–312.

He L, Hannon GJ. MicroRNAs: small RNAs with a big role in gene regulation. Nat Rev Genet. 2004;5:522–31.

Henschke CI, Yankelevitz DF, Libby DM, et al. Survival of patients with stage I lung cancer detected on CT screening. N Engl J Med. 2006;355:1763–71.

Hu Z, Chen X, Zhao Y, et al. Serum microRNA signatures identified in a genome-wide serum microRNA expression profiling predict survival of non-small-cell lung cancer. J Clin Oncol. 2010;28:1721–6.

Huang Z, Huang D, Ni S, et al. Plasma microRNAs are promising novel biomarkers for early detection of colorectal cancer. Int J Cancer. 2010;127:118–26.

Huber K, Kirchheimer JC, Ermler D, et al. Determination of plasma urokinase-type plasminogen activator antigen in patients with primary liver cancer: characterization as tumor-associated antigen and comparison with alpha-fetoprotein. Cancer Res. 1992;52:1717–20.

Hunter MP, Ismail N, Zhang X, et al. Detection of microRNA expression in human peripheral blood microvesicles. PLoS One. 2008;3:e3694.
Iorio MV, Ferracin M, Liu CG, et al. MicroRNA gene expression deregulation in human breast cancer. Cancer Res. 2005;65:7065–70.
Ji X, Takahashi R, Hiura Y, et al. Plasma miR-208 as a biomarker of myocardial injury. Clin Chem. 2009;55:1944–9.
Jones MR, Quinton LJ, Blahna MT, et al. Zcchc11-dependent uridylation of microRNA directs cytokine expression. Nat Cell Biol. 2009;11:1157–63.
Katoh T, Sakaguchi Y, Miyauchi K, et al. Selective stabilization of mammalian microRNAs by 3′ adenylation mediated by the cytoplasmic poly(A) polymerase GLD-2. Genes Dev. 2009;23:433–8.
Kosaka N, Izumi H, Sekine K, et al. microRNA as a new immune-regulatory agent in breast milk. Silence. 2010;1:7.
Kroh EM, Parkin RK, Mitchell PS, et al. Analysis of circulating microRNA biomarkers in plasma and serum using quantitative reverse transcription-PCR (qRT-PCR). Methods. 2010;50:298–301.
Laterza OF, Lim L, Garrett-Engele PW, et al. Plasma microRNAs as sensitive and specific biomarkers of tissue injury. Clin Chem. 2009;55:1977–83.
Lawrie CH, Gal S, Dunlop HM, et al. Detection of elevated levels of tumour-associated microRNAs in serum of patients with diffuse large B-cell lymphoma. Br J Haematol. 2008;141:672–5.
Li J, Smyth P, Flavin R, et al. Comparison of miRNA expression patterns using total RNA extracted from matched samples of formalin-fixed paraffin-embedded (FFPE) cells and snap frozen cells. BMC Biotechnol. 2007;7:36.
Lingel A, Simon B, Izaurralde E, et al. Structure and nucleic-acid binding of the Drosophila Argonaute 2 PAZ domain. Nature. 2003;426:465–9.
Liu CJ, Kao SY, Tu HF, et al. Increase of microRNA miR-31 level in plasma could be a potential marker of oral cancer. Oral Dis. 2010;16:360–4.
Lodes MJ, Caraballo M, Suciu D, et al. Detection of cancer with serum miRNAs on an oligonucleotide microarray. PLoS One. 2009;4:e6229.
Lu J, Getz G, Miska EA, et al. MicroRNA expression profiles classify human cancers. Nature. 2005;435:834–8.
Lu S, Sun YH, Chiang VL. Adenylation of plant miRNAs. Nucleic Acids Res. 2009;37:1878–85.
Lusi EA, Passamano M, Guarascio P, et al. Innovative electrochemical approach for an early detection of microRNAs. Anal Chem. 2009;81:2819–22.
Makawita S, Diamandis EP. The bottleneck in the cancer biomarker pipeline and protein quantification through mass spectrometry-based approaches: current strategies for candidate verification. Clin Chem. 2010;56:212–22.
Mascalchi M, Belli G, Zappa M, et al. Risk-benefit analysis of X-ray exposure associated with lung cancer screening in the Italung-CT trial. AJR Am J Roentgenol. 2006;187:421–9.
Mathew LK, Simon MC. MiR-210: a sensor for hypoxic stress during tumorigenesis. Mol Cell. 2009;35:737–8.
Meng FY, Henson R, Wehbe-Janek H, et al. MicroRNA-21 regulates expression of the PTEN tumor suppressor gene in human hepatocellular cancer. Gastroenterology. 2007;133:647–58.
Mitchell PS, Parkin RK, Kroh EM, et al. Circulating microRNAs as stable blood-based markers for cancer detection. Proc Natl Acad Sci USA. 2008;105:10513–8.
Ng EK, Chong WW, Jin H, et al. Differential expression of microRNAs in plasma of patients with colorectal cancer: a potential marker for colorectal cancer screening. Gut. 2009;58:1375–81.
Oken MM, Marcus PM, Hu P, et al. Baseline chest radiograph for lung cancer detection in the randomized prostate, lung, colorectal and ovarian cancer screening trial. J Natl Cancer Inst. 2005;97:1832–9.

Park NJ, Zhou H, Elashoff D, et al. Salivary microRNA: discovery, characterization, and clinical utility for oral cancer detection. Clin Cancer Res. 2009;15:5473–7.

Patz EF Jr., Goodman PC, Bepler G. Screening for lung cancer. N Engl J Med. 2000;343:1627–33.

Pegtel DM, Cosmopoulos K, Thorley-Lawson DA, et al. Functional delivery of viral miRNAs via exosomes. Proc Natl Acad Sci USA. 2010;107:6328–33.

Peltier HJ, Latham GJ. Normalization of microRNA expression levels in quantitative RT-PCR assays: identification of suitable reference RNA targets in normal and cancerous human solid tissues. RNA. 2008;14:844–52.

Resnick KE, Alder H, Hagan JP, et al. The detection of differentially expressed microRNAs from the serum of ovarian cancer patients using a novel real-time PCR platform. Gynecol Oncol. 2009;112:55–9.

Ribas J, Ni X, Haffner M, et al. MiR-21: an androgen receptor-regulated microRNA that promotes hormone-dependent and hormone-independent prostate cancer growth. Cancer Res. 2009;69:7165–9.

Rosenfeld N, Aharonov R, Meiri E, et al. MicroRNAs accurately identify cancer tissue origin. Nat Biotechnol. 2008;26:462–9.

Rossi A, Maione P, Colantuoni G, et al. Screening for lung cancer: new horizons? Crit Rev Oncol Hematol. 2005;56:311–20.

Schneider J, Velcovsky HG, Morr H, et al. Comparison of the tumor markers tumor M2-PK, CEA, CYFRA 21-1, NSE and SCC in the diagnosis of lung cancer. Anticancer Res. 2000;20: 5053–8.

Selaru FM, Olaru AV, Kan T, et al. MicroRNA-21 is overexpressed in human cholangiocarcinoma and regulates programmed cell death 4 and tissue inhibitor of metalloproteinase 3. Hepatology. 2009;49:1595–601.

Skog J, Wurdinger T, van Rijn S, et al. Glioblastoma microvesicles transport RNA and proteins that promote tumour growth and provide diagnostic biomarkers. Nat Cell Biol. 2008;10:1470–6.

Sorio C, Mauri P, Pederzoli P, et al. Non-invasive cancer detection: strategies for the identification of novel cancer markers. IUBMB Life. 2006;58:193–8.

Stenman UH, Leinonen J, Zhang WM, et al. Prostate-specific antigen. Semin Cancer Biol. 1999;9:83–93.

Tanaka M, Oikawa K, Takanashi M, et al. Down-regulation of miR-92 in human plasma is a novel marker for acute leukemia patients. PLoS One. 2009;4:e5532.

Tang F, Hajkova P, Barton SC, et al. MicroRNA expression profiling of single whole embryonic stem cells. Nucleic Acids Res. 2006;34:e9.

Taplin S, Abraham L, Barlow WE, et al. Mammography facility characteristics associated with interpretive accuracy of screening mammography. J Natl Cancer Inst. 2008;100:876–87.

Tarro G, Perna A, Esposito C. Early diagnosis of lung cancer by detection of tumor liberated protein. J Cell Physiol. 2005;203:1–5.

Thery C, Zitvogel L, Amigorena S. Exosomes: composition, biogenesis and function. Nat Rev Immunol. 2002;2:569–79.

Tijsen AJ, Creemers EE, Moerland PD, et al. MiR423-5p as a circulating biomarker for heart failure. Circ Res. 2010;106:1035–9.

Tsujiura M, Ichikawa D, Komatsu S, et al. Circulating microRNAs in plasma of patients with gastric cancers. Br J Cancer. 2010;102:1174–9.

Valadi H, Ekstrom K, Bossios A, et al. Exosome-mediated transfer of mRNAs and microRNAs is a novel mechanism of genetic exchange between cells. Nat Cell Biol. 2007;9:654–9.

Volinia S, Calin GA, Liu CG, et al. A microRNA expression signature of human solid tumors defines cancer gene targets. Proc Natl Acad Sci USA. 2006;103:2257–61.

Wang J, Chen J, Chang P, et al. MicroRNAs in plasma of pancreatic ductal adenocarcinoma patients as novel blood-based biomarkers of disease. Cancer Prev Res. 2009a;2:807–13.

Wang JF, Yu ML, Yu G, et al. Serum miR-146a and miR-223 as potential new biomarkers for sepsis. Biochem Biophys Res Commun. 2010a;394:184–8.

Wang K, Zhang S, Marzolf B, et al. Circulating microRNAs, potential biomarkers for drug-induced liver injury. Proc Natl Acad Sci USA. 2009b;106:4402–7.

Wang GK, Zhu JQ, Zhang JT, et al. Circulating microRNA: a novel potential biomarker for early diagnosis of acute myocardial infarction in humans. Eur Heart J. 2010b;31:659–66.

Ward DG, Suggett N, Cheng Y, et al. Identification of serum biomarkers for colon cancer by proteomic analysis. Br J Cancer. 2006;94:1898–905.

Xi Y, Nakajima G, Gavin E, et al. Systematic analysis of microRNA expression of RNA extracted from fresh frozen and formalin-fixed paraffin-embedded samples. RNA. 2007;13:1668–74.

Yanaihara N, Caplen N, Bowman E, et al. Unique microRNA molecular profiles in lung cancer diagnosis and prognosis. Cancer Cell. 2006;9:189–98.

Yu B, Yang Z, Li J, et al. Methylation as a crucial step in plant microRNA biogenesis. Science. 2005;307:932–5.

Yuan A, Farber EL, Rapoport AL, et al. Transfer of microRNAs by embryonic stem cell microvesicles. PLoS One. 2009;4:e4722.

Zhang Y, Liu D, Chen X, et al. Secreted monocytic miR-150 enhances targeted endothelial cell migration. Mol Cell. 2010;39:133–44.

Zhou Z, Licklider LJ, Gygi SP, et al. Comprehensive proteomic analysis of the human spliceosome. Nature. 2002;419:182–5.

Zhu W, Qin W, Atasoy U, et al. Circulating microRNAs in breast cancer and healthy subjects. BMC Res Notes. 2009;2:89.

Chapter 22
RNAi-based Approaches to the Treatment of Brain Tumors

Saroj P. Mathupala, Sandeep Mittal, Murali Guthikonda, and Andrew E. Sloan

Abstract The discovery of RNA interference (RNAi) as an alternate mode of gene regulation utilized by eukaryotes has greatly expanded the arsenal of molecular tools that a biomedical researcher can employ to target tumors. Thus, it will be natural to extend this methodology from bench to the bedside as a novel mode of clinical intervention against intracranial tumors, as these are frequently refractive to conventional therapies, i.e. surgery, radiotherapy, and chemotherapy. Induction of RNAi in the targeted tumor can be multifaceted, with two primary modes of therapeutic intervention postulated to date; delivery of in silico designed and in vitro synthesized non-coding RNA molecules to the tumor, or the pharmacological perturbation of non-coding RNA molecules that are endogenously expressed by the tumor. Both modes silence gene expression, but via somewhat different molecular pathways. Two key non-coding RNA molecules that are known to impart RNAi are small interfering RNA and microRNA. Both are currently under investigation in pre-clinical and clinical settings for their potential utility against tumors, with small interfering RNA already proving their efficacy in several small-scale clinical trials. However, despite demonstration of excellent potential as an anti-cancer therapy at the bench, the current inability to deliver therapeutically significant levels of RNAi into the diseased tissue remains a primary hurdle that needs to be overcome for widespread utility at the bedside. This chapter presents an update on both the latest methodologies that have been utilized for induction of RNAi in brain tumors, both in pre-clinical and clinical settings, as well as descriptions of the molecular pathways that have been targeted via RNAi. The pioneering RNAi-based clinical trials that have been tested against brain tumors are also described to illustrate both the utility and the impact of this novel mode of brain tumor therapy at the bedside.

S.P. Mathupala (✉)
Department of Neurological Surgery, Karmanos Cancer Institute, Wayne State University School of Medicine, Detroit, MI 48201, USA
e-mail: smathupala@med.wayne.edu

W.C.S. Cho (ed.), *MicroRNAs in Cancer Translational Research*,
DOI 10.1007/978-94-007-0298-1_22,

22.1 Introduction

Malignant brain tumors are frequently refractory to conventional clinical treatments, i.e. surgical intervention followed by combined radiation therapy and chemotherapy. Despite aggressive multimodal therapy, the prognosis still remains quite poor for these patients. Thus, translation of any novel molecular tool used in brain tumor research to the bedside as a clinical therapy will be of great significance for improving the clinical outcome for such patients.

Brain tumors are classified according to their cellular origin and are graded based on their histopathology [World Health Organization (WHO) grading system]. The grading takes into account atypical nuclear and cellular morphology, mitotic activity, endothelial proliferation, necrosis, and proliferation index (Louis et al. 2007). Brain tumors can be broadly divided into primary brain tumors (those with intracranial origin) and brain metastases (those that metastasize from elsewhere in the body). Approximately 40% of brain tumors are metastatic with origins in cancers of the lung, breast, melanoma, renal, or colon.

Primary brain tumors can arise from any cell type within the central nervous system (CNS). Of these, gliomas are neuroectodermal in origin and arise from the glia, the supporting cells of the CNS. They are the most common type of primary brain tumors, and harbor distinct clinical, histological, and molecular characteristics. They are grouped into low-grade gliomas (WHO grades I and II) and high-grade gliomas (WHO grades III and IV). Other primary brain tumors include meningiomas, ependymomas, choroid plexus papillomas, pituitary adenomas, and medulloblastomas.

It is these most malignant of these tumors (grades III and IV glioma) that are frequently refractory to conventional clinical treatments, with patients having a median survival of 12–15 months from the time of diagnosis. Thus, it is these patients that urgently need novel modes of clinical intervention, where interference RNA can play a role in identifying adjuvant therapies in support of current clinical interventional regimens.

Worldwide, 13% of all deaths occur due to cancer, which translates into 7.4 million deaths. The age-adjusted worldwide incidence rate for malignant brain tumors is 3.7 per 100,000 person-years for males and 2.6 for females (IARC; International Agency for Research on Cancer; www.iarc.fr), with rates being higher in the so-called developed countries than in less-developed areas (IARC). The detection rate of malignant brain tumors has increased over the past two decades, especially among the elderly, most likely due to improved diagnostic imaging regimens.

22.1.1 RNA Interference (RNAi)

RNAi had its origins in horticulture research. The existence of a mechanism for interfering with the cellular transcription machinery was first identified in 1990 by Napoli and co-workers (1990) during genetic engineering experiments on the petunia plant to engineer deeper colored flowers. When extra copies of the chalcone synthase gene (which encodes an enzyme crucial for plant pigment synthesis) were added to the plant's genetic makeup, white or variegated flowers developed instead of the expected deeply colored flowers. Although the added copies of the gene were being correctly transcribed, the chalcone synthase mRNA were somehow being destroyed, preventing production of the enzyme. This new form of gene suppression was termed post-transcriptional gene silencing (PTGS).

Similar gene-silencing effects were also observed in the nematode *Caenorhabditis elegans* (Guo and Kemphues 1995), where specific gene transcripts could be selectively silenced or inactivated via application of copies of sense RNA identical to the targeted mRNA sequence. Here too, the introduced sense RNA was acting in a manner similar to a classical antisense RNA strand, although the latter imparts its effect due to its sequence complementarity to the target messenger RNA. The mechanism behind this phenotype was identified in 1998 when Fire and co-workers (1998) showed that the transcriptional interference effect was due to the inadvertent presence of double-stranded RNA (dsRNA) molecules in the synthetic RNA preparations used in these experiments. Thus, application of both sense and antisense RNA had a far more potent silencing effect than either sense or antisense RNA alone. Fire and Mello (who were awarded the Nobel Prize in Physiology or Medicine in 2006 for their discovery) designated this mechanism as RNAi, which was later found to be active in mammals as well (Elbashir et al. 2001).

22.1.2 Non-coding RNAs that Impart RNAi

Since these early reports on RNAi by in vitro engineered RNA on the specific mRNA transcripts, several new classes of non-coding RNAs that impart RNAi on eukaryotic cells (including mammalian cells) have been identified. The exogenously applied RNAi molecules that led to the original discovery (described above) were referred to simply as small interfering RNA (siRNA). However, recent reports indicate that both insect and mammalian cells also generate similar siRNA in an endogenous manner, which has been termed endo-siRNA (Okamura and Lai 2008; Watanabe et al. 2008). Thus, the exogenously applied siRNA will need to be categorized as exo-siRNA to differentiate the in vitro from the in vivo. MicroRNA (miRNA) are another class of non-coding RNAs that are endogenously transcribed (Birchler and Kavi 2008; Kim and Rossi 2007; Liu et al. 2008; Mathupala et al. 2007; Pillai 2005).

While siRNA and miRNA populate most of the RNAi molecules studied to date, additional classes of RNAi molecules have been recently described. Piwi-interacting RNA (piRNA) is a class of non-coding RNAs present in germline and adjacent tissues (Lau et al. 2006). Another two classes have been described that influence the transcription-initiation RNA (tiRNA) or are associated with promoter regions promoter-associated small RNAs (pasRNA) (Kapranov et al. 2007; Taft et al. 2009).

Would RNAi impart only short-term changes to the transcriptome of a tumor cell? To date, studies in both non-mammalian and mammalian systems have indicated that RNAi can influence gene activity at the epigenetic level by additional mechanisms known as RNA-directed DNA methylation (RdDM) and RNAi mediated heterochromatin modification (Malecova and Morris 2010; Moazed 2009; Verdel et al. 2009). In contrast to PTGS commonly mediated by RNAi, this mode of longer term gene silencing, which also encompasses chromatin remodeling, is referred to as TGS or transcriptional gene silencing.

The present chapter will address RNAi studies that have been completed in vivo in model systems, or currently in clinical trials. With regard to miRNA, we will describe miRNA signatures that have been documented in primary glioma resections, as these will be of utility in clinical diagnosis. Thus, for clarity, we will refer to small RNAs of exogenous origins (in vitro synthesized) as siRNAs throughout this chapter, while small RNAs transcribed and processed off nuclear DNA templates will be referred to as miRNA.

22.1.3 MiRNA

As described above, miRNAs are transcribed from intergenic regions of chromosomes as well as non-coding regions within genes, i.e. introns and non-translated regions of protein-coding genes. However, it should be noted that miRNAs of intergenic origin are in fact, encoded by distinct transcription units (i.e. genes) (Kim and Nam 2006; Nelson et al. 2003), despite being non-coding in nature. MiRNAs are now known to mediate expression of target genes by base pairing with regions within the corresponding mRNA, which then result in either inhibition of translation, or cause direct destruction of the mRNA template. The latter is thought to closely parallel the mechanism for RNAi via siRNA. The fate of mRNA template is selected on the basis of perfect or imperfect base-pairing between miRNA and the target mRNA template, with a perfect match targeting the mRNA towards siRNA-type targeted destruction, while mismatches create conditions for inhibition of translation (Krol and Krzyzosiak 2004; Meltzer 2005; Nelson et al. 2003). MiRNAs that were expressed within introns were usually coordinately expressed with their host gene's mRNA, indicating that they were derived from introns of the same pre-mRNA transcript.

22.2 Mechanisms of Interference

Experimentally, the siRNA-mediated mRNA degradation cascade begins with cleavage of exogenously introduced synthetic dsRNA by an enzyme complex referred to as Dicer. The resultant, approximately 21 base pair (bp) dsRNA cleavage products (siRNA) are then incorporated into a protein multimer composed of two proteins referred to as Argonaute-2 (Ago-2; in the case of mammalian systems) (Kim and Rossi 2007; Liu et al. 2004) and RNA-induced silencing complex (RISC). When the pair of oligonucleotide strands of the dsRNA template loaded onto RISC will have perfect sequence complementarity as in the above situation, Ago-2 will cleave and discard the sense RNA strand to produce an active RISC complex containing the antisense strand only. The resultant siRNA/RISC/Ago-2 complex scans the mRNA milieu for target sites with perfect complementarity. Upon binding, the target mRNA is cleaved by the complex.

In contrast, the miRNA-based mRNA degradation cascade begins with endogenously (nuclear) encoded primary miRNA transcripts referred to as pri-miRNAs. As mentioned before, these may be transcribed from non-coding genes, or be components of introns of pre-mRNA templates of coding genes. The miRNA coding genes are also transcribed by RNA polymerase II, in a manner similar to conventional protein coding genes (Kim and Rossi 2007). During transcription, the pri-miRNA template is also spliced and polyadenylated and then processed by an enzyme complex referred to as Drosha to yield precursor miRNAs (pre-miRNAs). These are then exported to the cytoplasm across the nuclear envelop via the protein exportin-5 to be bound to the Dicer complex, which in turn processes the pre-miRNA template for loading onto the Ago-2/RISC complex in a manner similar to that for siRNA. The mature miRNA strand may, or may not have perfect sequence complementarity with target mRNA. When there is full complementarity, the target mRNA is directed for cleavage via the Ago-2/RISC in a mechanism similar to that for siRNA. When the complementarity is partial, translation off the targeted mRNA is first inhibited due to miRNA/RISC binding, or this step may be followed by degradation of the target mRNA in cytoplasmic vesicles referred to as processing bodies (P-bodies).

22.3 Rational Design of siRNA for Pre-clinical and Clinical Studies

Design algorithms for siRNA have evolved from design of 21-mer siRNA to 27-mer longer siRNA templates (Khvorova et al. 2003; Kim et al. 2005; Reynolds et al. 2004; Siolas et al. 2005). The design changes are due to greater understanding of the mechanisms behind the RNAi, with studies indicating that 21-mers are processed by Dicer (an RNAse III type endonuclease) from longer dsRNA templates, which may be necessary both for loading siRNA onto the RISC complex and for the assembly of RISC itself (Amarzguioui et al. 2006; Preall et al. 2006; Rose et al. 2005). Thus, most current algorithms outline the use of 27-mer or longer templates (known as 27-mer Dicer substrates) when designing siRNA.

siRNA can sometimes down-regulate non-targeted mRNA (Cullen 2006a) voiding experimental data interpretation. However, when RNAi experiments are conducted in vivo, this response is not observed (Heidel et al. 2004). More importantly, in the single clinical trial reported for RNAi against glioma (Zukiel et al. 2006), the authors report the absence of local inflammatory responses at the siRNA site-of-injection.

The 5′ region of the antisense strand (also called the guide strand) of the dsRNA loaded on the RISC complex is referred to as the seed region, due to its importance in targeting a specific mRNA for cleavage. This seed region consists of 6 nucleotides between positions 2–7 of the guide strand of the siRNA duplex.

Recent studies indicate that off-target siRNA effects may occur due to partial complementarity between this seed region with the 3′ untranslated regions (3′UTR) of non-specific mRNA targets (Jackson et al. 2006). Such partial matches can initiate down-regulation of non-specific mRNA due to siRNA mimicking the action of a miRNA (explained below). Thus, current algorithms analyze siRNA templates for global mRNA 3′UTR – seed region matches to minimize the potential for off-target effects.

With greater understanding of the mechanism employed by Dicer to process long dsRNA molecules to select the guide strand (referred to as functional polarity of the dsRNA template), and the method employed to load the RISC complex in mammalian model systems, 29 bp dsRNA templates with 2-nucleotide 3′ overhangs are now known to display better efficacy in inducing the siRNA response in comparison to shorter 21 bp dsRNA templates. Also, current algorithms design the two strands in a manner that directs Dicer to preferentially process the guide RNA strand for loading onto the RISC complex for optimal efficiency of targeting (Amarzguioui et al. 2006; Rose et al. 2005).

22.4 Pre-clinical Experimental Strategies

Three experimental methods are commonly employed for introduction of siRNA into a cell; (i) direct transfection of gene specific 21 bp RNA duplexes (dsRNA) (which may be chemically modified for better intracellular longevity) to directly load RISC complexes for short term induction of siRNA (Bumcrot et al. 2006); (ii) pre digestion of long dsRNA (that correspond to a full length mRNA template) in vitro with Dicer to produce a pool of 21-mer siRNA templates against a particular mRNA target, followed by transfection of the entire pool into cells; (iii) transfection of cells with an expression plasmid or PCR based expression cassette that can transcribe a short symmetric RNA strand that folds to form a 21-mer dsRNA (denoted a short hairpin RNA or shRNA) for loading the RISC complex (Brummelkamp et al. 2002; Cullen 2006a, b; Myslinski et al. 2001).

In contrast to siRNA, design and development of experimental strategies with miRNA poses a dilemma. First and foremost, miRNA are cellular products specific to a particular cell line or tissue. One can design synthetic miRNA or miRNA templates for in vitro synthesis in a manner similar to that for siRNA, with the caveat that imperfect complementarity needs to be built into the designed miRNA template. In silico analysis of miRNA sequences and the transcriptome hint at a given miRNA sometimes influencing multiple gene targets, sometimes into the hundreds. If so, it would throw a cell's intricate network of signaling and metabolic cascades into disarray. Thus, further studies are needed to understand the in vivo selectivity and efficacy displayed by miRNA on the cellular transcriptome.

Recent laboratory strategies have utilized miRNA-adapted shRNA hybrids (shRNAmiR), where 30-mer duplex shRNA are designed to mimic naturally transcribed miRNA transcripts in organism of interest (Chang et al. 2006). When transfected, the shRNAmiR will be processed first by the nuclear Drosha complex and then by the cytoplasmic Dicer and loaded onto the RISC complex. Since the method closely follows the endogenous miRNA processing pathways it has been shown to induce more effective suppression of translation from target mRNA and/or cleavage (Silva et al. 2005).

However, recent reports indicate that endogenous miRNA may, at most, regulate their target mRNA messages by approximately two-fold (Seitz 2009). Thus, while miRNA may be of use as a diagnostic or prognostic tool in the clinic, they may be of limited utility as a therapeutic tool.

22.5 siRNA Targets in Glioma – Pre-clinical Studies

Since the recognition of RNAi as a primary mode of post-transcriptional gene regulation in mammalian cells, over to 300 research reports have appeared in literature on targeting glioma via RNAi. In fact, the first publication on glioma and dsRNA was in 1991, where the authors investigated the anti-proliferative effect of mismatched dsRNA in a human glioma cell line (Hubbell et al. 1991). Since then, RNAi has been utilized to target glioma metabolism, signal transduction, cell-cycle progression, proliferation, invasion, angiogenesis, etc, both in vitro and in vivo (Fig. 22.1).

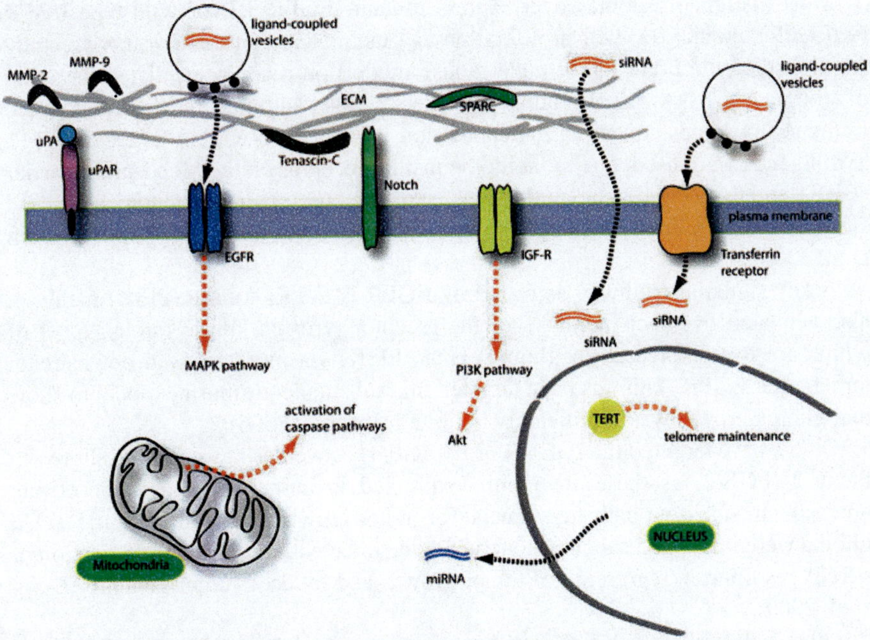

Fig. 22.1 Primary signaling pathways and extracellular matrix components that have been targeted by RNA interference (RNAi) in pre-clinical and clinical studies. Naked or encapsulated RNAi have been used to target glioma in vivo. Both the transferrin receptor and the epidermal growth factor receptor (EGFR) have been frequently used for ligand coupled homing of siRNA loaded vesicles. Both key signal transduction pathways and extracellular matrix proteins have been targeted to promote glioma apoptosis and growth inhibition, and to inhibit invasiveness. The only clinical trial on using RNAi in glioma therapy has targeted tenascin-C, an extracellular glycoprotein

Since the current chapter addresses the possible routes for using RNAi as a clinical tool, we will summarize published in vivo studies where rodent models were used to test the efficacy of select siRNA or miRNA molecules against implanted glioma cells (of both human and animal origin). The single clinical trial underway on using RNAi against glioma will also be described. Since miRNA may be of utility as a diagnostic tool, the miRNA signatures that have been identified to date from primary glioma tissues or other brain tumor types will be also described. In vitro studies (over 300 at present) that have used RNAi to target glioma, or to elucidate various cellular pathways, will not be discussed for sake of brevity as these in vitro studies have been reviewed elsewhere (Guo et al. 2010; Mathupala et al. 2006, 2007).

22.5.1 Targeting Cell Surface Receptors

Several receptors that are known to be highly expressed in glioma have been studied to either use as a therapeutic target [epidermal growth factor receptor (EGFR) or insulin-like growth factor-1 receptor (IGF-1 R)] or as a tool for targeted delivery of RNAi (transferrin receptor) (Cardoso et al. 2007).

Most malignant gliomas over-express mutant (EGFRvIII) or wild-type EGFR (wtEGFR), crucial for their proliferation and malignancy. In the first in vivo study (in an orthotopic brain tumor nude mouse model) on receptor-mediated targeting of glioma via RNAi, Zhang and co-workers (2004) administered ligand-coupled immuno-liposomes containing an expression vector (shRNA) that targeted EGFR. Two ligands were used, one to target the insulin receptor on the blood-brain barrier (BBB), and the second to target the transferrin receptor on glioma. Others have used HSV-1 based amplicons or siRNA to target EGFR in vivo (Kang et al. 2006; Saydam et al. 2005).

A key signaling pathway activated by EGFR is the PI3kinase (PI3K) pathway, also activated by other members of the receptor tyrosine kinase family, most of which are over-expressed in glioma. Thus, PI3K is a more focused downstream target, after EGFR. This has been targeted in a subcutaneous murine model to show that glioma growth was significantly inhibited (Pu et al. 2006).

Others have targeted the IGF1-R or the IGF-1, as well as basic fibroblast growth factor-2 (FGF-2) as these are highly expressed in human gliomas and activate downstream signaling pathways crucial for tumor growth and proliferation. Rat C6 glioma were implanted subcutaneously in nude mice with intra-tumoral injection of siRNA resulting in regression of tumor growth and reduced tumorigenicity (Dong et al. 2009).

Other cell-surface proteins that have been targeted by siRNA have been: (i) secreted protein acidic and rich in cysteine (SPARC), an extracellular glycoprotein switched-on in malignant gliomas and promote their invasion and cell-survival via Akt pathway activation. Intracranial glioma transfected with siRNA against SPARC were shown to form non-invasive tumors (Seno et al. 2009); (ii) protein tyrosine phosphatase zeta/receptor-type protein tyrosine phosphatase beta (PTPzeta/RPTPbeta) and its ligand pleiotrophin (PTN), which are normally involved in neuronal cell migration, but over-expressed in gliomas; tumor growth was abrogated when glioma engineered to express shRNA against above were implanted in an orthotopic brain tumor model (Ulbricht et al. 2006).

Notch receptors are normally involved in cellular homeostasis, including embryonic development and stem cell maintenance. It is also over-expressed in most gliomas. Thus, Notch-1 was targeted in a nude mouse model which resulted in not only cell-cycle arrest, but reduction in invasiveness and induction of apoptosis of the implanted glioma (Xu et al. 2010).

22.5.2 Targeting Invasive Nature of Glioma

Infiltration of surrounding normal brain tissue by malignant glioma is facilitated by proteolysis of extra-cellular matrix components, initiated by the urokinase plasminogen activator receptor (uPAR). This receptor, along with its agonist urokinase plasminogen activator (uPA) and matrix metalloproteinase-2 (MMP-2, a metalloproteinase over-expressed in glioma) have been targeted to affect the glioma infiltrative process via RNAi in vivo, to show that down-regulation these resulted in decreased cell invasion, angiogenesis, tumor growth, and in a meningioma model, increased radiosensitivity (Gogineni et al. 2010; Gondi et al. 2007, 2009; Kargiotis et al. 2008).

Other proteins that have been targeted in vivo in glioma via RNAi to investigate their effects on migration/invasiveness as well as proliferation and angiogenesis are (i) telomerase (hTERT) (Zhao et al. 2007); (ii) S100 calcium-binding protein A4 (S100A4), a member of the EF-hand family of calcium-binding proteins, which is up-regulated in glioma in contrast to white matter, and known to be involved in both invasion and metastasis. S100A4 was targeted via RNAi in C6 glioma cells (Takenaga et al. 2007) and shown to inhibit migration; (iii) Akt2, a serine/threonine kinase that regulates cell survival and proliferation. The expression of Akt2 was shown to closely correlate with malignancy of primary human glioma. Silencing of Akt2 resulted in decreased levels of MMP-9 and suppression of invasiveness (Zhang et al. 2009a).

22.5.3 Induction of Glioma Apoptosis

RNAi has been used to induce apoptosis in glioma by targeting survivin, an anti-apoptotic protein that belongs to the family of inhibitors of apoptosis (IAP). Survivin is highly expressed in glioma and inhibits caspase activation. Thus, when survivin was targeted in vivo with RNAi, it leads to glioma apoptosis (Uchida et al. 2004).

Another signaling molecule targeted for induction of apoptosis is Wnt, a molecule involved in embryogenesis as well as cancer. Several members of Wnt family are over-expressed in glioma and have been targeted by siRNA to show that silencing of Wnt proteins is associated with decreases in PI3-kinase/Akt signaling, with inhibition of tumor growth (Pu et al. 2009).

22.5.4 *Targeting Hypoxia-induced Glioma Angiogenesis Cascades via RNAi*

Hypoxia-inducible factor-1 (HIF-1) and vascular endothelial growth factor (VEGF) are both known to be over-expressed in the hypoxic regions of malignant glioma, and promote invasion and radioresistance. Both have been targeted (Gillespie et al. 2007) in vivo to show that silencing of either attenuates glioma migration and invasion.

22.5.5 *Modulation of Immuno-activity Against Glioma*

Malignant tumors contain microglia, which are known to promote glioma invasion. STAT-3 has been targeted in vivo via RNAi to activate anti-glioma immunity, where microglia and macrophage activation were studied. The latter resulted in glioma growth inhibition indicating that STAT-3 may serve as an adjuvant for immunotherapy against glioma (Zhang et al. 2009b).

22.6 MiRNA Targets in Glioma

An increasing number of investigators have turned to testing miRNA as a molecular tool for glioma therapy. However, in contrast to the extensive list of literature on targeting glioma via siRNA, only a few RNAi studies have been reported that harness the power of miRNA to target glioma even in vitro.

Oncogenic miRNA are classified as oncomirs (He et al. 2005), while those that act as tumor suppressors are termed anti-oncomirs (Goga and Benz 2007). Thus, miRNA therapies that will target oncomirs are referred to as antagomirs (Krutzfeldt et al. 2005). However, due to the homeostatic mechanisms active in mammalian cells, including tumor cells, it is becoming increasingly apparent that perturbing the miRNA to impart significant changes in the tumor proteome will be a difficult task at best. It is highly likely that miRNA will be of significant utility as a diagnostic tool, which has already been proven clinically.

In one expression profiling study, the miRNAs *miR-9, -15a, -16, -17, 19a, -20a, -21, -25, -28, -130b, -140, -210* were found to be over-expressed in glioma, while *miR-184* and *-328* were found to be under-expressed (Malzkorn et al. 2010). Others have located *miR-183, -367, -371* to be up-regulated in glioma (Lavon et al. 2010), similar to the profiles seen in neural precursor cells. In another study, *miR-10b* was found to be up-regulated with increasing grade of glioma, via RT-PCR analysis. The increase closely correlated with uPAR levels which is one of the receptors thought to be regulated by *miR-10b* (Sasayama et al. 2009). *MiR-221* expression was also found to be closely correlated with glioma malignancy in several other RT-PCR screens (Conti et al. 2009; Lu et al. 2009).

In the few in vivo studies that have examined the utility of miRNA, *miR-221* and *-222* have been shown to promote malignant progression by influencing the Akt signaling cascades (Zhang et al. 2010). Inhibition of *miR-21* was shown to suppress tumor growth in vivo, along with inhibition of EGFR, activation of Akt, cyclin D and Bcl-2 (Zhou et al. 2010). In another study *miR-21* was targeted for cytotoxic therapy with S-TRAIL, a pro-apoptotic ligand, where synergistic activity was seen between the two molecules (Corsten et al. 2007). *MiR-21* is also implicated in cell proliferation, apoptosis, and migration/invasion (Chan et al. 2005; Corsten et al. 2007; Gabriely et al. 2008) as well as regulation of inhibitors of MMP. Inhibition of *miR-21* was shown to reduce MMP activity in a mouse model and suppress glioma migration and invasion, indicating that *miR-21* contributes to glioma malignancy by down-regulation of MMP inhibitors leading to activation of MMPs, and thus promoting invasiveness of glioma. A detailed review of miRNA in glioma and the potential therapeutic efficacy is provided elsewhere (Mathupala et al. 2007).

22.7 Systemic Delivery of RNAi to Brain Tumors

Brain is a privileged organ, where the tissue is separated from systemic circulation and protected by the BBB, which forms a physical barrier to protect the brain tissue from invading pathogens and other toxins. Tight-junctions between the endothelial cells that line the cerebral capillaries ensure that the BBB will essentially be impenetrable to RNA molecules used in potential RNAi therapies (where the average molecular mass will be around 14 kDa) and to larger shRNA generating DNA vectors, or potential miRNA therapeutics.

Thus, the BBB needs to be traversed for any systemic application of RNAi to succeed. Both basic and potential clinical approaches for systemic application of RNAi have been reviewed in detail elsewhere (Boado 2007; Fountaine et al. 2005; Lesniak 2005; Pardridge 2004; 2007; Pirollo and Chang 2008; Xie et al. 2006).

Several strategies have been tested to deliver RNAi to the brain via the systemic route, by harnessing receptor-mediated transport across the BBB. As described previously, RNAi to target glioma EGFR have been encapsulated in pegylated immunoliposomes (poly-ethylene-glycol encapsulated liposomes) and tested in an orthotopic murine model. The liposomes were surface modified with monoclonal antibodies against the insulin receptor and against the transferrin receptor (Boado 2005; Zhang et al. 2004). The same group has also utilized direct conjugation of transferrin receptor-targeting antibodies to the siRNA, where the antibody was coupled to the RNA molecules via a biotin-streptavidin linker, again in an orthotopic brain tumor model (Xia et al. 2007).

Yet others have utilized coated nanoparticles for delivery of encapsulated siRNA or siRNA generating expression vectors through the BBB (Jain 2007). These nanoparticles are thought to mimic low-density lipoproteins (LDL) and thus interact with the LDL receptors on the endothelium resulting in their uptake across the BBB (Lesniak 2005). However, since the BBB is commonly compromised in patients with late-stage glioma, traversing the vascular endothelium may not be difficult in a systemic siRNA delivery strategy.

The most straightforward, non-obstructive clinical route to brain tumor may be direct application of RNAi therapeutics to the tumor, or the surgical cavity following tumor resection. In fact, the single clinical trial reported to date for glioma therapy with RNAi has followed this route (see Section 22.8), where siRNA was injected directly to infiltrative (inoperable) regions of glioma (Zukiel et al. 2006).

Others have used multifunctional carriers that are pH sensitive for peptide-targeted systemic delivery of siRNA in vivo, where stable nanoparticles were formed to administer HIF-1 targeting complexes in glioma bearing nude mice to show significant tumor inhibition (Wang et al. 2009).

22.8 RNAi-mediated Glioma Therapy Clinical Trials

Tenascin-C is an extracellular matrix glycoprotein that is also expressed during early development. It is also over-expressed in glioma. Tenascin-C expression in the stroma of tumors is known to be associated with poor prognosis. Most importantly, tenascin-C was targeted in the first (and only) clinical trial conducted thus far utilizing RNAi, where a significantly improved outcome was noted in a subset of patients with malignant gliomas (Zukiel et al. 2006).

In the initial study, a cohort of 11 low- and high-grade patients were treated after tumor resection, with at least two of the patients indicating absence of tumor recurrence. Tumor recurrence was observed in at least three other patients, but only at sites distal to the RNAi application point. Currently 46 patients have been treated in this clinical trial, with a median survival of 106.6 weeks for the cohort, whereas the median survival for patients with glioblastoma (grade IV) and anaplastic astrocytoma (grade III) was 48.2 weeks (Rolle et al. 2010; Wyszko et al. 2008). The success in this initial trial points to the efficacy of targeted RNAi-based therapy against glioma.

With regard to clinical trials based on interference RNA strategies, to date, over twenty trials have been initiated with some being completed or terminated, per descriptions outlined in the online database ClinicalTrials.gov. Of these, only five involve strategies against cancer (excluding the above described clinical trial against glioma). Of these, one (NCT00689065) has targeted the M2 subunit of ribonucleotide reductase, where siRNA was systemically delivered, embedded in nanoparticles for a prolonged half-life. In fact, initial results from this study on melanoma patients have provided the first proof-of-principle data to indicate the efficacy of systemic therapy via nanoparticle-mediated application of RNAi, in homing-in on the solid tumor to deliver the siRNA payload (Davis et al. 2010). Another formulation is being tested for advanced solid cancers, again by systemic delivery (NCT00938574). With liver being one of the first-pass organs during systemic siRNA delivery, a third trial targets hepatoma or those tumors with liver metastasis (NCT00882180). Two other trials are targeting melanoma via immunotherapeutic approaches, where siRNA-transfected (in one case, mRNA) dendritic cells are being utilized (NCT 00672542 and NCT00929019).

22.9 Conclusions

Given the hurdles faced by gene therapy in the past as an adjuvant therapeutic avenue against tumors, a somewhat different set of obstacles need to be overcome by RNAi drugs, primarily targeted delivery to the tumor (in a systemic application mode). However, the pre-clinical studies in animal models and results from the ongoing clinical trials (listed above) indicate the promise of RNAi as a potential adjuvant therapy against both malignant brain tumors as well as against other solid tumors. These initial success stories with systemic delivery clearly indicate that both research and clinical communities have bypassed the threshold problem of targeted delivery of therapeutic doses of RNAi to the tumor site. Studies with miRNA are already indicating their utility as a molecular tool for diagnostic screening and staging of tumors in the pathology laboratory, which are currently supplementing the traditional immune-histological methods. Thus, after a long pause after the discovery at the bench, these miniscule RNA entities are finally displaying their usefulness as a clinical tool for both diagnostic and therapeutic purposes.

Acknowledgments Research support for S.P. Mathupala was provided by a grant from the National Cancer Institute/National Institute of Health (CA 116257), the Fund for Medical Research and Education (FMRE), Wayne State University, and a gift from the Marvin E. Klein, M.D., Charitable Trust. A.E. Sloan is supported by grants from the National Cancer Institute/National Institute of Health (KO8 101954) and the Case Western Reserve University School of Medicine.

References

Amarzguioui M, Lundberg P, Cantin E, et al. Rational design and in vitro and in vivo delivery of Dicer substrate siRNA. Nat Protoc. 2006;1:508–17.

Birchler JA, Kavi HH. Molecular biology. Slicing and dicing for small RNAs. Science. 2008;320:1023–4.

Boado RJ. RNA interference and nonviral targeted gene therapy of experimental brain cancer. NeuroRx. 2005;2:139–50.

Boado RJ. Blood-brain barrier transport of non-viral gene and RNAi therapeutics. Pharm Res. 2007;24:1772–87.

Brummelkamp TR, Bernards R, Agami R. A system for stable expression of short interfering RNAs in mammalian cells. Science. 2002;296:550–3.

Bumcrot D, Manoharan M, Koteliansky V, et al. RNAi therapeutics: a potential new class of pharmaceutical drugs. Nat Chem Biol. 2006;2:711–9.

Cardoso ALC, Simoes S, de Almeida LP, et al. SiRNA detivery by a transferrin-associated lipid-based vector: a non-viral strategy to mediate gene silencing. J Gene Med. 2007;9:170–83.

Chan JA, Krichevsky AM, Kosik KS. MicroRNA-21 is an antiapoptotic factor in human glioblastoma cells. Cancer Res. 2005;65:6029–33.

Chang K, Elledge SJ, Hannon GJ. Lessons from Nature: microRNA-based shRNA libraries. Nat Methods. 2006;3:707–14.

Conti A, Aguennouz M, La Torre D, et al. MiR-21 and 221 upregulation and miR-181b downregulation in human grade II-IV astrocytic tumors. J Neurooncol. 2009;93:325–32.

Corsten MF, Miranda R, Kasmieh R, et al. MicroRNA-21 knockdown disrupts glioma growth in vivo and displays synergistic cytotoxicity with neural precursor cell delivered S-TRAIL in human gliomas. Cancer Res. 2007;67:8994–9000.

Cullen BR. Enhancing and confirming the specificity of RNAi experiments. Nat Methods. 2006a;3:677–81.

Cullen BR. Induction of stable RNA interference in mammalian cells. Gene Ther. 2006b;13:503–8.

Davis ME, Zuckerman JE, Choi CH, et al. Evidence of RNAi in humans from systemically administered siRNA via targeted nanoparticles. Nature. 2010;464:1067–70.

Dong WL, Hu J, Hu SY, et al. RNA interference affects tumorigenicity and expression of insulin-like growth factor-1, insulin-like growth factor-1 receptor, and basic fibroblast growth factor-2 in rat C6 glioma cells. Neural Regen Res. 2009;4:597–605.

Elbashir SM, Harborth J, Lendeckel W, et al. Duplexes of 21-nucleotide RNAs mediate RNA interference in cultured mammalian cells. Nature. 2001;411:494–8.

Fire A, Xu S, Montgomery MK, et al. Potent and specific genetic interference by double-stranded RNA in *Caenorhabditis elegans*. Nature. 1998;391:806–11.

Fountaine TM, Wood MJ, Wade-Martins R. Delivering RNA interference to the mammalian brain. Curr Gene Ther. 2005;5:399–410.

Gabriely G, Wurdinger T, Kesari S, et al. MicroRNA 21 promotes glioma invasion by targeting matrix metalloproteinase regulators. Mol Cell Biol. 2008;28:5369–80.

Gillespie DL, Whang K, Ragel BT, et al. Silencing of hypoxia inducible factor-1alpha by RNA interference attenuates human glioma cell growth in vivo. Clin Cancer Res. 2007;13:2441–8.

Goga A, Benz C. Anti-oncomir suppression of tumor phenotypes. Mol Interv. 2007;7:199–202.

Gogineni VR, Nalla AK, Gupta R, et al. Radiation-inducible silencing of uPA and uPAR in vitro and in vivo in meningioma. Int J Oncol. 2010;36:809–16.

Gondi CS, Dinh DH, Klopfenstein JD, et al. MMP-2 downregulation mediates differential regulation of cell death via ErbB-2 in glioma xenografts. Int J Oncol. 2009;35:257–63.

Gondi CS, Lakka SS, Dinh DH, et al. Intraperitoneal injection of a hairpin RNA-expressing plasmid targeting urokinase-type plasminogen activator (uPA) receptor and uPA retards angiogenesis and inhibits intracranial tumor growth in nude mice. Clin Cancer Res. 2007;13:4051–60.

Guo S, Kemphues KJ. Par-1, a gene required for establishing polarity in *C. elegans* embryos, encodes a putative Ser/Thr kinase that is asymmetrically distributed. Cell. 1995;81:611–20.

Guo D, Wang B, Han F, et al. RNA interference therapy for glioblastoma. Expert Opin Biol Ther. 2010;10:927–36.

He L, Thomson JM, Hemann MT, et al. A microRNA polycistron as a potential human oncogene. Nature. 2005;435:828–33.

Heidel JD, Hu S, Liu XF, et al. Lack of interferon response in animals to naked siRNAs. Nat Biotechnol. 2004;22:1579–82.

Hubbell HR, Boyer JE, Roane P, et al. Cyclic AMP mediates the direct antiproliferative action of mismatched double-stranded RNA. Proc Natl Acad Sci USA. 1991;88:906–10.

Jackson AL, Burchard J, Schelter J, et al. Widespread siRNA "off-target" transcript silencing mediated by seed region sequence complementarity. RNA. 2006;12:1179–87.

Jain KK. Use of nanoparticles for drug delivery in glioblastoma multiforme. Expert Rev Neurother. 2007;7:363–72.

Kang CS, Zhang ZY, Jia ZF, et al. Suppression of EGFR expression by antisense or small interference RNA inhibits U251 glioma cell growth in vitro and in vivo. Cancer Gene Ther. 2006;13:530–8.

Kapranov P, Cheng J, Dike S, et al. RNA maps reveal new RNA classes and a possible function for pervasive transcription. Science. 2007;316:1484–8.

Kargiotis O, Chetty C, Gondi CS, et al. Adenovirus-mediated transfer of siRNA against MMP-2 mRNA results in impaired invasion and tumor-induced angiogenesis, induces apoptosis in vitro and inhibits tumor growth in vivo in glioblastoma. Oncogene. 2008;27:4830–40.

Khvorova A, Reynolds A, Jayasena SD. Functional siRNAs and miRNAs exhibit strand bias. Cell. 2003;115:209–16.

Kim DH, Behlke MA, Rose SD, et al. Synthetic dsRNA Dicer substrates enhance RNAi potency and efficacy. Nat Biotechnol. 2005;23:222–6.

Kim VN, Nam JW. Genomics of microRNA. Trends Genet. 2006;22:165–73.

Kim DH, Rossi JJ. Strategies for silencing human disease using RNA interference. Nat Rev Genet. 2007;8:173–84.

Krol J, Krzyzosiak WJ. Structural aspects of microRNA biogenesis. IUBMB Life. 2004;56: 95–100.

Krutzfeldt J, Rajewsky N, Braich R, et al. Silencing of microRNAs in vivo with 'antagomirs'. Nature. 2005;438:685–9.

Lau NC, Seto AG, Kim J, et al. Characterization of the piRNA complex from rat testes. Science. 2006;313:363–7.

Lavon I, Zrihan D, Granit A, et al. Gliomas display a microRNA expression profile reminiscent of neural precursor cells. Neuro Oncol. 2010;12:422–33.

Lesniak MS. Novel advances in drug delivery to brain cancer. Technol Cancer Res Treat. 2005;4:417–28.

Liu J, Carmell MA, Rivas FV, et al. Argonaute2 is the catalytic engine of mammalian RNAi. Science. 2004;305:1437–41.

Liu X, Fortin K, Mourelatos Z. MicroRNAs: biogenesis and molecular functions. Brain Pathol. 2008;18:113–21.

Louis DN, Ohgaki H, Wiestler OD, et al. The 2007 WHO classification of tumours of the central nervous system. Acta Neuropathol. 2007;114:97–109.

Lu XM, Zhao P, Zhang CZ, et al. Analysis of miR-221 and p27 expression in human gliomas. Mol Med Rep. 2009;2:651–6.

Malecova B, Morris KV. Transcriptional gene silencing through epigenetic changes mediated by non-coding RNAs. Curr Opin Mol Ther. 2010;12:214–22.

Malzkorn B, Wolter M, Liesenberg F, et al. Identification and functional characterization of microRNAs involved in the malignant progression of gliomas. Brain Pathol. 2010;20: 539–50.

Mathupala SP, Guthikonda M, Sloan AE. RNAi based approaches to the treatment of malignant glioma. Technol Cancer Res Treat. 2006;5:261–9.

Mathupala SP, Mittal S, Guthikonda M, et al. MicroRNA and brain tumors: a cause and a cure? DNA Cell Biol. 2007;26:301–10.

Meltzer PS. Cancer genomics: small RNAs with big impacts. Nature. 2005;435:745–6.

Moazed D. Small RNAs in transcriptional gene silencing and genome defence. Nature. 2009;457:413–20.

Myslinski E, Ame JC, Krol A, et al. An unusually compact external promoter for RNA polymerase III transcription of the human H1RNA gene. Nucleic Acids Res. 2001;29:2502–9.

Napoli C, Lemieux C, Jorgensen R. Introduction of a chimeric chalcone synthase gene into petunia results in reversible co-suppression of homologous genes in trans. Plant Cell. 1990;2: 279–89.

Nelson P, Kiriakidou M, Sharma A, et al. The microRNA world: small is mighty. Trends Biochem Sci. 2003;28:534–40.

Okamura K, Lai EC. Endogenous small interfering RNAs in animals. Nat Rev Mol Cell Biol. 2008;9:673–8.

Pardridge WM. Intravenous, non-viral RNAi gene therapy of brain cancer. Expert Opin Biol Ther. 2004;4:1103–13.

Pardridge WM. shRNA and siRNA delivery to the brain. Adv Drug Deliv Rev. 2007;59:141–52.

Pillai RS. MicroRNA function: multiple mechanisms for a tiny RNA? RNA. 2005;11:1753–61.

Pirollo KF, Chang EH. Targeted delivery of small interfering RNA: approaching effective cancer therapies. Cancer Res. 2008;68:1247–50.

Preall JB, He Z, Gorra JM, et al. Short interfering RNA strand selection is independent of dsRNA processing polarity during RNAi in Drosophila. Curr Biol. 2006;16:530–5.

Pu PY, Kang CS, Zhang ZY, et al. Downregulation of PIK3CB by siRNA suppresses malignant glioma cell growth in vitro and in vivo. Technol Cancer Res. 2006;5:271–80.

Pu P, Zhang Z, Kang C, et al. Downregulation of Wnt2 and beta-catenin by siRNA suppresses malignant glioma cell growth. Cancer Gene Ther. 2009;16:351–61.

Reynolds A, Leake D, Boese Q, et al. Rational siRNA design for RNA interference. Nat Biotechnol. 2004;22:326–30.

Rolle K, Nowak S, Wyszko E, et al. Promising human brain tumors therapy with interference RNA intervention (iRNAi). Cancer Biol Ther. 2010;9:396–406.
Rose SD, Kim DH, Amarzguioui M, et al. Functional polarity is introduced by Dicer processing of short substrate RNAs. Nucleic Acids Res. 2005;33:4140–56.
Sasayama T, Nishihara M, Kondoh T, et al. MicroRNA-10b is overexpressed in malignant glioma and associated with tumor invasive factors, uPAR and RhoC. Int J Cancer. 2009;125:1407–13.
Saydam O, Glauser DL, Heid I, et al. Herpes simplex virus 1 amplicon vector-mediated siRNA targeting epidermal growth factor receptor inhibits growth of human glioma cells in vivo. Mol Ther. 2005;12:803–12.
Seitz H. Redefining microRNA targets. Curr Biol. 2009;19:870–3.
Seno T, Harada H, Kohno S, et al. Downregulation of SPARE expression inhibits cell migration and invasion in malignant gliomas. Int J Oncol. 2009;34:707–15.
Silva JM, Li MZ, Chang K, et al. Second-generation shRNA libraries covering the mouse and human genomes. Nat Genet. 2005;37:1281–8.
Siolas D, Lerner C, Burchard J, et al. Synthetic shRNAs as potent RNAi triggers. Nat Biotechnol. 2005;23:227–31.
Taft RJ, Glazov EA, Cloonan N, et al. Tiny RNAs associated with transcription start sites in animals. Nat Genet. 2009;41:572–8.
Takenaga K, Nygren J, Zelenina M, et al. Modified expression of Mts1/S100A4 protein in C6 glioma cells or surrounding astrocytes affects migration of tumor cells in vitro and in vivo. Neurobiol Dis. 2007;25:455–63.
Uchida H, Tanaka T, Sasaki K, et al. Adenovirus-mediated transfer of siRNA against survivin induced apoptosis and attenuated tumor cell growth in vitro and in vivo. Mol Ther. 2004;10:162–71.
Ulbricht U, Eckerich C, Fillbrandt R, et al. RNA interference targeting protein tyrosine phosphatase zeta/receptor-type protein tyrosine phosphatase beta suppresses glioblastoma growth in vitro and in vivo. J Neurochem. 2006;98:1497–506.
Verdel A, Vavasseur A, Le Gorrec M, et al. Common themes in siRNA-mediated epigenetic silencing pathways. Int J Dev Biol. 2009;53:245–57.
Wang XL, Xu RZ, Wu XM, et al. Targeted systemic delivery of a therapeutic siRNA with a multifunctional carrier controls tumor proliferation in mice. Mol Pharmaceut. 2009;6:738–46.
Watanabe T, Totoki Y, Toyoda A, et al. Endogenous siRNAs from naturally formed dsRNAs regulate transcripts in mouse oocytes. Nature. 2008;453:539–43.
Wyszko E, Rolle K, Nowak S, et al. A multivariate analysis of patients with brain tumors treated with atn-RNA. Acta Pol Pharm. 2008;65:677–84.
Xia CF, Zhang Y, Boado RJ, et al. Intravenous siRNA of brain cancer with receptor targeting and avidin-biotin technology. Pharm Res. 2007;24:2309–16.
Xie FY, Woodle MC, Lu PY. Harnessing in vivo siRNA delivery for drug discovery and therapeutic development. Drug Discov Today. 2006;11:67–73.
Xu P, Qiu MZ, Zhang ZY, et al. The oncogenic roles of Notch1 in astrocytic gliomas in vitro and in vivo. J Neurooncol. 2010;97:41–51.
Zhang LY, Alizadeh D, Van Handel M, et al. Stat3 Inhibition Activates Tumor Macrophages and Abrogates Glioma Growth in Mice. Glia. 2009b;57:1458–67.
Zhang BB, Gu F, She CH, et al. Reduction of Akt2 inhibits migration and invasion of glioma cells. Int J Cancer. 2009a;125:585–95.
Zhang JX, Han L, Ge YL, et al. MiR-221/222 promote malignant progression of glioma through activation of the Akt pathway. Int J Oncol. 2010;36:913–20.
Zhang Y, Zhang YF, Bryant J, et al. Intravenous RNA interference gene therapy targeting the human epidermal growth factor receptor prolongs survival in intracranial brain cancer. Clin Cancer Res. 2004;10:3667–77.
Zhao P, Wang C, Fu Z, et al. Lentiviral vector mediated siRNA knock-down of hTERT results in diminished capacity in invasiveness and in vivo growth of human glioma cells in a telomere length-independent manner. Int J Oncol. 2007;31:361–8.

Zhou X, Ren Y, Moore L, et al. Downregulation of miR-21 inhibits EGFR pathway and suppresses the growth of human glioblastoma cells independent of PTEN status. Lab Invest. 2010;90: 144–55.

Zukiel R, Nowak S, Wyszko E, et al. Suppression of human brain tumor with interference RNA specific for tenascin-C. Cancer Biol Ther. 2006;5:1002–7.

Index

A

Activator protein-1 (AP-1), 177
Acute lymphoblastic leukemia (ALL), 47, 270, 423, 426, 431, 434, 469, 470–471
Acute myeloid leukemia (AML), 14, 32, 35, 51, 270–273, 280, 316, 423, 435
Acute promyelocytic leukemia (APL), 270–271
Adenomatous polyposis coli (APC), 108–109, 115, 354
Adjuvant therapy, 50, 124, 182–183, 416
Adriamycin, 49
Akt signaling, 175, 251, 541
Alpha-fetoprotein (AFP), 172, 181, 378, 500
Androgen, 147–149, 155, 396, 406, 434
Antagomir, 84, 99, 102, 127, 141, 153, 156, 158, 165, 304, 351, 364, 488–489, 542
Anti-cancer agent, 452
Anti-estrogen, 98–99, 102
Anti-metastatic microRNA, 396, 403–407
Anti-miR, 50, 84, 96, 114, 182, 196, 210, 213–214, 300, 326, 335, 361–363, 383, 472, 475, 493
Anti-miRNA oligonucleotides, 52, 326, 335, 361–363
Antisense microRNA oligonucleotides (AMOs), 127
Apoptosis, 2–6, 19, 26, 33, 37, 45, 49–50, 80–81, 84, 93, 96–97, 110–113, 135, 138, 140, 154–155, 165, 169–170, 173–177, 183, 190, 203, 210–213, 227–228, 231–232, 234, 245–246, 249, 252, 257, 273, 276–277, 298, 300, 303, 315–318, 322, 349, 361, 375, 381, 390, 400, 405, 419–421, 423, 430–431, 433, 450–452, 465, 471–472, 474–475, 486, 492–494, 523, 539, 541, 543
Argonaute-2 (Ago-2), 536–537
Array comparative genomic hybridization (aCGH), 319
Arsenic, 425, 431, 466, 472
AUC (the areas under the ROC curve), 118, 522
Azacytidine, 435
5-Aza-deoxycytidine, 296

B

Barrett's esophagus (BE), 202, 204–208, 213
B cell, 2–3, 27, 32, 40, 47–48, 51, 240, 242–243, 245–246, 248–252, 254, 256–257, 273, 275–277, 280, 294, 300, 316–317, 325, 355, 357, 378, 418, 422, 507–508
B cell lymphoma, 3, 48, 51, 243, 248, 251, 256–257, 276, 300, 316, 325, 355, 357, 422, 507–508
BCR-ABL, 270, 273–275
Bead-based flow cytometric miRNA expression profiling, 272
Bead-based hybridization, 152
Bead-based microRNA profiling, 272, 519
Benign prostatic hyperplasia (BPH), 146, 148, 151, 500
Bevacizumab (Avastin), 434
Biological marker, 288
Biomarker, 50–51, 74, 76–77, 96, 108, 117–118, 123, 125, 128, 145–159, 168, 178–184, 203–206, 208–209, 226–228, 254–261, 272, 287–304, 310, 312–313, 318, 321, 325–326, 360–361, 395, 416–418, 421–423, 434, 475, 499–528
Bladder cancer, 212, 215, 223–235, 417, 422, 434, 436, 470, 516, 518
Blood-brain barrier (BBB), 363, 475, 540, 543
Bmi-1, 178, 323, 327, 346, 353, 355, 382–383
Body fluid, 361, 416, 501–507, 512–518, 522, 527–528

W.C.S. Cho (ed.), *MicroRNAs in Cancer Translational Research*,
DOI 10.1007/978-94-007-0298-1, © Springer Science+Business Media B.V. 2011

Bortezomib, 435
Brain tumor, 343–363, 533–545
Breast cancer, 48–49, 82–83, 91–103, 114, 137, 190, 196, 212–213, 316–317, 319, 324, 350, 358, 374, 376–378, 382–384, 391, 394–395, 397–402, 404–407, 417–418, 420–424, 428–429, 434–435, 452, 454, 457–459, 461–462, 466, 468–472, 474, 500, 505, 510
Burkitt lymphoma (BL), 243–251, 255–257

C
Caenorhabditis elegans (*C. elegans*), 1, 78, 203, 224, 313, 316, 378, 501, 519, 535
Camptothecin, 155, 427, 434, 450, 467
Cancer antigen 125 (CA125), 312
Cancer biomarker, 50–51, 421, 516, 521, 527
Cancer classification, 521
Cancer diagnosis, 178, 223, 318, 359, 390, 503, 509–510
Cancer prognosis, 195, 319, 326, 422
Cancer stem cell (CSC), 3, 6–7, 42, 93, 166–167, 172, 180–181, 183, 352–353, 373–385, 404, 418, 432, 473
Cancer therapy, 50, 102, 141, 181, 335, 362, 383, 385, 394–395, 404, 407–408, 486
Cancer therapy response, 109
Capecitabine, 50, 125, 427, 433, 453
Carbohydrate antigen 15–3 (CA 15–3), 500
Carbohydrate antigen 19–9 (CA19–9), 312, 500
Carboplatin, 434, 450
Carcinoembryonic antigen (CEA), 312, 500
Carcinogenesis, 82, 107, 109, 123, 136–138, 172, 183, 190–191, 193–195, 202–203, 206, 208, 214–216, 245, 293, 323, 489, 494, 501–502
Case-control study, 116, 151
β-Catenin, 100, 108–109, 171–172, 354, 357, 378, 421, 473
Cerebrospinal fluid (CSF), 331, 361
Cervical cancer, 189–196, 316, 469, 473, 500
Cetuximab (Erbitux), 126, 427, 434
Chemoradiotherapy, 50, 125
Chemoresistance, 97, 126, 159, 324, 326, 429, 433
Chemosensitivity, 49, 157, 196, 326, 423, 433–434
Chemotherapy (CT), 49–50, 80, 102, 125, 128, 147, 180, 182, 196, 250, 260, 262, 274, 288, 300, 303, 310, 312, 326, 344, 353, 360, 415–436, 449–450, 534
Chemotherapy resistance, 250
Chromogranin A (CgA), 500
Chromosomal abnormalities, 136
Chronic lymphocytic leukemia (CLL), 2, 13–15, 30–32, 34, 40, 44, 48–49, 139, 156, 243–258, 260–261, 270, 275–280, 315–317, 323, 419–420, 425, 430–431, 467, 470–471
Chronic myeloid leukemia (CML), 269–270, 273–275, 425, 431, 460, 466–467, 471
Circulating microRNA, 77, 183, 303, 325, 422, 501–516, 519–527
Cirrhosis, 166–167, 521
Cisplatin, 50, 196, 326, 425–426, 430, 432, 450, 452, 454–455, 466–473
Clinical trial, 84–85, 147, 158, 385, 416–418, 423, 434–435, 537, 539–540, 544
Cloning technology, 118
c-Myc, 3, 78–79, 81, 98, 112, 136–138, 178, 211, 214, 246–247, 250–253, 255, 257, 273, 295, 316–318, 328, 352, 354–355, 357, 404, 487–489, 493–494
Colorectal cancer (CRC), 107–118, 121–125, 127–128, 391, 417, 504–505, 507–509
Common fragile site (CFS), 202
Comparative genomic hybridization (CGH), 202, 243, 319
Copy number alterations, 243, 319
Cox regression analysis, 124, 150
CpG islands, 111, 123, 171, 226, 230–231
CXC chemokine receptor-4 (CXCR4), 38, 45, 102
Cyclin-dependent kinase (CDK), 2, 6, 19, 29–30, 32–34, 44, 108, 155, 173, 252, 297, 354–355, 357, 378, 419, 492
Cyclooxygenase-2 (COX-2), 31, 34, 113, 416
Cyclophosphamide, 49, 256, 260

D
Deep sequencing, 149, 152, 355
Dendritic cell, 544
Diagnostic marker, 301–302, 405
Dicer, 2, 19, 115, 122, 164, 171, 190, 233, 314–315, 318, 320–321, 335, 345, 403, 451–452, 536–538
Diffuse large B cell lymphoma (DLBCL), 48, 51, 243–244, 246–249, 251, 254–257, 259–260, 276, 316, 503, 507–508
Dihydrofolate reductase, 436
DNA methyltransferases (DNMTs), 36, 273, 280
DNA reparation, 113
Docetaxel, 147, 333–334, 423, 426, 430, 432, 434, 450, 456, 464, 467

Double-stranded RNA (dsRNA), 153, 164, 535–536
Doxorubicin, 154–155, 327–328, 425, 427, 429–430, 433, 450, 453, 457, 466–472
Drosha, 98, 164, 171, 178, 190, 233, 314–315, 318, 320–321, 335, 451, 494, 537–538
Drosophila, 26–27, 35, 38, 313, 361, 435
Drug resistance, 50, 394, 399, 431, 449–476
Drug sensitivity, 49, 76, 182–183, 450, 465, 474
Drug transport protein, 418, 429, 433

E
E2F family, 112, 154, 174
Early diagnosis, 50–51, 117, 178, 202, 331
E-cadherin, 21, 100–101, 114, 154, 156, 177, 210, 214–215, 299, 302, 324–325, 399–402, 421, 426, 473
EGFR signaling, 109–110, 349, 430
Electroporation, 384
Embryonal brain tumor, 344–345, 347, 353–359, 361
Embryonic stem (ES) cell, 6, 376–379
Endoscopic ultrasound (EUS), 295, 301
Enhancer of zeste homolog (EZH2), 31, 34, 156, 246, 251, 397, 399
Enzyme-linked immunosorbent assay (ELISA), 501
Epidermal growth factor receptor (EGFR), 36, 76–80, 91, 108–110, 126, 137, 299, 303, 328, 331, 344, 346–350, 360, 397, 417, 425, 428–430, 434, 467, 469, 539–540, 543
Epigenetics, 4, 49, 81, 83, 92, 107, 118, 140–141, 156, 168, 171–172, 244, 247–248, 251, 257, 271, 273, 288, 295–296, 311, 315, 318–320, 345, 348, 354, 450, 452, 500, 535
Epithelial-mesenchymal transition (EMT), 82, 95, 100, 108, 113–114, 154, 156, 211, 214, 225–226, 298, 324–325, 373, 380–381, 399, 418, 473
Epithelial ovarian cancer (EOC), 309–336, 454–455, 509
Epothilone-B, 426
Epstein-Barr virus (EBV), 81, 138, 244–249
Erlotinib (Tarceva), 80, 423, 427, 434, 450, 470
Esophageal adenocarcinoma (EAC), 202–214
Esophageal cancer (EC), 48, 201–216, 395, 402, 421, 471
Esophageal squamous cell carcinoma (ESCC), 202–206, 208–209, 212–215, 395
Estrogen receptor (ER), 23, 25, 37, 44, 91, 93, 95, 97–99, 174, 296, 424, 428, 474–475
Etoposide, 154, 425–426, 429–430, 433, 450, 458, 466, 469, 471
Event-free survival (EFS), 260, 272
Exosomes, 325, 523–524
Expression profiling, 8–9, 46–47, 82, 91, 109, 117–119, 122–123, 128, 139, 153, 159, 171–172, 191, 227, 254, 257, 259, 270–271, 278, 332, 353–354, 421, 423–424, 428–429, 431, 435, 519, 542
Extracellular matrix (ECM), 101, 108, 113–114, 177, 350, 395, 404, 407, 539, 544

F
Fine needle aspiration (FNA), 295, 301
Five-year survival, 233
Fluorouracil (5-FU), 125, 432, 450
Follicular lymphoma (FL), 248, 254–256, 276, 316
Formalin-fixed paraffin-embedded (FFPE), 46–47, 51, 76, 123–124, 204, 302–303, 455
Functional study, 9, 93, 102, 172, 190–191

G
Gain-of-function study, 159
Gastric cancer, 117, 135–141, 215, 419–420, 424, 464, 466–469, 506, 509–510
Gastro-esophageal reflux disease (GERD), 202
Gefitinib (Iressa), 80, 425, 429–430, 450, 466, 469, 470
Gemcitabine, 49, 298, 300, 303, 333, 426–427, 431–432, 434, 453, 459, 466, 467, 469–470, 472
Gene copy number, 79, 312, 319
Gene Ontology, 115
Genome-wide microRNA expression studies, 148, 272
Glioblastoma, 11, 13–15, 26–27, 38, 44, 109, 114, 137, 190, 317, 344, 346–347, 349–350, 374, 382, 391, 395, 397, 401, 406, 433, 462, 466–467, 469, 471, 492, 544
Glioblastoma multiforme (GBM), 433
Glioma, 34, 36, 212, 303, 344–353, 356–357, 359, 360, 362, 382, 406, 420, 424, 426, 430, 433, 435, 492, 536–537, 539–544
Glucocorticoids, 212, 426
Glycolysis, 356, 486–488, 490–493, 495

H
HBx, 171
Head and neck squamous cell carcinoma (HNSCC), 406, 468
Hedgehog signaling, 354
Helicobacter pylori (*H. pylori*), 40, 137–138, 140

Hematological cancer, 136, 210, 430–431, 502
Hepatitis B virus (HBV), 164, 166–167, 170–171, 179–180, 521
Hepatitis C virus (HCV), 84, 164, 166–167, 177, 180, 433, 475
Hepatocellular carcinoma (HCC), 9, 48, 50–51, 84, 110, 163–184, 190, 378, 383, 391, 397, 402, 404–405, 420, 433, 453, 466–469, 471, 474–475, 521
Hereditary non-polyposis colorectal cancer (HNPCC), 122
High-grade dysplasia (HGD), 201, 204–206
High-mobility group A protein 2 (HMGA2), 2–3, 6, 19, 78–79, 93–94, 177–178, 298, 316, 323, 327, 379, 397, 404, 419, 468
Histone deacetylase (HDAC), 140, 171, 226, 261, 296, 318, 320, 327, 333, 335, 418–419, 425, 430, 469
Hodgkin's lymphoma (HL), 47, 259
Homeobox protein A1 (HOXA1), 298, 327
Hormonal therapy, 428
Human chorionic gonadotropin (hCG), 500
Human epidermal growth factor receptor 2 (HER2 or ErbB2), 91
Human papillomavirus (HPV), 190–191, 193–196
Hypoxia, 4–5, 23–24, 42–43, 113, 211, 296, 348, 351–352, 407, 416–417, 421, 428, 489–491, 542
Hypoxia-inducible factor-1 (HIF-1), 4, 23–24, 42, 113, 348, 352, 407, 489–491, 493–495, 542, 544
Hypoxia responsive element (HRE), 296

I
Imatinib mesylate (IM) (Gleevec), 426, 431
Inflammation, 40, 202, 311, 323, 523–524
In situ hybridization (ISH), 76, 92, 124, 165, 209, 227, 258, 293–294, 301, 489
Insulin-like growth factor (IGF), 19, 96, 112, 331, 379, 540
Interferon-γ, 427, 433
International Agency for Research on Cancer (IARC), 534
International Union Against Cancer (UICC), 124, 128
Intracranial tumor, 534
Invasion, 2, 4, 17, 45, 48, 50, 82–83, 93–97, 99–103, 114, 128, 149, 154–156, 169–170, 172–173, 175–177, 183, 192, 196, 210–212, 214, 229, 298–299, 303, 323, 348–351, 353, 358, 383, 389–409, 419, 421, 472, 474, 524, 539–543
Irinotecan, 126

K
Kaplan-Meier, 47, 124–125, 150, 209
Kaposi's sarcoma-associated herpesvirus (KSHV), 245
Knock-out, 2, 159
KRAS, 18, 36, 76, 78–79, 83, 108–110, 125–126, 216, 229, 231, 293, 327, 331, 424–425

L
Lapatinib, 425, 429, 450, 470
Leukemia, 2, 8, 32–33, 35–36, 39–40, 42, 47, 51, 139, 156, 212, 243, 248, 256, 269–281, 294, 315–316, 356, 374–375, 391, 397, 418, 423, 430–431, 495, 505, 510
Liposome, 102, 384, 540, 543
LKB1, 352, 491–492
Locked nucleic acid (LNA), 83, 127, 153, 261, 384, 475
Logistic regression, 195
Loss-of-function study, 214
Loss of heterogeneity, 202
Loss of heterozygosity (LOH), 80, 92, 96, 136, 243
Low-density lipoproteins (LDL), 42, 543
Low-grade dysplasia (LGD), 204–206
Luciferase assay, 17, 211, 489
Lung cancer, 48, 51, 73–85, 109, 117, 137, 139, 146, 190, 214–216, 280, 316, 318, 336, 352, 376–377, 383, 385, 391, 394, 397–398, 404, 406, 419–420, 422–424, 429–430, 436, 471–472, 499, 507, 521, 526
Lymphoma, 3–4, 7, 13, 18, 21, 23, 32–33, 39–40, 47–48, 51, 78, 81, 99, 112, 239–262, 274, 276–277, 300, 316–317, 325, 355, 357, 377, 422, 487–488, 495, 507–508, 526

M
Mammalian target of rapamycin (mTOR), 173, 300
MammaPrint, 435
Mantle cell lymphoma (MCL), 211, 215, 243, 246–250, 252–253, 255–256, 258, 261, 276–277, 316
Matrix metalloproteinase (MMP), 113–114, 203, 211, 298–299, 406, 541, 543
Medulloblastoma (MB), 97, 344–346, 353–357, 359–362, 381
Mesenchymal-epithelial transition (MET), 6, 28, 41–42, 44, 101, 112–113, 156, 170, 211, 349, 382, 394, 397, 419, 425, 430, 466, 469

Metabolism, 113, 155, 177, 190, 203, 251, 303, 348, 351–352, 451, 485–495, 539
Metastasis, 4, 17, 21, 32, 45, 73–85, 92–95, 99–102, 110–111, 114, 123, 128–129, 150, 155–157, 164, 166–170, 173, 175–180, 182, 191, 195–196, 203, 211–214, 287, 297–300, 303, 310, 318, 321, 344, 351, 375, 382–383, 389–409, 416, 419–422, 428, 433, 450, 507, 523, 541, 544
Metastatic microRNA, 100, 396, 401–403
Methotrexate, 50, 126, 436, 467, 469–470
Microarray, 7, 9, 46–47, 51, 75–76, 91–92, 97, 99, 115, 119–123, 137, 140, 148, 152, 165, 173, 175, 191, 206, 208, 224, 229–230, 234, 270–272, 278–280, 288–289, 293, 301, 320, 322, 353, 376, 454–464, 487, 502, 505, 509–510, 512–513, 516, 518–519, 521–522
Microdissection, 289, 322
Microparticle, 523–524
MicroRNA biogenesis, 107, 117, 126, 164–165, 215, 229–230, 313–315, 318–321, 335, 383, 451, 490, 494
MicroRNA deregulation, 3–7, 12, 16, 46, 92, 168, 192, 317–322
MicroRNA diagnostics, 435–436
MicroRNA microarray, 91, 123, 137, 140, 173, 230, 278–280, 289, 301, 376, 516, 519
MicroRNA pharmacogenomics, 50
MicroRNA processing machinery, 114–115, 118
MicroRNA registry (miRBase), 1, 136, 165, 190, 203, 212, 375, 451, 465
MicroRNA replacement therapy, 102, 196
MicroRNA serial analysis of gene expression (miRAGE), 119, 122
MicroRNA therapeutics, 102, 543
Microsatellite instability (MSI), 47, 115, 119–122
Microsatellite stability (MSS), 119–122
Microvesicle (MV), 361, 523–524
Milk, 501, 516, 518
MiRBase, 1, 136, 165, 190, 203, 212, 375
Mitoxantrone, 425, 427, 429, 433, 471
Molecular epidemiology, 216
Molecular heterogeneity, 353
Multi-drug resistance (MDR), 138, 429
Multiple myeloma (MM), 47, 243, 246–249, 252–253, 255–256, 276, 423, 435
Multivariate analysis, 124–125, 302
Mutation, 2, 4, 7, 50, 77–80, 92, 115, 119, 124, 126, 168, 192, 244, 249–251, 271–273, 275–276, 278, 288, 297, 311–312, 314–315, 317, 321, 323, 344, 347, 355, 416, 432, 450, 491, 493
MYB, 240, 248, 277

N

Nasopharyngeal carcinoma (NPC), 392, 395, 398
Natural killer/T cell (NK/T) lymphoma, 256
NCI-60, 49, 76, 79, 423
Neural differentiation, 345, 353, 358
Neuroblastoma, 18, 23, 25, 36, 158, 362, 422–423
Neuron-specific enolase (NSE), 500
Never-smoker, 77–78
Next generation sequencing, 165, 273, 321–322
Non-coding RNA, 1, 74, 85, 135, 190, 313, 375, 501, 535–536
Non-Hodgkin's lymphoma (NHL), 47
Non-small cell lung cancer (NSCLC), 74, 214, 216, 383, 423, 471–472, 499, 503, 507
Normal squamous epithelium (NSE), 204–207, 211, 500
Northern blot, 293, 488
Notch receptors, 541
Notch signaling, 196, 353, 356
Novel therapeutics, 344, 351
Nuclear factor kappa B (NF-κB), 171, 173, 177, 182, 244, 249, 251, 299, 323, 327, 397, 400, 417, 427, 433, 473
Nucleoside analog S–1, 427, 433
Nucleotide excision repair (NER), 113

O

Odds ratio (OR), 117, 216, 230
Oligodendroglioma, 345, 347
Oligonucleotide array sequence analysis, 127
2′-*O*-methyl-antisense oligonucleotides, 153
Oncogene, 2–3, 6, 18, 22, 28, 30, 34–38, 40–42, 44, 46, 78, 93, 110, 137–138, 154, 156, 173, 193–194, 196, 212, 230, 244, 249, 252, 257, 274, 280, 288, 297–298, 316–317, 323, 362, 377, 394, 404, 418, 424, 431, 452, 472, 474, 487, 493
Oncomir, 81, 97, 246–248, 316, 394
Oncotype DX, 435
Oral squamous cell carcinoma, 422, 506, 510, 516–517
Osteosarcoma, 34, 418, 423, 469–471
Ovarian cancer, 50–51, 79, 117, 146, 309–336, 392, 402, 418, 432, 454–456, 458, 460–461, 463, 466–467, 468, 470–472, 500, 504, 509

Overall survival (OS), 47–48, 50, 75, 80, 126, 140, 181, 209, 253, 259–261, 279, 302, 321, 405, 507, 511
Oxaliplatin, 125, 450, 453
Oxidative phosphorylation, 486, 490

P

p53, 4–7, 27, 34, 38, 48–49, 80, 82, 93, 97–98, 108, 110–111, 119, 124, 138, 155, 171, 174, 177, 190, 193–195, 202, 212–214, 250, 253, 257, 280, 312, 321, 323, 329, 348–349, 354, 357, 362, 380–381, 395, 398, 405, 419, 430–434, 471, 492–495
Paclitaxel, 49, 159, 327–329, 335, 426–427, 432–433, 450, 461, 464, 468–469
Pancreatic cancer, 83, 287–304, 380–382, 392, 394, 396–397, 406, 418, 431–432, 453, 459–460, 469, 510
Pancreatic ductal adenocarcinoma (PDAC), 287, 295, 432, 505
Peptide nucleic acid (PNA), 384
Personalized medicine, 436
P-glycoprotein, 418, 429
Phase I clinical trial, 84, 158, 435
Phase II clinical trial, 84, 416–417, 423, 434
Phase III clinical trial, 300
Philadelphia chromosome, 269–270, 431
Phosphatidylinositol 3-kinase (PI3K), 108–110, 173, 176, 250, 300, 347–348, 472, 488, 493, 540
PI3K/Akt/mTOR pathway, 176, 488
PI3K/Akt/PTEN pathway, 176
Piwi-interacting RNA (piRNA), 535
Plasma, 47, 51, 76, 117–118, 121, 128, 176, 183, 216, 235, 242, 246, 254–255, 258, 301, 303, 417, 500–516, 519–522, 524, 526–528
Platelet-derived growth factor (PDGF), 156, 349, 351, 354
Polycistron, 7, 92, 205, 210–212
Polymorphism, 50, 168, 170, 179, 202, 229, 315, 436
Positron emission tomography (PET), 195, 417, 486
Post-transcriptional gene silencing (PTGS), 535–536
Post-transcriptional regulation, 17, 115, 165, 248–249, 254, 298, 349, 493
Post-translational modification, 230
Pre-clinical, 83, 537–542
Prediction, 17, 46–47, 49–50, 91, 110, 124–128, 136, 150, 159, 165, 179, 181, 191, 193, 195, 202, 210, 270, 272, 274, 280, 303, 451
Primary effusion lymphoma (PEL), 245, 256–257
Primary mediastinal B cell lymphoma (PMBCL), 248, 251, 254
Prognostic marker, 48, 50, 100, 158, 191, 195, 261, 279, 302, 318–319, 326, 402, 408
Progression-free survival (PFS), 125–126, 150, 260
Promoter-associated small RNAs (pasRNA), 535
Prostate cancer, 51, 75, 83, 117, 145–159, 213, 228, 276–277, 350, 392, 395–397, 399, 403, 406, 419–422, 434, 467–469, 488, 500, 503, 507–508, 510
Prostate-specific antigen (PSA), 146–147, 150, 155, 158, 500
Protein processing, 115, 230, 315
Proteomic analysis, 159, 347, 488

Q

Quantitative methylation specific PCR (qMS-PCR), 295
Quantitative reverse transcription-polymerase chain reaction (qRT-PCR), 46–48, 51, 76, 148–150, 153, 165, 205–206, 208, 272, 280, 288, 293, 301, 320, 376, 502–522

R

Radiation, 159, 182, 196, 274, 288, 300, 303, 344, 360, 401, 424, 428, 500, 534
Radioresistance, 353, 424–428, 542
Radiosensitivity, 157, 416–417, 424, 428, 541
Radiotherapy (RT), 128, 147, 344, 353, 415–436, 449–450, 453
Randomized clinical trial, 146
Randomized control trial, 183
Rb, 202, 230, 349–350, 377, 419
Reactive oxygen species (ROS), 416, 490–491, 493
Receiver operator characteristic (ROC) curve, 117–118, 228, 233, 509, 521–522
Reed-Sternberg cells (RSC), 254, 259
Renal cell carcinoma (RCC), 423
RNA-directed DNA methylation (RdDM), 536
RNA-induced silencing complex (RISC), 1, 9, 115, 164–165, 314, 451, 524, 526, 536–538
RNA interference (RNAi), 182, 194, 244, 274, 533–545
RNase protection assay, 151

S

Saliva, 422, 501, 506, 516–518
Sarcoma, 18, 22, 36–38, 44
Secreted microRNA, 527

Serum, 47, 51, 75–77, 84, 117–118, 128, 146–147, 151, 179, 183, 216, 257, 301, 303, 312, 325, 328–331, 361, 422, 500–516, 519–522, 524, 526–528
Serum marker, 312
Severe combined immune deficiency (SCID), 156, 181, 374
Short hairpin RNA (shRNA), 102, 538, 540–541, 543
Signaling pathway, 110, 112, 171, 173–175, 348, 353, 540
Signal transducer and activator of transcription 3 (STAT3), 18, 22–25, 97, 252, 298, 328, 349, 421
Single nucleotide polymorphism (SNP), 50, 92, 116–117, 126, 168, 202, 215–216, 229, 315, 331, 434
Small cell lung cancer (SCLC), 74, 425, 430, 469, 471
Small interfering RNA (siRNA), 99, 102, 384, 472, 474, 533, 535–544
Solexa sequencing, 503, 507–508, 510, 516, 519, 521
Sorafenib (Nexavar), 183, 427, 433, 450, 466, 468
Sputum, 422
Stemness, 3, 6, 38, 42, 93, 169, 172, 183, 377–378, 380, 382–383, 404
Stool, 118, 422
S-TRAIL, 426, 433, 435, 543
Sunitinib (Sutent), 450
Survivin, 97, 541
Systemic inflammatory response syndrome (SIRS), 511, 514–515

T

Tamoxifen, 44, 95, 99, 425, 428, 462, 470, 474–475
Targeted therapy, 102, 123, 401, 408–409, 435
Target gene, 79, 94–95, 98, 111, 210, 212, 214, 249, 252, 273, 323, 473, 488
Targetscan, 136, 165, 451
Taxol, 360, 362, 425, 429, 434, 461, 464, 471–472
T cell, 25–26, 33, 40, 240–242, 244–246, 248, 251, 256, 274, 276
T cell lymphoma, 33, 40, 248, 256, 276
TGF-β signaling, 171, 420
Therapeutic target, 50, 157, 182, 210, 281, 362, 540
Tight-junctions, 399–400, 543
Tobacco smoking, 77–78
Topoisomerase I, 434, 450
Topoisomerase II, 429, 450
Transcription-initiation RNA (tiRNA), 535
Transcriptomic, 17, 182
Transferrin, 385, 539, 540, 543
Trastuzumab (Herceptin), 102, 417
Tricarboxylic acid (TCA) cycle, 486, 488, 490
Trichostatin A (TSA), 171, 296, 430
Tumor classification, 91, 139, 325, 347, 363, 510–511
Tumorigenesis, 4, 17, 46, 74–75, 77–78, 82, 93, 113, 136, 148, 169, 171–173, 179, 181, 190–191, 195, 204, 210, 213, 315–316, 319, 322, 324, 326, 375, 377–378, 383–384, 403, 423, 435, 486, 488, 509
Tumor necrosis factor related apoptosis inducing ligand (TRAIL), 44, 228, 232, 362, 433, 463, 470–471, 474
Tumor suppressor gene, 2, 4, 97, 107, 110, 114, 137, 154, 166, 178, 182, 192–193, 196, 203, 210, 212, 230, 245–254, 260, 275, 280, 288, 295, 347, 351, 394–395, 402, 419, 487, 492
Twist, 4, 99, 397, 400–401, 421, 472–473

U

Univariate analysis, 150
Unknown primary cancer (UPC), 46
Urine, 228, 416, 422, 501, 516–517
Urothelial carcinoma (UC), 225–231, 233

V

Vascular endothelial growth factor (VEGF), 94, 97, 100, 211, 299, 351, 397, 407, 417, 421, 428, 434, 542
Vinblastine, 450, 463, 467, 471
Vincristine, 426, 450, 464, 466, 469, 471–472
Viral vector, 195, 384
Vogelstein's model, 108

W

Warburg effect, 155, 486, 493–494
Western blot, 17
Wnt/β-catenin signaling/pathway, 171–172
World Health Organization (WHO), 150, 344, 534

X

X-ray, 500

Z

ZEB, 299, 325, 397, 400–401, 473
Zebra fish, 345, 358